Asymptotic Statistical Inference

Shailaja Deshmukh • Madhuri Kulkarni

Asymptotic Statistical Inference

A Basic Course Using R

Shailaja Deshmukh
Department of Statistics
Savitribai Phule Pune University
Pune, Maharashtra, India

Madhuri Kulkarni
Department of Statistics
Savitribai Phule Pune University
Pune, Maharashtra, India

ISBN 978-981-15-9005-4   ISBN 978-981-15-9003-0   (eBook)
https://doi.org/10.1007/978-981-15-9003-0

© The Editor(s) (if applicable) and The Author(s), under exclusive license to Springer Nature
Singapore Pte Ltd. 2021
This work is subject to copyright. All rights are solely and exclusively licensed by the Publisher, whether
the whole or part of the material is concerned, specifically the rights of translation, reprinting, reuse of
illustrations, recitation, broadcasting, reproduction on microfilms or in any other physical way, and
transmission or information storage and retrieval, electronic adaptation, computer software, or by similar
or dissimilar methodology now known or hereafter developed.
The use of general descriptive names, registered names, trademarks, service marks, etc. in this
publication does not imply, even in the absence of a specific statement, that such names are exempt from
the relevant protective laws and regulations and therefore free for general use.
The publisher, the authors and the editors are safe to assume that the advice and information in this
book are believed to be true and accurate at the date of publication. Neither the publisher nor the
authors or the editors give a warranty, expressed or implied, with respect to the material contained
herein or for any errors or omissions that may have been made. The publisher remains neutral with regard
to jurisdictional claims in published maps and institutional affiliations.

This Springer imprint is published by the registered company Springer Nature Singapore Pte Ltd.
The registered company address is: 152 Beach Road, #21-01/04 Gateway East, Singapore 189721,
Singapore

Dedicated to

Our Respected Teachers and Beloved Students

Who Enriched Our Learning

Preface

Statistics as a scientific discipline deals with various methods of collection of data, a variety of tools for summarization and analysis of data to extract information from the data and logical techniques for meaningful interpretation of the analysis to convert information into knowledge. Numerous methods of analysis and their optimality properties are discussed in statistical inference literature. These differ depending on the size of the data. If the data size is relatively smaller, an optimal solution may not always exist. However, in many cases the scenario changes for the better as the sample size goes on increasing and the existence of an optimal solution can be ensured. Since statistics is concerned with accumulation of data, it is of prime interest to judge a variety of optimality properties of inference procedures, when we get an increasing amount of data. These optimality properties are investigated in asymptotic statistical inference theory.

In the present book, we study in detail some basic large sample optimality properties of estimators and some test procedures. A rigorous mathematical approach is directed towards the theoretical concepts and simultaneously supported by simulation studies, with the tool of R software, to absorb the notions easily.

The book begins with a brief introduction to basic framework of statistical inference. An overview of the concepts from parametric statistical inference for finite sample size and of the various modes of convergence of a sequence of random variables from probability theory is also provided. These notions form the foundation of the asymptotic statistical inference developed in subsequent chapters.

Chapters 2 and 3 form the core of the book. The basic concept of consistency of an estimator for a real parameter and a vector parameter is discussed in detail in Chap. 2. In Chap. 3 we present in depth the convergence in distribution of a suitably normalized estimator. In particular, the focus is on the consistent and asymptotically normal (CAN) estimators. The large sample optimality properties of an estimator are defined in terms of its limiting behaviour as sample size increases. Hence, the convergence of a sequence of random variables becomes the principal probability tool in the asymptotic investigation of an estimator. In Chap. 4, we discuss two families of distributions for which optimal estimators do exist for the parameters of interest. It is shown that for the probability models belonging to an exponential family and a Cramér family, the maximum likelihood estimators of the parameters are CAN.

Chapters 5 and 6 study various test procedures and their properties when the sample size is large. In Chap. 5, we introduce a likelihood ratio test procedure and prove results related to the asymptotic null distribution of the likelihood ratio test statistic. Chapter 6 addresses the applications of the likelihood ratio test procedure when the underlying probability model is a multinomial distribution. In particular, we study tests for goodness of fit, tests for validity of the model and a variety of tests for contingency tables. In Chap. 6, we also study a score test and Wald's test and examine their relationship with the likelihood ratio test and Karl Pearson's chi-square test. We have discovered an important result regarding a score test statistic and Karl Pearson's chi-square test statistic. While testing a hypothesis about a parameter of a multinomial distribution, these two statistics are identical.

Numerous illustrations and examples of differing difficulty level, from routine to challenging, are incorporated throughout each chapter to clarify the concepts. These illustrations and several remarks reveal the depth of the incorporated theory. For better assimilation of the notions contained in the book, various exercises are included at the end of each chapter. Solutions to almost all the conceptual exercises are given in Chap. 7, to motivate students towards solving these exercises and to enable digestion of the underlying concepts.

Over the years, we have noted that the concepts from asymptotic inference are crucial in modern statistics, but are difficult for the students to grasp due to their abstract nature. To overcome this difficulty, we have augmented the theory with R software as a tool for simulation and computation, which is a novel and a unique feature of our book. Nowadays R is a leading computational tool for statistics and data analysis. It can handle a variety of functions, such as data manipulation, statistical modeling, advanced statistical methods etc. It has numerous packages in the CRAN repository, which are constantly growing in number. R also has some excellent graphical facilities. Despite all these multiple advantages, not only R is free, but it is also platform-independent and hence can be used on any operating system. Hence, keeping up with the recent trend of using R software for statistical computations and data analysis, we too have used it extensively in this book for illustrating the concepts, verifying the properties of estimators and carrying out various test procedures.

Chapter 1 covers a brief introduction to R software. The last section of each of the Chapters from 2, 3, 4, 5 and 6 presents R codes for the verification of the concepts and procedures discussed in the respective chapters. The major benefit of these codes is to understand the complex notions with ease. The R codes also reveal the hidden aspects of different procedures and cater to the educational need of visual demonstration of the concepts and procedures. The asymptotic theory gives reasonable answers in many scenarios, which are found approximately valid. It may be theoretically very hard to ascertain whether the approximation errors involved are insignificant or not, but one can have recourse to simulation studies to empirically judge the accuracy of certain approximations. It has been demonstrated in the book using R codes. These codes are deliberately kept simple, so that readers can understand the underlying theory with the minimal effort. At the end of each chapter, computational exercises based on R software are included so as to provide a hands-on experience to students.

The book has evolved out of the instructional material prepared for teaching a course "Asymptotic Inference" for several years at Savitribai Phule Pune University, formerly known as University of Pune. To some extent, the topics coincide with what we used to cover in the course. No doubt, there are many excellent books on asymptotic inference. However, these books do not elaborate on the computational aspect. While teaching the course, we realized that the students need a simpler, user-friendly book. Students often come across terms such as "routine computations", "trivial", "obvious", when in fact, the underlying steps are not so obvious and trivial for them. Hence, we decided to compile the teaching material in the form of a customized book to fulfill the need of the students while trying to fill the gap. While the competing texts are often found substantially concise, contrarily, we have tried to well develop the subject with the help of a variety of nicely worked out examples. The main motive is to provide fairly thorough treatment of basic techniques, theoretically and computationally using R, so that the book will be suitable for self-study. The style of the book is purposely kept conversational, so that the reader may feel vicinity of a teacher. Hopefully, a better understanding can provide more insights and propel students towards a better appreciation of the beauty of the subject. We will be deeply rewarded, if the present book helps students to enhance their understanding and to enjoy the subject.

The mathematical prerequisites for this book are basic convergence concepts for a sequence of real numbers and familiarity with the properties of a variety of discrete and continuous distributions. It is assumed that the reader has the background knowledge of parametric inference for finite sample size. This includes concepts of sufficiency, information function, standard methods of estimation and finite sample optimality properties of estimators obtained using these methods. Some background to measure-theoretic probability theory would also be beneficial, since it forms the mathematical foundation for asymptotic inference. In particular, an awareness of concepts such as various modes of convergence, laws of large numbers and Lindeberg-Levy central limit theorem would be useful. In addition, a basic knowledge of R software is desirable. We have added three sections, devoted to a brief introduction to the basic concepts needed, in Chap. 1 for ready reference. For an interested reader, a list of reference books is given for an in-depth study of these concepts.

The intended target audience of the present book is mainly post graduate students in a quantitative program, such as Statistics, Bio-statistics or Econometrics and other disciplines, where inference for large sample size is needed for the data analysis. The book will be useful to data scientists and researchers in many areas in which data size is large and various analytical methods are to be deployed such as categorical data analysis, regression analysis, survival analysis etc. It will also provide sufficient background information for studying inference in stochastic processes. The book is designed primarily to serve as a text book for a one semester introductory course in asymptotic statistical inference in any post-graduate statistics program.

We wish to express our special thanks to all our teachers, in particular we are grateful to Prof. B. K. Kale and Prof. M. S. Prasad who have laid the strong foundation of statistical inference and influenced our understanding, appreciation, and taste for the subject. We sincerely thank Prof. M.S. Prasad, Prof. M. B. Rajarshi, Dr Vidyagouri Prayag, Dr Akanksha Kashikar and Namitha Pais for reading the entire manuscript very carefully and for their constructive inputs. Incorporation of their suggestions, and also the comments and criticism by a number of reviewers has definitely improved the presentation and the rigour of the book. We thank the Head, Department of statistics, Savitribai Phule Pune University, for providing the necessary facilities.

We wish to express our deep gratitude to the R core development team and the authors of contributed packages, who have invested a lot of time and effort in creating R as it is today. With the help of such a wonderful computational tool, it is possible to showcase the beauty of the theory of asymptotic statistical inference. We take this opportunity to acknowledge Nupoor Singh, editor of the Statistics section of Springer Nature and her team, for providing help from time to time and subsequent processing of the text to its present form.

We are deeply grateful to our family members for their constant support and encouragement. Last but not the least, we owe profound thanks to all our students whom we have taught during last several years and who have been the driving force to take up this immense task. Their reactions and doubts in the class and our urge to make the theory crystal clear to them, compelled us to pursue this activity and to prepare a variety of illustrations and exercises.

All mistakes and ambiguities in the book are exclusively our responsibility. We would love to know any mistakes that a reader comes across in the book. Feedback in the form of suggestions and comments from colleagues and readers is most welcome.

Pune, India   Shailaja Deshmukh
August 15, 2020   Madhuri Kulkarni

Contents

About the Authors

Shailaja Deshmukh is a visiting faculty at the Department of Statistics, Savitribai Phule Pune University (formerly known as University of Pune). She retired as a Professor of Statistics from Savitribai Phule Pune University. She has taught around twenty five different theoretical and applied courses. Her areas of interest are inference in stochastic processes, applied probability, actuarial statistics and analysis of microarray data. She has a number of research publications in various peer-reviewed journals, such as *Biometrika, Journal of Multivariate Analysis, J. R. Statist. Soc., Australian and New Zealand Journal of Statistics, Environmetrics, J. of Statistical Planning and Inference, Journal of Translational Medicine*. She has published four books, the last of which was 'Multiple Decrement Models in Insurance: An Introduction Using R', published by Springer. She has served as an executive editor and as a chief editor of the *Journal of Indian Statistical Association* and she is an elected member of the international Statistical Institute.

Madhuri Kulkarni has been working as an Assistant Professor at the Department of Statistics, Savitribai Phule Pune University since 2003. She has taught a variety of courses in the span of 17 years. The list includes programming languages like C and C++, core statistical courses like probability distributions, statistical inference, regression analysis, and applied statistical courses like actuarial statistics, Bayesian inference, reliability theory. She has been using R for teaching the practical and applied courses for more than a decade. She is a recipient of the prestigious U.S. Nair Young Statistician Award. She has completed research projects for Armament Research and Development Establishment (ARDE), Pune, and has also received core research grant for a research project on software reliability from DST-SERB, India in 2018. She writes regularly in English, Hindi and Marathi in her blog. She also shares the e-content developed by her.

List of Figures

List of Tables

Introduction

1

Contents

1.1 Introduction

Statistics is concerned with the collection of data, their analysis and interpretation. As a first step, the data are analyzed without any extraneous assumptions. The principal aim of such an analysis is the organization and summarization of the data to bring out their main features and clarify their underlying structure. The first step of such an analysis is known as an exploratory data analysis in which some graphs such as histogram, box plots, scatter plots are drawn and some interesting characteristics based on the given sample, such as sample mean, sample variance, coefficient of variation, correlation or regression, in case of multivariate data, are obtained. These sample characteristics are then used to estimate the corresponding population characteristics. It is the second step of analysis and is known as a confirmatory data analysis. Statistical inference plays a significant role in the confirmatory data analysis as it involves basically the inference procedures consisting of point estimation, interval estimation and testing of hypotheses. In these inference procedures it is essential to know the distribution of the sample characteristics. Probability theory and the distribution theory play the role of a bridge between the exploratory and the confirmatory data analysis. In many cases, it is difficult to find the distribution for a finite sample size and then one seeks to find it for a large sample size. To elaborate on all these issues we begin with the basic framework of parametric statistical inference.

© The Author(s), under exclusive license to Springer Nature Singapore Pte Ltd. 2021
S. Deshmukh and M. Kulkarni, *Asymptotic Statistical Inference*,
https://doi.org/10.1007/978-981-15-9003-0_1

The parametric approach to statistical modeling assumes a family of probability distributions. More specifically, suppose X is a random variable or a random vector under study, defined on a probability space $(\Omega, \mathbb{A}, P_\theta)$, with probability law $f(x, \theta)$, $\theta \in \Theta$. The probability law $f(x, \theta)$ is determined by the probability measure P_θ. The probability law is the probability mass function if X is a discrete random variable and it is a probability density function if X is a continuous random variable. The parameter θ may be a real parameter or a vector parameter. The set Θ is known as a parameter space. As θ varies over Θ, we get a family of probability distributions. An important condition on the probability measure P_θ in statistical inference is indexing of P_θ by a parameter $\theta \in \Theta$. The probability measure P_θ is said to be indexed by a parameter θ or labeled by a parameter θ if $P_{\theta_1}(\cdot) = P_{\theta_2}(\cdot)$ implies $\theta_1 = \theta_2$. A parameter θ is then known as an indexing parameter. In terms of probability law $f(x, \theta)$, it is stated as follows. Suppose a support S_f of $f(x, \theta)$ is defined as $S_f = \{x | f(x, \theta) > 0\}$. Then a parameter θ is known as an indexing parameter if $f(x, \theta_1) = f(x, \theta_2)$, $\forall x \in S_f$, implies that $\theta_1 = \theta_2$. A collection $\{P_\theta, \theta \in \Theta\}$ is known as a family of probability measures indexed by θ or $\{f(x, \theta), \theta \in \Theta\}$ is known as a family of probability distributions indexed by θ. Such an indexing condition of the probability measure is known as an identifiability condition as it uniquely identifies the member from the family $\{P_\theta, \theta \in \Theta\}$. For example, if X follows a binomial $B(n, p)$ distribution, then p is the indexing parameter; if X follows a normal $N(\mu, \sigma^2)$ distribution, then μ and σ^2 both are indexing parameters. Suppose X and Y denote the life lengths of the two components of a system in which the system fails if either of the components stops working. Thus, the life of a system is the same as the life of the component which fails first. Hence, the observable random variable is $Z = \min\{X, Y\}$. Suppose X and Y are independent random variables each having exponential distribution with failure rate θ and λ respectively. Then Z has exponential distribution with failure rate $\theta + \lambda$. In this situation, $(\theta, \lambda)'$ cannot be an indexing parameter of Z, as infinitely many pairs of $(\theta, \lambda)'$ will give rise to the same value of $\theta + \lambda$. Thus, an indexing parameter is $\theta + \lambda$ and not $(\theta, \lambda)'$. The problem of identifiability is basic to all statistical methods and data analysis, occurring in such diverse areas as reliability theory, survival analysis and econometrics, where stochastic modeling is widely used. For more details, one may refer to the book by Rao [1].

An important assumption in parametric inference is that the form of the probability law $f(x, \theta)$ is known and the only unknown quantities are the indexing parameters. The main aim of statistical inference is to have the best guess about θ or some parametric function $g(\theta)$, such as mean or variance of a distribution, on the basis of a sample $\underline{X} \equiv \{X_1, X_2, \dots, X_n\}$ of n observations from the distribution of X. It is to be noted the given data $\underline{X}$ is generated corresponding to some value of $\theta = \theta_0$, say, which is labeled as a true parameter. However, θ_0 is unknown and we wish to guess its value on the basis of the observations $\underline{X}$ generated under θ_0. A true parameter θ_0 is any member of Θ and hence usually θ_0 is simply referred to as θ. To have the best guess of θ we find a suitable statistic, that is a Borel measurable function $T_n(\underline{X})$ of observations $\underline{X}$. For example, suppose $\underline{X}$ is a random sample from a normal $N(\theta, 1)$ distribution, $\theta \in \mathbb{R}$. Then the sample mean $\overline{X}_n$ or the sample median are functions of sample observations and can be used to have a good guess about θ, as θ is a population mean as well as population median. Suppose $\Theta = [0, \infty)$, we do come across with such a restricted parameter space setup, particularly while developing the likelihood ratio test procedure to test $H_0 : \theta = 0$ against the alternative that $H_0 : \theta > 0$. In such a case, it is desirable to have a statistic whose value lies in the interval $[0, \infty)$ as a preliminary guess for θ. Observe that

$$P_\theta[\overline{X}_n < 0] = P_\theta[\sqrt{n}(\overline{X}_n - \theta) < -\sqrt{n}\theta] = \begin{cases} \Phi(-\sqrt{n}\theta) > 0, & \text{if } \theta > 0 \\ 1/2, & \text{if } \theta = 0. \end{cases}$$

Here $\Phi(\cdot)$ denotes the distribution function of the standard normal distribution. Thus, for any $\theta \in [0, \infty)$, $P_\theta[\overline{X}_n < 0] > 0$. As a consequence, for given data if $\overline{X}_n < 0$, one cannot use $\overline{X}_n$ as a possible value of θ which is non-negative. Further, if $\Theta = \{0, 1\}$, then $P_\theta[\overline{X}_n = 0] = 0$ and $P_\theta[\overline{X}_n = 1] = 0$. In such cases using the sample mean $\overline{X}_n$ to guess for θ does not seem to be reasonable, instead a suitable function of observations with range space as $\{0, 1\}$ should be used. In view of such limitations, any statistic cannot be used as an estimator of θ, it is essential that the range space of T_n is the same as the parameter space. We thus define an estimator and an estimate as follows. Suppose $\mathbb{X}$ is a set of all possible values of $\underline{X}$. It is known as a sample space.

▶ Definition 1.1.1

Estimator and Estimate: Suppose $\underline{X}$ is a sample from the probability distribution of X, indexed by a parameter $\theta \in \Theta$. A Borel measurable function $T_n(\underline{X})$ of $\underline{X}$ from $\mathbb{X}$ to Θ is known as an estimator of θ. For a given realization $\underline{x}$ of $\underline{X}$, the value $T_n(\underline{x})$ is known as an estimate of θ.

This approach of defining an estimator is followed by many, including Rohatgi and Saleh [2] and Shao [3]. In Example 2.2.3, we define a suitable statistic to estimate θ when the parameter space Θ for a normal $N(\theta, 1)$ distribution is either $[0, \infty)$ or $\{0, 1\}$. While defining consistency of an estimator T_n for a parameter θ in the next chapter, we study the limiting behavior of the probability that the distance between T_n and θ is small. This fact also indicates that the range space of T_n should be the same as that of θ, which may or may not be true for any statistic T_n.

It is to be noted that an estimator is a random variable or a random vector and an estimate is a specific value of the random variable or a random vector. An estimator $T_n(\underline{X})$ forms a basis of inference for the parameter θ. In all the inference procedures the important basic assumption is that, it is possible to suggest an estimator of θ by having observations on X; that is, we assume that observations on X provide information on θ. Such an assumption is valid as the probability law $f(x, \theta)$ changes as the value of an indexing parameter θ changes. To clarify this important point, in Fig. 1.1, we have plotted the probability mass function of binomial $B(5, p)$ distribution for $p = 0.1, 0.2, \ldots, 0.9$. From Fig. 1.1 we note that as the value of the parameter changes, the nature of the probability mass function changes. For some values of p, some values of X are more likely. On the other hand, if the observed value of X is 4, say, then from Fig. 1.1, we guess that it is more likely that $p \in \{0.7, 0.8, 0.9\}$.

Such a feature is also observed when we have more data. It is to be noted that the joint distribution of $\underline{X} = \{X_1, X_2, \ldots, X_n\}$, where each X_i is distributed as

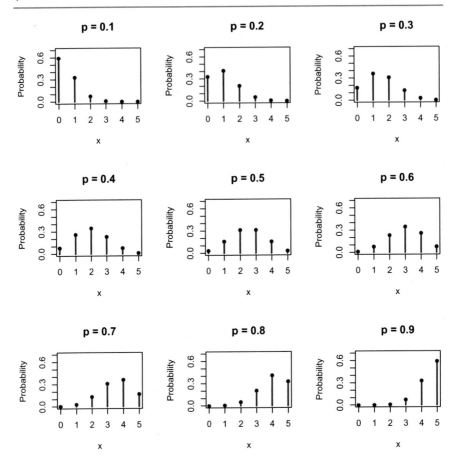

Fig. 1.1 Probability mass function of binomial $B(5, p)$ distribution

X, also changes as the underlying indexing parameter changes. Thus, the sample $\{X_1, X_2, \ldots, X_n\}$ does provide information on θ. Since $T_n(\underline{X})$ is a Borel measurable function of random variables $\underline{X}$, it is again a random variable and its probability distribution is determined by that of X. It then follows that the probability distribution of $T_n(\underline{X})$ is also indexed by θ and it also provides information on θ. For example, if $\underline{X}$ is a random sample from a normal $N(\theta, 1)$ distribution then the sample mean $\overline{X}_n$ again has a normal $N(\theta, 1/n)$ distribution. A random sample $\underline{X}$ from the distribution of X indicates that $\{X_1, X_2, \ldots, X_n\}$ are independent and identically distributed random variables, each having the same probability law $f(x, \theta)$ as that of X. The joint distribution of $\{X_1, X_2, \ldots, X_n\}$ is then given by $f(x_1, x_2, \ldots, x_n, \theta) = \prod_{i=1}^{n} f(x_i, \theta)$. For the given data $\underline{x} = \{x_1, x_2, \ldots, x_n\}$, $f(x_1, x_2, \ldots, x_n, \theta)$ is a function of θ and Fisher defines it as a likelihood function. We denote it by $L(\theta|\underline{x})$. Thus, $L(\theta|\underline{x})$ provides information on θ corresponding to the given observed data. If X is a discrete random variable, then $L(\theta|\underline{x})$ represents the probability of generating data $\underline{x} = \{x_1, x_2, \ldots, x_n\}$ when true parameter is θ. It varies as θ varies over Θ and hence provides information on θ corresponding to the given observed data. Most of the inference procedures are based on the likelihood function. The most popular is the method of maximum likelihood estimation to find an estimator of θ. This procedure has been proposed by Fisher in 1925. He proposed to estimate θ by that value of θ for which $L(\theta|\underline{x})$ is maximum corresponding to given data $\underline{x}$ and labeled it as a maximum likelihood estimate. Another heavily used inference procedure based on the likelihood function is the likelihood ratio test procedure for testing hypotheses. In the present book, we discuss both these procedures and their properties when the size of the random sample is large.

Another frequently used method of estimation is a method of moments. It has been proposed by Karl Pearson in nineteenth century. In this method, the estimator is obtained by solving the system of equations obtained by equating sample moments to the corresponding population moments and is labeled as a moment estimator. There are various other methods to obtain estimators of the parameter of interest on the basis of given data such as method of least squares, methods based on sample quantiles and methods based on estimating functions. In this book, we will not discuss these methods of estimation or the properties of these estimators for finite n as our focus is on the discussion of large sample optimality properties of estimators and test procedures. For details of these methods and properties of the estimators for finite n, for interval estimation and for testing of hypotheses for finite n, one may refer to the following books—Casella and Berger [4], Kale and Muralidharan [5], Lehmann and Casella [6], Lehmann and Romano [7], Rohatgi and Saleh [2]. However for ready reference, in the following section, we list various results from parametric statistical inference for finite sample size, as these form a foundation of the asymptotic statistical inference theory.

1.2 Basics of Parametric Inference

Point estimation is one of the most important branches of statistical inference. To study the optimality properties of an estimator, one needs to know its distribution. The probability distribution of an estimator is known as a sampling distribution . As discussed in Sect. 1.1, if $\underline{X}$ is a random sample from a normal $N(\theta, \sigma^2)$ distribution, $\theta \in \mathbb{R}$, $\sigma^2 > 0$, then the sample mean $\overline{X}_n$ is a Borel measurable function from $\mathbb{X}$ to Θ and hence is an estimator of θ. It has a normal $N(\theta, \sigma^2/n)$ distribution and it is the sampling distribution of $\overline{X}_n$. Sampling distribution of an estimator is useful to investigate the properties of an estimator. The first natural property of an estimator is unbiasedness, as defined below.

▶ Definition 1.2.1

Unbiased Estimator: An estimator $T_n(\underline{X})$ is an unbiased estimator of $g(\theta)$, if $E_\theta(T_n(\underline{X})) = g(\theta), \ \forall \ \theta \in \Theta$.

The concept of unbiasedness requires that the sampling distribution of an estimator is centered at $g(\theta)$. If $\underline{X}$ is a random sample from the normal $N(\theta, \sigma^2)$ distribution, $\theta \in \mathbb{R}$, $\sigma^2 > 0$, then the sample mean $\overline{X}_n$ is an unbiased estimator of θ. Here

$g(\theta) = \theta$. If we have two unbiased estimators for the same parametric function $g(\theta)$, then using sampling distribution of the estimators we can find their variances and choose that estimator which has smaller variance. If the two estimators are not unbiased then these can be compared using the mean squared error (MSE).

▶ **Definition 1.2.2**

Mean Squared Error of an Estimator: Mean squared error of $T_n(\underline{X})$ as an estimator of $g(\theta)$ is defined as $MSE(T_n(\underline{X})) = E_\theta(T_n(\underline{X}) - g(\theta))^2$.

An estimator with smaller MSE is always preferred. If $T_n(\underline{X})$ is an unbiased estimator of $g(\theta)$, then MSE of $T_n(\underline{X})$ is the same as the variance of $T_n(\underline{X})$. Within a class of unbiased estimators of a parameter $g(\theta)$, we seek for an estimator which has the smallest variance. Under certain regularity conditions, there exists a lower bound for the variance of an unbiased estimator of $g(\theta)$ and it is known as the Cramér-Rao lower bound for the variance of an unbiased estimator. To study this important and fairly general result we need a concept of information function, introduced by Fisher. In Sect. 1.1, we have discussed the concept of information and noted that the observations on X contain the information about the parameter. We now quantify this concept of information about θ in a sample or in any statistic, under the following two assumptions:

1. Suppose the support S_f of $f(x, \theta)$ is free from θ.
2. The identity $\int_{S_f} f(x, \theta)dx = 1$ can be differentiated with respect to θ at least twice under the integral sign.

As a consequence,

$$E_\theta\left(\frac{\partial}{\partial\theta}\log f(X, \theta)\right) = 0 \ \& \ E_\theta\left(\frac{\partial}{\partial\theta}\log f(X, \theta)\right)^2 = E_\theta\left(-\frac{\partial^2}{\partial\theta^2}\log f(X, \theta)\right).$$

▶ **Definition 1.2.3**

Information Function: The information function $I(\theta)$ which quantifies information about θ in a single observation on a random variable X is defined as

$$I(\theta) = E_\theta\left(\frac{1}{f(X, \theta)}\frac{\partial}{\partial\theta}f(X, \theta)\right)^2 = E_\theta\left(\frac{\partial}{\partial\theta}\log f(X, \theta)\right)^2$$

$$= E_\theta\left(-\frac{\partial^2}{\partial\theta^2}\log f(X, \theta)\right).$$

$I(\theta)$ is usually referred to as the Fisher information function. The function $\frac{1}{f(x,\theta)}\frac{\partial}{\partial\theta}f(x, \theta) = \frac{\partial}{\partial\theta}\log f(x, \theta)$ is interpreted as a relative rate of change in $f(x, \theta)$ as θ varies. Thus, it is similar to velocity while $\frac{\partial^2}{\partial\theta^2}\log f(x, \theta)$ is similar to acceleration. The function $\frac{\partial}{\partial\theta}\log f(X, \theta)$, viewed as a function of X for a fixed θ is known as a score function. Thus, for each fixed θ, it is a random variable. From the above expressions it is clear that its expectation is 0 and variance is $I(\theta)$.

We use these results in subsequent chapters. Observe that $I(\theta) \geq 0$ and it is 0 if and only if $E_\theta \left(\frac{\partial}{\partial \theta} \log f(X, \theta) \right)^2 = 0$ which is equivalent to the statement that $\frac{\partial}{\partial \theta} \log f(X, \theta) = 0$ with probability 1, that is, $f(X, \theta)$ does not depend on θ or the distribution of X does not change as θ changes.

The following theorem states a result about a lower bound for the variance of an unbiased estimator.

Theorem 1.2.1

Cramér-Rao inequality: *Suppose $\underline{X}$ is a random sample from the distribution of X with $\{f(x, \theta), \theta \in \Theta\}$ as a family of probability distributions of X. Suppose $\mathbb{U}$ is a class of all unbiased estimators T_n of $g(\theta)$ such that $E(T_n^2) < \infty$ for all $\theta \in \Theta$. Suppose the following conditions are satisfied.*

1. *The support S_f of $f(x, \theta)$ is free from θ.*
2. *The identity $\int_{S_f} f(x, \theta)dx = 1$ can be differentiated with respect to θ at least twice under the integral sign.*
3. *An estimator $T_n \in \mathbb{U}$ is such that*

$$g'(\theta) = \frac{d}{d\theta} E_\theta(T_n) = \frac{d}{d\theta} \int t \, h(t, \theta) \, dx = \int t \frac{\partial}{\partial \theta} \log h(t, \theta) \, h(t, \theta) dx,$$

where $h(t, \theta)$ is a probability law of T_n.

Then

$$Var(T_n) \geq \frac{(g'(\theta))^2}{I(\theta)}.$$

A function $(g'(\theta))^2 / I(\theta)$ is known as the Cramér-Rao lower bound for the variance of an unbiased estimator of $g(\theta)$. One would like to have an unbiased estimator which attains the lower bound as specified in the above theorem. Such an estimator is known as a minimum variance bound unbiased estimator (MVBUE). Such estimators exist for some models, but in general, it is difficult to find such an estimator. The next step is to find an unbiased estimator whose variance is smaller than that of any other unbiased estimator. It leads to the concept of a uniformly minimum variance unbiased estimator (UMVUE). It is as defined below.

▶ **Definition 1.2.4**

Suppose $\mathbb{U}$ is a class of all unbiased estimators T_n of θ such that $E(T_n^2) < \infty$ for all $\theta \in \Theta$. An estimator $T_n^* \in \mathbb{U}$ is called a UMVUE of θ if

$$E(T_n^* - \theta)^2 \leq E(T_n - \theta)^2 \quad \forall \, \theta \in \Theta \ \& \ \forall \, T_n \in \mathbb{U}.$$

If the family of distributions of an estimator satisfies certain properties, then the estimator is a UMVUE of its expectation. These properties involve concepts of sufficiency and completeness of a statistic. We define these below.

▶ **Definition 1.2.5**

Sufficient Statistic: Suppose $\underline{X} = \{X_1, X_2, \ldots, X_n\}$ is a random sample from the distribution of X with probability law $f(x, \theta)$. A statistic $U_n = U_n(\underline{X})$ is a sufficient statistic for the family $\{f(x, \theta), \theta \in \Theta\}$ if and only if the conditional distribution of $\underline{X}$, given U_n does not depend on θ.

If the conditional distribution of $\underline{X}$ given U_n does not depend on θ, then the conditional distribution of any statistic $V_n = V_n(\underline{X})$ given U_n also does not depend on θ. It means that a sufficient statistic $U_n(\underline{X})$ extracts all the information that the sample has about θ. If U_n is a sufficient statistic, then it can be shown that $I(\theta)$ corresponding to U_n is the same as $I(\theta)$ corresponding to a random sample $\underline{X}$. It implies that there is no loss of information if the inference procedures are based on the sufficient statistic. Thus, one of the desirable properties of an estimator $T_n(\underline{X})$ is that it should be a function of sufficient statistic. Well-known Neyman-Fisher factorization theorem gives a criterion for determining a sufficient statistic. We state it below.

Theorem 1.2.2

Neyman-Fisher factorization theorem: *Suppose the joint probability law* $f(x_1, x_2, \ldots, x_n, \theta)$ *is factorized as*

$$f(x_1, x_2, \ldots, x_n, \theta) = h(x_1, x_2, \ldots, x_n) g(T_n(\underline{x}), \theta),$$

where h *is a non-negative function of* $\{x_1, x_2, \ldots, x_n\}$ *only and does not depend on* θ *and* g *is a non-negative function of* θ *and* $\{x_1, x_2, \ldots, x_n\}$ *through* $T_n(\underline{x})$. *Then* $T_n(\underline{X})$ *is a sufficient statistic.*

In the statement of the above theorem, statistic $T_n(\underline{X})$ and parameter θ may be vector valued. The joint probability law $f(x_1, x_2, \ldots, x_n, \theta)$ viewed as a function of θ given data $\{x_1, x_2, \ldots, x_n\}$ is nothing but a likelihood function $L(\theta|\underline{x})$. Thus, according to the Neyman-Fisher factorization theorem, if $L(\theta|\underline{x})$ is factorized as $h(\underline{x})g(T_n(\underline{x}), \theta)$, then $T_n(\underline{x})$ is a sufficient statistic for the family of distributions or simply $T_n(\underline{x})$ is a sufficient statistic for θ.

The concept of sufficiency is used frequently with another concept, called completeness of the family of distributions. We define it below and also define what is meant by a complete statistic .

▶ **Definition 1.2.6**

Complete Statistic: A family $\{f(x, \theta), \theta \in \Theta\}$ of probability distributions of X is said to be complete, if for any function h

$$E_\theta(h(X)) = 0 \implies P_\theta[h(X) = 0] = 1 \ \forall \ \theta \in \Theta.$$

A statistic $U_n(\underline{X})$ based on a random sample $\underline{X}$ from the distribution of X is said to be complete if the family of distributions of $U_n(\underline{X})$ is complete.

Using Rao-Blackwell and Lemann-Scheffe theorem it can be shown that an unbiased estimator of $g(\theta)$, which is a function of complete sufficient statistic, is always the UMVUE of $g(\theta)$. We state these two theorems below.

Theorem 1.2.3

Rao-Blackwell theorem: *Suppose $\underline{X}$ is a random sample from the distribution of X with $\{f(x, \theta), \theta \in \Theta\}$ as a family of probability distributions. Suppose $\mathbb{U}$ is a class of unbiased estimators T_n of θ such that $E(T_n^2) < \infty$ for all $\theta \in \Theta$. Suppose $U_n = U_n(\underline{X})$ is a sufficient statistic for the family. Then the conditional expectation $E_\theta(T_n|U_n)$ is independent of θ and is an unbiased estimator of θ. Further, $E_\theta(E_\theta(T_n|U_n) - \theta)^2 \leq E_\theta(T_n - \theta)^2 \ \forall \ \theta \in \Theta$.*

Thus, according to Rao-Blackwell theorem, $E_\theta(T_n|U_n)$ is an unbiased estimator with variance smaller than that of any other unbiased estimator of $\theta \ \forall \ \theta \in \Theta$, that is, it is a UMVUE of θ. Lehmann-Scheffe theorem, stated below, conveys that under additional requirement of completeness of the sufficient statistic, $E_\theta(T_n|U_n)$ is a unique UMVUE of its expectation.

Theorem 1.2.4

Lehmann-Scheffe theorem: *Suppose $U_n = U_n(\underline{X})$ is a complete sufficient statistic and there exists an unbiased estimator T_n of θ. Then there exists a unique UMVUE of θ and it is given by $E_\theta(T_n|U_n)$.*

Rao-Blackwell theorem and Lehmann-Scheffe theorem together convey that an unbiased estimator, which is a function of a complete sufficient statistic, is a unique UMVUE of its expectation. In Chap. 2, we prove that it is also a consistent estimator of its expectation.

Testing of hypothesis is another fundamental branch of inference. It is different from estimation in some aspects, such as the accuracy measures and the appropriate asymptotic theory. We now present a brief introduction to a formal model for statistical hypothesis testing that was proposed by Neyman and Pearson in the late 1920. We are confronted with a hypothesis testing problem when we want to guess, which of two possible statements about a population is correct on the basis of observed data. Hypothesis is nothing but a statement about the population. When we are interested in studying a particular characteristic X of the population, with the assumption that the form of the probability law $f(x, \theta)$ of X is known and the only unknown quantities are the indexing parameters, a hypothesis reduces to a statement about the population parameter. On the basis of the observed data, one is interested in testing the validity of an assertion about the unknown parameter θ. For example, one may be interested in verifying if a proportion p of defectives in a lot of items is at most 5%. In such a situation the set of possible values of p are divided in two sets, one is $(0, 0.05]$ and the other is $(0.05, 1)$. One statement is $p \leq 0.05$ and the other statement is $p > 0.05$. It is necessary to distinguish between the two hypotheses under consideration. In each case, we declare one of the two hypotheses to be a null hypothesis, denote it by H_0, and the other to be an alternative hypothesis and denote it by H_1. Roughly speaking, the logic for determining which hypothesis is H_0 and which is H_1 is as follows. *A null hypothesis H_0 should be the hypothesis to which one defaults if the evidence given by data is doubtful or is insufficient and H_1 should be the hypothesis that one requires compelling evidence to embrace it.* Hence, a null hypothesis is always interpreted as hypothesis of "no difference". In general if Θ denotes the parameter space, then a null hypothesis corresponds to $H_0 : \theta \in \Theta_0$ and the alternative hypothesis corresponds to $H_1 : \theta \in \Theta_1$, where $\Theta_0 \cap \Theta_1 = \emptyset$ and $\Theta_0 \cup \Theta_1 = \Theta$. If Θ_0 contains only one point, we say that H_0 is a simple hypothesis, otherwise it is known as a composite hypothesis. Similarly if Θ_1 is a singleton set, then H_1 is a simple hypothesis, otherwise it is a composite hypothesis. Thus, if a hypothesis is simple, the probability distribution of X is specified completely under that hypothesis.

We now elaborate on the procedure of testing of hypotheses. Corresponding to given data $\underline{x} = (x_1, x_2, \ldots, x_n)$, we find a decision rule that will lead to a decision to accept or to reject the null hypothesis. Such a decision rule partitions the sample space $\mathbb{X}$ into two disjoint sets C and C' such that if $\underline{x} \in C$, we reject H_0, and if $\underline{x} \in C'$, we do not reject H_0. The set C is known as a critical region or a rejection region. The set C' is known as an acceptance region. There are two types of errors that can be made if one uses such a procedure. One may reject H_0 when in fact it is true. It is called as a type I error. Alternatively, one may accept H_0 when it is false. This error is called a type II error. Thus $P_\theta(C)$, for $\theta \in \Theta_0$ is the probability of type I error while $P_\theta(C')$, for $\theta \in \Theta_1$ is the probability of type II error. Ideally, one would like to find a critical region for which both these probabilities are 0. However, it is not possible. If a critical region is such that the probability of type I error is 0, then the probability of type II error will be 1. As a next step, we would like to devise procedures that minimize the probabilities of committing errors. Unfortunately, there is an inevitable tradeoff between type I and type II errors so that we cannot minimize the probabilities of both types of errors simultaneously.

The distinguishing feature of hypothesis testing is the manner in which it addresses the tradeoff between type I and type II errors. The Neyman-Pearson formulation of hypothesis testing offers a null hypothesis a privileged status. H_0 will be maintained unless there is compelling evidence against it. Such a status is equivalent to declaring type I error to be more serious than type II error. Hence in the Neyman-Pearson formulation, an upper bound is imposed on the maximum probability of type I error that will be tolerated. This bound is known as a level of significance, conventionally denoted by α. The level of significance is specified prior to examining the data. We consider test procedures for which the probability of type I error is not greater than α. Such tests are called level α tests. In 1925, Fisher suggested two values for α as $\alpha = 0.05$ and $\alpha = 0.01$ in his extremely influential book "Statistical Methods for Research Workers". These suggestions were intended as practical guidelines, but in view of the convenience of standardization in providing a common frame of reference, these values gradually became the conventional levels to use. Interested reader may refer to Lehmann and Romano [7] (p. 57) about the choice of significance level α.

In hypothesis testing, decisions typically are described in the language that acknowledges the privileged status of the null hypothesis and emphasizes that the decision criterion is based on the probability of committing a type I error. In describing the action of choosing H_0, many statisticians prefer the phrase "fail to reject the null hypothesis" to the phrase "accept the null hypothesis" because choosing H_0 does not imply an affirmation that H_0 is correct, it only means that the level of evidence against H_0 is not sufficiently compelling to warrant its rejection at significance level α.

To introduce some more concepts precisely, we proceed to define a test function as follows. A Borel measurable function $\phi : \mathbb{X} \to [0, 1]$ is known as a test function. A test function defined as

$$
\phi(\underline{X}) \;=\; \begin{cases} 1, & \text{if } \underline{X} \in C \\ \gamma(\underline{X}), & \text{if } \underline{X} \in B(C) \\ 0, & \text{if } \underline{X} \in C' \end{cases}
$$

is known as a randomized test function, where $B(C)$ denotes the boundary set of C. If $\gamma(\underline{X}) = 0$, it is known as a non-randomized test function. Thus, $\phi(\underline{X}) = 1$ implies that H_0 is rejected when the observed data are in the critical region. The function

$$
\beta_\phi(\theta) = E_\theta \phi(\underline{X}) = P_\theta[\underline{X} \in C] + E_\theta \gamma(\underline{X})
$$

is known as a power function of $\phi(\underline{X})$. A test is level α test if

$$
\beta_\phi(\theta) \le \alpha \;\; \forall \; \theta \in \Theta_0 \;\; \Leftrightarrow \;\; \sup_{\theta \in \Theta_0} \beta(\theta) = \alpha.
$$

$\sup_{\theta \in \Theta_0} \beta_\phi(\theta)$ is known as a size of the test. Within a class of all level α tests, we seek to find a test for which the probability of type II error is minimum. We thus get the most powerful (MP) test and uniformly most powerful (UMP) test. These are defined below.

▶ Definition 1.2.7

Most Powerful Test: Suppose U_α is a class of all level α tests to test $H_0 : \theta \in \Theta_0$ against the alternative $H_1 : \theta = \theta_1 \in \Theta_1$. A test $\phi_0 \in U_\alpha$ is said to be the most powerful test against an alternative H_1 if $\beta_{\phi_0}(\theta_1) \ge \beta_\phi(\theta_1) \; \forall \; \phi \in U_\alpha$.

It is to be noted that in the above definition, we have a fixed value θ_1 of a parameter in Θ_1. Thus, this definition is sufficient if Θ_1 is a singleton set. In general Θ_1 consists of more than one point. If a given test is an MP test for every point in Θ_1, then we get an UMP test. The precise definition is given below.

▶ **Definition 1.2.8**

Uniformly Most Powerful Test: Suppose U_α is a class of all level α tests to test $H_0 : \theta \in \Theta_0$ against the alternative $H_1 : \theta \in \Theta_1$. A test $\phi_0 \in U_\alpha$ is said to be an uniformly most powerful test against an alternative H_1 if

$$\beta_{\phi_0}(\theta) \geq \beta_\phi(\theta) \ \forall \ \phi \in U_\alpha \text{ uniformly in } \ \theta \in \Theta_1.$$

In most of the cases, UMP tests do not exist. These exist for one-sided null and alternative hypotheses if the underlying distribution belongs to an exponential family. For details, one may refer to Lehmann and Romano [7].

We state below the Neyman-Pearson lemma, which is a fundamental lemma giving a general method for finding the MP test of a simple null hypothesis against a simple alternative.

Theorem 1.2.5

Neyman-Pearson Lemma: *Suppose X is a random variable with probability law* $f(x, \theta)$, *where* $\theta \in \Theta = \{\theta_0, \theta_1\}$. *Suppose we want to test* $H_0 : \theta = \theta_0$ *against the alternative* $H_1 : \theta = \theta_1$.

1. Any test of the form

$$\phi(\underline{x}) = \begin{cases} 1, & \text{if } f(x, \theta_1) > k f(x, \theta_0) \\ \gamma(x), & \text{if } f(x, \theta_1) = k f(x, \theta_0) \\ 0, & \text{if } f(x, \theta_1) < k f(x, \theta_0), \end{cases}$$

for some $k \geq 0$ *and* $0 \leq \gamma(x) \leq 1$ *is the most powerful test of its size. If* $k = \infty$, *then the test*

$$\phi(\underline{x}) = \begin{cases} 1, & \text{if } f(x, \theta_0) = 0 \\ 0, & \text{if } f(x, \theta_0) > 0 \end{cases}$$

is the most powerful test of its size.
2. Given $\alpha \in (0, 1)$, *there exists a test* ϕ *of one of the two forms given in (1) with* $\gamma(x) = \gamma$, *(a constant) for which* $E_{\theta_0}(\phi(X)) = \alpha$.

It can be shown that the MP test as given by the Neyman-Pearson lemma is unique. This lemma is also useful to find a UMP test, if it exists. Suppose a null hypothesis $H_0 : \theta = \theta_0$ is simple and alternative is $H_1 : \theta \in \Theta_1$. We derive an MP test for testing $H_0 : \theta = \theta_0$ against the alternative $H_1 : \theta = \theta_1$, where $\theta_1 \in \Theta_1$. If the critical region remains invariant to fixed $\theta_1 \in \Theta_1$, then this an UMP test. A likelihood ratio test procedure discussed in Chap. 5 is an extension of the idea behind the Neyman-Pearson lemma.

In most of the examples the critical region C can be expressed as $[T_n < c]$ or $[T_n > c]$ or $[|T_n| > c]$, where T_n is a function of observed data $\underline{X}$ and fixed values of parameters under the hypotheses under study. T_n is known as a test statistic. If the family $\{f(x, \theta), \theta \in \Theta\}$ admits a sufficient statistic, then it is desirable to have a test statistic as a function of the sufficient statistic. In the critical region $[T_n < c]$, c is known as a cut-off point and it is determined so that the size of the test is α, that is, $\sup_{\theta \in \Theta_0} P[T_n < c] = \alpha$. Thus, to determine the cut-off point we need to know the distribution of the test statistic in the null set up, which is usually referred to as null distribution of a test statistic. It may not be always possible to find the null distribution for a finite sample size n. However, in most of the cases it is possible to obtain a large sample null distribution. We elaborate on this in Chaps. 5 and 6.

Testing at a fixed level α as described above is one of the two standard approaches in any testing procedure. The other approach is based on a concept of a p-value. Reporting "reject H_0" or "do not reject H_0" is not very informative. Instead, it is better to know for every α, whether the test rejects the null hypothesis at that level. Generally, if the test rejects H_0 at level α it will reject at level $\alpha' > \alpha$ also. Hence, there is a smallest α at which the test rejects and this number is known as the p-value. It is also known as significance probability or observed level of significance. It is defined below.

▶ Definition 1.2.9

p-value: Suppose for every $\alpha \in (0, 1)$ we have a size α test with rejection region C_α and $T(\underline{X})$ is the corresponding test statistic. Then, p-value is defined as

$$\text{p-value} = \inf\{\alpha | T(\underline{X}) \in C_\alpha\}.$$

Thus, the p-value is the smallest level at which we can reject H_0. We illustrate the evaluation of the p-value when the critical region is of the type $[T(\underline{X}) > c]$. Suppose $T(\underline{x})$ is the observed value of the test statistic corresponding to given data. Then p-value is $P_{H_0}[T(\underline{X}) > T(\underline{x})]$. Thus, it is the probability of observing under H_0, a sample outcome at least as extreme as the one observed. Informally, the p-value is a measure of the evidence against H_0. The smaller the p-value, the more extreme the outcome and the stronger the evidence against H_0. If p-value is large, H_0 is not rejected. However, a large p-value is not strong evidence in favor of H_0, because a large p-value can occur for two reasons: (i) H_0 is true or (ii) H_0 is false but the test has low power.

The approach of p-value is a good practice, as in this approach we determine not only whether the hypothesis is accepted or rejected at the given significance level, but also the smallest significance level at which the hypothesis would be rejected for the given observation. p-value gives an idea of how strongly the data contradict the hypothesis. It also enables to reach a verdict based on the significance level of the choice of the experimenter or the researcher. For example, p-value is equal to 0.07 is roughly interpreted as in 7 out of 100 simulations, we commit the error of rejecting H_0 when in fact it is true. An experimenter or the researcher can decide whether this much error is acceptable and can take the decision accordingly. Similarly, p-value 0.03 may be large or small depending on the experiment under study. In most of the software, p-values are reported and the decision about the acceptance or rejection of the null hypothesis is left to the experimenter.

In all these results and procedures, the sample size n is assumed to be a fixed finite number. It has been noticed that for a very few inference problems there exists an exact, optimal solution for finite n. In some cases for finite n, optimality theory may not exist or may not give satisfactory results due to intractability of a problem. Asymptotic optimality theory resolves these issues in many cases. There is a lot of literature related to asymptotic inference theory. For example, we mention a few books here such as Casella and Berger [4], DasGupta [8], Ferguson [9], Kale and Muralidharan [5], Lehmann and Casella [6], Lehmann [10], Lehmann and Romano [7], Rao [11], Rohatgi and Saleh [2], Shao [3], Silvey [12] and van der Vaart [13].

In the present book in Chaps. 2 to 4, we discuss large sample optimality properties of the estimators. Chapters 5 and 6 are devoted to the discussion on the test procedures when sample size is large.

In the next section we discuss some basic concepts from probability theory which form a foundation of the asymptotic statistical inference and list some results which are frequently used in the proofs of the theorems and in the solutions of the problems in the present book. For details, one may refer to Athreya and Lahiri [14], Bhat [15], Gut [16] and Loeve [17].

1.3   Basics of Asymptotic Inference

Large sample optimality properties of an estimator are defined in terms of its limiting behavior as sample size increases and hence are based on the various modes of convergence of a sequence of random variables. Thus, the principal probability tool in asymptotic investigation is the convergence of a sequence of random variables. As sample size increases, we study the limiting behavior of a sequence $\{T_n, n \geq 1\}$ of estimators of θ and examine how close it is to θ, in some sense to be defined appropriately.

Suppose $\{T_n, n \geq 1\}$ is a sequence of estimators of θ, that is, for every $n \geq 1$, T_n is a measurable function of sample observations with range space as the parameter space Θ, which we assume to be a real line to begin with. As a consequence, for each realization in $\mathbb{X}$, T_n is a real number for each $n \geq 1$ and hence a sequence $\{T_n, n \geq 1\}$ is equivalent to a sequence $\{a_n, n \geq 1\}$ of real numbers. Thus, all techniques of convergence of a sequence of real numbers can be used to study the convergence of a sequence of random variables. However, a sequence $\{T_n, n \geq 1\}$ of random variables is equivalent to a collection of sequences of real numbers. This collection is finite, countable or uncountable depending on whether Ω is finite, countable or uncountable. Thus, to discuss convergence of a sequence of random variables, one has to deal with convergence of a collection of sequences of real numbers. In various modes of convergence, a sequence of random variables is reduced to a collection of sequences of real numbers in some suitable way. The different ways lead to different types of convergence, such as point-wise convergence, almost sure convergence, convergence in probability, convergence in law or convergence in distribution and convergence in r-th mean. We define these modes of convergence below for a sequence of random variables and use these to study the optimality properties of the estimator for a large sample size.

Suppose a sequence $\{X_n, n \geq 1\}$ of random variables and X are defined on the same probability space $(\Omega, \mathbb{A}, P)$.

▶ **Definition 1.3.1**

Almost Sure Convergence: Suppose $N \in \mathbb{A}$ is such that $P(N) = 0$. Then $\{X_n, n \geq 1\}$ is said to converge almost surely to a random variable X, denoted by $X_n \overset{a.s.}{\to} X$, if $X_n(\omega) \to X(\omega) \; \forall \; \omega \in N^c$, such that $P(N) = 0$.

The set N is known as a P-null set.

▶ **Definition 1.3.2**

Convergence in Probability: A sequence $\{X_n, n \geq 1\}$ is said to converge in probability to a random variable X, denoted by $X_n \overset{P}{\to} X$, if $\forall \; \epsilon > 0$,

$$P\{\omega| \; |X_n(\omega) - X(\omega)| < \epsilon\} = P[|X_n - X| < \epsilon] \to 1 \;\; \text{as} \;\; n \to \infty.$$

▶ **Definition 1.3.3**

Convergence in Law: A sequence $\{X_n, n \geq 1\}$ is said to converge in law to a random variable X, denoted by $X_n \overset{L}{\to} X$, if

$$F_n(x) = P[X_n \leq x] \to P[X \leq x] = F(x), \;\; \forall \, x \in C_F(x) \;\; \text{as} \;\; n \to \infty,$$

where $C_F(x)$ is a set of points of continuity of the distribution function F of X.

In this mode of convergence, it is not necessary that a sequence of random variables $\{X_n, n \geq 1\}$ and X are defined on the same probability space $(\Omega, \mathbb{A}, P)$.

▶ **Definition 1.3.4**

Convergence in r-th Mean: A sequence $\{X_n, n \geq 1\}$ is said to converge in r-th mean to a random variable X, denoted by $X_n \overset{r}{\to} X$, if for any $r \geq 1$,

$$E(|X_n - X|^r) \to 0 \;\; \text{as} \;\; n \to \infty,$$

provided $E(|X_n - X|^r)$ is defined. If $r = 2$ then the convergence is referred to as convergence in quadratic mean and is denoted by $X_n \overset{q.m.}{\to} X$.

In asymptotic inference setup $X_n = T_n$, where T_n is an estimator whose properties are to be investigated and $X = \theta$, θ being a parameter under study. Thus, in all these modes of convergence we judge closeness or proximity of T_n to θ for large values of n. In general, limit of a sequence of random variables is a random variable. But when we are interested in the limiting behaviour of a sequence $\{T_n, n \geq 1\}$ of estimators of θ, the limit random variable is degenerate at θ, that is, it is a constant. In asymptotic inference, all the modes of convergence such as almost sure convergence, convergence in probability, convergence in law and convergence in r-th mean are heavily used. Many desirable properties of the estimators are defined in terms of these modes of convergence. For example, if $T_n \xrightarrow{P_\theta} \theta$, $\forall \ \theta \in \Theta$ then T_n is said to be weakly consistent for θ. If $T_n \xrightarrow{a.s.} \theta$, $\forall \ \theta \in \Theta$ then T_n is said to be strongly consistent for θ.

There are a number of implications among various modes of convergence. We state these results below. We also list some results, lemmas and theorems from probability theory, which help to verify the consistency of an estimator and to find the asymptotic distribution of the estimator with suitable normalization.

▶ Result 1.3.1
Almost sure convergence implies convergence in probability, but in general the converse is not true. It is true if the sequence $\{X_n, n \geq 1\}$ is a monotone sequence of random variables (Gut [16], p. 213).

▶ Result 1.3.2
Convergence in r-th mean implies convergence in probability, but the converse is not true.

▶ Result 1.3.3
Convergence in probability implies convergence in law. Convergence in probability and convergence in law are equivalent if the limit random variable is degenerate.

▶ Result 1.3.4
A limit random variable in convergence in probability and in almost sure convergence is almost surely unique, that is, if $X_n \xrightarrow{P} X$ and $X_n \xrightarrow{P} Y$, then X and Y are equivalent random variables, that is $X = Y$ a.s. Similarly if $X_n \xrightarrow{a.s.} X$ and $X_n \xrightarrow{a.s.} Y$, then $X = Y$ a.s.

▶ Result 1.3.5
If $X_n \xrightarrow{L} X$ and $X_n \xrightarrow{L} Y$, then X and Y are identically distributed random variables, that is $F_X(x) = F_Y(x) \ \forall \ x \in \mathbb{R}$.

▶ Result 1.3.6
Almost sure convergence and convergence in probability are closed under all arithmetic operations as stated below. Suppose $\{X_n, n \geq 1\}$, X, $\{Y_n, n \geq 1\}$ and Y are defined on the same probability space $(\Omega, \mathbb{A}, P)$. If $X_n \xrightarrow{a.s.} X$ and $Y_n \xrightarrow{a.s.} Y$, then
(i) $X_n \pm Y_n \xrightarrow{a.s.} X + Y$.
(ii) $X_n Y_n \xrightarrow{a.s.} XY$.
(iii) $X_n/Y_n \xrightarrow{a.s.} X/Y$, provided X_n/Y_n and X/Y are defined.
Same is true for convergence in probability.

▶ Result 1.3.7
If $X_n - Y_n \xrightarrow{P} 0$ and if $X_n \xrightarrow{P} X$, then $Y_n \xrightarrow{P} X$.

▶ **Result 1.3.8**

If $X_n - Y_n \overset{P}{\to} 0$ and if $X_n \overset{L}{\to} X$, then $Y_n \overset{L}{\to} X$.

▶ **Definition 1.3.5**

A sequence of random variables $\{X_n, n \geq 1\}$ is said to be bounded in probability if for any $\epsilon > 0$, there exists a constant K and an integer n_0 such that

$$P[|X_n| \leq K] \geq 1 - \epsilon \quad \forall \ n \geq n_0 .$$

It can be shown that a real random variable is always bounded in probability.

▶ **Result 1.3.9**

If $X_n \overset{P}{\to} X$, where X is a real random variable, then the sequence $\{X_n, n \geq 1\}$ is bounded in probability.

▶ **Result 1.3.10**

If $X_n \overset{L}{\to} X$, where X is a real random variable, then the sequence $\{X_n, n \geq 1\}$ is bounded in probability.

▶ **Result 1.3.11**

If $\{X_n, n \geq 1\}$ is bounded in probability and if $Y_n \overset{P}{\to} 0$, then $X_n Y_n \overset{P}{\to} 0$.

▶ **Result 1.3.12**

Slutsky's Theorem: Suppose $\{X_n, n \geq 1\}$ and $\{Y_n, n \geq 1\}$ are sequences of random variables defined on the same probability space $(\Omega, \mathbb{A}, P)$. If $X_n \overset{L}{\to} X$ and $Y_n \overset{L}{\to} C$ (or $Y_n \overset{P}{\to} C$), then

(i) $X_n Y_n \overset{P}{\to} 0$, if $C = 0$.

(ii) $X_n \pm Y_n \overset{L}{\to} X \pm C$.

(iii) $X_n Y_n \overset{L}{\to} XC$.

(iv) $X_n / Y_n \overset{L}{\to} X/C$, provided X_n/Y_n and X/C are defined.

▶ **Result 1.3.13**

Suppose g is a continuous function. Then

(i) If $X_n \overset{a.s.}{\to} X \implies g(X_n) \overset{a.s.}{\to} g(X)$.

(ii) If $X_n \overset{P}{\to} X \implies g(X_n) \overset{P}{\to} g(X)$.

(iii) If $X_n \overset{L}{\to} X \implies g(X_n) \overset{L}{\to} g(X)$.

The third result is known as a continuous mapping theorem.

▶ **Result 1.3.14**

Borel-Cantelli Lemma: Suppose $\{A_n, n \geq 1\}$ is a sequence of events defined on $(\Omega, \mathbb{A}, P)$. If $\sum_{n=1}^{\infty} P(A_n) < \infty$ then $P(\limsup A_n) = 0$.

▶ **Result 1.3.15**

Khintchine's Weak Law of Large Numbers (WLLN): Suppose $\{X_n, n \geq 1\}$ is a sequence of independent and identically distributed random variables with finite mean μ, then $S_n/n = \overline{X}_n \xrightarrow{P} \mu$.

▶ **Result 1.3.16**

Kolmogorov's Strong Law of Large Numbers (SLLN): Suppose $\{X_n, n \geq 1\}$ is a sequence of independent and identically distributed random variables with finite mean μ, then $S_n/n = \overline{X}_n \xrightarrow{a.s.} \mu$.

▶ **Result 1.3.17**

Lindeberg-Levy CLT: Suppose $\{X_n, n \geq 1\}$ is a sequence of independent and identically distributed random variables with mean μ and positive, finite variance σ^2, then

$$Y_n = \frac{\sum_{i=1}^{n} X_i - n\mu}{\sqrt{n}\sigma} = \frac{S_n - n\mu}{\sqrt{n}\sigma} = \frac{\sqrt{n}(\overline{X}_n - \mu)}{\sigma} \xrightarrow{L} Z \sim N(0,1).$$

From the above results we note that almost sure convergence implies convergence in probability which further implies convergence in law. Thus if $T_n \xrightarrow{P_\theta} \theta$, then $T_n \xrightarrow{L} \theta$, but then the limiting distribution is degenerate and hence is not informative. We study the convergence in distribution of a suitably normalized T_n to get a limiting non-degenerate distribution. Such a limiting non-degenerate distribution is useful to find a large sample interval estimator for θ and for testing hypotheses about θ. Lindeberg-Levy CLT, Slutsky's theorem, Khintchine's WLLN and Kolmogorov's SLLN are heavily used to find the asymptotic non-degenerate distribution of an estimator and the asymptotic null distribution of a test statistic.

Some of the most useful tools in probability theory and inference are moment inequalities. We state below some of these which are needed in the proofs of some theorems in Chaps. 2 to 6.

▶ **Inequality 1.3.1**

If $E(|X|^m) < \infty$, then $E(|X|^r) < \infty$, for $0 < r \leq m$. Thus, if a moment of a certain order is finite, then all the moments of lower order are also finite.

▶ **Inequality 1.3.2**

Schwarz Inequality: $E(|XY|) \leq \sqrt{E(|X|^2)E(|Y|^2)}$.

▶ **Inequality 1.3.3**

Jensen's Inequality: If $f(\cdot)$ is a convex function and if $E(X)$ is finite, then $f(E(X)) \leq E(f(X))$. If $f(\cdot)$ is a concave function, then $f(E(X)) \geq E(f(X))$.

▶ **Inequality 1.3.4**

Basic Inequality: Suppose X is an arbitrary random variable and $g(\cdot)$ is a non-negative Borel function on $\mathbb{R}$. If $g(\cdot)$ is even and non-decreasing on $[0, \infty)$, then

$$\forall\ a > 0, \quad \frac{E(g(X)) - g(a)}{M} \ \leq\ P[|X| \geq a] \ \leq\ \frac{E(g(X))}{g(a)},$$

where M denotes the almost sure supremum of $g(X)$.

▶ **Inequality 1.3.5**

Chebyshev's Inequality: $P(|X| \geq a) \ \leq\ E(X^2)/a^2$.

▶ **Inequality 1.3.6**

Markov Inequality: $P(|X| \geq a) \ \leq\ E(X^r)/a^r$, $r > 0$.

We now state the inverse function theorem from calculus. It is heavily used in the proofs of theorems in Chaps. 3 and 4, to examine whether inverse of some parametric function exists and has some desirable properties.

Theorem 1.3.1

Inverse Function Theorem: *Suppose D denotes the class of totally differentiable functions, that is a class of functions whose components have continuous partial derivatives. Suppose $\underline{f} = (f_1, f_2, \ldots, f_n) \in D$ is defined on an open set S in $\mathbb{R}^n$ and suppose $T = \underline{f}(S)$ is the range space. If the jacobian $J_{\underline{f}}(a) \neq 0$ for some $a \in S$, then there are two open sets $M \subset S$ and $N \subset T$ and a uniquely determined function $\underline{g}$ such that (i) $a \in M$ & $\underline{f}(a) \in N$, (ii) $N = \underline{f}(M)$, (iii) $\underline{f}$ is one to one on M, (iv) $\underline{g}$ is defined on N, $\underline{g}(N) = M$ and $\underline{g}(\underline{f}(x)) = x \ \forall\ x \in M$ and (v) $\underline{g} \in D$ on N.*

Inverse function theorem basically states that if the jacobian is non-zero then the unique inverse exists. In addition, if the given function is totally differentiable then the inverse function is also totally differentiable.

All the modes of convergence and related results and theorems listed above are heavily used in asymptotic statistical inference to establish the asymptotic optimality properties of the estimators. A major result in asymptotic inference theory is that for some smooth probability models, maximum likelihood estimators are asymptotically optimal, in the sense that these are consistent and asymptotically normal with suitable normalization. Moreover, the variance of a maximum likelihood estimator asymptotically attains the Cramér-Rao lower bound. Thus, asymptotic theory justifies the use of the method maximum likelihood estimation in certain situations and hence it is the most frequently used method. Wald's test procedure, a likelihood ratio test procedure and a score test procedure are the three major approaches of constructing tests of significance for parameters in statistical models. The asymptotic null distributions of these test statistics are heavily based on the results related to maximum likelihood estimation. We discuss all these results in detail in the following chapters.

An important feature of asymptotic inference is that it is non-parametric, in the sense that whatever may be the distribution of X, if its mean $E(X) = g(\theta)$, say, is finite, then a sample mean $\overline{X}_n$ based on a random sample from the distribution of X, converges almost surely and in probability to $g(\theta)$. In addition if the variance is positive and finite, then by the central limit theorem one can obtain the large sample distribution of a suitably normalized estimator of θ. All such limit theorems provide distribution-free approximations for statistical quantities such as significance levels, critical values, power, confidence coefficients, and so on. However, the accuracy of these approximations is not distribution-free, it very much depends both on the sample size, on the underlying distribution as well as on the values of parameters. These are some limitations of the asymptotic inference theory. Although asymptotic inference is both practically useful and of theoretical importance, it basically gives approximations. It is difficult to assess these approximations theoretically but can be judged by simulation. Thus, one of the ways to judge the approximation errors is to supplement the theoretical results by a simulation work. This is a crucial aspect of asymptotic inference theory.

The novelty of this book is use of R software (see R [18]) to illustrate such an important feature. The last section of every chapter is devoted to the application of R software to evaluate the performance of estimators and test procedures by simulation, to obtain solutions of the likelihood equations, to carry out the likelihood ratio test procedures, goodness of fit test procedures and tests for contingency tables. Moreover, it is also helpful to clarify the concepts of consistency and asymptotic distributions of the estimators.

Some readers may be familiar with R software as it has been introduced in the curriculum of many under-graduate and post-graduate statistics programs. In the following section we give a brief introduction to R, which will be useful to beginners. We have also tried to make the codes given in the last sections of Chaps. 2 to 6 to be self explanatory.

1.4   Introduction to R Software and Language

In statistical analysis phase one needs a good statistical software to carry out a variety of computations and draw different types of graphs. There are a number of software available for the computation such as Excel, Minitab, Matlab and SAS. In last two decades R software is strongly advocated and a large proportion of the world's leading statisticians use it for statistical analysis. It is a high-level language and an environment for data analysis and graphics, created by Ross Ihaka and Robert Gentleman in 1996. It is both a software and a programming language considered as a dialect of the S language developed by AT & T Bell Laboratories. The current R software is the result of a collaborative effort with contributions from all over the world. It has become very popular in academics and also in corporate world for variety reasons, such as its good computing performance, excellent built-in help system, flexibility in graphical environment, its vast coverage, availability of new, cutting edge applications in many fields and scripting and interfacing facilities. The most important advantage is that in spite of being the finest integrated software, it is freely available software from the site called CRAN (Comprehensive R Archive Network) with address http://cran.r-project.org/. From this site one need to "Download and Install R" by running the appropriate pre-compiled binary distributions. When R is installed properly, you will see the R icon on your desktop/laptop. To start R, one has to click on the R icon. The data analysis in R proceeds as an interactive dialogue with the interpreter. As soon as we type a command at the prompt (>), and press the enter key, the interpreter responds by executing the command. The session is ended by typing q(). The latest version of R is 4.0.2 released on June 22, 2020.

With this non-statistical part of the introduction to R, we now proceed to discuss how it is used for statistical analysis. The discussion is not exhaustive but we restrict to the functions or commands which are repeatedly used in this book. Like any other programming language, R contains data structures. Vectors are the basic data structures in R. The standard arithmetic functions and operators apply to vectors on an element-wise basis, with usual hierarchy. Below we state some functions which we need to write R codes for the concepts discussed in this book. The most useful R command for entering small data sets is the c ("combine") function. This function combines or concatenates terms together. In the following code, we specify some such basic functions with their output. One can use any variable names, but care is to be taken as R is case sensitive.

```
x=c(10, 23, 35, 49, 52, 67) ### c function to construct a vector with
                            given elements
x ### displays x, print(x) also displays the object x
length(x) ## specifies a number of elements in x
y=1:5; y ### constructs a vector with consecutive elements and
 prints it, two commands can be given on the same line
 with separator ";"
u=seq(10,25,5); u ## sequence function to create a vector with first
                element 10, last element 25 and with increment 5
v=c(rep(1,3),rep(2,2),rep(3,5)); v ## rep function to create a
                vector where 1 is repeated thrice, 2 twice and 3 five times
m=matrix(c(10, 23, 35, 49, 52, 67),nrow=2,ncol=3); m ### matrix with
 2 rows and 3 columns, with first two elements forming
                first column and so on
t(m) ### transpose of matrix m
m1=matrix(c(10, 23, 35, 49,52, 67),nrow=2,ncol=3,byrow=T);
 m1 ### with additional argument byrow=T, we get matrix with 2 rows
 and 3 columns, with first three elements forming first row and
                next three forming second row.
 #### Output
> x=c(10, 23, 35, 49, 52, 67)
> x
[1] 10 23 35 49 52 67
> length(x)
[1] 6
> y=1:5; y
> y
```

```
[1] 1 2 3 4 5
> u=seq(10,25,5); u
> u
[1] 10 15 20 25
> v=c(rep(1,3),rep(2,2),rep(3,5)); v
> v
 [1] 1 1 1 2 2 3 3 3 3 3
> m=matrix(c(10, 23, 35, 49, 52, 67),nrow=2,ncol=3); m
     [,1] [,2] [,3]
[1,]   10   35   52
[2,]   23   49   67
> t(m)
     [,1] [,2]
[1,]   10   23
[2,]   35   49
[3,]   52   67
> m1=matrix(c(10, 23, 35, 49, 52, 67),nrow=2,ncol=3,byrow=T); m1
     [,1] [,2] [,3]
[1,]   10   23   35
[2,]   49   52   67
```

For many probability distributions, to find the values of probability law or distribution function for specified values, to draw random samples from these distributions, R has excellent facility, specified in following four types of functions. The d function returns the probability law of the distribution, whereas the p function gives the distribution function of the distribution. The q function gives the quantiles, and the r function returns random samples from a distribution. Each family has a name and some parameters. The function name is found by combining either d, p, q or r with the name for the family. The parameter names vary from family to family but are consistent within a family. These functions are illustrated for the uniform $U(2,4)$ distribution in the following.

```
dunif(c(2.5,3.3,3.9),2,4) ###  probability density function at
                                          2.5,3.3,3.9
punif(c(2.5,3.3,3.9),2,4) ###  distribution function at 2.5,3.3,3.9
qunif(c(.25,.5,.75),2,4)  ###  first, second and third quartiles
r=runif(5,2,4)            ###  random  sample of size 5,
                               stored in object r
round(r,2)                ###  values in r rounded to
                               second decimal point

### Output
> dunif(c(2.5,3.3,3.9),2,4)
[1] 0.5 0.5 0.5
> punif(c(2.5,3.3,3.9),2,4)
[1] 0.25 0.65 0.95
> qunif(c(.25,.5,.75),2,4)
```

```
[1] 2.5 3.0 3.5
> r=runif(5,2,4)
> round(r,2)
[1] 2.62 3.56 2.80 2.45 2.30
```

We use functions `rbinom`, `rnorm` to draw random samples from binomial and normal distributions. Thus, we need to change the family name and add appropriate parameters. We can get the names for all probability distributions by following the path on R console as help → manuals (in pdf) → An Introduction to R → Probability distributions.

The function `round(r,2)` prints the values of r, rounded to the second decimal point. It is to be noted that rounding is mainly for printing purpose, original unrounded values of r are stored in the object r.

There are some useful built-in functions. We illustrate commonly used functions with a data set stored in variable x.

```
x=rnorm(25,3,2) ### random sample of size n=25 from normal distribution
                    with mean 3 and standard deviation 2
mean(x); median(x); max(x); min(x); sum(x); cumsum(x)### cumulative sum
var(x)### divisor is (n-1) and not n
quantile(x,c(.25,.5,.75)) ### three quartiles
summary(x) ### gives minimum, maximum, three quartiles and mean
shapiro.test(x) ###  Shapiro-Wilk test for normality, gives value of
                    statistic and p-value
### Partial output
> summary(x)
   Min. 1st Qu.  Median    Mean 3rd Qu.    Max.
  -1.387   2.250   3.426   3.356   4.360   7.978
> shapiro.test(x)
W = 0.97969, p-value = 0.8789
```

It is to be noted that we have drawn a random sample from normal distribution and from the p-value of Shapiro-Wilk test normality, normality is accepted, as is expected. We can carry out number of test procedures on similar lines, manual from help menu lists some of these. Apart from many built-in functions, one can write a suitable function as required in a specific situation. An illustration is given below after discussing commands to draw various types of plots.

Graphical features of R software have a remarkable variety. Each graphical function has a large number of options making production of graphics very flexible. Graphical device is a graphical window or a file. There are two kinds of graphical functions— the high-level plotting functions, which create a new graph, and low-level plotting functions, which add elements to an already existing graph. The standard high-level plotting functions are `plot()` function for scatter plot, `hist()` function for histogram and `boxplot()` function for box plot, etc. The lower level plotting functions are `lines()` to impose curves on the existing plot, `abline()` to add a line with given intercept and slope, `points()` to add points at appropriate places etc. These functions take extra arguments that control the graphic. The graphs are produced with respect to graphical parameters, which are defined by default and can be modified with the function "par". If we type `?par` on R console, we get the description on number of arguments for graphical functions, as documented in R. We explain one among these which is frequently used in the book. It is `par(mfrow=c(2,2))` or `par(mfcol=c(2,2))`. This command divides the graphical window invisibly in 2 rows and 2 columns to accommodate 4 graphs. A function legend() is usually added in plots to specify a list of symbols or colors used in the graphs. These features will be clear from a variety of graphs drawn in the last section of the subsequent chapters.

We now illustrate how to draw an histogram, a boxplot, how to impose additional curves on these plots using lines function. We use `rnorm(n,th,si)` function to generate a random sample of size n from normal distribution with mean θ and standard deviation σ. It is to be noted that the third argument of the function `rnorm(n,th,si)` is a value of the standard deviation and not the variance. We draw an histogram of $n = 120$ generated observations with $\theta = 1$ and $\sigma = 2$, using `hist` function, with relative frequency on y-axis. It is achieved by the argument `freq=FALSE` of the `hist` function. On this plot using `lines` function we impose the curve of probability density function of $N(1, 2^2)$ distribution. We use `dnorm(r,th,si)` function to obtain values of ordinates at r running from $\theta - 3\sigma$ to $\theta + 3\sigma$ as probability that observation lies in this interval is 0.9973. We use the function `seq` to generate a sequence of r values. The function `boxplot` draws the boxplot of 120 generated observations. We adopt the same steps for the random sample generated from χ_2^2 distribution to draw these plots. Normal distribution is a symmetric distribution while χ_2^2 distribution is an asymmetric distribution. Four plots drawn in one window using `par(mfrow= c(2,2))` function in Fig. 1.2 display these features.

```
n=120 ### sample size th = 1; si=2 ### mean 1 and standard deviation 2
x=rnorm(n,th,si) ### sample of size n from N(1,2^2) distribution
r = seq(th-3*si,th+3*si,.2) ### sequence of points on x axis at which
       to find ordinates
y = dnorm(r,th,si) ### ordinates of N(1,2^2) distribution at r
df=2 ### parameter in terms of degrees of freedom for
                             chi-square distribution
u=rchisq(n,df) ### sample of size n from chi-square distribution
                       with 2 df
v = seq(0,12,.3) ### sequence of points on x axis at which
                       to find ordinates
w = dchisq(v,df) ### ordinates of chi-square distribution with 2 df
par(mfrow= c(2,2)) ### divides the graphical window in four panels
hist(x,freq=FALSE,main="Histogram",xlab = "X", col="light blue")
lines(r,y,"o",pch=20,col="dark blue")
boxplot(x,main="Box Plot")
hist(u,freq=FALSE,main="Histogram",xlab = "X", col="light blue")
lines(v,w,"o",pch=20,col="dark blue")
boxplot(u,main="Box Plot")
```

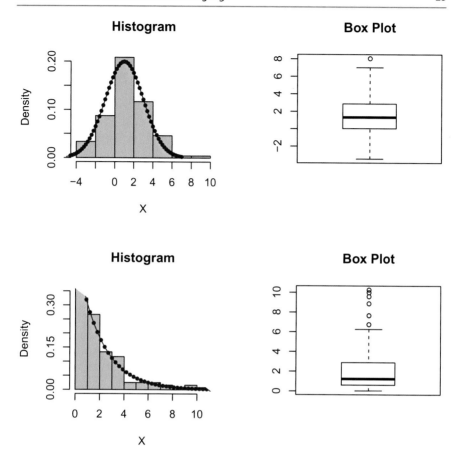

Fig. 1.2 Histogram and box plot

From the graph, we note that the curve of probability density of normal distribution is a close approximation to the histogram with relative frequencies. Box plot indicates the symmetry around the median 1 and range of the simulated values of X is approximately $(-4, 7)$. The curve of probability density is a close approximation to the histogram with relative frequencies for the chi-square distribution also. Further, the asymmetry of the chi-square distribution is reflected in the box plot. If we run the same code, we may not get exactly the same graphs, as the sample generated will be different. If we want to have the same sample to be generated each time we run the code, we have to fix the seed. It can be done by the function `set.seed(2)`, 2 is an arbitrary number and it can be replaced by any other number. We explain an important role of `set.seed` function in Sect. 2.7.

So far we discussed how to use built-in functions of R. In the following code we illustrate how to write our own functions, with the help of the R code used to draw Fig. 1.1. It includes a function written to find probability mass function of binomial $B(5, p)$ distribution for various values of p, plot function, points function and `par(mfrow=c(3,3))` function.

```
g=function(p)
{
x=0:5
g=dbinom(x,5,p)
return(g)
}
par(mfrow=c(3,3))
p=seq(.1,.9,.1)
pname=paste("p =",p,sep=" ")
for(i in 1:length(p))
{
x=0:5
plot(x,g(p[i]),type="h", xlab="x", ylab = "Probability",main= pname[i],
     ylim=c(0,0.7),col="blue",lwd=2)
points(x,g(p[i]),pch=16,col="dark blue")
}
```

Observe the use of `seq` function to create a vector of p-values from 0.1 to 0.9 with an increment of 0.1. The command `pname` in the above code is used to assign a title giving the value of p for each of the nine graphs. Within a loop `plot` function and `points` function are used to have a panel of nine graphs of probability mass function of binomial distribution. Note the arguments in a `plot` function, `type=h` produces vertical lines with height proportional to the probability at that point, `xlab="x"` and `ylab = "Probability"` assign labels on x-axis and y-axis, `main=` assigns title to the graph, `ylim=c(0,0.7)` specifies the lower and upper limit on the y-axis, `col="blue"` specifies the color of the vertical lines and `lwd=2` determines the width of the line. In `points` function, `pch=16` decides the point characteristic, that is, type of points, there are such 25 types specified by the numbers 1–25.

There are a number of excellent books on introduction to statistics using R, such as Crawley [19], Dalgaard [20], Purohit et al. [21] and Verzani [22]. There is a tremendous amount of information about R on the web at `http://cran.r-project.org/` with a variety of R manuals. Following are some links useful for beginners to learn R software.

1. https://www.datacamp.com/courses/free-introduction-to-r
2. http://www.listendata.com/p/r-programming-tutorials.html
3. http://www.r-tutor.com/r-introduction
4. https://www.r-bloggers.com/list-of-free-online-r-tutorials/
5. https://www.tutorialspoint.com/r/
6. https://www.codeschool.com/courses/try-r

As for any software or programming language, best way to learn R is to use it for understanding the concepts and solving problems. We hope that this brief introduction will definitely be useful to a reader to be comfortable with R and follow the codes written in subsequent chapters.

In the next chapter we discuss the concept of consistency of an estimator in a real and a vector parameter setup, along with some methods to generate consistent estimators.

References

1. Rao, B. L. S. P. (1992). *Identifiability in stochastic models: Characterization of probability distributions*. Cambridge: Academic Press.
2. Rohatgi, V. K., & Saleh, A. K. Md. E. (2001). *Introduction to probability and statistics*. New York: Wiley.
3. Shao, J. (2003). *Mathematical statistics* (2nd ed.). New York: Springer.
4. Casella, G., & Berger, R. L. (2002). *Statistical inference* (2nd ed.). USA: Duxbury.
5. Kale, B. K., & Muralidharan, K. (2016). *Parametric inference: An introduction*. Delhi: Narosa.
6. Lehmann, E. L., & Casella, G. (1998). *Theory of point estimation* (2nd ed.). New York: Springer.
7. Lehmann, E. L., & Romano, J. P. (2005). *Testing of statistical hypothesis* (3rd ed.). New York: Springer.
8. DasGupta, A. (2008). *Asymptotic theory of statistics and probability*. New York: Springer.
9. Ferguson, T. S. (1996). *A course in large sample theory*. London: Chapman and Hall.
10. Lehmann, E. L. (1999). *Elements of large sample theory*. New York: Springer.
11. Rao, C. R. (1978). *Linear statistical inference and its applications*. New York: Wiley.
12. Silvey, S. D. (1975). *Statistical inference*. London: Chapman and Hall.
13. van der Vaart, A. (1998). *Asymptotic statistics*. Cambridge: Cambridge University Press.
14. Athreya, K. B., & Lahiri, S. N. (2006). *Measure theory and probability theory*. New York: Springer.
15. Bhat, B. R. (1999). *Modern probability theory* (3rd ed.). New Delhi: New Age International.
16. Gut, A. (2005). *Probability: A graduate course*. New York: Springer.
17. Loeve, M. (1978). *Probability theory I* (4th ed.). New York: Springer.
18. R Core Team. (2019). *R: A language and environment for statistical computing*. Vienna, Austria: R Foundation for Statistical Computing. https://www.R-project.org/.
19. Crawley, M. J. (2007). *The R book*. London: Wiley.
20. Dalgaard, P. (2008). *Introductory statistics with R* (2nd ed.). New York: Springer.
21. Purohit, S. G., Gore, S. D., & Deshmukh, S. R. (2008). *Statistics using R* (2nd ed.). New Delhi: Narosa Publishing House.
22. Verzani, J. (2005). *Using R for introductory statistics*. New York: Chapman and Hall/CRC Press.

Consistency of an Estimator

2

Contents

> ━ **Learning Objectives** After going through this chapter, the readers should be able
> - to comprehend the concept of consistency of an estimator for a real and vector valued parameter
> - to compare performance of consistent estimators based on different criteria such as mean squared error and coverage probability
> - to verify consistency of an estimator using R

2.1 Introduction

As discussed in Chap. 1, in asymptotic inference theory, we study the limiting behavior of a sequence $\{T_n, n \geq 1\}$ of estimators of θ and examine how close it is to θ using various modes of convergence. The most frequently investigated large sample property of an estimator is weak consistency. Weak consistency of an estimator is defined in terms of convergence in probability. We examine how close the estimator is to the true parameter value in terms of probability of proximity. Weak consistency

© The Author(s), under exclusive license to Springer Nature Singapore Pte Ltd. 2021
S. Deshmukh and M. Kulkarni, *Asymptotic Statistical Inference*,
https://doi.org/10.1007/978-981-15-9003-0_2

is always referred to as consistency in literature. In the next section, we define it for a real parameter and illustrate by a variety of examples. We study some properties of consistent estimators, the most important being the invariance of consistency under continuous transformation. Strong consistency and uniform consistency of an estimator are discussed briefly in Sects. 2.3 and 2.4. In Sect. 2.5, we define consistency when the distribution of a random variable or a random vector is indexed by a vector parameter. It is defined in two ways as marginal consistency and joint consistency, the two approaches are shown to be equivalent. This result is heavily used in applications. Thus, to obtain a consistent estimator for a vector parameter, one can proceed marginally and use all the tools discussed in Sect. 2.2. From examples in Sects. 2.2 and 2.5, we note that, for a given parameter, one can have an uncountable family of consistent estimators and hence one has to deal with the problem of selecting the best from the family. It is discussed in Sect. 2.6. Within a family of consistent estimators of θ, the performance of a consistent estimator is judged by the rate of convergence of a true coverage probability to 1 and of MSE to 0 for a consistent estimator whose MSE exists, faster the rate better is the estimator. Section 2.7 is devoted to the verification of the consistency of an estimator by simulation. It is illustrated through some examples and R software.

2.2   Consistency: Real Parameter Setup

Suppose X is a random variable or a random vector defined on a probability space $(\Omega, \mathbb{A}, P_\theta)$, where the probability measure P_θ is indexed by a parameter $\theta \in \Theta \subset \mathbb{R}$. Suppose $\underline{X} \equiv \{X_1, X_2, \ldots, X_n\}$ is a random sample from the distribution of X and $T_n \equiv T_n(\underline{X})$ is an estimator of θ. Weak consistency of T_n is defined below.

▶ **Definition 2.2.1**
Weakly Consistent Estimator: A sequence $\{T_n, n \geq 1\}$ of estimators of θ is said to be weakly consistent for θ if for each $\theta \in \Theta$, $T_n \overset{P_\theta}{\to} \theta$, that is, given $\epsilon > 0$ and δ in $(0, 1)$, $\exists\ n_0(\epsilon, \delta, \theta)$ such that

$$P_\theta[|T_n - \theta| > \epsilon] < \delta, \quad \Leftrightarrow \quad P_\theta[|T_n - \theta| < \epsilon] \geq 1 - \delta, \quad \forall\ n \geq n_0(\epsilon, \delta, \theta).$$

Hence onwards, weakly consistent estimator will be simply referred to as a consistent estimator and, instead of saying that a sequence of estimators is consistent, we will say that an estimator T_n is consistent for θ. $P_\theta[|T_n - \theta| < \epsilon]$ is known as a coverage probability as it is the probability of the event that a random interval $(T_n - \epsilon, T_n + \epsilon)$ covers the true but unknown parameter θ. If it converges to 1 as $n \to \infty$, $\forall\ \theta \in \Theta$ and $\forall\ \epsilon > 0$, then T_n is a consistent estimator of θ. In other words, T_n is consistent for θ, if with very high chance, for large n, T_n and θ are close to each other. It is to be noted that in the definition of consistency, the probability of the event $[|T_n - \theta| > \epsilon]$ is obtained under P_θ probability measure, that is two θ's involved in $P_\theta[|T_n - \theta| < \epsilon]$ must be the same. We elaborate on this issue after Example 2.2.2.

An important feature to be noted is that there is a subtle difference between convergence in probability in probability theory and consistency in inference setup, although consistency of an estimator is essentially convergence in probability. When one says that $X_n \xrightarrow{P} X$, one deals with one probability measure specified by P. In the definition of consistency, T_n is a consistent estimator of θ if $T_n \xrightarrow{P_\theta} \theta$, $\forall\ \theta \in \Theta$ as $n \to \infty$. Thus, the definition of consistency deals with the entire family of probability measures indexed by θ. For each value of θ, the probability structure associated with the sequence $\{T_n, n \geq 1\}$ is different. The definition of consistency requires that for each possible value of θ, the probability structure is such that the sequence converges in probability to that value of θ. To emphasize such an important requirement, we use P_θ instead of P when we express consistency of T_n as an estimator of θ. Hence, we write $T_n \xrightarrow{P_\theta} \theta$ and not $T_n \xrightarrow{P} \theta$ as $n \to \infty$.

We now present a variety of examples to elaborate on the concept of consistency of an estimator based on a random sample $\underline{X} = \{X_1, X_2, \ldots, X_n\}$ from the distribution under study. In this and all the subsequent chapters, we denote the likelihood of θ given data $\underline{X}$ by $L_n(\theta|\underline{X})$ instead of $L_n(\theta|\underline{X} = \underline{x})$, where $\underline{x}$ denotes an observed realization of $\underline{X}$.

✐ Example 2.2.1

Suppose $\underline{X} = \{X_1, X_2, \ldots, X_n\}$ is a random sample of size n from a uniform $U(0, \theta)$ distribution, $\theta \in \Theta = (0, \infty)$. If $X \sim U(0, \theta)$, then its probability density function $f_X(x, \theta)$ is given by

$$f_X(x, \theta) = \begin{cases} 1/\theta, & \text{if } 0 \leq x \leq \theta \\ 0, & \text{otherwise.} \end{cases}$$

Hence, the likelihood function $L_n(\theta|\underline{X})$ is given by,

$$L_n(\theta|\underline{X}) = 1/\theta^n, \quad \text{if} \quad X_i \leq \theta, \ \forall\ i = 1, 2, \ldots, n \ \Leftrightarrow\ X_{(n)} \leq \theta \quad \text{and}$$
$$L_n(\theta|\underline{X}) = 0, \quad \text{if} \quad X_{(n)} > \theta.$$

Observe that the likelihood function is not a continuous function of θ and hence is not a differentiable function of θ. Thus, the routine calculus theory is not applicable. However, it is strictly decreasing over the interval $[X_{(n)}, \infty)$. Hence, it attains maximum at the smallest possible value of θ. The smallest possible value of θ given the data is $X_{(n)}$. Hence, the maximum likelihood estimator $\hat{\theta}_n$ of θ is given by $\hat{\theta}_n = X_{(n)}$. From the likelihood, it is clear that $X_{(n)}$ is a sufficient statistic for the family of uniform $U(0, \theta)$ distributions for $\theta > 0$. We establish the consistency of $\hat{\theta}_n$ by showing that its coverage probability converges to 1. To find the coverage probability, we need to find the distribution function of $X_{(n)}$. If $X \sim U(0, \theta)$ distribution, then it is easy to verify that the distribution function $F_{X_{(n)}}(x, \theta)$ of $X_{(n)}$ is given by,

$$F_{X_{(n)}}(x, \theta) = \begin{cases} 0, & \text{if} \quad x < 0 \\ (x/\theta)^n, & \text{if} \quad 0 \leq x < \theta \\ 1, & \text{if} \quad x \geq \theta. \end{cases}$$

Now for $\epsilon > 0$, the coverage probability is given by,

$$P_\theta[|X_{(n)} - \theta| < \epsilon] = P_\theta[\theta - \epsilon < X_{(n)} < \theta + \epsilon] = P_\theta[\theta - \epsilon < X_{(n)} < \theta]$$
$$= 1 \text{ if } \epsilon \geq \theta,$$

as $P_\theta[0 < X_{(n)} < \theta] = 1 \; \forall \; \theta \in \Theta$. For $\epsilon < \theta$, we have

$$P_\theta[|X_{(n)} - \theta| < \epsilon] = P_\theta[\theta - \epsilon < X_{(n)} < \theta] = 1 - F_{X_{(n)}}(\theta - \epsilon, \theta)$$
$$= 1 - \left(\frac{\theta - \epsilon}{\theta}\right)^n = 1 - \left(1 - \frac{\epsilon}{\theta}\right)^n$$
$$\to 1 \quad \text{as} \quad \frac{\epsilon}{\theta} < 1 \, .$$

Thus, we conclude that $X_{(n)}$ is a consistent estimator of θ. Using the expression for the coverage probability, we can obtain the minimum sample size n_0 so that $P_\theta[|X_{(n)} - \theta| < \epsilon] \geq 1 - \delta$, where $\epsilon > 0$ and $\delta \in (0, 1)$ are specified constants. Thus, for $\epsilon < \theta$,

$$P_\theta[|X_{(n)} - \theta| < \epsilon] \geq 1 - \delta \Rightarrow 1 - \left(\frac{\theta - \epsilon}{\theta}\right)^n \geq 1 - \delta$$
$$\Rightarrow \left(\frac{\theta - \epsilon}{\theta}\right)^n \leq \delta$$
$$\Rightarrow n \geq \log \delta / \log((\theta - \epsilon)/\theta) \, .$$

Hence, the minimum sample size is $n_0 = \left[\log \delta / \log((\theta - \epsilon)/\theta)\right] + 1$. If $\epsilon \geq \theta$, coverage probability is 1 and hence $n_0 = 1$. We now examine whether $X_{(1)}$ is consistent for θ. Instead of finding the coverage probability $P_\theta[|X_{(1)} - \theta| < \epsilon]$, we show that $X_{(1)} \xrightarrow{P_\theta} 0, \; \forall \; \theta \in \Theta$ and appeal to Result 1.3.4, which states that the limit random variable in convergence in probability is almost surely unique to arrive at the conclusion that $X_{(1)}$ is not consistent for θ. Observe that the distribution function $F_{X_{(1)}}(x, \theta)$ of $X_{(1)}$ is given by

$$F_{X_{(1)}}(x, \theta) = 1 - [1 - F_X(x, \theta)]^n = \begin{cases} 0, & \text{if} \quad x < 0 \\ 1 - (1 - x/\theta)^n, & \text{if} \quad 0 \leq x < \theta \\ 1, & \text{if} \quad x \geq \theta. \end{cases}$$

Now, for $\epsilon > 0$,

$$P_\theta[|X_{(1)} - 0| < \epsilon] = P_\theta[-\epsilon < X_{(1)} < \epsilon] = P_\theta[0 < X_{(1)} < \epsilon] = 1 \text{ if } \epsilon \geq \theta.$$

For $\epsilon < \theta$, we have

$$P_\theta[|X_{(1)} - 0| < \epsilon] = P_\theta[0 < X_{(1)} < \epsilon] = F_{X_{(1)}}(\epsilon, \theta) - F_{X_{(1)}}(0, \theta)$$
$$= 1 - \left(1 - \frac{\epsilon}{\theta}\right)^n \to 1 \quad \text{as} \quad \frac{\epsilon}{\theta} < 1.$$

Thus, $X_{(1)} \xrightarrow{P_\theta} 0$, $\forall$ $\theta \in \Theta$. From Result 1.3.4, $X_{(1)}$ cannot converge in probability to θ. Thus, $X_{(1)}$ cannot be a consistent estimator of θ. However, $X_{(n)} + cX_{(1)} \xrightarrow{P_\theta} \theta$, as convergence in probability is closed under all arithmetic operations and hence $X_{(n)} + cX_{(1)}$ is also a consistent estimator of θ, where c is any real number such that range space of $X_{(n)} + cX_{(1)}$ is same as the parameter space $(0, \infty)$.

If $X \sim U(0, \theta)$, then $E(X) = \theta/2 < \infty$, hence the moment estimator $\tilde{\theta}_n$ of θ is given by, $\tilde{\theta}_n = 2\overline{X}_n$. By Khintchine's WLLN $\overline{X}_n \xrightarrow{P_\theta} E(X) = \theta/2$, $\forall$ $\theta \in \Theta$ and hence $\tilde{\theta}_n = 2\overline{X}_n \xrightarrow{P_\theta} \theta$, $\forall$ $\theta \in \Theta$ which proves that $\tilde{\theta}_n$ is a consistent estimator of θ. $\square$

In the following example, we examine the consistency of an estimator using different approaches.

✒ Example 2.2.2

Suppose $\{X_1, X_2, \ldots, X_n\}$ is a random sample of size n from a normal $N(\theta, 1)$ distribution, $\theta \in \mathbb{R}$. In this example, we illustrate various approaches to examine the consistency of the sample mean $\overline{X}_n$ as an estimator of a population mean θ. We use the result that

$$\overline{X}_n \sim N(\theta, 1/n) \text{ distribution } \Rightarrow Z = \sqrt{n}(\overline{X}_n - \theta) \sim N(0, 1) \text{ distribution.}$$

(i) The first approach is verification of consistency by the definition. For given $\epsilon > 0$,

$$P_\theta[|\overline{X}_n - \theta| < \epsilon] = P_\theta[\sqrt{n}|\overline{X}_n - \theta| < \sqrt{n}\epsilon]$$
$$= \Phi(\sqrt{n}\epsilon) - \Phi(-\sqrt{n}\epsilon) \rightarrow 1, \text{ as } n \rightarrow \infty, \forall \theta \in \mathbb{R}.$$

Thus, the coverage probability converges to 1 as $n \rightarrow \infty$, $\forall$ $\theta \in \Theta$ and $\forall$ $\epsilon > 0$, hence the sample mean $\overline{X}_n$ is a consistent estimator of θ.

(ii) Since $\overline{X}_n \sim N(\theta, 1/n)$ distribution,

$$E(\overline{X}_n - \theta)^2 = Var(\overline{X}_n) = 1/n \rightarrow 0, \text{ as } n \rightarrow \infty, \forall \theta \in \mathbb{R}.$$

Thus, $\overline{X}_n$ converges in quadratic mean to θ and hence converges in probability to θ.

(iii) Suppose $F_n(x)$, $x \in \mathbb{R}$ denotes the distribution function of $\overline{X}_n - \theta$. Then

$$F_n(x) = P_\theta[\overline{X}_n - \theta \leq x] = P_\theta[\sqrt{n}(\overline{X}_n - \theta) \leq \sqrt{n}x] = \Phi(\sqrt{n}x), \ x \in \mathbb{R}.$$

The limiting behavior of $F_n(x)$ as $n \rightarrow \infty$ is as follows:

$$F_n(x) \rightarrow \begin{cases} 0, & \text{if } x < 0 \\ 1/2, & \text{if } x = 0 \\ 1, & \text{if } x > 0. \end{cases}$$

Suppose F is a distribution function of a random variable which is degenerate at 0, then it is given by

$$F(x) \;=\; \begin{cases} 0, & \text{if } \; x < 0 \\ 1, & \text{if } \; x \geq 0. \end{cases}$$

It is to be noted that $F_n(x) \to F(x)$, $\forall \; x \in C_F(x) = \mathbb{R} - \{0\}$, where $C_F(x)$ is a set of points of continuity of F. It implies that $(\overline{X}_n - \theta) \overset{L}{\to} 0$, where the limit law is degenerate and hence, $(\overline{X}_n - \theta) \overset{P_\theta}{\to} 0$, for all $\theta \in \mathbb{R}$, which proves that $\overline{X}_n$ is consistent for θ.

(iv) Observe that $\{X_1, X_2, \ldots, X_n\}$ are independent and identically distributed random variables with finite mean θ, hence by Khintchine's WLLN, $\overline{X}_n \overset{P_\theta}{\to} \theta$, for all $\theta \in \mathbb{R}$.

As stated in Result 1.3.4, limit random variable in convergence in probability is almost surely unique, thus if $\overline{X}_n \overset{P_\theta}{\to} \theta$, then $\overline{X}_n$ cannot converge in probability to any other parametric function $g(\theta)$. Hence, $\overline{X}_n$ cannot be consistent for any other parametric function $g(\theta)$.   □

It is to be noted that in the above example, the first approach uses definition based on coverage probability, the second uses the result that convergence in r-th mean implies convergence in probability. In the third approach, we use the result that if a limit random variable is degenerate, then convergence in law implies convergence in probability. The last approach uses well known Khintchine's WLLN.

___

✍ Remark 2.2.1

In the first approach of the above example, we have shown that the coverage probability $P_\theta[|\overline{X}_n - \theta| < \epsilon]$ converges to 1 as $n \to \infty$. Here, θ is a true, but unknown, parameter value. Suppose we label $P_\theta[|\overline{X}_n - \theta| < \epsilon]$ as a probability of true coverage for each fixed ϵ. The probability $P_\theta[|\overline{X}_n - \theta_1| < \epsilon]$ is then labeled as a probability of false coverage as θ is a true parameter value and we compute the probability $\overline{X}_n$ is in a ϵ neighborhood of θ_1. Now consider

$$P_\theta[|\overline{X}_n - \theta_1| < \epsilon] = P_\theta[\sqrt{n}(-\epsilon+\theta_1 - \theta) < \sqrt{n}(\overline{X}_n - \theta) < \sqrt{n}(\epsilon+\theta_1 - \theta)]$$
$$= \Phi(\sqrt{n}(\epsilon + \theta_1 - \theta)) - \Phi(\sqrt{n}(-\epsilon + \theta_1 - \theta)).$$

Suppose $\theta_1 < \theta$, then for ϵ such that $\epsilon + \theta_1 - \theta < 0$, both the terms in the above expression converge to 0, while if $\theta_1 > \theta$, for $\epsilon > 0$, the first term converges to 1 and for ϵ such that $-\epsilon + \theta_1 - \theta > 0$, the second term also goes to 1 and hence the probability of false coverage $P_\theta[|\overline{X}_n - \theta_1| < \epsilon]$ converges to 0. Thus for $\theta = \theta_1$ only the probability converges to 1 for all $\epsilon > 0$. In the definition of a consistency of an estimator, it is expected that a probability of true coverage converges to 1.

If $\{X_1, X_2, \ldots, X_n\}$ is a random sample of size n from either a Bernoulli $B(1, \theta)$ or a Poisson $Poi(\theta)$ or a uniform $U(0, 2\theta)$ or an exponential distribution with mean θ, then the sample mean $\overline{X}_n$ can be shown to be consistent for θ using the four approaches discussed in Example 2.2.2 and the central limit theorem for independent and identically distributed random variables with positive finite variance.

The next example is again related to the consistency of θ when we have a random sample of size n from a normal $N(\theta, 1)$ distribution. However, the parameter space is not the entire real line, but a subset of real line, which may not be open. We come across such a parameter space while deriving likelihood ratio test procedures for testing $H_0 : \theta = \theta_0$ against $H_1 : \theta > \theta_0$ or for testing $H_0 : \theta \in [a, b]$ against $H_1 : \theta \notin [a, b]$. This example illustrates the basic concepts of consistency very well.

✐ Example 2.2.3

Suppose $\underline{X} \equiv \{X_1, X_2, \ldots, X_n\}$ is a random sample of size n from a normal $N(\theta, 1)$ distribution, $\theta \in \Theta$, when Θ is either (i) $\Theta = [0, \infty)$ or (ii) $\Theta = [a, b]$ or (iii) $\Theta = \{a, b\}$, $a < b \in \mathbb{R}$ or (iv) $\Theta = I$, the set of all integers. We find the maximum likelihood estimator of θ in each case and examine whether it is consistent for θ. Corresponding to a random sample $\underline{X}$ from normal $N(\theta, 1)$ distribution, the likelihood function of θ is given by

$$L_n(\theta|\underline{X}) = \prod_{i=1}^{n} \frac{1}{\sqrt{2\pi}} \exp\left\{-\frac{1}{2}(X_i - \theta)^2\right\}$$

$$= \left(\sqrt{2\pi}\right)^{-n} \exp\left\{-\frac{1}{2}\sum_{i=1}^{n}(X_i - \theta)^2\right\}.$$

The log likelihood function $Q(\theta) = \log L_n(\theta|\underline{X})$ and its first and second derivatives are given by,

$$Q(\theta) = c - \frac{1}{2}\sum_{i=1}^{n}(X_i - \theta)^2 \Rightarrow Q'(\theta) = \sum_{i=1}^{n}(X_i - \theta)$$

$$= n(\overline{X}_n - \theta) \text{ and } Q''(\theta) = -n,$$

where c is a constant free from θ. Thus, the solution of the likelihood equation $Q'(\theta) = 0$ is given by, $\theta = \overline{X}_n$. The second derivative is negative for all θ.
(i) We first find the maximum likelihood estimator $\hat{\theta}_n$ of θ when $\Theta = [0, \infty)$. If $\overline{X}_n \geq 0$, then it is an estimator and the likelihood is maximum at $\overline{X}_n$. As discussed in Sect. 1.1,

$$P_\theta[\overline{X}_n < 0] = P_\theta[\sqrt{n}(\overline{X}_n - \theta) < -\sqrt{n}\theta] = \begin{cases} \Phi(-\sqrt{n}\theta) > 0, & \text{if } \theta > 0 \\ 1/2, & \text{if } \theta = 0. \end{cases}$$

Thus, it is possible that $\overline{X}_n < 0$ and hence it cannot be an estimator of θ. Observe that

$$\overline{X}_n < 0 \leq \theta \Rightarrow \overline{X}_n - \theta < 0 \Rightarrow Q'(\theta) = n(\overline{X}_n - \theta) < 0$$

which further implies that $Q(\theta)$ is a decreasing function of θ. Thus, $Q(\theta)$ attains maximum at the smallest possible value of θ which is 0. Thus, the maximum likelihood estimator $\hat{\theta}_n$ of θ is given by

$$\hat{\theta}_n = \begin{cases} \overline{X}_n, & \text{if } \overline{X}_n \geq 0 \\ 0, & \text{if } \overline{X}_n < 0. \end{cases}$$

To verify the consistency, we proceed as follows. By WLLN $\overline{X}_n \overset{P_\theta}{\to} \theta$, for all $\theta \geq 0$. Now, for $\epsilon > 0$,

$$P_\theta[|\hat{\theta}_n - \overline{X}_n| < \epsilon] \geq P_\theta[\hat{\theta}_n = \overline{X}_n] = P_\theta[\overline{X}_n \geq 0]$$
$$= 1 - \Phi(-\sqrt{n}\theta) \to 1 \text{ if } \theta > 0 .$$

As a consequence, if $\theta > 0$, then $\hat{\theta}_n - \overline{X}_n \overset{P_\theta}{\to} 0$ and $\overline{X}_n \overset{P_\theta}{\to} \theta$ implies that $\hat{\theta}_n \overset{P_\theta}{\to} \theta$ when $\theta > 0$. Suppose $\theta = 0$. Then using the fact that $\hat{\theta}_n \geq 0$, for $\epsilon > 0$, as $n \to \infty$,

$$P_0[|\hat{\theta}_n| > \epsilon] = P_0[\hat{\theta}_n > \epsilon] = P_0[\overline{X}_n > \epsilon] = 1 - \Phi(\sqrt{n}\epsilon) \to 0 \Rightarrow \hat{\theta}_n \overset{P_0}{\to} 0.$$

Thus, it is proved that $\hat{\theta}_n \overset{P_\theta}{\to} \theta$ for all $\theta \geq 0$ and hence $\hat{\theta}_n$ is a consistent estimator of θ.

(ii) In this case, the parameter space Θ is $[a, b] \subset \mathbb{R}$. As in the case (i), if $\overline{X}_n \in [a, b]$ then only it can be labeled as an estimator. Note that in this case, the likelihood attains maximum at $\overline{X}_n$. However, for any $\theta \in [a, b]$, it is possible that $\overline{X}_n < a$ and $\overline{X}_n > b$. It is shown below.

$$P_\theta[\overline{X}_n < a] = P_\theta[\sqrt{n}(\overline{X}_n - \theta) < \sqrt{n}(a - \theta)]$$
$$= \begin{cases} \Phi(\sqrt{n}(a - \theta)) > 0 , & \text{if } a < \theta \leq b \\ 1/2 , & \text{if } \theta = a . \end{cases}$$

Along similar lines,

$$P_\theta[\overline{X}_n > b] = P_\theta[\sqrt{n}(\overline{X}_n - \theta) > \sqrt{n}(b - \theta)]$$
$$= \begin{cases} 1 - \Phi(\sqrt{n}(b - \theta)) > 0 , & \text{if } a \leq \theta < b \\ 1/2 , & \text{if } \theta = b . \end{cases}$$

Thus, if $\overline{X}_n \notin [a, b]$, then according to the definition of an estimator, it cannot be an estimator of θ. Suppose

$$\overline{X}_n < a \leq \theta \Rightarrow \overline{X}_n - \theta < 0 \Rightarrow Q'(\theta) = n(\overline{X}_n - \theta) < 0,$$

which further implies that $Q(\theta)$ is a decreasing function of θ and thus attains maximum at the smallest possible value of θ which is a. Similarly,

$$\overline{X}_n > b \geq \theta \Rightarrow \overline{X}_n - \theta > 0 \Rightarrow Q'(\theta) = n(\overline{X}_n - \theta) > 0,$$

implying that $Q(\theta)$ is an increasing function of θ and thus attains maximum at the largest possible value of θ which is b. Thus, the maximum likelihood estimator $\hat{\theta}_n$ of θ is given by

$$
\hat{\theta}_n = \begin{cases} a, & \text{if} \quad \overline{X}_n < a \\ \overline{X}_n, & \text{if} \quad \overline{X}_n \in [a, b] \\ b, & \text{if} \quad \overline{X}_n > b. \end{cases}
$$

To verify the consistency, we proceed on similar lines as in (i). By WLLN $\overline{X}_n \overset{P_\theta}{\to} \theta$, for all $\theta \in [a, b]$. Now for $\epsilon > 0$ and $\theta \in (a, b)$,

$$
P_\theta[|\hat{\theta}_n - \overline{X}_n| < \epsilon] \geq P_\theta[\hat{\theta}_n = \overline{X}_n] = P_\theta[a \leq \overline{X}_n \leq b]
$$
$$
= \Phi(\sqrt{n}(b - \theta)) - \Phi(\sqrt{n}(a - \theta)) \quad \to \quad 1.
$$

Hence, $\forall \ \theta \in (a, b)$, $\hat{\theta}_n \overset{P_\theta}{\to} \theta$. Now to examine convergence in probability at the boundary points a and b, consider for $\theta = a$,

$$
P_a[|\hat{\theta}_n - a| > \epsilon] = P_a[\hat{\theta}_n - a > \epsilon]
$$
$$
= \begin{cases} P_a[\hat{\theta}_n > a + \epsilon] = 0, & \text{if} \quad \epsilon > b - a \\ P_a[\overline{X}_n > a + \epsilon] = 1 - \Phi(\sqrt{n}\epsilon) \to 0, & \text{if} \quad \epsilon \leq b - a. \end{cases}
$$

Thus, $\hat{\theta}_n \overset{P_a}{\to} a$. Further for the boundary point b,

$$
P_b[|\hat{\theta}_n - b| > \epsilon] = P_b[b - \hat{\theta}_n > \epsilon]
$$
$$
= \begin{cases} P_b[\hat{\theta}_n < b - \epsilon] = 0, & \text{if} \quad \epsilon > b - a \\ P_b[\overline{X}_n < b - \epsilon] = \Phi(-\sqrt{n}\epsilon) \to 0, & \text{if} \quad \epsilon \leq b - a. \end{cases}
$$

Hence, $\hat{\theta}_n \overset{P_b}{\to} b$. Thus, we have shown that $\forall \ \theta \in [a, b]$, $\hat{\theta}_n \overset{P_\theta}{\to} \theta$ and hence $\hat{\theta}_n$ is a consistent estimator of θ.

(iii) Suppose the parameter space is $\Theta = \{a, b\}$. Thus, it consists of only two points where a, b are any fixed real numbers and we assume that $b > a$. It is to be noted that in this case the likelihood is not even a continuous function of θ and hence to find the maximum likelihood estimator of θ, we compare $L_n(a|\underline{X})$ with $L_n(b|\underline{X})$. Observe that, for $b > a$,

$$
\frac{L_n(b|\underline{X})}{L_n(a|\underline{X})} = \exp\left\{ -\frac{1}{2} \sum_{i=1}^{n}(X_i - b)^2 + \frac{1}{2} \sum_{i=1}^{n}(X_i - a)^2 \right\}
$$
$$
= \exp\left\{ -\frac{1}{2}\left(-2(b - a)\sum_{i=1}^{n} X_i + n(b^2 - a^2) \right) \right\}
$$
$$
= \exp\left\{ n(b - a)\left(\overline{X}_n - \frac{a + b}{2} \right) \right\} \tag{2.1}
$$
$$
\Rightarrow \quad L_n(b|\underline{X}) > L_n(a|\underline{X}) \ \text{if} \ \overline{X}_n > (a + b)/2
$$

& $L_n(b|\underline{X}) \leq L_n(a|\underline{X})$ if $\overline{X}_n \leq (a+b)/2$.

Hence, the maximum likelihood estimator $\hat{\theta}_n$ of θ is given by

$$\hat{\theta}_n = \begin{cases} b, & \text{if } \overline{X}_n > (a+b)/2 \\ a, & \text{if } \overline{X}_n \leq (a+b)/2 . \end{cases}$$

To verify consistency of $\hat{\theta}_n$, we have to check whether $\hat{\theta}_n \overset{P_a}{\to} a$ and $\hat{\theta}_n \overset{P_b}{\to} b$. Observe that, for all $\epsilon > 0$,

$$P_a[|\hat{\theta}_n - a| < \epsilon] \geq P_a[\hat{\theta}_n = a] = P_a\left[\overline{X}_n \leq (a+b)/2\right]$$
$$= \Phi\left(\sqrt{n}(b-a)/2\right) \to 1$$

as $n \to \infty$ since $(b-a) > 0$. On similar lines, for all $\epsilon > 0$,

$$P_b[|\hat{\theta}_n - b| < \epsilon] \geq P_b[\hat{\theta}_n = b] = P_b\left[\overline{X}_n > (a+b)/2\right]$$
$$= 1 - \Phi\left(\sqrt{n}(a-b)/2\right) \to 1$$

as $n \to \infty$ since $(a-b) < 0$. Thus, $\hat{\theta}_n \overset{P_a}{\to} a$ and $\hat{\theta}_n \overset{P_b}{\to} b$ implying that $\hat{\theta}_n$ is a consistent estimator of θ. In particular, if $a = 0$ and $b = 1$, the maximum likelihood estimator $\hat{\theta}_n$ of θ is given by

$$\hat{\theta}_n = \begin{cases} 1, & \text{if } \overline{X}_n > 1/2 \\ 0, & \text{if } \overline{X}_n \leq 1/2 . \end{cases}$$

It is to be noted that for $0 < \epsilon \leq 1$,
$P_0[|\hat{\theta}_n - 0| < \epsilon] = P_1[|\hat{\theta}_n - 1| < \epsilon] = \Phi\left(\sqrt{n}/2\right)$. For $n = 1$, the probability is 0.69 and for $n = 36$, it is almost 1. Thus, the coverage probability is close to 1 even for a small sample size. In Sect. 2.7, while verifying consistency by simulation, we discuss this feature in more detail.

(iv) In this case, the parameter space is the set of integers. As in the above cases, the log likelihood is given by $Q(\theta) = c - \frac{1}{2}\sum_{i=1}^{n}(X_i - \theta)^2$, which is maximum with respect to the variations in θ if $\theta = \overline{X}_n$. However, $P_\theta[\overline{X}_n = k] = 0$ for any integer k, hence $\overline{X}_n$ cannot be an estimator of θ. To find the maximum likelihood estimator for θ, we compare the values of $L_n(\theta|\underline{X})$ at $\theta - 1$, θ and at $\theta + 1$. Proceeding on similar lines as in Eq. (2.1), we get

$$\frac{L_n(\theta|\underline{X})}{L_n(\theta - 1|\underline{X})} = \exp\left\{n(\overline{X}_n - (\theta - 1/2))\right\} \geq 1 \text{ if } \overline{X}_n \geq (\theta - 1/2).$$

Similarly, it can show that $L_n(\theta + 1|\underline{X}) \geq L_n(\theta|\underline{X})$ if $\overline{X}_n \geq (\theta + 1/2)$. As a consequence, we conclude that the likelihood at θ is larger than or equal to that at $\theta - 1$ and $\theta + 1$ if $\overline{X}_n \in [\theta - 1/2, \theta + 1/2)$. Hence, the maximum likelihood estimator $\hat{\theta}_n$ of θ is given by

$$\hat{\theta}_n = k \quad \text{if} \quad \overline{X}_n \in [k - 1/2, k + 1/2), \quad \text{where} \quad k \in I.$$

Suppose for a given random sample $\overline{X}_n = 5.6$, then observe that $6 - 1/2 < 5.6 < 6 + 1/2$ and hence $\hat{\theta}_n = 6$, if for a given random sample $\overline{X}_n = 5.5$, then $6 - 1/2 \leq 5.5 < 6 + 1/2$ and hence $\hat{\theta}_n = 6$, if for given random sample $\overline{X}_n = 3.3$, then $3 - 1/2 < 3.3 < 3 + 1/2$ and hence $\hat{\theta}_n = 3$, if $\overline{X}_n = -7.8$, then $-8 - 1/2 < -7.8 < -8 + 1/2$ and hence $\hat{\theta}_n = -8$. Thus, $\hat{\theta}_n$ is the nearest integer to $\overline{X}_n$. It can also be expressed as $\hat{\theta}_n = [\overline{X}_n + 1/2]$, where $[x]$ is an integer part of x. To examine the consistency of $\hat{\theta}_n$, for any $\epsilon > 0$ and for $\theta \in I$,

$$P_\theta[|\hat{\theta}_n - \theta| < \epsilon] \geq P_\theta[\hat{\theta}_n = \theta] = P_\theta[\theta - 1/2 \leq \overline{X}_n < \theta + 1/2]$$
$$= \Phi(\sqrt{n}/2) - \Phi(-\sqrt{n}/2) \to 1, \quad \text{as} \quad n \to \infty, \quad \forall \ \theta \in I.$$

Hence, $\hat{\theta}_n$ is a consistent estimator of θ.

In all the above cases, the equation to find a moment estimator is given by $\overline{X}_n = \theta$, and if $\overline{X}_n$ belongs to the parameter space, then we can call it as a moment estimator. Whenever it exists, it is consistent for θ.   □

✒ Example 2.2.4

Suppose X follows a Poisson $Poi(\theta)$ distribution, $\theta \in \Theta = (0, \infty)$ and $\underline{X} = \{X_1, X_2, \ldots, X_n\}$ is a random sample of size n from it. The probability mass function of X is given by $P[X = x] = e^{-\theta}\theta^x/x!$, $x = 0, 1, \ldots$. The log-likelihood function of θ corresponding to $\underline{X}$ and its first and second derivatives are given by

$$\log L_n(\theta|\underline{X}) = c + (\log \theta)\sum_{i=1}^{n} X_i - n\theta,$$

$$\frac{\partial}{\partial \theta}\log L_n(\theta|\underline{X}) = \frac{\sum_{i=1}^{n} X_i}{\theta} - n \ \& \ \frac{\partial^2}{\partial \theta^2}\log L_n(\theta|\underline{X}) = -\frac{\sum_{i=1}^{n} X_i}{\theta^2},$$

where c is a constant free from θ. The solution of the likelihood equation is $\theta = \overline{X}_n$ and, at this solution, the second derivative is negative, provided $\overline{X}_n > 0$. It is to be noted that if the parameter space is $\Theta = (0, \infty)$, then $\overline{X}_n$ is an estimator provided $\overline{X}_n > 0$. However, it is possible that $\overline{X}_n = 0 \Leftrightarrow X_i = 0$ $\forall \ i = 1, 2, \ldots, n$, the probability of which is $\exp(-n\theta) > 0$. In this case, the likelihood of θ is given by $\exp(-n\theta)$. It is a decreasing function of θ and attains supremum at $\theta = 0$. However, 0 is not included in the parameter space. Hence, the maximum likelihood estimator of θ does not exist. To examine whether the moment estimator is consistent for θ, observe that mean of the $Poi(\theta)$ distribution is θ. The equation to find a moment estimator is given by $\overline{X}_n = \theta$. If $\overline{X}_n > 0$, then the moment estimator will be $\tilde{\theta}_n = \overline{X}_n$ and it will be consistent for θ.   □

✑ Remark 2.2.2

If in the above example, the parameter space is $[0, \infty)$, then the maximum likelihood estimator is $\overline{X}_n$ and it will be consistent for θ. However, if we define $0^0 = 1$, then for $\theta = 0$

$$P[X = x] = \frac{e^{-\theta}\theta^x}{x!} = \begin{cases} 1, & \text{if} & x = 0 \\ 0, & \text{if} & x = 1, 2, \dots . \end{cases}$$

Thus, at $\theta = 0$, X is degenerate at 0.

✒ Example 2.2.5

Suppose $X \sim Poi(\theta)$, $\theta \in \Theta = (0, \infty)$. Suppose an estimator T_n based on a random sample of size n from the distribution of X is defined as follows:

$$T_n = \begin{cases} \overline{X}_n, & \text{if} \ \overline{X}_n > 0 \\ 0.05, & \text{if} \ \overline{X}_n = 0. \end{cases}$$

To examine whether it is consistent for θ, observe that $\forall \ \epsilon > 0$ and $\forall \ \theta > 0$,

$$P[|T_n - \overline{X}_n| < \epsilon] \geq P[T_n = \overline{X}_n] = P[\overline{X}_n > 0]$$
$$= 1 - \exp(-n\theta) \ \rightarrow \ 1, \ \text{as } n \rightarrow \infty.$$

Thus, $(T_n - \overline{X}_n) \overset{P_\theta}{\rightarrow} 0$, $\forall \ \theta \in \Theta$, but by the WLLN, $\overline{X}_n \overset{P_\theta}{\rightarrow} \theta$ and hence, $T_n \overset{P_\theta}{\rightarrow} \theta$, $\forall \ \theta \in \Theta$, which proves that T_n is a consistent estimator of θ. □

✐ Remark 2.2.3

In the above example, it is to be noted that the consistency of T_n follows from the fact that $T_n = \overline{X}_n$ on a set whose probability converges to 1 as $n \rightarrow \infty$ and $\overline{X}_n \overset{P_\theta}{\rightarrow} \theta$. It is to be noted that the value of T_n on the set $[\overline{X}_n = 0]$ does not matter, it can be any arbitrary small positive number. Thus, in general, if an estimator T_n is defined as

$$T_n = \begin{cases} U_n, & \text{with probability} \quad p_n \\ c, & \text{with probability} \quad 1 - p_n , \end{cases}$$

where $U_n \overset{P_\theta}{\rightarrow} \theta$, $\forall \ \theta \in \Theta$, $p_n \rightarrow 1$ as $n \rightarrow \infty$ and c is any arbitrary number so that the range space of T_n is the parameter space, then T_n is consistent for θ.

We now proceed to establish an important property of a consistent estimator which is well known as an invariance property of consistency under continuous transformation. It is in contrast with the unbiasedness property which is invariant under only linear transformation and not in general under any other transformations. In Chap. 1, we have noted that if

$$X_n \overset{P}{\rightarrow} X \ \Rightarrow \ g(X_n) \overset{P}{\rightarrow} g(X) \ \text{where } g \text{ is a continuous function.}$$

The next theorem is the same result in the inference setup.

Theorem 2.2.1

Suppose T_n is a consistent estimator of θ and $g : \Theta \to \mathbb{R}$ is a continuous function. Then $g(T_n)$ is a consistent estimator of $g(\theta)$.

Proof It is to be noted that g being a continuous function, is a Borel function and hence $g(T_n)$ is a random variable. Continuity of g implies that, for any x and y in domain of g, given $\epsilon > 0$, $\exists\ \delta > 0$ such that if $|x - y| < \delta$, then $|g(x) - g(y)| < \epsilon$. Thus, given $\epsilon > 0$, $\exists\ \delta > 0$ such that $|T_n - \theta| < \delta$ implies $|g(T_n) - g(\theta)| < \epsilon$. As a consequence,

$$[|g(T_n)(\omega) - g(\theta)| < \epsilon] \supset [|T_n(\omega) - \theta| < \delta]$$
$$\Rightarrow\ P[|g(T_n) - g(\theta)| < \epsilon] \geq P[|T_n - \theta| < \delta]\ \to\ 1,\ \forall\ \delta > 0$$
$$\Rightarrow\ P[|g(T_n) - g(\theta)| < \epsilon] \to\ 1,\ \forall\ \epsilon > 0$$

and it is true $\forall\ \theta \in \Theta$. Hence, $g(T_n)$ is a consistent estimator of $g(\theta)$.   □

This theorem is one of the most frequently used theorems to obtain a consistent estimator for a parametric function of interest, as is evident from the following examples.

✐ Example 2.2.6

Suppose $\{X_1, X_2, \ldots, X_n\}$ is a random sample of size n from a Poisson $Poi(\theta)$ distribution, $\theta \in \Theta = (0, \infty)$. It is shown in Example 2.2.5 that T_n is a consistent estimator for θ. Further, $P[X_1 = 0] = \exp(-\theta)$ is a continuous function of θ. Hence, by Theorem 2.2.1, $\exp(-T_n)$ is a consistent estimator for $\exp(-\theta)$. It to be noted that T_n is a biased estimator of θ and $\exp(-T_n)$, although consistent, is also a biased estimator for $\exp(-\theta)$. To find a consistent and unbiased estimator for $P[X_1 = 0] = \exp(-\theta)$, we define random variables Y_i, $i = 1, 2, \ldots, n$ as follows:

$$Y_i = \begin{cases} 1, & \text{if} \quad X_i = 0 \\ 0, & \text{otherwise} . \end{cases}$$

Thus, $Y_i = I_{[X_i=0]}$ is a Borel function of X_i, $i = 1, 2, \ldots, n$. Hence, $\{X_1, X_2, \ldots, X_n\}$ are independent and identically distributed random variables implies that $\{Y_1, Y_2, \ldots, Y_n\}$ are also independent and identically distributed random variables, with $E(Y_i) = P[X_i = 0]$, $i = 1, 2, \ldots, n$. Hence, the sample mean $\overline{Y}_n$ is an unbiased estimator of $E(Y_1) = P[X_1 = 0]$ and by the WLLN it is also consistent for $P[X_1 = 0]$.   □

✐ Example 2.2.7

Suppose $\{X_1, X_2, \ldots, X_n\}$ is a random sample of size n from a normal $N(\theta, 1)$ distribution, $\theta \in \Theta = \mathbb{R}$, then by WLLN, $\overline{X}_n$ is a consistent estimator of θ. Further, $P[X_1 \leq a] = P[(X_1 - \theta) \leq (a - \theta)] = \Phi(a - \theta)$, where $\Phi(\cdot)$ is a distribution function of the standard normal distribution. It is a continuous function.

Hence, by Theorem 2.2.1, $\Phi(a - \overline{X}_n)$ is a consistent estimator for $\Phi(a - \theta)$. To find a consistent and unbiased estimator for $P[X_1 \leq a]$, we adopt the same procedure as in Example 2.2.6. We define random variables Y_i, $i = 1, 2, \ldots, n$ as follows:

$$Y_i = \begin{cases} 1, & \text{if} \quad X_i \leq a \\ 0, & \text{otherwise} . \end{cases}$$

Thus, $\{Y_1, Y_2, \ldots, Y_n\}$ are independent and identically distributed random variables, with $E(Y_i) = P[X_i \leq a] = \Phi(a - \theta)$, $i = 1, 2, \ldots, n$. Hence, the sample mean $\overline{Y}_n$ is an unbiased estimator of $\Phi(a - \theta)$ and by the WLLN it is also consistent for $\Phi(a - \theta)$. $\qquad\square$

The following example illustrates the application of Theorem 2.2.1 to generate a consistent estimator for the indexing parameter using a moment estimator.

☑ Example 2.2.8

Suppose $\{X_1, X_2, \ldots, X_n\}$ is a random sample of size n from a distribution with probability density function $f(x, \theta) = \theta x^{\theta-1}$, $0 < x < 1$, $\theta \in \Theta = (0, \infty)$. We find a consistent estimator for θ based on a sample mean. We have $E(X) = \theta/(\theta + 1) < \infty$ $\forall$ $\theta \in (0, \infty)$. Thus, by the WLLN, $\overline{X}_n \overset{P_\theta}{\to} \theta/(\theta + 1)$, $\forall \theta > 0$. Suppose $\theta/(\theta + 1) = \phi$, say. Then $\phi/(1 - \phi) = \theta$. Hence, if a function g is defined as $g(\phi) = \phi/(1 - \phi)$, $0 < \phi < 1$, then $g(\phi) = \theta$. It is clear that g is a continuous function. Hence, by Theorem 2.2.1, $g(\overline{X}_n) = \overline{X}_n/(1 - \overline{X}_n)$ is consistent for $g(\phi) = \theta$. One more approach to show that $\overline{X}_n/(1 - \overline{X}_n)$ is consistent for θ, is based on the fact that convergence in probability is closed under the arithmetic operations. Thus,

$$\overline{X}_n \overset{P_\theta}{\to} \frac{\theta}{\theta + 1} \quad \Rightarrow \quad \frac{1}{\overline{X}_n} \overset{P_\theta}{\to} \frac{\theta + 1}{\theta} = \frac{1}{\theta} + 1$$

$$\Rightarrow \quad \frac{1}{\overline{X}_n} - 1 \overset{P_\theta}{\to} \frac{1}{\theta} \quad \Rightarrow \quad \frac{\overline{X}_n}{1 - \overline{X}_n} \overset{P_\theta}{\to} \theta.$$

We now find a consistent estimator for θ based on a sufficient statistic. The likelihood of θ given the random sample $\underline{X} = \{X_1, X_2, \ldots, X_n\}$ is

$$L_n(\theta | \underline{X}) = \prod_{i=1}^{n} \theta X_i^{\theta-1}$$

$$\Leftrightarrow \quad \log L_n(\theta | \underline{X}) = n \log \theta - \theta \sum_{i=1}^{n} (-\log X_i) - \sum_{i=1}^{n} \log X_i.$$

By the Neyman-Fisher factorization theorem, $\sum_{i=1}^{n}(-\log X_i)$ is a sufficient statistic. Suppose a random variable Y is defined as $Y = -\log X$, then the probability density function $f_Y(y, \theta)$ of Y is given by $f_Y(y, \theta) = \theta e^{-\theta y}$, $y > 0$. Thus, the distribution of Y is exponential with mean $1/\theta$. Hence, the moment estimator

$\tilde{\theta}_n$ of θ based on a sufficient statistic is given by the equation $\overline{Y}_n = S_n/n = E(Y) = 1/\theta$ and $\tilde{\theta}_n = n/S_n = 1/\overline{Y}_n$. By the WLLN,

$$S_n/n \overset{P_\theta}{\to} E(Y) = 1/\theta \implies \tilde{\theta}_n \overset{P_\theta}{\to} \theta \ \forall \ \theta \in \Theta.$$

Thus, $\tilde{\theta}_n$ is a consistent estimator of θ. It can be easily verified that $\tilde{\theta}_n$ is the maximum likelihood estimator of θ. □

Using the WLLN, one can show that an empirical distribution function is a consistent estimator of the distribution function for a fixed real number, from which the sample is drawn. Further, by Theorem 2.2.1, we can obtain the consistent estimator of the indexing parameter based on an empirical distribution function. The following example elaborates on it. We first define an empirical distribution function, also known as a sample distribution function.

▶ **Definition 2.2.2**
Empirical Distribution Function: Suppose $\underline{X} = \{X_1, X_2, \ldots, X_n\}$ is a random sample from the distribution of X with distribution function $F(x)$, $x \in \mathbb{R}$. For each fixed $x \in \mathbb{R}$, the empirical distribution function $F_n(x)$ corresponding to the given random sample $\underline{X}$ is defined as

$$F_n(x) = \frac{\text{number of } X_i \leq x}{n} = \frac{1}{n} \sum_{i=1}^{n} Y_i ,$$

where for $i = 1, 2, \ldots, n$, Y_i is defined as

$$Y_i = \begin{cases} 1, & \text{if } X_i \leq x \\ 0, & \text{if } X_i > x, \end{cases}$$

It is clear from the definition that $F_n(x)$ is non-decreasing, right continuous with $F_n(-\infty) = 0$ and $F_n(\infty) = 1$. Thus, it satisfies all the properties of a distribution function. Moreover, it is a step function with discontinuities at n points. However, an important point to be noted is that it is not a deterministic function. From the definition, we note that the empirical distribution function is a Borel measurable function of $\{X_1, X_2, \ldots, X_n\}$, and hence it is a random variable for each fixed x. Observe that for each fixed x, $Y_i \sim B(1, F(x))$ distribution. Further, Y_i is a Borel function of X_i, $i = 1, 2, \ldots, n$ and hence $\{Y_1, Y_2, \ldots, Y_n\}$ are also independent and identically distributed random variables. It then follows that $n F_n(x) \sim B(n, F(x))$ distribution, with mean $n F(x)$ and variance $n F(x)(1 - F(x))$. By the WLLN,

$$F_n(x) = \frac{1}{n} \sum_{i=1}^{n} Y_i \overset{P}{\to} E(Y_i) = F(x), \quad \text{for each fixed } x \in \mathbb{R}.$$

Using the result that an empirical distribution function is a consistent estimator of the distribution function, for each fixed $x \in \mathbb{R}$, we can obtain a consistent estimator of the indexing parameter, when the distribution function $F(x, \theta)$ is indexed by a parameter θ. We illustrate it in the following example.

✐ Example 2.2.9

Suppose $F_n(x)$ is an empirical distribution function corresponding to a random sample $\underline{X} = \{X_1, X_2, \ldots, X_n\}$ from the distribution of X, with the distribution function $F(x, \theta)$, $x \in \mathbb{R}$. We have noted that by the WLLN,

$$F_n(x) = \frac{1}{n} \sum_{i=1}^{n} Y_i \overset{P_\theta}{\to} E(Y_i) = F(x, \theta), \quad \forall \ \theta \in \Theta.$$

If F^{-1} exists and is continuous, then we can find a consistent estimator of θ based on $F_n(x)$. We illustrate the procedure for two distributions. Suppose X follows an exponential distribution with scale parameter θ. Then its distribution function $F(x, \theta)$ is given by

$$F(x, \theta) = \begin{cases} 0, & \text{if } x < 0 \\ 1 - \exp\{-\theta x\}, & \text{if } x \geq 0. \end{cases}$$

For $x < 0$, $F_n(x) = 0$ and $F(x) = 0$. For fixed $x > 0$, $1 - \exp\{-\theta x\} = y$ $\Rightarrow \theta = -\frac{1}{x} \log(1 - y)$. Thus,

$$F_n(x) \overset{P_\theta}{\to} F(x, \theta) \quad \Rightarrow \quad -\frac{1}{x} \log(1 - F_n(x)) \overset{P_\theta}{\to} -\frac{1}{x} \log(1 - F(x, \theta))$$
$$= \theta, \forall \ \theta \in \Theta.$$

Thus, for any fixed $x > 0$, $-\frac{1}{x} \log(1 - F_n(x))$ is a consistent estimator of θ, in fact, we have an uncountable family of consistent estimators of θ. Suppose $\{X_1, X_2, \ldots, X_n\}$ is a random sample from a Weibull distribution with probability density function

$$f(x, \theta) = \theta \, x^{\theta - 1} \exp\{-x^\theta\}, \quad x > 0, \quad \theta > 0.$$

Then its distribution function $F(x, \theta)$ for $x > 0$ is given by, $F(x, \theta) = 1 - \exp\{-x^\theta\}$. Hence, for fixed $x > 0$,

$$F_n(x) \overset{P_\theta}{\to} F(x, \theta), \quad \forall \ \theta \in \Theta$$
$$\Rightarrow \quad -\log(1 - F_n(x)) \overset{P_\theta}{\to} x^\theta, \quad \forall \ \theta \in \Theta$$
$$\Rightarrow \quad \frac{\log(-\log(1 - F_n(x)))}{\log x} \overset{P_\theta}{\to} \theta, \quad \forall \ \theta \in \Theta.$$

Thus, a consistent estimator of θ can be obtained from the empirical distribution function for each fixed $x > 0$. ☐

Theorem 2.2.1 states that if g is a continuous function, then T_n is a consistent estimator of θ implies that $g(T_n)$ is a consistent estimator for $g(\theta)$. However, if g is not continuous, still $g(T_n)$ can be a consistent estimator of $g(\theta)$ as shown in the following example. We need following theorem in the solution.

Theorem 2.2.2

If $W_n \xrightarrow{P} C < 0$, then $P[W_n \leq 0] \to 1$ as $n \to \infty$.

Proof It is known that $W_n \xrightarrow{P} C \Rightarrow W_n \xrightarrow{L} C$ and hence $F_{W_n}(x)$ converges to the distribution function of $W \equiv C$ at all real numbers except C, it being a point of discontinuity. Thus,

$$F_{W_n}(x) = P[W_n \leq x] \to \begin{cases} 0, & \text{if } x < C \\ 1, & \text{if } x > C . \end{cases}$$

Hence, $P[W_n \leq 0] \to 1$ as $C < 0$. □

Theorem 2.2.2 is useful in proving some results in Cramér-Huzurbazar theory in Chap. 4.

✐ Example 2.2.10

Suppose T_n is a consistent estimator of $\theta \in \mathbb{R}$. It is assumed that $P_\theta[T_n = 1] = 0 \ \forall \ \theta$. Suppose a function $g : \mathbb{R} \to \mathbb{R}$ is defined as

$$g(x) = \begin{cases} -1, & \text{if } x < 1 \\ 1, & \text{if } x \geq 1. \end{cases}$$

It is clear that g is not a continuous function, 1 being a point of discontinuity, and hence we cannot use Theorem 2.2.1 to claim consistency of $g(T_n)$ for $g(\theta)$. We use the definition of consistency to verify whether $g(T_n)$ is consistent for $g(\theta)$. Suppose $\theta \geq 1$, then $g(\theta) = 1$. Now,

$$P_\theta[|g(T_n) - g(\theta)| < \epsilon] = P_\theta[|g(T_n) - 1| < \epsilon]$$
$$= P_\theta[1 - \epsilon < g(T_n) < 1 + \epsilon] = 1, \quad \text{if } \epsilon > 2 ,$$

as possible values of $g(T_n)$ are -1 and 1.

For $0 < \epsilon \leq 2$, $P_\theta[1 - \epsilon < g(T_n) < 1 + \epsilon] = P_\theta[g(T_n) = 1] = P_\theta[T_n \geq 1]$.

Suppose now $\theta < 1$, then $g(\theta) = -1$. Further,

$$P_\theta[|g(T_n) - g(\theta)| < \epsilon] = P_\theta[|g(T_n) - (-1)| < \epsilon]$$
$$= P_\theta[-1 - \epsilon < g(T_n) < -1 + \epsilon] = 1, \quad \text{if } \epsilon > 2 .$$

For $0 < \epsilon \leq 2$, $P_\theta[-1 - \epsilon < g(T_n) < -1 + \epsilon] = P_\theta[g(T_n) = -1]$
$$= P_\theta[T_n < 1].$$

It is given that T_n is consistent for θ, that is $T_n \overset{P_\theta}{\to} \theta \ \forall \ \theta \in \mathbb{R}$. To examine the limiting behavior of $P_\theta[T_n < 1]$ and of $P_\theta[T_n \geq 1]$, we use Theorem 2.2.2. Suppose $\theta < 1$, then

$$T_n \overset{P_\theta}{\to} \theta \Rightarrow T_n - 1 \overset{P_\theta}{\to} \theta - 1 < 0 \Rightarrow P[T_n - 1 \leq 0] \to 1$$
$$\Rightarrow P[T_n \leq 1] \to 1 \text{ as } n \to \infty.$$

Now with the assumption that $P[T_n = 1] = 0$, $[T_n \leq 1] = [T_n < 1] \to 1$. Thus, if $\theta < 1$ then $P_\theta[|g(T_n) - g(\theta)| < \epsilon] \to 1$. Suppose $\theta \geq 1$, then

$$T_n \overset{P_\theta}{\to} \theta \Rightarrow 1 - T_n \overset{P_\theta}{\to} 1 - \theta \leq 0 \Rightarrow P[1 - T_n \leq 0] \to 1$$
$$\Rightarrow P[T_n \geq 1] \to 1 \text{ as } n \to \infty.$$

Thus, if $\theta \geq 1$, then $P_\theta[|g(T_n) - g(\theta)| < \epsilon] \to 1$. Hence, we claim that $g(T_n) \overset{P_\theta}{\to} g(\theta)$ for all $\theta \in \mathbb{R}$. Hence, $g(T_n)$ is a consistent estimator of $g(\theta)$, even if g is not a continuous function. Observe that for $\epsilon = 2$ and $\theta = 8$, $P[|g(T_n) - \theta| < \epsilon] = P[6 < g(T_n) < 10] = 0$ and hence $g(T_n)$ is not consistent for θ. It cannot be consistent for θ in view of the fact that the limit random variable in convergence in probability is almost surely unique. □

✐ Remark 2.2.4

It is to be noted that in the above example the assumption that $P[T_n = 1] = 0$, that is, the probability assigned to a discontinuity point is 0, plays a crucial role.

The following theorem states the most useful result for verifying the consistency of an estimator provided its MSE exists.

Theorem 2.2.3

An estimator T_n is a consistent estimator of θ if $MSE_\theta(T_n) \to 0$ as $n \to \infty$.

Proof By Chebyshev's inequality, for any $\epsilon > 0$ and for any $\theta \in \Theta$,

$$P_\theta[|T_n - \theta| > \epsilon] \leq E(T_n - \theta)^2/\epsilon^2 = MSE_\theta(T_n)/\epsilon^2 \to 0$$

and hence T_n is consistent for θ. □

✍ Remark 2.2.5

By the definition of MSE of T_n as an estimator of θ, we have

$$
\begin{aligned}
MSE_\theta(T_n) &= E(T_n - \theta)^2 = E(T_n - E(T_n) + E(T_n) - \theta)^2 \\
&= E(T_n - E(T_n))^2 + (E(T_n) - \theta)^2 \\
&= Var_\theta(T_n) + (b_\theta(T_n))^2 \\
&\to 0 \quad \text{if } b_\theta(T_n) \to 0 \text{ and } Var_\theta(T_n) \to 0,
\end{aligned}
$$

where $b_\theta(T_n) = (E(T_n) - \theta)$ is a bias of T_n as an estimator of θ. Thus, Theorem 2.2.3 can be restated as follows. If $b_\theta(T_n) = (E(T_n) - \theta) \to 0$ and $Var_\theta(T_n) \to 0$, then T_n is consistent for θ. Such a consistent estimator is referred to as a MSE consistent estimator of θ.

▶ Definition 2.2.3

MSE Consistent Estimator: Suppose T_n is an estimator of θ such that MSE of T_n exists. If MSE of T_n converges to 0 as $n \to \infty$, then T_n is called as a MSE consistent estimator of θ.

Theorem 2.2.3 is nothing but the well-known result from probability theory which states that convergence in r-th mean implies convergence in probability. However, if $MSE_\theta(T_n)$ does not converge to zero then we cannot conclude that T_n is not consistent for θ. It is in view of the result that convergence in probability does not imply convergence in quadratic mean. The following example illustrates that the converse of Theorem 2.2.3 is not true.

✐ Example 2.2.11

Suppose $\{X_n, n \geq 1\}$ is a sequence of random variables defined as $X_n = \mu + \epsilon_n$, where $\{\epsilon_n, n \geq 1\}$ is a sequence of independent random variables such that

$$
\epsilon_n = \begin{cases} 0, & \text{with probability } 1 - 1/n \\ n, & \text{with probability } 1/n. \end{cases}
$$

Suppose $\mu \in \mathbb{R}$. With the given distribution of ϵ_n, we have $E(\epsilon_n) = 1$ and $Var(\epsilon_n) = n - 1$. Observe that

$$
\text{For } 0 < \epsilon < n, \quad P[|\epsilon_n| < \epsilon] = P[|\epsilon_n| = 0] = 1 - 1/n \to 1 \text{ as } n \to \infty
$$
$$
\text{For } \epsilon > n, P[|\epsilon_n| < \epsilon] = 1
$$
$$
\Rightarrow \epsilon_n \xrightarrow{P} 0 \Rightarrow X_n \xrightarrow{P} \mu.
$$

Thus, X_n as an estimator of μ is consistent for μ. But

$$
\begin{aligned}
MSE_\mu(X_n) = E(X_n - \mu)^2 &= E(X_n^2) - 2\mu E(X_n) + \mu^2 \\
&= E((\mu + \epsilon_n)^2) - 2\mu(\mu + 1) + \mu^2 = n + 1 \\
&\to \infty \text{ as } n \to \infty.
\end{aligned}
$$

Thus, X_n as an estimator of μ is consistent for μ but its MSE does not converge to 0. □

✎ Example 2.2.12

Suppose $\{X_1, X_2, \ldots, X_n\}$ is a random sample of size n from a uniform $U(0, \theta)$ distribution, $\theta \in \Theta = (0, \infty)$.

$$X \sim U(0, \theta) \Rightarrow E(X_{(n)}) = \frac{n\theta}{n + 1} \ \& \ E(X_{(n)}^2) = \frac{n\theta^2}{n + 2}$$

$$\Rightarrow MSE_\theta(X_{(n)}) = E(X_{(n)} - \theta)^2 = \frac{n\theta^2}{n + 2} - 2\theta\frac{n\theta}{n + 1} + \theta^2$$

$$= \frac{2\theta^2}{(n + 1)(n + 2)} \ \rightarrow \ 0,$$

as $n \to \infty$. Hence, $X_{(n)}$ is a MSE consistent estimator of θ. Similarly,

$$E(X_{(n-1)}) = \frac{(n - 1)\theta}{(n + 1)} \ \& \ E(X_{(n-1)}^2) = \frac{n(n - 1)\theta^2}{(n + 1)(n + 2)}$$

$$\Rightarrow \ MSE_\theta(X_{(n-1)}) = E(X_{(n-1)} - \theta)^2 = \frac{n(n - 1)\theta^2}{(n + 1)(n + 2)} - 2\theta\frac{(n - 1)\theta}{(n + 1)} + \theta^2$$

$$= \frac{6\theta^2}{(n + 1)(n + 2)} \ \rightarrow \ 0,$$

as $n \to \infty$. Hence, $X_{(n-1)}$ is also MSE consistent for θ. □

Using Theorem 2.2.3, we now prove that the UMVUE, as defined in Sect. 1.2 is always a consistent estimator of its expectation.

Theorem 2.2.4

Suppose an estimator T_n based on a random sample $\{X_1, X_2, \ldots, X_n\}$ is an UMVUE of its expectation $g(\theta)$. Then T_n is a MSE consistent estimator of $g(\theta)$.

Proof It is given that T_n is an UMVUE of its expectation $g(\theta)$. Hence, if U_n is any other unbiased estimator of $g(\theta)$ based on $\{X_1, X_2, \ldots, X_n\}$, we have

$$\sigma_n^2 = Var(T_n) \leq Var(U_n), \ \forall \ n \geq 1 \ \& \ \forall \ \theta \in \Theta \ \Rightarrow \ \sigma_{n+1}^2 \leq Var(U_{n+1})$$

where $\sigma_{n+1}^2 = Var(T_{n+1})$ and T_{n+1} is UMVUE of $g(\theta)$ based on a random sample $\{X_1, X_2, \ldots, X_{n+1}\}$. In particular, suppose $U_{n+1} = T_n$, where T_n is viewed as a function of $\{X_1, X_2, \ldots, X_{n+1}\}$. Hence, we have

$$\sigma_{n+1}^2 \leq \sigma_n^2, \ \forall \ n \geq 1 \ \Rightarrow \ \{\sigma_n^2, n \geq 1\} \text{ is a non-increasing sequence.}$$

Further, it is bounded below by 0 and hence is convergent. Consequently, every subsequence of $\{\sigma_n^2, n \geq 1\}$ is convergent with the same limit. To find the limit, we find an appropriate subsequence of $\{\sigma_n^2, n \geq 1\}$ and show that it converges to 0. Suppose $n = mk$ and T_{mk} is an UMVUE of $g(\theta)$ based on a random sample $\{X_1, X_2, \ldots, X_{mk}\}$. Suppose an estimator U_{mk} is defined as

$$U_{mk} = \frac{1}{k} \{T(X_1, X_2, \ldots, X_m) + T(X_{m+1}, \ldots, X_{2m})$$
$$+ \cdots + T(X_{(m-1)k+1}, X_{(m-1)k+2}, \ldots, X_{mk})\}.$$

It is clear that $E(U_{mk}) = g(\theta)$ and thus U_{mk} is an unbiased estimator of $g(\theta)$. Further, $Var(U_{mk}) = \sigma_m^2/k$. Now T_{mk} is an UMVUE of $g(\theta)$ and hence

$$Var(T_{mk}) \leq Var(U_{mk}) \quad \Rightarrow \quad \sigma_{mk}^2 \leq \sigma_m^2/k \quad \to \quad 0 \text{ as } k \to \infty \quad \Leftrightarrow \quad n \to \infty.$$

Thus, a subsequence $\{\sigma_{mk}^2, k \geq 1\}$ of the sequence $\{\sigma_n^2, n \geq 1\}$ converges to 0 and hence the sequence $\{\sigma_n^2, n \geq 1\}$ also converges to 0. Thus, T_n is an unbiased estimator of $g(\theta)$ with variance converging to 0 and hence it is a MSE consistent estimator of $g(\theta)$.   □

Following two examples illustrate Theorem 2.2.4.

✐ Example 2.2.13

Suppose $\{X_1, X_2, \ldots, X_n\}$ is a random sample from a Bernoulli $B(1, p)$ distribution, $0 < p < 1$. Then $S_n = \sum_{i=1}^{n} X_i$ is a complete sufficient statistic. Further, sample mean $\overline{X}_n$ is an unbiased estimator of p and it is a function of a complete sufficient statistic. Hence, it is the UMVUE of p and hence a consistent estimator of p. Consistency also follows from WLLN. Observe that

$$E(nS_n) = n^2 p \ \& \ E(S_n^2) = np - np^2 + n^2 p^2 \quad \Rightarrow \quad E(nS_n - S_n^2)$$
$$= n(n-1)p(1-p).$$

Thus, $(nS_n - S_n^2)/n(n-1)$ is an unbiased estimator of $p(1-p)$ and it is a function of a complete sufficient statistic. Hence, it is the UMVUE of $p(1-p)$ and hence a consistent estimator of $p(1-p)$. Consistency of $(nS_n - S_n^2)/n(n-1)$ can also be established as follows:

$$\frac{nS_n - S_n^2}{n(n-1)} = \frac{n}{n-1}\frac{S_n}{n} - \frac{n}{n-1}\frac{S_n^2}{n^2} \xrightarrow{P_p} p - p^2 = p(1-p).$$

□

✐ Example 2.2.14

Suppose $\{X_1, X_2, \ldots, X_n\}$ is a random sample from a normal $N(\theta, 1)$ distribution $\theta \in \mathbb{R}$. Then $\overline{X}_n$ is a complete sufficient statistic and its distribution is normal $N(\theta, 1/n)$. Hence, for any $t \in \mathbb{R}$,

$$E(e^{t\overline{X}_n}) = e^{t\theta + t^2/2n} \quad \Rightarrow \quad E(e^{t\overline{X}_n - t^2/2n}) = e^{t\theta}.$$

Thus, $T_n = e^{t\overline{X}_n - t^2/2n}$ is the UMVUE of $e^{t\theta}$ and hence a consistent estimator of $e^{t\theta}$. Consistency of T_n also follows by noting that $\overline{X}_n \xrightarrow{P_\theta} \theta$ implies

$$T_n = e^{t\overline{X}_n - t^2/2n} \xrightarrow{P_\theta} e^{t\theta} \; \forall \; \theta \in \Theta. \qquad \square$$

Following theorem proves that sample raw moments are consistent for corresponding population raw moments.

Theorem 2.2.5
Suppose $\{X_1, X_2, \ldots, X_n\}$ is random sample from the distribution of X, with indexing parameter $\theta \in \Theta$. (i) Suppose g is a Borel function, such that $E_\theta(g(X)) = h(\theta) < \infty$. Then $\sum_{i=1}^{n} g(X_i)/n \xrightarrow{P_\theta} h(\theta)$, $\forall \; \theta \in \Theta$. (ii) Suppose a population raw moment $\mu'_r(\theta) = E_\theta(X^r)$ of order r is finite, $r \geq 1$. Then sample raw moment $m'_r = \sum_{i=1}^{n} X_i^r/n$ of order r is a consistent estimator of $\mu'_r(\theta)$

Proof (i) Since $\{X_1, X_2, \ldots, X_n\}$ is random sample and g is a Borel function, $\{g(X_1), g(X_2), \ldots, g(X_n)\}$ are also independent and identically distributed random variables with finite mean $h(\theta)$. Hence, by Khintchine's WLLN, $\sum_{i=1}^{n} g(X_i)/n \xrightarrow{P_\theta} h(\theta)$, $\forall \; \theta \in \Theta$.
(ii) In particular, suppose $g(X) = X^r$, $r \geq 1$, then $\mu'_r(\theta) = E(X^r)$ is the r-th population raw moment, which is assumed to be finite. Then by (i) the sample raw moment of order r given by $m'_r = \sum_{i=1}^{n} X_i^r/n$ is a consistent estimator of r-th population raw moment $\mu'_r(\theta)$. $\qquad \square$

As a consequence of Theorem 2.2.5 and the fact that the convergence in probability is closed under the arithmetic operations, we get that

$$m_2 = m'_2 - (m'_1)^2 \xrightarrow{P_\theta} \mu'_2 - (\mu'_1)^2 = \mu_2 \quad \&$$

$$m_3 = m'_3 - 3m'_2 m'_1 + 2(m'_1)^3 \xrightarrow{P_\theta} \mu'_3 - 3\mu'_2\mu'_1 + 2(\mu'_1)^3 = \mu_3.$$

In general, the sample central moment of order r given by $m'_r = \sum_{i=1}^{n}(X_i - \overline{X}_n)^r/n$ is consistent for r-th population central moment $\mu_r(\theta)$, provided it is finite. This result is proved in Sect. 2.5 using a different approach. Further, by taking $g(X) = e^{tX}$, we get that the sample moment generating function

$$M_n(t) = \sum_{i=1}^{n} e^{tX_i}/n \xrightarrow{P_\theta} E(e^{tX}) = M_X(t),$$

for a fixed t for which the moment generating function $M_X(t)$ of X exists. On similar lines, we can show that the sample probability generating function

$$P_n(t) = \sum_{i=1}^{n} t^{X_i}/n \xrightarrow{P_\theta} E(t^X) = P_X(t),$$

for a fixed $t \in (0, 1)$, where $P_X(t)$ is the probability generating function of a positive integer valued random variable X.

In Theorem 2.2.5, it is proved that the sample raw moments are consistent for corresponding population raw moments, provided these exist. There are certain distributions for which moments do not exist, for example, Cauchy distribution. In such situations, one can find a consistent estimator for the parameter of interest based on sample quantiles. We now discuss this approach. Suppose X is an absolutely continuous random variable with distribution function $F(x, \theta)$ and probability density function $f(x, \theta)$, where θ is an indexing parameter. Suppose $a_p(\theta)$ is such that

$$P[X \leq a_p(\theta)] \geq p \quad \& \quad P[X \geq a_p(\theta)] \geq 1 - p.$$

Then $a_p(\theta)$ is known as the p-th population quantile or fractile. There may be multiple values of $a_p(\theta)$ unless the distribution function $F(x, \theta)$ is strictly monotone. We assume that the distribution function $F(x, \theta)$ is strictly monotone, hence p-th population quantile $a_p(\theta)$ is a unique solution of $F(a_p(\theta), \theta) = p$, $0 < p < 1$. We assume that the solution of the equation $F(a_p(\theta), \theta) = p$ exists. For example, suppose X follows an exponential distribution with scale parameter θ. Its distribution function $F(x, \theta)$ is given by, $F(x, \theta) = 1 - \exp(-\theta x)$ for $x > 0$. Hence, the solution of the equation $F(a_p(\theta), \theta) = p$ is given by, $a_p(\theta) = -\log(1 - p)/\theta$, $0 < p < 1$ and it is a p-th population quantile of the distribution of X.

To define the corresponding sample quantile, suppose $\{X_1, X_2, \ldots, X_n\}$ is a random sample from the distribution of X with distribution function $F(x, \theta)$. Suppose $\{X_{(1)}, X_{(2)}, \ldots, X_{(n)}\}$ is the corresponding order statistics. Then $X_{(r_n)}$ is defined as a p-th sample quantile, where $r_n = [np] + 1$, $0 < p < 1$. In the following theorem, we prove that the p-th sample quantile is consistent for the p-th population quantile. We can then use the invariance property of consistency to get a consistent estimator for the desired parametric function.

Theorem 2.2.6

Suppose $\{X_1, X_2, \ldots, X_n\}$ is a random sample from the distribution of X which is an absolutely continuous random variable with distribution function $F(x, \theta)$ where θ is an indexing parameter. Suppose $F(x, \theta)$ is strictly increasing, and p-th population quantile $a_p(\theta)$ is a unique solution of $F(a_p(\theta), \theta) = p$, $0 < p < 1$. Then the p-th sample quantile $X_{(r_n)}$, where $r_n = [np] + 1$, $0 < p < 1$, is a consistent estimator of the p-th population quantile $a_p(\theta)$.

Proof It is given that the distribution of X is absolutely continuous with distribution function $F(x, \theta)$, hence by the probability integral transformation it follows that $U = F(X, \theta) \sim U(0, 1)$ distribution. As a consequence, $\{F(X_{(1)}), F(X_{(2)}), \ldots, F(X_{(n)})\}$ can be treated as an order statistics corresponding to a random sample of size n from $U(0, 1)$. Thus, the distribution of $U_{(r_n)} = F(X_{(r_n)})$ is same as that of the r_n-th-order statistics from uniform $U(0, 1)$ distribution. The probability density function of $U_{(r_n)}$ is given by,

$$g_{r_n}(u) = \frac{n!}{(r_n - 1)!(n - r_n)!} u^{r_n - 1}(1 - u)^{n - r_n}, \quad 0 < u < 1.$$

It then follows from the definition of beta function that

$$E(U_{(r_n)}) = \frac{r_n}{n + 1} \quad \& \quad E(U_{(r_n)}^2) = \frac{r_n(r_n + 1)}{(n + 1)(n + 2)}$$

$$\Rightarrow E(U_{(r_n)} - p)^2 = E(U_{(r_n)}^2) - 2pE(U_{(r_n)}) + p^2$$

$$= \frac{r_n(r_n + 1)}{(n + 1)(n + 2)} - 2p\frac{r_n}{n + 1} + p^2.$$

To find the limit of $E(U_{(r_n)} - p)^2$ observe that,

$$r_n = [np] + 1 \Rightarrow np < r_n \leq np + 1$$

$$\Rightarrow \frac{np}{n + 1} < \frac{r_n}{n + 1} \leq \frac{np + 1}{n + 1}$$

$$\Rightarrow \lim_{n \to \infty} \frac{np}{n + 1} \leq \lim_{n \to \infty} \frac{r_n}{n + 1} \leq \lim_{n \to \infty} \frac{np + 1}{n + 1}$$

$$\Rightarrow p \leq \lim_{n \to \infty} \frac{r_n}{n + 1} \leq p$$

$$\Rightarrow \lim_{n \to \infty} \frac{r_n}{n + 1} = p \quad \text{and}$$

$$\lim_{n \to \infty} \frac{r_n + 1}{n + 2} = \lim_{n \to \infty} \frac{r_n}{n + 1} \frac{n + 1}{n + 2} + \frac{1}{n + 2} = p$$

Hence,

$$\lim_{n \to \infty} E(U_{(r_n)} - p)^2 = \lim_{n \to \infty} \left\{ \frac{r_n(r_n + 1)}{(n + 1)(n + 2)} - 2p\frac{r_n}{n + 1} + p^2 \right\} = 0$$

$$\Rightarrow U_{(r_n)} \overset{q.m.}{\to} p \Rightarrow U_{(r_n)} \overset{P}{\to} p \Rightarrow F(X_{(r_n)}) \overset{P_\theta}{\to} p$$

$$\Rightarrow F^{-1}(F(X_{(r_n)})) \overset{P_\theta}{\to} F^{-1}(p)$$

$$\Rightarrow X_{(r_n)} \overset{P_\theta}{\to} a_p(\theta), \quad \forall \; \theta \in \Theta.$$

Thus, the p-th sample quantile $X_{(r_n)}$ is a consistent estimator of the p-th population quantile $a_p(\theta)$. $\quad\square$

☑ Example 2.2.15

For normal $N(\theta, 1)$ and Cauchy $C(\theta, 1)$ distributions, θ is a population median. Hence, in both the cases, by Theorem 2.2.6, the sample median $X_{([n/2]+1)}$ is consistent for θ. For a uniform $U(0, \theta)$ distribution, $\theta/2$ is a population median. Hence, $2X_{([n/2]+1)}$ is consistent for θ. More generally, for a normal $N(\theta, 1)$ distribution, the p-th quantile is $a_p(\theta) = \theta + \Phi^{-1}(p)$, hence $X_{([np]+1)} - \Phi^{-1}(p)$

is consistent for θ, $0 < p < 1$. For a uniform $U(0, \theta)$ distribution, the p-th quantile is given by, $a_p(\theta) = p\theta$, hence $X_{([np]+1)}/p$ is consistent for θ, $0 < p < 1$. For a Cauchy $C(\theta, 1)$ distribution, the p-th quantile is given by, $a_p(\theta) = \theta + \tan(\pi(p - 1/2))$, hence $X_{([np]+1)} - \tan(\pi(p - 1/2))$ is consistent for θ, $0 < p < 1$. Thus, we have an uncountable family of consistent estimators for θ as p varies over $(0, 1)$ for normal $N(\theta, 1)$, uniform $U(0, \theta)$ and Cauchy $C(\theta, 1)$ distributions. We have to choose p appropriately to get a better estimator. $\qquad\qquad\square$

The following example illustrates how to obtain consistent estimators based on sample moments and sample quantiles.

✐ Example 2.2.16

Suppose X follows an exponential distribution with location parameter (also known as a threshold parameter) θ and scale parameter 1. The probability density function of X is given by, $f_X(x, \theta) = \exp\{-(x - \theta)\}$, $x \geq \theta$ and its distribution function $F_X(x, \theta)$ is given by,

$$F_X(x, \theta) = \begin{cases} 0, & \text{if} \quad x < \theta \\ 1 - \exp\{-(x - \theta)\}, & \text{if} \quad x \geq \theta. \end{cases}$$

The solution of the equation $F_X(x, \theta) = 1/2$ gives the population median $a_{1/2}(\theta) = \theta + \log_e 2$. Suppose $\underline{X} \equiv \{X_1, X_2, \ldots, X_n\}$ is a random sample of size n from the distribution of X. Then by Theorem 2.2.6, the sample median is a consistent estimator of the population median. Hence, $X_{([n/2]+1)} - \log_e 2$ is consistent for θ. If a random variable Y is defined as $Y = X - \theta$ then it follows that Y has an exponential distribution with scale parameter 1. Hence, we immediately get that $E(X) = \theta + 1 < \infty$. By the WLLN the sample mean $\overline{X}_n$ is consistent for $E(X) = \theta + 1$ and hence the consistent estimator for θ based on the sample mean is $\overline{X}_n - 1$. It is an unbiased estimator of θ. Further, $Var(\overline{X}_n - 1) = Var(\overline{X}_n) = 1/n \to 0$. Hence, $\overline{X}_n - 1$ is a MSE consistent estimator of θ. To obtain a maximum likelihood estimator of θ, observe that the likelihood function $L_n(\theta|\underline{X})$ is given by

$$L_n(\theta|\underline{X}) = \prod_{i=1}^{n} \exp\{-(X_i - \theta)\}, \text{ if } X_i > \theta, \ \forall \ i = 1, 2, \ldots, n \ \Leftrightarrow \ X_{(1)} > \theta.$$

It is to be noted that the likelihood function is not a continuous function of θ. However, it is strictly increasing over the interval $(-\infty, X_{(1)})$ and hence attains supremum at the largest possible value of θ given the data, which is $X_{(1)}$. Hence, the maximum likelihood estimator of θ is given by, $\hat{\theta}_n = X_{(1)}$. We examine the consistency of $\hat{\theta}_n$ by showing that its coverage probability converges to 1. From the distribution function of X, the distribution function $F_{X_{(1)}}(x, \theta)$ of $X_{(1)}$ is

given by

$$F_{X_{(1)}}(x, \theta) \ = \ \begin{cases} 0, & \text{if } \ x < \theta \\ 1 - \exp\{-n(x - \theta)\}, & \text{if } \ x \geq \theta. \end{cases}$$

Thus, the distribution of $X_{(1)}$ is again exponential with scale parameter n and location parameter θ. Now the coverage probability is given by

$$\begin{aligned} P_\theta[|X_{(1)} - \theta| < \epsilon] &= P_\theta[\theta - \epsilon < X_{(1)} < \theta + \epsilon] \\ &= P_\theta[\theta < X_{(1)} < \theta + \epsilon] \quad \text{as } X_{(1)} \geq \theta \\ &= F_{X_{(1)}}(\theta + \epsilon, \theta) - F_{X_{(1)}}(\theta, \theta) \\ &= 1 - \exp(-n\epsilon) \\ &\to 1 \text{ as } \quad n \to \infty, \ \forall \ \epsilon > 0 \text{ and } \forall \ \theta \in \Theta. \end{aligned}$$

Hence, $X_{(1)}$ is a consistent estimator of θ. It is to be noted that the coverage probability does not depend on θ. Further, the distribution of $X_{(1)}$ is exponential with scale parameter n and location parameter θ, hence $E(X_{(1)}) = \theta + 1/n$ and $Var(X_{(1)}) = 1/n^2$. Thus, the bias of $X_{(1)}$ as an estimator of θ and variance of $X_{(1)}$ both converge to 0 and hence $X_{(1)}$ is MSE consistent for θ. It is known that the first sample quartile is consistent for the first population quartile which is a solution of $F(x, \theta) = 1 - \exp\{-(x - \theta)\} = 1/4 \ \Rightarrow \ a_{1/4}(\theta) = \theta + \log(4/3)$. Thus, $X_{([n/4]+1)}$ is consistent for $\theta + \log(4/3)$ and hence $X_{([n/4]+1)} - \log(4/3)$ is consistent for θ. ☐

Following example illustrates a different approach to verify the consistency of an estimator.

☑ Example 2.2.17

Suppose X is a random variable with mean μ and known variance σ^2, $0 < \sigma^2 < \infty$. Suppose $\{X_1, X_2, \ldots, X_n\}$ is a random sample from the distribution of X. A parametric function g is defined as $g(\mu) = 0$ if $\mu \neq 0$ and $g(0) = 1$. Suppose for $-1/2 < \delta < 0$, an estimator T_n is defined as

$$T_n \ = \ \begin{cases} 1, & \text{if } \ |\overline{X}_n| < n^\delta \\ 0, & \text{if } \ |\overline{X}_n| \geq n^\delta. \end{cases}$$

We examine whether T_n is a consistent estimator of $g(\mu)$. When $\mu = 0$, by the central limit theorem, $\sqrt{n}\overline{X}_n \overset{L}{\to} Z_1 \sim N(0, \sigma^2)$, since $0 < \sigma^2 < \infty$. Hence, for $\mu = 0$ and $\epsilon > 0$,

$$\begin{aligned} P_0[|T_n - g(\mu)| < \epsilon] &= P_0[|T_n - 1| < \epsilon] \\ &\geq P_0[|T_n = 1] = P_0[|\overline{X}_n| < n^\delta] \\ &= \Phi\left(n^{1/2+\delta}/\sigma\right) - \Phi\left(-n^{1/2+\delta}/\sigma\right) \quad \to \quad 1 \text{ as } n \to \infty. \end{aligned}$$

Suppose $\mu \neq 0$ and $\epsilon > 1$. Then $P_\mu[|T_n - g(\mu)| < \epsilon] = P_\mu[|T_n - 0| < \epsilon] = 1$. Now suppose $\mu \neq 0$ and $0 < \epsilon \leq 1$. By the WLLN,

$$\overline{X}_n \overset{P}{\to} \mu \;\Rightarrow\; 1/|\overline{X}_n| \overset{P}{\to} 1/|\mu| \;\Rightarrow\; n^\delta/|\overline{X}_n| \overset{P}{\to} 0$$
$$\Rightarrow\; P_\mu\left[n^\delta/|\overline{X}_n| < \epsilon\right] \to 1 \;\forall\; \epsilon > 0.$$

Hence, for $0 < \epsilon \leq 1$

$$P_\mu[|T_n - g(\mu)| < \epsilon] = P_\mu[|T_n - 0| < \epsilon] = P_\mu[T_n = 0]$$
$$= P_\mu[|\overline{X}_n| \geq n^\delta] = P_\mu\left[n^\delta/|\overline{X}_n| < 1\right] \to 1\,.$$

As a consequence, T_n is a consistent estimator of $g(\mu)$.   □

In the following two sections, we briefly discuss strong consistency and uniform consistency.

2.3   Strong Consistency

As a pre-requisite to the concept of the strongly consistent estimator, we have given the definition of almost sure convergence of a sequence of random variables in Sect. 1.3. Using it, we define a strongly consistent estimator as follows:

▶ Definition 2.3.1
Strongly Consistent Estimator: A sequence of estimators $\{T_n, n \geq 1\}$ is said to be strongly consistent for θ if as $n \to \infty$,

$$T_n \overset{a.s.}{\to} \theta, \;\forall\; \theta \in \Theta, \;\; \text{that is}\;\; T_n(\omega) \to \theta, \;\forall\; \omega \in N^c, \;\text{with}\; P_\theta(N) = 0, \;\forall\; \theta \in \Theta,$$

where set N is a p-null set. Equivalently, a sequence of estimators $\{T_n, n \geq 1\}$ is said to be strongly consistent for θ if

$$P_\theta[\lim_{n \to \infty} T_n = \theta] = 1, \;\forall\; \theta \in \Theta.$$

Kolmogorov's SLLN stated in Chap. 1, is useful to examine the strong consistency of an estimator which is in the form of an average. Suppose $\{X_1, X_2, \ldots, X_n\}$ is random sample from the distribution of X, with indexing parameter $\theta \in \Theta$. Suppose g is a Borel function, such that $E_\theta(g(X)) = h(\theta) < \infty$. Then by Kolmogorov's SLLN, $T_n = \sum_{i=1}^n g(X_i)/n \overset{a.s.}{\to} h(\theta), \;\forall\; \theta \in \Theta$. If h^{-1} exists and is continuous, then $h^{-1}(T_n)$ is a strongly consistent estimator of θ. Here, we use the result stated in Sect. 1.3 that, if f is a continuous function, then $X_n \overset{a.s.}{\to} X \;\Rightarrow\; f(X_n) \overset{a.s.}{\to} f(X)$. In particular if $\{X_1, X_2, \ldots, X_n\}$ is a random sample from a normal $N(0, \theta)$ distribution or a Bernoulli $B(1, \theta)$ distribution or a Poisson $Poi(\theta)$ distribution, then by Kolmogorov's SLLN, $\overline{X}_n \overset{a.s.}{\to} \theta \;\forall\; \theta \in \Theta$. Thus, $\overline{X}_n$ is a strongly consistent estimator of θ. Further, if $g(x) = x^r$, $r \geq 1$, then the sample raw moment $m'_r = \sum_{i=1}^{n} X_i^r/n$ of order, $r \geq 1$, is a strongly consistent estimator of population raw moment $\mu'_r(\theta) = E_\theta(X^r)$ of order r, provided it is finite.

In the following theorem, we state two sufficient conditions for the almost sure convergence of $\{X_n, n \geq 1\}$ to X, which follow from the Borel-Cantelli lemma.

Theorem 2.3.1

Suppose a sequence $\{X_n, n \geq 1\}$ of random variables and X are defined on the same probability space.

1.   *If $\forall \; \epsilon > 0$, $\sum_{n \geq 1} P[|X_n - X| > \epsilon] < \infty$, then $X_n \xrightarrow{a.s.} X$.*

2.   *If for some $r > 0$, $\sum_{n \geq 1} E(|X_n - X|^r) < \infty$, then $X_n \xrightarrow{a.s.} X$.*

Following examples illustrate how these sufficient conditions are useful to examine strong consistency of an estimator.

✒ Example 2.3.1

Suppose $\{X_1, X_2, \ldots, X_n\}$ is a random sample of size n from a uniform $U(0, \theta)$ distribution. Hence, the distribution function $F_{X_{(n)}}(x, \theta)$ of $X_{(n)}$ is given by

$$F_{X_{(n)}}(x, \theta) = \begin{cases} 0, & \text{if} \quad x < 0 \\ (x/\theta)^n, & \text{if} \quad 0 \leq x < \theta \\ 1, & \text{if} \quad x \geq \theta. \end{cases}$$

For $\epsilon \geq \theta$, $P_\theta[|X_{(n)} - \theta| > \epsilon] = 1 - P_\theta[|X_{(n)} - \theta| < \epsilon] = 0$

$$\Rightarrow \sum_{n \geq 1} P_\theta[|X_{(n)} - \theta| > \epsilon] < \infty, \; \forall \; \theta.$$

For $\epsilon < \theta$, as derived in Example 2.2.1,

$$P_\theta[|X_{(n)} - \theta| > \epsilon] = (1 - \epsilon/\theta)^n \Rightarrow \sum_{n \geq 1} P_\theta[|X_{(n)} - \theta| > \epsilon]$$

$$= \sum_{n \geq 1} (1 - \epsilon/\theta)^n < \infty \text{ as } \epsilon/\theta < 1.$$

Thus, by the sufficient condition stated in Theorem 2.3.1, it follows that $X_{(n)}$ is a strongly consistent estimator of θ.   □

✒ Example 2.3.2

Suppose $\{X_1, X_2, \ldots, X_n\}$ is a random sample of size n from a normal $N(\theta, 1)$ distribution. For $N(\theta, 1)$ distribution, θ is the population mean, hence by Kolmogorov's SLLN, it immediately follows that the sample mean $\overline{X}_n$ is strongly consistent for θ. In what follows, we use an alternative method to establish the same. By the Markov inequality for $\epsilon > 0$,

$$\sum_{n \geq 1} P[|\overline{X}_n - \theta| > \epsilon] \leq \sum_{n \geq 1} E|\overline{X}_n - \theta|^r / \epsilon^r.$$

If $r = 2$, $E|\overline{X}_n - \theta|^2 = Var(\overline{X}_n) = 1/n$, but the series $\sum_{n \geq 1} 1/n$ is divergent and we cannot draw any conclusion from sufficient condition (ii). However, in the same condition, it is required that the series be convergent for some $r > 0$. Suppose $r = 4$. To find $E|\overline{X}_n - \theta|^4$, it is to be noted that,

$$X \sim N(\theta, 1) \Rightarrow \sqrt{n}(\overline{X}_n - \theta) \sim N(0, 1) \Rightarrow Y_n = (\sqrt{n}(\overline{X}_n - \theta))^2 \sim \chi_1^2$$
$$\Rightarrow E(Y_n^2) = n^2 E(\overline{X}_n - \theta)^4 = Var(Y_n) + (E(Y_n))^2 = 2 + 1 = 3$$
$$\Rightarrow \sum_{n \geq 1} E(\overline{X}_n - \theta)^4 = \sum_{n \geq 1} 3/n^2 < \infty.$$

Thus, by the sufficient condition (ii), $\overline{X}_n$ is a strongly consistent estimator of θ.   □

2.4   Uniform Weak and Strong Consistency

As discussed in Sect. 2.2, an estimator T_n is said to be consistent for θ if T_n converges in probability to θ for all $\theta \in \Theta$. If the convergence is uniform in θ, then we get uniform consistency. A precise definition is given below.

▶ Definition 2.4.1
Uniformly Consistent Estimator: Suppose $T_n \xrightarrow{P_\theta} \theta$, $\forall \; \theta \in \Theta$, that is, for given $\epsilon > 0$ and $\delta \in (0, 1)$, for each θ, $\exists \; n_0(\epsilon, \delta, \theta)$ such that

$$\forall \; n \geq n_0(\epsilon, \delta, \theta), \; P_\theta[|T_n - \theta| < \epsilon] \geq 1 - \delta, \; \forall \; \theta \in \Theta$$
$$\Leftrightarrow \; P_\theta[|T_n - \theta| > \epsilon] \leq \delta, \; \forall \; \theta \in \Theta.$$

T_n is said to be uniformly consistent if $n_0(\epsilon, \delta, \theta)$ does not depend on θ.

    If $\sup_\Theta n_0(\epsilon, \delta, \theta)$ is finite then the convergence is uniform in θ. Following examples illustrate uniform consistency. Using the WLLN, sample averages are consistent for corresponding population averages, but the WLLN does not provide any information about the rate of convergence and it is also not useful to find out $n_0(\epsilon, \delta, \theta)$. Chebyshev's inequality comes out to be handy to deal with the rate of convergence and determination of $n_0(\epsilon, \delta, \theta)$.

✒ Example 2.4.1

Suppose $\{X_1, X_2, \ldots, X_n\}$ is a random sample of size n from a normal $N(\theta, 1)$ distribution. Then $\overline{X}_n \sim N(\theta, 1/n)$. Hence, by Chebyshev's inequality, the bound on coverage probability is given by,

$$P_\theta[|\overline{X}_n - \theta| < \epsilon] \geq 1 - E(\overline{X}_n - \theta)^2/\epsilon^2$$
$$= 1 - 1/n\epsilon^2 \to 1 \text{ as } n \to \infty \ \forall \ \theta \in \Theta.$$

Hence, $\overline{X}_n$ is consistent for θ. We select $n_0(\epsilon, \delta, \theta)$ such that

$$1 - 1/n\epsilon^2 \geq 1 - \delta \ \Rightarrow \ n \geq 1/\epsilon^2\delta \ \Rightarrow \ n_0(\epsilon, \delta, \theta) = \left[1/\epsilon^2\delta\right] + 1,$$

thus, $n_0(\epsilon, \delta, \theta)$ does not depend on θ and hence $\overline{X}_n$ is a uniformly consistent estimator of θ. □

✒ Example 2.4.2

Suppose X follows an exponential distribution with scale parameter 1 and location parameter θ, then its probability density function is $f_X(x, \theta) = \exp\{-(x - \theta)\}, \ x \geq \theta$. It is shown in Example 2.2.16 that corresponding to a random sample of size n from this distribution, the distribution of $X_{(1)}$ is again exponential with scale parameter n and location parameter θ. Hence, $E(X_{(1)}) = \theta + 1/n$ and $Var(X_{(1)}) = 1/n^2$, from which MSE of $X_{(1)}$ as an estimator of θ is,

$$MSE_\theta(X_{(1)}) = E((X_{(1)} - \theta)^2) = E(X_{(1)}^2) - 2\theta E(X_{(1)}) + \theta^2$$
$$= 2/n^2 \to 0 \text{ as } n \to \infty.$$

Thus, $X_{(1)}$ is a consistent estimator of θ. To examine uniform consistency, we use Chebyshev's inequality as in the previous example. By Chebyshev's inequality,

$$P_\theta[|X_{(1)} - \theta| < \epsilon] \geq 1 - E(X_{(1)} - \theta)^2/\epsilon^2 = 1 - 2/n^2\epsilon^2, \ \forall \ \theta \in \Theta.$$

We select $n_0(\epsilon, \delta, \theta)$ such that

$$1 - 2/n^2\epsilon^2 \geq 1 - \delta \ \Rightarrow \ n^2 \geq 2/\epsilon^2\delta \ \Rightarrow \ n_0(\epsilon, \delta, \theta) = \left[\sqrt{2/\epsilon^2\delta}\right] + 1,$$

thus, $n_0(\epsilon, \delta, \theta)$ does not depend on θ and hence $X_{(1)}$ is a uniformly consistent estimator θ. □

✒ Example 2.4.3

Suppose X follows an exponential distribution with scale parameter $1/\theta$ and location parameter 0, then its probability density function is $f_X(x, \theta) = (1/\theta) \exp\{-x/\theta\}, \ x \geq 0, \ \theta > 0$. Then by the WLLN, the sample mean $\overline{X}_n$ based on a random sample of size n from the distribution of X is consistent for θ. To examine its uniform consistency, we use Chebyshev's inequality as in the previous example. Thus,

$$P_\theta[|\overline{X}_n - \theta| < \epsilon] \geq 1 - E(\overline{X}_n - \theta)^2/\epsilon^2 = 1 - \theta^2/n\epsilon^2, \ \forall \ \theta > 0.$$

If $n_0(\epsilon, \delta, \theta)$ is such that $1 - \theta^2/n\epsilon^2 \geq 1 - \delta$, then $n_0(\epsilon, \delta, \theta)$ depends on θ and hence $\overline{X}_n$ is not uniformly consistent for θ. We get the same result if X follows a Poisson distribution with mean θ.   □

✒ Example 2.4.4

In Example 2.2.9, it is shown that the empirical distribution function $F_n(x)$, corresponding to a random sample $\{X_1, X_2, \ldots, X_n\}$ drawn from the distribution with distribution function $F(x)$, is consistent for $F(x)$ for fixed $x \in \mathbb{R}$. Further defining the random variables Y_i as in Example 2.2.9, it follows that $nF_n(x)$ has binomial $B(n, F(x))$ distribution. Hence, $Var(F_n(x)) = \frac{F_n(x)(1-F_n(x))}{n} \leq \frac{1}{4n}$. Thus, by Chebyshev's inequality, it follows that the convergence in probability is uniform in x.   □

Now, we introduce the concept of a uniformly strongly consistent estimator. It is defined on similar lines as the uniformly consistent estimator.

▶ Definition 2.4.2

Uniformly Strongly Consistent Estimator: Suppose T_n is a strongly consistent estimator of θ, that is, given $\epsilon > 0$, $\exists \ n_0(\epsilon, \omega, \theta)$ such that $\ \forall \ n \geq n_0(\epsilon, \omega, \theta)$, $|T_n(\omega) - \theta| < \epsilon, \forall \ \omega \in N^c$, with $P_\theta(N) = 0, \ \forall \ \theta \in \Theta$. If $n_0(\epsilon, \omega, \theta)$ does not depend on θ, then T_n is said to be a uniformly strongly consistent estimator of θ.

In Example 2.4.4, it is shown that the sample distribution function is uniformly weakly consistent for the population distribution function. One can appeal to the SLLN to claim the strong consistency of the sample distribution function for the population distribution function for each fixed $x \in \mathbb{R}$. It can further be proved that the strong convergence is also uniform in x, which is a well-known Glivenko-Cantelli theorem. The statement is given below. For proof, refer to Gut [1] (p. 306).

Theorem 2.4.1

Glivenko-Cantelli Theorem: Suppose $F_n(x)$ is an empirical distribution function, corresponding to a random sample $\{X_1, X_2, \ldots, X_n\}$ drawn from the distribution with distribution function $F(x)$. Then

$$\sup_{x \in \mathbb{R}} |F_n(x) - F(x)| \overset{a.s.}{\to} 0 \ \text{ as } \ n \to \infty.$$

The theorem states that the sample distribution function is uniformly strongly consistent for the distribution function from which we have drawn the random sample. It is an important theorem as it forms a basis of non-parametric inference.

In the next section, we extend the concept of consistency and all the related results when the distribution of a random variable or a random vector is indexed by a vector parameter.

2.5   Consistency: Vector Parameter Setup

Suppose X is a random variable or a random vector defined on a probability space $(\Omega, \mathbb{A}, P_\theta)$, where the probability measure P_θ is indexed by a vector parameter $\underline{\theta} \in \Theta \subset \mathbb{R}^k$. Suppose $\underline{\theta} = (\theta_1, \theta_2, \ldots, \theta_k)'$. Given a random sample $\underline{X} = \{X_1, X_2, \ldots, X_n\}$ of size n from the distribution of X, suppose $\underline{T}_n \equiv \underline{T}_n(\underline{X}) = (T_{1n}, T_{2n}, \ldots, T_{kn})'$ is an estimator of $\underline{\theta}$, that is $\underline{T}_n$ is a random vector with range space as the parameter space $\Theta \subset \mathbb{R}^k$ and T_{in} is an estimator of θ_i for $i = 1, 2, \ldots, k$. Consistency of $\underline{T}_n$ as an estimator of $\underline{\theta}$ is defined in two ways as joint consistency and marginal consistency. These are defined below.

▶ **Definition 2.5.1**
Jointly Weakly Consistent Estimator: An estimator $\underline{T}_n$ of $\underline{\theta}$ is said to be jointly weakly consistent for $\underline{\theta}$ if

$$\forall \; \underline{\theta} \in \Theta \; \& \; \forall \; \epsilon > 0, \quad \lim_{n \to \infty} P_{\underline{\theta}}\left[\underline{T}_n \in N_\epsilon(\underline{\theta})\right] = 1,$$

where $N_\epsilon(\underline{\theta})$ is an ϵ neighbourhood of $\underline{\theta}$ with respect to Euclidean or square Euclidean norm, or in particular,

$$\forall \; \underline{\theta} \in \Theta \; \& \; \forall \; \epsilon > 0, \quad \lim_{n \to \infty} P_{\underline{\theta}}[\max_{i=1,2,\ldots,k} |T_{in} - \theta_i| < \epsilon] = 1.$$

▶ **Definition 2.5.2**
Marginally Weakly Consistent Estimator: An estimator $\underline{T}_n$ of $\underline{\theta}$ is said to be marginally weakly consistent for $\underline{\theta}$ if T_{in} is consistent for θ_i, $\forall \; i = 1, 2, \ldots, k$.

As in the real parameter setup, a weakly consistent estimator will be referred to as simply a consistent estimator. In the following theorem, we establish the equivalence of two definitions. Such an equivalence plays an important role in examining the consistency of a vector estimator, as one can simply proceed marginally and use all the tools discussed in Sect. 2.2.

Theorem 2.5.1

Suppose $\underline{T}_n = (T_{1n}, T_{2n}, \ldots, T_{kn})'$ is an estimator of $\underline{\theta} = (\theta_1, \theta_2, \ldots, \theta_k)'$. Then $\underline{T}_n$ is jointly consistent for $\underline{\theta}$ if and only if $\underline{T}_n$ is marginally consistent for $\underline{\theta}$.

Proof Part(i) - In this part, we prove that joint consistency implies marginal consistency. Suppose $\underline{T}_n$ is jointly consistent for $\underline{\theta}$, then $\forall \quad \epsilon > 0$,

$\lim_{n \to \infty} P_{\underline{\theta}} \left[\max_{i=1,2,\dots,k} |T_{in} - \theta_i| < \epsilon \right] = 1$.   Suppose   the   events E and E_i, $i = 1, 2, \dots, k$ are defined as follows.

$$E = [\max_{i=1,2,\dots,k} |T_{in} - \theta_i| < \epsilon] \ \& \ E_i = [|T_{in} - \theta_i| < \epsilon]$$

$$\Rightarrow \ E = \bigcap_{i=1}^{k} E_i \Leftrightarrow E^c = \bigcup_{i=1}^{k} E_i^c.$$

Now to prove marginal consistency observe that

$$E = \bigcap_{i=1}^{k} E_i \Rightarrow E_i \supset E, \ \forall \ i = 1, 2, \dots, k$$

$$\Rightarrow P_{\underline{\theta}}(E_i) \ \geq \ P_{\underline{\theta}}(E), \ \forall \ i = 1, 2, \dots, k$$

$$\Rightarrow P_{\underline{\theta}}(E_i) \to 1, \ \forall \ i = 1, 2, \dots, k \ \text{ as } \ P_{\underline{\theta}}(E) \to 1$$

$$\Rightarrow T_{in} \ \overset{P_{\underline{\theta}}}{\to} \ \theta_i \ \forall \ i = 1, 2, \dots, k$$

$$\Rightarrow \underline{T}_n \ \text{ is marginally consistent for } \ \underline{\theta}.$$

Thus, joint consistency implies marginal consistency of $\underline{T}_n$.

Part(ii) - In this part, we prove that marginal consistency implies joint consistency. Suppose $\underline{T}_n$ is marginally consistent for $\underline{\theta}$, then T_{in} is consistent for θ_i, $\forall \ i = 1, 2, \dots, k$. Thus as $n \to \infty$,

$$P_{\underline{\theta}}[|T_{in} - \theta_i| < \epsilon] = P_{\underline{\theta}}(E_i) \to 1 \ \Rightarrow \ P_{\underline{\theta}}(E_i^c) \to 0 \ \forall \ i = 1, 2, \dots, k$$

$$\Rightarrow \ P_{\underline{\theta}}(E^c) = P_{\underline{\theta}} \left(\bigcup_{i=1}^{k} E_i^c \right) \ \leq \ \sum_{i=1}^{k} P_{\underline{\theta}}(E_i^c) \to 0$$

$$\Rightarrow \underline{T}_n \ \text{ is jointly consistent for } \ \underline{\theta}. \qquad \square$$

Once the equivalence between joint and marginal consistency is established, it is of interest to find out which of the results from real parameter setup can be extended to vector parameter setup. In the next theorem, we prove that the invariance property of consistency under continuous transformation remains valid in vector setup as well.

Theorem 2.5.2

Suppose $\underline{T}_n$ is a consistent estimator of $\underline{\theta}$.

(i)   *Suppose $g : \mathbb{R}^k \to \mathbb{R}$ is a continuous function. Then $g(\underline{T}_n)$ is consistent for $g(\underline{\theta})$.*

(ii)   *Suppose $\underline{g} : \mathbb{R}^k \to \mathbb{R}^l$, $l \leq k$ is a continuous function. Then $\underline{g}(\underline{T}_n)$ is consistent for $\underline{g}(\underline{\theta})$.*

Proof (i) Continuity of $g : \mathbb{R}^k \to \mathbb{R}$ implies that given $\epsilon > 0$, $\exists \; \delta > 0$ such that when $\underline{T}_n \in N_\delta(\underline{\theta})$, $|g(\underline{T}_n) - g(\underline{\theta})| < \epsilon$. Thus, $\forall \; \epsilon > 0$, $\forall \; \underline{\theta} \in \Theta$,

$$[|g(\underline{T}_n) - g(\underline{\theta})| < \epsilon] \supset [\underline{T}_n \in N_\delta(\underline{\theta})]$$
$$\Rightarrow \quad P_{\underline{\theta}}[|g(\underline{T}_n) - g(\underline{\theta})| < \epsilon] \geq P_{\underline{\theta}}[\underline{T}_n \in N_\delta(\underline{\theta})] \to 1 \quad \forall \; \delta > 0$$
$$\Rightarrow \quad P_{\underline{\theta}}[|g(\underline{T}_n) - g(\underline{\theta})| < \epsilon] \to 1 \; \forall \; \epsilon > 0 \;\; n \to \infty$$
$$\Rightarrow \quad g(\underline{T}_n) \overset{P_{\underline{\theta}}}{\to} g(\underline{\theta}), \; \forall \; \underline{\theta} \in \Theta$$
$$\Rightarrow \quad g(\underline{T}_n) \text{ is consistent for } g(\underline{\theta}).$$

(ii) A function $\underline{g} : \mathbb{R}^k \to \mathbb{R}^l$ can be expressed as $\underline{g}(\underline{x}) = (g_1(\underline{x}), g_2(\underline{x}), \ldots, g_l(\underline{x}))'$, where $\underline{x} = (x_1, x_2, \ldots, x_k)$ and $g_i(\underline{x})$, $\forall \; i = 1, 2, \ldots, l$ is a function from $\mathbb{R}^k$ to $\mathbb{R}$. It is given that $\underline{g}$ is a continuous function, hence $g_i(\underline{x})$, $\forall \; i = 1, 2, \ldots, l$ is a continuous function from $\mathbb{R}^k$ to $\mathbb{R}$. Hence

$$g_i(\underline{T}_n) \overset{P_{\underline{\theta}}}{\to} g_i(\underline{\theta}), \forall \underline{\theta} \in \Theta \; \& \; \forall \; i = 1, 2, \ldots, l \text{ by (i)}$$
$$\Rightarrow \; (g_1(\underline{T}_n), g_2(\underline{T}_n), \ldots, g_l(\underline{T}_n))' \overset{P_{\underline{\theta}}}{\to} (g_1(\underline{\theta}), g_2(\underline{\theta}), \ldots, g_l(\underline{\theta})) \text{ by Theorem 2.5.1}$$
$$\Rightarrow \; \underline{g}(\underline{T}_n) \overset{P_{\underline{\theta}}}{\to} \underline{g}(\underline{\theta}), \; \forall \; \underline{\theta} \in \Theta$$
$$\Rightarrow \; \underline{g}(\underline{T}_n) \text{ is consistent for } \underline{g}(\underline{\theta}).$$

$\square$

In Theorem 2.2.5, it has been proved that the sample raw moments are consistent for the corresponding population raw moments. Using the invariance property of consistency under continuous transformations in vector parameter setup, in the next theorem, we prove that sample central moments are consistent for the corresponding population central moments.

Theorem 2.5.3

Suppose $\{X_1, X_2, \ldots, X_n\}$ is a random sample from the distribution of X, with indexing parameter $\theta \in \Theta$. Then sample central moment $m_r = \frac{1}{n} \sum_{i=1}^{n}(X_i - \overline{X}_n)^r$ of order, $r \geq 1$, is a consistent estimator of population central moment $\mu_r(\theta) = E_\theta(X - E(X))^r$ of order r, provided it is finite.

Proof By the binomial theorem, we have,

$$\mu_r(\theta) = E_\theta(X - E(X))^r = \sum_{j=0}^{r} \binom{r}{j} E(X^j)(-1)^{r-j}(E(X))^{r-j}$$

$$= \sum_{j=0}^{r} \binom{r}{j} \mu'_j (-1)^{r-j} (\mu'_1)^{r-j},$$

and it can be presented as a continuous function $g = g(\mu'_1, \mu'_2, \ldots, \mu'_r)$ from $\mathbb{R}^r$ to $\mathbb{R}$. On similar lines, $m_r = \sum_{i=1}^{n}(X_i - \overline{X}_n)^r/n$ can be expressed as,

$$m_r = \frac{\sum_{i=1}^{n}(X_i - \overline{X}_n)^r}{n} = \frac{\sum_{i=1}^{n}\sum_{j=0}^{r}\binom{r}{j}X_i^j(-1)^{r-j}(\overline{X}_n)^{r-j}}{n}$$

$$= \sum_{j=0}^{r}\binom{r}{j}m'_j(-1)^{r-j}(m'_1)^{r-j} = g(m'_1, m'_2, \ldots, m'_r).$$

In Theorem 2.2.5, it has been proved that m'_i is consistent for μ'_i, $i = 1, 2, \ldots, r$ provided μ'_r is finite. Since marginal consistency is equivalent to joint consistency we get that $(m'_1, m'_2, \ldots, m'_r)'$ is consistent for $(\mu'_1, \mu'_2, \ldots, \mu'_r)'$ and hence by Theorem 2.5.2,

$$m_r = g(m'_1, m'_2, \ldots, m'_r)' \overset{P_\theta}{\to} g(m'_1, m'_2, \ldots, m'_r)' = \mu_r(\theta).$$

Thus, the sample central moment m_r is a consistent estimator of the population central moment $\mu_r(\theta)$. $\qquad\square$

In the next theorem, we extend Theorem 2.5.3 to product moments.

Theorem 2.5.4

Suppose $\underline{Z} = (X, Y)'$ has a bivariate probability distribution with $E(X^2) < \infty$ and $E(Y^2) < \infty$. Then a sample correlation coefficient between X and Y based on a random sample of size n from the distribution of $\underline{Z}$ is a consistent estimator of a population correlation coefficient between X and Y.

Proof The population correlation coefficient ρ between X and Y is given by $\rho = Cov(X, Y)/\sigma_X\sigma_Y$ and the sample correlation coefficient R_n based on a random sample of size n is defined as,

$$R_n = \frac{S_{XY}^2}{S_X S_Y} \text{ where } S_{XY}^2 = \frac{1}{n}\sum_{i=1}^{n}(X_i - \overline{X}_n)(Y_i - \overline{Y}_n),$$

$$S_X^2 = \frac{1}{n}\sum_{i=1}^{n}(X_i - \overline{X}_n)^2 \ \& \ S_Y^2 = \frac{1}{n}\sum_{i=1}^{n}(Y_i - \overline{Y}_n)^2.$$

By Theorem 2.5.3, S_X and S_Y are consistent for σ_X and σ_Y respectively. The population covariance between X and Y is $Cov(X, Y) = E(XY) - E(X)E(Y)$. By Khintchine's WLLN, $\overline{X}_n$ and $\overline{Y}_n$ are consistent for $E(X)$ and $E(Y)$ respectively. To find a consistent estimator for $E(XY)$ we define $U = XY$, being Borel function it is a random variable. A random sample of size n from the distribution of $\underline{Z}$, gives a random sample of size n from the distribution of U and again by Khintchine's WLLN,

$\overline{U}_n = \sum_{i=1}^n U_i/n = \sum_{i=1}^n X_i Y_i/n$ is consistent for $E(U) = E(XY)$. Hence, a consistent estimator for covariance between X and Y is given by

$$\frac{1}{n} \sum_{i=1}^n X_i Y_i - \overline{X}_n \overline{Y}_n = \frac{1}{n} \sum_{i=1}^n (X_i - \overline{X}_n)(Y_i - \overline{Y}_n) = S_{XY}^2.$$

Convergence in probability is closed under all arithmetic operations. Hence, R_n is a consistent estimator of ρ. $\qquad\square$

📖 Remark 2.5.1

From Theorem 2.2.5 and Theorem 2.5.3, it is clear that the sample mean is consistent for population mean, sample variance is consistent for population variance. It is in contrast with the result that sample variance is not unbiased for population variance. Further, convergence in probability is closed under all arithmetic operations. Hence, from these two theorems along with Theorem 2.5.4, we get that sample regression coefficients, sample multiple correlation coefficient and sample partial correlation coefficients are consistent for corresponding population coefficients.

Following examples illustrate the results established in the above theorems.

✒️ Example 2.5.1

Suppose T_{in}, $i = 1, 2, \ldots, l$ are consistent estimators for θ. The convex combination $T_n = \sum_{i=1}^l \alpha_i T_{in}$ can be expressed as $T_n = g(T_{1n}, T_{2n}, \ldots, T_{ln})'$ where $g(x_1, x_2, \ldots, x_l) = \sum_{i=1}^l \alpha_i x_i$, with $\sum_{i=1}^l \alpha_i = 1$, is a continuous function from $\mathbb{R}^l \to \mathbb{R}$. Hence, consistency of T_n follows from Theorem 2.5.2. Thus, a convex combination of consistent estimators of θ is again a consistent estimator of θ. $\square$

✒️ Example 2.5.2

Suppose $\{X_1, X_2, \ldots, X_n\}$ is a random sample from a distribution of a random variable X, which is absolutely continuous with support $[\theta_1, \theta_2]$, where $\theta_1 < \theta_2 \in \mathbb{R}$, and distribution function F. As discussed in Theorem 2.2.6, if the distribution of X is absolutely continuous with distribution function $F(\cdot)$, then by the probability integral transformation $U = F(X) \sim U(0, 1)$ distribution. As a consequence, $\{F(X_{(1)}), F(X_{(n)})\}$ can be treated as $\{U_{(1)}, U_{(n)}\}$, the minimum and the maximum order statistics corresponding to a random sample of size n from $U(0, 1)$ distribution. If $U \sim U(0, 1)$, then the distribution functions $F_{U_{(1)}}(x)$ of $U_{(1)}$ and $F_{U_{(n)}}(x)$ of $U_{(n)}$ are given by,

$$F_{U_{(1)}}(x) = \begin{cases} 0, & \text{if} \quad x < 0 \\ 1 - (1 - x)^n, & \text{if} \quad 0 \le x < 1 \\ 1, & \text{if} \quad x \ge 1. \end{cases}$$

$$\& \quad F_{U_{(n)}}(x) \;=\; \begin{cases} 0, & \text{if} \quad x < 0 \\ x^n, & \text{if} \quad 0 \le x < 1 \\ 1, & \text{if} \quad x \ge 1. \end{cases}$$

Now $P[|U_{(1)}| < \epsilon] = P[-\epsilon < U_{(1)} < \epsilon] = P[0 < U_{(1)} < \epsilon] = 1$ if $\epsilon > 1$, as $P[0 < U_{(1)} < 1] = 1$. For $\epsilon \le 1$, we have

$$P[|U_{(1)}| < \epsilon] = P[0 < U_{(1)} < \epsilon] = F_{U_{(1)}}(\epsilon) = 1 - (1 - \epsilon)^n \;\to\; 1 .$$

Thus, $U_{(1)} \xrightarrow{P} 0$. Similarly,

$$P[|U_{(n)} - 1| < \epsilon] = P[1 - \epsilon < U_{(n)} < 1 + \epsilon]$$
$$= P[1 - \epsilon < U_{(n)} < 1] = 1 \text{ if } \epsilon > 1,$$

as $P[0 < U_{(n)} < 1] = 1$. For $\epsilon \le 1$, we have

$$P[|U_{(n)} - 1| < \epsilon] = P[1 - \epsilon < U_{(n)} < 1] = F_{U_{(n)}}(1) - F_{U_{(n)}}(1 - \epsilon)$$
$$= 1 - (1 - \epsilon)^n \;\to\; 1 .$$

Hence, we conclude that $U_{(n)} \xrightarrow{P} 1$. Thus,

$$\text{if} \quad U \sim U(0, 1), \quad \text{then} \quad U_{(1)} \xrightarrow{P} 0 \;\&\; U_{(n)} \xrightarrow{P} 1 .$$

Suppose $F^{-1}(\cdot)$ exists. It is continuous, as F is continuous, hence by the invariance property of consistency under continuous transformation, it follows that

$$U_{(1)} = F(X_{(1)}) \xrightarrow{P} 0 \;\Rightarrow\; F^{-1}(F(X_{(1)})) = X_{(1)} \xrightarrow{P} F^{-1}(0) = \theta_1$$

$$\& \quad U_{(n)} = F(X_{(n)}) \xrightarrow{P} 1 \;\Rightarrow\; F^{-1}(F(X_{(n)})) = X_{(n)} \xrightarrow{P} F^{-1}(1) = \theta_2 ,$$

for all θ_1 and θ_2. Thus, $(X_{(1)}, X_{(n)})'$ is consistent for $(\theta_1, \theta_2)'$. $\qquad\square$

✒ Example 2.5.3

Suppose $\{X_1, X_2, \ldots, X_n\}$ is a random sample of size n from a normal $N(\mu, \sigma^2)$ distribution. For normal $N(\mu, \sigma^2)$ distribution, $\mu_1' = \mu$ and $\mu_2 = \sigma^2$. Hence, by the WLLN, $\overline{X}_n$ is consistent for μ and by Theorem 2.5.3, m_2 is consistent for σ^2. $\qquad\square$

✒ Example 2.5.4

Suppose X follows a gamma distribution with scale parameter α and shape parameter λ. Then its probability density function, mean and variance are given by

$$f(x, \alpha, \lambda) = \frac{\alpha^\lambda}{\Gamma(\lambda)} e^{-\alpha x} x^{\lambda - 1}, \; x > 0, \; \alpha > 0, \; \lambda > 0, \; \mu_1' = \lambda/\alpha \;\&\; \mu_2 = \lambda/\alpha^2.$$

Suppose $\{X_1, X_2, \ldots, X_n\}$ is a random sample from the distribution of X. By the WLLN, $m_1' \xrightarrow{P_\theta} \mu_1' = \lambda/\alpha$ and by Theorem 2.5.3, $m_2 \xrightarrow{P_\theta} \mu_2 = \lambda/\alpha^2$. Convergence in probability is closed under all arithmetic operations, hence $m_1'/m_2 \xrightarrow{P_\theta} \alpha$ and $m_1'^2/m_2 \xrightarrow{P_\theta} \lambda$. By Theorem 2.5.1, joint consistency is equivalent to marginal consistency, hence the moment estimator $\underline{T}_n = \left(m_1'/m_2, m_1'^2/m_2\right)'$ of $\underline{\theta} = (\alpha, \lambda)'$ is a consistent estimator of $\underline{\theta}$.   □

✍ Remark 2.5.2

Consistency of $\underline{T}_n$ in Example 2.5.4 can also be shown by finding an appropriate transformation $\underline{g} : \mathbb{R}^2 \to \mathbb{R}^2$ and using the result established in Theorem 2.5.2. We adopt such an approach in the next example.

✏ Example 2.5.5

Suppose $\{X_1, X_2, \ldots, X_n\}$ is a random sample from a lognormal $LN(\mu, \sigma^2)$ distribution. The first and the second population raw moments are $\mu_1' = \exp(\mu + \sigma^2/2)$ and $\mu_2' = \exp(2\mu + 2\sigma^2)$. We examine whether a moment estimator of $\underline{\theta} = (\mu, \sigma^2)'$ is a consistent estimator. By the WLLN,

$$m_1' \xrightarrow{P_\theta} \mu_1' \ \& \ m_2' \xrightarrow{P_\theta} \mu_2' \ \Rightarrow \ (m_1', m_2')' \xrightarrow{P_\theta} (\mu_1', \mu_2')'$$

by Theorem 2.5.1. Now we find a continuous function $\underline{g} : \mathbb{R}^2 \to \mathbb{R}^2$ such that $\underline{g}(m_1', m_2')' \xrightarrow{P_\theta} \underline{g}(\mu_1', \mu_2')' = (\mu, \sigma^2)'$. Suppose $\underline{g}$ is defined as $\underline{g}(x_1, x_2) = (g_1(x_1, x_2), g_2(x_1, x_2))'$, where

$$g_1(x_1, x_2) = 2 \log x_1 - (\log x_2)/2 \ \& \ g_2(x_1, x_2) = \log x_2 - 2 \log x_1.$$

It is easy to see that $\underline{g}$ is a continuous function and

$$g_1(\mu_1', \mu_2') = 2 \log \mu_1' - (\log \mu_2')/2 = \mu$$
$$\& \ g_2(\mu_1', \mu_2') = \log \mu_2' - 2 \log \mu_1' = \sigma^2.$$

Hence, by the invariance property of consistency under continuous transformation,

$$\underline{T}_n = \underline{g}(m_1', m_2') = (2 \log m_1' - (\log m_2')/2, \ \log m_2' - 2 \log m_1') \xrightarrow{P_\theta} \underline{\theta} = (\mu, \sigma^2).$$
Observe that $\underline{T}_n$ is a moment estimator of $\underline{\theta} = (\mu, \sigma^2)'$ and is a consistent estimator of $\underline{\theta}$.   □

📑 Remark 2.5.3

There is one more approach to solve Example 2.5.5. It is known that if $X \sim LN(\mu, \sigma^2)$ distribution, then $Y = \log X \sim N(\mu, \sigma^2)$ distribution. A random sample $\{X_1, X_2, \ldots, X_n\}$ from the distribution of X is equivalent to a random sample $\{Y_1, Y_2, \ldots, Y_n\}$ from the distribution of Y. In Example 2.5.3, we have shown that for $N(\mu, \sigma^2)$ distribution the sample mean and the sample variance are consistent for μ and σ^2 respectively. Hence, $(\overline{Y}_n, S_Y^2)'$ is consistent for $\underline{\theta} = (\mu, \sigma^2)'$, where

$$\overline{Y}_n = \frac{1}{n} \sum_{i=1}^{n} Y_i = \frac{1}{n} \sum_{i=1}^{n} \log X_i \text{ and } S_Y^2 = \frac{1}{n} \sum_{i=1}^{n} (Y_i - \overline{Y}_n)^2.$$

✏️ Example 2.5.6

Suppose $\underline{Z} = (X, Y)'$ has a bivariate normal $N_2(\mu_1, \mu_2, \sigma_1^2, \sigma_2^2, \rho)$ distribution and $\{\underline{Z}_1, \underline{Z}_2, \ldots, \underline{Z}_n\}$ is a random sample of size n from the distribution of $\underline{Z}$. Since $\underline{Z} = (X, Y)'$ has a bivariate normal distribution, $X \sim N(\mu_1, \sigma_1^2)$ distribution and $Y \sim N(\mu_2, \sigma_2^2)$ distribution. A random sample $\{\underline{Z}_1, \underline{Z}_2, \ldots, \underline{Z}_n\}$ gives a random sample $\{X_1, X_2, \ldots, X_n\}$ from the distribution of X and a random sample $\{Y_1, Y_2, \ldots, Y_n\}$ from the distribution of Y. Hence, by Example 2.5.3, $(\overline{X}_n, S_X^2)'$ is consistent for $(\mu_1, \sigma_1^2)'$ and $(\overline{Y}_n, S_Y^2)'$ is consistent for $(\mu_2, \sigma_2^2)'$, where $S_X^2 = \sum_{i=1}^{n}(X_i - \overline{X}_n)^2/n$ and $S_Y^2 = \sum_{i=1}^{n}(Y_i - \overline{Y}_n)^2/n$. To find a consistent estimator for $\rho = \text{Cov}(X, Y)/\sigma_1\sigma_2$, we note that from Theorem 2.5.4, the sample correlation coefficient R_n is a consistent estimator of ρ, where R_n is given by,

$$R_n = \frac{S_{XY}^2}{S_X S_Y} = \left(\frac{1}{n} \sum_{i=1}^{n} (X_i - \overline{X}_n)(Y_i - \overline{Y}_n) \right) \times$$

$$\left(\frac{1}{n} \sum_{i=1}^{n} (X_i - \overline{X}_n)^2 \frac{1}{n} \sum_{i=1}^{n} (Y_i - \overline{Y}_n)^2 \right)^{-1/2}.$$

Thus, $\underline{T}_n = (\overline{X}_n, \overline{Y}_n, S_X^2, S_Y^2, R_n)'$ is consistent for $\underline{\theta} = (\mu_1, \mu_2, \sigma_1^2, \sigma_2^2, \rho)'$. It is to be noted that $\underline{T}_n$ is a moment as well as a maximum likelihood estimator of $\underline{\theta}$. □

The following example illustrates that a maximum likelihood estimator need not be consistent. It was the first example of a maximum likelihood estimator being inconsistent and was given by Neyman and Scott in 1948 and hence it is known as a Neyman-Scott example.

✎ Example 2.5.7

Suppose $X_{ij} = \mu_i + \epsilon_{ij}$, where $\{\epsilon_{ij}, \ i = 1, 2, \ldots, n, \ j = 1, 2\}$ are independent and identically distributed random variables such that $\epsilon_{ij} \sim N(0, \sigma^2)$ distribution. It is a balanced one-way ANOVA design with two observations in each of the n groups. Thus, the random variables $X_{ij} \sim N(\mu_i, \sigma^2)$ distribution. The likelihood of $\mu_i \ i = 1, 2, \ldots, n$ and σ^2 given the data $\underline{X} = \{X_{ij}, \ i = 1, 2, \ldots, n, \ j = 1, 2\}$ is given by

$$L_n(\mu_1, \mu_2, \ldots, \mu_n, \sigma^2 | \underline{X}) = \left(\sqrt{2\pi\sigma^2}\right)^{-2n} \exp\left\{-\frac{1}{2\sigma^2} \sum_{i=1}^{n} \sum_{j=1}^{2} (X_{ij} - \mu_i)^2\right\}.$$

The maximum likelihood estimators of the parameters are given by,

$$\hat{\mu}_i = \frac{1}{2} \sum_{j=1}^{2} X_{ij} = \overline{X}_i, \ i = 1, 2, \ldots, n, \quad \& \quad \hat{\sigma}_n^2 = \frac{1}{2n} \sum_{i=1}^{n} T_i,$$

$$\text{where} \ T_i = \sum_{j=1}^{2} (X_{ij} - \overline{X}_i)^2.$$

It is to be noted that for each $i = 1, 2, \ldots, n$, $\hat{\mu}_i$ is an average of two observations and it does not depend on n at all. Observe that $\frac{T_i}{\sigma^2} \sim \chi_1^2$ distribution, $\forall \ i = 1, 2, \ldots, n$. Hence, by the WLLN,

$$\frac{1}{n} \sum_{i=1}^{n} \frac{T_i}{\sigma^2} \xrightarrow{P} 1 \ \Rightarrow \ \hat{\sigma}_n^2 = \frac{1}{2n} \sum_{i=1}^{n} T_i \xrightarrow{P} \frac{\sigma^2}{2}.$$

Thus, $\hat{\sigma}_n^2$ is the maximum likelihood estimator of σ^2, but it is not consistent for σ^2. □

✐ Remark 2.5.4

It is to be noted that in the model of Neyman-Scott example, there are two observations in each of the n groups, and as n increases the number of groups increases. We face the same problem even if each group has a finite number k of observations. Thus, inconsistency of $\hat{\sigma}_n^2$ is not due to the method of maximum likelihood estimation, but it is due to the model in which number of observations in each group remains the same but number of groups increases and hence the number of parameters also increases. The problem of inconsistency arises because the number of observations and the number of parameters grow at the same rate. On the other hand, if we fix the number of groups and if the number of observations in each group increases, then the scenario changes. Thus, if $\{X_{ij}, \ i = 1, 2, \ j = 1, 2, \ldots, n\}$ are independent random variables such that $X_{ij} \sim N(\mu_i, \sigma^2)$ distribution. Then the maximum likelihood estimators of $\mu_i \ i = 1, 2$ and σ^2 are consistent estimators for μ_i and σ^2 respectively (see solution of Exercise 2.8.28).

The Limit random variable in convergence in probability is almost surely unique, hence a given estimator cannot be consistent for two different parametric functions, but from the various examples discussed above, it is to be noted that for a given parametric function $g(\theta)$, there are a number of consistent estimators. Hence, we must have some criterion to choose the best estimator from the family of consistent estimators for $g(\theta)$. In the following section, we discuss two such criteria, one is based on the coverage probability and the other is based on the mean squared error.

2.6   Performance of a Consistent Estimator

Within a family of consistent estimators of θ, the performance of a consistent estimator is judged by the rate of convergence of a true coverage probability to 1 and of MSE to 0, provided the MSE of the estimator exists, faster the rate better is the estimator. We discuss this concept below.

Criterion based on true coverage probability: Suppose T_{1n} and T_{2n} are two consistent estimators of θ. Then for $\epsilon > 0$ the true coverage probabilities are given by

$$p_1(\epsilon, \theta, n) = P_\theta[|T_{1n} - \theta| < \epsilon] \quad \text{and} \quad p_2(\epsilon, \theta, n) = P_\theta[|T_{2n} - \theta| < \epsilon].$$

Since both T_{1n} and T_{2n} are consistent estimators of θ, both the true coverage probabilities converge to 1 as $n \to \infty$ and hence the criterion for the preference between the two is based on the rate of convergence to 1. Thus, if $p_1(\epsilon, \theta, n) \to 1$ faster than $p_2(\epsilon, \theta, n) \to 1$, then T_{1n} is preferred to T_{2n}.

In general, it is difficult to find the coverage probabilities and hence the second criterion based on the mean squared error is more useful.

Criterion based on mean squared error: Suppose T_{1n} and T_{2n} are two consistent estimators of θ such that the mean squared errors of both exist. Then the mean squared errors of both will converge to 0. The criterion for the preference between the two is again based on the rate of convergence to 0. Thus, if $MSE_\theta(T_{1n}) \to 0$ faster than that of T_{2n} then T_{1n} is preferred to T_{2n}.

In the class of unbiased estimators, the estimator with smallest variance is the best estimator, analogously within the class of MSE consistent estimators, the estimator whose mean squared error converges to 0 faster is a preferable estimator.

Following examples illustrate how to judge the performance of a consistent estimator on the basis of the coverage probability and the MSE.

✒ Example 2.6.1

Suppose $\{X_1, X_2, \ldots, X_n\}$ is a random sample of size n from a uniform $U(0, \theta)$ distribution. We examine whether $T_{1n} = 2\overline{X}_n$ or $T_{2n} = X_{(n)}$ is a better consistent estimator for θ. If $X \sim U(0, \theta)$, then $E(X) = \theta/2$ and $Var(X) = \theta^2/12$. Hence,

$$MSE_\theta(T_{1n}) = E_\theta(2\overline{X}_n - \theta)^2 = 4Var(\overline{X}_n) = \frac{4\,\theta^2}{n\,12}$$

$$= \frac{\theta^2}{3n} \;\to\; 0 \text{ as } n \to \infty \text{ at the rate of } 1/n.$$

To find MSE of $T_{2n} = X_{(n)}$, we have $E_\theta(X_{(n)}) = n\theta/(n+1)$ and $E_\theta(X_{(n)})^2 = n(n+1)\theta^2/((n+1)(n+2))$. Hence,

$$MSE_\theta(T_{2n}) = E_\theta(X_{(n)} - \theta)^2$$

$$= \frac{2\theta^2}{(n+1)(n+2)} \;\to\; 0 \text{ as } n \to \infty \text{ at the rate of } 1/n^2.$$

Thus, $T_{2n} = X_{(n)}$ is a better consistent estimator of θ than $T_{1n} = 2\overline{X}_n$ as its MSE converges to 0 faster than that of $T_{1n} = 2\overline{X}_n$. □

✔ Example 2.6.2

Suppose $\{X_1, X_2, \ldots, X_n\}$ is a random sample of size n from an exponential distribution with location parameter θ and scale parameter 1. In Example 2.2.14, we have obtained the maximum likelihood estimator of θ and it is $X_{(1)}$, which is shown to be consistent. Similarly the consistent estimator based on sample mean is $\overline{X}_n - 1$. Further,

$$MSE_\theta(X_{(1)}) = \frac{2}{n^2} \quad \& \quad MSE_\theta(\overline{X}_n - 1) = \frac{1}{n}.$$

Thus, $MSE_\theta(X_{(1)})$ converges to 0 faster than $MSE_\theta(\overline{X}_n - 1)$. Hence, $X_{(1)}$ is a better consistent estimator of θ than the consistent estimator-based sample mean. In fact

$$\forall\, n > 2, \;\; \frac{2}{n^2} - \frac{1}{n} = \frac{2-n}{n^2} < 0 \;\Rightarrow\; \forall\, n > 2, \; MSE_\theta(X_{(1)}) < MSE_\theta(\overline{X}_n - 1).$$

□

✐ Remark 2.6.1

In both the above examples, the maximum likelihood estimator, which is a function of a sufficient statistic, is a better consistent estimator than the moment estimator.

✔ Example 2.6.3

Suppose $\{X_1, X_2, \ldots, X_n\}$ is a random sample of size n from a normal $N(\theta, 1)$ distribution, $\theta \in \Theta = \{0, 1\}$. In Example 2.2.3 we have obtained the maximum likelihood estimator $\hat{\theta}_n$ of θ. It is given by

$$\hat{\theta}_n = \begin{cases} 1, & \text{if } \overline{X}_n > 1/2 \\ 0, & \text{if } \overline{X}_n \le 1/2. \end{cases}$$

It is shown in the same example that for both $\theta = 0$ and $\theta = 1$, the coverage probability is 1 when $\epsilon > 1$. For $0 < \epsilon \le 1$, to find the rate of convergence to 1, we use the result that for $x > 0$ and sufficiently large, $\Phi(-x) \approx \frac{1}{x}\phi(-x)$, where $\Phi(\cdot)$ and $\phi(\cdot)$ denote the distribution function and the probability density function of the standard normal distribution respectively. Thus, for $0 < \epsilon \le 1$,

$$\begin{aligned} P_0[|\hat{\theta}_n - 0| < \epsilon] = P_0[\hat{\theta}_n = 0] = P_0\left[\overline{X}_n \le 1/2\right] &= \Phi\left(\sqrt{n}/2\right) \\ &= 1 - \Phi\left(-\sqrt{n}/2\right) \\ &= 1 - \frac{2}{\sqrt{n}}\phi\left(-\sqrt{n}/2\right) = 1 - \frac{2}{\sqrt{n}}\frac{1}{\sqrt{2\pi}}\exp(-n/8) \to 1, \end{aligned}$$

exponentially fast as $n \to \infty$. On similar lines,

$$\begin{aligned} P_1[|\hat{\theta}_n - 1| < \epsilon] = P_1[\hat{\theta}_n = 1] = P_1\left[\overline{X}_n > 1/2\right] &= 1 - \Phi\left(-\sqrt{n}/2\right) \\ &= 1 - \frac{2}{\sqrt{n}}\phi\left(-\sqrt{n}/2\right) = 1 - \frac{2}{\sqrt{n}}\frac{1}{\sqrt{2\pi}}\exp(-n/8) \to 1, \end{aligned}$$

exponentially fast as $n \to \infty$. Thus, for both $\theta = 0$ and $\theta = 1$, the coverage probability $p(\epsilon, \theta, n)$ is the same and thus the rate of convergence of coverage probability to 1 is the same. To find the rate of convergence of MSE to 0, we find MSE from the probability mass function of $\hat{\theta}_n$. Thus,

$$P_\theta[\hat{\theta}_n = 0] = P_\theta[\overline{X}_n < 1/2] = \begin{cases} \Phi\left(\sqrt{n}/2\right), & \text{if } \theta = 0 \\ \Phi\left(-\sqrt{n}/2\right), & \text{if } \theta = 1 \end{cases}$$

$$\& \ P_\theta[\hat{\theta}_n = 1] = P_\theta[\overline{X}_n \ge 1/2] = \begin{cases} 1 - \Phi\left(\sqrt{n}/2\right), & \text{if } \theta = 0 \\ 1 - \Phi\left(-\sqrt{n}/2\right), & \text{if } \theta = 1. \end{cases}$$

Hence, the MSE of $\hat{\theta}_n$ as an estimator of θ is given by

$$\begin{aligned} MSE_\theta(\hat{\theta}_n) = E_\theta(\hat{\theta}_n - \theta)^2 &= (1 - \theta)^2 P_\theta[\hat{\theta}_n = 1] + (0 - \theta)^2 P_\theta[\hat{\theta}_n = 0] \\ &= (1 - \theta)^2 P_\theta[\overline{X}_n \ge 1/2] + \theta^2 P_\theta[\overline{X}_n < 1/2] \\ &= 1 - \Phi\left(\sqrt{n}/2\right) \ \text{if } \theta = 0 \ \text{ and} \\ &= \Phi\left(-\sqrt{n}/2\right) = 1 - \Phi\left(\sqrt{n}/2\right) \ \text{if } \theta = 1. \end{aligned}$$

Thus, for both $\theta = 0$ and $\theta = 1$,

$$\begin{aligned} MSE_\theta(\hat{\theta}_n) = 1 - \Phi\left(\frac{\sqrt{n}}{2}\right) &= \Phi\left(\frac{-\sqrt{n}}{2}\right) \\ &= \frac{2}{\sqrt{n}}\frac{1}{\sqrt{2\pi}}\exp(-n/8) \to 0 \ \text{exponentially fast,} \end{aligned}$$

as $n \to \infty$. Thus, the rate of convergence of coverage probability to 1 and of MSE to 0 is the same for $\theta = 0$ and $\theta = 1$. Further observe that $p(\epsilon, \theta, n) = 1 - MSE_\theta(\hat{\theta}_n)$. Now we consider a family $\{T_k(\overline{X}_n), 0 < k < 1\}$ of estimators for θ, where an estimator $T_k(\overline{X}_n)$ is defined as follows. For $0 < k < 1$,

$$T_k(\overline{X}_n) = \begin{cases} 0, & \text{if } \overline{X}_n < k \\ 1, & \text{if } \overline{X}_n \geq k. \end{cases}$$

We verify whether $T_k(\overline{X}_n)$ is consistent for θ. Observe that

$$P_0[|T_k(\overline{X}_n) - 0| < \epsilon] = \begin{cases} 1, & \text{if } \epsilon > 1 \\ P_0[T_k(\overline{X}_n) = 0], & \text{if } 0 < \epsilon \leq 1 \end{cases}$$

& $P_0[T_k(\overline{X}_n) = 0] = P_0\left[\overline{X}_n \leq k\right] = \Phi\left(\sqrt{n}k\right) \to 1$ as $n \to \infty$, as $k > 0$.

On similar lines,

$$P_1[|T_k(\overline{X}_n) - 1| < \epsilon] = P_1\left[1 - \epsilon < T_k(\overline{X}_n) < 1 + \epsilon\right]$$
$$= \begin{cases} 1, & \text{if } \epsilon > 1 \\ P_1[T_k(\overline{X}_n) = 1], & \text{if } 0 < \epsilon \leq 1 \end{cases}$$

& $P_1[T_k(\overline{X}_n) = 1] = P_1\left[\overline{X}_n > k\right]$
$$= 1 - \Phi\left(\sqrt{n}(k - 1)\right) \to 1 \text{ as } n \to \infty, \text{ as } k < 1.$$

Thus, if $0 < k < 1$ then $T_k(\overline{X}_n)$ is consistent for θ. We obtain MSE of $T_k(\overline{X}_n)$ on similar lines as those for $\hat{\theta}_n$ Thus,

$$P_\theta[T_k(\overline{X}_n) = 0] = P_\theta[\overline{X}_n < k] = \begin{cases} \Phi\left(\sqrt{n}k\right), & \text{if } \theta = 0 \\ \Phi\left(\sqrt{n}(k - 1)\right), & \text{if } \theta = 1 \end{cases}$$

and

$$P_\theta[T_k(\overline{X}_n) = 1] = P_\theta[\overline{X}_n \geq k] = \begin{cases} 1 - \Phi\left(\sqrt{n}k\right), & \text{if } \theta = 0 \\ 1 - \Phi\left(\sqrt{n}(k - 1)\right), & \text{if } \theta = 1. \end{cases}$$

Hence, the MSE of $T_k(\overline{X}_n)$ as an estimator of θ is given by

$$\begin{aligned} MSE_\theta(T_k(\overline{X}_n)) &= E_\theta(T_k(\overline{X}_n) - \theta)^2 \\ &= (1 - \theta)^2 P_\theta[T_k(\overline{X}_n) = 1] + (0 - \theta)^2 P_\theta[T_k(\overline{X}_n) = 0] \\ &= 1 - \Phi\left(\sqrt{n}k\right) \text{ if } \theta = 0 \text{ and} \\ &= \Phi\left(\sqrt{n}(k - 1)\right) \text{ if } \theta = 1. \end{aligned}$$

Observe that $MSE_\theta(T_k(\overline{X}_n)) \to 0$ as $n \to \infty$, thus $T_k(\overline{X}_n)$ is MSE consistent for θ. For the estimator $T_k(\overline{X}_n)$ also observe that the coverage probability $p_k(\epsilon, \theta, n) = 1 - MSE_\theta(T_k(\overline{X}_n))$. Thus, we have an uncountable family of consistent estimators of θ and it is of interest to choose $k \in (0, 1)$ to have a better estimator. Note that, for $k = 1/2$, expressions for the coverage probability are same for $\theta = 0$ and 1. Similarly, expressions for the MSE are same for $\theta = 0$ and 1. However, for $0 < k < 1/2$ and for $1/2 < k < 1$, their behavior is in the opposite direction, as shown below.

For $0 < k < 1/2$, $p_k(\epsilon, 0, n) = \Phi\left(\sqrt{n}k\right) < \Phi\left(\sqrt{n}/2\right)$

$$p_k(\epsilon, 1, n) = 1 - \Phi\left(\sqrt{n}(k - 1)\right) > 1 - \Phi\left(-\sqrt{n}/2\right)$$
$$= \Phi\left(\sqrt{n}/2\right)$$
$$\Rightarrow \quad p_k(\epsilon, 0, n) < p_k(\epsilon, 1, n)$$
$$\Rightarrow \quad MSE_0(T_k(\overline{X}_n)) > MSE_1(T_k(\overline{X}_n))$$

For $1/2 < k < 1$, $p_k(\epsilon, 0, n) = \Phi\left(\sqrt{n}k\right) > \Phi\left(\sqrt{n}/2\right)$

$$p_k(\epsilon, 1, n) = 1 - \Phi\left(\sqrt{n}(k - 1)\right) < 1 - \Phi\left(-\sqrt{n}/2\right)$$
$$= \Phi\left(\sqrt{n}/2\right)$$
$$\Rightarrow \quad p_k(\epsilon, 0, n) > p_k(\epsilon, 1, n)$$
$$\Rightarrow \quad MSE_0(T_k(\overline{X}_n)) < MSE_1(T_k(\overline{X}_n)).$$

Thus, if we have to choose $k \in (0, 1)$ so that coverage probability and MSE have the same nature, the only choice is $k = 1/2$. It is to be noted that the choice of k is not dictated by the rate of convergence of coverage probability or of MSE. In the next section, using R code we show that $T_{1/2}(\overline{X}_n) = \hat{\theta}_n$ performs better than any other estimator, as is expected since it is the maximum likelihood estimator of θ.   □

In the next section, we discuss how to verify the consistency of an estimator by simulation, using R software.

2.7   Verification of Consistency Using R

Suppose $T_n = T_n(X_1, ..., X_n)$ is a consistent estimator of θ, based on a random sample from the distribution of a random variable or a random vector X. According to the definition, an estimator T_n is consistent for θ if for all $\theta \in \Theta$ and for all $\epsilon > 0$, the coverage probability $p_n(\epsilon, \theta) = P_\theta[|T_n - \theta| < \epsilon] \to 1$ as $n \to \infty$. The consistency of T_n is verified by simulating m random samples from the distribution of X and finding the estimate r_n of the coverage probability. The estimator r_n is defined as $r_n = \sum_{i=1}^{m} I_{[|T_{ni} - \theta| < \epsilon]}/m$, where I_A is an indicator function of event A and T_{ni} is the estimator T_n based on i-th simulated sample. Thus, r_n is the proportion of times $|T_{ni} - \theta| < \epsilon$ in m samples or the relative frequency of the event $[|T_n - \theta| < \epsilon]$ in *m* samples. In other words, we estimate the coverage probability $p_n(\epsilon, \theta)$ by the relative frequency r_n. We then examine whether r_n approaches 1 as *n* increases for various values of θ and ϵ. The stepwise procedure to verify that T_n is a consistent estimator of θ is as follows.

1. Generate *m* random samples each of size *n* from the distribution of *X* with a fixed value of θ.
2. Compute T_{ni}, $i = 1, 2, \ldots, m$ using the samples generated in Step 1.
3. Find the estimate r_n of the coverage probability $p_n(\epsilon, \theta)$.
4. Increase the sample size *n* and repeat the steps 1 to 3 and examine whether $r_n \to 1$ as *n* increases.
5. Repeat steps 1 to 4 for various values of ϵ and θ.

An important point to be noted is that in step 4 as we increase sample size from *n* to $n + 50$ say, initial *n* observations should remain the same when we compute r_n. It is achieved by setting seed while generating a random sample. Following code and the table given below illustrate how the initial set of observations remain the same while generating the sample of larger size in every simulation, when we fix the seed. We illustrate it for three simulations.

```
n = c(4,8,12) ## sample sizes;
nsim = 3 ## number of simulations
x1=matrix(nrow=nsim, ncol=n[1])
x2=matrix(nrow=nsim, ncol=n[2])
x3=matrix(nrow=nsim, ncol=n[3])
for(i in 1:nsim)
{
    set.seed(i)
    x1[i,] = runif(n[1],0,1)
    set.seed(i)
    x2[i,] = runif(n[2],0,1)
    set.seed(i)
    x3[i,] = runif(n[3],0,1)
}
x1=round(x1,2); x2=round(x2,2); x3=round(x3,2); x1; x2; x3
u1=1:4; u2=1:8; u3=1:12
par(mfrow=c(3,1))
p=c(1,2,3)
pname=paste("Simulation",p,sep=" ")
for(i in 1:length(p))
{
plot(u3,x3[i,],"o", main= pname[i],
                ylab="Observations",xlab="Sample Size",
        yaxt="n",xaxt="n",col="blue",lwd=1,lty=1,xlim=c(0,14))
axis(2,at=sort(unique(x3[i,])),
   labels = sort(unique(x3[i,])),cex.axis = 0.7)
```

```
axis(1,at = u3,las = 2,cex.axis = 0.7)
lines(u2,x2[i,],"o",col="green",lwd=2,lty=2)
lines(u1,x1[i,],"o",col="red",lwd=3,lty=3)
legend("topright",legend=c(n[1],n[2],n[3]),lty=c(3,2,1),lwd=c(3,2,1),
       col=c("red","green","blue"),title=expression(paste(n)))
}
```

Observe that for simulation 1, while generating samples of size 4, 8 and 12, the seed is same as 1, while for simulation 2, it is 2 and for simulation 3, it is 3 for all the sample sizes. The matrix $x1$ presents three simulations each of size 4, the matrix $x2$ corresponds to the sample size 8 and the matrix $x3$ corresponds to the sample size 12. To clearly display, the desired feature the output is organized in Table 2.1 and Fig. 2.1.

It is to be noted that when sample size is 8, the initial 4 observations remain the same in all the three simulations, same is observed when sample size is increased to 12. Figure 2.1 also conveys that in all the three simulations as we increase the sample size from 4 to 8, the initial 4 observations remain the same and as we increase the sample size from 8 to 12, the initial 8 observations remain the same. The red, green and blue lines coincide for the initial four observations, green and blue lines coincide for the initial eight observations.

Now, we discuss a variety of examples to verify consistency by simulation. An important point to be noted from these illustrative examples is that *sample size required for* $r_n \to 1$, depends on ϵ, *the distribution of X, the value of the parameter* θ *and the estimator* T_n.

Table 2.1 Random Samples from Uniform $U(0, 1)$ Distribution

	1	2	3	4	5	6	7	8	9	10	11	12
Sim 1	0.27	0.37	0.57	0.91								
	0.27	0.37	0.57	0.91	0.20	0.90	0.94	0.66				
	0.27	0.37	0.57	0.91	0.20	0.90	0.94	0.66	0.63	0.06	0.21	0.18
Sim 2	0.18	0.70	0.57	0.17								
	0.18	0.70	0.57	0.17	0.94	0.94	0.13	0.83				
	0.18	0.70	0.57	0.17	0.94	0.94	0.13	0.83	0.47	0.55	0.55	0.24
Sim 2	0.17	0.81	0.38	0.33								
	0.17	0.81	0.38	0.33	0.60	0.60	0.12	0.29				
	0.17	0.81	0.38	0.33	0.60	0.60	0.12	0.29	0.58	0.63	0.51	0.51

Simulation 1

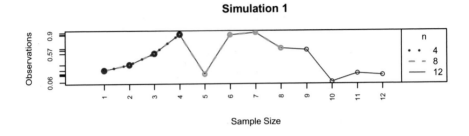

Simulation 2

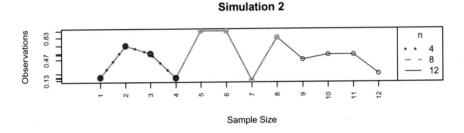

Simulation 3

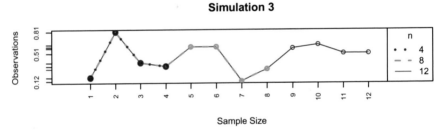

Fig. 2.1 Random samples from uniform U(0,1) distribution

✏ Example 2.7.1

We have shown in Example 2.2.1 that based on a random sample of size n from a uniform $U(0, \theta)$ distribution, $X_{(n)}$ is a consistent estimator of θ, but $X_{(1)}$ is not consistent for θ. Further by the WLLN, it follows that $\overline{X}_n$ is not consistent for θ but $2\overline{X}_n$ is consistent for θ. Using the following R code we verify these results. To have some guess for the initial sample size, if possible, we find the minimum sample size needed to attain the given level of accuracy specified by ϵ and δ, either from the the expression for coverage probability, or by approximating the distribution of T_n or by Chebyshev's inequality. We have shown in Example 2.6.1 that MSE of $2\overline{X}_n$ converges to 0 at the rate $1/n$ while that of $X_{(n)}$ converges to 0 at the rate $1/n^2$. It is reflected in the computation of minimum sample size. In Example 2.2.1, using the expression for the coverage probability for $X_{(n)}$, we have obtained the minimum sample size n_0 so that $P_\theta[|X_{(n)} - \theta| < \epsilon] \geq 1 - \delta$ where $\epsilon > 0$ and

$\delta \in (0, 1)$ are specified constants. It is given by $n_0 = \lceil \log \delta / \log(\theta - \epsilon)/\theta \rceil + 1$.
Using the CLT, $\sqrt{n}(2\overline{X}_n - \theta) \overset{L}{\rightarrow} Z_1 \sim N(0, \theta^2/3)$. Thus,

$$P_\theta[|2\overline{X}_n - \theta| < \epsilon] \geq 1 - \delta \quad \Leftrightarrow \quad 2\Phi\left(\frac{\epsilon\sqrt{3n}}{\theta}\right) \geq 1 - \delta$$

$$\Rightarrow n_0 = \left[\frac{\theta^2}{3\epsilon^2}\Phi^{-2}\left(1 - \frac{\delta}{2}\right)\right] + 1 .$$

For $2\overline{X}_n$ from Chebyshev's inequality we get $n_0 = \lceil \theta^2/3\delta\epsilon^2 \rceil + 1$. In the
following code using R software, we compute the values of n_0 for $\theta = 2$,
$\epsilon = 0.02, 0.03, 0.04$ and $\delta = 0.02$ when $T_n = X_{(n)}$ and $T_n = 2\overline{X}_n$. We use the
function $\texttt{floor(x)}$ which gives the integer part of x.

```
th=2; ep=c(0.02,0.03,0.04); del=0.02
n0=floor(log(del)/log((2-ep)/2)) + 1;
n0 ## Minimum sample size for X_(n)
n1=floor(th^2*(qnorm(1-del/2))^2/(3*ep^2)) + 1;
n1 ## Minimum sample size for
    # 2*sample mean by normal approximation
n2=floor(th^2/(3*del*ep^2)) + 1;
n2 ## Minimum sample size for 2*sample mean
    # by Chebyshev's inequality
```

From the output, we note that corresponding to $\epsilon = 0.02, 0.03, 0.04$ and
$\delta = 0.02$, the minimum sample sizes for $X_{(n)}$ are $390, 259, 194$, for $2\overline{X}_n$ using
normal approximation, the minimum sample sizes are $18040, 8018, 4510$ and
using Chebyshev's inequality, these $166667, 74075, 41667$. It is to be noted that
for $\epsilon = 0.03$, minimum sample sizes for $X_{(n)}$ is 259 while for $2\overline{X}_n$, it is 8018,
almost 30 times more.

Following code gives one more approach to find the minimum sample size for
$2\overline{X}_n$ corresponding to the given precision in terms of ϵ and δ.

```
th = 2; ep = 0.04; del = 0.02; n = 100; nsim = 1000;
ind = 0; t2 = c(); pr = 0
while(ind==0)
{
for(i in 1:nsim)
{
set.seed(i)
x = runif(n,0,th)
t2[i] = 2*mean(x)
}
```

```
pr = length(which(abs(t2-th)<ep))/nsim
if(pr > 1-del) ind = 1
n = n + 50
}
n0 = n - 50;  n0
```

The minimum sample size for $2\overline{X}_n$ corresponding to $\epsilon = 0.4$ and $\delta = 0.02$ is 4650.

✍ Remark 2.7.1

It is to be noted that such an approach in the above code to decide the minimum sample size is quite general and is not based on any formula. It can be used for any distribution and any estimator. In this setup, instead of using any formula from theory, we rely purely on computations. We keep computing the desired probability for increasing values of sample size n and stop when the threshold in terms of $1 - \delta$ is crossed. The corresponding value is the required minimum sample size. One more feature to be noted in such a purely computational approach is that the minimum sample size may change if we change the seed while generating the samples. More precisely, if we change the seed, the minimum sample size will change by the increment in the sample size n, it is 50 in the above code. For example, if in the above code we set the seed as $2i$, then the minimum sample size corresponding to the same ϵ and δ is 4600. In the theoretical formulae based on normal approximation or based on Chebyshev's inequality, the minimum sample size does not depend on the generated samples at all. Using normal approximation the minimum sample size for $2\overline{X}_n$ corresponding to $\epsilon = 0.4$ and $\delta = 0.02$ is 4510. Thus, using the formula and using the purely computational approach we get approximately the similar results. In the above code, if we replace the estimator $2\overline{X}_n$ by $X_{(n)}$, the minimum sample size corresponding to $\epsilon = 0.4$ and $\delta = 0.02$ is 200. With the formula it is 194, thus again two approaches give comparable results.

In view of the major difference in minimum sample size for $X_{(n)}$ and $2\overline{X}_n$, in the following code, we have taken different initial sample sizes to verify the consistency of $X_{(n)}$ and $2\overline{X}_n$.

```
### Consistency of X(n)
th = 2; eps = c(0.02,0.03,0.04); tmax = c()
init = 100; incr = 100; nmax = 700; nsim = 1000
N = seq(init,nmax,incr); p1 = matrix(nrow=3,ncol=length(N))
for(i in 1:3)
{
for(j in 1:length(N))
{
n = N[j]
for(m in 1:nsim)
{
set.seed(m)
x = runif(n,0,th)
tmax[m] = max(x)      ## Estimator for m-th sample
}
p1[i,j] = length(which(abs(tmax-th)<eps[i]))/nsim
    ##  Estimate of coverage probability for ith epsilon
                               and jth sample size
}
}
par(mfrow= c(1,2))
plot(N,p1[1,],xlab="Sample Size",type="l",
               ylab="Estimate of Coverage Probability",
    main=expression(paste("X"[(n)])),lty=1,pch=20,
                              ylim=range(.4,1.02),col="red")
lines(N,p1[2,],lwd=2,col="maroon")
lines(N,p1[3,],lwd=3,col="dark red")
legend("bottomright",legend=c(eps[1],eps[2],eps[3]),lwd= c(1,2,3),
       col = c("red","maroon","dark red"),
      title=expression(paste(epsilon)))
###Consistency of 2* sample mean
t =c(); init = 2000; incr = 1000; nmax = 8000; nsim = 1000
N = seq(init,nmax,incr); p2 = matrix(nrow=3,ncol=length(N))
for(i in 1:3)
{
for(j in 1:length(N))
{
n = N[j]
for(m in 1:nsim)
{
set.seed(m)
x = runif(n,0,th)
t[m] = 2*mean(x)
}
p2[i,j] = length(which(abs(t-th)<eps[i]))/nsim
}
}
```

```
plot (N,p2 [1,],xlab="Sample Size",
               ylab="Estimate of Coverage Probability",
type="1",main=expression(paste(2*bar(X))),lty=1,
         pch=20,ylim=range(.4,1.02),col="blue")
lines (N,p2 [2,],lwd=2,col="purple") lines (N,p2 [3,],lwd=3,col="dark
blue") legend("bottomright",legend=c(eps[1],eps[2],eps[3]),lty =
c(1,1,1), lwd= c(1,2,3),col = c("blue","purple","dark blue"),
title=expression(paste(epsilon)))
```

From Fig. 2.2, we observe that as n increases, the coverage probability increases to 1 for both $X_{(n)}$ and $2\overline{X}_n$. Further, rate of convergence of coverage probability for $X_{(n)}$ is much higher than that for $2\overline{X}_n$. For smaller ϵ, the rate is slower, as expected. From this figure, we also note that when $T_n = X_{(n)}$, corresponding to all ϵ, the estimate of coverage probability is above 0.95 when n is above 400, supporting the computation of minimum sample sizes. Similar picture is revealed for $T_n = 2\overline{X}_n$.

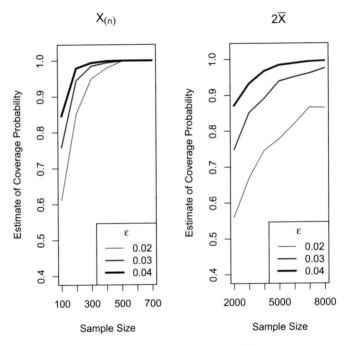

Fig. 2.2 Uniform $U(0, \theta)$ distribution: consistency of $X_{(n)}$ and $2\overline{X}_n$

In Example 2.2.1, we have shown that $X_{(1)}$ and $\overline{X}_n$ are not consistent estimators of θ. Following is a code to verify the same.

```
th = 2; t2 = c(); t3 = c(); eps = 0.1
init = 1000; incr = 1000; nmax = 20000; nsim = 1000
N = seq(init,nmax,incr); length(N); p2 = c(); p3 = c()
for(j in 1:length(N))
{
n = N[j]
for(m in 1:nsim)
{
set.seed(m)
x = runif(n,0,th)
t2[m] = min(x)
t3[m] = mean(x)
}
p2[j] = length(which(abs(t2-th)<eps))/1000
p3[j] = length(which(abs(t3-th)<eps))/1000
}
summary(p2);summary(p3)
```

The summary of both $p2$ and $p3$ displays that minimum observation, maximum observation, three quartiles and mean is 0, thus the estimate of coverage probability for $X_{(1)}$ and $\overline{X}_n$ are 0 for 20 values of n ranging from 1000 to 20000 with the increment of 1000 and for $\epsilon = 0.1$. Thus, the inconsistency of these estimators is reflected in simulation results.   □

All these codes and the entire procedure can be repeated for various values of $\theta \in \Theta$.

☑ Example 2.7.2

Suppose $X \sim N(\theta, 1)$ distribution, where $\theta \in \{0, 1\}$. On the basis of a random sample from the distribution X, in Example 2.2.3, we have shown that $\hat{\theta}_n$ as given below, is a consistent estimator of θ.

$$\hat{\theta}_n = \begin{cases} 1, & \text{if } \overline{X}_n > 1/2 \\ 0, & \text{if } \overline{X}_n \leq 1/2. \end{cases}$$

We verify these results by simulating samples from $N(0, 1)$ and $N(1, 1)$ and computing the estimate of coverage probability in both the cases and for three values of ϵ. Following is the R code.

```
t0 = c(); t1 = c(); th0=0; th1=1; eps = c(0.001,0.005,0.010)
init = 20; incr = 10; nmax = 60; nsim=1000; N = seq(init,nmax,incr)
p0 = matrix(nrow=3,ncol=length(N));
p1 = matrix(nrow=3,ncol=length(N))
for(i in 1:3)
{
for(j in 1:length(N))
{
n = N[j]
for(m in 1:nsim)
{
set.seed(m)
x0 = rnorm(n,th0,1)
t0[m] = ifelse(mean(x0)<0.5,0,1)
x1 = rnorm(n,th1,1)
t1[m] = ifelse(mean(x1)<0.5,0,1)
}
p0[i,j] = length(which(abs(t0-th0)<eps[i]))/nsim
p1[i,j] = length(which(abs(t1-th1)<eps[i]))/nsim
}
}
p0; p1
```

$p0$ and $p1$ give the estimates of coverage probability when $\theta = 0$ and $\theta = 1$ and these are displayed in Table 2.1 and Table 2.2 respectively.

For both the values of θ, and three values of ϵ, coverage probability is almost 1 for $n = 40$. It is in contrast with the results of Example 2.7.1 where the estimate of coverage probability for $X_{(n)}$ is close to 1 for $n > 800$ and for $2\overline{X}_n$, sample size needed is much larger. Thus, the minimum sample size required to attain the given level of accuracy depends on the underlying probability model. In Example 2.2.3, we have shown that for $N(\theta, 1)$ distribution with $\theta = 0, 1$ and for $0 < \epsilon \leq 1$,

$$P_0[|\hat{\theta}_n - 0| < \epsilon] = P_1[|\hat{\theta}_n - 1| < \epsilon] = \Phi\left(\sqrt{n}/2\right),$$

which can be obtained by the function $\mathtt{pnorm}\left(\sqrt{n}/2\right)$ in R. For $n=20, 30, 40, 50$ and 60, the values of coverage probabilities are $0.987, 0.997, 0.999, 0.998$ and 0.999 respectively. Using the expression for the coverage probability, we can find the minimum sample size n_0 from $\Phi\left(\sqrt{n}/2\right) \geq 1 - \delta$, for various values of δ. It is $n_0 = \left[4\Phi^{-2}(1 - \delta)\right] + 1$, which can be obtained as $n_0 = 4(\mathtt{qnorm}(1 - \delta))^2 + 1$ in R. Thus, for any $\epsilon \in (0, 1]$ and $\delta = 0.001, 0.003, 0.005, 0.007, 0.009$, $n_0 = 39, 31, 27, 25, 23$ respectively and these minimum sample sizes are similar to those obtained by simulation as is clear from values of n in Tables 2.2 and 2.3. $\square$

Table 2.2 Estimate of coverage probability when $\theta = 0$

ϵ	n				
	20	30	40	50	60
0.001	0.989	0.999	1	1	1
0.005	0.989	0.999	1	1	1
0.010	0.989	0.999	1	1	1

Table 2.3 Estimate of coverage probability when $\theta = 1$

ϵ	n				
	20	30	40	50	60
0.001	0.987	0.996	1	0.999	1
0.005	0.987	0.996	1	0.999	1
0.010	0.987	0.996	1	0.999	1

✒ Example 2.7.3

Suppose $\{X_1, X_2, \ldots, X_n\}$ is a random sample of size n from a gamma distribution with scale parameter α and shape parameter λ, with probability density function,

$$f(x, \alpha, \lambda) = \frac{\alpha^\lambda}{\Gamma(\lambda)} e^{-\alpha x} x^{\lambda-1}, \quad x > 0, \ \alpha > 0, \ \lambda > 0.$$

In Example 2.5.4, it is shown that $\underline{T}_n = \left(m_1'/m_2, m_1'^2/m_2\right)'$ is a consistent estimator of $\underline{\theta} = (\alpha, \lambda)'$. Using the following code, we verify both the marginal and joint consistency by simulation.

```
lambda = 4; alpha = 3; eps = 0.1; alest = laest = E = c();
nsim = 1000; init = 2000; incr = 1000; nmax = 15000
N = seq(init,nmax,incr); length(N); prob1 = prob2 = prob3 = c()
for(s in 1:length(N))
{
n = N[s]
for(i in 1:nsim)
{
set.seed(i)
x = rgamma(n,shape = lambda,scale = 1/alpha )
m1d = mean(x)
m2 = sum((x-m1d)^2)/n
alest[i] = m1d/m2
```

```
laest[i] = m1d^2/m2
E[i]=max(abs(alest[i]-alpha),abs(laest[i]-lambda))
}
prob1[s] = length(which(abs(alest-alpha)<eps))/nsim
           # consistency of alpha
prob2[s] = length(which(abs(laest-lambda)<eps))/nsim
           # consistency of lambda
prob3[s] = length(which(E <eps))/nsim
           # consistency of (alpha, lambda)'

}
d=data.frame(N, prob1, prob2, prob3); d
```

Estimates of coverage probabilities as obtained in data frame d are presented in Table 2.4.

In the table, the first two columns correspond to the estimate of coverage probability for α and λ respectively, while the third column gives the estimate of coverage probability $P[\max\{|\hat{\alpha}_n - \alpha|, |\hat{\lambda}_n - \lambda|\} < \epsilon]$. As the sample size n increases, the estimates of coverage probability increase in all the three columns, supporting the marginal and joint consistency. It is to be noted that estimates in the second and the third columns are almost the same. The result may not hold for all the distributions such as when the underlying model is a Cauchy distribution. It is clear from the next example. It is to be noted that for a gamma distribution, we need a large sample size and a large value of ϵ for coverage probability to be close to 1.   □

Table 2.4 Estimates of coverage probabilities: $G(\alpha, \lambda)$ distribution

n	For α	For λ	For α & λ
2000	0.643	0.538	0.514
3000	0.739	0.620	0.604
4000	0.799	0.685	0.669
5000	0.849	0.732	0.728
6000	0.882	0.777	0.771
7000	0.903	0.818	0.815
8000	0.919	0.839	0.837
9000	0.936	0.865	0.860
10000	0.954	0.887	0.886
11000	0.962	0.907	0.907
12000	0.978	0.919	0.919
13000	0.984	0.934	0.934
14000	0.985	0.945	0.945
15000	0.990	0.954	0.953

✏ Example 2.7.4

Suppose $\{X_1, X_2, \ldots, X_n\}$ is a random sample of size n from a Cauchy $C(\theta, \lambda)$ distribution with location parameter θ and scale parameter λ, with probability density function,

$$f(x, \theta, \lambda) = \frac{\lambda}{\pi} \frac{1}{\lambda^2 + (x - \theta)^2}, \quad x \in \mathbb{R}, \ \theta \in \mathbb{R}, \ \lambda > 0.$$

Since sample quantiles are consistent for corresponding population quantiles, we have $X_{([n/2]+1)} \xrightarrow{P_{\theta,\lambda}} \theta$ and $X_{([n/4]+1)} \xrightarrow{P_{\theta,\lambda}} \theta - \lambda$. Thus, $(\hat{\theta}_n, \hat{\lambda}_n)' = (X_{([n/2]+1)}, X_{([n/2]+1)} - X_{([n/4]+1)})'$ is consistent for $(\theta, \lambda)'$. Using the following code, which is similar to that in Example 2.7.3, we verify marginal and joint consistency of $(\hat{\theta}_n, \hat{\lambda}_n)'$.

```
lambda = 2; theta = 3; eps = 0.1; thest = laest = E = c()
nsim = 1000; init = 2000; incr = 1000; nmax = 10000
N = seq(init,nmax,incr); length(N); prob1 = prob2 = prob3 = c()
for(s in 1:length(N))
{
n = N[s]
for(i in 1:nsim)
{
set.seed(i)
x = rcauchy(n,location = theta,scale = lambda )
m1=median(x)
m2=quantile(x,.25)
thest[i] = m1
laest[i] = m1-m2
E[i]=max(abs(thest[i]-theta),abs(laest[i]-lambda))
}
prob1[s] = length(which(abs(thest-theta)<eps))/nsim
prob2[s] = length(which(abs(laest-lambda)<eps))/nsim
prob3[s] = length(which(E <eps))/nsim
}
d=data.frame(N, prob1, prob2, prob3); d
```

Estimates of coverage probabilities as obtained in data frame d are presented in Table 2.5.

Estimates in all the three columns converge to 1 as n increases, supporting the joint and marginal consistency. According to the notation in Theorem 2.5.1, we have,

$$P(E) = P(E_1 \cap E_2) \implies P(E^c) = P(E_1^c \cup E_2^c) \le P(E_1^c) + P(E_2^c)$$
$$\implies P(E) \ge P(E_1) + P(E_2) - 1.$$

Table 2.5 Estimates of coverage probabilities: $C(\theta, \lambda)$ distribution

N	For θ	For λ	For θ & λ
2000	0.844	0.708	0.598
3000	0.915	0.792	0.729
4000	0.953	0.825	0.787
5000	0.974	0.886	0.862
6000	0.983	0.924	0.909
7000	0.991	0.947	0.939
8000	0.991	0.967	0.958
9000	0.996	0.975	0.971
10000	0.998	0.980	0.978

We observe such a relationship, $prob3 \geq prob1 + prob2 - 1$, in three columns of the above output. Such a relationship is also observed in the previous example for a gamma distribution. □

✎ Example 2.7.5

Suppose $\{X_1, X_2, \ldots, X_n\}$ is a random sample from a normal $N(\theta, 1)$ distribution, $\theta \in \Theta = \{0, 1\}$. In Example 2.6.3, it is shown that for $0 < k < 1$,

$$T_k(\overline{X}_n) = \begin{cases} 0, & \text{if } \overline{X}_n < k \\ 1, & \text{if } \overline{X}_n \geq k \end{cases}$$

is consistent for θ as its coverage probability converges to 1 and its MSE converges to 0. Thus, we have an uncountable family of consistent estimator of θ. It has been argued in Example 2.6.3, that $T_{1/2}(\overline{X}_n) = \hat{\theta}_n$ performs better than any other estimator in the family. Using following R code, we verify the results established in Example 2.6.3.

```
n=c(20,30,40); k=seq(0.1,.9,.1)
x1=matrix(nrow=length(k),ncol=length(n))
 ## coverage probability at theta=0
x2=matrix(nrow=length(k),ncol=length(n))
 ## coverage probability at theta=1
x3=matrix(nrow=length(k),ncol=length(n)) ## MSE at theta=0
x4=matrix(nrow=length(k),ncol=length(n)) ## MSE at theta=1
for(i in 1:length(k))
{
for(j in 1:length(n))
{
x1[i,j]=pnorm(sqrt(n[j])*k[i])
x2[i,j]=1-pnorm(sqrt(n[j])*(k[i]-1))
```

```
x3[i,j]=1-pnorm(sqrt(n[j])*k[i])
x4[i,j]=pnorm(sqrt(n[j])*(k[i]-1))
}
}
x1=round(x1,2);x2=round(x2,2);x3=round(x3,2);x4=round(x4,2)
d=data.frame(k,x1[,1],x2[,1],x3[,1],x4[,1],
             x1[,2],x2[,2],x3[,2],x4[,2],
             x1[,3],x2[,3],x3[,3],x4[,3]); d
```

In Table 2.6, p_0, p_1, m_0, m_1 denote the coverage probability for $\theta = 0$ and $\theta = 1$ and MSE for $\theta = 0$ and $\theta = 1$ respectively. For $k \leq 0.5$, as sample size increases p_0 approaches to 1, p_1 is 1 for all the three sample sizes. while m_0 approaches to 0 and again $m_1 = 0$, for all the three sample sizes, as expected since $p_1 + m_1 = 1$. For $k > 0.5$, as sample size increases p_1 approaches to 1, p_0 is 1 for all the three sample sizes, while m_1 approaches to 0 and again $m_0 = 0$, for all the three sample sizes. As discussed in Example 2.6.3, we observe that as k increases from 0.1 to 0.9, for all the three sample sizes, p_0 increases, while p_1 decreases and m_0 decreases while m_1 increases. These are same for $k = 1/2$. Hence, $k = 1/2$ is the best choice for selecting an estimator from the family $\{T_k(\overline{X}_n), 0 < k < 1\}$. The estimator $T_k(\overline{X}_n)$ is a test function of the MP test as derived by the Neyman-Pearson lemma for testing $H_0 : \theta = 0$ against the alternative $H_1 : \theta = 1$ based on a random sample of size n from a normal $N(\theta, 1)$ distribution. It is to be noted that the probability of type I error is same as $1 - p_0$ and p_1 is the power of the test, that is, $1 - p_1$ is the probability of type II error. It is known that as the probability of type I error increases, the probability of type II error decreases. It is reflected in Table 2.6. Suppose we fix the probability of type I error α say as 0.05. Then $\Phi(\sqrt{n}k) = 0.95 \Rightarrow k = 0.3678$. Thus, $[\overline{X}_n > 0.3678]$ is the critical region of the MP test with level of significance $\alpha = 0.05$. Power of the test which is p_1 corresponding to $k = 0.3678$ is 0.9976. □

Table 2.6 Performance of $T_k(\overline{X}_n)$

k	$n = 20$				$n = 30$				$n = 40$			
	p_0	p_1	m_0	m_1	p_0	p_1	m_0	m_1	p_0	p_1	m_0	m_1
0.1	0.67	1.00	0.33	0.00	0.71	1.00	0.29	0.00	0.74	1.00	0.26	0.00
0.2	0.81	1.00	0.19	0.00	0.86	1.00	0.14	0.00	0.90	1.00	0.10	0.00
0.3	0.91	1.00	0.09	0.00	0.95	1.00	0.05	0.00	0.97	1.00	0.03	0.00
0.4	0.96	1.00	0.04	0.00	0.99	1.00	0.01	0.00	0.99	1.00	0.01	0.00
0.5	**0.99**	**0.99**	**0.01**	**0.01**	**1.00**	**1.00**	**0.00**	**0.00**	**1.00**	**1.00**	**0.00**	**0.00**
0.6	1.00	0.96	0.00	0.04	1.00	0.99	0.00	0.01	1.00	0.99	0.00	0.01
0.7	1.00	0.91	0.00	0.09	1.00	0.95	0.00	0.05	1.00	0.97	0.00	0.03
0.8	1.00	0.81	0.00	0.19	1.00	0.86	0.00	0.14	1.00	0.90	0.00	0.10
0.9	1.00	0.67	0.00	0.33	1.00	0.71	0.00	0.29	1.00	0.74	0.00	0.26

In Sect. 2.6, we discussed two criteria for comparing consistent estimators. Both the criteria reduce to a single criterion if we restrict the class of all consistent estimators of θ to a subclass of consistent estimators whose asymptotic distribution with a suitable normalization is a normal distribution. Such a class of estimators is known as a class of consistent and asymptotically normal (CAN) estimators and is always referred to as a class of CAN estimators. The next chapter is devoted to the discussion of CAN estimators.

2.8   Conceptual Exercises

2.8.1   Suppose T_n is a consistent estimator of θ. Obtain conditions on the sequence $\{a_n, n \geq 1\}$ such that the following are also consistent estimators of θ.
(i) $a_n T_n$,
(ii) $a_n + T_n$ and
(iii) $(a_n + nT_n)/(n + 1)$.

2.8.2   Suppose $\{X_1, X_2, \ldots, X_n\}$ is a random sample of size n from a distribution of X where $E(X) = \theta$ and $E(X^2) = V < \infty$. Show that $T_n = \frac{2}{n(n+1)} \sum_{i=1}^{n} i X_i$ is a consistent estimator of θ.

2.8.3   Suppose $\{X_1, X_2, \ldots, X_n\}$ is a random sample from a distribution with mean θ and variance σ^2. Suppose $T_n = \sum_{i=1}^{n} a_i X_i$, where $\sum_{i=1}^{n} a_i \to 1$ and $\sum_{i=1}^{n} a_i^2 \to 0$. Show that T_n is consistent for θ.

2.8.4   Suppose $g(x)$ is an even, non-decreasing and non-negative function on $[0, \infty)$. Then show that T_n is consistent for $\eta(\theta)$ if $E(g(T_n - \eta(\theta)) \to 0 \ \forall \ \theta$.

2.8.5   Suppose $\{X_1, X_2, \ldots, X_n\}$ is a random sample of size n from the following distributions. (i) Bernoulli $B(1, \theta)$, (ii) Poisson $Poi(\theta)$, (iii) uniform $U(0, 2\theta)$ and exponential distribution with mean θ. Show that the sample mean $\overline{X}_n$ is consistent for θ using the four approaches discussed in Example 2.2.2.

2.8.6   Suppose $\{X_1, X_2, \ldots, X_n\}$ is a random sample of size n from a Poisson $Poi(\theta)$ distribution. Find the maximum likelihood estimator of θ and examine whether it is consistent for θ when (i) $\theta \in [a, b] \subset (0, \infty)$ and (ii) $\theta \in \{1, 2\}$.

2.8.7   Suppose $\{X_1, X_2, \ldots, X_n\}$ is a random sample from a uniform $U(\theta - 1, \theta + 1)$ distribution, $\theta \in \mathbb{R}$. (i) Examine whether $T_{1n} = \overline{X}_n$, $T_{2n} = X_{(1)} + 1$ and $T_{3n} = X_{(n)} - 1$ are consistent estimators for θ.

Which one is better? Why? (ii) Find an uncountable family of consistent estimators of θ based on sample quantiles.

2.8.8   Suppose $\{X_1, X_2, \ldots, X_n\}$ are independent random variables where X_i follows a uniform $U(i(\theta - 1), i(\theta + 1))$ distribution, $\theta \in \mathbb{R}$. Find the maximum likelihood estimator of θ and examine whether it is a consistent estimator of θ.

2.8.9   Suppose $\{X_1, X_2, \ldots, X_n\}$ are independent random variables where $X_i \sim U(0, i\theta)$ distribution, $\theta \in \Theta = \mathbb{R}^+$. (i) Find the maximum likelihood estimator of θ and examine whether it is consistent for θ. (ii) Find the moment estimator of θ and examine whether it is consistent for θ.

2.8.10   Suppose $\{X_1, X_2, \ldots, X_n\}$ is a random sample from a uniform $U(-\theta, \theta)$ distribution. Examine whether $-X_{(1)}$ and $X_{(n)}$ are both consistent for θ. Is $(X_{(n)} - X_{(1)})/2$ consistent for θ? Justify your answer.

2.8.11   Suppose $\{X_1, X_2, \ldots, X_{2n+1}\}$ is a random sample from a uniform $U(\theta - 1, \theta + 1)$ distribution, $\theta \in \mathbb{R}$. Examine whether $X_{(n)} - 1$ and $X_{([n/4]+1)}$ are consistent for θ.

2.8.12   Suppose $\{X_1, X_2, \ldots, X_n\}$ is a random sample of size n from a binomial $B(1, \theta)$ distribution, $\theta \in \Theta = (0, 1)$. (i) Find the maximum likelihood estimator of θ and examine whether it is consistent for θ. (ii) Find the moment estimator of θ and examine whether it is consistent for θ. (iii) Find the maximum likelihood estimator of θ and examine whether it is consistent for θ, if $\Theta = (a, b) \subset (0, 1)$.

2.8.13   Suppose $\{X_1, X_2, \ldots, X_n\}$ is a random sample of size n from a normal $N(\theta, 1)$ distribution, $\theta \in \Theta = \{0, 1\}$. An estimator $T_k(\overline{X}_n)$ is defined as follows. Prove that it is consistent for θ if and only if $0 < k < 1$.

$$T_k(\overline{X}_n) = \begin{cases} 0, & \text{if } \overline{X}_n < k \\ 1, & \text{if } \overline{X}_n \geq k. \end{cases}$$

2.8.14   Suppose $\{X_1, X_2, \ldots, X_n\}$ is a random sample of size n from a normal $N(\theta, 1)$ distribution, $\theta \in \Theta = \{-1, 0, 1\}$. (i) Find the maximum likelihood estimator of θ and examine whether it is consistent for θ. (ii) Examine whether it is unbiased for θ. Examine whether it is asymptotically unbiased for θ.

2.8.15   Suppose $\{X_1, X_2, \ldots, X_n\}$ is a random sample from a Cauchy $C(\theta, 1)$ distribution, where $\theta \in \mathbb{R}$. Examine whether the sample mean is consistent for θ.

2.8.16   Suppose $X_1, \ldots, X_n$ is a random sample from X with probability density
          function $f(x, \theta) = \theta/x^2$, $x \geq \theta$, $\theta > 0$. (i) Find the maximum likeli-
          hood estimator of θ and examine its consistency for θ by computing the
          coverage probability and also the MSE. (ii) Examine if $X_{(n)}$ is consistent
          for θ.

2.8.17   Suppose X follows a Laplace distribution with probability density func-
          tion $f(x, \theta) = \exp\{-|x - \theta|\}/2$, $x \in \mathbb{R}$, $\theta \in \mathbb{R}$. A random sample of
          size n is drawn from the distribution of X. Examine whether $\overline{X}_n$ is con-
          sistent for θ. Examine if it is MSE consistent. Is sample median consistent
          for θ? Justify. Find the maximum likelihood estimator of θ and examine
          whether it is consistent for θ. Find a family of consistent estimators of θ
          based on sample quantiles.

2.8.18   Suppose $\{X_1, X_2, \ldots, X_n\}$ is a random sample from an exponential dis-
          tribution with mean θ. Show that $\overline{X}_n$ is MSE consistent for θ. Find a
          constant $c \in \mathbb{R}$ such that $T_n = n\overline{X}_n/(n + c)$ has MSE smaller than that
          of $\overline{X}_n$. Is T_n consistent for θ? Justify.

2.8.19   Suppose $\{X_1, X_2, \ldots, X_n\}$ is a random sample from a Laplace distribu-
          tion with probability density function $f(x, \theta)$ given by

$$f(x, \theta) = (1/2\theta) \exp\{-|x|/\theta\}, \quad x \in \mathbb{R}, \quad \theta > 0.$$

          Examine whether following estimators are consistent for θ.
          (i) Sample mean, (ii) sample median, (iii) $\sum_{i=1}^{n} |X_i|/n$ and
          (iv) $(\sum_{i=1}^{n} X_i^2/n)^{1/2}$.

2.8.20   Suppose $\{X_1, X_2, \ldots, X_n\}$ is a random sample of size n from a distr-
          ibution of X with probability density function $f(x, \theta) = \theta/x^{\theta+1}$,
          $x > 1, \theta > 0$. Examine whether $\overline{X}_n$ is consistent for θ for $\theta > 1$. What
          happens if $0 < \theta \leq 1$? Obtain consistent estimator of θ based on the
          transformations $\log x$ and $1/x$.

2.8.21   Suppose $\{X_1, X_2, \ldots, X_n\}$ is a random sample of size n from an exponen-
          tial distribution with scale parameter 1 and location parameter θ. Examine
          whether $X_{(1)}$ is strongly consistent for θ.

2.8.22   Suppose $\{X_1, X_2, \ldots, X_n\}$ is a random sample from a normal $N(\theta, 1)$
          distribution, where $\theta \in \{0, 1\}$. Examine whether the maximum likelihood
          estimator $\hat{\theta}_n$ of θ is strongly consistent for θ. Examine whether $\hat{\theta}_n = \theta$
          almost surely for large n.

2.8.23   Suppose $\{X_1, X_2, \ldots, X_n\}$ is a random sample of size n from a normal $N(\theta, 1)$ distribution, $\theta \in \Theta = I$, the set of integers. Examine whether the maximum likelihood estimator $\hat{\theta}_n$ of θ is strongly consistent for θ. Examine whether $\hat{\theta}_n = \theta$ almost surely for large n.

2.8.24   Suppose $\{X_1, X_2, \ldots, X_n\}$ is a random sample of size n from a Bernoulli $B(1, \theta)$ distribution. Examine whether $\overline{X}_n$ is uniformly consistent for $\theta \in \Theta = (0, 1)$.

2.8.25   Suppose X follows an exponential distribution with location parameter μ and scale parameter σ, with probability density function $f(x, \mu, \sigma)$ as,

$$f(x, \mu, \sigma) = (1/\sigma) \exp\{-(x - \mu)/\sigma\} \quad x \geq \mu, \quad \sigma > 0, \quad \mu \in \mathbb{R}.$$

Suppose $\{X_1, X_2, \ldots, X_n\}$ is a random sample from the distribution of X. (i) Verify whether $\overline{X}_n$ is consistent for μ or σ. (ii) Find a consistent estimator for $\underline{\theta} = (\mu, \sigma)'$ based on the sample median and the sample mean.

2.8.26   Suppose $\{(X_1, Y_1)', (X_2, Y_2)', \ldots, (X_n, Y_n)'\}$ is a random sample from a bivariate Cauchy $C_2(\theta_1, \theta_2, \lambda)$ distribution, with probability density function given by Kotz et al. [2],

$$f(x, y, \theta_1, \theta_2, \lambda) = \frac{\lambda}{2\pi}\{\lambda^2 + (x - \theta_1)^2 + (y - \theta_2)^2\}^{-3/2} \ (x, y)' \in \mathbb{R}^2,$$
$$\theta_1, \theta_2 \in \mathbb{R}, \ \lambda > 0.$$

Using marginal sample quartiles obtain two distinct consistent estimators of $(\theta_1, \theta_2, \lambda)'$. Hence, obtain a family of consistent estimators of $(\theta_1, \theta_2, \lambda)'$.

2.8.27   Suppose $\{X_1, X_2, \ldots, X_n\}$ is a random sample from an exponential distribution with probability density function $f(x, \theta)$ given by, $f(x, \theta) = (1/\alpha) \exp\{-(x - \theta)/\alpha\}, \quad x \geq \theta, \ \theta \in \mathbb{R}, \alpha > 0$. Show that $(X_{(1)}, (\sum_{i=2}^{n}(X_{(i)} - X_{(1)})/(n - 1))'$ is consistent for $(\theta, \alpha)'$. Obtain a consistent estimator of $(\theta, \alpha)'$ based on sample moments.

2.8.28   Suppose $\{X_{ij}, i = 1, 2, \ j = 1, 2, \ldots, n\}$ are independent random variables such that $X_{ij} \sim N(\mu_i, \sigma^2)$ distribution. Find maximum likelihood estimator of μ_i $i = 1, 2$ and σ^2. Examine whether these are consistent.

2.8.29   Suppose $\{X_1, X_2, \ldots, X_n\}$ is a random sample from a normal $N(\theta, \sigma^2)$ distribution. Suppose $S_n^2 = \sum_{i=1}^{n}(X_i - \overline{X}_n)^2$. (i) Examine whether $T_{1n} = S_n^2/n$ and $T_{2n} = S_n^2/(n-1)$ are consistent for θ. (ii) Show that $MSE_\theta(T_{1n}) < MSE_\theta(T_{2n})$, $\forall\, n \geq 2$. (iii) Show that $T_{3n} = S_n^2/(n+k)$ is consistent for θ. Determine k such that $MSE_\theta(T_{3n})$ is minimum.

2.8.30   An electronic device is such that the probability of its instantaneous failure is θ, that is, if X denotes the life length random variable of the device, then $P[X = 0] = \theta$. Given that $X > 0$, the conditional distribution of life length is exponential with mean α. In a random sample of size n, it is observed that r items failed instantaneously and remaining $n - r$ items had life times $\{X_{i_1}, X_{i_2}, \ldots, X_{i_{n-r}}\}$. On the basis of these data, find consistent estimator of $(\theta, \alpha)'$.

2.8.31   A linear regression model is $Y = a + bX + \epsilon$ where $E(\epsilon) = 0$ & $Var(\epsilon) = \sigma^2$. Suppose $\{(X_i, Y_i)', \; i = 1, 2, \ldots, n\}$ is a random sample from the distribution of $(X, Y)'$. Examine whether the least square estimators of a and b are consistent for a and b respectively.

2.8.32   Suppose $\{X_1, X_2, \ldots, X_n\}$ is a random sample of size n from a normal $N(\mu, \sigma^2)$ distribution. Find a consistent estimator of $P[X_1 < a]$ where a is any real number.

2.8.33   Suppose $\{\underline{Z}_1, \underline{Z}_2, \ldots, \underline{Z}_n\}$ is a random sample of size n from a multivariate normal $N_p(\underline{\mu}, \Sigma)$ distribution. Find a consistent estimator of $\underline{\theta} = (\underline{\mu}, \Sigma)$. Also find a consistent estimator of $\underline{l}'\underline{\mu}$ where $\underline{l}$ is a vector in $\mathbb{R}^p$.

2.8.34   On the basis of a random sample of size n from a multinomial distribution in k cells with cell probabilities $(p_1, p_2, \ldots, p_k)$, with $\sum_{i=1}^{k} p_i = 1$, find a consistent estimator for $\underline{p} = (p_1, p_2, \ldots, p_{k-1})$.

2.8.35   Suppose $\{X_1, X_2, \ldots, X_n\}$ is a random sample of size n from a uniform $U(\theta_1, \theta_2)$ distribution, $-\infty < \theta_1 < x < \theta_2 < \infty$. Examine whether $(X_{(1)}, X_{(n)})'$ is consistent for $(\theta_1, \theta_2)'$. Obtain consistent estimator for $(\theta_1 + \theta_2)/2$ and $(\theta_2 - \theta_1)^2/12$ based on $(X_{(1)}, X_{(n)})'$ and also based on sample moments.

2.8.36   Suppose $\{X_1, X_2, \ldots, X_n\}$ is a random sample of size n from a Laplace distribution with probability density function given by,

$$f(x, \theta, \lambda) = (1/2\lambda)\exp\{-|x - \theta|/\lambda\}, \quad x \in \mathbb{R}, \quad \theta \in \mathbb{R}, \quad \lambda > 0.$$

Using stepwise maximization procedure find the maximum likelihood estimator of θ and λ and examine if those are consistent for θ and λ respectively.

2.9   Computational Exercises

Verify the results established in the following examples and exercises by simulation using R software.

2.9.1   Exercise 2.8.7(i) (Hint: Use code similar to Example 2.7.1).

2.9.2   Exercise 2.8.25 (Hint: Use code similar to Examples 2.7.1 and 2.7.4).

2.9.3   Exercise 2.8.27 (Hint: Use code similar to Example 2.7.4).

2.9.4   Exercise 2.8.29((i)&(ii)) (Hint: Use code similar to Example 2.7.1).

2.9.5   Example 2.2.6

2.9.6   Example 2.2.8

2.9.7   Example 2.2.15, for $p = 0.25, 0.50, 0.75$

2.9.8   Example 2.5.3

2.9.9   Example 2.5.4

2.9.10   Example 2.6.1

2.9.11   Example 2.6.2

References

1. Gut, A. (2005). *Probability: A graduate course*. New York: Springer.
2. Kotz, S., Balakrishnan, N., & Johnson, N. L. (2000). *Continuous multivariate distributions: Models and applications* (Vol. I, 2nd ed.). New York: Wiley.

Consistent and Asymptotically Normal Estimators

3

Contents

> **Learning Objectives** After going through this chapter, the readers should be able
> - to understand the concept of a consistent and asymptotically normal (CAN) estimator of a real and vector valued parameter
> - to generate a CAN estimator using different methods such as method of moments, method based on the sample quantiles and the delta method
> - to judge the asymptotic normality of estimators using R

3.1   Introduction

In Chap. 2 we discussed a basic large sample property of consistency of an estimator. The present chapter is devoted to a study of an additional property of a consistent estimator, which involves its asymptotic distribution with a suitable normalization. Suppose T_n is a consistent estimator of θ. In view of the fact that convergence

© The Author(s), under exclusive license to Springer Nature Singapore Pte Ltd. 2021
S. Deshmukh and M. Kulkarni, *Asymptotic Statistical Inference*,
https://doi.org/10.1007/978-981-15-9003-0_3

in probability implies convergence in law, $T_n \xrightarrow{L} \theta$, $\forall$ θ. Thus, the asymptotic distribution of T_n is degenerate at θ. Such a degenerate distribution is not helpful to find the rate of convergence or to find an interval estimator of θ. Hence, our aim is to find a blowing factor a_n such that the asymptotic distribution of $a_n(T_n - \theta)$ is non-degenerate.

Suppose $a_n(T_n - \theta) \xrightarrow{L} U$, where U is a real random variable and $a_n \to \infty$ as $n \to \infty$, then $a_n(T_n - \theta)$ is bounded in probability, by Result 1.3.10 which states that if $X_n \xrightarrow{L} X$ then the sequence $\{X_n, n \geq 1\}$ is bounded in probability. Hence,

$$T_n - \theta = \frac{1}{a_n} a_n(T_n - \theta) \xrightarrow{P_\theta} 0, \ \forall \ \theta.$$

Hence, T_n is a consistent estimator for θ. Thus, if $a_n(T_n - \theta)$ converges in law, then T_n is consistent for θ. In particular, if $a_n = \sqrt{n}$, then an estimator T_n is said to be a $\sqrt{n}$ consistent estimator of θ.

It is particularly of interest to find a sequence $\{a_n, n \geq 1\}$ of real numbers tending to ∞ as $n \to \infty$, so that the asymptotic distribution of $a_n(T_n - \theta)$ is normal. Estimators for which large sample distribution of $a_n(T_n - \theta)$ is normal are known as consistent and asymptotically normal (CAN) estimators. These play a key role in large sample inference theory. In Sects. 3.2 and 3.3, we investigate various properties of CAN estimators for a real as well as a vector parameter respectively. Section 3.4 is devoted to verification of CAN property by simulation using R.

3.2   CAN Estimator: Real Parameter Setup

We begin with a precise definition of a CAN estimator for a real parameter setup.

▶ **Definition 3.2.1**
Consistent and Asymptotically Normal Estimator: Suppose T_n is an estimator of θ and suppose there exists a sequence $\{a_n, n \geq 1\}$ of real numbers tending to ∞ as $n \to \infty$, such that $a_n(T_n - \theta)/\sqrt{v(\theta)} \xrightarrow{L} Z$ as $n \to \infty$, where $Z \sim N(0, 1)$ distribution and $0 < v(\theta) < \infty$. Then the estimator T_n is said to be a consistent and asymptotically normal estimator with approximate variance $v(\theta)/a_n^2$.

As mentioned in the last chapter, two criteria for comparing consistent estimators reduce to a single criterion for comparing approximate variances, if we restrict the class of all consistent estimators to a subclass of CAN estimators. If T_n is a CAN estimator of θ, it is asymptotically unbiased and hence MSE of T_n as an estimator of θ is the approximate variance $v(\theta)/a_n^2$. Thus, comparison based on MSE reduces to comparison based on the approximate variance. Suppose T_{1n} and T_{2n} are two CAN estimators of θ such that

$$a_n(T_{1n} - \theta) \xrightarrow{L} U_1 \sim N(0, v_1(\theta)) \ \text{ and } \ a_n(T_{2n} - \theta) \xrightarrow{L} U_2 \sim N(0, v_2(\theta)).$$

If $v_1(\theta) < v_2(\theta) \ \forall \ \theta$, then T_{1n} is preferred to T_{2n} for large n.

Using the asymptotic normal distribution of T_n, we can obtain the rate of convergence of the coverage probability, as shown below. Suppose for $\epsilon > 0$, $p(\epsilon, n)$ denotes the coverage probability of a CAN estimator T_n of θ. Then,

$$
\begin{aligned}
p(\epsilon, n) &= P_\theta[|T_n - \theta| < \epsilon] = P_\theta\left[\frac{-\epsilon a_n}{\sqrt{v(\theta)}} < \frac{a_n(T_n - \theta)}{\sqrt{v(\theta)}} < \frac{\epsilon a_n}{\sqrt{v(\theta)}}\right] \\
&\approx \Phi\left(\frac{\epsilon a_n}{\sqrt{v(\theta)}}\right) - \Phi\left(\frac{-\epsilon a_n}{\sqrt{v(\theta)}}\right) \quad \text{as } T_n \text{ is CAN for } \theta \\
&= 2\Phi\left(\frac{\epsilon a_n}{\sqrt{v(\theta)}}\right) - 1 \\
&\to 1 \quad \text{at the rate of } a_n \to \infty.
\end{aligned}
$$

Thus, if T_n is CAN, then the rate of convergence of true coverage probability to 1 is the same as the rate of convergence of a_n to ∞. We now examine the preference rule based on the coverage probability. For $\epsilon > 0$, the coverage probabilities are given by

$$
p_1(\epsilon, n) = 2\Phi\left(\frac{\epsilon a_n}{\sqrt{v_1(\theta)}}\right) - 1 \quad \text{and} \quad p_2(\epsilon, n) = 2\Phi\left(\frac{\epsilon a_n}{\sqrt{v_2(\theta)}}\right) - 1.
$$

Suppose $v_1(\theta) < v_2(\theta) \ \forall \ \theta$. Then for any $\epsilon > 0$ and for large n,

$$
v_1(\theta) < v_2(\theta) \ \Rightarrow \ \frac{\epsilon a_n}{\sqrt{v_1(\theta)}} > \frac{\epsilon a_n}{\sqrt{v_2(\theta)}} \ \Rightarrow \ p_1(\epsilon, n) > p_2(\epsilon, n)
$$

and T_{1n} is preferred to T_{2n} for large n. If the sequence of norming constants is $\{a_n, n \geq 1\}$ for T_{1n} and $\{b_n, n \geq 1\}$ for T_{2n} then the comparison is based on the approximate variances $v_1(\theta)/a_n^2$ and $v_2(\theta)/b_n^2$. If $v_1(\theta)/a_n^2 < v_2(\theta)/b_n^2 \ \forall \ \theta$ then T_{1n} is preferred to T_{2n} for large n.

From the large sample expression of $p(\epsilon, n) = 2\Phi\left(\epsilon a_n/\sqrt{v(\theta)}\right) - 1$, we can find the minimum sample size $n_0(\epsilon, \delta, \theta)$ required for a consistent estimator T_n to achieve the degree of accuracy specified by (ϵ, δ), $\epsilon > 0$, $0 < \delta < 1$. Thus, in particular with $a_n^2 = n$,

$$
p(\epsilon, n) = 2\Phi\left(\frac{\epsilon \sqrt{n}}{\sqrt{v(\theta)}}\right) - 1 > 1 - \delta
$$

$$
\Rightarrow \ n_0(\epsilon, \delta, \theta) = \left[\frac{v(\theta)}{\epsilon^2}\left[\Phi^{-1}\left(1 - \frac{\delta}{2}\right)\right]^2\right] + 1.
$$

If T_{1n} and T_{2n} are two CAN estimators of θ with approximate variances $v_1(\theta)/n$ and $v_2(\theta)/n$ respectively, then corresponding to the degree of accuracy specified by (ϵ, δ), minimum sample sizes are given by

$$n_{01}(\epsilon, \delta, \theta) = \left[\frac{v_1(\theta)}{\epsilon^2} \left[\Phi^{-1}\left(1 - \frac{\delta}{2}\right) \right]^2 \right] + 1$$

$$\&\ n_{02}(\epsilon, \delta, \theta) = \left[\frac{v_2(\theta)}{\epsilon^2} \left[\Phi^{-1}\left(1 - \frac{\delta}{2}\right) \right]^2 \right] + 1 .$$

Hence, if $v_1(\theta) < v_2(\theta)$ $\forall$ θ, then $n_{01}(\epsilon, \delta, \theta) < n_{02}(\epsilon, \delta, \theta)$ $\forall$ θ and T_{1n} is preferred to T_{2n} for large n.

Following example illustrates how to verify that a given estimator is a CAN estimator.

☑ Example 3.2.1

(i) Suppose $\underline{X} \equiv \{X_1, X_2, \ldots, X_n\}$ is a random sample from a $N(\theta, 1)$, $\theta \in \mathbb{R}$. Then the maximum likelihood estimator $\hat{\theta}_n$ of θ is $\hat{\theta}_n = \overline{X}_n$. In Example 2.2.2, it is shown that $\overline{X}_n$ is consistent for θ. Further, $\overline{X}_n$ has normal $N(\theta, 1/n)$ distribution, that is, for each n, $\sqrt{n}(\overline{X}_n - \theta)$ has $N(0, 1)$ distribution, hence its asymptotic distribution is also standard normal. Thus, $\overline{X}_n$ is CAN for θ with approximate variance $1/n$. It may be noted that for $N(\theta, 1)$ distribution, the approximate variance of $\overline{X}_n$ is $I^{-1}(\theta)/n$, where $I(\theta)$ is the information function.

(ii) If $\theta \in \{0, 1\}$, it is shown in Example 2.2.3 that the maximum likelihood estimator $\hat{\theta}_n$ of θ is given by

$$\hat{\theta}_n = \begin{cases} 1, & \text{if } \overline{X}_n > \frac{1}{2} \\ 0, & \text{if } \overline{X}_n \leq \frac{1}{2} \end{cases}$$

and it is consistent for θ. To examine if it is CAN, we find the limit law of $\sqrt{n}(\hat{\theta}_n - \theta)$ for $\theta = 0$ and $\theta = 1$. For $\theta = 0$,

$$P_0[\sqrt{n}(\hat{\theta}_n - 0) \leq x] = P_0[\hat{\theta}_n \leq x/\sqrt{n}] = 0 \text{ if } x < 0 \text{ as } \hat{\theta}_n = 0 \text{ or } 1.$$

For $x = 0$, $P_0[\sqrt{n}(\hat{\theta}_n - 0) \leq 0] = P_0[\hat{\theta}_n = 0] = P_0[\overline{X}_n \leq 1/2] = \Phi(\sqrt{n}/2)$.

For $x > 0$,

$$P_0[\hat{\theta}_n \leq x/\sqrt{n}] = \begin{cases} P_0[\hat{\theta}_n = 0] = \Phi(\sqrt{n}/2), & \text{if } 0 < x/\sqrt{n} < 1 \\ 1, & \text{if } x/\sqrt{n} \geq 1 . \end{cases}$$

Thus,

$$P_0[\sqrt{n}(\hat{\theta}_n - 0) \leq x] = \begin{cases} 0, & \text{if } x < 0 \\ \Phi(\sqrt{n}/2), & \text{if } x = 0 \\ \Phi(\sqrt{n}/2), & \text{if } 0 < x < \sqrt{n} \\ 1, & \text{if } x \geq \sqrt{n} . \end{cases}$$

Consequently as $n \to \infty$,

$$P_0[\sqrt{n}(\hat{\theta}_n - 0) \leq x] \quad \to \quad \begin{cases} 0, & \text{if } x < 0 \\ 1, & \text{if } x \geq 0 . \end{cases}$$

Thus, for $\theta = 0$, $\sqrt{n}(\hat{\theta}_n - 0) \overset{L}{\to} U \equiv 0$. Suppose $\theta = 1$,

$$P_1[\sqrt{n}(\hat{\theta}_n - 1) \leq x] = P_1[\hat{\theta}_n \leq 1 + x/\sqrt{n}] = 0$$
$$\text{if } 1 + x/\sqrt{n} < 0 \quad \Leftrightarrow \quad x < -\sqrt{n}.$$

Suppose $-\sqrt{n} < x < 0 \quad \Leftrightarrow \quad 1 + x/\sqrt{n} < 1$. Then

$$\begin{aligned} P_1[\hat{\theta}_n \leq 1 + x/\sqrt{n}] &= P_1[\hat{\theta}_n = 0] = P_1[\overline{X}_n \leq 1/2] \\ &= P_1[\sqrt{n}(\overline{X}_n - 1) \leq \sqrt{n}(1/2 - 1)] = \Phi(-\sqrt{n}/2). \end{aligned}$$

For $x = 0$, $P_1[\sqrt{n}(\hat{\theta}_n - 1) \leq 0] = P_1[\hat{\theta}_n \leq 1] = 1$.

For $x > 0 \quad \Leftrightarrow \quad 1 + x/\sqrt{n} > 1$, then $P_1[\hat{\theta}_n \leq 1 + x/\sqrt{n}] = 1$. Thus,

$$P_1[\sqrt{n}(\hat{\theta}_n - 1) \leq x] \quad = \quad \begin{cases} 0, & \text{if } x < -\sqrt{n} \\ \Phi(-\sqrt{n}/2), & \text{if } -\sqrt{n} < x < 0 \\ 1, & \text{if } x \geq 0 . \end{cases}$$

Hence, as $n \to \infty$,

$$P_1[\sqrt{n}(\hat{\theta}_n - 1) \leq x] \quad \to \quad \begin{cases} 0, & \text{if } x < 0 \\ 1, & \text{if } x \geq 0 . \end{cases}$$

Thus, for $\theta = 1$ also, $\sqrt{n}(\hat{\theta}_n - 1) \overset{L}{\to} U \equiv 0$. Hence, the asymptotic distribution of $\sqrt{n}(\hat{\theta}_n - \theta)$ is degenerate at 0 for $\theta \in \{0, 1\}$. Thus, with the norming factor $\sqrt{n}$, the asymptotic distribution of $Y_n = \sqrt{n}(\hat{\theta}_n - \theta)$ is not normal, but we cannot claim that $\hat{\theta}_n$ is not CAN. From the definition of the CAN estimator, one has to check whether the asymptotic distribution of $(\hat{\theta}_n - \theta)$ is normal with some other norming factor. Hence, we consider a collection of sequences $\{a_n, n \geq 1\}$ of real numbers tending to ∞ as $n \to \infty$. Such a collection can be classified in three groups depending on the rate of convergence of a_n as follows: (i) a_n and $\sqrt{n}$ converge to ∞ at the same rate, (ii) $a_n \to \infty$ faster than $\sqrt{n} \to \infty$ and (iii) $a_n \to \infty$ slower than $\sqrt{n} \to \infty$. Now,

$$U_n = a_n(\hat{\theta}_n - \theta) = \frac{a_n}{\sqrt{n}}Y_n \overset{L}{\to} \begin{cases} U, & \text{if } a_n \text{ \& } n \to \infty \text{ at the same rate} \\ \infty, & \text{if } a_n \to \infty \text{ faster than } \sqrt{n} \to \infty \\ 0, & \text{if } a_n \to \infty \text{ slower than } \sqrt{n} \to \infty. \end{cases}$$

Thus, there exists no sequence $\{a_n, n \geq 1\}$ of real numbers tending to ∞ as $n \to \infty$ such that the asymptotic distribution of $a_n(\hat{\theta}_n - \theta)$ is normal. Hence we claim that $\hat{\theta}_n$ is not a CAN estimator of θ.

(iii) Suppose the parameter space is $\Theta = [a, b]$. It is shown in Example 2.2.3 that the maximum likelihood estimator $\hat{\theta}_n$ of θ is

$$\hat{\theta}_n = \begin{cases} a, & \text{if} \quad \overline{X}_n < a \\ \overline{X}_n, & \text{if} \quad \overline{X}_n \in [a, b] \\ b, & \text{if} \quad \overline{X}_n > b \end{cases}$$

and is consistent for θ. To investigate the asymptotic distribution of $\sqrt{n}(\hat{\theta}_n - \theta)$, it is to be noted that $\sqrt{n}(\hat{\theta}_n - \theta) - \sqrt{n}(\overline{X}_n - \theta) = \sqrt{n}(\hat{\theta}_n - \overline{X}_n)$ and

$$\forall \epsilon > 0, \ P_\theta[\sqrt{n}|\hat{\theta}_n - \overline{X}_n| < \epsilon] \geq P_\theta[\hat{\theta}_n = \overline{X}_n] \ \to \ 1 \text{ as } n \to \infty, \ \forall \theta \in (a, b)$$

as shown in Example 2.2.3. Thus, the asymptotic distribution of $\sqrt{n}(\hat{\theta}_n - \theta)$ and of $\sqrt{n}(\overline{X}_n - \theta)$ is the same $\forall \theta \in (a, b)$. But $\sqrt{n}(\overline{X}_n - \theta)$ has standard normal distribution for all $\theta \in [a, b]$. Hence, for all $\theta \in (a, b)$ the asymptotic distribution of $\sqrt{n}(\hat{\theta}_n - \theta)$ is standard normal. To find the asymptotic distribution of $\sqrt{n}(\hat{\theta}_n - \theta)$ at $\theta = a$, we study the limit of $P_a[\sqrt{n}(\hat{\theta}_n - a) \leq x]$ for all $x \in \mathbb{R}$. Since $\hat{\theta}_n \geq a$, for $x < 0$, $P_a[\sqrt{n}(\hat{\theta}_n - a) \leq x] = 0$. Suppose $x = 0$. Then

$$P_a[\sqrt{n}(\hat{\theta}_n - a) \leq 0] = P_a[\sqrt{n}(\hat{\theta}_n - a) = 0] = P_a[\hat{\theta}_n = a] = P_a[\overline{X}_n < a]$$
$$= P_a[\sqrt{n}(\overline{X}_n - a) < 0] = \Phi(0) = 1/2.$$

From the expression of $\hat{\theta}_n$, we have $\hat{\theta}_n \leq b \ \Rightarrow \ \sqrt{n}(\hat{\theta}_n - a) \leq \sqrt{n}(b - a)$. Suppose $0 < x < \sqrt{n}(b - a)$. Then

$$P_a[\sqrt{n}(\hat{\theta}_n - a) \leq x] = P_a[\sqrt{n}(\hat{\theta}_n - a) \leq 0] + P_a[0 < \sqrt{n}(\hat{\theta}_n - a) \leq x]$$
$$= 1/2 + P_a[0 < \sqrt{n}(\overline{X}_n - a) \leq x]$$
$$= 1/2 + \Phi(x) - 1/2 = \Phi(x).$$

If $x \geq \sqrt{n}(b - a)$, then $P_a[\sqrt{n}(\hat{\theta}_n - a) \leq x] = 1$. Thus,

$$P_a[\sqrt{n}(\hat{\theta}_n - a) \leq x] = \begin{cases} 0, & \text{if} & x < 0 \\ 1/2 = \Phi(x), & \text{if} & x = 0 \\ \Phi(x), & \text{if} & 0 < x < \sqrt{n}(b - a) \\ 1, & \text{if} & x \geq \sqrt{n}(b - a) . \end{cases}$$

Hence,

$$P_a[\sqrt{n}(\hat{\theta}_n - a) \leq x] \ \to \ \begin{cases} 0, & \text{if} \ x < 0 \\ \Phi(x), & \text{if} \ x \geq 0 . \end{cases}$$

Thus, for $\theta = a$, the asymptotic distribution of $\sqrt{n}(\hat{\theta}_n - \theta)$ is not normal and 0 is a point of discontinuity. We show that it is a mixture of discrete and continuous distributions and the continuous distribution is related to the standard normal distribution. Suppose U_1 is a random variable with a distribution degenerate at 0. Then its distribution function is given by

$$F_{U_1}(x) = \begin{cases} 0, \text{ if } x < 0 \\ 1, \text{ if } x \geq 0. \end{cases}$$

Suppose a random variable U_2 is defined as $U_2 = |U|$ where $U \sim N(0, 1)$. Then $P[U_2 \leq x] = 0$ if $x < 0$. Suppose $x \geq 0$, then

$$P[U_2 \leq x] = P[|U| \leq x] = P[-x \leq U \leq x] = \Phi(x) - \Phi(-x) = 2\Phi(x) - 1.$$

Thus, the distribution function of U_2 is given by

$$F_{U_2}(x) = \begin{cases} 0, & \text{if } x < 0 \\ 2\Phi(x) - 1, & \text{if } x \geq 0. \end{cases}$$

It is easy to verify that

$$P_a[\sqrt{n}(\hat{\theta}_n - a) \leq x] \to 0.5 F_{U_1}(x) + 0.5 F_{U_2}(x).$$

To find the asymptotic distribution of $\sqrt{n}(\hat{\theta}_n - \theta)$ at $\theta = b$, we adopt the similar procedure as that for $\theta = a$ and study the limit of $P_b[\sqrt{n}(\hat{\theta}_n - b) \leq x]$ for all $x \in \mathbb{R}$. By definition $\hat{\theta}_n \geq a$, so that $\sqrt{n}(\hat{\theta}_n - b) \geq \sqrt{n}(a - b) < 0$. Hence, for $x < \sqrt{n}(a - b)$, $P_b[\sqrt{n}(\hat{\theta}_n - b) \leq x] = 0$. Suppose $\sqrt{n}(a - b) \leq x < 0$, then

$$P_b[\sqrt{n}(\hat{\theta}_n - b) \leq x] = P_b[\sqrt{n}(\overline{X}_n - b) \leq x] = \Phi(x).$$

Suppose $x = 0$. Then $P_b[\sqrt{n}(\hat{\theta}_n - b) \leq 0] = P_b[\hat{\theta}_n \leq b] = 1$. Suppose $x > 0$. By definition, $\hat{\theta}_n \leq b$ implying that $\sqrt{n}(\hat{\theta}_n - b) \leq 0$, hence $P_b[\sqrt{n}(\hat{\theta}_n - b) \leq x] = 1$, $\forall x > 0$. Thus,

$$P_b[\sqrt{n}(\hat{\theta}_n - b) \leq x] = \begin{cases} 0, & \text{if } x < \sqrt{n}(a - b) \\ \Phi(x), & \text{if } \sqrt{n}(a - b) \leq x < 0 \\ 1, & \text{if } x \geq 0. \end{cases}$$

Consequently, as $n \to \infty$,

$$P_b[\sqrt{n}(\hat{\theta}_n - b) \leq x] \to \begin{cases} \Phi(x), \text{ if } x < 0 \\ 1, \text{ if } x \geq 0. \end{cases}$$

It is to be noted that as in the case of $\theta = a$, the asymptotic distribution of $\sqrt{n}(\hat{\theta}_n - \theta)$ is not normal at $\theta = b$, and 0 is a point of discontinuity. Further, the asymptotic distribution is a mixture of discrete and continuous distributions as shown below. Suppose a random variable U_3 is defined as $U_3 = -|U|$ where $U \sim N(0, 1)$. Since $U_3 \leq 0$, $P[U_3 \leq x] = 1$ for all $x \geq 0$. Suppose $x < 0$, then

$$P[U_3 \leq x] = P[|U| \geq -x] = 1 - P[x \leq U \leq -x]$$
$$= 1 - \Phi(-x) + \Phi(x) = 2\Phi(x).$$

Thus, the distribution function of U_3 is given by

$$F_{U_3}(x) = \begin{cases} 2\Phi(x), & \text{if } x < 0 \\ 1, & \text{if } x \geq 0. \end{cases}$$

It then follows that $P_b[\sqrt{n}(\hat{\theta}_n - b) \leq x] \to 0.5 F_{U_1}(x) + 0.5 F_{U_3}(x)$. Thus, for $\theta = a$ and $\theta = a$, the asymptotic distribution of $\sqrt{n}(\hat{\theta}_n - \theta)$ is not normal. As in (ii), we can show that there exists no sequence $\{a_n, n \geq 1\}$ of real numbers tending to ∞ as $n \to \infty$ such that the asymptotic distribution of $a_n(\hat{\theta}_n - \theta)$ is normal for $\theta = a, b$. Hence, we conclude that $\hat{\theta}_n$ is not a CAN estimator of θ when $\theta \in [a, b]$. $\square$

🖎 Remark 3.2.1

In Example 3.2.1, it is to be noted that when $\Theta = \{0, 1\}$, the maximum likelihood estimator of θ is consistent but not CAN. Similarly, when $\Theta = [a, b]$, the maximum likelihood estimator of θ is consistent but not CAN as the asymptotic distribution is not normal at the boundary points a and b. In both the cases, we do not get asymptotic normality at the parametric points which are not the interior points. In the next chapter, we discuss Cramér Huzurbazar theory in which one of the regularity conditions is that the parameter θ is an interior point of the parameter space. Then for large n, its maximum likelihood estimator exists and is CAN with approximate variance $1/nI(\theta)$.

The most frequently used method to generate a CAN estimator for θ is based on the WLLN and the CLT, provided the underlying assumptions are satisfied. Suppose $\{X_1, X_2, \ldots, X_n\}$ is a random sample from the distribution of X with indexing parameter θ. Further, suppose $E(X) = h(\theta)$ and $Var(X) = v(\theta)$. It is assumed that $v(\theta)$ is positive and finite which implies that $E(X) = h(\theta) < \infty$.

By the WLLN, $\overline{X}_n \xrightarrow{P_\theta} h(\theta)$ and by the CLT,

$$\sqrt{n}(\overline{X}_n - h(\theta)) \xrightarrow{L} U \sim N(0, v(\theta)) \ \forall \ \theta \in \Theta.$$

Hence, $\overline{X}_n$ is CAN for $h(\theta)$ with approximate variance $v(\theta)/a_n^2$. From $\overline{X}_n$ one can find a CAN estimator for θ provided the function h satisfies certain assumptions. In the following theorem, we discuss how the CAN property remains invariant under differentiable transformation. This result is usually referred to in literature as the delta method. The proof is based on Result 1.3.10 which states that if $U_n \xrightarrow{L} U$, where U is a real random variable, then U_n is bounded in probability.

Theorem 3.2.1

Delta method: *Suppose T_n is a CAN estimator of θ with approximate variance $v(\theta)/a_n^2$. Suppose g is a differentiable function such that $g'(\theta) \neq 0$ and $g'(\theta)$ is continuous. Then $g(T_n)$ is a CAN estimator for $g(\theta)$ with approximate variance $(g'(\theta))^2 v(\theta)/a_n^2$.*

Proof It is given that T_n is a CAN estimator of θ, thus,

$$T_n \xrightarrow{P_\theta} \theta \text{ and } a_n(T_n - \theta) \xrightarrow{L} U \sim N(0, v(\theta)), \ \forall \ \theta \in \Theta.$$

It is given that g is differentiable, hence g is continuous and by the invariance of consistency under continuous transformation, $g(T_n)$ is consistent for $g(\theta)$. Since g is a differentiable function, by the Taylor series expansion,

$$g(T_n) = g(\theta) + (T_n - \theta)g'(\theta) + R_n, \text{ where } |R_n| \leq M|T_n - \theta|^{1+\delta}, \ \delta > 0.$$

Thus,

$$a_n(g(T_n) - g(\theta)) = a_n(T_n - \theta)g'(\theta) + a_n R_n \xrightarrow{L} U_1 \sim N(0, (g'(\theta))^2 v(\theta))$$

$$\text{provided } a_n R_n \xrightarrow{P_\theta} 0. \tag{3.2.1}$$

Now, $|a_n R_n| \leq M a_n |T_n - \theta||T_n - \theta|^\delta$. Since T_n is consistent for θ, $|T_n - \theta|^\delta \xrightarrow{P_\theta} 0$. Further, $a_n(T_n - \theta) \xrightarrow{L} U \Rightarrow a_n|T_n - \theta| \xrightarrow{L} |U|$, by continuous mapping theorem. Thus, $a_n|T_n - \theta|$ is bounded in probability. Hence by Slutsky's theorem, $a_n|T_n - \theta||T_n - \theta|^\delta \xrightarrow{P_\theta} 0$ implying that $a_n R_n \xrightarrow{P_\theta} 0$. Thus, from (3.2.1) we get that $a_n(g(T_n) - g(\theta)) \xrightarrow{L} U_1 \sim N(0, (g'(\theta))^2 v(\theta))$ and hence $g(T_n)$ is a CAN estimator of $g(\theta)$ with approximate variance $(g'(\theta))^2 v(\theta)/a_n^2$. $\qquad \square$

✍ Remark 3.2.2

By the continuous mapping theorem, if $X_n \xrightarrow{L} X$ then $g(X_n) \xrightarrow{L} g(X)$ if g is a continuous function. In particular if $a_n(T_n - \theta) \xrightarrow{L} U \sim N(0, v(\theta))$, then $g(a_n(T_n - \theta)) \xrightarrow{L} g(U)$, if g is a continuous function. Distribution of $g(U)$ will again be normal if g is a linear function as normality is preserved under linear transformations, but not in general. Thus if $g(x) = \exp(x)$ or $g(x) = a_0 + a_1 x + a_2 x^2$, $a_2 \neq 0$ or $g(x) = 1/x$, $x \neq 0$, then the distribution of $g(X)$ is not normal. In view of such situations, the result of Theorem 3.2.1 seems surprising, the explanation of which is found in its proof. If instead of continuity we impose the additional condition that g is differentiable, which implies continuity, then for large n in a small neighborhood of θ, $g(T_n)$ is approximately a linear function of T_n and a linear function of a random variable having normal distribution again has a normal distribution. Further, T_n is CAN for θ and hence the remainder term converges to 0 in probability. Thus, in delta method, $a_n(g(T_n) - g(\theta))$ is approximated by a linear function of $a_n(T_n - \theta)$ and hence the normality is preserved.

The following examples illustrate the application of delta method to generate CAN estimators.

✐ Example 3.2.2

Suppose X follows a beta distribution with parameter $(\theta, 1)'$ having probability density function $f(x, \theta) = \theta x^{\theta-1}$, $0 < x < 1$, $\theta > 0$. Hence,

$$E(X) = \frac{\theta}{\theta + 1}, \quad E(X^2) = \frac{\theta}{\theta + 2} \quad \text{and}$$

$$Var(X) = \frac{\theta}{(\theta + 2)(\theta + 1)^2} = v(\theta), \quad \text{say.}$$

For $\theta > 0$, $v(\theta)$ is positive and finite. Suppose $\{X_1, X_2, \ldots, X_n\}$ is a random sample from the distribution of X. Then by the WLLN and by the CLT,

$$\overline{X}_n \xrightarrow{P_\theta} \frac{\theta}{\theta + 1} \quad \& \quad \sqrt{n}\left(\overline{X}_n - \frac{\theta}{\theta + 1}\right) \xrightarrow{L} Z_1 \sim N(0, v(\theta)), \quad \forall \; \theta > 0.$$

Thus, $\overline{X}_n$ is CAN for $\theta/(\theta + 1) = \phi$, say, with approximate variance $v(\theta)/n = v_1(\phi)/n$, say. To get a CAN estimator for θ, we use delta method and define a function g such that $g(\phi) = \theta$. Suppose $g(\phi) = \phi/(1 - \phi)$, $0 < \phi < 1$, then $g(\phi) = \theta$. Further, $g'(\phi) = 1/(1 - \phi)^2 \neq 0$. Hence by the delta method, $g(\overline{X}_n) = \overline{X}_n/(1 - \overline{X}_n)$ is CAN for $g(\phi) = \theta$ with approximate variance $v_1(\phi)(g'(\phi))^2/n = \theta(1 + \theta)^2/n(\theta + 2)$. We find some more CAN estimators of θ as follows. In Example 2.2.8 we have shown that $S_n = -\sum_{i=1}^{n} \log X_i$ is a sufficient statistic. Further, it is shown that if a random variable Y is defined as $Y = -\log X$, then the distribution of Y is exponential and that of S_n is gamma $G(\theta, n)$ with scale parameter θ and shape parameter n. Thus, the moment estimator $\tilde{\theta}_n$ of θ based on a sufficient statistic is given by the equation $\overline{Y}_n = S_n/n = E(Y) = 1/\theta$ and hence $\tilde{\theta}_n = n/S_n = 1/\overline{Y}_n$. By the WLLN, $S_n/n \xrightarrow{P_\theta} E(Y) = 1/\theta$, hence $\tilde{\theta}_n$ is consistent for θ. Further, $Var(Y) = 1/\theta^2 < \infty$. Hence by the CLT,

$$\sqrt{n}(\overline{Y}_n - 1/\theta) \xrightarrow{L} Z_2 \sim N(0, 1/\theta^2) \; \forall \; \theta > 0.$$

Thus, $\overline{Y}_n$ is CAN for $1/\theta = \phi$ with approximate variance $1/n\theta^2$. Suppose a function g is defined as $g(\phi) = 1/\phi = \theta$, then $g'(\phi) = -1/\phi^2 \neq 0$. Hence by the delta method, $g(\overline{Y}_n) = 1/\overline{Y}_n$ is CAN for $g(\phi) = 1/\phi = \theta$, with approximate variance $(1/n\theta^2)\theta^4 = \theta^2/n$. Thus, $\overline{X}_n/(1 - \overline{X}_n)$ and $1/\overline{Y}_n$ are both CAN for θ with the same norming factor. We compare their approximate variances to examine which is better. It is to be noted that

$$\frac{\theta(1+\theta)^2}{\theta+2} - \theta^2 = \frac{\theta}{\theta+2} > 0 \; \forall \; \theta > 0,$$

and hence $1/\overline{Y}_n$ is better than $\overline{X}_n/(1 - \overline{X}_n)$. We now find the maximum likelihood estimator of θ. The likelihood of θ given the random sample $\underline{X} = \{X_1, X_2, \ldots, X_n\}$ is

$$L_n(\theta|\underline{X}) = \prod_{i=1}^{n} \theta X_i^{\theta-1}$$

$$\Leftrightarrow \quad \log L_n(\theta|\underline{X}) = n\log\theta + \theta\sum_{i=1}^{n}\log X_i - \sum_{i=1}^{n}\log X_i.$$

From the log likelihood we have

$$\frac{\partial}{\partial\theta}\log L_n(\theta|\underline{X}) = \frac{n}{\theta} + \sum_{i=1}^{n}\log X_i = 0 \; \Rightarrow \; \theta = \frac{1}{\overline{Y}_n}.$$

Further, $\frac{\partial^2}{\partial\theta^2}\log L_n(\theta|\underline{X}) = -n/\theta^2 < 0 \; \forall \; \theta > 0$ and hence at the solution of the likelihood equation also. Hence, the maximum likelihood estimator $\hat{\theta}_n$ of θ is given by $\hat{\theta}_n = 1/\overline{Y}_n$, which is CAN with approximate variance θ^2/n. For a random variable X with probability density function $f(x, \theta) = \theta x^{\theta-1}$, the information function is $I(\theta) = 1/\theta^2$ and thus the approximate variance $\theta^2/n = 1/nI(\theta)$. Further, the maximum likelihood estimator $\hat{\theta}_n$ is the same as the moment estimator of θ based on the sufficient statistic and is better than the estimator based on the sample mean.   □

✒ Example 3.2.3

Suppose $\{X_1, X_2, \ldots, X_n\}$ is a random sample from a Poisson distribution with mean $\theta > 0$. An estimator T_n is defined as

$$T_n = \begin{cases} \overline{X}_n, & \text{if } \overline{X}_n > 0 \\ 0.01, & \text{if } \overline{X}_n = 0 \end{cases}$$

Consistency of T_n follows using the same arguments as in Example 2.2.5. For a Poisson distribution with mean $\theta > 0$, $E(X) = Var(X) = \theta < \infty$. Hence by the WLLN and by the CLT, $\overline{X}_n$ is CAN for θ with approximate variance θ/n. Observe that for $\epsilon > 0$

$$P[|\sqrt{n}(T_n - \theta) - \sqrt{n}(\overline{X}_n - \theta)| < \epsilon] = P[\sqrt{n}|T_n - \overline{X}_n| < \epsilon] \geq P[T_n = \overline{X}_n]$$
$$= P[\overline{X}_n > 0]$$
$$= 1 - \exp(-n\theta) \quad \to \quad 1, \ \forall \ \theta > 0$$

$$\Rightarrow \quad \sqrt{n}(T_n - \theta) - \sqrt{n}(\overline{X}_n - \theta) \xrightarrow{P_\theta} 0, \ \forall \ \theta > 0$$

$$\Rightarrow \quad \text{If} \ \sqrt{n}(\overline{X}_n - \theta) \xrightarrow{L} U \ \text{then}$$

$$\sqrt{n}(T_n - \theta) \xrightarrow{L} U \ \text{by Result 1.3.8}$$

$$\text{Now} \ \sqrt{n}(\overline{X}_n - \theta) \xrightarrow{L} Z_1 \sim N(0, \theta) \ \text{by CLT}$$

$$\Rightarrow \quad \sqrt{n}(T_n - \theta) \xrightarrow{L} Z_1 \sim N(0, \theta) \ \forall \ \theta > 0,$$

which proves that T_n is CAN for θ with approximate variance θ/n. Suppose $g(\theta) = e^{-\theta}$, then it is a differentiable function with $g'(\theta) = -e^{-\theta} \neq 0$. Hence by the delta method, e^{-T_n} is CAN for $e^{-\theta} = P[X_1 = 0]$ with approximate variance $\theta e^{-2\theta}/n$.   □

☑ Example 3.2.4

Suppose $\{X_1, X_2, \ldots, X_n\}$ is a random sample from a normal $N(\theta, 1)$ distribution, where $\theta \in \mathbb{R}$. In Example 3.2.1, we have proved that $\overline{X}_n$ is CAN for θ with approximate variance $1/n$. Suppose $g : \mathbb{R} \to \mathbb{R}$ is a function defined as $g(x) = x^2$. It is a differentiable function with $g'(x) = 2x$ and it is $\neq 0$, $\forall \ x \neq 0$. Hence, from Theorem 3.2.1, it follows that $\sqrt{n}(\overline{X}_n^2 - \theta^2) \xrightarrow{L} Z_1$ which has $N(0, 4\theta^2)$ distribution for all $\theta \in \mathbb{R} - \{0\}$. At $\theta = 0$, we cannot apply Theorem 3.2.1. When $\theta = 0$, $\overline{X}_n \sim N(0, 1/n)$, that is, $\sqrt{n}\overline{X}_n \sim N(0, 1)$ and hence $n\overline{X}_n^2 \sim \chi_1^2$. As a consequence $n\overline{X}_n^2$ is bounded in probability. Thus, $\sqrt{n}(\overline{X}_n^2 - 0) = (1/\sqrt{n})n\overline{X}_n^2 \xrightarrow{P_0} 0$ and hence $\sqrt{n}(\overline{X}_n^2 - 0) \xrightarrow{L} U \equiv 0$. As in Example 3.2.1, it can be shown that there exists no sequence $\{a_n, n \geq 1\}$ of real numbers tending to ∞ as $n \to \infty$ such that the asymptotic distribution of $a_n\overline{X}_n^2$ is normal. Thus, $\overline{X}_n^2$ is not CAN for θ^2 if $\theta \in \mathbb{R}$. It is CAN if the parameter space is taken as $\mathbb{R} - \{0\}$. It is to be noted that with norming factor n, $n\overline{X}_n^2 \sim \chi_1^2$.   □

Example 3.2.4 conveys that if $g'(\theta) = 0$, then $g(\overline{X}_n)$ is not CAN if the parameter space is $\mathbb{R}$. With the norming factor n, the asymptotic distribution of $g(\overline{X}_n) - g(\theta)$ is chi-square. In the following theorem, we prove that such a result is true in general for any distribution.

Theorem 3.2.2

Suppose T_n is a CAN estimator of θ with approximate variance $v(\theta)/a_n^2$. Suppose g is a differentiable function such that $g'(\theta) = 0$ and $g''(\theta) \neq 0$. Then

$$\frac{2}{g''(\theta)v(\theta)} a_n^2(g(T_n) - g(\theta)) \xrightarrow{L} U \sim \chi_1^2 \text{ and } a_n(g(T_n) - g(\theta)) \xrightarrow{L} 0.$$

Proof It is given that T_n is a CAN estimator of θ, thus,

$$T_n \xrightarrow{P_\theta} \theta \text{ and } a_n(T_n - \theta) \xrightarrow{L} W \sim N(0, v(\theta)), \ \forall \ \theta \in \Theta.$$

Since g is differentiable, by the Taylor series expansion

$$g(T_n) = g(\theta) + (T_n - \theta)g'(\theta) + \frac{1}{2}(T_n - \theta)^2 g''(\theta)$$
$$+ R_n, \text{ where } |R_n| \leq M|T_n - \theta|^{2+\delta}, \ \delta > 0.$$

Hence,

$$a_n^2(g(T_n) - g(\theta)) = \frac{1}{2}v(\theta)\frac{a_n^2(T_n - \theta)^2}{v(\theta)}g''(\theta) + a_n^2 R_n. \qquad (3.2.2)$$

Since T_n is consistent for θ, $|T_n - \theta|^\delta \xrightarrow{P_\theta} 0$. Now

$$a_n(T_n - \theta) \xrightarrow{L} W \sim N(0, v(\theta))$$

$$\Rightarrow \frac{a_n^2(T_n - \theta)^2}{v(\theta)} \xrightarrow{L} U \sim \chi_1^2$$

$$\Rightarrow a_n^2(T_n - \theta)^2 \text{ is bounded in probability}$$

$$\Rightarrow |a_n^2 R_n| \leq M a_n^2(T_n - \theta)^2 |T_n - \theta|^\delta \xrightarrow{P_\theta} 0$$

$$\Rightarrow a_n^2 R_n \xrightarrow{P_\theta} 0$$

$$\Rightarrow (2/g''(\theta)v(\theta))a_n^2(g(T_n) - g(\theta)) \xrightarrow{L} U \sim \chi_1^2 \text{ from (3.2.2)}$$

$$\Rightarrow a_n(g(T_n) - g(\theta))$$

$$= \frac{1}{a_n}\frac{g''(\theta)v(\theta)}{2}\left(\frac{2}{g''(\theta)v(\theta)}a_n^2(g(T_n) - g(\theta))\right) \xrightarrow{P_\theta} 0$$

$$\Rightarrow a_n(g(T_n) - g(\theta)) \xrightarrow{L} 0,$$

the second last step follows from Slutsky's theorem. $\qquad \square$

☑ Example 3.2.5

Suppose $\{X_1, X_2, \ldots, X_n\}$ is a random sample from a distribution of a random variable X with mean 0, variance σ^2 and finite fourth central moment μ_4. Since $E(X) = 0$, $Var(X) = E(X^2)$. In Theorem 2.5.3, it is proved that sample central moments are consistent for corresponding population central moments, hence $m_2 = S_n^2 \xrightarrow{P_{\sigma^2}} \mu_2 = \sigma^2$. Now,

$$\sqrt{n}(S_n^2 - \sigma^2) = \sqrt{n}\left(\frac{1}{n}\sum_{i=1}^{n} X_i^2 - \overline{X}_n^2 - \sigma^2\right)$$

$$= \frac{1}{\sqrt{n}}\left(\sum_{i=1}^{n} X_i^2 - n\sigma^2\right) - (\sqrt{n}\overline{X}_n)\overline{X}_n.$$

Observe that by the CLT

$$\sqrt{n}\overline{X}_n \xrightarrow{L} Z_1 \sim N(0, \sigma^2) \Rightarrow \sqrt{n}\overline{X}_n \text{ is bounded in probability}$$

By the WLLN $\overline{X}_n \xrightarrow{P_{\sigma^2}} 0 \Rightarrow (\sqrt{n}\overline{X}_n)\overline{X}_n \xrightarrow{P_{\sigma^2}} 0.$

Now, $\{X_1, X_2, \ldots, X_n\}$ are independent and identically distributed random variables implies that $\{X_1^2, X_2^2, \ldots, X_n^2\}$ are also independent and identically distributed random variables with mean σ^2 and variance $\mu_4 - \mu_2^2$. Hence, by the CLT

$$(1/\sqrt{n})(\sum_{i=1}^{n} X_i^2 - n\sigma^2) \xrightarrow{L} Z_2 \sim N(0, \mu_4 - \mu_2^2)$$

$$\Rightarrow \sqrt{n}(S_n^2 - \sigma^2) = \frac{1}{\sqrt{n}}(\sum_{i=1}^{n} X_i^2 - n\sigma^2) - (\sqrt{n}\overline{X}_n)\overline{X}_n \xrightarrow{L} Z_2 \sim N(0, \mu_4 - \mu_2^2),$$

by Slutsky's theorem. The unbiased estimator of σ^2 is given by

$$U_n^2 = \sum_{i=1}^{n}(X_i - \overline{X}_n)^2/(n-1) = (n/(n-1))S_n^2$$

$$= a_n S_n^2 \text{ where } a_n = n/(n-1) \to 1$$

and the consistency of U_n^2 follows from the consistency of S_n^2. To examine whether it is CAN, consider

$$\sqrt{n}(U_n^2 - \sigma^2) - \sqrt{n}(S_n^2 - \sigma^2) = \sqrt{n}(a_n S_n^2 - \sigma^2) - \sqrt{n}(S_n^2 - \sigma^2)$$

$$= \sqrt{n}S_n^2(a_n - 1) = \frac{\sqrt{n}}{n-1}S_n^2 \xrightarrow{P_{\sigma^2}} 0,$$

as $S_n^2 \overset{P_{\sigma^2}}{\to} \sigma^2$ and hence is bounded in probability. But $\sqrt{n}(S_n^2 - \sigma^2) \overset{L}{\to} Z_2 \sim$ $N(0, \mu_4 - \mu_2^2)$ and hence $\sqrt{n}(U_n^2 - \sigma^2) \overset{L}{\to} Z_2 \sim N(0, \mu_4 - \mu_2^2)$. These results remain valid even if $E(X) \neq 0$, observe that

$$S_n^2 = \frac{1}{n}\sum_{i=1}^{n}(X_i - \overline{X}_n)^2 = \frac{1}{n}\sum_{i=1}^{n}((X_i - E(X)) - (\overline{X}_n - E(X)))^2$$

$$= \frac{1}{n}\sum_{i=1}^{n}(Y_i - \overline{Y}_n)^2,$$

where $Y_i = X_i - E(X)$, $i = 1, 2, \ldots, n$ and $E(Y_i) = 0$, $i = 1, 2, \ldots, n$. We then proceed exactly on similar lines as in (i) and (ii) and show that even if $E(X) \neq 0$, sample variance S_n^2 is CAN for σ^2 and the unbiased estimator of σ^2 is also CAN for σ^2. □

In Sect. 2.2, we have discussed a method based on sample quantiles to generate consistent estimators for the parameter of interest. Thus, for a Cauchy $C(\theta, 1)$ distribution, with location parameter θ, the sample median is consistent for the population median which is θ. The population first quartile is $\theta - 1$ while the population third quartile is $\theta + 1$. Thus, $X_{([n/4]+1)} + 1$ and $X_{([3n/4]+1)} - 1$ are also consistent for θ. It is of interest to see whether these are CAN for θ and which among these is the best estimator. Below we state a theorem which is useful to find CAN estimators based on the sample quantiles. For proof, we refer to Serfling [1] and DasGupta [2].

Theorem 3.2.3

Suppose $\{X_1, X_2, \ldots, X_n\}$ is a random sample from the distribution of X which is absolutely continuous with distribution function $F(x, \theta)$ and probability density function $f(x, \theta)$, where θ is an indexing parameter. Suppose F^{-1} exists and $f(a_p(\theta), \theta) \neq 0$. Suppose $r_n = [np] + 1$, $0 < p < 1$. Then (i) the p-th sample quantile $X_{(r_n)}$ is consistent for the p-th population quantile $a_p(\theta)$ and (ii) $\sqrt{n}(X_{(r_n)} - a_p(\theta)) \overset{L}{\to} Z_1 \sim N(0, v(\theta))$, where $v(\theta) = p(1 - p)/(f(a_p(\theta), \theta))^2$.

✎ Example 3.2.6

Suppose $\{X_1, X_2, \ldots, X_n\}$ is a random sample from each of the following distributions. (i) $N(\theta, 1)$, (ii) $C(\theta, 1)$ and (iii) $U(\theta - 1, \theta + 1)$, $\theta \in \mathbb{R}$. For a normal $N(\theta, 1)$ distribution, the p-th population quantile is given by $a_p(\theta) = \theta + \Phi^{-1}(p)$, hence $X_{([np]+1)} - \Phi^{-1}(p)$ is CAN for θ, with approximate variance $(1/n)2\pi p(1 - p)\exp((\Phi^{-1}(p))^2)$. For a uniform $U(\theta - 1, \theta + 1)$ distribution, the p-th population quantile is given by $a_p(\theta) = 2p + (\theta - 1)$, hence $X_{([np]+1)} - 2p + 1$ is CAN for θ, with approximate variance $4p(1 - p)/n$. For a Cauchy $C(\theta, 1)$ distribution, the p-th population quantile is given by $a_p(\theta) = \theta + \tan(\pi(p - 1/2))$, hence $X_{([np]+1)} - \tan(\pi(p - 1/2))$ is CAN for

θ, with approximate variance $(p(1 - p)/n)\, \pi^2(1 + \tan(\pi(p - 1/2))^2)$. Thus, we have an uncountable family of CAN estimators for θ for each distribution. For all the three distributions, θ is the population median and hence the sample median $X_{([n/2]+1)}$ is CAN for θ with approximate variance $v(\theta)/n$ where $v(\theta) = 1/4(f(a_p(\theta), \theta))^2$. For $N(\theta, 1)$, $v(\theta) = \pi/2$, for $C(\theta, 1)$, $v(\theta) = \pi^2/4$ and for $U(\theta - 1, \theta + 1)$, $v(\theta) = 1$. For both normal $N(\theta, 1)$ and uniform $U(\theta - 1, \theta + 1)$ distributions, θ is the population mean and hence the sample mean $\overline{X}_n$ is CAN for θ with approximate variance $1/n$ for $N(\theta, 1)$ and $1/3n$ for $U(\theta - 1, \theta + 1)$ distribution. In both the cases, the sample mean is better than the sample median as the approximate variance is less for the sample mean than that for the sample median. For a $C(\theta, 1)$ distribution, the sample median is CAN for θ with approximate variance $\pi^2/4n$. The population first quartile is $\theta - 1$ while the population third quartile is $\theta + 1$. Thus, $X_{([n/4]+1)} + 1$ and $X_{([3n/4]+1)} - 1$ are also CAN for θ with the same approximate variance $3\pi^2/4n$. Thus, among the CAN estimators of θ based on the sample median, the sample first quartile and the sample third quartile, the one based on the sample median is the best. In fact, within the family of CAN estimators based on sample quantiles and for distributions symmetric around θ, the CAN estimator based on the sample median is the best, its approximate variance being the smallest, which follows from the fact that θ is the mode of the distribution.   □

✍ Example 3.2.7

Suppose a random variable X has an exponential distribution with location parameter θ and scale parameter 1. Its probability density function is given by

$$f_X(x, \theta) = \exp\{-(x - \theta)\}, \quad x \geq \theta, \ \theta \in \mathbb{R}.$$

Suppose $\{X_1, X_2, \ldots, X_n\}$ is a random sample from the distribution of X. In Example 2.2.16, we have obtained the maximum likelihood estimator of θ. It is $X_{(1)}$ and we have also verified that it is consistent for θ. In the same example, we derived the distribution function $F_{X_{(1)}}(x, \theta)$ of $X_{(1)}$ and it is given by

$$F_{X_{(1)}}(x, \theta) = \begin{cases} 0, & \text{if } x < \theta \\ 1 - \exp\{-n(x - \theta)\}, & \text{if } x \geq \theta. \end{cases}$$

It thus follows that for each n, $X_{(1)}$ has an exponential distribution with location parameter θ and scale parameter n, which further implies that for each n, $Y_n = n(X_{(1)} - \theta)$ has the exponential distribution with location parameter 0 and scale parameter 1. Hence, its asymptotic distribution is the same. Thus, with norming factor n, the asymptotic distribution of $X_{(1)}$ is not normal, but we cannot conclude that $X_{(1)}$ is not CAN. As in Example 3.2.1, it can be shown that there exists no sequence $\{a_n, n \geq 1\}$ of real numbers tending to ∞ as $n \to \infty$ such that the asymptotic distribution of $a_n(X_{(1)} - \theta)$ is normal. Hence, we conclude that $X_{(1)}$ is not CAN for θ. We now find CAN estimator of θ based on a sample mean.

If a random variable X has an exponential distribution with location parameter θ and scale parameter 1, then $E(X) = \theta + 1$ and $Var(X) = 1$. Hence, by the WLLN, $\overline{X}_n \xrightarrow{P_\theta} \theta + 1$, $\forall \ \theta$, which implies that $\overline{X}_n - 1$ is consistent for θ. Further by the CLT,

$$\sqrt{n}(\overline{X}_n - (\theta + 1)) = \sqrt{n}((\overline{X}_n - 1) - \theta) \xrightarrow{L} Z \sim N(0, 1).$$

Hence, $\overline{X}_n - 1$ is also CAN for θ with approximate variance $1/n$. We now find a CAN estimator of θ based on the sample median. From the distribution function of X, we find the median of X to be $a_{1/2}(\theta) = \theta + \log_e 2$. By Theorem 3.3.3, the sample median $X_{(r_n)}$, where $r_n = [n/2] + 1$, is CAN for $\theta + \log_e 2$ with approximate variance $1/4n(\exp(-\log_e 2))^2 = 1/n$. Hence, $X_{(r_n)} - \log_e 2$ is CAN for θ with approximate variance $1/n$. Thus, both $\overline{X}_n - 1$ and the sample median are CAN with the same approximate variance. $\qquad \square$

✒ Example 3.2.8

Suppose X is a random variable with probability density function $f(x, \theta) = 2\theta^2/x^3$, $x \geq \theta$, $\theta \in \mathbb{R}$. Suppose $\underline{X} \equiv \{X_1, X_2, \ldots, X_n\}$ is a random sample from the distribution of X. Corresponding to a random sample $\underline{X}$, the likelihood of θ is given by

$$L_n(\theta|\underline{X}) = \prod_{i=1}^{n} 2\theta^2/X_i^3 = 2^n \theta^{2n} \prod_{i=1}^{n} X_i^{-3}, \quad X_i \geq \theta, \ \forall \ i \ \Leftrightarrow \ X_{(1)} \geq \theta.$$

Thus, the likelihood is an increasing function of θ on $(-\infty, X_{(1)}]$ and attains maximum at the maximum possible value of θ given the data $\underline{X}$. The maximum possible value of θ given data is $X_{(1)}$ and hence the maximum likelihood estimator $\hat{\theta}_n$ of θ is given by $X_{(1)}$. To verify the consistency of $X_{(1)}$ as an estimator of θ, we find the coverage probability using the distribution function of $X_{(1)}$. The distribution function $F_X(x)$ of X is given by

$$F_X(x) = \begin{cases} 0, & \text{if } x < \theta \\ 1 - \theta^2/x^2, & \text{if } x \geq \theta. \end{cases}$$

Hence, the distribution function of $X_{(1)}$ is given by

$$F_{X_{(1)}}(x) = 1 - [1 - F_X(x)]^n = \begin{cases} 0, & \text{if } x < \theta \\ 1 - \theta^{2n}/x^{2n}, & \text{if } x \geq \theta. \end{cases}$$

For $\epsilon > 0$, the coverage probability is given by

$$
\begin{aligned}
P_\theta[|X_{(1)} - \theta| < \epsilon] &= P_\theta[\theta - \epsilon < X_{(1)} < \theta + \epsilon] \\
&= P_\theta[\theta < X_{(1)} < \theta + \epsilon] \quad \text{as} \quad X_{(1)} \geq \theta \\
&= F_{X_{(1)}}(\theta + \epsilon) - F_{X_{(1)}}(\theta) \\
&= 1 - \frac{\theta^{2n}}{(\theta + \epsilon)^{2n}} - 0 = 1 - \left(\frac{\theta}{\theta + \epsilon}\right)^{2n} \\
&\to 1 \quad \forall \ \epsilon > 0 \ \text{and} \ \forall \ \theta \ \text{as} \ n \to \infty.
\end{aligned}
$$

Hence, $X_{(1)}$ is consistent for θ. To derive the asymptotic distribution of $X_{(1)}$, with suitable norming, as in Example 3.2.7, we define $Y_n = n(X_{(1)} - \theta)$ and derive its distribution function $G_{Y_n}(y)$ for $y \in \mathbb{R}$. Since $X_{(1)} \geq \theta$, $Y_n \geq 0$, hence for $y < 0$, $G_{Y_n}(y) = 0$. Suppose $y \geq 0$, then

$$
\begin{aligned}
G_{Y_n}(y) &= P_\theta[n(X_{(1)} - \theta) \leq y] = P_\theta[X_{(1)} \leq \theta + y/n] = F_{X_{(1)}}(\theta + y/n) \\
&= 1 - \left(\frac{\theta}{\theta + y/n}\right)^{2n} \to 1 - \exp(-2y/\theta).
\end{aligned}
$$

Thus, the asymptotic distribution of $Y_n = n(X_{(1)} - \theta)$ is exponential with location parameter 0 and scale parameter $2/\theta$. Thus, with norming factor n, the asymptotic distribution of $X_{(1)}$ is not normal. Proceeding on similar lines as in Example 3.2.1, we claim that there exists no sequence $\{a_n, n \geq 1\}$ of real numbers tending to ∞ as $n \to \infty$ such that the asymptotic distribution of $a_n(X_{(1)} - \theta)$ is normal, hence $X_{(1)}$ is not CAN for θ. We find CAN estimator for θ based on p-th sample quantile. From the distribution function of X, the p-th population quantile $a_p(\theta) = \theta/\sqrt{(1-p)}$. Hence, the p-th sample quantile $X_{([np]+1)}$ is CAN for $a_p(\theta)$ with approximate variance $\theta^2 p/4n(1-p)^2$. Thus, the family of CAN estimators of θ based on p-th sample quantile is given by $T_n = \sqrt{(1-p)}X_{([np]+1)}$ with approximate variance $\theta^2 p/4n(1-p)$, $0 < p < 1$.

It is to be noted that $E(X) = 2\theta < \infty$. Hence, by the WLLN,

$\overline{X}_n \xrightarrow{P_\theta} E(X) = 2\theta$, $\forall \ \theta$. Hence, the moment estimator $\overline{X}_n/2$ is consistent for θ. Now $E(X^2) = 2\theta^2 \int_\theta^\infty x^2/x^3 dx$ is not a convergent integral, as the integral $\int_a^\infty \frac{p_m(x)}{q_n(x)} dx$ is convergent if $n - m \geq 2$ where $p_m(x)$ and $q_n(x)$ are polynomials of degree m and n respectively. Hence for this distribution, the second raw moment and hence the variance does not exist and we cannot appeal to the CLT to claim normality.

□

The main motive to find an asymptotic non-degenerate distribution of a suitably normalized estimator for a parameter θ is (i) to find an asymptotic null distribution of a test statistic for testing certain hypotheses about θ and (ii) to find an interval estimator of θ, as an interval estimator is more informative than the point estimator. In Chaps. 5 and 6, we study how to derive the asymptotic null distribution of a test statistic on the basis of an asymptotic non-degenerate distribution of a suitably normalized estimator. We now discuss how the asymptotic distribution is useful to find an asymptotic confidence interval for a parameter of interest.

Suppose T_n is a CAN estimator for θ. Then the asymptotic normal distribution is useful to find large sample confidence interval for θ. Following examples illustrate the procedure.

☑ Example 3.2.9

Suppose $\{X_1, X_2, \ldots, X_n\}$ is a random sample from an exponential distribution with mean θ. Then variance of X_1 is θ^2. By the WLLN and the CLT we immediately get that $\overline{X}_n$ is CAN for θ with approximate variance θ^2/n. Hence,

$$Q_n = \frac{\sqrt{n}}{\theta}\left(\overline{X}_n - \theta\right) = \sqrt{n}\left(\frac{\overline{X}_n}{\theta} - 1\right) \xrightarrow{L} Z \sim N(0, 1)$$

and for large n, Q_n can be treated as a pivotal quantity to construct a confidence interval for θ. Thus, given the confidence coefficient $(1 - \alpha)$, we find $a_{(1-\alpha/2)}$ such that $P[-a_{(1-\alpha/2)} < Q_n < a_{(1-\alpha/2)}] = 1 - \alpha$, where $a_{(1-\alpha/2)}$ is $(1 - \alpha/2)$ th quantile of the standard normal distribution. Now,

$$-a_{(1-\alpha/2)} < Q_n < a_{(1-\alpha/2)} \iff \frac{\overline{X}_n}{1 + a_{(1-\alpha/2)}/\sqrt{n}} < \theta < \frac{\overline{X}_n}{1 - a_{(1-\alpha/2)}/\sqrt{n}}.$$

Thus, asymptotic confidence interval for θ with confidence coefficient $(1 - \alpha)$ is given by

$$\left(\frac{\overline{X}_n}{1 + a_{(1-\alpha/2)}/\sqrt{n}}, \frac{\overline{X}_n}{1 - a_{(1-\alpha/2)}/\sqrt{n}}\right).$$

□

☑ Example 3.2.10

Suppose $\{X_1, X_2, \ldots, X_n\}$ is a random sample from a normal $N(\theta, \theta^2)$ distribution, $\theta \in \mathbb{R} - \{0\}$. Then

$$Q_n = \frac{\sqrt{n}}{\theta}\left(\overline{X}_n - \theta\right) = \sqrt{n}\left(\frac{\overline{X}_n}{\theta} - 1\right) \sim N(0, 1)$$

for each n. Hence, Q_n is a pivotal quantity to construct a confidence interval for θ. Thus, proceeding as in Example 3.2.9, for each n, the confidence interval for θ with confidence coefficient $(1 - \alpha)$ is given by

$$\left(\frac{\overline{X}_n}{1 + a_{(1-\alpha/2)}/\sqrt{n}}, \frac{\overline{X}_n}{1 - a_{(1-\alpha/2)}/\sqrt{n}}\right).$$

□

Example 3.2.9 and Example 3.2.10 convey that once we have a CAN estimator for θ, it is very easy to find an asymptotic confidence interval for θ. However, such an easy procedure may not work always. For example, suppose $\{X_1, X_2, \ldots, X_n\}$ is a random sample from Poisson $Poi(\theta)$ distribution. To find asymptotic confidence interval for θ, note that by the WLLN and the CLT,

$$\sqrt{n}(\overline{X}_n - \theta) \xrightarrow{L} Z_1 \sim N(0, \theta) \;\Rightarrow\; Q_n = (\sqrt{n}/\sqrt{\theta})(\overline{X}_n - \theta) \xrightarrow{L} Z \sim N(0, 1)$$

and Q_n can be treated as a pivotal quantity to construct a confidence interval for θ. Thus, given the confidence coefficient $(1 - \alpha)$, we find $a_{(1-\alpha/2)}$ such that $P[-a_{(1-\alpha/2)} < Q_n < a_{(1-\alpha/2)}]$ is $1 - \alpha$. Now,

$$-a_{(1-\alpha/2)} < Q_n < a_{(1-\alpha/2)} \;\Leftrightarrow\; \overline{X}_n - \frac{\sqrt{\theta}}{\sqrt{n}}a_{(1-\alpha/2)} < \theta < \overline{X}_n + \frac{\sqrt{\theta}}{\sqrt{n}}a_{(1-\alpha/2)}.$$

Thus, θ is involved in both the lower and upper bounds of the interval and hence this procedure does not give us the desired confidence interval. Such a problem is very common and arises because θ is involved in the approximate variance $v(\theta)/a_n^2$ of the CAN estimator T_n of θ. There are two approaches to resolve this problem, one is a studentization procedure and the other is a variance stabilization technique. In a studentization procedure, the variance function $v(\theta)$ is replaced by its consistent estimator, whereas in a variance stabilization technique we find a transformation g so that variance of $g(T_n)$ is free from θ. Using $g(T_n)$ we first form a confidence interval for $g(\theta)$ and hence find a confidence interval for θ. We discuss below these procedures in detail.

Studentization procedure: Suppose T_n is CAN for θ with approximate variance $v(\theta)/a_n^2$. Then $Q_n = (a_n/\sqrt{v(\theta)})(T_n - \theta) \xrightarrow{L} Z \sim N(0, 1)$. However, such a pivotal quantity may not be useful to get the asymptotic confidence interval, as θ is involved in the variance function. Suppose $v(\theta)$ is a continuous function of θ. Then by the invariance property of consistency under continuous transformation, $v(T_n)$ is consistent for $v(\theta)$. Using Slutsky's theorem,

$$\tilde{Q}_n = \frac{a_n}{\sqrt{v(T_n)}}(T_n - \theta) = \left(\frac{\sqrt{v(T_n)}}{\sqrt{v(\theta)}}\right)\left(\frac{a_n}{\sqrt{v(\theta)}}(T_n - \theta)\right) \xrightarrow{L} Z \sim N(0, 1).$$

Thus, for large n, $\tilde{Q}_n$ is also a pivotal quantity and hence given the confidence coefficient $(1 - \alpha)$, we find $a_{(1-\alpha/2)}$ such that $P[-a_{(1-\alpha/2)} < \tilde{Q}_n < a_{(1-\alpha/2)}] = 1 - \alpha$. Now,

$$-a_{(1-\alpha/2)} < \tilde{Q}_n < a_{(1-\alpha/2)}$$
$$\Leftrightarrow\; T_n - \frac{\sqrt{v(T_n)}}{a_n}a_{(1-\alpha/2)} < \theta < T_n + \frac{\sqrt{v(T_n)}}{a_n}a_{(1-\alpha/2)}.$$

Thus, asymptotic confidence interval for θ with confidence coefficient $(1 - \alpha)$ is given by

$$\left(T_n - \frac{\sqrt{v(T_n)}}{a_n} a_{(1-\alpha/2)}, \ T_n + \frac{\sqrt{v(T_n)}}{a_n} a_{(1-\alpha/2)} \right)$$

which can also be expressed as

$$\left(T_n - s.e.(T_n) a_{(1-\alpha/2)}, \ T_n + s.e.(T_n) a_{(1-\alpha/2)} \right) ,$$

where $s.e.(T_n)$ is the standard error of T_n for large n. We get a symmetric confidence interval as the limiting distribution of the pivotal quantity is a symmetric distribution.

This procedure is known as a studentization procedure in view of the fact that to get the pivotal quantity $\tilde{Q}_n$ from Q_n we replace the variance function by its consistent estimator. We adopt similar procedure to get Student's t distribution. Suppose $\{X_1, X_2, \ldots, X_n\}$ are independent and identically distributed random variables each having normal $N(0, \sigma^2)$ distribution. Then $U_n = \sqrt{n}\bar{X}_n/\sigma$ has standard normal distribution. If σ is replaced by S_n where S_n^2 is an unbiased estimator of σ^2, then the distribution of U_n, apart from some constants, is t distribution with $(n - 1)$ degrees of freedom.

Variance stabilization technique: The technique is heavily based on the invariance property of CAN estimators under differentiable transformation as proved in Theorem 3.2.1. The delta method states that if T_n is CAN for θ with approximate variance $v(\theta)a_n^2$ and g is a differentiable function with $g'(\theta) \neq 0$, then $g(T_n)$ is CAN for $g(\theta)$ with approximate variance $(g'(\theta))^2 v(\theta)/a_n^2$. Variance function of $g(T_n)$ thus depends on θ. It is noted in the examples of constructing confidence intervals, for the parameter of exponential and Poisson distributions and in the studentization procedure, that it is better to have a variance function to be free from θ so that the associated pivotal quantity is useful to construct an asymptotic confidence interval. In a variance stabilization technique, as the nomenclature indicates, we try to find a function g so that the variance function of $g(T_n)$ does not depend on θ. More specifically we try to find g such that

$$g'(\theta)\sqrt{v(\theta)} = c \quad \Leftrightarrow \quad g(\theta) = \int \frac{c\,d\theta}{\sqrt{v(\theta)}} + k,$$

where c is any positive real number and k is a constant of integration. With such a choice of g, the approximate variance of $g(T_n)$ will be c^2/a_n^2. Using the pivotal quantity based on the large sample distribution of $g(T_n)$, we obtain the confidence interval for $g(\theta)$. By the inverse function theorem, the condition $g'(\theta) \neq 0$ imposed on g assures that a unique inverse of g exists and hence from the confidence interval for $g(\theta)$ we can get the confidence interval for θ, assuming g is a one-to-one function of θ. Of course this technique works only for those functions g, for which the indefinite integral $\int \frac{c\,d\theta}{\sqrt{v(\theta)}}$ can be explicitly found out.

Following examples illustrate the procedure.

☑ Example 3.2.11

Suppose $\{X_1, X_2, \ldots, X_n\}$ is a random sample from a Poisson distribution with mean θ. Then $Var(X) = \theta < \infty$ and hence by the WLLN and by the CLT, $\overline{X}_n$ is CAN for θ with asymptotic variance θ/n. Thus,

$Q_n = (\sqrt{n}/\sqrt{\theta})(\overline{X}_n - \theta) \overset{L}{\to} Z \sim N(0, 1)$ and Q_n is a pivotal quantity. However, as discussed above Q_n is not useful to construct a confidence interval for θ. Hence, adopting studentization procedure we define a pivotal quantity $\tilde{Q}_n$ as $\tilde{Q}_n = (\sqrt{n}/\sqrt{\overline{X}_n})(\overline{X}_n - \theta)$. By Slutsky's theorem, $\tilde{Q}_n$ has the standard normal distribution. Hence, given the confidence coefficient $(1 - \alpha)$, we find $a_{(1-\alpha/2)}$ such that $P[-a_{(1-\alpha/2)} < \tilde{Q}_n < a_{(1-\alpha/2)}] = 1 - \alpha$. Now,

$$-a_{(1-\alpha/2)} \;\; < \;\; \tilde{Q}_n < a_{(1-\alpha/2)}$$
$$\Leftrightarrow \;\; \overline{X}_n - \frac{\sqrt{\overline{X}_n}}{\sqrt{n}} a_{(1-\alpha/2)} \;\; < \;\; \theta \;\; < \;\; \overline{X}_n + \frac{\sqrt{\overline{X}_n}}{\sqrt{n}} a_{(1-\alpha/2)}.$$

Thus, asymptotic confidence interval for θ with confidence coefficient $(1 - \alpha)$ is given by

$$\left(\overline{X}_n - \frac{\sqrt{\overline{X}_n}}{\sqrt{n}} a_{(1-\alpha/2)}, \;\; \overline{X}_n + \frac{\sqrt{\overline{X}_n}}{\sqrt{n}} a_{(1-\alpha/2)} \right)$$

which can be expressed as

$$\left(\overline{X}_n - s.e.(\overline{X}_n)a_{(1-\alpha/2)}, \;\; \overline{X}_n + s.e.(\overline{X}_n)a_{(1-\alpha/2)} \right),$$

where $\sqrt{\overline{X}_n/n}$ is the standard error of $\overline{X}_n$. In variance stabilization technique, we find a differentiable function g such that $(g'(\theta))^2 \neq 0$ and $Var(g(\overline{X}_n))$ is free from θ, that is
$(g'(\theta))^2\theta = c^2 \;\; \Leftrightarrow \;\; g(\theta) = \int \frac{c\,d\theta}{\sqrt{\theta}}$. Hence, $g(\theta) = 2c\sqrt{\theta}$. Thus, we have a pivotal quantity $\tilde{Q}_n$ as

$$\tilde{Q}_n = \frac{\sqrt{n}}{c}\left(2c\sqrt{\overline{X}_n} - 2c\sqrt{\theta}\right) = 2\sqrt{n}\left(\sqrt{\overline{X}_n} - \sqrt{\theta}\right) \sim N(0, 1).$$

Hence corresponding to the confidence coefficient $(1 - \alpha)$, asymptotic confidence interval for $\sqrt{\theta}$ is given by $\left(\sqrt{\overline{X}_n} - \frac{1}{2\sqrt{n}}a_{(1-\alpha/2)}, \;\; \sqrt{\overline{X}_n} + \frac{1}{2\sqrt{n}}a_{(1-\alpha/2)} \right)$.
Consequently, asymptotic confidence interval for θ is given by

$$\left(\left(\sqrt{\overline{X}_n} - \frac{1}{2\sqrt{n}}a_{(1-\alpha/2)}\right)^2, \;\; \left(\sqrt{\overline{X}_n} + \frac{1}{2\sqrt{n}}a_{(1-\alpha/2)}\right)^2 \right).$$

□

✏ Example 3.2.12

Suppose $\{X_1, X_2, \ldots, X_n\}$ is a random sample from a normal $N(\theta, \theta^2)$ distribution, $\theta > 0$. Then $T_n = \sqrt{n}(\overline{X}_n - \theta) \sim N(0, \theta^2)$ distribution. We now find a differentiable function g such that $(g'(\theta))^2 \neq 0$ and $Var(g(\overline{X}_n))$ are free from θ, that is $(g'(\theta))^2\theta^2 = c^2$, that is, $g(\theta) = \int \frac{c\, d\theta}{\theta} = c \log \theta$. Hence,

$$Q_n = \sqrt{n}(\log \overline{X}_n - \log \theta) \xrightarrow{L} Z \sim N(0, 1) \;\Rightarrow\; Q_n \text{ is a pivotal quantity.}$$

Thus, we find $a_{(1-\alpha/2)}$ such that $P[-a_{(1-\alpha/2)} < Q_n < a_{(1-\alpha/2)}] = 1 - \alpha$, the given confidence coefficient. Now,

$$-a_{(1-\alpha/2)} \;<\; Q_n < a_{(1-\alpha/2)}$$
$$\Leftrightarrow\quad \log \overline{X}_n - a_{(1-\alpha/2)}/\sqrt{n} \;<\; \log \theta \;<\; \log \overline{X}_n + a_{(1-\alpha/2)}/\sqrt{n}.$$

Hence, the asymptotic confidence interval for θ with confidence coefficient $(1 - \alpha)$ is given by

$$\left(\exp\left(\log \overline{X}_n - a_{(1-\alpha/2)}/\sqrt{n}\right), \quad \exp\left(\log \overline{X}_n + a_{(1-\alpha/2)}/\sqrt{n}\right)\right) .$$

□

✐ Remark 3.2.3

It is to be noted that $\overline{X}_n \sim N(\theta, \theta^2/n)$ distribution for each n, but $\log \overline{X}_n \sim N(\log \theta, 1/n)$ distribution for large n.

✏ Example 3.2.13

Suppose $\{X_1, X_2, \ldots, X_n\}$ is a random sample of size n from an exponential distribution with mean θ. We Obtain $100(1 - \alpha)\%$ asymptotic confidence interval for the survival function $e^{-t/\theta}$, where t is a fixed positive real number, by two methods. In the first method, we use CAN estimator of $e^{-t/\theta}$ and studentization procedure. In the second method, we use the fact that $e^{-t/\theta}$ is a monotone function of θ and using the confidence interval for θ based on $\overline{X}_n$, we obtain $100(1 - \alpha)\%$ asymptotic confidence interval for $e^{-t/\theta}$. If X follows an exponential distribution with mean θ, then $Var(X) = \theta^2$. Corresponding to a given random sample, by the WLLN and the CLT $\overline{X}_n$ is CAN for θ with approximate variance θ^2/n.

(i) Suppose $g(x) = e^{-t/x}$, then g is a differentiable function and $g'(x) = \frac{t}{x^2}e^{-t/x} \neq 0$ for all $x > 0$. Hence by the delta method, $g(\overline{X}_n) = e^{-t/\overline{X}_n}$ is CAN for $g(\theta) = e^{-t/\theta} = P[X > t]$ with approximate variance $t^2 e^{-2t/\theta}/n\theta^2$.

The consistent estimator of the approximate variance is $t^2 e^{-2t/\overline{X}_n}/n\,\overline{X}_n^2$. Hence by Slutsky's theorem,

$$Q_n = \frac{\overline{X}_n\sqrt{n}}{t e^{-t/\overline{X}_n}}\left(e^{-t/\overline{X}_n} - e^{-t/\theta}\right) \xrightarrow{L} Z \sim N(0,1)\,.$$

Thus for large n, Q_n is a pivotal quantity and is useful to find asymptotic confidence interval for the survival function $e^{-t/\theta}$. Given a confidence coefficient $(1-\alpha)$, we find the quantile $a_{1-\alpha/2}$ of the standard normal distribution so that $P[-a_{1-\alpha/2} < Q_n < a_{1-\alpha/2}] = 1 - \alpha$. Inverting the inequality $-a_{1-\alpha/2} < Q_n < a_{1-\alpha/2}$, we get

$$e^{-t/\overline{X}_n} - a_{1-\alpha/2}\frac{t e^{-t/\overline{X}_n}}{\overline{X}_n\sqrt{n}} < e^{-t/\theta} < e^{-t/\overline{X}_n} + a_{1-\alpha/2}\frac{t e^{-t/\overline{X}_n}}{\overline{X}_n\sqrt{n}}\,.$$

Hence using studentization technique, $100(1-\alpha)\%$ large sample confidence interval for $e^{-t/\theta}$ is given by

$$\left(e^{-t/\overline{X}_n}\left(1 - a_{1-\alpha/2}\frac{t}{\overline{X}_n\sqrt{n}}\right),\;\; e^{-t/\overline{X}_n}\left(1 + a_{1-\alpha/2}\frac{t}{\overline{X}_n\sqrt{n}}\right)\right)\,.$$

(ii) Since $\overline{X}_n$ is CAN for θ with approximate variance θ^2/n,

$$Q_n = \frac{\sqrt{n}}{\theta}\left(\overline{X}_n - \theta\right) = \sqrt{n}\left(\frac{\overline{X}_n}{\theta} - 1\right) \xrightarrow{L} Z \sim N(0,1)\,.$$

Thus for large n, Q_n is a pivotal quantity. Given a confidence coefficient $(1-\alpha)$, we can find the quantile $a_{1-\alpha/2}$ of the standard normal distribution so that $P[-a_{1-\alpha/2} < Q_n < a_{1-\alpha/2}] = 1 - \alpha$. Inverting the inequality $-a_{1-\alpha/2} < Q_n < a_{1-\alpha/2}$, we get

$$\frac{\overline{X}_n}{1 + \frac{a_{1-\alpha/2}}{\sqrt{n}}} < \theta < \frac{\overline{X}_n}{1 - \frac{a_{1-\alpha/2}}{\sqrt{n}}}\,.$$

Now, if $g(\theta) = e^{-t/\theta}$ then $g'(\theta) = \frac{t}{\theta^2}e^{-t/\theta} > 0$ and hence $g(\theta) = e^{-t/\theta}$ is a monotone increasing function of θ. Thus, $a < \theta < b \Rightarrow e^{-t/a} < e^{-t/\theta} < e^{-t/b}$. Hence, $100(1-\alpha)\%$ large sample confidence interval for $e^{-t/\theta}$ is given by

$$\left(\exp\left\{\frac{-t\left(1 + \frac{a_{1-\alpha/2}}{\sqrt{n}}\right)}{\overline{X}_n}\right\},\;\; \exp\left\{\frac{-t\left(1 - \frac{a_{1-\alpha/2}}{\sqrt{n}}\right)}{\overline{X}_n}\right\}\right)\,.$$

It is to be noted that we cannot compare the two confidence intervals by comparing the approximate variance of the CAN estimators of $e^{-t/\theta}$, since in the second method we use the fact that $e^{-t/\theta}$ is a monotone function of θ to construct the confidence interval. We compare the lower and upper limits and arrive at the conclusion that the two methods essentially lead to the same large sample confidence intervals. The lower limit L_n of the $100(1-\alpha)\%$ asymptotic confidence interval for $e^{-t/\theta}$, by using CAN estimator of $e^{-t/\theta}$ and by estimating its variance can be expressed as follows:

$$
\begin{aligned}
L_{1n} &= e^{-t/\overline{X}_n}\left(1 - a_{1-\alpha/2}\,\frac{t}{\overline{X}_n\sqrt{n}}\right)\\
&= \left(1 - \frac{t}{\overline{X}_n} + \frac{t^2}{2!\overline{X}_n^2} - \frac{t^3}{3!\overline{X}_n^3} - \cdots\right)\left(1 - a_{1-\alpha/2}\,\frac{t}{\overline{X}_n\sqrt{n}}\right)\\
&= 1 - \frac{t}{\overline{X}_n} + \frac{t^2}{2!\overline{X}_n^2} - a_{1-\alpha/2}\,\frac{t}{\overline{X}_n\sqrt{n}}\\
&\quad + a_{1-\alpha/2}\,\frac{t^2}{\overline{X}_n^2\sqrt{n}} - a_{1-\alpha/2}\,\frac{t^3}{2\overline{X}_n^3\sqrt{n}} + \cdots.
\end{aligned}
$$

Similarly, the lower limit L_n of the $100(1-\alpha)\%$ asymptotic confidence interval for $e^{-t/\theta}$, using the fact that $e^{-t/\theta}$ is a monotone function of θ, can be expressed as follows:

$$
\begin{aligned}
L_{2n} &= \exp\left\{\frac{-t\left(1 + \frac{a_{1-\alpha/2}}{\sqrt{n}}\right)}{\overline{X}_n}\right\}\\
&= 1 - \frac{-t\left(1 + \frac{a_{1-\alpha/2}}{\sqrt{n}}\right)}{\overline{X}_n} + \frac{t^2\left(1 + \frac{a_{1-\alpha/2}}{\sqrt{n}}\right)^2}{2\overline{X}_n^2} - \cdots\\
&= 1 - \frac{t}{\overline{X}_n} + \frac{t^2}{2!\overline{X}_n^2} + a_{1-\alpha/2}\,\frac{t^2}{\overline{X}_n^2\sqrt{n}} - \frac{t a_{1-\alpha/2}}{\overline{X}_n\sqrt{n}} + \frac{t^2 a_{1-\alpha/2}^2}{2n\overline{X}_n^2} - \cdots.
\end{aligned}
$$

If we compare L_{1n} with L_{2n}, then we note that the first four terms are identical in both the expressions. In rest of the terms of L_{1n}, the denominator involves $\sqrt{n}$, while in rest of the terms of L_{2n}, the denominator involves $n, n^2, n^3, \ldots$. Similar scenario is observed in upper limits. It is to be noted that these comparisons are for limits which are obtained for large n. Hence, in general, there would not be any difference between the two. We verify it by simulation in Sect. 3.4.   □

In the next section, we extend the results of a CAN estimator for a real parameter setup to a CAN estimator for a vector parameter setup.

3.3   CAN Estimator: Vector Parameter Setup

Suppose X is a random variable or a random vector defined on a probability space $(\Omega, \mathbb{A}, P_\theta)$, where the probability measure P_θ is indexed by a vector parameter $\underline{\theta} \in \Theta \subset \mathbb{R}^k$. Suppose $\underline{\theta} = (\theta_1, \theta_2, \ldots, \theta_k)'$. Given a random sample $\{X_1, X_2, \ldots, X_n\}$ of size n from the distribution of X, suppose $\underline{T}_n = (T_{1n}, T_{2n}, \ldots, T_{kn})'$ is an estimator of $\underline{\theta}$, that is, $\underline{T}_n$ is a random vector with range space as the parameter space $\Theta \subset \mathbb{R}^k$ and T_{in} is an estimator of θ_i for $i = 1, 2, \ldots, k$. Consistency of $\underline{T}_n$ as an estimator of $\underline{\theta}$ is defined in Sect. 2.5 using the two approaches, marginal and joint consistency, and it is proved that the two approaches are equivalent. However, such an equivalence is not valid for a CAN estimator in a vector setup, and it is essential to treat real and vector parameter setups separately. We define below a CAN estimator for a vector parameter.

▶ **Definition 3.3.1**
Consistent and Asymptotically Normal Estimator for a Vector Parameter: Suppose $\underline{T}_n$ is an estimator of $\underline{\theta}$ and suppose there exists a sequence $\{a_n, n \geq 1\}$ of real numbers tending to ∞ as $n \to \infty$, such that $a_n(\underline{T}_n - \underline{\theta}) \overset{L}{\to} \underline{U} \sim N_k(0, \Sigma(\theta))$ distribution as $n \to \infty$, where $\Sigma(\theta)$ is a positive definite matrix. Then $\underline{T}_n$ is a CAN estimator of $\underline{\theta}$ with approximate dispersion matrix $\Sigma(\theta)/a_n^2$.

As for a real parameter, if $a_n(\underline{T}_n - \underline{\theta}) \overset{L}{\to} \underline{U}$, then $\underline{T}_n \overset{P_\theta}{\to} \underline{\theta}$ for all $\underline{\theta}$ and $\underline{T}_n$ is consistent for $\underline{\theta}$. It is to be noted that if $a_n(\underline{T}_n - \underline{\theta}) \overset{L}{\to} \underline{U} \sim N_k(0, \Sigma(\theta))$ distribution, then each component of $a_n(\underline{T}_n - \underline{\theta})$ has asymptotically normal distribution. However, it is known from the theory of multivariate normal distribution that though each component of random vector $\underline{X}$ has normal distribution, the distribution of $\underline{X}$ need not be multivariate normal. As a consequence, though each of the components of $\underline{T}_n = (T_{1n}, T_{2n}, \ldots, T_{kn})'$ is CAN for the corresponding component of $\underline{\theta} = (\theta_1, \theta_2, \ldots, \theta_k)'$, the vector estimator $\underline{T}_n$ may not be a CAN estimator of $\underline{\theta}$. Thus, we need to deal with multivariate setup to obtain a CAN estimator for a vector parameter. If an estimator is in the form of an average, then the standard tool to generate a CAN estimator in multiparameter setup is the multivariate CLT and the extension of delta method. For the consistent estimators based on sample quantiles, the asymptotic normality can be established using the asymptotic joint distribution of the order statistics. In the following theorems, we state all these results and illustrate using examples.

Theorem 3.3.1

Multivariate CLT: *Suppose $\underline{X}$ is a k-dimensional random vector with mean vector $E(\underline{X}) = \underline{\mu}$ and dispersion matrix Σ, which is a positive definite matrix. Suppose $\{\underline{X}_1, \underline{X}_2, \ldots, \underline{X}_n\}$ is a random sample from the distribution of $\underline{X}$. Suppose $\overline{\underline{X}}_n$ denotes the sample mean vector, then $\sqrt{n}(\overline{\underline{X}}_n - \underline{\mu}) \overset{L}{\to} \underline{U} \sim N_k(0, \Sigma)$ distribution as $n \to \infty$.*

In Theorem 2.2.5, it has been proved that the sample raw moments are consistent for the corresponding population raw moments. Further, the joint consistency and the marginal consistency are equivalent. Hence, $\underline{T}_n = (m'_1, m'_2, \ldots, m'_k)'$ is consistent for $\underline{\mu} = (\mu'_1, \mu'_2, \ldots, \mu'_k)'$, provided $\mu'_k < \infty$. In the following theorem, using the multivariate CLT, we prove that $\underline{T}_n$ is also a CAN estimator of $\underline{\mu}$.

Theorem 3.3.2

Suppose $\{X_1, X_2, \ldots, X_n\}$ is a random sample from the distribution of X whose raw moments up to order $2k$ are finite. Then a random vector $\underline{T}_n = (m'_1, m'_2, \ldots, m'_k)'$ of first k sample raw moments is a CAN estimator of a vector $\underline{\mu} = (\mu'_1, \mu'_2, \ldots, \mu'_k)'$ of corresponding population raw moments with approximate dispersion matrix Σ/n, where $\Sigma = [\sigma_{ij}]$ and $\sigma_{ij} = Cov(X^i, X^j)$, $i, j = 1, 2, \ldots, k$.

Proof Consistency of $\underline{T}_n$ for the parameter $\underline{\mu}$ follows from Theorem 2.2.5 and the equivalence of joint and marginal consistency. To show that it is CAN, suppose a random vector $\underline{Z}$ is defined as $\underline{Z} = (X, X^2, \ldots, X^k)$. Then $E(\underline{Z}) = \underline{\mu}$ and the dispersion matrix Σ of $\underline{Z}$ is given by $\Sigma = [\sigma_{ij}]$, where $\sigma_{ij} = Cov(X^i, X^j)$, $i, j = 1, 2, \ldots, k$. To examine whether Σ is positive definite, suppose a random variable Y is defined as $Y = \sum_{i=1}^{k} a_i X^i$. Without loss of generality, we assume that X is a non-degenerate random variable which implies that Y is also a non-degenerate random variable and hence $Var(Y) > 0$. Observe that for any non-zero vector $\underline{a} = (a_1, a_2, \ldots, a_k)'$ of real numbers

$$0 < Var(Y) = Var\left(\sum_{i=1}^{k} a_i X^i\right) = \sum_{i=1}^{k}\sum_{j=1}^{k} a_i a_j Cov(X^i, X^j) = \underline{a}'\Sigma\underline{a},$$

which proves that Σ is a positive definite matrix. Thus $\underline{Z}$ is a random vector with mean vector $\underline{\mu}$ and the positive definite dispersion matrix Σ. Now a random sample $\{X_1, X_2, \ldots, X_n\}$ from the distribution of X gives a corresponding random sample $\{\underline{Z}_1, \underline{Z}_2, \ldots, \underline{Z}_n\}$ from the distribution of $\underline{Z}$. Suppose $\overline{Z}_n = \sum_{i=1}^{n} \underline{Z}_i/n$ denotes the sample mean vector. Then by the multivariate CLT,
$\sqrt{n}(\overline{Z}_n - \underline{\mu}) \overset{L}{\to} \underline{U} \sim N_k(0, \Sigma)$ distribution as $n \to \infty$. But $\overline{Z}_n = T_n$ and hence we have proved that $\underline{T}_n = (m'_1, m'_2, \ldots, m'_k)'$, a random vector of first k sample raw moments is a CAN estimator of a vector $\underline{\mu} = (\mu'_1, \mu'_2, \ldots, \mu'_k)'$ of corresponding population raw moments with approximate dispersion matrix Σ/n. □

Multivariate CLT is useful to verify whether the estimator $\underline{T}_n$ is CAN, provided it is in the form of an average. If an estimator is based on sample quantiles, then the following theorem, which states the asymptotic joint distribution of sample quantiles, is useful to verify whether the given estimator is a CAN estimator in a vector parameter setup. We state the theorem for $k = 2$. It can be easily generalized for higher dimensions.

Theorem 3.3.3

Suppose $\{X_1, X_2, \ldots, X_n\}$ is a random sample from the distribution of X which is absolutely continuous random variable with probability density function $f(x, \underline{\theta})$, where $\underline{\theta}$ is an indexing parameter. Suppose $\{X_{(1)}, X_{(2)}, \ldots, X_{(n)}\}$ is a corresponding order statistic. Suppose $Y_{1n} = X_{([np_1]+1)}$ and $Y_{2n} = X_{([np_2]+1)}$ are the p_1-th and p_2-th sample quantiles respectively and $a_{p_1}(\underline{\theta})$ and $a_{p_2}(\underline{\theta})$ are the p_1-th and p_2-th population quantiles respectively, $0 < p_1 < p_2 < 1$. Suppose $0 < f(a_{p_1}(\underline{\theta}), \underline{\theta}) < \infty$ and $0 < f(a_{p_2}(\underline{\theta}), \underline{\theta}) < \infty$. Then, as $n \to \infty$,

$$\sqrt{n}((Y_{1n} - a_{p_1}(\underline{\theta})), (Y_{2n} - a_{p_2}(\underline{\theta})))' \xrightarrow{L} \underline{U} \sim N_2(\underline{0}, \Sigma(\underline{\theta})),$$
$$\forall \; \underline{\theta} \; for \; which \; \Sigma(\underline{\theta}) \; is \; positive \; definite,$$

where $\Sigma(\underline{\theta}) = [\sigma_{ij}]$ with

$$\sigma_{11} = \frac{p_1(1 - p_1)}{(f(a_{p_1}(\underline{\theta}), \underline{\theta}))^2}, \quad \sigma_{22} = \frac{p_2(1 - p_2)}{(f(a_{p_2}(\underline{\theta}), \underline{\theta}))^2} \quad \&$$

$$\sigma_{12} = \sigma_{21} = \frac{p_1(1 - p_2)}{f(a_{p_1}(\underline{\theta}), \underline{\theta}) f(a_{p_2}(\underline{\theta}), \underline{\theta})}.$$

Theorems 3.3.2 and 3.3.3 together with the invariance property of CAN estimators under differentiable transformation is useful to find CAN estimators for parameters of interest. Invariance property of CAN estimators under differentiable transformation, also known as the delta method, is stated in the following theorem. The proof is given for part(i).

Theorem 3.3.4

Delta method: *Suppose $\underline{T}_n = (T_{1n}, T_{2n}, \ldots, T_{kn})'$ is a CAN estimator of $\underline{\theta} = (\theta_1, \theta_2, \ldots, \theta_k)'$ with approximate dispersion matrix $\Sigma(\underline{\theta})/a_n^2$, where $\Sigma(\underline{\theta}) = [\sigma_{ij}(\underline{\theta})]$ is a positive definite matrix.*
(i) Suppose $g : \mathbb{R}^k \to \mathbb{R}$ is a totally differentiable function. Then

$$a_n(g(\underline{T}_n) - g(\underline{\theta})) \xrightarrow{L} U \sim N(0, v(\underline{\theta})), \quad as \; n \to \infty, \; \forall \; \underline{\theta} \; for \; which \; v(\underline{\theta}) > 0,$$

where $v(\underline{\theta}) = \Delta'\Sigma(\underline{\theta})\Delta$ and Δ is a gradient vector of order $k \times 1$ of g, with i-th component given by $\frac{\partial g}{\partial \theta_i}$.
(ii) Suppose $\underline{g} : \mathbb{R}^k \to \mathbb{R}^l$, $l \le k$ is such that $\underline{g}(\underline{x}) = (g_1(\underline{x}), g_2(\underline{x}), \ldots, g_l(\underline{x}))'$ and $\underline{x} \in \mathbb{R}^k$. Suppose all partial derivatives of the type $\partial g_i/\partial x_j$, $i = 1, 2, \ldots, l$ and $j = 1, 2, \ldots, k$ exist and are continuous, that is, $g_1, g_2, \ldots, g_l$ are totally differentiable functions. Suppose M is a matrix of order $l \times k$ with (i, j)-th element $\frac{\partial g_i}{\partial \theta_j}$, $i = 1, 2, \ldots, l$ and $j = 1, 2, \ldots, k$. Then as $n \to \infty$,

$$\sqrt{n}(\underline{g}(\underline{T}_n) - \underline{g}(\underline{\theta})) \xrightarrow{L} \underline{U} \sim N_l(\underline{0}, M\Sigma(\underline{\theta})M') \; for \; all \; \underline{\theta} \; for \; which \; M\Sigma(\underline{\theta})M' \; is$$
a positive definite matrix.

Proof (i) It is given that $\underline{T}_n$ is a CAN estimator for $\underline{\theta}$. Thus,

$$\underline{T}_n \xrightarrow{P_\theta} \underline{\theta} \text{ and } a_n(\underline{T}_n - \underline{\theta}) \xrightarrow{L} U \sim N_k(0, \Sigma(\underline{\theta})) \ \forall \ \theta \in \Theta.$$

$g : \mathbb{R}^k \to \mathbb{R}$ is a totally differentiable function which implies that g is continuous and hence by invariance of consistency under continuous transformation, $g(\underline{T}_n)$ is consistent for $g(\underline{\theta})$. Now g is a totally differentiable function, hence by the Taylor series expansion,

$$g(\underline{T}_n) = g(\underline{\theta}) + \sum_{i=1}^{k}(T_{in} - \theta_i)\frac{\partial g}{\partial T_{in}}\Big|_{\underline{\theta}} + R_n,$$

$$\text{where } R_n = \frac{1}{2}\sum_{i=1}^{k}\sum_{j=1}^{k}(T_{in} - \theta_i)(T_{jn} - \theta_j)\frac{\partial^2 g}{\partial T_{in}\partial T_{in}}\Big|_{\underline{\theta}_n^*}$$

and $\underline{\theta}_n^* = \alpha\underline{T}_n + (1-\alpha)\underline{\theta}$, $0 < \alpha < 1$. Since $\underline{T}_n \xrightarrow{P_\theta} \underline{\theta}$, we have $\underline{\theta}_n^* \xrightarrow{P_\theta} \underline{\theta}$. Hence,

$\frac{\partial^2 g}{\partial T_{in}\partial T_{in}}\Big|_{\underline{\theta}_n^*} \xrightarrow{P_\theta} \frac{\partial^2 g}{\partial \theta_i \partial \theta_j} = d_{ij}$, say. Further, $a_n(T_{in} - \theta_i) \xrightarrow{L} U_1 \sim N(0, \sigma_{ii}(\underline{\theta}))$ hence $a_n(T_{in} - \theta_i)$ is bounded in probability for all $i = 1, 2, \ldots k$ and $(T_{jn} - \theta_j) \xrightarrow{P_\theta} 0$ for all $j = 1, 2, \ldots k$. Thus,

$$a_n R_n = \frac{1}{2}\sum_{i=1}^{k}\sum_{j=1}^{k}\{a_n(T_{in} - \theta_i)\}(T_{jn} - \theta_j)\frac{\partial^2 g}{\partial T_{in}\partial T_{jn}}\Big|_{\underline{\theta}_n^*} \xrightarrow{P_\theta} 0.$$

As a consequence,

$$a_n(g(\underline{T}_n) - g(\underline{\theta})) = \Delta'\{a_n(\underline{T}_n - \underline{\theta})\} + a_n R_n \xrightarrow{L} U_2 \sim N(0, v(\underline{\theta}))$$
$$\text{where } v(\underline{\theta}) = \Delta'\Sigma(\underline{\theta})\Delta.$$

Hence $g(\underline{T}_n)$ is a CAN estimator for $g(\underline{\theta})$ with approximate variance $v(\underline{\theta})/a_n^2$. □

In the following examples, we illustrate how delta method is useful to obtain a CAN estimator.

✐ Example 3.3.1

Suppose $\{X_1, X_2, \ldots, X_n\}$ is a random sample from X following an exponential distribution with mean θ_1 and $\{Y_1, Y_2, \ldots, Y_n\}$ is a random sample from Y following an exponential distribution with mean θ_2. Suppose X and Y are independent. By the WLLN, $\overline{X}_n$ is consistent for θ_1 and $\overline{Y}_n$ is consistent for θ_2. Thus, $\underline{T}_n = (\overline{X}_n, \overline{Y}_n)'$ is consistent for $\underline{\theta} = (\theta_1, \theta_2)'$. To examine whether it is CAN for $(\theta_1, \theta_2)'$, define $\underline{Z} = (X, Y)'$. Then $E(\underline{Z}) = \underline{\theta} = (\theta_1, \theta_2)'$ and dispersion matrix Σ of $\underline{Z}$ is $\Sigma = \text{diag}(\theta_1^2, \theta_2^2)$ as X and Y are independent random variables. A random sample $\{X_1, X_2, \ldots, X_n\}$ from X and a random sample $\{Y_1, Y_2, \ldots, Y_n\}$

from Y are equivalent to a random sample $\{\underline{Z}_1, \underline{Z}_2, \dots, \underline{Z}_n\}$ from $\underline{Z}$. Hence, by the multivariate CLT, we have $\sqrt{n}(\overline{Z}_n - \underline{\theta}) \xrightarrow{L} \underline{U} \sim N_2(\underline{0}, \Sigma)$, $\forall\ \underline{\theta}$, that is, $\underline{T}_n = (\overline{X}_n, \overline{Y}_n)'$ is CAN for $\underline{\theta} = (\theta_1, \theta_2)'$ with approximate dispersion matrix Σ/n. As in Example 3.2.2, the approximate dispersion matrix Σ/n is related to the information matrix. For this probability model $I(\underline{\theta}) = \mathrm{diag}(1/\theta_1^2, 1/\theta_2^2)$. Thus, $I(\underline{\theta}) = \Sigma^{-1}$. Suppose we want to find a CAN estimator for $P[X < Y]$. First we obtain its expression in terms of $\underline{\theta} = (\theta_1, \theta_2)'$ and then use the delta method. Since X and Y are independent,

$$
\begin{aligned}
P[X < Y] &= E(I_{[X<Y]}) = E(E(I_{[X<Y]}|Y)) = E(P[X < Y]|Y) \\
&= \frac{1}{\theta_1\theta_2} \int_0^\infty \left[\int_0^y \exp(-x/\theta_1)dx\right] \exp(-y/\theta_2)dy \\
&= \frac{1}{\theta_2} \int_0^\infty \left[1 - \exp(-y/\theta_1)\right] \exp(-y/\theta_2)dy \\
&= \frac{1}{\theta_2} \left[\int_0^\infty \exp(-y/\theta_2)dy - \int_0^\infty \exp\{-(1/\theta_1 + 1/\theta_2)y\}dy\right] \\
&= 1 - \frac{\theta_1}{\theta_1 + \theta_2} = \frac{\theta_2}{\theta_1 + \theta_2} \ .
\end{aligned}
$$

Thus, $P[X < Y] = \theta_2/(\theta_1 + \theta_2) = g(\theta_1, \theta_2)$, say, where g is a function from $\mathbb{R}^2 \to \mathbb{R}$. Its partial derivatives exist and are continuous, thus g is a totally differentiable function, hence by Theorem 3.3.4, $g(\underline{T}_n) = \overline{Y}_n/(\overline{X}_n + \overline{Y}_n)$ is CAN for $g(\theta_1, \theta_2) = P[X < Y] = \theta_2/(\theta_1 + \theta_2)$ with the approximate variance $\Delta'\Sigma\Delta/n$, where $\Delta' = \left(-\theta_2/(\theta_1 + \theta_2)^2, \theta_1/(\theta_1 + \theta_2)^2\right)$ and $\Delta'\Sigma\Delta = 2\theta_1^2\theta_2^2/(\theta_1 + \theta_2)^4$. There is one more approach to find a CAN estimator for $P[X < Y]$. We define random variables Z_i, $i = 1, 2, \dots, n$ as

$$
Z_i = \begin{cases} 1, & \text{if } X_i < Y_i \\ 0, & \text{if } X_i \ge Y_i. \end{cases}
$$

Then $\{Z_1, Z_2, \dots, Z_n\}$ are independent and identically distributed random variables, each having Bernoulli $B(1, p(\theta_1, \theta_2))$ distribution, where $p(\theta_1, \theta_2) = P[X_i < Y_i] = \theta_2/(\theta_1 + \theta_2)$. Hence by the WLLN and the CLT, $\overline{Z}_n$ is CAN for $p(\theta_1, \theta_2) = \theta_2/(\theta_1 + \theta_2)$ with approximate variance $p(\theta_1, \theta_2)(1 - p(\theta_1, \theta_2))/n = \theta_1\theta_2/n(\theta_1 + \theta_2)^2$. We compare the approximate variances of the two CAN estimators of $P[X < Y]$ to examine which is better. Observe that

$$
\begin{aligned}
\frac{2\theta_1^2\theta_2^2}{(\theta_1 + \theta_2)^4} - \frac{\theta_1\theta_2}{(\theta_1 + \theta_2)^2} &= \frac{\theta_1\theta_2}{(\theta_1 + \theta_2)^2}\left[\frac{2\theta_1\theta_2}{(\theta_1 + \theta_2)^2} - 1\right] \\
&= \frac{-\theta_1\theta_2(\theta_1 - \theta_2)^2}{(\theta_1 + \theta_2)^4} < 0 \ .
\end{aligned}
$$

Thus, $\overline{Y}_n/(\overline{X}_n + \overline{Y}_n)$ is a better CAN estimator of $P[X < Y]$ than $\overline{Z}_n$. It sounds reasonable, as while defining $\overline{Z}_n$ we ignore actual magnitudes of X and Y. Further, $\overline{Y}_n/(\overline{X}_n + \overline{Y}_n)$ is function of a sufficient statistic while $\overline{Z}_n$ is not. □

✎ Example 3.3.2

Suppose $\underline{X} = \{X_1, X_2, \ldots, X_n\}$ is a random sample from a normal distribution with mean μ and variance σ^2. Then the likelihood of $\underline{\theta} = (\mu, \sigma^2)'$ is given by

$$L_n(\underline{\theta}|\underline{X}) = \prod_{i=1}^{n} \frac{1}{\sqrt{2\pi}\sigma} \exp\left\{-\frac{1}{2\sigma^2}(X_i - \mu)^2\right\}$$

$$\Leftrightarrow \quad \log L_n(\underline{\theta}|\underline{X}) = c - \frac{n}{2}\log\sigma^2 - \frac{1}{2\sigma^2}\sum_{i=1}^{n}(X_i - \mu)^2,$$

where c is a constant free from $\underline{\theta}$. The system of likelihood equations is given by

$$\frac{\partial}{\partial\mu}\log L_n(\underline{\theta}|\underline{X}) = \frac{1}{\sigma^2}\sum_{i=1}^{n}(X_i - \mu) = 0$$

$$\text{and} \quad \frac{\partial}{\partial\sigma^2}\log L_n(\underline{\theta}|\underline{X}) = -\frac{n}{2\sigma^2} + \frac{1}{2\sigma^4}\sum_{i=1}^{n}(X_i - \mu)^2 = 0$$

with solution $\hat{\mu}_n = \overline{X}_n$ and $\hat{\sigma}_n^2 = \frac{1}{n}\sum_{i=1}^{n}(X_i - \overline{X}_n)^2$. The matrix D of second order partial derivatives at the solution of the system of likelihood equations is $D = \text{diag}(-n/\hat{\sigma}_n^2, -n/2\hat{\sigma}_n^4)$. The first principal minor of D is negative and the second is positive, hence the matrix D is negative definite a.s. Hence, $\hat{\underline{\theta}}_n = (\hat{\mu}_n, \hat{\sigma}_n^2)'$ is the maximum likelihood estimator of $\underline{\theta} = (\mu, \sigma^2)'$. By the WLLN $\hat{\mu}_n = \overline{X}_n \xrightarrow{P_\theta} \mu$ and in Theorem 2.5.3, it is proved that the sample central moments are consistent for corresponding population central moments, hence $\hat{\sigma}_n^2 = \sum_{i=1}^{n}(X_i - \overline{X}_n)^2/n$ is consistent for σ^2. Thus, $\hat{\underline{\theta}}_n = (\hat{\mu}_n, \hat{\sigma}_n^2)'$ is consistent for $\underline{\theta}$. To examine whether it is CAN, we adopt the procedure as in the proof of Theorem 3.3.2 and further use invariance property of CAN estimators under differentiable transformation as stated in Theorem 3.3.4. Suppose a random vector $\underline{Z}$ is defined as $\underline{Z} = (X, X^2)'$. Then $E(\underline{Z}) = (\mu, \sigma^2 + \mu^2)'$ and the dispersion matrix Σ of $\underline{Z}$ is given by

$$\Sigma = \begin{pmatrix} Var(X) & Cov(X, X^2) \\ Cov(X, X^2) & Var(X^2) \end{pmatrix},$$

where $Var(X) = \sigma^2$, $Cov(X, X^2) = \mu'_3 - \mu\mu'_2$ and $Var(X^2) = \mu'_4 - (\mu'_2)^2$, where $\mu'_2 = \mu^2 + \sigma^2$. To find μ'_3 and μ'_4, we use the well-known result that the normal distribution is symmetric implying that $\mu_3 = 0$. Thus,

$$0 = \mu_3 = \mu'_3 - 3\mu'_2\mu'_1 + 2\mu'^3_1 \Rightarrow \mu'_3 = 3\mu(\mu^2 + \sigma^2) - 2\mu^3 = \mu^3 + 3\mu\sigma^2.$$

Further, coefficient of kurtosis is 3 which gives $\mu_4 = 3\sigma^4$ and $\mu_4 = \mu'_4 - 4\mu'_3\mu'_1 + 6\mu'_2\mu'^2_1 - 3\mu'^4_1$ gives $\mu'_4 = 3\sigma^4 + \mu^4 + 6\mu^2\sigma^2$. From these expressions for μ'_3 and μ'_4, we get $Cov(X, X^2) = 2\mu\sigma^2$ and $Var(X^2) = 2\sigma^4 + 4\mu^2\sigma^2$. Thus, the dispersion matrix Σ of $\underline{Z}$ is

$$\Sigma = \begin{pmatrix} \sigma^2 & 2\mu\sigma^2 \\ 2\mu\sigma^2 & 2\sigma^4 + 4\mu^2\sigma^2 \end{pmatrix}.$$

A random sample $\{X_1, X_2, \ldots, X_n\}$ is equivalent to a random sample $\{\underline{Z}_1, \underline{Z}_2, \ldots, \underline{Z}_n\}$ from $\underline{Z}$. Suppose $\overline{Z}_n = \sum_{i=1}^{n} Z_i/n$ denotes the sample mean vector. Then by the multivariate CLT, $\sqrt{n}(\overline{Z}_n - E(\underline{Z})) \xrightarrow{L} \underline{U} \sim N_2(0, \Sigma)$ distribution as $n \to \infty$. But $\overline{Z}_n = (m'_1, m'_2)'$ and hence $\underline{T}_n = (m'_1, m'_2)'$ is CAN for $(\mu, \mu^2 + \sigma^2)' = \underline{\phi}$, say, with approximate dispersion matrix Σ/n. We now find a transformation $\underline{g} : \mathbb{R}^2 \to \mathbb{R}^2$ such that $\underline{g}(\underline{T}_n)$ is CAN for $\underline{g}(\underline{\phi}) = \underline{\theta} = (\mu, \sigma^2)'$. Suppose $\underline{g} = (g_1, g_2)' : \mathbb{R}^2 \to \mathbb{R}^2$ is defined as $g_1(x_1, x_2) = x_1$ and $g_2(x_1, x_2) = x_2 - x_1^2$. Then

$$\frac{\partial}{\partial x_1} g_1(x_1, x_2) = 1, \quad \frac{\partial}{\partial x_2} g_1(x_1, x_2) = 0,$$

$$\frac{\partial}{\partial x_1} g_2(x_1, x_2) = -2x_1 \quad \& \quad \frac{\partial}{\partial x_2} g_2(x_1, x_2) = 1.$$

These partial derivatives are continuous and hence g_1 and g_2 are totally differentiable functions. The matrix M of partial derivatives evaluated at $\underline{\theta}$ is given by

$$M = \begin{pmatrix} 1 & 0 \\ -2\mu & 1 \end{pmatrix}.$$

Hence by Theorem 3.3.4,

$$\underline{g}(\underline{T}_n) = (g_1(m'_1, m'_2), g_2(m'_1, m'_2))' = (m'_1, m'_2 - m'^2_1)'$$
$$= (\hat{\mu}_n, \hat{\sigma}^2_n)' \text{ is CAN for } \underline{g}(\underline{\phi}) = \underline{\theta} = (\mu, \sigma^2)',$$

with approximate dispersion matrix $M\Sigma M'/n$, where $M\Sigma M' = \text{diag}[\sigma^2, 2\sigma^4]$. It indicates that $\hat{\mu}_n$ and $\hat{\sigma}^2_n$ are asymptotically independent. It is consistent with the result, that for a random sample from a normal distribution, the sample mean and the sample variance are statistically independent. Now we find a moment estimator of $\underline{\theta} = (\mu, \sigma^2)'$ based on the sufficient statistic and examine whether it is CAN. From the likelihood, it is clear that $(\sum_{i=1}^{n} X_i, \sum_{i=1}^{n} X_i^2)$ is jointly sufficient for the $N(\mu, \sigma^2)$ family. Hence, the system of moment equations based on the sufficient statistic is given by

$$\overline{X}_n = m_1' = \mu \quad \text{and} \quad \frac{1}{n} \sum_{i=1}^{n} X_i^2 = m_2' = \mu^2 + \sigma^2$$

and its solution is given by $\mu = m_1'$ and $\sigma^2 = m_2' - m_1'^2$. Hence, the moment estimator $\underline{\tilde{\theta}}_n$ of $\underline{\theta}$ based on the sufficient statistics is given by

$$\underline{\tilde{\theta}}_n = (\tilde{\mu}_n, \tilde{\sigma}_n^2)' = (m_1', m_2' - m_1'^2)' = (\hat{\mu}_n, \hat{\sigma}_n^2)',$$

which is the same as the maximum likelihood estimator of $\underline{\theta}$ and hence is CAN with approximate dispersion matrix $M \Sigma M'/n$. As in Example 3.3.1, we investigate whether the approximate dispersion matrix $M \Sigma M'/n$ is related to the information matrix $I(\underline{\theta})$. To find the information matrix, we find the second order partial derivatives of $\log f(x, \mu, \sigma^2) = c - (1/2)\log \sigma^2 - (1/2\sigma^2)(x - \mu)^2$. These are as follows:

$$\frac{\partial}{\partial \mu} \log f(x, \mu, \sigma^2) = \frac{(x - \mu)}{\sigma^2}, \quad \frac{\partial^2}{\partial \mu^2} \log f(x, \mu, \sigma^2) = \frac{-1}{\sigma^2}$$

$$\frac{\partial}{\partial \sigma^2} \log f(x, \mu, \sigma^2) = \frac{-1}{2\sigma^2} + \frac{(x - \mu)^2}{2\sigma^4}$$

$$\frac{\partial^2}{\partial \sigma^2 \partial \mu} \log f(x, \mu, \sigma^2) = -\frac{(x - \mu)}{\sigma^4},$$

$$\frac{\partial^2}{\partial (\sigma^2)^2} \log f(x, \mu, \sigma^2) = \frac{1}{2\sigma^4} - \frac{(x - \mu)^2}{\sigma^6}.$$

Hence, the information matrix $I(\underline{\theta})$ is $\text{diag}[1/\sigma^2, 1/2\sigma^4]$ and $M \Sigma M' = I^{-1}(\underline{\theta})$. $\square$

✍ Example 3.3.3

Suppose $\{(X_1, Y_1), (X_2, Y_2), \ldots, (X_n, Y_n)\}$ is a random sample from a distribution of $(X, Y)'$ which is bivariate normal $BN(\mu_1, \mu_2, \sigma_1^2, \sigma_2^2, \rho)$. Suppose a random variable Z is defined as $Z = X - Y$, then $Z \sim N(\mu_1 - \mu_2, \sigma^2)$, where $\sigma^2 = \sigma_1^2 + \sigma_2^2 - 2\rho\sigma_1\sigma_2$. A random sample $\{(X_1, Y_1), (X_2, Y_2), \ldots (X_n, Y_n)\}$ from the distribution of $(X, Y)'$ gives a random sample $\{Z_1, Z_2, \ldots, Z_n\}$ from the distribution of Z. Hence, from Example 3.3.2 it follows that $\overline{Z}_n = \overline{X}_n - \overline{Y}_n$ is CAN for $\mu_1 - \mu_2$ with approximate variance σ^2/n and $S_n^2 = \sum_{i=1}^{n}(Z_i - \overline{Z}_n)^2/n$ is CAN for σ^2 with approximate variance $2\sigma^4/n$. Using these results, we can find an asymptotic confidence interval for $\mu_1 - \mu_2$. $\square$

☑ Example 3.3.4

Suppose the number X of telephone calls arriving at an exchange in a unit time period has a Poisson distribution with mean λ. Given that $X = x$ calls have been arrived, the number of calls Y that lasts more than a unit time period is distributed as binomial $B(x, p)$. Thus, the probability mass function of $(X, Y)'$ is

$$P[X = x, Y = y] = \frac{e^{-\lambda}\lambda^x}{x!}\binom{x}{y}p^y(1-p)^{x-y},$$

$$y = 0, 1, \ldots, x; \quad x = 0, 1, 2, \ldots .$$

The unconditional distribution of Y is given by

$$P[Y = y] = \sum_{x=y}^{\infty} P[X = x, Y = y] = \frac{e^{-\lambda}p^y\lambda^y}{y!}\sum_{x=y}^{\infty}\frac{\lambda^{x-y}(1-p)^{x-y}}{(x-y)!}$$

$$= \frac{e^{-\lambda}p^y\lambda^y}{y!}\sum_{r=0}^{\infty}\frac{(\lambda(1-p))^r}{r!} = \frac{e^{-\lambda}p^y\lambda^y}{y!}e^{\lambda(1-p)}$$

$$= \frac{e^{-\lambda p}(\lambda p)^y}{y!}, \quad y = 0, 1, \ldots, .$$

Thus, the unconditional distribution of Y is Poisson $Poi(\lambda p)$. Now the likelihood of $\underline{\theta} = (\lambda, p)'$ corresponding to a random sample from the distribution of $(X, Y)'$ is

$$L_n(\underline{\theta}|\underline{X}, \underline{Y}) = \prod_{i=1}^{n}\binom{X_i}{Y_i}e^{-\lambda}\lambda^{X_i}p^{Y_i}(1-p)^{X_i-Y_i}/X_i!$$

$$= c + e^{-n\lambda}(\lambda(1-p))^{\sum_{i=1}^{n}X_i}\left(\frac{p}{1-p}\right)^{\sum_{i=1}^{n}Y_i},$$

where c is a constant free from the parameters. Thus, by the Neyman-Fisher factorization theorem, a sufficient statistic for the family is $\left(\sum_{i=1}^{n}X_i, \sum_{i=1}^{n}Y_i\right)'$. Hence, the system of equations to find the moment estimator of $\underline{\theta}$ based on the sufficient statistic and its solution are given by

$$\frac{1}{n}\sum_{i=1}^{n}X_i = \overline{X}_n = E(X) = \lambda \text{ and } \frac{1}{n}\sum_{i=1}^{n}Y_i = \overline{Y}_n = E(Y) = \lambda p$$

$$\Rightarrow \lambda = \overline{X}_n \text{ and } p = \overline{Y}_n/\overline{X}_n.$$

Hence, the moment estimator $\underline{\tilde{\theta}}_n$ of $\underline{\theta}$ based on the sufficient statistic is given by

$$\underline{\tilde{\theta}}_n = (\tilde{\lambda}_n, \tilde{p}_n)' = (\overline{X}_n, \overline{Y}_n/\overline{X}_n)'.$$

It is to be noted that if $\overline{X}_n = 0$, then $\overline{Y}_n$ is also 0 and estimator of $(\lambda, p)'$ is not defined. We assume that for a given random sample $\overline{X}_n > 0$ and $\overline{Y}_n > 0$. By the WLLN $\tilde{\lambda}_n = \overline{X}_n \overset{P_\theta}{\to} \lambda$ and $\overline{Y}_n \overset{P_\theta}{\to} \lambda p$. Convergence in probability is closed under all arithmetic operations, hence $\tilde{p}_n = \overline{Y}_n / \overline{X}_n \overset{P_\theta}{\to} p$. Hence, $\underline{\tilde{\theta}}_n = (\tilde{\lambda}_n, \tilde{p}_n)'$ is a consistent estimator of $\underline{\theta} = (\lambda, p)'$. To examine whether it is CAN, a random vector $\underline{Z}$ is defined as $\underline{Z} = (X, Y)'$. Then $E(\underline{Z}) = (\lambda, \lambda p)'$ and the dispersion matrix Σ of $\underline{Z}$ is given by

$$\Sigma = \begin{pmatrix} Var(X) & Cov(X, Y) \\ Cov(X, Y) & Var(Y) \end{pmatrix},$$

where $Var(X) = \lambda$, $Var(Y) = \lambda p$ and $Cov(X, Y) = E(XY) - E(X)E(Y)$. Now

$$E(XY) = E(E(XY)|X) = E(XE(Y|X)) = E(XXp)$$
$$= p(Var(X) + (E(X))^2) = p(\lambda + \lambda^2),$$

hence $Cov(X, Y) = p(\lambda + \lambda^2) - \lambda^2 p = \lambda p$. Thus, the dispersion matrix Σ of $\underline{Z}$ is

$$\Sigma = \begin{pmatrix} \lambda & \lambda p \\ \lambda p & \lambda p \end{pmatrix}.$$

A random sample from the distribution of $(X, Y)'$ is the same as a random sample $\{\underline{Z}_1, \underline{Z}_2, \ldots, \underline{Z}_n\}$ from $\underline{Z}$. Suppose $\overline{\underline{Z}}_n = \sum_{i=1}^{n} \underline{Z}_i / n$ denotes the sample mean vector. Then by the multivariate CLT, $\sqrt{n}(\overline{\underline{Z}}_n - E(\underline{Z})) \overset{L}{\to} \underline{U} \sim N_2(0, \Sigma)$ distribution as $n \to \infty$. But $\overline{\underline{Z}}_n = (\overline{X}_n, \overline{Y}_n)'$ and hence $\underline{T}_n = (\overline{X}_n, \overline{Y}_n)'$ is CAN for $(\lambda, \lambda p)' = \underline{\phi}$, say, with approximate dispersion matrix Σ/n. We now find a transformation $\underline{g} : \mathbb{R}^2 \to \mathbb{R}^2$ such that $\underline{g}(\underline{T}_n)$ is CAN for $\underline{g}(\underline{\phi}) = \underline{\theta} = (\lambda, p)'$. Suppose $\underline{g} = (g_1, g_2)' : \mathbb{R}^2 \to \mathbb{R}^2$ is defined as $g_1(x_1, x_2) = x_1$ and $g_2(x_1, x_2) = x_2/x_1$. Then

$$\frac{\partial}{\partial x_1} g_1(x_1, x_2) = 1, \quad \frac{\partial}{\partial x_2} g_1(x_1, x_2) = 0,$$
$$\frac{\partial}{\partial x_1} g_2(x_1, x_2) = -\frac{x_2}{x_1^2} \quad \& \quad \frac{\partial}{\partial x_2} g_2(x_1, x_2) = \frac{1}{x_1}.$$

These partial derivatives are continuous and hence g_1 and g_2 are totally differentiable functions. The matrix M of partial derivatives evaluated at $(\lambda, p)'$ is given by

$$M = \begin{pmatrix} 1 & 0 \\ -p/\lambda & 1/\lambda \end{pmatrix}.$$

Hence by Theorem 3.3.4,

$$
\underline{g}(\underline{T}_n) = (g_1(\overline{X}_n, \overline{Y}_n), g_2(\overline{X}_n, \overline{Y}_n))'
$$

$$
= \left(\overline{X}_n, \frac{\overline{Y}_n}{\overline{X}_n} \right)' = (\tilde{\lambda}_n, \tilde{p}_n)' \text{ is CAN for } \underline{g}(\underline{\phi}) = \underline{\theta} = (\lambda, p)',
$$

with the approximate dispersion matrix $M \Sigma M'/n$, which is diag$[\lambda, p(1-p)/\lambda]$. Now we check whether $M \Sigma M' = I^{-1}(\theta)$. From the likelihood we have

$$
\frac{\partial}{\partial \lambda} \log L_n(\underline{\theta}|\underline{X}) = -n + \frac{\sum\limits_{i=1}^{n} X_i}{\lambda},
$$

$$
\frac{\partial^2}{\partial \lambda^2} \log L_n(\underline{\theta}|\underline{X}) = -\frac{n\overline{X}_n}{\lambda^2}, \quad \frac{\partial^2}{\partial p \partial \lambda} \log L_n(\underline{\theta}|\underline{X}) = 0
$$

and $\displaystyle \frac{\partial}{\partial p} \log L_n(\underline{\theta}|\underline{X}) = \frac{\sum\limits_{i=1}^{n} Y_i - \sum\limits_{i=1}^{n} X_i}{1-p} + \frac{\sum\limits_{i=1}^{n} Y_i}{p}$,

$$
\frac{\partial^2}{\partial p^2} \log L_n(\underline{\theta}|\underline{X}) = \frac{\sum\limits_{i=1}^{n} Y_i - \sum\limits_{i=1}^{n} X_i}{(1-p)^2} - \frac{\sum\limits_{i=1}^{n} Y_i}{p^2}.
$$

Thus, $I(\theta)$ is given by

$$
nI(\underline{\theta}) = \begin{pmatrix} n/\lambda & 0 \\ 0 & n\lambda/p(1-p) \end{pmatrix}.
$$

Thus, the approximate dispersion matrix $M \Sigma M'/n = I^{-1}(\underline{\theta})/n$, as in Example 3.3.2. To obtain the maximum likelihood estimator, the system of likelihood equations is given by

$$
-n + \frac{\sum\limits_{i=1}^{n} X_i}{\lambda} = 0, \text{ and } \frac{\sum\limits_{i=1}^{n} Y_i - \sum\limits_{i=1}^{n} X_i}{1-p} + \frac{\sum\limits_{i=1}^{n} Y_i}{p} = 0
$$

and its solution is given as $\lambda = \overline{X}_n$ and $p = \overline{Y}_n/\overline{X}_n$. The matrix D of second order partial derivatives is given by

$$
D = \begin{pmatrix} -\frac{n\overline{X}_n}{\lambda^2} & 0 \\ 0 & \frac{\sum\limits_{i=1}^{n} Y_i - \sum\limits_{i=1}^{n} X_i}{(1-p)^2} - \frac{\sum\limits_{i=1}^{n} Y_i}{p^2} \end{pmatrix}.
$$

At the solution of the system of likelihood equations it is given by

$$D_{sol} = \begin{pmatrix} -\frac{n}{\overline{X}_n} & 0 \\ 0 & -\frac{n\overline{X}_n^3}{(\overline{X}_n - \overline{Y}_n)} \end{pmatrix}.$$

Its first principal minor is negative and the second is positive, as $(\overline{X}_n - \overline{Y}_n) > 0$, hence D_{sol} is an almost surely negative definite matrix and hence at the solution, the likelihood attains its maximum. Thus, the maximum likelihood estimator $\hat{\underline{\theta}}_n = (\hat{\lambda}_n, \hat{p}_n)'$ of $\underline{\theta}$ is given by

$$\hat{\underline{\theta}}_n = \left(\overline{X}_n, \overline{Y}_n/\overline{X}_n\right)' = (\hat{\lambda}_n, \, \hat{p}_n)'.$$

Thus, the maximum likelihood estimator $\hat{\underline{\theta}}_n = (\hat{\lambda}_n, \hat{p}_n)'$ of $\underline{\theta}$ is the same as the moment estimator of $\underline{\theta}$ based on the sufficient statistics and hence is CAN with the approximate dispersion matrix $M\Sigma M'/n$, which is a diagonal matrix, implying that for large n, $\overline{X}_n$ and $\overline{Y}_n/\overline{X}_n$ are independent.

$\square$

✒ Example 3.3.5

Suppose X follows an exponential distribution with location parameter μ and scale parameter $1/\sigma$ with probability density function

$$f(x, \mu, \sigma) = (1/\sigma) \exp\{-(x - \mu)/\sigma\}, \quad x \geq \mu, \quad \sigma > 0, \quad \mu \in \mathbb{R}.$$

Then

$$E(X) = \mu + \sigma \ \& \ Var(X) = \sigma^2 \ \Rightarrow \ E(X^2) = \sigma^2 + (\mu + \sigma)^2.$$

Suppose $\{X_1, X_2, \ldots, X_n\}$ is a random sample from a distribution of X. The moment estimator $\tilde{\underline{\theta}}_n$ of $\underline{\theta} = (\mu, \sigma)'$ is given by

$$\tilde{\underline{\theta}}_n = (\tilde{\mu}_n, \tilde{\sigma}_n)' = (m_1' - \sqrt{m_2}, \sqrt{m_2})' = \left(m_1' - \sqrt{m_2' - m_1'^2}, \sqrt{m_2' - m_1'^2}\right)'.$$

By the WLLN, $m_1' = \overline{X}_n \xrightarrow{P_\theta} \mu + \sigma$ and by Theorem 2.5.3, $m_2 \xrightarrow{P_\theta} \sigma^2$. Convergence in probability is closed under all arithmetic operations, hence $\sqrt{m_2} \xrightarrow{P_\theta} \sigma$ and $m_1' - \sqrt{m_2} \xrightarrow{P_\theta} \mu$. Thus, the moment estimator $\tilde{\underline{\theta}}_n$ is consistent for $\underline{\theta}$. To examine whether it is CAN, using the same the procedure as in Example 3.3.2 and Example 3.3.4, we get that $\underline{T}_n = (m_1', m_2')'$ is CAN for

$(\mu + \sigma, \sigma^2 + (\mu + \sigma)^2)' = \underline{\phi}$ with approximate dispersion matrix as Σ/n where Σ is given by

$$\Sigma = \begin{pmatrix} Var(X) & Cov(X, X^2) \\ Cov(X, X^2) & Var(X^2) \end{pmatrix}.$$

To find the elements of matrix Σ, we need to find the third and fourth raw moments of X. If we define $Y = X - \mu$, then Y has an exponential distribution with location parameter 0 and mean σ. Hence, the r-th raw moment of Y is given by $E(Y^r) = r!\sigma^r$, $r \geq 1$. From the raw moments of Y, we find the raw moments of X. Thus,

$$E(Y^3) = 6\sigma^3 \Rightarrow E(X^3) = 6\sigma^3 + \mu^3 + 6\mu\sigma^2 + 3\sigma\mu^2$$
$$\Rightarrow Cov(X, X^2) = 4\sigma^3 + 2\mu\sigma^2$$
$$\& \ E(Y^4) = 24\sigma^4 \Rightarrow E(X^4) = 24\sigma^4 + \mu^4 + 24\sigma^3\mu + 12\sigma^2\mu^2 + 4\sigma\mu^3$$
$$\Rightarrow Var(X^2) = 4\sigma^2(\mu^2 + 4\mu\sigma + 5\sigma^2).$$

Thus, the dispersion matrix Σ is given by

$$\Sigma = \begin{pmatrix} \sigma^2 & 4\sigma^3 + 2\mu\sigma^2 \\ 4\sigma^3 + 2\mu\sigma^2 & 4\sigma^2(\mu^2 + 4\mu\sigma + 5\sigma^2) \end{pmatrix}.$$

To find a CAN estimator for $(\mu, \sigma)'$, suppose $\underline{g} = (g_1, g_2)' : \mathbb{R}^2 \to \mathbb{R}^2$ is defined as $g_1(x_1, x_2) = x_1 - \sqrt{x_2 - x_1^2}$ and $g_2(x_1, x_2) = \sqrt{x_2 - x_1^2}$ so that $\underline{g}(\underline{\phi}) = \underline{\theta}$. Further,

$$\frac{\partial}{\partial x_1} g_1(x_1, x_2) = 1 + \frac{x_1}{\sqrt{x_2 - x_1^2}}, \quad \frac{\partial}{\partial x_2} g_1(x_1, x_2) = -\frac{1}{2\sqrt{x_2 - x_1^2}}$$

$$\frac{\partial}{\partial x_1} g_2(x_1, x_2) = -\frac{x_1}{\sqrt{x_2 - x_1^2}} \quad \& \quad \frac{\partial}{\partial x_2} g_2(x_1, x_2) = \frac{1}{2\sqrt{x_2 - x_1^2}}.$$

These partial derivatives are continuous and hence g_1 and g_2 are totally differentiable functions. The matrix M of partial derivatives evaluated at $\underline{\phi}$ is

$$M = \begin{pmatrix} \frac{\mu + 2\sigma}{\sigma} & -\frac{1}{2\sigma} \\ -\frac{\mu + \sigma}{\sigma} & \frac{1}{2\sigma} \end{pmatrix}.$$

Hence, by Theorem 3.3.4,

$$\underline{g}(\underline{T}_n) = (g_1(m_1', m_2'), g_2(m_1', m_2'))' = (m_1' - \sqrt{m_2}, \sqrt{m_2})' = (\tilde{\mu}_n, \tilde{\sigma}_n)'$$

is CAN for $g(\underline{\phi}) = \underline{\theta} = (\mu, \sigma)'$, with the approximate dispersion matrix $M \Sigma M'/n$, where

$$M \Sigma M' = \begin{pmatrix} \sigma^2 & -\sigma^2 \\ -\sigma^2 & 2\sigma^2 \end{pmatrix}.$$

□

In Example 3.3.5, the distribution of X involves two parameters μ and σ. Further, $E(X)$ is a function of both the parameters but $Var(X)$ is a function of only σ. In Example 3.2.5, we have shown that the sample variance is CAN for population variance. Thus, it is of interest to examine whether a random vector of sample mean and sample variance is CAN for a vector of population mean and population variance. In Theorem 3.3.2, we have proved that a random vector of raw moments is CAN for a vector of corresponding population raw moments. In the following example, we show that a random vector $(\overline{X}_n, S_n^2)'$ of sample mean and sample variance is CAN for $(E(X), Var(X))'$.

✔ Example 3.3.6

Suppose $\{X_1, X_2, \ldots, X_n\}$ is a random sample from the distribution of X with mean μ, variance σ^2 and finite raw moments up to order 4. Then from Theorem 3.3.2, $\underline{T}_n = (m_1', m_2')'$ is CAN for $\underline{\phi} = (\mu_1', \mu_2')'$ with approximate dispersion matrix Σ/n where Σ is

$$\Sigma = \begin{pmatrix} Var(X) & Cov(X, X^2) \\ Cov(X, X^2) & Var(X^2) \end{pmatrix} = \begin{pmatrix} \mu_2' - \mu_1'^2 & \mu_3' - \mu_1'\mu_2' \\ \mu_3' - \mu_1'\mu_2' & \mu_4' - \mu_2'^2 \end{pmatrix}.$$

Suppose $\underline{g} = (g_1, g_2)' : \mathbb{R}^2 \to \mathbb{R}^2$ is defined as $g_1(x_1, x_2) = x_1$ and $g_2(x_1, x_2) = x_2 - x_1^2$ so that $\underline{g}(\underline{\phi}) = \underline{\theta} = (\mu, \sigma^2)'$. Further, $\frac{\partial}{\partial x_1} g_1(x_1, x_2) = 1$, $\frac{\partial}{\partial x_2} g_1(x_1, x_2) = 0$, $\frac{\partial}{\partial x_1} g_2(x_1, x_2) = -2x_1$ and $\frac{\partial}{\partial x_2} g_2(x_1, x_2) = 1$. These partial derivatives are continuous and hence g_1 and g_2 are totally differentiable functions. The matrix M of partial derivatives evaluated at $\underline{\phi}$ is given by

$$M = \begin{pmatrix} 1 & 0 \\ -2\mu_1' & 1 \end{pmatrix}.$$

Hence, by Theorem 3.3.4,

$$\underline{g}(\underline{T}_n) = (g_1(m_1', m_2'), \ g_2(m_1', m_2'))' = (m_1', m_2)'$$
$$= (\overline{X}_n, S_n^2)' \text{ is CAN for } \underline{g}(\underline{\phi}) = \underline{\theta} = (\mu, \sigma^2)'$$

with the approximate dispersion matrix $M \Sigma M'/n$, where

$$M \Sigma M' = \begin{pmatrix} \mu_2 & \mu_3 \\ \mu_3 & \mu_4 - \mu_2^2 \end{pmatrix}.$$

We assume here that $\mu_4 \neq \mu_2^2$.   □

📝 Remark 3.3.1

In Example 3.3.2 when $X \sim N(\mu, \sigma^2)$ distribution, we have shown that $(\overline{X}_n, S_n^2)'$ is CAN for $(\mu, \sigma^2)'$ with approximate dispersion matrix D/n, where

$$D = \begin{pmatrix} \sigma^2 & 0 \\ 0 & 2\sigma^4 \end{pmatrix}.$$

When $X \sim N(\mu, \sigma^2)$ distribution, it is known that $\mu_3 = 0$ and $\mu_4 = 3\sigma^4$. Thus, $\mu_4 - \mu_2^2 = 2\sigma^4$. Thus, results of Example 3.3.2 can be derived from Example 3.3.6.

✏️ Example 3.3.7

Suppose X follows a Laplace distribution, also known as a double exponential distribution, with probability density function

$$f(x, \mu, \alpha) = \frac{1}{2\alpha} \exp\left(-\frac{|x - \mu|}{\alpha}\right), \quad x, \mu \in \mathbb{R}, \alpha > 0.$$

We obtain a CAN estimator of $(\mu, \alpha)'$ based on (i) the sample quantiles and (ii) the sample moments. The distribution function $F_X(x)$ of X is given by

$$F_X(x) = \begin{cases} \frac{1}{2} \exp\{(x - \mu)/\alpha\}, & \text{if } x < \mu \\ 1 - \frac{1}{2} \exp\{-(x - \mu)/\alpha\}, & \text{if } x \geq \mu. \end{cases}$$

Hence,

$$F_X(x) = \frac{1}{4} \implies \frac{1}{2} \exp(x - \mu)/\alpha = \frac{1}{4} \implies a_{1/4}(\underline{\theta}) = \mu - \alpha \log_e 2 \text{ and}$$

$$F_X(x) = \frac{3}{4} \implies 1 - \frac{1}{2} \exp\{-(x - \mu)\}/\alpha = \frac{3}{4} \implies a_{3/4}(\underline{\theta}) = \mu + \alpha \log_e 2.$$

It is to be noted that in view of symmetry of the double exponential distribution around μ, the first and the third quartile are equidistant from μ. Further,

$$f(a_{1/4}(\underline{\theta}), \mu, \alpha) = \frac{1}{2\alpha} \exp\left(-\frac{|-\alpha \log_e 2|}{\alpha}\right)$$

$$= \frac{1}{2\alpha} \exp(\log \frac{1}{2}) = \frac{1}{4\alpha}, \quad \& \quad f(a_{3/4}(\underline{\theta}), \mu, \alpha) = \frac{1}{4\alpha}.$$

Suppose $\{X_1, X_2, \ldots, X_n\}$ is a random sample from the distribution of X. From Theorem 3.3.3 it follows that $\underline{T}_n = \left(X_{([n/4]+1)},\ X_{([3n/4]+1)}\right)'$ is CAN for $\underline{\phi} = (\mu - \alpha \log_e 2,\ \mu + \alpha \log_e 2)'$ with approximate dispersion matrix

$$\Sigma = \begin{pmatrix} 3\alpha^2 & \alpha^2 \\ \alpha^2 & 3\alpha^2 \end{pmatrix} = \alpha^2 \begin{pmatrix} 3 & 1 \\ 1 & 3 \end{pmatrix}.$$

We now find a transformation $\underline{g} : \mathbb{R}^2 \to \mathbb{R}^2$ such that $\underline{g}(\underline{T}_n)$ is CAN for $\underline{g}(\underline{\phi}) = \underline{\theta} = (\mu, \alpha)'$. Suppose $\underline{g} = (g_1, g_2)' : \mathbb{R}^2 \to \mathbb{R}^2$ is defined as $g_1(x_1, x_2) = (x_1 + x_2)/2$ and $g_2(x_1, x_2) = (x_2 - x_1)/2 \log_e 2$. Then

$$\frac{\partial}{\partial x_1} g_1(x_1, x_2) = \frac{1}{2}, \quad \frac{\partial}{\partial x_2} g_1(x_1, x_2) = \frac{1}{2},$$
$$\frac{\partial}{\partial x_1} g_2(x_1, x_2) = -\frac{1}{2 \log_e 2} \quad \& \quad \frac{\partial}{\partial x_2} g_2(x_1, x_2) = \frac{1}{2 \log_e 2}.$$

These partial derivatives are continuous and hence g_1 and g_2 are totally differentiable functions. The matrix M of partial derivatives evaluated at $\underline{\phi}$ is given by

$$M = \begin{pmatrix} 1/2 & 1/2 \\ -1/2 \log_e 2 & 1/2 \log_e 2 \end{pmatrix} = \begin{pmatrix} 0.5 & 0.5 \\ -0.7213 & 0.7213 \end{pmatrix}.$$

Hence by Theorem 3.3.4,

$$\underline{g}(\underline{T}_n) = \left(\frac{X_{([n/4]+1)} + X_{([3n/4]+1)}}{2},\ \frac{X_{([3n/4]+1)} - X_{([n/4]+1)}}{2 \log_e 2} \right)' = (\hat{\mu}_n, \hat{\alpha}_n)'$$

is CAN for $\underline{g}(\underline{\phi}) = \underline{\theta} = (\mu, \alpha)'$ with the approximate dispersion matrix $M \Sigma M'/n$, which is $\mathrm{diag}[2\alpha^2, \alpha^2/(\log 2)^2] = \mathrm{diag}[2\alpha^2, 2.0814\alpha^2]$. It implies that for large n, $\hat{\mu}_n$ and $\hat{\alpha}_n$ are independent. To obtain a CAN estimator for $(\mu, \alpha)'$ based on the sample moments, note that the double exponential distribution is symmetric around μ, hence $E(X) = \mu$. To find $Var(X)$, we define $Y = X - \mu$, then $Var(X) = Var(Y)$. Further, Y is distributed as $U_1 - U_2$, where U_1 and U_2 are independent random variables each having an exponential distribution with mean α and $E(U_i)^r = r!\alpha^r$, $i = 1, 2$. Hence, $Var(X) = Var(Y) = Var(U_1 - U_2) = 2\alpha^2$. Thus, for a double exponential distribution, $E(X)$ is a function of μ only and $Var(X)$ is a function of α only. Hence, instead of using Theorem 3.3.2, it is better to use the result established in Example 3.3.6, which states that $(m_1', m_2)'$ is CAN for $(\mu_1', \mu_2)' = (\mu, 2\alpha^2)' = \underline{\phi}$, say with approximate dispersion matrix Σ/n, where Σ is

$$\Sigma = \begin{pmatrix} \mu_2 & \mu_3 \\ \mu_3 & \mu_4 - \mu_2^2 \end{pmatrix}.$$

Now to find the central moments μ_3 and μ_4 of X, we use the fact that $Y = X - \mu$ is distributed as $U_1 - U_2$, where U_1 and U_2 are independent random variables each having an exponential distribution with mean α. Hence the common characteristic functions of U_j, $j = 1, 2$ is $\phi(t) = (1 - it\alpha)^{-1}$. Thus, the characteristic function of Y is given by

$$\phi_Y(t) = E(\exp\{itY\}) = E(\exp\{itU_1\})E(\exp\{-itU_1\})$$
$$= (1 - it\alpha)^{-1}(1 + it\alpha)^{-1} = (1 + t^2\alpha^2)^{-1}.$$

It is now clear that all odd ordered raw moments of Y, which are same as the central moments, are 0 and $\mu'_{2r} = \mu_{2r} = (2r)!\alpha^{2r}$. Thus, $\mu_3 = 0$ and $\mu_4 - \mu_2^2 = 24\alpha^4 - 4\alpha^4 = 20\alpha^4$ and hence Σ is diag$[2\alpha^2, 20\alpha^4]$. To find CAN estimator for $(\mu, \alpha)'$, suppose $g = (g_1, g_2)' : \mathbb{R}^2 \to \mathbb{R}^2$ is defined as $g_1(x_1, x_2) = x_1$ and $g_2(x_1, x_2) = \sqrt{x_2/2}$ so that $g(\phi) = (\mu, \alpha)'$. Now,

$$\frac{\partial}{\partial x_1}g_1(x_1, x_2) = 1, \quad \frac{\partial}{\partial x_2}g_1(x_1, x_2) = 0,$$

$$\frac{\partial}{\partial x_1}g_2(x_1, x_2) = 0 \ \& \ \frac{\partial}{\partial x_2}g_2(x_1, x_2) = -\frac{1}{2\sqrt{2x_2}}.$$

These partial derivatives are continuous and hence g_1 and g_2 are totally differentiable functions. The matrix M of partial derivatives, evaluated at ϕ, is given by

$$M = \begin{pmatrix} 1 & 0 \\ 0 & -1/4\alpha \end{pmatrix}.$$

Hence, by Theorem 3.3.4,

$$g(\underline{T}_n) = (g_1(m'_1, m_2), \ g_2(m'_1, m_2))'$$
$$= \left(m'_1, \ \sqrt{m_2/2}\right)' \ \text{is CAN for} \ g(\underline{\phi}) = (\mu, \alpha)'$$

with the approximate dispersion matrix $M\Sigma M'/n$, where $M\Sigma M' = \text{diag}[2\alpha^2, 1.25\alpha^2]$. $\qquad\square$

✍ Remark 3.2.2

In a multiparameter setup, the estimators are compared on the basis of generalized variance, which is nothing but the determinant of the approximate dispersion matrix. In the above example, we have obtained two CAN estimators of $(\mu, \alpha)'$, one based on the sample quantiles and the other based on the sample moments. We compare their generalized variances to examine which one is better. Generalized variance of the CAN estimator based on the sample quantiles is

$2\alpha^4/(log2)^2 = 4.1627\alpha^4$ and generalized variance of the CAN estimator based on the sample moments is $2.5\alpha^4$. Thus, CAN estimator based on the sample moments is better than that based on the sample quantiles.

✒ Example 3.3.8

Suppose $X \sim C(\mu, \sigma)$ distribution with location parameter μ and scale parameter σ. The probability density function $f_X(x, \underline{\theta})$ and the distribution function $F_X(x, \underline{\theta})$ of X are given by

$$f_X(x, \underline{\theta}) = \frac{\sigma}{\pi} \frac{1}{\sigma^2 + (x - \mu)^2} \quad \text{and} \quad F_X(x, \underline{\theta}) = \frac{1}{2} + \frac{1}{\pi} \tan^{-1}\left(\frac{x - \mu}{\sigma}\right),$$

$$x \in \mathbb{R}, \ \mu \in \mathbb{R}, \sigma > 0.$$

From the distribution function, we have $a_{1/4}(\underline{\theta}) = \mu - \sigma$ and $a_{3/4}(\underline{\theta}) = \mu + \sigma$. Suppose $\{X_1, X_2, \ldots, X_n\}$ is a random sample from Cauchy $C(\mu, \sigma)$ distribution. From Theorem 2.2.6, $X_{([n/4]+1)}$ is consistent for $\mu - \sigma$ and $X_{([3n/4]+1)}$ is consistent for $\mu + \sigma$. Hence,
$\left((X_{([n/4]+1)} + X_{([3n/4]+1)})/2, \ (X_{([3n/4]+1)} - X_{([n/4]+1)})/2\right)'$ is consistent for $\underline{\theta} = (\mu, \sigma)'$. We examine if it is CAN for $\underline{\theta}$. From Theorem 3.3.3, it follows that $\underline{T}_n = \left(X_{([n/4]+1)}, \ X_{([3n/4]+1)}\right)'$ is CAN for $\underline{\phi} = (\mu - \sigma, \mu + \sigma)'$ with approximate dispersion matrix,

$$\Sigma = \begin{pmatrix} 3\sigma^2\pi^2/4 & \sigma^2\pi^2/4 \\ \sigma^2\pi^2/4 & 3\sigma^2\pi^2/4 \end{pmatrix} = \frac{\sigma^2\pi^2}{4} \begin{pmatrix} 3 & 1 \\ 1 & 3 \end{pmatrix}.$$

We now find a transformation $g : \mathbb{R}^2 \to \mathbb{R}^2$ such that $g(\underline{T}_n)$ is CAN for $g(\underline{\phi}) = \underline{\theta} = (\mu, \sigma)'$. Suppose $g = (g_1, g_2)' : \mathbb{R}^2 \to \mathbb{R}^2$ is defined as $g_1(x_1, x_2) = (x_1 + x_2)/2$ and $g_2(x_1, x_2) = (x_2 - x_1)/2$. Then

$$\frac{\partial}{\partial x_1} g_1(x_1, x_2) = \frac{1}{2}, \quad \frac{\partial}{\partial x_2} g_1(x_1, x_2) = \frac{1}{2},$$

$$\frac{\partial}{\partial x_1} g_2(x_1, x_2) = -\frac{1}{2} \quad \& \quad \frac{\partial}{\partial x_2} g_2(x_1, x_2) = \frac{1}{2}.$$

These partial derivatives are continuous and hence g_1 and g_2 are totally differentiable functions. The matrix M of partial derivatives evaluated at $\underline{\phi}$ is given by

$$M = \begin{pmatrix} 1/2 & 1/2 \\ -1/2 & 1/2 \end{pmatrix}.$$

Hence, by Theorem 3.3.4,

$$\underline{g}(\underline{T}_n) = \left(\frac{X_{([n/4]+1)} + X_{([3n/4]+1)}}{2}, \quad \frac{X_{([3n/4]+1)} - X_{([n/4]+1)}}{2} \right)' \quad \text{is CAN for}$$

$$\underline{g}(\underline{\phi}) = \underline{\theta} = (\mu, \sigma)'$$

with the approximate dispersion matrix $M \Sigma M'/n$, where
$M \Sigma M' = \text{diag} \left[\sigma^2 \pi^2/2, \sigma^2 \pi^2/4 \right]$. We know that for $C(\mu, \sigma)$ distribution, the
second quartile, that is, median is μ. From Theorem 3.3.3, it follows that
$\underline{T}_n = \left(X_{([n/4]+1)} X_{([n/2]+1)} \right)'$ is CAN for $\underline{\phi} = (\mu - \sigma, \mu)'$ with approximate dispersion matrix,

$$\Sigma = \begin{pmatrix} 3\sigma^2 \pi^2/4 & \sigma^2 \pi^2/4 \\ \sigma^2 \pi^2/4 & \sigma^2 \pi^2/4 \end{pmatrix} = \frac{\sigma^2 \pi^2}{4} \begin{pmatrix} 3 & 1 \\ 1 & 1 \end{pmatrix}.$$

We now find a transformation $g : \mathbb{R}^2 \to \mathbb{R}^2$ such that $\underline{g}(\underline{T}_n)$ is CAN for
$\underline{g}(\underline{\phi}) = \underline{\theta} = (\mu, \sigma)'$. Suppose $\underline{g} = (g_1, g_2)' : \mathbb{R}^2 \to \mathbb{R}^2$ is defined as
$g_1(x_1, x_2) = x_2$ and $g_2(x_1, x_2) = x_2 - x_1$. Then $\frac{\partial}{\partial x_1} g_1(x_1, x_2) = 0$,
$\frac{\partial}{\partial x_2} g_1(x_1, x_2) = 1$, $\frac{\partial}{\partial x_1} g_2(x_1, x_2) = -1$ and $\frac{\partial}{\partial x_2} g_2(x_1, x_2) = 1$. These partial
derivatives are continuous and hence g_1 and g_2 are totally differentiable functions.
The matrix M of partial derivatives evaluated at ϕ is given by

$$M = \begin{pmatrix} 0 & 1 \\ -1 & 1 \end{pmatrix}.$$

Hence, by Theorem 3.3.4, $\underline{g}(\underline{T}_n) = \left(X_{([n/2]+1)}, \quad X_{([n/2]+1)} - X_{([n/4]+1)} \right)'$ is
CAN for $\underline{g}(\underline{\phi}) = \underline{\theta} = (\mu, \sigma)'$ with the approximate dispersion matrix $M \Sigma M'/n$,
where $M \Sigma M' = \text{diag} \left[\sigma^2 \pi^2/4, \sigma^2 \pi^2/2 \right]$. Now, we proceed and find a CAN
estimator based on the second and the third quartiles, from Theorem 3.3.3,
$\underline{T}_n = \left(X_{([n/2]+1)}, \quad X_{([3n/4]+1)} \right)'$ is CAN for $\underline{\phi} = (\mu, \mu + \sigma)'$ with approximate
dispersion matrix,

$$\Sigma = \begin{pmatrix} \sigma^2 \pi^2/4 & \sigma^2 \pi^2/4 \\ \sigma^2 \pi^2/4 & 3\sigma^2 \pi^2/4 \end{pmatrix} = \frac{\sigma^2 \pi^2}{4} \begin{pmatrix} 1 & 1 \\ 1 & 3 \end{pmatrix}.$$

We search for a transformation $g : \mathbb{R}^2 \to \mathbb{R}^2$ such that $\underline{g}(\underline{T}_n)$ is CAN for
$\underline{g}(\underline{\phi}) = \underline{\theta} = (\mu, \sigma)'$. Suppose $\underline{g} = (g_1, g_2)' : \mathbb{R}^2 \to \mathbb{R}^2$ is defined as
$g_1(x_1, x_2) = x_1$ and $g_2(x_1, x_2) = x_2 - x_1$. Then $\frac{\partial}{\partial x_1} g_1(x_1, x_2) = 1$,
$\frac{\partial}{\partial x_2} g_1(x_1, x_2) = 0$, $\frac{\partial}{\partial x_1} g_2(x_1, x_2) = -1$ and $\frac{\partial}{\partial x_2} g_2(x_1, x_2) = 1$. These partial
derivatives are continuous and hence g_1 and g_2 are totally differentiable functions.
The matrix M of partial derivatives evaluated at ϕ is given by

$$M = \begin{pmatrix} 1 & 0 \\ -1 & 1 \end{pmatrix}.$$

Hence, by Theorem 3.3.4,

$$\underline{g}(\underline{T}_n) = \left(X_{([n/2]+1)}, \quad X_{([3n/4]+1)} - X_{([n/2]+1)} \right)'$$

is CAN for $\underline{g}(\underline{\phi}) = \underline{\theta} = (\mu, \sigma)'$ with the approximate dispersion matrix $M \Sigma M' / n$, where $M \Sigma M'$ is again a diagonal matrix given by $M \Sigma M' = \text{diag}\left[\sigma^2 \pi^2 / 4, \sigma^2 \pi^2 / 2 \right]$. Thus, CAN estimator based on the second and third quartiles has the same approximate dispersion matrix as that based on the first and the second quartiles. Further, CAN estimators of $(\mu, \sigma)'$ based on (i) the first and the third quartiles, (ii) the first and the second quartiles and (iii) the second and the third quartiles have the same generalized variance $\sigma^4 \pi^4 / 8$. In this sense, all the three estimators are equivalent. $\square$

In the following example using multivariate CLT and the delta method as discussed in Theorem 3.3.4, we find famous arctan transformation of Fisher, popularly known as Fisher's Z transformation, for the correlation parameter in the bivariate normal distribution.

✒ Example 3.3.9

Suppose $(X, Y)'$ has a bivariate normal distribution with zero mean vector and dispersion matrix Σ given by

$$\Sigma = \begin{pmatrix} 1 & \rho \\ \rho & 1 \end{pmatrix},$$

$\rho \in (-1, 1)$. Suppose $\{(X_i, Y_i)', \ i = 1, 2, \ldots, n\}$ is a random sample from the distribution of $(X, Y)'$. The sample correlation coefficient R_n is defined as

$$R_n = \frac{S_{XY}^2}{S_X S_Y} \quad \text{where} \quad S_{XY}^2 = \frac{1}{n} \sum_{i=1}^{n} (X_i - \overline{X}_n)(Y_i - \overline{Y}_n),$$

$$S_X^2 = \frac{1}{n} \sum_{i=1}^{n} (X_i - \overline{X}_n)^2 \ \& \ S_Y^2 = \frac{1}{n} \sum_{i=1}^{n} (Y_i - \overline{Y}_n)^2.$$

From Example 2.5.4, it follows that R_n is consistent for ρ. We now examine whether R_n is CAN for ρ. Note that $X \sim N(0, 1)$, $Y \sim N(0, 1)$ and the conditional distribution of X given $Y = y$ is also normal $N(\rho y, 1 - \rho^2)$. Suppose a random vector $\underline{U}$ is defined as $\underline{U} = (X^2, Y^2, XY)'$. Then $\underline{\mu} = E(\underline{U}) = (1, 1, \rho)'$. To find the dispersion matrix V of $\underline{U}$, note that

$$Var(X^2) = 2,\ Var(Y^2) = 2\ \&\ \ Var(XY) = E(X^2Y^2) - (E(XY))^2$$
$$= E(X^2Y^2) - \rho^2.$$

Using the conditional distribution of X given $Y = y$, we find $E(X^2Y^2)$ and the remaining elements of V as follows:

$$\begin{aligned}
E(X^2Y^2) &= E(E(X^2Y^2)|Y) = E(Y^2E(X^2|Y)) \\
&= E(Y^2(Var(X|Y) + (E(X|Y))^2)) \\
&= E(Y^2(1 - \rho^2 + \rho^2Y^2)) = 1 - \rho^2 + \rho^2E(Y^4) \\
&= 1 - \rho^2 + 3\rho^2 = 1 + 2\rho^2 \\
\Rightarrow\ Var(XY) &= 1 + \rho^2\ \&\ Cov(X^2, Y^2) = 1 + 2\rho^2 - 1 = 2\rho^2 \\
Cov(X^2, XY) &= E(X^3Y) - E(X^2)E(XY) \\
&= E(E(X^3Y|Y)) - \rho = E(YE(X^3|Y)) - \rho \\
E((X - \rho Y)^3|Y) &= 0\ \text{since}\ X|Y = y \sim N(\rho y, 1 - \rho^2) \\
\Rightarrow\ E(X^3|Y) &= 3\rho Y E(X^2|Y) - 3\rho^2Y^2E(X|Y) + \rho^3Y^3 \\
&= 3\rho(1 - \rho^2)Y + \rho^3Y^3 \\
\Rightarrow\ E(X^3Y) &= E(YE(X^3|Y)) = 3\rho(1 - \rho^2)E(Y^2) + \rho^3E(Y^4) = 3\rho \\
\Rightarrow\ Cov(X^2, XY) &= 2\rho\ \text{similarly}\ \ Cov(Y^2, XY) = 2\rho.
\end{aligned}$$

Hence the dispersion matrix V of $\underline{U}$ is given by

$$V = \begin{pmatrix} 2 & 2\rho^2 & 2\rho \\ 2\rho^2 & 2 & 2\rho \\ 2\rho & 2\rho & 1+\rho^2 \end{pmatrix}.$$

From the principal minors of V it follows that it is a positive definite matrix with determinant $4(1 - \rho^2)^2 \neq 0$. A random sample $\{(X_i, Y_i)',\ i = 1, 2, \ldots, n\}$ gives a random sample $\{\underline{U}_1, \underline{U}_2, \ldots, \underline{U}_n\}$ from the distribution of $\underline{U}$. Hence by the multivariate CLT,

$$\sqrt{n}(\overline{\underline{U}}_n - \underline{\mu}) \xrightarrow{L} W \sim N_3(\underline{0}, V)\ \text{where}$$

$$\overline{\underline{U}}_n = \left(\sum_{i=1}^{n} X_i^2/n,\ \sum_{i=1}^{n} Y_i^2/n,\ \sum_{i=1}^{n} X_iY_i/n \right)'.$$

It is to be noted that

$$\sqrt{n}((S_X^2, S_Y^2, S_{XY})' - \underline{\mu}) = \sqrt{n}(\overline{\underline{U}}_n - \underline{\mu}) - \sqrt{n}(\overline{X}_n^2, \overline{Y}_n^2, \overline{X}_n\overline{Y}_n)'.$$

Now,

$$X \sim N(0, 1) \Rightarrow \sqrt{n}\,\overline{X}_n \sim N(0, 1) \ \& \ \overline{X}_n \overset{P}{\to} 0 \text{ by WLLN}$$

$$Y \sim N(0, 1) \Rightarrow \sqrt{n}\,\overline{Y}_n \sim N(0, 1) \ \& \ \overline{Y}_n \overset{P}{\to} 0 \text{ by WLLN}$$

$$\Rightarrow (\sqrt{n}\,\overline{X}_n)\overline{X}_n \overset{P}{\to} 0 \text{ as } \sqrt{n}\overline{X}_n \text{ is bounded in probability}$$

$$\Rightarrow (\sqrt{n}\,\overline{Y}_n)\overline{Y}_n \overset{P}{\to} 0 \text{ as } \sqrt{n}\overline{Y}_n \text{ is bounded in probability}$$

$$\Rightarrow (\sqrt{n}\,\overline{X}_n)\overline{Y}_n \overset{P}{\to} 0 \text{ as } \sqrt{n}\overline{X}_n \text{ is bounded in probability}$$

$$\Rightarrow \sqrt{n}((S_X^2, S_Y^2, S_{XY})' - \underline{\mu}) - \sqrt{n}(\underline{U}_n - \underline{\mu}) \overset{P}{\to} \underline{0}$$

$$\Rightarrow \sqrt{n}((S_X^2, S_Y^2, S_{XY})' - \underline{\mu}) \overset{L}{\to} W \sim N_3(\underline{0}, V) \text{ as}$$

$$\sqrt{n}(\underline{U}_n - \underline{\mu}) \overset{L}{\to} W.$$

To examine whether the sample correlation coefficient R_n is CAN for ρ, observe that

$$R_n = \frac{S_{XY}^2}{S_X S_Y} = g(S_X^2, S_Y^2, S_{XY}) \text{ where } g(x_1, x_2, x_3) = x_3/(x_1 x_2)^{1/2},$$

a function from $\mathbb{R}^3 - \{(0, 0, 0)\} \to (-1, 1)$. Hence, the vector of partial derivatives of g is given by

$$\frac{\partial}{\partial x_1} g(x_1, x_2, x_3) = \frac{-x_3}{2x_1^{3/2} x_2^{1/2}},$$

$$\frac{\partial}{\partial x_2} g(x_1, x_2, x_3) = \frac{-x_3}{2x_1^{1/2} x_2^{3/2}}, \quad \frac{\partial}{\partial x_3} g(x_1, x_2, x_3) = \frac{1}{x_1^{1/2} x_2^{1/2}}.$$

By Theorem 3.3.4,

$$\sqrt{n}(g(S_X^2, S_Y^2, S_{XY}) - g(\underline{\mu})) = \sqrt{n}(R_n - \rho) \overset{L}{\to} W_1 \sim N(0, \sigma^2) \text{ where}$$

$$\sigma^2 = \Delta' V \Delta$$

and $\Delta' = (-\rho/2, -\rho/2, 1)$. Hence $\sigma^2 = (1 - \rho^2)^2$. Thus, R_n is a CAN estimator of ρ with approximate variance $(1 - \rho^2)^2/n$. To obtain the asymptotic confidence interval for ρ, we use variance stabilization technique and find a function g so that the approximate variance of $g(R_n)$ is free from ρ, that is, $(g'(\rho))^2(1 - \rho^2)^2 = c$. With $c = 1$, $g(\rho) = \int \frac{1}{1-\rho^2}\, d\rho$. Now,

$$g(\rho) = \int \frac{1}{1 - \rho^2}\, d\rho = \int \left[\frac{1/2}{1 - \rho} + \frac{1/2}{1 + \rho} \right] d\rho$$

$$= \frac{1}{2} \log \frac{1 + \rho}{1 - \rho} = \tanh^{-1}(\rho).$$

The transformation $g(\rho) = \frac{1}{2} \log \frac{1+\rho}{1-\rho}$ is known as Fisher's Z transformation or arctan transformation. Thus,

$$\sqrt{n} \left[\frac{1}{2} \log \frac{1 + R_n}{1 - R_n} - \frac{1}{2} \log \frac{1 + \rho}{1 - \rho} \right] \xrightarrow{L} Z \sim N(0, 1).$$

Using this transformation and routine calculations, we get the asymptotic confidence interval for $\theta = \frac{1}{2} \log \frac{1+\rho}{1-\rho}$, and further obtain the asymptotic confidence interval for ρ from it. With $W_n = \frac{1}{2} \log \frac{1+R_n}{1-R_n}$, $Q_n = \sqrt{n}(W_n - \theta)$ is a pivotal quantity having the standard normal distribution. Hence, we find $a_{(1-\alpha/2)}$ such that $P[-a_{(1-\alpha/2)} < Q_n < a_{(1-\alpha/2)}] = 1 - \alpha$, the given confidence coefficient. Now,

$$-a_{(1-\alpha/2)} < Q_n < a_{(1-\alpha/2)}$$
$$\Leftrightarrow W_n - a_{(1-\alpha/2)}/\sqrt{n} < \theta < W_n + a_{(1-\alpha/2)}/\sqrt{n}$$
$$\Leftrightarrow 2(W_n - a_{(1-\alpha/2)}/\sqrt{n}) < \log \frac{1 + \rho}{1 - \rho} < 2(W_n + a_{(1-\alpha/2)}/\sqrt{n})$$
$$\Leftrightarrow e^{\{2(W_n - a_{(1-\alpha/2)}/\sqrt{n})\}} < \frac{1 + \rho}{1 - \rho} < e^{\{2(W_n + a_{(1-\alpha/2)}/\sqrt{n})\}}$$
$$\Leftrightarrow \frac{e^{\{2(W_n - a_{(1-\alpha/2)}/\sqrt{n})\}} - 1}{e^{\{2(W_n - a_{(1-\alpha/2)}/\sqrt{n})\}} + 1} < \rho < \frac{e^{\{2(W_n + a_{(1-\alpha/2)}/\sqrt{n})\}} - 1}{e^{\{2(W_n + a_{(1-\alpha/2)}/\sqrt{n})\}} + 1}.$$

Thus, the asymptotic confidence interval for ρ with confidence coefficient $(1 - \alpha)$ is given by

$$\left(\frac{\exp\{2(W_n - a_{(1-\alpha/2)}/\sqrt{n})\} - 1}{\exp\{2(W_n - a_{(1-\alpha/2)}/\sqrt{n})\} + 1}, \quad \frac{\exp\{2(W_n + a_{(1-\alpha/2)}/\sqrt{n})\} - 1}{\exp\{2(W_n + a_{(1-\alpha/2)}/\sqrt{n})\} + 1} \right).$$

Another approach to obtain an asymptotic confidence interval for ρ is the studentization procedure. We know that R_n is a consistent estimator of ρ. Hence,

$$\frac{\sqrt{n}(R_n - \rho)}{(1 - \rho^2)} \xrightarrow{L} Z \sim N(0, 1) \Rightarrow Q_n = \frac{\sqrt{n}(R_n - \rho)}{(1 - R_n^2)} \xrightarrow{L} Z \sim N(0, 1).$$

With Q_n as a pivotal quantity and using routine method, the asymptotic confidence interval for ρ with confidence coefficient $(1 - \alpha)$ is given by

$$\left(R_n - \frac{a_{(1-\alpha/2)}(1 - R_n^2)}{\sqrt{n}}, \quad R_n + \frac{a_{(1-\alpha/2)}(1 - R_n^2)}{\sqrt{n}} \right).$$

$\square$

We compare the asymptotic confidence intervals for ρ obtained using the variance stabilization technique and the studentization procedure in Example 3.4.7 in the next section using R.

✍ Remark 3.3.3

An important property of Fisher's Z transformation is that it attains normality much quicker than R_n (DasGupta [2], p. 53 and Anderson [3], p. 134). We verify this property using R in Example 3.4.6 in the next section.

Various examples discussed in the present section are useful to find an asymptotic non-degenerate distribution of a suitably normalized estimator for a parameter $\underline{\theta}$. We elaborate in Chaps. 5 and 6 how these are useful to decide the asymptotic null distribution of a test statistic in large sample tests of hypothesis.

3.4   Verification of CAN Property Using R

This section is devoted to a verification of CAN property of an estimator, by simulation using R software. Suppose $T_n = T_n(X_1, ..., X_n)$ is a CAN estimator of θ, based on a random sample from the given distribution. Then according to Definition 3.2.1, there exists a sequence $\{a_n, n \geq 1\}$ of real numbers tending to ∞ as $n \to \infty$, such that $U_n = a_n(T_n - \theta)/\sqrt{v(\theta)} \xrightarrow{L} U \sim N(0, 1)$ distribution as $n \to \infty$, where $0 < v(\theta) < \infty$. It then follows that T_n is a consistent estimator of θ. To verify whether T_n is a consistent estimator, we adopt the procedure as discussed in Sect. 2.7. To examine whether it is CAN, we use Shapiro-Wilk test of normality for the simulated values of U_n. Normality can also be judged visually using various plots. For example, we may draw a histogram of simulated values of U_n, with relative frequencies on Y-axis and impose a curve of probability density function of the standard normal distribution on it. We observe whether the two are close to each other as sample size increases. We may also draw a box plot and Q-Q plot with Q-Q line to assess normality of U_n. In addition, we can plot a graph of empirical distribution function of simulated values of U_n and impose the graph of distribution function of the standard normal distribution on it. The visual impression produced by these four types of plots, judged analytically by the Shapiro-Wilk test, will help us to conclude whether T_n is a CAN estimator of θ. The stepwise procedure to verify whether T_n is CAN is as follows:

1. Generate m random samples of size n from the distribution of X with $\theta = \theta_0$, say.
2. Compute T_{ni} and $U_{ni} = a_n(T_{ni} - \theta_0)\sqrt{v(\theta_0)}$, for $i = 1, 2, \ldots, m$.
3. Find the estimate of the coverage probability $p_n(\epsilon, \theta)$ as $r_n = \frac{1}{m}\sum_{i=1}^{m} I_{[|T_{ni} - \theta_0| < \epsilon]}$.
4. Corresponding to m values of U_{ni}, carry out the Shapiro-Wilk test. Further, draw a histogram with relative frequencies on Y-axis and impose a curve of probability density function of the standard normal distribution on it, also draw a box plot, a QQ plot and the graph of empirical distribution function.

5. Increase the sample size n and repeat the steps 1 to 4.
6. Repeat steps 1 to 5 for various values of ϵ and θ_0.

We illustrate the procedure by some examples and R code. The code does not include code for verification of consistency.

✎ Example 3.4.1

Suppose $\{X_1, X_2, \ldots, X_n\}$ is a random sample from a Cauchy $C(\theta, 1)$ distribution. In Example 3.2.6, it is shown that for $C(\theta, 1)$ distribution the sample median $T_n = X_{([n/2]+1)}$ is CAN for θ with approximate variance $\pi^2/4n$, thus for large n the distribution of $U_n = (2\sqrt{n}/\pi)(X_{([n/2]+1)} - \theta)$ is approximately normal. We verify this result by simulation. Using the procedure given in Sect. 2.7, one can verify the consistency of $X_{([n/2]+1)}$ for θ. Following is the R code to verify whether the distribution of U_n can be approximated by the standard normal distribution. We verify it for four sample sizes 20, 40, 60, 80 and $\theta = 3$.

```
th = 3; n = c(20,40,60,80); nsim = 1000
t=matrix(nrow=length(n),ncol=nsim); p = c()
for(m in 1:length(n))
{
for(i in 1:nsim)
{
  set.seed(i)
 x=rcauchy(n[m],th,1)
me = median(x)
t[m,i] = (2*sqrt(n[m])/pi)*(me-th)
}
p[m]=shapiro.test(t[m,])$p.value
}
d=round(data.frame(n,p),4);d; r = seq(-4,4,.1); y = dnorm(r)
par(mfrow= c(2,2))
nname=paste("n =",n,sep=" ")
for(i in 1:length(n))
{
hist(t[i,],freq=FALSE,main=nname[i],ylim=c(0,0.4),
     xlab = expression(paste("U"[n])),col="light blue")
lines(r,y,"o",pch=20,col="dark blue")
}
u=t[4,]
hist(u,freq=FALSE,main="Histogram",ylim=c(0,0.4),
     xlab = expression(paste("T"[n])),col="light blue")
lines(r,y,"o",pch=20,col="dark blue") boxplot(u,main="Box Plot")
qqnorm(u); qqline(u); plot(ecdf(u),main="Empirical Distribution
Function",xlab=expression(paste("T"[n])),
     ylab = expression(paste("F"[n](t))))
lines(r,pnorm(r),col="blue")
```

The p-values of Shapiro-Wilk test for normality for sample sizes 20, 40, 60, 80 are 0.0014, 0.6081, 0.7511 and 0.6135 respectively. Thus, normality is not accepted for $n = 20$, and it sounds reasonable, as we wish to examine whether the large sample distribution is normal. Normality is acceptable for $n = 40, 60, 80$. It is to be noted that p-value is not a monotone function of sample size, as it does not just depend on the sample size, but also on the realized sample. Figure 3.1 displays histograms of 1000 simulated values of U_n for sample sizes $n = 20, 40, 60, 80$. We note that the curve of probability density of the standard normal distribution is not a close approximation to the histogram for $n = 20$, indicating that for $n = 20$ the distribution of U_n cannot be approximated by the standard normal distribution. For other sample sizes, the curve of probability density of the standard normal distribution is a close approximation to the histogram with relative frequencies, indicating that for $n = 40, 60, 80$ the distribution of U_n is approximately normal. These results are consistent with the p-values of Shapiro-Wilk test. Figure 3.2 displays four plots for sample size $n = 80$. The first is the histogram of 1000 simulated values of U_n with the probability density of the standard normal dis-

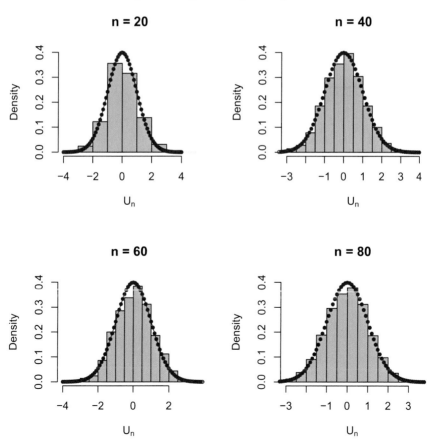

Fig. 3.1 Cauchy $C(\theta, 1)$ distribution: histograms of normalized sample median

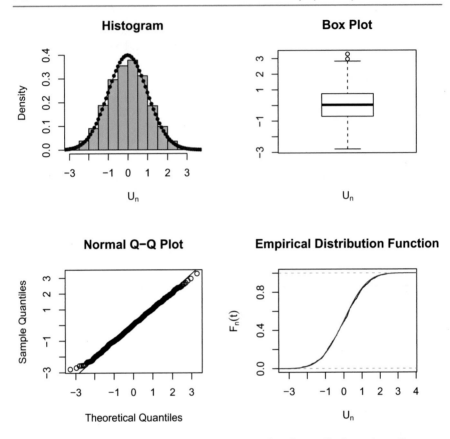

Fig. 3.2 Cauchy $C(\theta, 1)$ distribution: approximate normality of normalized sample median

tribution imposed on it. As already noted, the curve of probability density of
the standard normal distribution is a close approximation to the histogram with
relative frequencies. Box plot indicates symmetry around 0 and range of the simulated values of U_n is approximately $(-3, 3)$. The QQ plot of simulated 1000
values of U_n also shows that sample quantiles and theoretical quantiles are in
close agreement and thus supports the approximation of the distribution of U_n
to that of the standard normal. In the fourth plot, we see close approximation of
empirical distribution function of 1000 values of U_n with the distribution function
of the standard normal distribution. The visual impression of these four plots is
further supported by p-value 0.6135 of Shapiro-Wilk test. The above procedure
may be repeated for various values of $\theta \in \Theta$.   □

✏ Example 3.4.2

Suppose $\{X_1, X_2, \ldots, X_n\}$ is a random sample from a Weibull distribution with probability density function

$$f(x, \theta) = \theta \, x^{\theta-1} \exp\{-x^\theta\}, \quad x > 0, \quad \theta > 0 .$$

Then its distribution function $F(x, \theta)$ for $x > 0$ is given by $F(x, \theta) = 1 - \exp\{-x^\theta\}$, which implies that the population median is $a_{1/2}(\theta) = (\log 2)^{1/\theta} = (0.6931)^{1/\theta}$. Hence by the delta method, $\tilde{\theta}_n = \log(\log 2)/(\log X_{([n/2]+1)}) = -0.3666/(\log X_{([n/2]+1)})$ is CAN for θ with approximate variance

$$\frac{v(\theta)}{n} = \frac{\theta^2}{n(\log 2)^2(\log(\log 2))^2} = \frac{\theta^2}{0.06456n} .$$

Thus for large n, the distribution of

$$T_n = \frac{\sqrt{0.06456n}}{\theta}\left(\frac{-0.3666}{\log X_{([n/2]+1)}} - \theta\right) = -\frac{0.2541\sqrt{n}}{\theta}\left(\frac{0.3666}{\log X_{([n/2]+1)}} + \theta\right)$$

is approximately normal. We verify the same by simulation using following R code. Consistency of $\tilde{\theta}_n$ can be verified using the procedure described in Sect. 2.7. We take $\theta = 2$ and sample sizes from 500 to 7500 with the increment of 1000.

```
th = 2; n = seq(500,7500,1000); length(n); nsim = 1000
t = matrix(nrow=length(n),ncol=nsim); p = c()
for(m in 1:length(n))
{
for(i in 1:nsim)
{
set.seed(i)
x = rweibull(n[m],th,1)
me = median(x)
t[m,i] = -(0.2541*sqrt(n[m])/th)*((0.3666/log(me))+th)
}
p[m] = shapiro.test(t[m,])$p.value
}
d = round(data.frame(n,p),4); d
u = t[length(n),]; r = seq(-4,4,.1); y = dnorm(r)
par(mfrow=c(2,2))
hist(u,freq=FALSE,main="Histogram",ylim=c(0,0.4),
     xlab = expression(paste("T"[n])),col="light blue")
lines(r,y,"o",pch=20,col="dark blue")
boxplot(u,main="Box Plot",xlab = expression(paste("T"[n])))
```

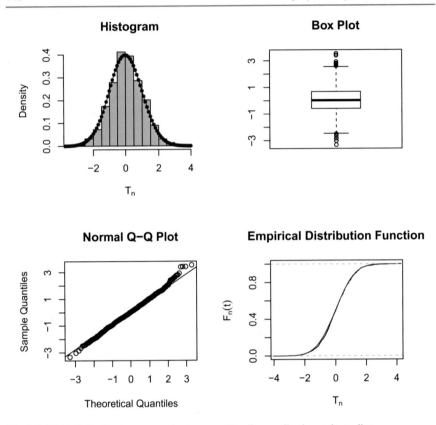

Fig. 3.3 Weibull distribution: approximate normality of normalized sample median

```
qqnorm(u);qqline(u)
plot(ecdf(u),main="Empirical Distribution Function",
xlab=expression(paste("T"[n])),ylab = expression(paste("F"[n](t))))
lines(r,pnorm(r),col="blue")
```

The *p*-values of Shapiro-Wilk test for normality for sample sizes 500, 1500, 2500, 3500 are 0, for sample sizes 4500, 5500, 6500, 7500 are 0.0033, 0.0242, 0.0107 and 0.1104 respectively. Thus, for sample size $n = 7500$, normality of sample median is acceptable according to Shapiro-Wilk test with the p-value 0.1104. It is to be noted that the sample size required to achieve similar results in Example 3.4.1 for Cauchy distribution is 80. Thus, the requirement of sample size depends on the underlying probability model and even the values of the indexing parameter. Figure 3.3 displays four plots for $n = 7500$, which indicate that the distribution of T_n is approximately normal. The same procedure can be repeated for various values of $\theta \in \Theta$.   □

☑ Example 3.4.3

Suppose $\{X_1, X_2, \ldots, X_n\}$ is a random sample from an exponential distribution, with location parameter $\theta \in \mathbb{R}$ and scale parameter 1. In Example 3.2.7, it is proved that $X_{(1)}$ is not a CAN estimator of θ, that is, there does not exist any sequence $\{a_n, n \geq 1\}$ of real numbers, such that $a_n \to \infty$ as $n \to \infty$ for which the asymptotic distribution of $a_n(X_{(1)} - \theta)$ is normal. By taking $a_n = n$, we verify it by simulation. Further, it is known that for each n and hence asymptotically, distribution of $T_n = n(X_{(1)} - \theta)$ is exponential with mean 1. We verify this result by simulation, graphically and with goodness of fit test, using following R code. We also use one built-in function for goodness of fit test.

```
th = 1.5; n = 60; nsim = 2000; t = c()
for(i in 1:nsim)
{
set.seed(i)
x = th + rexp(n,rate=1)
t[i] = min(x)
}
tn = n*(t-th); summary(tn); v=var(tn)*(nsim-1)/nsim; v
r = seq(-4,4,.2); y = dnorm(r)
par(mfrow= c(2,2))
hist(tn,freq=FALSE,main="Histogram",ylim=c(0,0.4),xlim=c(-4,5),
     xlab = expression(paste("T"[n])), col="light blue")
lines(r,y,"o",pch=20,col="dark blue")
boxplot(tn,main="Box Plot",xlab = expression(paste("T"[n])));
qqnorm(tn); qqline(tn)
plot(ecdf(tn),main="Empirical Distribution Function",
xlab=expression(paste("T"[n])),ylab = expression(paste("F"[n](t))))
lines(r,pnorm(r),"o",pch=20,col="blue")
shapiro.test(tn)
O = hist(tn,plot=FALSE)$counts; sum(O)
bk = hist(tn,plot=FALSE)$breaks; bk
M = max(bk); u = seq(0,M,.1); y1 = dexp(u, rate=1)
par(mfrow= c(1,1))
hist(tn,freq=FALSE,main="Histogram of T_n",ylim=c(0,1),xlim=c(0,M),
     xlab = expression(paste("T"[n])),col="light blue")
lines(u,y1,"o",pch=20,col="dark blue")
e = exp(1); ep = c()
for(i in 1:(length(bk)-1))
{
ep[i] = e^(-bk[i]) - e^(-bk[i+1])
}
a = 1-sum(ep); a; ep1 = c(ep, a); ef = sum(O)*ep1; O1 = c(O,0)
d = data.frame(O1,round(ef,2)); d
```

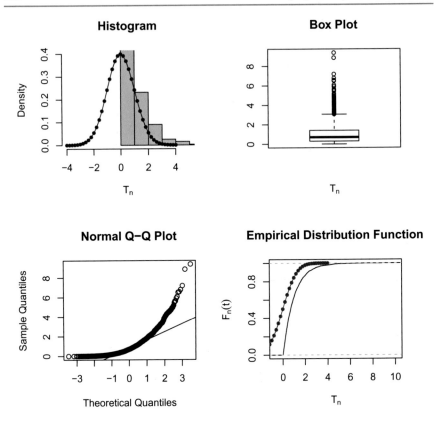

Fig. 3.4 Exponential $Exp(\theta, 1)$ distribution: MLE is not CAN

```
ts = sum((O1-ef)^2/ef);ts
df = length(ef)-1; df; b = qchisq(.95,df); b;
                 p1 = 1-pchisq(ts,df); p1
chisq.test(O1,p=ep1)
### pooling of frequencies less than 5 is ignored
```

We know that the distribution of $T_n = n(X_{(1)} - \theta)$ is exponential with mean 1. Thus, $Var(T_n) = 1$. From summary statistic and variance, we note that based on 2000 simulations, the mean of T_n is 1.0055, very close to theoretical mean 1 and variance is 1.0371, again close to theoretical variance 1. From all the four graphs in Fig. 3.4, it is clear that the asymptotic distribution of T_n is not normal. Shapiro-Wilk test also rejects the null hypothesis of normality of T_n as its p-value is 0 (less than $2.2e^{-16}$). All these results are for sample size 60, but even for the large sample size we get similar results and normality of T_n is not acceptable. The first graph of histogram in Fig. 3.4, superimposed by the curve of probability density function of the standard normal distribution indicates that an exponential distribution will be a better fit. Figure 3.5 supports such a hunch

Fig. 3.5 Exponential
$Exp(\theta, 1)$ distribution:
asymptotic distribution of
MLE

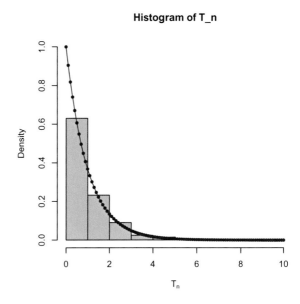

and the theory that the distribution of T_n is exponential with mean 1. To verify
the visual impression analytically, we carry out goodness of fit test by using Karl
Pearson's chi-square test statistic. We obtain observed frequencies O by extracting `counts` from `hist` and class intervals by extracting `breaks` from `hist`.
We take the last interval as (M, L), where M is the maximum value of vector
`breaks` and L denotes the maximum value of support of the distribution in
null setup. In this example, $L = \infty$. Expected frequencies are obtained by computing the probabilities of the class intervals using the exponential distribution
with mean 1. The data frame d shows the close agreement between observed and
expected frequencies. With these data, we compute Karl Pearson's test statistic
and the corresponding p-value. It is 0.5668, which supports the theoretical result
that T_n has the exponential distribution with mean 1. There is a built-in function
`chisq.test(O1,p=ep1)` for Karl Pearson's goodness of fit test, where `O1`
is a vector of observed frequencies, when data are grouped in class intervals and
`ep1` is a vector specifying probabilities of these class intervals under the null
hypothesis. The built-in function `chisq.test` also gives the same results. It
is to be noted that we get these results for the sample size 60 as for finite n also
distribution of $T_n = n(X_{(1)} - \theta)$ is exponential.   □

   In the next example we illustrate how to verify a CAN property of a vector
estimator for a vector parameter.

✒ Example 3.4.4

Suppose X follows an exponential distribution with location parameter μ and scale parameter $1/\sigma$ with probability density function

$$f(x, \mu, \sigma) = \frac{1}{\sigma} \exp\left\{-\frac{x - \mu}{\sigma}\right\}, \quad x \geq \mu, \quad \sigma > 0, \quad \mu \in \mathbb{R}.$$

In Example 3.3.5, it is shown that $(\tilde{\mu}_n, \tilde{\sigma}_n)' = (m_1' - \sqrt{m_2}, \sqrt{m_2})'$ is CAN for $(\mu, \sigma)'$, with the approximate dispersion matrix D/n, where

$$D = \begin{pmatrix} \sigma^2 & -\sigma^2 \\ -\sigma^2 & 2\sigma^2 \end{pmatrix}.$$

There are two approaches to verify that $(\tilde{\mu}_n, \tilde{\sigma}_n)' = (m_1' - \sqrt{m_2}, \sqrt{m_2})'$ is CAN for $(\mu, \sigma)'$. One is based on the Cramér-Wold device which states that

$$\underline{X}_n \xrightarrow{L} \underline{X} \text{ if and only if } \underline{L}'\underline{X}_n \xrightarrow{L} \underline{L}'\underline{X} \ \forall \ \underline{L} \in \mathbb{R}^k, \ \underline{L} \neq \underline{0}.$$

Cramér-Wold device allows higher dimensional problems to reduce to the one-dimensional case. Thus, $\underline{X}_n$ converges to k-multivariate normal distribution, if and only if $l'\underline{X}_n$ converges to univariate normal distribution for any k-dimensional vector l of real numbers. In the following code, we adopt this approach to verify normality of $(\tilde{\mu}_n, \tilde{\sigma}_n)'$ for five vectors.

```
mu = 3; si = 2; n = 1000; nsim = 1500; m1 = m2 = m3 = m4 = c()
for(i in 1:nsim)
{
set.seed(i)
u = runif(n,0,1)
x = mu-si*log(1-u)
m1[i] = mean(x)
m2[i] = mean(x^2)
m3[i] = m2[i]-m1[i]^2
m4[i] = sqrt(m3[i])
}
x1 = m1-m4; summary(x1)
x2 = m4; summary(x2)
y = matrix(nrow=nsim,ncol=2)
y = cbind(sqrt(n)*(x1-mu),sqrt(n)*(x2-si))
v = matrix(c(si^2,-si^2,-si^2,2*si^2),nrow=2,byrow=TRUE); v
l = matrix(nrow=5,ncol=2)
l[,1] = c(2,5,7,8,10); l[,2] = c(4,8,12,14,18)
ll = t(l); pv = wi = c()
```

```
for(i in 1:5)
{
di = l[i,]%*%v%*%l1[,i]
di = as.numeric(di)
wi = (y%*%l1[,i])/sqrt(di)
pv[i] = shapiro.test(wi)$p.value
}
round(pv,3)
```

For five two-dimensional vectors as specified in l, p-values of Shapiro-Wilk test are 0.2530.1710.1840.1930.208. These convey that the distribution of $T_n = \underline{l}' \left(\sqrt{n}(\tilde{\mu}_n - \mu), \sqrt{n}(\tilde{\sigma}_n - \sigma) \right) / \sqrt{\underline{l}' V \underline{l}}$ can be approximated by the standard normal distribution, which further implies that $(\tilde{\mu}_n, \tilde{\sigma}_n)'$ is CAN for $(\mu, \sigma)'$, with the approximate dispersion matrix D/n.

In the second approach, we use the result that if $(\tilde{\mu}_n, \tilde{\sigma}_n)' = (m'_1 - \sqrt{m_2}, \sqrt{m_2})'$ is CAN for $(\mu, \sigma)'$, then
$T_n = n(m'_1 - \sqrt{m_2} - \mu, \sqrt{m_2} - \sigma)' D^{-1}(m'_1 - \sqrt{m_2} - \mu, \sqrt{m_2} - \sigma) \sim \chi^2_2$.
We verify whether T_n has χ^2_2 distribution graphically and by the chi-square test using the following R code.

```
mu = 3; si = 2; n = 500; nsim = 1200; m1 = m2 = m3 = m4 = w = c()
for(i in 1:nsim)
{
set.seed(i)
u = runif(n,0,1)
x = mu-si*log(1-u)
m1[i] = mean(x)
m2[i] = mean(x^2)
m3[i] = m2[i]-m1[i]^2
m4[i] = sqrt(m3[i])
}
x1 = m1-m4; x2 = m4; summary(x1);
summary(x2); y = cbind(x1-mu,x2-si);summary(y);
v = matrix(c(si^2,-si^2,-si^2,2*si^2),nrow=2,byrow=TRUE);v
v1 = solve(v); y1 = t(y)
for(i in 1:nsim)
{
w[i] = n*y[i,]%*%v1%*%y1[,i]
}
summary(w)
O = hist(w,plot=FALSE)$counts; sum(O)
bk = hist(w,plot=FALSE)$breaks; bk
M = max(bk); u = seq(0,M,.2); y2 = dchisq(u,2)
hist(w,freq=FALSE,main="Histogram of T_n",xlim=c(0,M),ylim=c(0,.5),
     xlab = expression(paste("T"[n])),col="light blue")
```

Fig. 3.6 Exponential
$Exp(\mu, \sigma)$ distribution:
CAN estimator based on
moments

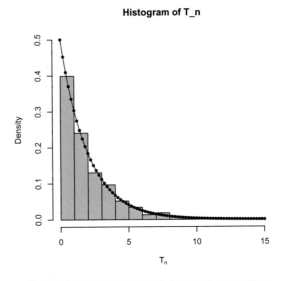

Histogram of T_n

```
lines(u,y2,"o",pch=20,col="dark blue")
ep = c()
for(i in 1:(length(bk)-1))
{
ep[i] = pchisq(bk[i+1],2) - pchisq(bk[i],2)
}
a = 1-sum(ep); a; ep1 = c(ep, a); ef = sum(O)*ep1; O1 = c(0,0)
d = data.frame(O1,round(ef,2)); d
ts = sum((O1-ef)^2/ef); ts; df = length(ef)-1; df
b = qchisq(.95,df); b; p1 = 1-pchisq(ts,df); p1;
                     chisq.test(O1,p=ep1)
```

In the output, $x1$ is a vector of simulated values of $\tilde{\mu}_n$, $x2$ is a vector of simulated values of $\tilde{\sigma}_n$ and w is a vector of simulated values of T_n. Summary statistics (not shown) of $x1$, $x2$ and of w indicate that mean of simulated values of $\tilde{\mu}_n$ is close to the assumed value of $\mu = 3$, the mean of simulated values of $\tilde{\sigma}_n$ is close to $\sigma = 2$ and the mean of w is close to 2, which is a mean of χ^2_2 distribution. The p-value 0.1758 of Karl Pearson's chi-square test statistic supports the theory that the distribution of T_n is χ^2_2. The built-in function chisq.test also gives the same results. Figure 3.6 shows that the histogram of simulated values of w is closely approximated by the probability density function of χ^2_2 distribution. It is to be noted that the sample size required in this example is 500, while in the previous example it is 60. $\square$

✐ Example 3.4.5

In Example 3.2.13, we have obtained $100(1 - \alpha)\%$ asymptotic confidence interval for the survival function $e^{-t/\theta}$, where t is a fixed positive real number. We have used following two methods—(i) based on the studentization procedure and (ii) using the fact that $e^{-t/\theta}$ is a monotone function of θ. By method (i), $100(1 - \alpha)\%$ large sample confidence interval for $e^{-t/\theta}$ is given by

$$\left(e^{-t/\overline{X}_n} \left(1 - a_{1-\alpha/2} \frac{t}{\overline{X}_n \sqrt{n}} \right), \quad e^{-t/\overline{X}_n} \left(1 + a_{1-\alpha/2} \frac{t}{\overline{X}_n \sqrt{n}} \right) \right)$$

and by method (ii) it is

$$\left(\exp \left\{ \frac{-t \left(1 + \frac{a_{1-\alpha/2}}{\sqrt{n}} \right)}{\overline{X}_n} \right\}, \quad \exp \left\{ \frac{-t \left(1 - \frac{a_{1-\alpha/2}}{\sqrt{n}} \right)}{\overline{X}_n} \right\} \right).$$

We compare these two asymptotic confidence intervals by simulation using the R code given below. For comparison, we obtain the empirical confidence coefficient (ECC), estimate of average length of the interval (AL) and its variance (VL).

```
th = 2; t = 0.4; par = exp(-t/th); nsim = 2000;
n = c(50,100,150,200,250); alpha = 0.05; a = qnorm(1-alpha/2,0,1)
mth_est = mest = cc1 = cc2 = alc1 = alc2 = vlc1 = vlc2 = c()
for(s in 1:5)
{
L1 = U1 = L2 = U2 = est = th_est = c()
for(j in 1:nsim)
{
set.seed(j)
x = rexp(n[s],rate = 1/th)
th_est[j] = mean(x)
est[j] = exp(-t/th_est[j])
L1[j] = est[j]-a*t*est[j]/(sqrt(n[s])*th_est[j])
U1[j] = est[j]+a*t*est[j]/(sqrt(n[s])*th_est[j])
L2[j] = exp(-t*(1+a/sqrt(n[s]))/th_est[j])
U2[j] = exp(-t*(1-a/sqrt(n[s]))/th_est[j])
}
cc1[s] = length(which(L1<par&U1>par))/nsim
cc2[s] = length(which(L2<par&U2>par))/nsim
length1 = U1-L1
length2 = U2-L2
alc1[s] = mean(length1)
alc2[s] = mean(length2)
```

Table 3.1 Comparison of two methods of constructing an asymptotic confidence interval

n	$\hat{\theta}_n$	$e^{-t/\hat{\theta}_n}$	ECC1	ECC2	AL1	AL2	VL1	VL2
50	2.0055	0.8160	0.9440	0.9450	0.0918	0.0918	0.0001	0.0001
100	2.0047	0.8186	0.9545	0.9540	0.0645	0.0645	0	0
150	2.0052	0.8182	0.9550	0.9565	0.0525	0.0525	0	0
200	2.0042	0.8183	0.9580	0.9565	0.0455	0.0455	0	0
250	2.0032	0.8184	0.9510	0.9520	0.0406	0.0406	0	0

```
vlc1[s] = var(length1)
vlc2[s] = var(length2)
mth_est[s]=mean(th_est)
mest[s]=mean(est)
}
d = cbind(mth_est,mest,cc1,cc2,alc1,alc2,vlc1,vlc2)
d = round(d,4); d1 = data.frame(n,d); d1
```

The output of data frame d1 is displayed in Table 3.1.

The first and the second column present the mean of 2000 values of $\hat{\theta}_n$ and $e^{-t/\hat{\theta}_n}$ respectively for sample sizes $n = 50, 100, \ldots, 250$. From these columns, we note that for all the sample sizes, the estimate of θ is close to 2 and of $e^{-t/\theta}$ is close to $e^{-4/2} = 0.8187$. Further, the average length of the confidence interval is same for both the methods and all the sample sizes. It decreases as the sample size increases, as is expected. The variance of the length is almost 0 for both the methods and all the sample sizes. The empirical confidence coefficients for both the methods are more or less the same for all the sample sizes and approach 0.95 as the sample size increases. Thus, there is not much variation in the properties of large sample confidence intervals corresponding to two methods.   □

In Example 3.3.9, we have shown that the asymptotic distribution of $\sqrt{n}(R_n - \rho)/(1 - \rho^2)$ and also of Fisher's Z transformation of R_n is standard normal, where R_n is the sample correlation coefficient. However, as pointed out in Remark 3.3.3, the rates of convergence to normality are different. In the next example, we compare the rate of convergence to normality of R_n and Fisher's Z transformation of R_n. To generate a random sample from a bivariate normal distribution, we use the package mvtnorm. For details of this package, refer to Genz and Bretz [4] and Genz et al. [5].

☞ Example 3.4.6

Suppose $(X, Y)'$ has a bivariate normal distribution with zero mean vector and dispersion matrix Σ given by

$$\Sigma = \begin{pmatrix} 1 & \rho \\ \rho & 1 \end{pmatrix},$$

$\rho \in (-1, 1)$. In this example, we compare the rate of convergence to normality of R_n and Fisher's Z transformation of R_n. For various sample sizes, we simulate from the bivariate normal distribution and judge the rate of convergence to normality based on p-values of Shapiro-Wilk test. We also compare the performance based on the coefficient of skewness and coefficient of kurtosis. The R code is given below.

```
library(mvtnorm)
rho = .5; a =.5*(log((1+rho)/(1-rho))); mu = c(0,0);
sig=matrix(c(1,rho,rho,1),nrow=2)
n = c(100,140,180,220); nsim = 1200; pt = pz = c()
t = matrix(nrow=length(n),ncol=nsim);
z=matrix(nrow=length(n),ncol=nsim)
for(m in 1:length(n))
{
for(i in 1:nsim)
{
set.seed(i)
x = rmvnorm(n[m],mu,sig)
v = cor(x)
R = v[1,2]
s = 0.5*(log((1+R)/(1-R)))
t[m,i] = (sqrt(n[m])/(1-rho^2))*(R-rho)
z[m,i] = sqrt(n[m])*(s-a)
}
pt[m] = shapiro.test(t[m,])$p.value
pz[m] = shapiro.test(z[m,])$p.value
}
d = round(data.frame(n,pt,pz),4); d
ct = matrix(nrow=length(n),ncol=3)
cz = matrix(nrow=length(n),ncol=3)
skt = kut = c(); skz = kuz = c(); r = 2:4
for(m in 1:length(n))
{
for(l in 1:length(r))
{
ct[m,l] = mean((t[m,] - mean(t[m,]))^r[l])
cz[m,l] = mean((z[m,] - mean(z[m,]))^r[l])
}
}
ct = round(ct,4); cz = round(cz,4); ct; cz
for(m in 1:length(n))
{
skt[m] = (ct[m,2])^2/(ct[m,1])^3
skz[m] = (cz[m,2])^2/(cz[m,1])^3
kut[m] = ct[m,3]/(ct[m,1])^2
kuz[m] = cz[m,3]/(cz[m,1])^2
}
d1 = round(data.frame(n,skt,skz,kut,kuz),4); d1
```

Table 3.2 $N_2(0, 0, 1, 1, \rho)$ distribution: p-values of Shapiro-Wilk test

n	r_n	Z
100	0.0000	0.9611
140	0.0009	0.7433
180	0.0144	0.4470
220	0.0755	0.9388

The p-values of Shapiro-Wilk test are displayed in Table 3.2.

It is noted that for a sample of size 100, 140 and 180 normality of $\sqrt{n}(R_n - \rho)/(1 - \rho^2)$ is not accepted but normality of corresponding Z transformation is accepted. For a sample of size 220, normality of $\sqrt{n}(R_n - \rho)$ is acceptable with p-value 0.0755, while the p-value of corresponding Z transformation is 0.9388. Observe that the p-values corresponding to Z transformation are not monotone functions of sample size, these decrease initially and increase at a later stage. Such a feature is in view of the fact that the p-value is not just a function of the sample size, but it is also a function of observed value of the test statistics, which changes as observed data change. To summarize, normality of Z transformation is acceptable for small sizes but not for $\sqrt{n}(R_n - \rho)/(1 - \rho^2)$. Hence, we conclude that rate of convergence to normality of R_n is much slower than that of its Z transformation. Following Table 3.3 displays the values of coefficient of skewness $\beta_1 = \mu_3^2/\mu_2^3$ and of coefficient of kurtosis $\gamma_1 = \mu_4/\mu_2^2$ of $\sqrt{n}(R_n - \rho)/(1 - \rho^2)$ and of normalized Fisher's Z transformation for four sample sizes.

From the table, we note that there is a difference between the values of the coefficient of skewness β_1, but the values of coefficient of kurtosis are more or less the same for all the sample sizes. The coefficient of skewness of normalized Z is closer to 0 than that of normalized R_n, which again implies that the rate of convergence to normality of Z is faster than that of R_n. From the definitions of β_1 and of γ_1, we note that the difference between the values of the coefficient of skewness and similarity between the values of the coefficient of kurtosis, should depend on the sample central moments. Table 3.4 displays the values of second, third and fourth sample central moments, m_2, m_3, m_4 respectively, both for normalized R_n and normalized Z.

From the table, we observe that the second central moment for both R_n and Z is close to 1 as expected, because R_n and Z are normalized to have variance

Table 3.3 Coefficient of skewness and kurtosis of normalized R_n and normalized Z

n	$\beta_1(R_n)$	$\beta_1(Z)$	$\gamma_1(R_n)$	$\gamma_1(Z)$
100	0.1122	0.0016	3.1402	3.0091
140	0.0652	0.0000	3.1788	3.0364
180	0.0350	0.0012	3.0858	2.9764
220	0.0324	0.0007	3.1598	3.1012

Table 3.4 Sample central moments of normalized R_n and normalized Z

n	$m_2(R_n)$	$m_2(Z)$	$m_3(R_n)$	$m_3(Z)$	$m_4(R_n)$	$m_4(Z)$
100	1.0238	1.0228	-0.3470	-0.0411	3.2913	3.1481
140	0.9882	0.9940	-0.2508	-0.0021	3.1045	3.0003
180	1.0119	1.0210	-0.1905	0.0359	3.1595	3.1028
220	0.9682	0.9723	-0.1715	0.0262	2.9621	2.9316

1. Further, the fourth central moment for both R_n and Z is close to 3, which is reflected in the values of the coefficient of kurtosis to be close to 3. The third central moments for R_n and Z are different. The third central moment of Z is closer to 0, which is reflected in the values of the coefficient of skewness.

□

Fisher's Z transformation, being a variance stabilizing transformation, is useful to find an asymptotic confidence interval for ρ. In the next example, we discuss how to compute an asymptotic confidence interval for ρ, using Z transformation and using the studentization procedure. We compare the performance of these two procedures by empirical confidence coefficient, average length of the confidence interval and variance of the length of the confidence interval.

✏ Example 3.4.7

Suppose $(X, Y)'$ has a bivariate normal distribution with zero mean vector and dispersion matrix Σ given by

$$\Sigma = \begin{pmatrix} 1 & \rho \\ \rho & 1 \end{pmatrix},$$

$\rho \in (-1, 1)$. In Example 3.3.9, we have shown that the asymptotic distribution of $\sqrt{n}(W_n - \theta)$ is standard normal, where W_n is Fisher's Z transformation of R_n given by $W_n = \frac{1}{2} \log \frac{1+R_n}{1-R_n}$, where R_n is the sample correlation coefficient and $\theta = \frac{1}{2} \log \frac{1+\rho}{1-\rho}$. In the same example, we have obtained the asymptotic confidence interval for ρ using Fisher's Z transformation, that is, using variance stabilization technique and also the asymptotic confidence interval for ρ using the studentization procedure. The asymptotic confidence interval with confidence coefficient $(1 - \alpha)$ for ρ using variance stabilization technique is given by

$$\left(\frac{\exp\{2(W_n - a_{(1-\alpha/2)}/\sqrt{n})\} - 1}{\exp\{2(W_n - a_{(1-\alpha/2)}/\sqrt{n})\} + 1}, \frac{\exp\{2(W_n + a_{(1-\alpha/2)}/\sqrt{n})\} - 1}{\exp\{2(W_n + a_{(1-\alpha/2)}/\sqrt{n})\} + 1} \right),$$

where $W_n = \frac{1}{2} \log \frac{1+R_n}{1-R_n}$. The asymptotic confidence interval with confidence coefficient $(1 - \alpha)$ for ρ using the studentization procedure is given by

$$\left(R_n - \frac{a_{(1-\alpha/2)}(1 - R_n^2)}{\sqrt{n}}, R_n + \frac{a_{(1-\alpha/2)}(1 - R_n^2)}{\sqrt{n}} \right).$$

Following R code computes asymptotic confidence interval for ρ for a sample of size 250 from the bivariate normal distribution. The performance of the two procedures is compared on the basis of 1500 simulations.

```
library(mvtnorm)
n = 250; nsim = 1500
rho = .5; a=.5*(log((1+rho)/(1-rho)));
           alpha = 0.05; b = qnorm(1-alpha/2)
mu=c(0,0); sig=matrix(c(1,rho,rho,1),nrow=2)
R = s = LCI = UCI = LR = UR = c()
for(i in 1:nsim)
{
set.seed(i)
x = rmvnorm(n,mu,sig)
v = cor(x)
R[i] = v[1,2]
s[i] = .5*(log((1+R[i])/(1-R[i])))
LCI[i] = (exp(2*(s[i]-b/sqrt(n)))-1)/(exp(2*(s[i]-b/sqrt(n)))+1)
UCI[i] = (exp(2*(s[i]+b/sqrt(n)))-1)/(exp(2*(s[i]+b/sqrt(n)))+1)
LR[i] =  R[i]-b*(1-R[i]^2)/sqrt(n)
UR[i] =  R[i]+b*(1-R[i]^2)/sqrt(n)
}
mean(R); shapiro.test(s); shapiro.test(R)
ecc = length(which(LCI<rho&UCI>rho))/nsim;
        ecc ##empirical conf. coeff.
eccR = length(which(LR<rho&UR>rho))/nsim; eccR
d = data.frame(head(LCI),head(UCI),head(LR),
                        head(UR)); d1 = round(d,4); d1
length_CI = UCI-LCI; AVG_L_CI = mean(length_CI);
                     V_L_CI = var(length_CI)
AVG_L_CI;V_L_CI
length_R = UR-LR; AVG_L_R = mean(length_R); V_L_R = var(length_R)
AVG_L_R;V_L_R
```

The mean of 1500 simulated values of R_n comes out to be 0.4998, very close to the value of $\rho = 0.5$. The p-values of Shapiro-Wilk test for normality are 0.6128 and 0.1896 for Fisher's Z transformation and R_n respectively, supporting the normality of both. It is to be noted that as in the previous example, the p-value corresponding to Fisher's Z transformation is larger than that of R_n. Based on 1500 simulations, the empirical confidence coefficient is 0.946 for Fisher's Z transformation and 0.9413 for R_n, both of them being close to the confidence coefficient 0.95. The average length of the confidence interval is 0.1852 for Fisher's Z transformation with the variance of the length being 0.00013, very small. The average length of the confidence interval is 0.1854 for R_n, with the variance of the length being 0.00014. It is also very small. Thus, on the basis of these three criteria, the performance of the two techniques is almost the same. The first 6 confidence intervals obtained using both the techniques are displayed in

Table 3.5 Confidence intervals for ρ based on R_n and Fisher's Z transformation

Number	Variance stabilization technique	Studentization technique
1	$(0.3707, 0.5630)$	$(0.3762, 0.5687)$
2	$(0.3891, 0.5775)$	$(0.3946, 0.5833)$
3	$(0.3522, 0.5483)$	$(0.3575, 0.5540)$
4	$(0.4279, 0.6076)$	$(0.4336, 0.6135)$
5	$(0.4189, 0.6007)$	$(0.4246, 0.6066)$
6	$(0.3931, 0.5806)$	$(0.3986, 0.5864)$

Table 3.5. From the table, we note that the confidence intervals corresponding to two techniques are comparable. More precisely, the confidence intervals obtained using studentization technique are slightly shifted to the right of those obtained using variance stabilization technique. □

3.5 Conceptual Exercises

3.5.1 Suppose $\{X_1, X_2, \ldots, X_n\}$ is a random sample from a uniform $U(0, \theta)$ distribution, $\theta > 0$. (i) Examine whether the maximum likelihood estimator of θ is a CAN estimator of θ.
(ii) Examine whether the moment estimator of θ is a CAN estimator of θ.
(iii) Solve (i) and (ii) if $\{X_1, X_2, \ldots, X_n\}$ are independent random variables where X_i has uniform $U(0, i\theta)$ distribution, $\theta > 0$.

3.5.2 Suppose $\underline{X} \equiv \{X_1, X_2, \ldots, X_n\}$ is a random sample from a uniform $U(0, \theta)$ distribution. Obtain $100(1 - \alpha)\%$ asymptotic confidence interval for θ based on a sufficient statistic.

3.5.3 Suppose $\{X_1, X_2, \ldots, X_n\}$ is a random sample from a uniform $U(\theta, 1)$ distribution, $0 < \theta < 1$. Find the maximum likelihood estimator of θ and the moment estimator of θ. Examine whether these are CAN estimators of θ.

3.5.4 Suppose $\{X_1, X_2, \ldots, X_n\}$ is a random sample from a Bernoulli $B(1, \theta)$ distribution, $\theta \in (0, 1)$. (i) Suppose an estimator $\hat{\theta}_n$ is defined as follows:

$$\hat{\theta}_n = \begin{cases} 0.01, & \text{if } \overline{X}_n = 0 \\ \overline{X}_n, & \text{if } 0 < \overline{X}_n < 1 \\ 0.98, & \text{if } \overline{X}_n = 1. \end{cases}$$

Examine whether it is a CAN estimator of θ. (ii) Examine whether the maximum likelihood estimator of θ is a CAN estimator of θ if $\theta \in [a, b] \subset (0, 1)$.

3.5.5 Suppose $\{X_1, X_2, \ldots, X_n\}$ is a random sample from a normal $N(\theta, 1)$ distribution. Find the maximum likelihood estimator of θ and examine if it is CAN for θ if $\theta \in [0, \infty)$. Identify the limiting distribution at $\theta = 0$.

3.5.6 Suppose $\{X_1, X_2, \ldots, X_n\}$ is a random sample from a distribution of a random variable X with probability density function

$f(x, \theta) = k\theta^k / x^{k+1}$, $x \geq \theta$, $k \geq 3$.

(i) Find the maximum likelihood estimator of θ and examine whether it is CAN for θ.

(ii) Find the moment estimator of θ and examine whether it is CAN for θ.

(iii) Find 95% asymptotic confidence interval for θ based on the maximum likelihood estimator.

3.5.7 Suppose $\{X_1, X_2, \ldots, X_n\}$ is a random sample from an exponential distribution with mean θ. (i) Find the CAN estimator for the mean residual life $E(X - t | X > t)$, $t > 0$. (ii) Show that for some constant $c(p)$, $\sqrt{n}(c(p)X_{([np]+1)} - \theta)$ converges in law to $N(0, \sigma^2(p))$. Find the constant $c(p)$ and $\sigma^2(p)$.

3.5.8 Suppose $\{X_1, X_2, \ldots, X_n\}$ are independent and identically distributed random variables each having a Poisson distribution with mean θ, $\theta > 0$. Find a CAN estimator of $P[X_1 = 1]$. Is it necessary to impose any condition on the parameter space? Under this condition, using the CAN estimator of $P[X_1 = 1]$, obtain a large sample confidence interval for $P[X_1 = 1]$.

3.5.9 Suppose $\{X_1, X_2, \ldots, X_n\}$ is a random sample from $f(x, \theta) = \theta x^{\theta - 1}$, $0 < x < 1$, $\theta > 0$. Find a CAN estimator of $e^{-\theta}$ based on the sample mean and also based on a sufficient statistic. Compare the two estimators.

3.5.10 Suppose $\{X_1, X_2, \ldots, X_n\}$ is a random sample from a Bernoulli $B(1, \theta)$. Find a CAN estimator of $\theta(1 - \theta)$ when $\theta \in (0, 1) - \{1/2\}$. What is the limiting distribution of the estimator when $\theta = 1/2$ and when the norming factor is $\sqrt{n}$ and n?

3.5.11 Suppose $\{X_1, X_2, \ldots, X_n\}$ is a random sample from a geometric distribution, with probability mass function

$$p(x, \theta) = \theta(1 - \theta)^x, \quad x = 0, 1, \ldots.$$

However, $X_1, X_2, \ldots, X_n$ are not directly observable, but one can note whether $X_i \geq 2$ or not. (i) Find a CAN estimator for θ, based on the observed data. (ii) Find a CAN estimator for θ, if $X_i \geq 2$ is replaced by $X_i > 2$.

3.5.12 Suppose $\{X_1, X_2, \ldots, X_n\}$ is a random sample from a normal $N(\theta, 1)$ distribution. However, $X_1, X_2, \ldots, X_n$ are not directly observable, but one can note whether $X_i > 2$ or not. Find a CAN estimator for θ, based on the observed data.

3.5.13 Suppose $\{X_1, X_2, \ldots, X_n\}$ is a random sample from a distribution of X, with probability density function $f(x, \alpha, \theta)$ as given by

$$f(x, \alpha, \theta) = \begin{cases} \frac{2x}{\alpha\theta}, & \text{if} \quad 0 < x \leq \theta \\ \frac{2(\alpha - x)}{\alpha(\alpha - \theta)}, & \text{if} \quad \theta < x \leq \alpha. \end{cases}$$

Find a CAN estimator of θ when α is known.

3.5.14 Suppose $\{X_1, X_2, \ldots, X_n\}$ is a random sample from a negative binomial distribution with probability mass function given by

$$P_\theta[X = x] = \binom{x + k - 1}{x} p^k (1 - p)^x, \quad x = 0, 1, \ldots, \quad 0 < p < 1, \quad k > 0.$$

Obtain a CAN estimator for p assuming k to be known.

3.5.15 Suppose $\{X_1, X_2, \ldots, X_n\}$ is a random sample from an exponential distribution with mean θ. Suppose $T_{1n} = \sum_{i=1}^n X_i/n$ and $T_{2n} = \sum_{i=1}^n X_i/(n+1)$.
(i) Examine whether T_{1n} and T_{2n} are consistent for θ. Prove that T_{1n} is CAN for θ.
(ii) Prove that $\sqrt{n}(T_{2n} - T_{1n}) \overset{P_\theta}{\to} 0$ and hence both T_{1n} and T_{2n} are CAN for θ with the same approximate variance, but $MSE_\theta(T_{2n}) < MSE_\theta(T_{1n})$ $\forall\ n \geq 1$.
(iii) Find a CAN estimator for $P[X_1 > t]$ where t is a positive real number.

3.5.16 Suppose $\{X_1, X_2, \ldots, X_n\}$ is a random sample from a Poisson $Poi(\theta)$ distribution. Examine whether the sample variance is CAN for θ.

3.5.17 Show that the empirical distribution function $F_n(a)$ is CAN for $F(a)$, where a is a fixed real number. Hence obtain a asymptotic confidence interval for $F(a)$.

3.5.18 Suppose $\{X_1, X_2, \ldots, X_n\}$ is a random sample from a normal $N(\theta, a\theta^2)$ distribution, $\theta > 0$ and a is a known positive real number. Find the maximum likelihood estimator of θ. Examine whether it is CAN for θ.

3.5.19 Suppose $\{X_1, X_2, \ldots, X_n\}$ is a random sample from a uniform $U(-\theta, \theta)$ distribution. Find CAN estimator for θ based on $\sum_{i=1}^n |X_i|$. Are sample mean and sample median CAN for θ? Justify your answer. Find a consistent estimator for θ based on $X_{(1)}$ and find a consistent estimator for θ based on $X_{(n)}$. Examine if these are CAN for θ.

3.5.20 Suppose $\{X_1, X_2, \ldots, X_n\}$ is a random sample from a uniform $U(0, \theta)$ distribution, $\theta > 0$. Examine whether $S_n = (\prod_{i=1}^n X_i)^{1/n}$ is CAN for θe^{-1}.

3.5.21 Suppose $\{X_1, X_2, \ldots, X_{2n+1}\}$ is a random sample from a uniform $U(\theta - 1, \theta + 1)$ distribution. (i) Show that $\overline{X}_{2n+1}$ and $X_{(n+1)}$ are both CAN for θ. Compare the two estimators. (ii) Using the large sample distribution, obtain the minimum sample size n_0 required for both the estimators to attain a given level of accuracy specified by ϵ and δ such that $P[|T_n - \theta| < \epsilon] \geq 1 - \delta$, $\forall\ n \geq n_0$, where T_n is either $\overline{X}_{2n+1}$ or $X_{(n+1)}$.

3.5.22 Suppose $\{X_1, X_2, \ldots, X_n\}$ is a random sample from X with probability density function $f(x, \theta) = \theta/x^2$, $x \geq \theta$, $\theta > 0$. (i) Find the maximum likelihood estimator of θ and examine if it is CAN for θ. (ii) Find a CAN estimator of θ based on the sample quantiles.

3.5.23 Suppose $\{X_1, X_2, \ldots, X_n\}$ is a random sample from a Weibull distribution with probability density function

$$f(x, \theta) = \theta\, x^{\theta-1} \exp\{-x^\theta\}, \quad x > 0, \quad \theta > 0.$$

Obtain the estimator of θ based on the sample quantiles. Is it CAN? Justify your answer.

3.5.24 Suppose $\{X_1, X_2, \ldots, X_n\}$ is a random sample from a Laplace $(\theta, 1)$ distribution. Find a family of CAN estimators of θ based on the sample quantiles. Also find the CAN estimator of θ based on the sample mean and the sample median. Which one is better? Why?

3.5.25 Suppose $\{X_1, X_2, \ldots, X_n\}$ is a random sample from a gamma $G(\theta, \lambda)$ distribution. Find $100(1 - \alpha)\%$ asymptotic confidence interval for the mean of the distribution.

3.5.26 Suppose $\{X_1, X_2, \ldots, X_n\}$ is a random sample from a Bernoulli distribution with mean θ. Find an asymptotic confidence interval for θ with confidence coefficient $(1 - \alpha)$ using both the studentization procedure and the variance stabilization technique.

3.5.27 Suppose $\{X_1, X_2, \ldots, X_n\}$ is a random sample from a normal $N(\mu, \sigma^2)$ distribution. Find a CAN estimator of the p-th population quantile based on p-th sample quantile as well as based on a CAN estimator of $\underline{\theta} = (\mu, \sigma^2)'$. Which is better? Why? Obtain the asymptotic confidence interval for the p-th population quantile using the estimator which is better between the two.

3.5.28 Suppose $\{X_1, X_2, \ldots, X_n\}$ is a random sample from an exponential distribution with location parameter μ and scale parameter $1/\sigma$. (i) Obtain an asymptotic confidence interval for μ when σ is known and when it is unknown. (ii) Obtain an asymptotic confidence interval for σ when μ is known and when it is unknown. (iii) Obtain an asymptotic confidence interval for the p-th population quantile when both μ and σ are unknown.

3.5.29 Suppose $\{X_1, X_2, \ldots, X_n\}$ is a random sample of size n from a uniform $U(\theta_1, \theta_2)$ distribution, $-\infty < \theta_1 < \theta_2 < \infty$. Obtain a CAN estimator of p-th population quantile and hence based on it obtain the asymptotic confidence interval for p-th population quantile.

3.5.30 Suppose $\{X_1, X_2, \ldots, X_n\}$ is a random sample from a Poisson $Poi(\theta)$ distribution, $\theta > 0$. (i) Obtain a CAN estimator of the coefficient of variation $cv(\theta)$ of X when it is defined as
$cv(\theta) = \text{standard deviation}/\text{mean} = 1/\sqrt{\theta}$. (ii) If the estimator of $cv(\theta)$ is proposed as $\tilde{cv}(\theta) = S_n/\overline{X}_n$, where $\overline{X}_n$ is the sample mean and S_n is the sample standard deviation, examine if it is CAN for θ. Compare the two estimators.

3.5.31 Suppose $\{X_1, X_2, \ldots, X_n\}$ is a random sample from a log-normal distribution with parameters μ and σ^2. Find a CAN estimator of $(\mu'_1, \mu'_2)'$. Hence obtain a CAN estimator for $\underline{\theta} = (\mu, \sigma^2)'$ and its approximate variance-covariance matrix.

3.5.32 Suppose $\{X_1, X_2, \ldots, X_n\}$ is a random sample from a gamma distribution with scale parameter α and shape parameter λ. Find a moment estimator of $(\alpha, \lambda)'$ and examine whether it is CAN. Find its approximate variance-covariance matrix.

3.5.33 Suppose $X_{ij} = \mu_i + \epsilon_{ij}$ where $\{\epsilon_{ij}, i = 1, 2, 3, \ j = 1, 2, \ldots, n\}$ are independent and identically distributed random variables each having a normal

$N(0, \sigma^2)$ distribution. (i) Obtain a CAN estimator for $\theta = \mu_1 - 2\mu_2 + \mu_3$. (ii) Suppose $\{\epsilon_{ij}, i = 1, 2, 3, \ j = 1, 2, \ldots, n\}$ are independent and identically distributed random variables with $E(\epsilon_{ij}) = 0$ and $Var(\epsilon_{ij}) = \sigma^2$. Is the estimator of θ obtained in (i) still a CAN estimator of θ? Justify your answer.

3.5.34 Suppose $\{X_1, X_2, \ldots, X_n\}$ is a random sample from a uniform $U(\theta_1, \theta_2)$ distribution, where $\theta_1 < \theta_2 \in \mathbb{R}$. (i) Find the maximum likelihood estimator of $(\theta_1, \theta_2)'$. Show that it is consistent but not CAN. (ii) Find a CAN estimator of $(\theta_1 + \theta_2)/2$.

3.5.35 Suppose $\{X_1, X_2, \ldots, X_n\}$ are independent and identically distributed random variables with finite fourth order moment. Suppose $E(X_1) = \mu$ and $Var(X_1) = \sigma^2$. Find a CAN estimator of the coefficient of variation σ/μ.

3.5.36 Suppose $\{X_1, X_2, \ldots, X_n\}$ is a random sample from a distribution of X with distribution function F. Suppose random variables Z_1 and Z_2 are defined as follows. For $a < b$,

$$Z_1 = \begin{cases} 1, & \text{if} \quad X \le a \\ 0, & \text{if} \quad X > a. \end{cases}$$

$$Z_2 = \begin{cases} 1, & \text{if} \quad X \le b \\ 0, & \text{if} \quad X > b. \end{cases}$$

Show that for large n, the distribution of $(\overline{Z}_{1n}, \overline{Z}_{2n})'$ is bivariate normal. Hence obtain a CAN estimator for $(F(a), F(b))'$.

3.5.37 Suppose $\{X_1, X_2, \ldots, X_n\}$ is a random sample from the following distributions—(i) Normal $N(\mu, \sigma^2)$ and (ii) exponential distribution with location parameter θ and scale parameter λ. Find the maximum likelihood estimators of the parameters using stepwise maximization procedure and examine whether these are CAN.

3.6   Computational Exercises

Verify the results of the following exercises and examples by simulation using R code.

3.6.1 Exercise 3.5.1 (i) and (ii) (Hint: Code will be similar to that of Example 3.4.4).

3.6.2 Exercise 3.5.2 (Hint: Code will be similar to that of Example 3.4.4).

3.6.3 Exercise 3.5.4 (Hint: Code will be similar to that of Example 3.4.1).

3.6.4 In Exercise 3.5.5, compare the performance of the asymptotic confidence intervals based on the maximum likelihood estimator and moment estimator of θ.(Hint: Use the approach similar to that of Example 3.4.5 and Example 3.4.7).

3.6.5 In Exercise 3.5.7, find the value of p for which $\sigma^2(p)$ is minimum. (Hint: In the solution of Exercise 3.5.7, you will find the expression of $\sigma^2(p)$.)

3.6.6 Exercise 3.5.15, (*i*) & (*ii*) (Hint: Code will be similar to that of Example 3.4.1. See the solution of Exercise 3.5.15 for the expressions of MSE).

3.6.7 Exercise 3.5.21 (Hint: Code will be similar to that of Example 3.4.1. See the solution of Exercise 3.5.21 for the expressions of the minimum sample size).

3.6.8 Exercise 3.5.26. Display the first 6 confidence intervals for both the procedures. Also compute the empirical confidence coefficient for both the procedures.(Hint: Use the approach similar to that of Example 3.4.7).

3.6.9 Exercise 3.5.27. Take $p = 0.40$. Display the first 6 confidence intervals for both the procedures. Also compute the empirical confidence coefficient for both the procedures.

3.6.10 Exercise 3.5.30 (Hint: See the solution of Exercise 3.5.30)

3.6.11 Exercise 3.5.31 (Hint: See the solution of Exercise 3.5.31)

3.6.12 Example 3.2.2

3.6.13 Example 3.2.6. Take any value of p.

3.6.14 Example 3.2.8 (Hint: Code will be similar to that of Example 3.4.4).

3.6.15 Example 3.2.11 (Hint: Use the approach similar to that of Example 3.4.7).

3.6.16 Example 3.3.1 (Hint: Code will be similar to that of Example 3.4.1).

References

1. Serfling, R. J. (1980). *Approximation theorems of mathematical statistics*. New York: Wiley.
2. DasGupta, A. (2008). *Asymptotic theory of statistics and probability*. New York: Springer.
3. Anderson, T. W. (2003). *An introduction to multivariate statistical analysis*. New York: Wiley.
4. Genz, A., & Bretz, F. (2009). *Computation of multivariate normal and t probabilities: Lecture notes in statistics* (Vol. 195). Heidelberg: Springer.
5. Genz, A., Bretz, F., Miwa, T., Mi, X., Leisch, F., Scheipl, F., Hothorn, T. (2020). *mvtnorm: Multivariate normal and t distributions*. R package version 1.1-0. http://CRAN.R-project.org/package=mvtnorm.

CAN Estimators in Exponential and Cramér Families

4

Contents

━ **Learning Objectives** After going through this chapter, the readers should be able
- to derive CAN estimators of parameters of the probability models belonging to one-parameter and multiparameter exponential family
- to appreciate the Cramér-Huzurbazar theory of maximum likelihood estimation of parameters of the probability models belonging to one-parameter and multiparameter Cramér family
- to compare CAN estimators on the basis of asymptotic relative efficiency
- to implement the iterative procedures for computation of maximum likelihood estimators using R

4.1 Introduction

In Chap. 2, we discussed the concept of consistency of an estimator based on a random sample from the distribution of X, which is either a random variable or a

---

random vector with probability law $f(x, \theta)$, indexed by a parameter $\theta \in \Theta$. It may be a real parameter or a vector parameter. Chapter 3 was devoted to the discussion of a CAN estimator of a parameter. The present chapter is concerned with the study of a CAN estimator of a parameter, when a probability distribution of X belongs to a specific family of distributions such as an exponential family or a Cramér family. An exponential family is a subclass of a Cramér family. In Sect. 4.2, we prove that in a one-parameter exponential family and in a multiparameter exponential family, the maximum likelihood estimator and the moment estimator based on a sufficient statistic are the same and these are CAN estimators. Section 4.3 presents the Cramér-Huzurbazar theory for the distributions belonging to a Cramér family. Cramér-Huzurbazar theory, which is usually referred to as standard large sample theory of maximum likelihood estimation, asserts that for large sample size with high probability, the maximum likelihood estimator of a parameter is a CAN estimator. These results are heavily used in Chaps. 5 and 6 to find the asymptotic null distribution of the likelihood ratio test statistic, Wald's test statistic and the score test statistic. In many models, the system of likelihood equations cannot be solved explicitly and we need some numerical procedures. In Sect. 4.4, we discuss frequently used numerical procedures to solve the system of likelihood equations, such as Newton-Raphson procedure and method of scoring. The last section illustrates the results established in Sects. 4.2 to 4.4 using R software.

4.2   Exponential Family

An exponential family is a practically convenient and widely used unified family of distributions indexed either by a real parameter or by a vector parameter. It contains most of the standard discrete and continuous distributions that are used for modeling, such as normal, Poisson, binomial, exponential, gamma, and multivariate normal. The reason for the special status of an exponential family is that a number of important and useful results regarding estimation and testing of hypotheses can be unified within the framework of an exponential family. This family also forms the basis for an important class of models, known as generalized linear models. A fundamental treatment of the general exponential family is available in Lehmann and Romano [1] and other books cited therein.

Suppose X is a random random variable or a random vector with probability law $f(x, \theta)$, which is either a probability density function or a probability mass function, $\theta \in \Theta \subset \mathbb{R}$ and it is an indexing parameter. The distribution of X belongs to a one-parameter exponential family if the following conditions are satisfied.

(i) The support S_f of $f(x, \theta)$ defined as $S_f = \{x | f(x, \theta) > 0\}$ is free from θ.
(ii) The parameter space Θ is such that $\int_{S_f} f(x, \theta)\, dx = 1$ and it is an open set.
(iii) The probability law $f(x, \theta)$ is expressible as

$$f(x, \theta) = \exp\{U(\theta)K(x) + V(\theta) + W(x)\},$$

where U and V are functions of θ only and K and W are functions of x only.

(iv) U is a twice differentiable function of θ with $U'(\theta) \neq 0$.

(v) $K(x)$ and 1 are linearly independent, that is, $a + bK(x) = 0 \Rightarrow a = 0$ and $b = 0$.

In condition (ii), it is enough to assume that the true parameter θ_0 is an interior point. If Θ is assumed to be an open set, it is always satisfied. It has been proved that Θ is a convex set (see Lehmann and Romano [1]).

It is easy to verify that all standard distributions, such as $N(\theta, 1)$, $\theta \in \mathbb{R}$, exponential distribution with mean $\theta > 0$, Binomial $B(n, \theta)$ where n is known and $\theta \in (0, 1)$, Poisson $Poi(\theta)$, $\theta > 0$, geometric with success probability $\theta \in (0, 1)$ and the truncated versions of these distributions, constitute a one-parameter exponential family of distributions. $N(\theta, \theta)$ distribution where $\theta > 0$, belongs to a one-parameter exponential family. However, $N(\theta, \theta^2)$ distribution, where $\theta > 0$ does not belong to a one-parameter exponential family, but belongs to a curved exponential family, refer to van der Vaart [2] for details. In Sect. 4.3, we show that it belongs to a Cramér family. A family of normal $N(\theta, 1)$ distributions, with $\theta \in I^+$ where $I^+ = \{1, 2, \ldots\}$ or $\theta \in \{-1, 0, 1\}$, is not an exponential family as the parameter space is not open. Similarly, a family of gamma distributions with a known scale parameter and shape parameter belonging to I^+, also known as Erlang distribution, is not an exponential family, as the parameter space is not open. A uniform $U(0, \theta)$ or a uniform $U(\theta - 1, \theta + 1)$ distribution, an exponential distribution with scale parameter 1 and location parameter θ, do not belong to an exponential family as the support of the distribution depends on the parameter θ. A Cauchy distribution with scale parameter 1 and location parameter θ does not belong to the exponential family, as its probability density function $f(x, \theta)$ cannot be expressed in a required form. For the same reason, a Laplace distribution does not belong to an exponential family.

Some of the properties of the distributions belonging to an exponential family are as follows:

(i) If $\{X_1, X_2, \ldots, X_n\}$ is a random sample from the distribution belonging to a one-parameter exponential family, then $\sum_{r=1}^{n} K(X_r)$ is a minimal sufficient statistic.

(ii) A function V is differentiable any number of times with respect to θ.

(iii) The identity $\int_{S_f} f(x, \theta) dx = 1$ can be differentiated with respect to θ any number of times under the integral sign. As a consequence, $E_\theta \left(\frac{\partial}{\partial \theta} \log f(X, \theta) \right) = 0$ and information function $I(\theta)$ is given by

$$I(\theta) = E_\theta \left(\frac{\partial}{\partial \theta} \log f(X, \theta) \right)^2 = E_\theta \left(-\frac{\partial^2}{\partial \theta^2} \log f(X, \theta) \right)$$

$$= Var_\theta \left(\frac{\partial}{\partial \theta} \log f(X, \theta) \right).$$

Observe that $I(\theta) = Var_\theta \left(\frac{\partial}{\partial \theta} \log f(X, \theta) \right) \geq 0$. If $Var_\theta \left(\frac{\partial}{\partial \theta} \log f(X, \theta) \right) = 0$, then $\frac{\partial}{\partial \theta} \log f(X, \theta) = U'(\theta)K(X) + V'(\theta)$ is a degenerate random variable; it may be a function of θ. It further implies that $K(X)$ is some function of θ, which is contrary to the assumption that $K(X)$ is a function of X only. Thus, it follows that $I(\theta) = Var_\theta \left(\frac{\partial}{\partial \theta} \log f(X, \theta) \right) > 0$. Further,

$$E_\theta \left(\frac{\partial}{\partial \theta} \log f(X, \theta) \right) = 0 \Rightarrow E_\theta(K(X)) = \frac{-V'(\theta)}{U'(\theta)} = \eta(\theta), \text{ say } \quad (4.2.1)$$

It is to be noted that $E_\theta(K(X)) < \infty$ as $U'(\theta) \neq 0$. We find the variance of $K(X)$ using the following formula for $I(\theta)$. We have

$$I(\theta) = E_\theta \left(\frac{\partial}{\partial \theta} \log f(X, \theta) \right)^2 = E_\theta \left(U'(\theta)K(X) + V'(\theta) \right)^2$$

$$= (U'(\theta))^2 E_\theta \left(K(X) - \left(\frac{-V'(\theta)}{U'(\theta)} \right) \right)^2$$

$$= (U'(\theta))^2 E_\theta \left(K(X) - E(K(X)) \right)^2$$

$$= (U'(\theta))^2 Var_\theta(K(X))$$

$$\Rightarrow Var_\theta(K(X)) = \frac{I(\theta)}{(U'(\theta))^2}, \quad \text{as } U'(\theta) \neq 0. \quad (4.2.2)$$

Since $I(\theta) = (U'(\theta))^2 Var_\theta(K(X))$ and $U'(\theta) \neq 0$, if $Var_\theta(K(X)) < \infty$ then $I(\theta) < \infty$. Thus, if the variance of $K(X)$ is finite, then $0 < I(\theta) < \infty$. We find one more expression of $I(\theta)$ as follows:

$$I(\theta) = E_\theta \left(-\frac{\partial^2}{\partial \theta^2} \log f(X, \theta) \right) = E_\theta \left(-U''(\theta)K(X) - V''(\theta) \right)$$

$$= \left(U''(\theta) \frac{V'(\theta)}{U'(\theta)} - V''(\theta) \right) = U'(\theta) \left(\frac{U''(\theta)V'(\theta) - V''(\theta)U'(\theta)}{(U'(\theta))^2} \right)$$

$$= U'(\theta) \frac{d}{d\theta} \left(\frac{-V'(\theta)}{U'(\theta)} \right) = U'(\theta)\eta'(\theta). \quad (4.2.3)$$

We use these expressions of $E(K(X))$, $Var(K(X))$ and $I(\theta)$ in the following Theorem 4.2.1, which proves an important result that for the distributions belonging to an exponential family, method of moment estimation and maximum likelihood estimation lead to the best CAN estimator of an indexing parameter θ. It is true when the moment estimator is a function of a sufficient statistic. We use the inverse function theorem, stated in Chap. 1, in the proof of Theorem 4.2.1 and of Theorem 4.2.2 to examine whether inverse of some parametric function exists and has some desirable properties.

Theorem 4.2.1

Suppose the distribution of a random variable or a random vector X belongs to a one-parameter exponential family with indexing parameter θ. Suppose $\{X_1, X_2, \ldots, X_n\}$ is a random sample from the distribution of X. Then the moment estimator of θ based on a sufficient statistic is the same as the maximum likelihood estimator of θ and it is CAN for θ with approximate variance $1/nI(\theta)$.

Proof Since the distribution of X belongs to a one-parameter exponential family with indexing parameter θ, its probability law is given by $f(x, \theta) = \exp\{U(\theta)K(x) + V(\theta) + W(x)\}$, where U and V are differentiable functions of θ with $U'(\theta) \neq 0$. Further, 1 and $K(x)$ are linearly independent, the parameter space Θ is an open set and the support S_f of $f(x, \theta)$ is free from θ. Corresponding to a random sample $\underline{X} = \{X_1, X_2, \ldots, X_n\}$ of size n from the distribution of X, the likelihood of θ is given by

$$L_n(\theta|\underline{X}) = \prod_{r=1}^{n} \exp\{U(\theta)K(X_r) + V(\theta) + W(X_r)\}$$

$$= \exp\left\{U(\theta)\sum_{r=1}^{n} K(X_r) + nV(\theta) + \sum_{r=1}^{n} W(X_r)\right\}. \qquad (4.2.4)$$

From the Neyman-Fisher factorization theorem, it follows that $\sum_{r=1}^{n} K(X_r)$ is a sufficient statistic for the family. Thus, the moment estimator of θ based on a sufficient statistic is a solution of the equation $T_n = \sum_{r=1}^{n} K(X_r)/n = E_\theta(K(X)) = \eta(\theta) = -V'(\theta)/U'(\theta)$ by Eq. (4.2.1). To investigate whether this equation has a solution, we note that $I(\theta) = U'(\theta)\eta'(\theta)$ from Eq. (4.2.3). Further, $I(\theta) > 0$ and $U'(\theta) \neq 0 \ \forall \ \theta \in \Theta$, thus if $U'(\theta) > 0$, then $\eta'(\theta) > 0$ and if $U'(\theta) < 0$, then $\eta'(\theta) < 0$ which implies that $\eta'(\theta) \neq 0 \ \forall \ \theta \in \Theta$. Hence, by the inverse function theorem, unique η^{-1} exists and the moment equation $T_n = \eta(\theta)$ has a unique solution. Thus, the moment estimator $\tilde{\theta}_n$ based on a sufficient statistic is given by $\tilde{\theta}_n = \eta^{-1}(T_n)$. To find the maximum likelihood estimator, from the likelihood as given in Eq. (4.2.4), we have

$$\frac{\partial}{\partial \theta} \log L_n(\theta|\underline{X}) = U'(\theta)nT_n + nV'(\theta) \text{ and}$$

$$\frac{\partial^2}{\partial \theta^2} \log L_n(\theta|\underline{X}) = U''(\theta)nT_n + nV''(\theta).$$

Thus, the likelihood equation is given by $T_n = \eta(\theta)$ with its solution as $\theta = \eta^{-1}(T_n)$. To claim it to be the maximum likelihood estimator, we examine whether the second derivative of the log-likelihood is negative at the solution of the likelihood equation. Observe that at $T_n = \eta(\theta)$, that is at $\theta = \eta^{-1}(T_n)$

$$\frac{\partial^2}{\partial \theta^2} \log L_n(\theta|\underline{X})|_{\theta=\eta^{-1}(T_n)} = n\left(U''(\theta)\eta(\theta) + V''(\theta)\right)$$

$$= n\left(U''(\theta)\frac{-V'(\theta)}{U'(\theta)} + V''(\theta)\right)$$

$$= n \frac{(-U''(\theta)V'(\theta) + U'(\theta)V''(\theta))}{U'(\theta)}$$

$$= -nI(\theta) = -nI(\eta^{-1}(T_n)) < 0 \ a.s.$$

Hence, $\hat{\theta}_n = \eta^{-1}(T_n)$ is the maximum likelihood estimator of θ and it is the same as the moment estimator $\tilde{\theta}_n = \eta^{-1}(T_n)$ based on a sufficient statistic. To establish that $\hat{\theta}_n = \tilde{\theta}_n$ is CAN, observe that $E(K(X)) = \eta(\theta) < \infty$ as $U'(\theta) \neq 0$ and hence by Khintchine's WLLN, $T_n = \frac{1}{n}\sum_{r=1}^{n} K(X_r) \overset{P_\theta}{\to} E_\theta(K(X)) = \eta(\theta)$ as $n \to \infty$, $\forall \ \theta \in \Theta$. Thus, T_n is consistent for $\eta(\theta)$. To find its asymptotic distribution with suitable normalization, it is to be noted that $\{K(X_1), K(X_2), \ldots, K(X_n)\}$ are independent and identically distributed random variables with $E_\theta(K(X)) = \eta(\theta)$ and $Var_\theta(K(X)) = I(\theta)/(U'(\theta))^2$, which is positive and finite. Hence, by the CLT applied to $\{K(X_1), K(X_2), \ldots, K(X_n)\}$, we have

$$\frac{\sum_{r=1}^{n} K(X_r) - n\eta(\theta)}{\sqrt{nI(\theta)}/U'(\theta)} \overset{L}{\to} Z \sim N(0,1) \quad \Leftrightarrow \quad \sqrt{n}(T_n - \eta(\theta)) \overset{L}{\to} Z_1 \sim N(0, \sigma(\theta)),$$

where $\sigma(\theta) = I(\theta)/(U'(\theta))^2$. Thus, T_n is CAN for $\eta(\theta) = \phi$, say with approximate variance $\sigma(\theta)/n = \sigma_1(\phi)/n$. To find the CAN estimator for θ, we use the delta method. Suppose $g(\phi) = \eta^{-1}(\phi)$ then $g(\phi) = \eta^{-1}(\phi) = \eta^{-1}(\eta(\theta)) = \theta$. Further, $\eta(\theta) = -V'(\theta)/U'(\theta)$. Since U and V are differentiable twice, $\eta(\theta)$ is differentiable and by the inverse function theorem $g(\phi) = \eta^{-1}(\phi)$ is a differentiable function and hence a continuous function. Hence, by the invariance property of consistency under continuous transformation,

$$T_n \overset{P_\theta}{\to} \eta(\theta) \ \Rightarrow \ \eta^{-1}(T_n) = \hat{\theta}_n = \tilde{\theta}_n \overset{P_\theta}{\to} \eta^{-1}(\eta(\theta)) = \theta, \ \forall \ \theta \in \Theta.$$

Thus, $\hat{\theta}_n = \tilde{\theta}_n$ is consistent for θ. Now, g is differentiable and

$$g'(\phi) = \frac{d}{d\phi}\eta^{-1}(\phi) = \frac{d}{d\eta(\theta)}\eta^{-1}(\eta(\theta)) = \frac{d\theta}{d\eta(\theta)}.$$

To find the expression for $\frac{d\theta}{d\eta(\theta)}$ observe that,

$$\eta^{-1}(\eta(\theta)) = \theta \Rightarrow \frac{d}{d\theta}\eta^{-1}(\eta(\theta)) = \frac{d}{d\theta}\theta = 1$$

$$\Rightarrow \frac{d\eta^{-1}(\eta(\theta))}{d\eta(\theta)}\frac{d\eta(\theta)}{d\theta} = 1, \quad \text{by chain rule}$$

$$\Rightarrow \frac{d\theta}{d\eta(\theta)}\eta'(\theta) = 1$$

$$\Rightarrow \frac{d\theta}{d\eta(\theta)} = (\eta'(\theta))^{-1} = \frac{U'(\theta)}{I(\theta)} \neq 0,$$

as $U'(\theta) \neq 0$ and $I(\theta) < \infty$. Thus, $g'(\phi) = \frac{d\theta}{d\eta(\theta)} = \frac{U'(\theta)}{I(\theta)} \neq 0 \ \forall \ \theta \in \Theta$ and hence for all ϕ. Hence, by the delta method, $g(T_n) = \eta^{-1}(T_n) = \hat{\theta}_n = \tilde{\theta}_n$ is CAN for $g(\phi) = \theta$ with approximate variance given by

$$\frac{1}{n}V_1(\phi)(g'(\phi))^2 = \frac{1}{n}\frac{I(\theta)}{(U'(\theta))^2}\left(\frac{U'(\theta)}{I(\theta)}\right)^2 = \frac{1}{nI(\theta)}.$$

Thus, for the distributions belonging to a one-parameter exponential family with indexing parameter θ, the moment estimator of θ based on a sufficient statistic is the same as the maximum likelihood estimator of θ and it is a CAN estimator of θ with approximate variance $1/nI(\theta)$. □

It is to be noted that $\hat{\theta}_n$ is only a local maximum likelihood estimator of θ. Theorem 4.2.1 can be easily verified for all standard distributions, such as normal $N(\theta, 1)$, $\theta \in \mathbb{R}$, exponential distribution with mean $\theta > 0$, gamma distribution with scale 1 and shape parameter $\lambda > 0$, Binomial $B(n, \theta)$ where n is known and $\theta \in (0, 1)$, Poisson $Poi(\theta)$, $\theta > 0$, geometric with success probability $\theta \in (0, 1)$ and even the truncated versions of these distributions. In the following examples we illustrate the results of Theorem 4.2.1 for some distributions.

✎ Example 4.2.1

The number of insects per leaf is believed to follow a Poisson $Poi(\theta)$ distribution. Many leaves have no insects, because those are unsuitable for feeding. For the given situation if X denotes the number of insects per leaf, then the distribution of X is modeled as a truncated Poisson $Poi(\theta)$ distribution, truncated at 0, $\theta > 0$. The probability mass function of the Poisson $Poi(\theta)$ distribution truncated at 0 is given by

$$f(x, \theta) = P_\theta[X = x] = \frac{e^{-\theta}}{(1 - e^{-\theta})}\frac{\theta^x}{x!}$$
$$\Rightarrow \ \log f(x, \theta) = U(\theta)K(x) + V(\theta) + W(x),$$

where $x = 1, 2, \ldots,$ and $U(\theta) = \log \theta$, $K(x) = x$, $V(\theta) = -\theta - \log(1 - e^{-\theta})$ and $W(x) = -\log x!$. Further, U and V are differentiable functions of θ and can be differentiated any number of times and $U'(\theta) = 1/\theta \neq 0$. 1 and $k(x) = x$ are linearly independent. The parameter space is an open set and support of X, which is a set of natural numbers, and is free from θ. Thus, Poisson $Poi(\theta)$ distribution truncated at 0 is a member of a one-parameter exponential family. Hence, by Theorem 4.2.1, the moment estimator of θ based on a sufficient statistic is the same as the maximum likelihood estimator of θ and it is CAN with approximate variance $I^{-1}(\theta)/n$. We now proceed to find the expressions for the estimator and $I(\theta)$. Corresponding to a random sample of size n from the distribution of X, the likelihood of θ is given by

$$L_n(\theta|\underline{X}) \equiv L_n(\theta|X_1, X_2, \ldots, X_n) = \prod_{i=1}^{n} \frac{e^{-\theta}}{(1 - e^{-\theta})} \frac{\theta^{X_i}}{X_i!}$$

$$= \left(\frac{e^{-\theta}}{1 - e^{-\theta}}\right)^n \frac{\theta^{\sum_{i=1}^{n} X_i}}{\prod_{i=1}^{n} X_i!}.$$

By the Neyman-Fisher factorization theorem, it follows that $\sum_{i=1}^{n} X_i$ is a sufficient statistic. The moment estimator of θ based on the sufficient statistic is then given by the equation, $\overline{X}_n = E(X) = \theta/(1 - e^{-\theta}) = \eta(\theta)$ say. It is to be noted that

$$\eta'(\theta) = \frac{1 - e^{-\theta} - \theta e^{-\theta}}{(1 - e^{-\theta})^2} = \frac{P_\theta[Y > 1]}{(1 - e^{-\theta})^2} > 0, \quad \forall \ \theta > 0,$$

where $Y \sim Poi(\theta)$ with support $\{0, 1, \ldots, \}$. Hence, by the inverse function theorem, η^{-1} exists and by using numerical methods, which are discussed in Sect. 4.4, we get the moment estimator $\tilde{\theta}_n$ of θ based on the sufficient statistic as $\tilde{\theta}_n = \eta^{-1}(\overline{X}_n)$. We now proceed to find the maximum likelihood estimator. From the likelihood of θ, as specified above we get the following likelihood equation:

$$-n - \frac{ne^{-\theta}}{1 - e^{-\theta}} + \frac{\sum_{i=1}^{n} X_i}{\theta} = 0 \quad \Leftrightarrow \quad \overline{X}_n = \frac{\theta}{1 - e^{-\theta}}.$$

Further,

$$\frac{\partial^2}{\partial \theta^2} \log L_n(\theta|\underline{X}) = \frac{ne^{-\theta}}{(1 - e^{-\theta})^2} - \frac{\sum_{i=1}^{n} X_i}{\theta^2}$$

and at the solution of the likelihood equation it is $-n(1 - e^{-\theta} - \theta e^{-\theta})/\theta(1 - e^{-\theta})^2 < 0$ for all $\theta > 0$. Thus, the maximum likelihood estimator $\hat{\theta}_n$ of θ is given by $\hat{\theta}_n = \eta^{-1}(\overline{X}_n)$, which is the same as the moment estimator based on the sufficient statistic. The information function $I(\theta)$ is given by

$$nI(\theta) = E\left(-\frac{\partial^2}{\partial \theta^2} \log L_n(\theta|\underline{X})\right) = \frac{-ne^{-\theta}}{(1 - e^{-\theta})^2} + \frac{n\theta}{\theta^2}$$

$$= \frac{n(1 - e^{-\theta} - \theta e^{-\theta})}{\theta(1 - e^{-\theta})^2} = nU'(\theta)\eta'(\theta).$$

Thus, $\tilde{\theta}_n = \hat{\theta}_n$ is CAN for θ with approximate variance $1/nI(\theta) = \theta(1 - e^{-\theta})^2/n(1 - e^{-\theta} - \theta e^{-\theta})$. $\qquad \square$

✒ Example 4.2.2

Suppose $X \sim N(\theta, \theta)$, $\theta > 0$. Then its probability density function is given by

$$f(x, \theta) = \frac{1}{\sqrt{2\pi\theta}} \exp\left\{-\frac{1}{2\theta}(x - \theta)^2\right\}$$

$$= \frac{1}{\sqrt{2\pi\theta}} \exp\left\{-\frac{1}{2\theta}x^2 + x - \frac{\theta}{2}\right\}, \quad x \in \mathbb{R} \quad \theta > 0.$$

Hence, $\log f(x, \theta) = U(\theta)K(x) + V(\theta) + W(x)$, where $U(\theta) = -1/2\theta$, $K(x) = x^2$, $W(x) = x$ and $V(\theta) = -\log(2\pi\theta)/2 - \theta/2$. Further U and V are differentiable functions of θ and can be differentiated any number of times and $U'(\theta) = 1/2\theta^2 \neq 0$. To prove that 1 and $k(x) = x^2$ are linearly independent, consider $a + bx^2 = 0$. Taking derivative with respect to x, we get $2bx = 0$, with one more derivative $2b = 0$, that is $b = 0$ and then from $a + bx^2 = 0$, we have $a = 0$ which implies that 1 and $k(x) = x^2$ are linearly independent. The parameter space $(0, \infty)$ is an open set and support of X, which is a real line, is free from θ. Thus, a family of normal $N(\theta, \theta)$ distributions with $\theta > 0$ is a one-parameter exponential family. Suppose $\{X_1, X_2, \ldots, X_n\}$ is a random sample from the distribution of X. By Theorem 4.2.1, the moment estimator of θ based on a sufficient statistic is the same as the maximum likelihood estimator of θ and it is CAN with approximate variance $1/nI(\theta)$. We now proceed to find the expressions for the estimator and $I(\theta)$. Corresponding to a random sample of size n from the distribution of X, the likelihood of θ is given by

$$L_n(\theta|\underline{X}) \equiv L_n(\theta|X_1, X_2, \ldots, X_n) = \prod_{i=1}^n \frac{1}{\sqrt{2\pi\theta}} \exp\left\{-\frac{1}{2\theta}(X_i - \theta)^2\right\}$$

$$= \left(\frac{1}{\sqrt{2\pi\theta}}\right)^n \exp\left\{-\frac{1}{2\theta}\sum_{i=1}^n X_i^2 + \sum_{i=1}^n X_i - \frac{n\theta}{2}\right\}.$$

By the Neyman-Fisher factorization theorem, it follows that $\sum_{i=1}^n X_i^2$ is a sufficient statistic. The moment estimator of θ based on the sufficient statistic is the solution of the equation, $m_2' = E(X^2)$. It is a quadratic equation given by $\theta^2 + \theta - m_2' = 0$ with solution $\theta = \left(-1 \pm \sqrt{1 + 4m_2'}\right)/2$. Since $\theta > 0$, we discard the root $\theta = \left(-1 - \sqrt{1 + 4m_2'}\right)/2$ and hence the moment estimator $\tilde{\theta}_n$ of θ based on the sufficient statistic is $\tilde{\theta}_n = \left(-1 + \sqrt{1 + 4m_2'}\right)/2$. To find the maximum likelihood estimator, from the likelihood of θ as specified above we get the likelihood equation and the second derivative of the log likelihood at the solution as follows:

$$\frac{\partial}{\partial\theta} \log L_n(\theta|\underline{X}) = -\frac{n}{2\theta} + \frac{1}{2\theta^2}\sum_{i=1}^n X_i^2 - \frac{n}{2} = 0 \Leftrightarrow \theta^2 + \theta - m_2' = 0$$

$$\& \quad \frac{\partial^2}{\partial \theta^2} \log L_n(\theta | \underline{X}) = \frac{n}{2\theta^2} - \frac{\sum\limits_{i=1}^{n} X_i^2}{\theta^3} = -\frac{1}{\theta}\left[-\frac{n}{2\theta} + \frac{\sum\limits_{i=1}^{n} X_i^2}{2\theta^2} + \frac{\sum\limits_{i=1}^{n} X_i^2}{2\theta^2} \right]$$

$$= -\frac{1}{\theta}\left[\frac{n}{2} + \frac{\sum\limits_{i=1}^{n} X_i^2}{2\theta^2} \right] \quad \text{at the solution of likelihood equation}$$

$$< 0, \quad \forall \; \theta > 0.$$

Thus, the maximum likelihood estimator $\hat{\theta}_n$ of θ is given by
$\hat{\theta}_n = (-1 + \sqrt{1 + 4m_2'})/2$, which is the same as the moment estimator based on the sufficient statistic. The information function $I(\theta)$ is given by

$$n I(\theta) = E\left(-\frac{\partial^2}{\partial \theta^2} \log L_n(\theta | \underline{X}) \right) = -\frac{n}{2\theta^2} + \frac{n(\theta + \theta^2)}{\theta^3} = \frac{n(1 + 2\theta)}{2\theta^2}.$$

Thus, $\tilde{\theta}_n = \hat{\theta}_n$ is CAN for θ with approximate variance
$1/n I(\theta) = 2\theta^2/n(1 + 2\theta)$. We can also prove that $\tilde{\theta}_n = \hat{\theta}_n$ is CAN for θ using the expression for $\hat{\theta}_n$ and the delta method. From Theorem 2.2.5, we have $m_2' \xrightarrow{P_\theta} \mu_2'(\theta) = \theta + \theta^2$. Hence,

$$\hat{\theta}_n = \frac{-1 + \sqrt{1 + 4m_2'}}{2} \xrightarrow{P_\theta} \frac{-1 + \sqrt{1 + 4(\theta + \theta^2)}}{2} = \theta.$$

To prove that it is CAN, we use the delta method. From Example 3.3.2, we have $\underline{T}_n = (m_1', m_2')'$ is CAN for $(\theta, \theta^2 + \theta)' = \phi$, say, with approximate dispersion matrix Σ/n, where Σ is given by

$$\Sigma = \begin{pmatrix} \theta & 2\theta^2 \\ 2\theta^2 & 2\theta^2 + 4\theta^3 \end{pmatrix}.$$

We now define a function $g : \mathbb{R}^2 \to \mathbb{R}$ such that $g(x_1, x_2) = (-1 + \sqrt{1 + 4x_2})/2$. Its partial derivatives are given by $\frac{\partial}{\partial x_1} g(x_1, x_2) = 0$ and $\frac{\partial}{\partial x_2} g(x_1, x_2) = 1/\sqrt{1 + 4x_2}$. Thus, partial derivatives exist and are continuous, hence g is a totally differentiable function. Hence by Theorem 3.3.4, $g(\underline{T}_n) = (-1 + \sqrt{1 + 4m_2'})/2 = \hat{\theta}_n$ is CAN for $g(\theta, \theta^2 + \theta) = \theta$ with approximate variance $\Delta' \Sigma \Delta/n$, where $\Delta' = (0, 1/(1 + 2\theta))$ and $\Delta' \Sigma \Delta = 2\theta^2/(1 + 2\theta)$.

$\square$

📖 Remark 4.2.1

In Example 4.2.2, we have seen that a normal $N(\theta, \theta)$ distribution with $\theta > 0$ forms a one-parameter exponential family. In the next example, we note that a normal $N(\theta, \theta^2)$ distribution $\theta > 0$ does not belong to a one-parameter exponential family, but using the WLLN and the CLT along with the delta method, it can be shown that the maximum likelihood estimator is a CAN estimator of θ.

✍️ Example 4.2.3

Suppose $X \sim N(\theta, \theta^2)$, where $\theta > 0$. Then its probability density function is given by

$$f(x, \theta) = \frac{1}{\sqrt{2\pi\theta^2}} \exp\left\{-\frac{1}{2\theta^2}(x - \theta)^2\right\} = \frac{1}{\sqrt{2\pi\theta^2}} \exp\left\{-\frac{x^2}{2\theta^2} + \frac{x}{\theta} - \frac{1}{2}\right\}.$$

In view of the two terms $-\frac{x^2}{2\theta^2}$ and $\frac{x}{\theta}$ in the exponent, both involving θ and x, $f(x, \theta)$ cannot be expressed as $f(x, \theta) = \exp\{U(\theta)K(x) + V(\theta) + W(x)\}$ and hence the family of $N(\theta, \theta^2)$ distributions, where $\theta > 0$ is not a one-parameter exponential family. Suppose $\{X_1, X_2, \ldots, X_n\}$ is a random sample from the distribution of X. To find the maximum likelihood estimator of θ, the log-likelihood of θ corresponding to the data $\underline{X} \equiv \{X_1, X_2, \ldots, X_n\}$ is given by

$$\log L_n(\theta|\underline{X}) = c - \frac{n}{2}\log\theta^2 - \frac{1}{2\theta^2}\sum_{i=1}^{n}(X_i - \theta)^2$$

$$= c - \frac{n}{2} - n\log\theta - \frac{1}{2\theta^2}\sum_{i=1}^{n}X_i^2 + \frac{1}{\theta}\sum_{i=1}^{n}X_i,$$

where c is a constant, free from θ. It is a differentiable function of θ. Hence the likelihood equation and its solution are given by

$$-\frac{n}{\theta} + \frac{\sum_{i=1}^{n}X_i^2}{\theta^3} - \frac{\sum_{i=1}^{n}X_i}{\theta^2} = 0 \Leftrightarrow \theta^2 + \theta m_1' - m_2' = 0$$

$$\Rightarrow \theta = \frac{-m_1' \pm \sqrt{m_1'^2 + 4m_2'}}{2}.$$

Since $\theta > 0$, we discard the negative root. The second order partial derivative of the log-likelihood and its value at the positive root of the likelihood equation are given by

$$\frac{\partial^2}{\partial \theta^2} \log L_n(\theta|\underline{X}) = \frac{n}{\theta^2} - \frac{3 \sum_{i=1}^n X_i^2}{\theta^4} + \frac{2 \sum_{i=1}^n X_i}{\theta^3}$$

$$= \frac{n}{\theta^4}(\theta^2 + 2m_1'\theta - 3m_2')$$

$$= \frac{n}{\theta^4}(\theta^2 + m_1'\theta - m_2' + m_1'\theta - 2m_2')$$

$$= \frac{n}{\theta^4}(0 + m_1'\theta - m_2' - m_2') \quad \text{from the likelihood equation}$$

$$= \frac{n}{\theta^4}(-\theta^2 - m_2') \quad \text{from the likelihood equation}$$

$$< 0.$$

Thus, at the positive root of the likelihood equation, the second order partial derivative of the log-likelihood is negative. Hence, $\hat{\theta}_n = (-m_1' + \sqrt{m_1'^2 + 4m_2'})/2$ is the maximum likelihood estimator of θ. It is to be noted that form of this estimator is similar to the maximum likelihood estimator of θ for $N(\theta, \theta)$, $\theta > 0$ family in Example 4.2.2. We now examine whether it is CAN. Consistency of $\hat{\theta}_n$ follows immediately from the consistency of raw moments for the corresponding population raw moments and the fact that convergence in probability is closed under all arithmetic operations. To examine whether $\hat{\theta}_n = (-m_1' + \sqrt{m_1'^2 + 4m_2'})/2$ is asymptotically normal, we use Theorem 3.3.2 and an appropriate transformation. From Theorem 3.3.2, we have $\underline{T}_n = (m_1', m_2')'$ is CAN for $\underline{\phi} = (\mu_1', \mu_2')' = (\theta, 2\theta^2)'$ with approximate dispersion matrix Σ/n where Σ is given by

$$\Sigma = \begin{pmatrix} \mu_2' - (\mu_1')^2 & \mu_3' - \mu_1'\mu_2' \\ \mu_3' - \mu_1'\mu_2' & \mu_4' - (\mu_2')^2 \end{pmatrix} = \begin{pmatrix} \theta^2 & 2\theta^3 \\ 2\theta^3 & 6\theta^4 \end{pmatrix},$$

with $(\mu_1', \mu_2')' = (\theta, 2\theta^2)'$. We have obtained the third and fourth raw moments for $N(\mu, \sigma^2)$ distribution in Example 3.3.2. From those expressions we have $\mu_3' = \mu^3 + 3\mu\sigma^2 = \theta^3 + 3\theta^3 = 4\theta^3$ and $\mu_4' = 3\sigma^4 + \mu^4 + 6\mu^2\sigma^2 = 10\theta^4$. We further define a transformation $g : \mathbb{R}^2 \to \mathbb{R}$ as $g(x_1, x_2)=(-x_1 + \sqrt{x_1^2 + 4x_2})/2$. Then

$$\frac{\partial}{\partial x_1} g(x_1, x_2) = \frac{1}{2}\left(-1 + \frac{x_1}{\sqrt{x_1^2 + 4x_2}}\right) \quad \& \quad \frac{\partial}{\partial x_2} g(x_1, x_2) = \frac{1}{\sqrt{x_1^2 + 4x_2}}.$$

These partial derivatives are continuous and hence g is a totally differentiable function. The gradient vector Δ evaluated at $(\theta, 2\theta^2)'$ is given by $\Delta = [-1/3, 1/3\theta]'$. Hence, by Theorem 3.3.4,

$g(m_1', m_2')' = (-m_1' + \sqrt{m_1'^2 + 4m_2'})/2 = \hat{\theta}_n$ is CAN for $g(\theta, 2\theta^2)' = \theta$ with approximate variance $\Delta'\Sigma\Delta/n$, where $\Delta'\Sigma\Delta = \theta^2/3 > 0$ $\forall$ $\theta > 0$. It is to noted that

$$nI(\theta) = E_\theta\left(-\frac{\partial^2}{\partial\theta^2}\log L_n(\theta|\underline{X})\right) = n\left(-\frac{1}{\theta^2} + \frac{6\theta^2}{\theta^4} - \frac{2\theta}{\theta^3}\right) = \frac{3n}{\theta^2}.$$

Thus, $\hat{\theta}_n$ is CAN for $\theta \in (0, \infty)$ with approximate variance $\theta^2/3n = 1/nI(\theta)$. $\square$

✒ Example 4.2.4

Suppose a random variable X has a power series distribution with probability mass function as $p(x, \theta) = P_\theta[X = x] = a_x\theta^x/A(\theta)$, $x = 0, 1, 2, \ldots$, where $A(\theta) \neq 0$ is a norming constant and the parameter space Θ is such that $A(\theta)$ is a convergent series and $p(x, \theta) \geq 0$. Suppose $\underline{X} \equiv \{X_1, X_2, \ldots, X_n\}$ is a random sample from the distribution of X. To find a sufficient statistic, the likelihood of θ corresponding to the given random sample is given by

$$L_n(\theta|\underline{X}) = \left(\prod_{i=1}^n a_{X_i}\right)\theta^{\sum_{i=1}^n X_i}/(A(\theta))^n .$$

Hence by the Neyman-Fisher factorization theorem, $\sum_{i=1}^n X_i$ is a sufficient statistic for the family of power series distributions. Thus, the moment estimator of θ based on a sufficient statistic is a solution of the equation $\overline{X}_n = \sum_{i=1}^n X_i/n = E(X)$. However, we do not have the explicit form of the probability mass function and hence $E(X)$ cannot be evaluated using the formula $E(X) = \sum_{i=1}^\infty x_i p(x_i, \theta)$. To find $E(X)$, we use the following theorem for the power series.

Theorem: If the power series expansion $f(x) = \sum_{n\geq 1} a_n x^n$ is valid in an open interval $(-r, r)$ then for every $x \in (-r, r)$ the derivative $f'(x)$ exists and is given by the power series expansion $f'(x) = \sum_{n\geq 1} n a_n x^{n-1}$. (Apostol [3], p. 448). As a corollary to this theorem, f has derivatives of every order and these can be obtained by repeated differentiation, term by term of the power series. Thus,

$$\sum_{i=1}^\infty p(x_i, \theta) = 1 \quad \Leftrightarrow \quad \sum_{i=1}^\infty a_x\theta^x = A(\theta)$$

is a power series, hence is differentiable any number of times with respect to θ. As a consequence,

$$\sum_{i=1}^\infty p(x_i, \theta) = 1 \Rightarrow \frac{\partial}{\partial\theta}\sum_{i=1}^\infty p(x_i, \theta) = 0 \Rightarrow \sum_{i=1}^\infty \frac{\partial}{\partial\theta}p(x_i, \theta) = 0$$

$$\Rightarrow \sum_{i=1}^\infty \left(\frac{\partial}{\partial\theta}\log p(x_i, \theta)\right)p(x_i, \theta) = 0$$

$$\Rightarrow E\left(\frac{\partial}{\partial\theta}\log p(X,\theta)\right) = 0$$

$$\Rightarrow E\left(\frac{-A'(\theta)}{A(\theta)} + \frac{X}{\theta}\right) = 0$$

$$\Rightarrow E(X) = \frac{\theta A'(\theta)}{A(\theta)} = \eta(\theta), \quad \text{say}.$$

Thus, the moment equation is given by $\overline{X}_n = \eta(\theta)$. To investigate whether this equation has a solution, we proceed as follows. We have

$$I(\theta) = E\left(-\frac{\partial^2}{\partial\theta^2}\log p(X,\theta)\right)$$

$$= E\left(\frac{X}{\theta^2} + \frac{A(\theta)A''(\theta) - (A'(\theta))^2}{(A(\theta))^2}\right)$$

$$= \frac{\theta A'(\theta)}{A(\theta)\theta^2} + \frac{A(\theta)A''(\theta) - (A'(\theta))^2}{(A(\theta))^2}$$

$$= \frac{A(\theta)A'(\theta) + \theta A(\theta)A''(\theta) - \theta(A'(\theta))^2}{\theta(A(\theta))^2}.$$

Now observe that $\eta(\theta) = \theta A'(\theta)/A(\theta)$. Since $A(\theta)$ is differentiable any number of times with respect to θ, $\eta(\theta)$ is also differentiable any number of times. We have

$$\eta'(\theta) = \frac{A(\theta)A'(\theta) + \theta A(\theta)A''(\theta) - \theta(A'(\theta))^2}{(A(\theta))^2} = \theta I(\theta).$$

Further, $I(\theta) > 0$ and we can assume $\theta \neq 0$, as if $\theta = 0$ then the distribution of X is degenerate at 0. Thus if $\theta > 0$, then $\eta'(\theta) > 0$ and if $\theta < 0$, then $\eta'(\theta) < 0$ which implies that $\eta'(\theta) \neq 0 \ \forall \ \theta \in \Theta$. Hence, by the inverse function theorem, a unique η^{-1} exists. Since $\eta(\theta)$ is differentiable, η^{-1} is also differentiable and hence continuous. Thus, the moment equation $\overline{X}_n = \eta(\theta)$ has a unique solution and the moment estimator $\tilde{\theta}_n$ based on the sufficient statistic is given by $\tilde{\theta}_n = \eta^{-1}(\overline{X}_n)$. To examine whether $\tilde{\theta}_n$ is CAN, observe that $E(X) = \eta(\theta) < \infty$ as $A(\theta) \neq 0$ and hence by Khintchine's WLLN, $\overline{X}_n = \sum_{i=1}^{n} X_i/n \to E(X) = \eta(\theta)$ as $n \to \infty$, $\forall \ \theta \in \Theta$. Thus, $\overline{X}_n$ is consistent for $\eta(\theta)$ and by the invariance of consistency under continuous transformation $\tilde{\theta}_n = \eta^{-1}(\overline{X}_n)$ is consistent for θ. To find its asymptotic distribution with suitable normalization, we first find the variance of X using the following formula for $I(\theta)$. We have

$$I(\theta) = E\left(\frac{\partial}{\partial\theta}\log p(X,\theta)\right)^2 = E\left(\frac{X}{\theta} - \frac{A'(\theta)}{A(\theta)}\right)^2$$

$$= \frac{1}{\theta^2}E\left(X - \frac{\theta A'(\theta)}{A(\theta)}\right)^2 = \frac{1}{\theta^2}E(X - E(X))^2 = \frac{1}{\theta^2}Var(X).$$

Thus, $Var(X) = \theta^2 I(\theta)$ and it is positive and finite. Hence, by the CLT

$$\frac{\sum_{i=1}^{n} X_i - n\eta(\theta)}{\sqrt{n\theta^2 I(\theta)}} \xrightarrow{L} Z \sim N(0,1) \quad \Leftrightarrow \quad \sqrt{n}(\overline{X}_n - \eta(\theta)) \xrightarrow{L} Z_1 \sim N(0, \sigma(\theta)) \,,$$

where $\sigma(\theta) = \theta^2 I(\theta)$. Thus, $\overline{X}_n$ is CAN for $\eta(\theta) = \phi$, say with approximate variance $\sigma(\theta)/n = \sigma_1(\phi)/n$, say. To find a CAN estimator for θ, we use the delta method. Suppose $g(\phi) = \eta^{-1}(\phi)$ then $g(\phi) = \eta^{-1}(\phi) = \eta^{-1}(\eta(\theta)) = \theta$. Further, $\eta(\theta)$ is differentiable any number of times and by the inverse function theorem $g(\phi) = \eta^{-1}(\phi)$ is a differentiable function. Now,

$$g'(\phi) = \frac{d}{d\phi}\eta^{-1}(\phi) = \frac{d}{d\eta(\theta)}\eta^{-1}(\eta(\theta)) = \frac{d\theta}{d\eta(\theta)} \,.$$

Now, as shown in Theorem 4.2.1, we have

$$\frac{d\theta}{d\eta(\theta)} = (\eta'(\theta))^{-1} = \frac{1}{\theta I(\theta)} \neq 0 \text{ as } \theta \neq 0 \ \& \ I(\theta) < \infty$$

$$\Rightarrow \quad g'(\phi) = \frac{d\theta}{d\eta(\theta)} \neq 0 \ \forall \ \theta \in \Theta$$

and hence for all ϕ. Hence, by the delta method, $g(\overline{X}_n) = \eta^{-1}(\overline{X}_n) = \tilde{\theta}_n$ is CAN for $g(\phi) = \theta$ and its approximate variance is given by

$$\frac{1}{n}\sigma_1(\phi)(g'(\phi))^2 = \frac{1}{n}\theta^2 I(\theta)\left(\frac{1}{\theta I(\theta)}\right)^2 = \frac{1}{nI(\theta)} \,.$$

To find the maximum likelihood estimator, from the likelihood as specified above, we have

$$\frac{\partial}{\partial\theta}\log L_n(\theta|\underline{X}) = \frac{-nA'(\theta)}{A(\theta)} + \frac{\sum_{i=1}^{n} X_i}{\theta} \text{ and}$$

$$\frac{\partial^2}{\partial\theta^2}\log L_n(\theta|\underline{X}) = \frac{-nA(\theta)A''(\theta) + n(A'(\theta))^2}{(A(\theta))^2} - \frac{\sum_{i=1}^{n} X_i}{\theta^2} \,.$$

Thus, the likelihood equation is given by $\overline{X}_n = \eta(\theta)$ with its solution as $\theta = \eta^{-1}(\overline{X}_n)$. To claim it to be the maximum likelihood estimator, we examine whether the second derivative of the log-likelihood is negative at the solution of the likelihood equation. Observe that at $\overline{X}_n = \eta(\theta)$, that is at $\theta = \eta^{-1}(\overline{X}_n)$

$$\frac{\partial^2}{\partial\theta^2}\log L_n(\theta|\underline{X}) = \frac{-nA(\theta)A''(\theta) + n(A'(\theta))^2}{(A(\theta))^2} - \frac{n\theta A'(\theta)}{\theta^2 A(\theta)}$$

$$= \frac{-nA(\theta)A'(\theta) - n\theta A(\theta)A''(\theta) + n\theta(A'(\theta))^2}{\theta(A(\theta))^2}$$

$$= -nI(\theta) = -nI(\eta^{-1}(\overline{X}_n)) < 0 \ a.s.$$

Hence, $\hat{\theta}_n = \eta^{-1}(\overline{X}_n)$ is the maximum likelihood estimator of θ. It is the same as the moment estimator $\tilde{\theta}_n = \eta^{-1}(\overline{X}_n)$ based on the sufficient statistic and hence is CAN for θ with approximate variance $1/nI(\theta)$.   □

Thus, for the power series distribution with indexing parameter θ, the moment estimator of θ based on a sufficient statistic is the same as the maximum likelihood estimator of θ and it is CAN for θ with approximate variance $1/nI(\theta)$. This result is the same as that for a one-parameter exponential family. In fact, the family of power series distributions, which are discrete distributions, is also a one-parameter exponential family as shown below. Suppose the random variable X has a power series distribution. Then its probability mass function is given by $p(x, \theta) = P_\theta[X = x] = a_x \theta^x / A(\theta), \ x = 0, 1, 2, \ldots$, where $A(\theta) \neq 0$ is a norming constant and Θ is such that $A(\theta)$ is a convergent series and $p(x, \theta) \geq 0$. Some of the a_x may be zero. The probability mass function $p(x, \theta)$ can be rewritten as follows:

$$\log p(x, \theta) = \log a_x + x \log \theta - \log A(\theta) = U(\theta)K(x) + V(\theta) + W(x),$$

where $U(\theta) = \log \theta$, $K(x) = x$, $W(x) = \log a_x$ and $V(\theta) = -\log A(\theta)$. Further U is differentiable function of θ and can be differentiated any number of times and $U'(\theta) = 1/\theta \neq 0$. Differentiability of $V(\theta)$ follows from the theorem for power series which states that the power series has derivatives of every order and those can be obtained by repeated differentiation, term by term of the power series. To prove that 1 and $k(x) = x$ are linearly independent, consider $a + bx = 0$, if $x = 0$ then $a = 0$ which implies $bx = 0$. If we take $x = 1$, then $b = 0$. Thus, 1 and $k(x) = x$ are linearly independent. The parameter space is an open set $(-r, r)$, where r is a radius of convergence of the series $A(\theta)$. Support of X is $0, 1, 2, \ldots$, which is free from θ. Thus, the family of power series distributions is a one-parameter exponential family.

Many standard discrete distributions, such as Poisson, binomial, geometric, negative binomial, logarithmic series distribution and their truncated versions belong to the class of power series distributions. For example, suppose X follows truncated Poisson $Poi(\theta)$ distribution, truncated at 0, $\theta > 0$. The probability mass function of the Poisson $Poi(\theta)$ distribution, truncated at 0 is given by

$$f(x, \theta) = P_\theta[X = x] = \frac{e^{-\theta}}{(1 - e^{-\theta})} \frac{\theta^x}{x!} = \frac{a_x \theta^x}{A(\theta)}, \quad x = 1, 2, \ldots,$$

where $a_x = 0$ for $x = 0$ and $a_x = 1/x!$ for $x = 1, 2, \ldots$, and $A(\theta) = (1 - e^{-\theta})/e^{-\theta} = e^\theta - 1$. Thus, Poisson $Poi(\theta)$ distribution truncated at 0 is a power series distribution. In Example 4.2.1, it is shown that the maximum likelihood estimator of θ is the same as the moment estimator of θ based on a sufficient statistic and is CAN for θ with approximate variance $1/nI(\theta)$.

We now proceed to extend the results of a one-parameter exponential family to a multiparameter exponential family. We first define it and state some of its important properties. For the proof of these we refer to Lehmann and Romano [1].

Suppose X is a random variable or a random vector with probability law $f(x, \underline{\theta})$, which is either a probability density function or a probability mass function. It is indexed by a vector parameter $\underline{\theta} \in \Theta \subset \mathbb{R}^k$. The distribution of X is said to belong to a k-parameter exponential family if the following conditions are satisfied.

(i) The parameter space Θ contains an open rectangle of dimension k, which is satisfied if Θ is an open set.
(ii) The support S_f of $f(x, \underline{\theta})$ is free from $\underline{\theta}$.
(iii) The probability law $f(x, \underline{\theta})$ is expressible as

$$f(x, \underline{\theta}) = \exp\left\{\sum_{i=1}^k U_i(\underline{\theta})K_i(x) + V(\underline{\theta}) + W(x)\right\},$$

where U_i and V are functions of $\underline{\theta}$ only and K_i and W are functions of x only, $i = 1, 2, \ldots, k$.
(iv) U_i, $i = 1, 2, \ldots, k$ have continuous partial derivatives with respect to $\theta_1, \theta_2, \ldots, \theta_k$ and

$$|J| = \left\|\left[\frac{dU_i}{d\theta_j}\right]\right\| \neq 0.$$

(v) The functions 1 and $K_i(x)$ $i = 1, 2, \ldots, k$ are linearly independent, that is, $l_0 + l_1 K_1(x) + l_2 K_2(x) + \cdots + l_k K(x) = 0 \Rightarrow l_i = 0, \forall i = 0, 1, 2, \ldots, k$. This condition implies that $1, K_1(x), K_2(x), \ldots, K_k(x)$ are not functionally related to each other and it further implies that $V(\underline{\theta}), U_1(\underline{\theta}), \ldots, U_k(\underline{\theta})$ are also not functionally related to each other.

The fifth condition which states that $V(\underline{\theta}), U_1(\underline{\theta}), \ldots, U_k(\underline{\theta})$ are also not functionally related to each other is useful in the proof of Theorem 4.2.2.

The distributions belonging to a multiparameter exponential family satisfy the following properties. Suppose the dimension of the parameter space is k.

(i) If $\{X_1, X_2, \ldots, X_n\}$ is a random sample from the distribution belonging to a k-parameter exponential family, then $\{\sum_{r=1}^n K_i(X_r) \ i = 1, 2, \ldots, k\}$ is a minimal sufficient statistic for $\underline{\theta}$.
(ii) U_i, $i = 1, 2, \ldots, k$ and V have partial derivatives up to second order with respect to $\theta_i's$.
(iii) The identity $\int_{S_f} f(x, \underline{\theta})dx = 1$ can be differentiated with respect to $\theta_i's$ under the integral sign at least twice. As a consequence,

$E_{\underline{\theta}} \left(\frac{\partial}{\partial \theta_i} \log f(X, \underline{\theta}) \right) = 0$, $i = 1, 2, \ldots, k$ and the information matrix
$I(\underline{\theta}) = \left[I_{ij}(\underline{\theta}) \right]$ is given by

$$I_{ij}(\underline{\theta}) = E_{\underline{\theta}} \left(\frac{\partial}{\partial \theta_i} \log f(X, \underline{\theta}) \frac{\partial}{\partial \theta_j} \log f(X, \underline{\theta}) \right)$$

$$= E_{\underline{\theta}} \left(-\frac{\partial^2}{\partial \theta_i \partial \theta_j} \log f(X, \underline{\theta}) \right), \quad i, j = 1, 2, \ldots, k.$$

Further, $I(\underline{\theta})$ is also a dispersion matrix as

$$I_{ij}(\underline{\theta}) = Cov \left(\frac{\partial}{\partial \theta_i} \log f(X, \underline{\theta}), \frac{\partial}{\partial \theta_j} \log f(X, \underline{\theta}) \right)$$

and it is a positive definite matrix.

By the inverse function theorem, the condition that $|J| = \left| \left[\frac{dU_i}{d\theta_j} \right] \right| \neq 0$ implies that
$U_i(\underline{\theta})$, $i = 1, 2, \ldots, k$ are one-to-one functions of $\{\theta_1, \theta_2, \ldots, \theta_k\}$ and are invertible. Hence, if we relabel $U_i(\underline{\theta}) = \phi_i$, $i = 1, 2, \ldots, k$, then $\{\theta_1, \theta_2, \ldots, \theta_k\}$ can be uniquely expressed in terms of $\underline{\phi} = \{\phi_1, \phi_2, \ldots, \phi_k\}$. With such relabeling, the probability law $f(x, \underline{\theta}) = \exp \left\{ \sum_{i=1}^{k} U_i(\underline{\theta}) K_i(x) + V(\underline{\theta}) + W(x) \right\}$ is expressible as

$$f(x, \underline{\phi}) = \exp \left\{ \sum_{i=1}^{k} \phi_i K_i(x) + V_1(\underline{\phi}) + W(x) \right\} = \beta(\underline{\phi}) \exp \left\{ \sum_{i=1}^{k} \phi_i K_i(x) \right\} g(x),$$
$$(4.2.5)$$

where $\beta(\underline{\phi}) = \exp(V_1(\underline{\phi}))$ and $g(x) = \exp(W(x))$. The representation of $f(x, \underline{\phi})$ as in (4.2.5) is known as a canonical representation of a k-parameter exponential family and $\{\phi_1, \phi_2, \ldots, \phi_k\}$ are known as natural parameters. The condition that 1 and $K_i(x) \, i = 1, 2, \ldots, k$ are linearly independent implies that there is no functional relation among $\{V_1(\underline{\phi}), \phi_1, \phi_2, \ldots, \phi_k\}$, in particular there is no functional relation between ϕ_i and ϕ_j. As a consequence,

$$\frac{\partial \phi_i}{\partial \phi_j} = \begin{cases} 1, & \text{if} \quad i = j \\ 0, & \text{if} \quad i \neq j \end{cases}$$

We use this result in the proof of Theorem 4.2.2.

There is an interesting relation between information matrices $I(\underline{\theta})$ and $I(\underline{\phi})$, when the probability law $f(x, \underline{\theta})$ is expressed in a general form and when the same is expressed in a canonical form. We derive it below. We first find it for $k = 1$. Suppose

$$f(x, \phi) = \beta(\phi) \exp\{\phi K(x)\} g(x) \quad \& \quad f(x, \theta) = \beta_1(\theta) \exp\{U(\theta) K(x)\} W_1(x),$$
$$\text{where} \quad \phi = U(\theta).$$

By the chain rule,

$$\frac{\partial \log f(x, \theta)}{\partial \theta} = \frac{\partial \log f(x, \phi)}{\partial \phi} \frac{\partial \phi}{\partial \theta}$$

$$\Rightarrow E_\theta \left(\frac{\partial \log f(X, \theta)}{\partial \theta} \right)^2 = \left(\frac{\partial \phi}{\partial \theta} \right)^2 E_\phi \left(\frac{\partial \log f(x, \phi)}{\partial \phi} \right)^2$$

$$\Rightarrow I(\theta) = (U'(\theta))^2 I(\phi) .$$

The identity $I(\theta) = (U'(\theta))^2 I(\phi)$ has following nice interpretation. The Cramér-Rao lower bound for the variance of an unbiased estimator of ϕ is $1/nI(\phi)$ whereas the Cramér-Rao lower bound for the variance of an unbiased estimator of $U(\theta)$ is $(U'(\theta))^2/nI(\theta)$. Here $\phi = U(\theta)$ and hence,

$$\frac{1}{nI(\phi)} = \frac{(U'(\theta))^2}{nI(\theta)} \quad \Leftrightarrow \quad I(\theta) = (U'(\theta))^2 I(\phi).$$

The identity $I(\theta) = (U'(\theta))^2 I(\phi)$ is extended for vector parameter $\underline{\theta}$ and $\underline{\phi}$ as follows. Suppose in a k-parameter exponential family, the probability law

$$f(x, \underline{\theta}) = \exp \left\{ \sum_{i=1}^{k} U_i(\underline{\theta}) K_i(x) + V(\underline{\theta}) + W(x) \right\} \quad \&$$

$$f(x, \underline{\phi}) = \beta(\underline{\phi}) \exp \left\{ \sum_{i=1}^{k} \phi_i K_i(x) \right\} g(x) ,$$

where $\underline{\phi} = (U_1(\underline{\theta}), U_2(\underline{\theta}), \ldots, U_k(\underline{\theta}))'$. By the chain rule,

$$\frac{\partial \log f(x, \underline{\theta})}{\partial \theta_i} = \sum_{r=1}^{k} \frac{\partial \log f(x, \underline{\phi})}{\partial \phi_r} \frac{\partial \phi_r}{\partial \theta_i} \quad \& \quad \frac{\partial \log f(x, \underline{\theta})}{\partial \theta_j} = \sum_{s=1}^{k} \frac{\partial \log f(x, \underline{\phi})}{\partial \phi_s} \frac{\partial \phi_s}{\partial \theta_j} .$$

Suppose a matrix $J_{k \times k}$ is defined as $J = \left[\frac{\partial \phi_i}{\partial \theta_j} \right] = \left[\frac{\partial U_i(\underline{\theta})}{\partial \theta_j} \right]$. From the conditions of a k-parameter exponential family we know that $|J| \neq 0$. Thus,

$$I_{ij}(\underline{\theta}) = E_{\underline{\theta}} \left(\frac{\partial \log f(X, \underline{\theta})}{\partial \theta_i} \frac{\partial \log f(X, \underline{\theta})}{\partial \theta_j} \right)$$

$$= \sum_{r=1}^{k} \sum_{s=1}^{k} \frac{\partial \phi_r}{\partial \theta_i} \frac{\partial \phi_s}{\partial \theta_j} E_\phi \left(\frac{\partial \log f(X, \underline{\phi})}{\partial \phi_r} \frac{\partial \log f(X, \underline{\phi})}{\partial \phi_s} \right)$$

$$= \sum_{r=1}^{k} \sum_{s=1}^{k} \frac{\partial \phi_r}{\partial \theta_i} \frac{\partial \phi_s}{\partial \theta_j} I_{rs}(\underline{\phi})$$

$$= \sum_{r=1}^{k} \sum_{s=1}^{k} (i, r)\text{-th element of } J' \times (r, s)\text{-th element of}$$

$I(\underline{\phi}) \times (s, j)\text{-th element of } J$

$= (i, j)\text{-th element of } J'I(\underline{\phi})J .$

Thus, we have the identity $I(\underline{\theta}) = J'I(\underline{\phi})J$, which is analogous to that for $k = 1$.

In the following example, we illustrate how to express the two-parameter exponential family in a canonical form and also verify the relation derived above between the information matrices.

✒ Example 4.2.5

Suppose $(X, Y)'$ has a bivariate normal distribution with zero mean vector and dispersion matrix Σ given by

$$\Sigma = \sigma^2 \begin{pmatrix} 1 & \rho \\ \rho & 1 \end{pmatrix},$$

$\sigma^2 > 0,\ -1 < \rho < 1$. We examine whether the distribution belongs to a two-parameter exponential family and then express the probability law in a canonical form. The probability density function $f(x, y, \sigma^2, \rho)$ of $\underline{Z} = (X, Y)'$ is given by

$$f(x, y, \sigma^2, \rho) = \frac{1}{2\pi\sigma^2\sqrt{1-\rho^2}} \exp\left\{-\frac{1}{2\sigma^2(1-\rho^2)}(x^2 - 2\rho xy + y^2)\right\},$$

$$(x, y)' \in \mathbb{R}^2,$$

$\sigma^2 > 0,\ -1 < \rho < 1$. (i) It is to be noted that the parameter space Θ given by $\Theta = \{(\sigma^2, \rho)' | \sigma^2 > 0, -1 < \rho < 1\}$ is an open set. (ii) The support of $\underline{Z} = (X, Y)'$ is $\mathbb{R}^2$ which does not depend on the parameters. (iii) The probability density function $f(x, y, \sigma^2, \rho)$ can be rewritten as follows:

$$\begin{aligned}
\log f(x, y, \sigma^2, \rho) &= -\log 2\pi - \log \sigma^2 \\
&\quad - \frac{1}{2}\log(1 - \rho^2) - \frac{1}{2\sigma^2(1-\rho^2)}\left((x^2 + y^2) - 2\rho xy\right) \\
&= U_1(\sigma^2, \rho)K_1(x, y) + U_2(\sigma^2, \rho)K_2(x, y) \\
&\quad + V(\sigma^2, \rho) + W(x, y),
\end{aligned}$$

where $U_1(\sigma^2, \rho) = -\frac{1}{2\sigma^2(1-\rho^2)}$, $K_1(x, y) = x^2 + y^2$, $U_2(\sigma^2, \rho) = \frac{\rho}{\sigma^2(1-\rho^2)}$, $K_2(x, y) = xy$, $V(\sigma^2, \rho) = -\log \sigma^2 - \frac{1}{2}\log(1-\rho^2)$ and $W(x, y) = -\log 2\pi$.
(iv) The matrix $J = \left[\frac{dU_i}{d\theta_j}\right]$ of partial derivatives, where $\theta_1 = \sigma^2$ and $\theta_2 = \rho$, is given by

$$J = \begin{pmatrix} \frac{1}{2\sigma^4(1-\rho^2)} & -\frac{\rho}{\sigma^2(1-\rho^2)^2} \\ -\frac{\rho}{\sigma^4(1-\rho^2)} & \frac{1+\rho^2}{\sigma^2(1-\rho^2)^2} \end{pmatrix}.$$

It is clear that U_1 and U_2 have continuous partial derivatives with respect to σ^2 and ρ and $|J| = \left\| \left[\frac{dU_i}{d\theta_j} \right] \right\| = \frac{1}{2\sigma^6(1-\rho^2)^2} \neq 0$. (v) To examine whether functions $1, x^2 + y^2$ and xy are linearly independent, suppose $g(x, y) = l_1(x^2 + y^2) + l_2 xy + l_3 = 0$. Then

$$\frac{\partial}{\partial x} g(x, y) = 2xl_1 + l_2 y = 0 \quad \& \quad \frac{\partial^2}{\partial x^2} g(x, y) = 2l_1 = 0$$

$$\Rightarrow \quad l_1 = 0 \quad \Rightarrow \quad l_2 = 0$$

from $\frac{\partial}{\partial x} g(x, y) = 2xl_1 + l_2 = 0$. Now $g(x, y) = l_1(x^2 + y^2) + l_2 xy + l_3 = 0$ implies $l_3 = 0$. Thus, a bivariate normal distribution with zero mean vector and dispersion matrix Σ, satisfies all the requirements of a two-parameter exponential family and hence belongs to a two-parameter exponential family. The condition that $|J| \neq 0$ implies that $U_i(\sigma^2, \rho)$, $i = 1, 2$ are one-to-one functions of $(\sigma^2, \rho)'$ and are invertible. Hence, we define

$$U_1(\sigma^2, \rho) = \phi_1 = \frac{-1}{2\sigma^2(1 - \rho^2)} \quad \&$$

$$U_2(\sigma^2, \rho) = \phi_2 = \frac{\rho}{\sigma^2(1 - \rho^2)} \quad \Leftrightarrow \quad \sigma^2 = \frac{-2\phi_1}{4\phi_1^2 - \phi_2^2} \quad \& \quad \rho = -\frac{\phi_2}{2\phi_1}.$$

With such a relabeling, the probability law $f(x, y, \sigma^2, \rho)$ is expressible as

$$f(x, y, \phi_1, \phi_2) = \beta(\phi_1, \phi_2) \exp\{\phi_1 K_1(x, y) + \phi_2 K_2(x, y)\} g(x, y) \quad (4.2.6)$$

where $\beta(\phi) = 1/\sigma^2\sqrt{1 - \rho^2}$ and $g(x, y) = 1/2\pi$. Thus, the probability law of a bivariate normal distribution, as expressed in Eq. (4.2.6), is a canonical representation of a two-parameter exponential family and $\{\phi_1, \phi_2\}$ are natural parameters. We now find the information matrices $I(\sigma^2, \rho)$ and $I(\phi_1, \phi_2)$ and verify the relation $I(\sigma^2, \rho) = J'I(\phi_1, \phi_2)J$ where J is as derived above. To find the information matrix $I(\sigma^2, \rho)$, we note that $E(K_1(X, Y)) = E(X^2 + Y^2) = 2\sigma^2$ and $E(K_2(X, Y)) = E(XY) = \sigma^2\rho$. The derivatives of $\log f(x, y, \sigma^2, \rho)$ are given below.

$$\frac{\partial}{\partial \sigma^2} \log f(x, y, \sigma^2, \rho) = -\frac{1}{\sigma^2} + \frac{K_1(X, Y)}{2(1 - \rho^2)\sigma^4} - \frac{\rho K_2(X, Y)}{(1 - \rho^2)\sigma^4}$$

$$\frac{\partial^2}{\partial (\sigma^2)^2} \log f(x, y, \sigma^2, \rho) = \frac{1}{\sigma^4} - \frac{K_1(X, Y)}{(1 - \rho^2)\sigma^6} + \frac{2\rho K_2(X, Y)}{(1 - \rho^2)\sigma^6}$$

$$\frac{\partial^2}{\partial \rho \partial \sigma^2} \log f(x, y, \sigma^2, \rho) = \frac{\rho K_1(X, Y)}{(1 - \rho^2)^2 \sigma^4} - \frac{(1 + \rho^2) K_2(X, Y)}{(1 - \rho^2)^2 \sigma^4}$$

$$\frac{\partial}{\partial \rho} \log f(x, y, \sigma^2, \rho) = \frac{\rho}{(1 - \rho^2)} - \frac{\rho K_1(X, Y)}{(1 - \rho^2)^2 \sigma^2} + \frac{(1 + \rho^2) K_2(X, Y)}{(1 - \rho^2)^2 \sigma^2}$$

$$\frac{\partial^2}{\partial \rho^2} \log f(x, y, \sigma^2, \rho) = \frac{(1 + \rho^2)}{(1 - \rho^2)^2} - \frac{K_1(X, Y)(1 + 3\rho^2)}{(1 - \rho^2)^3 \sigma^2}$$

$$+ \frac{2\rho(3 + \rho^2) K_2(X, Y)}{(1 - \rho^2)^3 \sigma^2}.$$

Hence by the definition, the information matrix $I(\sigma^2, \rho)$ is given by

$$I(\sigma^2, \rho) = \begin{pmatrix} \frac{1}{\sigma^4} & \frac{-\rho}{\sigma^2(1-\rho^2)} \\ \frac{-\rho}{\sigma^2(1-\rho^2)} & \frac{1+\rho^2}{(1-\rho^2)^2} \end{pmatrix}.$$

To find $I(\phi_1, \phi_2)$ we have
$$\log f(x, y, \phi_1, \phi_2) = \log \beta(\phi_1, \phi_2) + \phi_1 K_1(x, y) + \phi_2 K_2(x, y) + \log g(x, y)$$
where $\log \beta(\phi_1, \phi_2) = \log \left(1/\sigma^2 \sqrt{1 - \rho^2} \right) = (\log(4\phi_1^2 - \phi_2^2))/2$. Hence,

$$\frac{\partial}{\partial \phi_1} \log f(x, y, \phi_1, \phi_2) = \frac{\partial}{\partial \phi_1} \log \beta(\phi_1, \phi_2) + K_1(x, y)$$

$$= \frac{4\phi_1}{4\phi_1^2 - \phi_2^2} + K_1(x, y)$$

$$\frac{\partial^2}{\partial \phi_1^2} \log f(x, y, \phi_1, \phi_2) = \frac{\partial^2}{\partial \phi_1^2} \log \beta(\phi_1, \phi_2) = -\frac{4(4\phi_1^2 - \phi_2^2)}{(4\phi_1^2 - \phi_2^2)^2}$$

$$\frac{\partial^2}{\partial \phi_2 \partial \phi_1} \log f(x, y, \phi_1, \phi_2) = \frac{\partial^2}{\partial \phi_2 \partial \phi_1} \log \beta(\phi_1, \phi_2) = \frac{8\phi_1 \phi_2}{(4\phi_1^2 - \phi_2^2)^2}$$

$$\frac{\partial}{\partial \phi_2} \log f(x, y, \phi_1, \phi_2) = \frac{\partial}{\partial \phi_2} \log \beta(\phi_1, \phi_2) + K_2(x, y)$$

$$= \frac{-\phi_2}{4\phi_1^2 - \phi_2^2} + K_2(x, y)$$

$$\frac{\partial^2}{\partial \phi_2^2} \log f(x, y, \phi_1, \phi_2) = \frac{\partial^2}{\partial \phi_2^2} \log \beta(\phi_1, \phi_2) = -\frac{4\phi_1^2 - \phi_2^2}{(4\phi_1^2 - \phi_2^2)^2}.$$

Hence the Information matrix $I(\phi_1, \phi_2)$ is given by

$$I(\phi_1, \phi_2) = \frac{1}{(4\phi_1^2 - \phi_2^2)^2} \begin{pmatrix} 4(4\phi_1^2 - \phi_2^2) & -8\phi_1 \phi_2 \\ -8\phi_1 \phi_2 & 4\phi_1^2 - \phi_2^2 \end{pmatrix}$$

$$= \sigma^4 \begin{pmatrix} 4(1 + \rho^2) & 4\rho \\ 4\rho & 1 + \rho^2 \end{pmatrix}.$$

Now matrix J is given by

$$J = \frac{1}{\sigma^2(1-\rho^2)} \begin{pmatrix} \frac{1}{2\sigma^2} & -\frac{\rho}{1-\rho^2} \\ -\frac{\rho}{\sigma^2} & \frac{1+\rho^2}{1-\rho^2} \end{pmatrix}.$$

It then follows that $I(\sigma^2, \rho) = J'I(\phi_1, \phi_2)J$. □

Theorem 4.2.1 is extended to a multiparameter exponential family in two steps. First, we prove it for a multiparameter exponential family with a canonical representation of its probability law and later for a general form. Following theorem is an extension of Theorem 4.2.1 to a k-parameter exponential family expressed in a canonical form.

Theorem 4.2.2

Suppose the distribution of a random variable or a random vector X belongs to a k-parameter exponential family with probability law
$f(x, \underline{\phi}) = \beta(\underline{\phi}) \exp\left\{\sum_{i=1}^{k} \phi_i K_i(x)\right\} g(x)$ with indexing parameter $\underline{\phi}$. Suppose $\underline{X} = \{X_1, X_2, \ldots, X_n\}$ is a random sample from the distribution of X. Then the moment estimator of $\underline{\phi}$ based on a sufficient statistic is the same as the maximum likelihood estimator of $\underline{\phi}$ and it is CAN for $\underline{\phi}$ with approximate dispersion matrix $I^{-1}(\underline{\phi})/n$.

Proof The distribution of X belongs to a k-parameter exponential family with indexing parameter $\underline{\phi} = (\phi_1, \phi_2, \ldots, \phi_k)'$, where $\{\phi_1, \phi_2, \ldots, \phi_k\}$ are natural parameters. Hence its probability law is given by $f(x, \underline{\phi}) = \beta(\underline{\phi}) \exp\left\{\sum_{i=1}^{k} \phi_i K_i(x)\right\} g(x)$. Corresponding to a random sample $\underline{X}$ from the distribution of X, the likelihood of $\underline{\phi}$ is given by

$$L_n(\underline{\phi}|\underline{X}) = \prod_{r=1}^{n} \beta(\underline{\phi}) \exp\left\{\sum_{i=1}^{k} \phi_i K_i(x_r)\right\} g(x_r)$$

$$= (\beta(\underline{\phi}))^n \exp\left\{\sum_{r=1}^{n}\sum_{i=1}^{k} \phi_i K_i(x_r)\right\} \prod_{r=1}^{n} g(x_r). \qquad (4.2.7)$$

From the Neyman-Fisher factorization theorem, it follows that $\{\sum_{r=1}^{n} K_i(X_r), i = 1, 2, \ldots, k\}$ is jointly sufficient for the family. Thus, the system of moment equations to find moment estimator of $\underline{\phi}$ based on a sufficient statistic is given by $T_{in} = \sum_{r=1}^{n} K_i(X_r)/n = E_{\underline{\phi}}(K_i(X))$, $i = 1, 2, \ldots, k$. To find $E_{\underline{\phi}}(K_i(X))$ we use the result that $E_{\underline{\phi}}(\frac{\partial}{\partial \phi_i} \log f(X, \underline{\phi})) = 0$, $i = 1, 2, \ldots, k$. Observe that

$$\log f(x, \underline{\phi}) = \log \beta(\underline{\phi}) + \sum_{i=1}^{k} \phi_i K_i(x) + \log g(x)$$

$$\Rightarrow \frac{\partial}{\partial \phi_i} \log f(X, \underline{\phi}) = \frac{\partial}{\partial \phi_i} \log \beta(\underline{\phi}) + K_i(X).$$

Hence,

$$E_{\underline{\phi}}\left(\frac{\partial}{\partial \phi_i} \log f(X, \underline{\phi})\right) = 0$$

$$\Rightarrow E_{\underline{\phi}}(K_i(X)) = -\frac{\partial}{\partial \phi_i} \log \beta(\underline{\phi}) = h_i(\underline{\phi}), \text{ say, } i = 1, 2, \ldots, k.$$

Thus, the system of moment equations is given by $T_{in} = h_i(\underline{\phi})$, $i = 1, 2, \ldots, k$. To ensure that this system of equations has a unique solution, we verify whether $|H| = \left|\left[\frac{\partial h_i(\underline{\phi})}{\partial \phi_j}\right]\right| \neq 0$. We have

$$\frac{\partial}{\partial \phi_i} \log f(X, \underline{\phi}) = \frac{\partial}{\partial \phi_i} \log \beta(\underline{\phi}) + K_i(X)$$

$$\Rightarrow \frac{\partial^2}{\partial \phi_j \partial \phi_i} \log f(X, \underline{\phi}) = \frac{\partial^2}{\partial \phi_j \partial \phi_i} \log \beta(\underline{\phi}).$$

Now,

$$I_{ij}(\underline{\phi}) = E_{\underline{\phi}}\left(-\frac{\partial^2}{\partial \phi_j \partial \phi_i} \log f(X, \underline{\phi})\right) = -\frac{\partial^2}{\partial \phi_j \partial \phi_i} \log \beta(\underline{\phi}) = \frac{\partial}{\partial \phi_j} h_i(\underline{\phi}).$$

Thus, $H = I(\underline{\phi})$, which is positive definite and hence $|H| \neq 0$. Thus, by the inverse function theorem the system of moment equations given by $T_{in} = h_i(\underline{\phi})$, $i = 1, 2, \ldots, k$ has a unique solution given by $\phi_i = q_i(T_{1n}, T_{2n}, \ldots, T_{kn})$, say, for $i = 1, 2, \ldots, k$ and hence the moment estimator $\underline{\tilde{\phi}}_n$ of $\underline{\phi}$ based on the sufficient statistic is given by

$$\underline{\tilde{\phi}}_n = \underline{q}(\underline{T}_n), \quad \text{where} \quad \underline{T}_n = (T_{1n}, T_{2n}, \ldots, T_{kn})' \ \& \ \underline{q} = (q_1, q_2, \ldots, q_k)'.$$

To find the maximum likelihood estimator of $\underline{\phi}$, from the likelihood as given in Eq. (4.2.7), we have the system of likelihood equations as

$$\frac{\partial}{\partial \phi_i} \log L_n(\underline{\phi}|\underline{X}) = n\frac{\partial}{\partial \phi_i} \log \beta(\underline{\phi}) + \sum_{r=1}^{n} K_i(X_r) = 0$$

$$\Leftrightarrow T_{in} = h_i(\underline{\phi}), \ i = 1, 2, \ldots, k.$$

Thus, the system of likelihood equations is the same as the system of moment equations and hence it has a unique solution given by $\phi_i = q_i(T_{1n}, T_{2n}, \ldots, T_{kn})$ $i = 1, 2, \ldots, k$. To examine whether this solution gives the maximum likelihood estimator, we verify whether the matrix of second order partial derivatives of the log-likelihood is negative definite almost surely at the solution. Now,

$$\frac{\partial^2}{\partial \phi_j \partial \phi_i} \log L_n(\underline{\phi}|\underline{X}) = n\frac{\partial^2}{\partial \phi_j \partial \phi_i} \log \beta(\underline{\phi}) = -n\frac{\partial}{\partial \phi_j}\left(-\frac{\partial}{\partial \phi_i} \log \beta(\underline{\phi})\right)$$

$$= -n\frac{\partial}{\partial \phi_j}h_i(\underline{\phi}) = -nI_{ij}(\underline{\phi}) \ .$$

Hence, the matrix of second order partial derivatives of the log-likelihood is $-nI(\underline{\phi})$, which is a negative definite matrix for any $\underline{\phi}$ and hence at the solution of the system of likelihood equations. Thus, the maximum likelihood estimator $\hat{\underline{\phi}}_n$ of $\underline{\phi}$ is given by $\hat{\phi}_{in} = q_i(T_{1n}, T_{2n}, \ldots, T_{kn}), i = 1, 2, \ldots, k$. Thus, the maximum likelihood estimator of $\underline{\phi}$ is the same as the moment estimator based on a sufficient statistic and is given by

$$\tilde{\underline{\phi}}_n = \hat{\underline{\phi}}_n = \underline{q}(\underline{T}_n), \quad \text{where} \ \ \underline{T}_n = (T_{1n}, T_{2n}, \ldots, T_{kn})' \ \&$$
$$\underline{q}(\underline{T}_n) = (q_1(\underline{T}_n), q_2(\underline{T}_n), \ldots, q_k(\underline{T}_n))' \ .$$

To establish that $\tilde{\underline{\phi}}_n = \hat{\underline{\phi}}_n$ is CAN, observe that $\{X_1, X_2, \ldots, X_n\}$ are independent and identically distributed random variables implies that $\{K_i(X_1), K_i(X_2), \ldots, K_i(X_n)\}$ are also independent and identically distributed random variables, being Borel functions, for all $i = 1, 2, \ldots, k$. Hence by Khintchine's WLLN,

$$T_{in} = \frac{1}{n}\sum_{r=1}^{n} K_i(X_r) \overset{P}{\to} E_{\underline{\phi}}(K_i(X)) = h_i(\underline{\phi}), \quad \text{as } n \to \infty, \ i = 1, 2, \ldots, k, \ \forall \ \underline{\phi} \ .$$

Thus, $\underline{T}_n$ is consistent for $\underline{h} \equiv \underline{h}(\underline{\phi}) = (h_1(\underline{\phi}), h_2(\underline{\phi}), \ldots, h_k(\underline{\phi}))'$. It is known that $\beta(\underline{\phi})$ has partial derivatives up to order 2, hence $\frac{\partial}{\partial \phi_j \partial \phi_i} \log \beta(\underline{\phi}) = \frac{\partial}{\partial \phi_j}h_i(\underline{\phi})$ exists and is continuous. Thus, $\{h_1(\underline{\phi}), h_2(\underline{\phi}), \ldots, h_k(\underline{\phi})\}$ are totally differentiable functions. Hence, by the inverse function theorem, $\{q_1, q_2, \ldots, q_k\}$ are also totally differentiable functions and hence are continuous. Now we have proved that $\underline{T}_n \overset{P}{\to} \underline{h}$, hence by the invariance property of consistency under continuous transformation, we get $q_i(\underline{T}_n) \overset{P}{\to} q_i(\underline{h}) = \phi_i, \ i = 1, 2, \ldots, k$. Since marginal consistency and joint consistency are equivalent,

$$\tilde{\underline{\phi}}_n = \hat{\underline{\phi}}_n = \underline{q}(\underline{T}_n) \overset{P}{\to} \underline{q}(\underline{h}) = \underline{\phi} \ \forall \ \underline{\phi} \ .$$

Thus, $\tilde{\underline{\phi}}_n = \hat{\underline{\phi}}_n$ is a consistent estimator of $\underline{\phi}$. To find its asymptotic distribution with suitable normalization, we first find the dispersion matrix D of $\underline{U} = (K_1(X), K_2(X), \ldots, K_k(X))'$ using the following formula for $I(\underline{\phi})$. We have

$$I_{ij}(\underline{\phi}) = E_{\underline{\phi}}\left(\frac{\partial}{\partial \phi_i} \log f(X, \underline{\phi})\frac{\partial}{\partial \phi_j} \log f(X, \underline{\phi})\right)$$
$$= E_{\underline{\phi}}\left\{\left(\frac{\partial}{\partial \phi_i} \log \beta(\underline{\phi}) + K_i(X)\right)\left(\frac{\partial}{\partial \phi_j} \log \beta(\underline{\phi}) + K_j(X)\right)\right\}$$
$$= E_{\underline{\phi}}\left\{(-h_i + K_i(X))(-h_j + K_j(X))\right\}$$

$$= E_\phi \left\{ (K_i(X) - E(K_i(X)))(K_j(X) - E(K_j(X))) \right\}$$
$$= Cov(K_i(X), K_j(X)) \quad i, \ j = 1, 2, \ldots, k \ .$$

Hence, $D = I(\underline{\phi})$, which is a positive definite matrix. Thus,
$\underline{U}_r = (K_1(X_r), K_2(X_r), \ldots, K_k(X_r))', r = 1, 2, \ldots, n$ are independent and identically distributed random vectors with $E_\phi(\underline{U}) = \underline{h}$ and dispersion matrix $I(\underline{\phi})$, which is positive definite. Hence, by the multivariate CLT applied to $\{\underline{U}_1, \underline{U}_2, \ldots, \underline{U}_n\}$ we have

$$\sqrt{n}(\underline{\overline{U}}_n - \underline{h}) = \sqrt{n}(\underline{T}_n - \underline{h}) \xrightarrow{L} Z_1 \sim N_k(\underline{0}, I(\underline{\phi})) \ .$$

Thus, $\underline{T}_n$ is CAN for $\underline{h}$ with approximate dispersion matrix $I(\underline{\phi})/n$. To find the CAN estimator for $\underline{\phi}$ we use the delta method. It is known that $\{q_1, q_2, \ldots, q_k\}$ are totally differentiable functions, hence $\underline{\tilde{\phi}}_n = \underline{\hat{\phi}}_n = \underline{q}(\underline{T}_n)$ is CAN for $\underline{q}(\underline{h}) = \underline{\phi}$ with approximate dispersion matrix $MI(\underline{\phi})M'/n$, where $M = \left[\frac{\partial q_i}{\partial h_j}\right]$. Now to find the matrix $MI(\underline{\phi})M'$, note that $\phi_i = q_i(h_1, h_2, \ldots, h_k)$, hence

$$\frac{\partial \phi_i}{\partial \phi_j} = \frac{\partial q_i(h_1, h_2, \ldots, h_k)}{\partial \phi_j}$$

$$= \sum_{m=1}^{k} \frac{\partial q_i(h_1, h_2, \ldots, h_k)}{\partial h_m} \frac{\partial h_m}{\partial \phi_j} \quad \text{by chain rule}$$

$$= \sum_{m=1}^{k} (i, m)\text{-th element of } M \ \times (m, j)\text{-th element of } I(\underline{\phi})$$

$$= (i, j)\text{-th element of } MI(\underline{\phi}) \ .$$

To find the value of $\frac{\partial \phi_i}{\partial \phi_j}$, we use the property that $K_1(X), K_2(X), \ldots, K_k(X)$ and 1 are linearly independent which implies that $\{\phi_1, \phi_2, \ldots, \phi_k, 1\}$ are not functionally related to each other. Hence,

$$\frac{\partial \phi_i}{\partial \phi_j} = \begin{cases} 1, & \text{if} \quad i = j \\ 0, & \text{if} \quad i \neq j \end{cases}$$

Thus, $MI(\underline{\phi})$ is an identity matrix which implies that $M = I^{-1}(\underline{\phi})$ and hence $MI(\underline{\phi})M' = I^{-1}(\underline{\phi})$, as $I(\underline{\phi})$ is a symmetric matrix. Thus, $\underline{\tilde{\phi}}_n = \underline{\hat{\phi}}_n = \underline{q}(\underline{T}_n)$ is CAN for $\underline{q}(\underline{h}) = \underline{\phi}$ with approximate dispersion matrix $MI(\underline{\phi})M'/n = I^{-1}(\underline{\phi})/n$.
$\qquad\square$

Using the relation $I(\underline{\theta}) = J'I(\underline{\phi})J$ between information matrices, we extend Theorem 4.2.2 to a multiparameter exponential family with the probability law $f(x, \underline{\theta})$ in a general form.

Theorem 4.2.3

Suppose the distribution of a random variable or a random vector X belongs to a k-parameter exponential family with the probability law
$f(x, \underline{\theta}) = \exp\left\{\sum_{i=1}^{k} U_i(\underline{\theta})K_i(x) + V(\underline{\theta}) + W(x)\right\}$, *where $\underline{\theta}$ is an indexing parameter. Suppose $\{X_1, X_2, \ldots, X_n\}$ is a random sample from the distribution of X. Then the moment estimator of $\underline{\theta}$ based on a sufficient statistic is the same as the maximum likelihood estimator of $\underline{\theta}$ and it is CAN for $\underline{\theta}$ with approximate dispersion matrix $I^{-1}(\underline{\theta})/n$.*

Proof Since the distribution of X belongs to a k-parameter exponential family with the probability law $f(x, \underline{\theta}) = \exp\left\{\sum_{i=1}^{k} U_i(\underline{\theta})K_i(x) + V(\underline{\theta}) + W(x)\right\}$, we have $\left|\left[\frac{dU_i}{d\theta_j}\right]\right| \neq 0$. Hence by the inverse function theorem, $U_i(\underline{\theta})$ can be uniquely relabeled as $U_i(\underline{\theta}) = \phi_i$, $i = 1, 2, \ldots, k$ and the inverse exists which is continuous and totally differentiable. Suppose

$$\theta_i = p_i(\phi_1, \phi_2, \ldots, \phi_k), \ i = 1, 2, \ldots, k$$
$$\Leftrightarrow \quad \underline{\theta} = \underline{p}(\underline{\phi}) = (p_1(\underline{\phi}), p_2(\underline{\phi}), \ldots, p_k(\underline{\phi}))'.$$

Then each of $\{p_1, p_2, \ldots, p_k\}$ are continuous and totally differentiable functions from $\mathbb{R}^k$ to $\mathbb{R}$. With such a relabeling of the parameters, the probability law $f(x, \underline{\theta})$ can be expressed as $f(x, \underline{\phi}) = \beta(\underline{\phi})\exp\{\sum_{i=1}^{k}\phi_i K_i(x)\}g(x)$, which is a canonical representation of the exponential family and $\{\phi_1, \phi_2, \ldots, \phi_k\}$ are natural parameters. Hence by Theorem 4.2.2, the moment estimator $\underline{\tilde{\phi}}_n$ of $\underline{\phi}$ based on a sufficient statistic is the same as the maximum likelihood estimator $\underline{\hat{\phi}}_n$ of $\underline{\phi}$ and it is CAN for $\underline{\phi}$ with approximate dispersion matrix $I^{-1}(\underline{\phi})/n$. Now,

$$\tilde{\phi}_{in} = \hat{\phi}_{in} \ \ \forall \ i = 1, 2, \ldots, k$$
$$\Rightarrow p_i(\tilde{\phi}_{1n}, \tilde{\phi}_{2n}, \ldots, \tilde{\phi}_{kn}) = p_i(\hat{\phi}_{1n}, \hat{\phi}_{2n}, \ldots, \hat{\phi}_{kn}) \ \ \forall \ i = 1, 2, \ldots, k$$
$$\Rightarrow \tilde{\theta}_{in} = \hat{\theta}_{in} \ \ \forall \ i = 1, 2, \ldots, k$$
$$\Rightarrow \underline{\tilde{\theta}}_n = \underline{\hat{\theta}}_n.$$

$\underline{\hat{\theta}}_n$ is a maximum likelihood estimator of $\underline{\theta}$, being a function of maximum likelihood estimators $\hat{\phi}_{in}$ $i = 1, 2, \ldots, k$. Further, the system of moment equations based on a sufficient statistic is given by $U_i(\underline{\theta}) = \phi_i = q_i(\underline{T}_n)$, $i = 1, 2, \ldots, k$. Thus, the moment estimator based on a sufficient statistic is given by

$$U_i(\underline{\tilde{\theta}}_n) = \tilde{\phi}_{in} = q_i(\underline{T}_n), \ i = 1, 2, \ldots, k \ \Rightarrow \ \underline{U}(\underline{\tilde{\theta}}_n) = \underline{\tilde{\phi}}_n \ \Leftrightarrow \ \underline{\tilde{\theta}}_n = \underline{p}(\underline{\tilde{\phi}}_n)$$

is a moment estimator of $\underline{\theta}$ based on a sufficient statistic. On the other hand, we can obtain a system of moment equations and a system of likelihood equations in terms of $\underline{\theta}$ and show that the two are the same. Now $\underline{\tilde{\phi}}_n = \underline{\hat{\phi}}_n$ is consistent for $\underline{\phi}$. Further, $\underline{p} = (p_1, p_2, \ldots, p_k)'$ is a totally differentiable and hence continuous function from

$\mathbb{R}^k$ to $\mathbb{R}^k$ and hence by the invariance property of consistency under continuous transformation we have

$$\underline{p}(\hat{\underline{\phi}}_n) = (p_1(\hat{\underline{\phi}}_n), p_2(\hat{\underline{\phi}}_n), \ldots, p_k(\hat{\underline{\phi}}_n))' = (\hat{\theta}_{1n}, \hat{\theta}_{2n}, \ldots, \hat{\theta}_{kn})'$$

$$= \hat{\underline{\theta}}_n = \tilde{\underline{\theta}}_n \xrightarrow{P_\theta} \underline{\theta} \ \forall \ \underline{\theta} \in \Theta.$$

From Theorem 4.2.2, we know that $\hat{\underline{\phi}}_n$ is CAN for ϕ with approximate dispersion matrix $I^{-1}(\underline{\phi})/n$. Further $\underline{p} = (p_1, p_2, \ldots, p_k)'$ is a totally differentiable function. Hence by the delta method, $\underline{p}(\hat{\underline{\phi}}_n)$ is CAN for $\underline{p}(\underline{\phi}) = \underline{\theta}$ with approximate dispersion matrix $MI^{-1}(\underline{\phi})M'/n$ where $M = \left[\frac{\partial p_i}{\partial \phi_j}\right]$. To find the matrix $MI^{-1}(\underline{\phi})M'$, we use the identity $I(\underline{\theta}) = J'I(\underline{\phi})J$ where $J = \left[\frac{\partial \phi_i}{\partial \theta_j}\right]$. Observe that $\phi_i = U_i(\underline{\theta})$ $i = 1, 2, \ldots, k$. Hence,

$$\frac{\partial \phi_i}{\partial \phi_j} = \sum_{r=1}^{k} \frac{\partial U_i(\underline{\theta})}{\partial \theta_r} \frac{\partial \theta_r}{\partial \phi_j} = \sum_{r=1}^{k} \frac{\partial \phi_i}{\partial \theta_r} \frac{\partial p_r(\underline{\phi})}{\partial \phi_j}$$

$$= \sum_{r=1}^{k}\sum_{s=1}^{k} (i, r)\text{-th element of } J \times (r, j)\text{-th element of } M$$

$$= (i, j)\text{-th element of } JM.$$

It is known that $\{\phi_1, \phi_2, \ldots, \phi_k\}$ are not functionally related to each other. Hence,

$$\frac{\partial \phi_i}{\partial \phi_j} = \begin{cases} 1, & \text{if} \quad i = j \\ 0, & \text{if} \quad i \neq j \end{cases}$$

Hence, JM is an identity matrix which implies that $M = J^{-1}$. Thus,

$$MI^{-1}(\underline{\phi})M' = J^{-1}I^{-1}(\underline{\phi})(J^{-1})' = (J'I(\underline{\phi})J)^{-1} = I^{-1}(\underline{\theta}).$$

Hence, we have proved that if the distribution of X belongs to a multiparameter exponential family, then the moment estimator of $\underline{\theta}$ based on a sufficient statistic is the same as the maximum likelihood estimator of $\underline{\theta}$ and it is CAN for $\underline{\theta}$ with approximate dispersion matrix $I^{-1}(\underline{\theta})/n$.

In the following example, we verify that normal $N(\mu, \sigma^2)$ distribution with $\mu \in \mathbb{R}$ and $\sigma^2 > 0$, which belongs to a two-parameter exponential family.

✐ Example 4.2.6

Suppose $X \sim N(\mu, \sigma^2)$ with $\mu \in \mathbb{R}$ and $\sigma^2 > 0$. Then, (i) the parameter space $\Theta = \{(\mu, \sigma^2) | \mu \in \mathbb{R}, \ \sigma^2 \in \mathbb{R}^+\}$ is an open set. (ii)The support of X is the real line which does not depend on the parameters. (iii) The probability density function $f(x, \mu, \sigma^2)$ is given by

$$f(x, \mu, \sigma^2) = \frac{1}{\sqrt{2\pi\sigma^2}} \exp\left\{-\frac{1}{2\sigma^2}(x - \mu)^2\right\}$$

$$= \frac{1}{\sqrt{2\pi\sigma^2}} \exp\left\{-\frac{1}{2\sigma^2}x^2 + \frac{\mu}{\sigma^2}x - \frac{\mu^2}{2\sigma^2}\right\}.$$

Hence,

$$\log f(x, \mu, \sigma^2) = -\frac{1}{2\sigma^2}x^2 + \frac{\mu}{\sigma^2}x - \frac{1}{2}\log 2\pi\sigma^2 - \frac{\mu^2}{2\sigma^2}$$

$$= U_1(\underline{\theta})K_1(x) + U_2(\underline{\theta})K_2(x) + V(\underline{\theta}) + W(x),$$

where $U_1(\underline{\theta}) = -1/2\sigma^2$, $K_1(x) = x^2$, $U_2(\underline{\theta}) = \frac{\mu}{\sigma^2}$, $K_2(x) = x$, $V(\underline{\theta}) = -(1/2)\log 2\pi\sigma^2 - \mu^2/2\sigma^2$ and $W(x) = 0$. (iv) The matrix $J = \left[\frac{dU_i}{d\theta_j}\right]$ of partial derivatives is given by

$$J = \begin{pmatrix} 0 & 1/2\sigma^4 \\ 1/\sigma^2 & -\mu/\sigma^4 \end{pmatrix}.$$

It is clear that U_1 and U_2 have continuous partial derivatives with respect to μ and σ^2 and $|J| = \left|\left[\frac{dU_i}{d\theta_j}\right]\right| = -\frac{1}{2\sigma^6} \neq 0$. (v) Using routine procedure, the functions 1, x and x^2 can be shown to be linearly independent. Thus, a normal $N(\mu, \sigma^2)$ distribution with $\mu \in \mathbb{R}$ and $\sigma^2 > 0$ satisfies all the requirements of a two-parameter exponential family. In Example 3.3.2, we have noted that the moment estimator of $\underline{\theta}$ based on a sufficient statistic is the same as the maximum likelihood estimator of $\underline{\theta}$ and it is CAN for $\underline{\theta}$ with approximate dispersion matrix $I^{-1}(\underline{\theta})/n$. □

✐ Example 4.2.7

Suppose $(Y_1, Y_2)'$ has a multinomial distribution in three cells with cell probabilities $(\theta + \phi)/2$, $(1 - \theta)/2$ and $(1 - \phi)/2, 0 < \theta, \ \phi < 1$. It is to be noted that (i) the parameter space is $\Theta = \{(\theta, \phi)' | 0 < \theta, \phi < 1\}$ and it is an open set. (ii)The support of $(Y_1, Y_2)'$ is $\{(0, 0), (0, 1), (1, 0)\}$, which does not depend on the parameters. (iii) The joint probability mass function of $(Y_1, Y_2)'$ is given by

$$p_{y_1 y_2} = P[Y_1 = y_1, Y_2 = y_2] = \left(\frac{\theta+\phi}{2}\right)^{y_1} \left(\frac{1-\theta}{2}\right)^{y_2} \left(\frac{1-\phi}{2}\right)^{1-y_1-y_2},$$

$$y_1, \ y_2 = 0, 1 \ \& \ y_1 + y_2 \leq 1.$$

Hence,

$$\log p_{y_1 y_2} = y_1 \log \left(\frac{\theta + \phi}{2} \right) + y_2 \log \left(\frac{1 - \theta}{2} \right) + (1 - y_1 - y_2) \log \left(\frac{1 - \phi}{2} \right)$$

$$= y_1 [\log(\theta + \phi) - \log(1 - \phi)]$$
$$+ y_2 [\log(1 - \theta) - \log(1 - \phi)] - \log 2 + \log(1 - \phi)$$
$$= K_1(y_1, y_2) U_1(\theta, \phi) + K_2(y_1, y_2) U_2(\theta, \phi) + W(y_1, y_2) + V(\theta, \phi),$$

where $K_1(y_1, y_2) = y_1$, $K_2(y_1, y_2) = y_2$, $W(y_1, y_2) = -\log 2$,
$U_1(\theta, \phi) = \log(\theta + \phi) - \log(1 - \phi)$, $U_2(\theta, \phi) = \log(1 - \theta) - \log(1 - \phi)$ and
$V(\theta, \phi) = \log(1 - \phi)$. (iv) Now the matrix $J = \left[\frac{dU_i}{d\theta_j} \right]$ of partial derivatives is
given by

$$J = \begin{pmatrix} \frac{1}{\theta + \phi} & \frac{1}{\theta + \phi} + \frac{1}{1 - \phi} \\ \frac{-1}{1 - \theta} & \frac{1}{1 - \phi} \end{pmatrix}.$$

It is clear that U_1 and U_2 have continuous partial derivatives with respect to θ
and ϕ and $|J| = \left\| \left[\frac{dU_i}{d\theta_j} \right] \right\| = -\frac{-2\theta}{(\theta + \phi)(1 - \theta)(1 - \phi)} \neq 0$. (v) To examine whether the
functions 1, $K_1(y_1, y_2) = y_1$ and $K_2(y_1, y_2) = y_2$ are linearly independent, sup-
pose $l_1 y_1 + l_2 y_2 + l_3 = 0$. Then $y_1 = y_2 = 0$ implies $l_3 = 0$. In $l_1 y_1 + l_2 y_2 = 0$,
$y_1 = 1$, $y_2 = 0$ implies $l_1 = 0$ and in $l_2 y_2 = 0$ if $y_2 = 1$, then $l_2 = 0$. Thus, multinomial distribution when the cell probabilities are $(\theta + \phi)/2$, $(1 - \theta)/2$ and
$(1 - \phi)/2$ satisfies all the requirements of a two-parameter exponential family
and hence belongs to a two-parameter exponential family. Thus, by Theorem
4.2.3, based on a random sample of size n, the moment estimator of $(\theta, \phi)'$ based
on a sufficient statistic is the same as the maximum likelihood estimator of $(\theta, \phi)'$
and it is CAN for $(\theta, \phi)'$ with approximate dispersion matrix $I^{-1}(\theta, \phi)/n$. Suppose $(X_1, X_2, X_3)'$ denote the cell frequencies corresponding to the random sample of size n from the given trinomial distribution, $X_1 + X_2 + X_3 = n$. Then the
likelihood of $(\theta, \phi)'$ corresponding to the observed data (X_1, X_2, X_3) is given by

$$\log L_n(\theta, \phi | X_1, X_2, X_3) = X_1 [\log(\theta + \phi) - \log(1 - \phi)]$$
$$+ X_2 [\log(1 - \theta) - \log(1 - \phi)] - n \log 2 + n \log(1 - \phi).$$

Hence,

$$\frac{\partial}{\partial \theta} \log L_n(\theta, \phi | X_1, X_2, X_3) = \frac{X_1}{\theta + \phi} - \frac{X_2}{1 - \theta}$$

$$\frac{\partial}{\partial \phi} \log L_n(\theta, \phi | X_1, X_2, X_3) = X_1 \left(\frac{1}{\theta + \phi} + \frac{1}{1 - \phi} \right) + \frac{X_2 - n}{1 - \phi}$$

$$= \frac{X_1}{\theta + \phi} + \frac{1}{1 - \phi} (X_1 + X_2 - n)$$

$$= \frac{X_1}{\theta + \phi} - \frac{X_3}{1 - \phi}.$$

Thus, the system of likelihood equations is given by

$$\frac{\partial}{\partial\theta}\log L_n(\theta,\phi|X_1,X_2,X_3) = \frac{X_1}{\theta+\phi} - \frac{X_2}{1-\theta} = 0 \qquad (4.2.8)$$

$$\frac{\partial}{\partial\phi}\log L_n(\theta,\phi|X_1,X_2,X_3) = \frac{X_1}{\theta+\phi} - \frac{X_3}{1-\phi} = 0 . \qquad (4.2.9)$$

From the Eqs. (4.2.8) and (4.2.9) we get

$$\frac{X_2}{1-\theta} = \frac{X_3}{1-\phi} \quad\Leftrightarrow\quad X_3\theta - X_2\phi = X_3 - X_2 . \qquad (4.2.10)$$

Further from Eq. (4.2.8) we have

$$(1-\theta)X_1 = (\theta+\phi)X_2 \quad\Leftrightarrow\quad (X_1+X_2)\theta + X_2\phi = X_1 . \qquad (4.2.11)$$

From Eqs. (4.2.10) and (4.2.11) we get

$$\theta = \frac{X_1+X_3-X_2}{n} = \frac{X_1+X_2+X_3-2X_2}{n} = 1 - \frac{2X_2}{n}.$$

From Eq. (4.2.10) we have

$$X_2\phi = X_3\theta - X_3 + X_2 = X_3\left(1 - \frac{2X_2}{n}\right) - X_3 + X_2$$

$$= X_2 - \frac{2X_2X_3}{n} \quad\Rightarrow\quad \phi = 1 - \frac{2X_3}{n}.$$

To verify that the solution of the system of likelihood equations leads to a maximum likelihood estimator, we examine whether the matrix of second order partial derivatives of the log-likelihood is negative definite almost surely, at the solution. The second order partial derivatives of the log-likelihood are given by

$$\frac{\partial^2}{\partial\theta^2}\log L_n(\theta,\phi|X_1,X_2,X_3) = \frac{-X_1}{(\theta+\phi)^2} - \frac{X_2}{(1-\theta)^2}$$

$$\frac{\partial^2}{\partial\phi^2}\log L_n(\theta,\phi|X_1,X_2,X_3) = \frac{-X_1}{(\theta+\phi)^2} - \frac{X_3}{(1-\phi)^2}$$

$$\frac{\partial^2}{\partial\theta\partial\phi}\log L_n(\theta,\phi|X_1,X_2,X_3) = \frac{-X_1}{(\theta+\phi)^2}.$$

If M denotes the matrix of second order partial derivatives of the log-likelihood, then it is clear that the first principal minor is negative and the second is positive for all θ and ϕ. Hence, M is negative definite almost surely, at the solution. Thus, the maximum likelihood estimators $\hat{\theta}_n$ and $\hat{\phi}_n$ of θ and ϕ are given by

$$\hat{\theta}_n = 1 - \frac{2X_2}{n} \quad\&\quad \hat{\phi}_n = 1 - \frac{2X_3}{n}.$$

Now we find the moment estimators based on the sufficient statistic. From the likelihood equation, it is clear that (X_1, X_2) is jointly sufficient for the family. Since $X_1 + X_2 + X_3 = n$, (X_2, X_3) is also jointly sufficient for the family. Hence, to find the moment estimators based on the sufficient statistic, we have following two equations.

$$\frac{X_2}{n} = \frac{1}{n} \sum_{r=1}^{n} Y_{2r} = E(Y_2) = \frac{1-\theta}{2} \quad \Rightarrow \quad \theta = 1 - \frac{2X_2}{n}$$

$$\& \quad \frac{X_3}{n} = \frac{1}{n} \sum_{r=1}^{n} Y_{3r} = E(Y_3) = \frac{1-\phi}{2} \quad \Rightarrow \quad \phi = 1 - \frac{2X_3}{n} .$$

Thus, the moment estimators based on the sufficient statistic are given by

$$\tilde{\theta}_n = 1 - \frac{2X_2}{n} \quad \& \quad \tilde{\phi}_n = 1 - \frac{2X_3}{n} ,$$

which are the same as the maximum likelihood estimators. From Theorem 4.2.3, $(\hat{\theta}_n, \hat{\phi}_n)'$ is CAN for $(\theta, \phi)'$ with approximate dispersion matrix $I^{-1}(\theta, \phi)/n$. Now, from the second order partial derivatives of the log-likelihood as given above we find the information matrix $I(\theta, \phi)$. We have $E(X_1) = n(\theta + \phi)/2$, $E(X_2) = n(1 - \theta)/2$ and $E(X_3) = n(1 - \phi)/2$. Hence, the information matrix $I(\theta, \phi)$ is given by

$$I(\theta, \phi) = \begin{pmatrix} \frac{1+\phi}{2(\theta+\phi)(1-\theta)} & \frac{1}{2(\theta+\phi)} \\ \frac{1}{2(\theta+\phi)} & \frac{1+\theta}{2(\theta+\phi)(1-\phi)} \end{pmatrix},$$

which is a positive definite matrix.   □

In the following section, we discuss CAN estimators for the parameters of a distribution belonging to a Cramér family which includes an exponential family.

4.3   Cramér Family

Suppose X is a random variable or a random vector with a probability law $f(x, \theta)$, indexed by a real parameter $\theta \in \Theta \subset \mathbb{R}$. Suppose the probability law $f(x, \theta)$ satisfies the following conditions in an open interval $N_\rho(\theta_0) = (\theta_0 - \rho, \theta_0 + \rho) \subset \Theta$, where $\rho > 0$ and θ_0 is a true parameter value, that is, θ_0 is the value of the parameter which generated a random sample $\{X_1, X_2, \ldots, X_n\}$ from the distribution of X.

C-1 The support S_f is free from the parameter θ.
C-2 The parameter space is an open set.

C-3 The partial derivatives $\frac{\partial}{\partial\theta} \log f(x,\theta)$, $\frac{\partial^2}{\partial\theta^2} \log f(x,\theta)$ and $\frac{\partial^3}{\partial\theta^3} \log f(x,\theta)$ exist for almost all values of $x \in S_f$.

C-4 The identity $\int_{S_f} f(x,\theta)dx = 1$ can be differentiated with respect to θ under the integral sign at least twice. Thus, the information function $I(\theta)$ is given by

$$I(\theta) = E_\theta \left(\frac{\partial}{\partial\theta} \log f(X,\theta) \right)^2 = E_\theta \left(-\frac{\partial^2}{\partial\theta^2} \log f(X,\theta) \right) \quad \& \quad 0 < I(\theta) < \infty.$$

C-5 $\left| \frac{\partial^3}{\partial\theta^3} \log f(x,\theta) \right| < M(x)$, where $M(x)$ may depend on θ_0 and ρ and $E(M(X)) < \infty$. Thus, the third derivative of $\log f(x,\theta)$ is bounded by an integrable function.

A family of distributions satisfying conditions C-1 to C-5 is known as a Cramér family of distributions and the five conditions are known as the Cramér regularity conditions. If a distribution belongs to a one-parameter exponential family, then the Cramér regularity conditions are satisfied. Thus a one-parameter exponential family is a sub class of a Cramér family. The Cauchy distribution with location parameter θ and scale parameter 1 does not belong to a one-parameter exponential family but it belongs to a Cramér family as the probability density function of the Cauchy distribution satisfies the Cramér regularity conditions, it is shown in Example 4.3.1. Uniform distributions $U(0,\theta)$ or $U(\theta-1,\theta+1)$ or $U(\theta,1)$ do not belong to a Cramér family, since the support of the distributions depends on θ. For the same reason, exponential distribution with location parameter θ and scale parameter 1 does not belong to a Cramér family. A Laplace distribution with location θ is not a member of a Cramér family, since the third condition gets violated. More precisely, suppose X follows a Laplace distribution with probability density function as given by $f(x,\theta) = (1/2) \exp\{-|x-\theta|\}$. Then $\log f(x,\theta) = -\log 2 - |x-\theta|$ and $|x-\theta|$ is not a differentiable function of θ for fixed x at $\theta = x$. As θ varies over $N_\rho(\theta_0)$, the set of values for which $\frac{\partial}{\partial\theta} \log f(x,\theta)$ does not exist is $(\theta_0 - \rho, \theta_0 + \rho)$ and $P_\theta[X \in (\theta_0 - \rho, \theta_0 + \rho)] > 0$ and hence the third condition gets violated. However, a Laplace distribution with probability density function given by $f(x,\theta) = (2\theta)^{-1} \exp\{-|x|/\theta\}$, $x \in \mathbb{R}, \theta > 0$ belongs to a Cramér family. A gamma distribution with a known scale parameter and shape parameter $\lambda \in I^+$ does not belong to a Cramér family as the parameter space is not an open set. Suppose $\{X_1, X_2, \ldots, X_n\}$ is a random sample from the distribution of X with the probability law $f(x,\theta)$ and θ_0 is a true parameter. For the distributions belonging to a Cramér family following four results are true.

▶ **Result 4.3.1**

With probability approaching 1 as $n \to \infty$, the likelihood equation $\frac{\partial}{\partial\theta} \log L_n(\theta|\underline{X}) = 0$ admits a solution $\hat{\theta}_n(\underline{X})$ and it is consistent for θ_0.

▶ **Result 4.3.2**

For large n, the distribution of $\hat{\theta}_n(\underline{X})$ can be approximated by the normal distribution $N(\theta_0, 1/nI(\theta_0))$, that is,

$$\sqrt{n}(\hat{\theta}_n(\underline{X}) - \theta_0) \xrightarrow{L} Z_1 \sim N(0, 1/I(\theta_0)).$$

▶ **Result 4.3.3**

With probability approaching 1 as $n \to \infty$, there is a relative maximum of the likelihood function at $\hat{\theta}_n(\underline{X})$, that is,

$$P\left[\frac{\partial^2}{\partial \theta^2} \log L_n(\theta|\underline{X})|_{\hat{\theta}_n(\underline{X})} < 0\right] \to 1, \quad \text{as } n \to \infty.$$

▶ **Result 4.3.4**

With probability approaching 1 as $n \to \infty$, the consistent solution of the likelihood equation is unique.

The first two results were established by Cramér [4]. However, issues such as whether there is a relative maximum at the solution of the likelihood equation and whether the consistent solution is unique were not addressed by Cramér [4]. Huzurbazar [5] established these results, which are the last two results listed above. The four results collectively are known as Cramér-Huzurbazar theorem (Kale and Muralidharan [6]). These are usually referred to as the standard theory of maximum likelihood estimation when the estimation is based on a random sample $\{X_1, X_2, \ldots, X_n\}$. Now we prove these results. We first prove a lemma, which is heavily used to prove the Cramér-Huzurbazar theorem. Suppose f and g are two probability density functions, then the Kullback Leibler distance between f and g is defined as $D(f, g) = \int \log\left(\frac{f(x)}{g(x)}\right) f(x)dx$. It can be shown that $D(f, g) \geq 0$. In the following lemma, we prove its particular case.

▶ **Lemma 4.3.1**

Suppose $f(x, \theta)$ is the probability density function of X and it is indexed by θ. Then

$$I(\theta_1, \theta_0) = E_{\theta_0}\left(-\log\left(\frac{f(X, \theta_1)}{f(X, \theta_0)}\right)\right) \geq 0,$$

and the equality holds if and only if $\theta_1 = \theta_0$.

Proof It is clear that

$$E_{\theta_0}\left(\frac{f(X, \theta_1)}{f(X, \theta_0)}\right) = \int_{S_f}\left(\frac{f(x, \theta_1)}{f(x, \theta_0)}\right) f(x, \theta_0)dx$$

$$= \int_{S_f} f(x, \theta_0)dx = 1 \quad \Rightarrow \quad \log E_{\theta_0}\left(\frac{f(X, \theta_1)}{f(X, \theta_0)}\right) = 0.$$

Suppose a function g is defined as

$$g(u) = -\log u, \ u > 0, \ \Rightarrow \ g'(u) = -1/u \ \&$$
$$g''(u) = 1/u^2 > 0 \Rightarrow \ g \ \text{is a convex function.}$$

Thus by Jensen's inequality,

$$E_{\theta_0}\left(-\log\left(\frac{f(X, \theta_1)}{f(X, \theta_0)}\right)\right) \geq -\log E_{\theta_0}\left(\left(\frac{f(X, \theta_1)}{f(X, \theta_0)}\right)\right) = 0 \ \Rightarrow \ I(\theta_1, \theta_0) \geq 0.$$

Equality holds if and only if $\forall \ x \in S_f$

$$\frac{f(x, \theta_1)}{f(x, \theta_0)} = 1 \ \Leftrightarrow \ f(x, \theta_1) = f(x, \theta_0) \ \Leftrightarrow \ \theta_1 = \theta_0,$$

as θ is an indexing parameter. $\qquad\qquad\qquad\qquad\qquad\qquad\qquad\qquad\qquad\square$

📖 Remark 4.3.1

$I(\theta_1, \theta_0)$ is known as the Kullback-Leibler information per unit of observation. It is a measure of the ability of the likelihood ratio to distinguish between $f(x, \theta_1)$ and $f(x, \theta_0)$ when θ_0 is the true parameter value. The inequality $I(\theta_1, \theta_0) \geq 0$ is known as the Shanon-Kolmogorov information inequality. Suppose $L_n(\theta|\underline{X})$ denotes the likelihood of θ given data $\underline{X}$. Then by Kolmogorov's SLLN and Lemma 4.3.1, we have

$$\frac{1}{n}\log\left(\frac{L_n(\theta|\underline{X})}{L_n(\theta_0|\underline{X})}\right) = \frac{1}{n}\sum_{i=1}^{n}\log\left(\frac{f(X_i, \theta)}{f(X_i, \theta_0)}\right) \overset{a.s.}{\to} -I(\theta, \theta_0) < 0. \quad (4.3.1)$$

Thus for large n, the likelihood function has higher value at θ_0, than at any specific value of θ, provided different θ correspond to different distributions, that is, provided θ is a labeling parameter. The inequality (4.3.1) specifies the rate of convergence of the likelihood ratio. If θ_0 is the true parameter value, then the likelihood ratio $\frac{L_n(\theta|\underline{X})}{L_n(\theta_0|\underline{X})}$ converges to 0 exponentially fast at the rate $\exp(-nI(\theta, \theta_0))$.

Another result needed in the proofs of the Cramér-Huzurbazar theorem is Theorem 2.2.2, proved in Sect. 2.2 which states that if $W_n \overset{P}{\to} C < 0$, then $P[W_n \leq 0] \to 1$ as $n \to \infty$.

In the following theorem, we prove Result 4.3.1 and Result 4.3.3 together. Proof of Result 4.3.3 is also given separately later.

Theorem 4.3.1

Suppose $\underline{X} = \{X_1, X_2, \ldots, X_n\}$ is a random sample from the distribution of X which belongs to a Cramér family and θ_0 is a true parameter value. With probability approaching 1 as $n \to \infty$, the likelihood equation $\frac{\partial}{\partial \theta} \log L_n(\theta|\underline{X}) = 0$ admits a solution $\hat{\theta}_n(\underline{X})$ and $\hat{\theta}_n(\underline{X})$ is a consistent estimator of θ_0. Further, the likelihood function has a relative maximum at $\hat{\theta}_n(\underline{X})$.

Proof The likelihood of θ corresponding to the given sample $\underline{X}$ is given by

$$L_n(\theta|\underline{X}) = \prod_{i=1}^{n} f(X_i, \theta) \quad \Leftrightarrow \quad \log L_n(\theta|\underline{X}) = \sum_{i=1}^{n} \log f(X_i, \theta).$$

For $\delta > 0$, the logarithm of the likelihood ratio is expressed as follows:

$$\log \left[\frac{L_n(\theta_0 + \delta|\underline{X})}{L_n(\theta_0|\underline{X})} \right] = \log L_n(\theta_0 + \delta|\underline{X}) - \log L_n(\theta_0|\underline{X})$$

$$= \sum_{i=1}^{n} \log f(X_i, \theta_0 + \delta) - \sum_{i=1}^{n} \log f(X_i, \theta_0)$$

$$= \sum_{i=1}^{n} \log \left[\frac{f(X_i, \theta_0 + \delta)}{f(X_i, \theta_0)} \right]$$

$$= \sum_{i=1}^{n} Y_i \quad \text{where} \quad Y_i = \log \left[\frac{f(X_i, \theta_0 + \delta)}{f(X_i, \theta_0)} \right].$$

Note that Y_i is a Borel function of X_i $i = 1, 2, \ldots, n$. Since $\{X_1, X_2, \ldots, X_n\}$ are independent and identically distributed random variables, being Borel functions, $\{Y_1, Y_2, \ldots, Y_n\}$ are also independent and identically distributed random variables with $E_{\theta_0}(Y_i) = -I(\theta_0 + \delta, \theta_0) < \infty$. Hence, by Khintchine's WLLN

$$\frac{1}{n} \log \left[\frac{L_n(\theta_0 + \delta|\underline{X})}{L_n(\theta_0|\underline{X})} \right] = \frac{1}{n} \sum_{i=1}^{n} Y_i \xrightarrow{P_{\theta_0}} -I(\theta_0 + \delta, \theta_0) < 0. \qquad (4.3.2)$$

On similar lines,

$$\frac{1}{n} \log \left[\frac{L_n(\theta_0 - \delta|\underline{X})}{L_n(\theta_0|\underline{X})} \right] \xrightarrow{P_{\theta_0}} -I(\theta_0 - \delta, \theta_0) < 0. \qquad (4.3.3)$$

It is to be noted that θ_0 being an indexing parameter, $L_n(\theta_0 \pm \delta|\underline{X}) \neq L_n(\theta_0|\underline{X})$ for any realization of $\underline{X}$. Suppose events E_n and F_n are defined as follows:

$$E_n = \{\omega| \log L_n(\theta_0 + \delta|\underline{X}(\omega)) < \log L_n(\theta_0|\underline{X}(\omega))\}$$
$$F_n = \{\omega| \log L_n(\theta_0 - \delta|\underline{X}(\omega)) < \log L_n(\theta_0|\underline{X}(\omega))\}$$
$$\Rightarrow \quad \log L_n(\theta_0 \pm \delta|\underline{X}(\omega)) < \log L_n(\theta_0|\underline{X}(\omega)) \quad \forall \; \omega \in E_n \cap F_n.$$

Further from the Cramér regularity conditions, it follows that the likelihood function is differentiable and hence continuous over the interval $(\theta_0 - \delta, \theta_0 + \delta)$. Hence, there exists a point $\hat{\theta}_n(\underline{X}) \in (\theta_0 - \delta, \theta_0 + \delta)$ at which log-likelihood attains its maximum. Again from the Cramér regularity conditions, the first and the second derivative of log-likelihood exist. Thus, using the theory of calculus we have

$$\frac{\partial}{\partial\theta} \log L_n(\theta_0|\underline{X})|_{\hat{\theta}_n(\underline{X})} = 0 \text{ and } \frac{\partial^2}{\partial\theta^2} \log L_n(\theta_0|\underline{X})|_{\hat{\theta}_n(\underline{X})} < 0$$

$$\text{for } \hat{\theta}_n(\underline{X}) \in (\theta_0 - \delta, \theta_0 + \delta).$$

Suppose an event H_n is defined as $H_n = A_n \cap B_n \cap C_n$, where

$$A_n = \left\{\omega | \frac{\partial}{\partial\theta} \log L_n(\theta_0|\underline{X}(\omega))|_{\hat{\theta}_n(\underline{X}(\omega))} = 0\right\},$$

$$B_n = \left\{\omega | \hat{\theta}_n(\underline{X}(\omega)) \in (\theta_0 - \delta, \theta_0 + \delta)\right\}$$

$$\& \quad C_n = \left\{\omega | \frac{\partial^2}{\partial\theta^2} \log L_n(\theta_0|\underline{X}(\omega))|_{\hat{\theta}_n(\underline{X}(\omega))} < 0\right\}.$$

It is to be noted that

$$\omega \in E_n \cap F_n \Rightarrow \omega \in H_n \quad \Rightarrow \quad E_n \cap F_n \subset H_n \quad \Rightarrow \quad P(H_n) \geq P(E_n \cap F_n).$$

From (4.3.2), we have

$W_n = \frac{1}{n} \log \left[L_n(\theta_0 + \delta|\underline{X})/L_n(\theta_0|\underline{X})\right] \xrightarrow{P_{\theta_0}} -I(\theta_0 + \delta, \theta_0) < 0$. Hence by Theorem 2.2.2, $P_{\theta_0}[W_n \leq 0] \to 1$ as $n \to \infty$. However, θ_0 being an indexing parameter $W_n \neq 0$ a.s. Thus, $P(E_n) \to 1$. Similarly, from Eq. (4.3.3), $P(F_n) \to 1$. Now

$$P(E_n) \to 1 \ \& \ P(F_n) \to 1 \Rightarrow P(E_n^c) \to 0 \ \& \ P(F_n^c) \to 0$$

$$\Rightarrow P(E_n^c \cup F_n^c) \leq P(E_n^c) + P(F_n^c) \to 0$$

$$\Rightarrow P(E_n \cap F_n) \to 1 \quad \Rightarrow \quad P(H_n) \to 1$$

$$\Rightarrow P(A_n) \geq P(H_n) \to 1, P(B_n) \geq P(H_n) \to 1$$

$$\& \ P(C_n) \geq P(H_n) \to 1, P(A_n \cap B_n) \geq P(H_n) \to 1.$$

Now $P(A_n) \to 1$ implies that with probability approaching 1, there is a solution to the likelihood equation. Similarly, $P(B_n) \to 1$ implies that $\hat{\theta}_n(\underline{X})$ is consistent for θ_0 as δ is arbitrary. The fact that $P(A_n \cap B_n) \to 1$ states that with probability approaching 1, there is a consistent solution of the likelihood equation and the Result 4.3.1 is proved. From the statement $P(C_n) \to 1$, we conclude that with probability approaching 1, there is a relative maximum of the likelihood at $\hat{\theta}_n(\underline{X})$ and the Result 4.3.3 is proved. $\qquad\square$

In the following theorem, we give an alternative proof of Result 4.3.3 and it is as given by Huzurbazar [5].

Theorem 4.3.2

With probability approaching 1 as $n \to \infty$, there is a relative maximum of the likelihood function at $\hat{\theta}_n(\underline{X})$, that is,

$$P\left[\frac{\partial^2}{\partial \theta^2} \log L_n(\theta|\underline{X})|_{\hat{\theta}_n(\underline{X})} < 0\right] \to 1, \text{ as } n \to \infty.$$

Proof Under the Cramér regularity conditions, the likelihood function is twice differentiable and the information function $I(\theta)$ is positive and finite. Observe that by Khintchine's WLLN

$$\frac{1}{n}\frac{\partial^2}{\partial \theta^2} \log L_n(\theta|\underline{X})|_{\theta_0} = \frac{1}{n}\frac{\partial^2}{\partial \theta^2} \sum_{i=1}^{n} \log f(X_i, \theta)|_{\theta_0} = \frac{1}{n}\sum_{i=1}^{n}\frac{\partial^2}{\partial \theta^2} \log f(X_i, \theta)|_{\theta_0}$$

$$\overset{P_{\theta_0}}{\to} -I(\theta_0). \tag{4.3.4}$$

Suppose $\frac{\partial^2}{\partial \theta^2} \log L_n(\theta|\underline{X})|_{\hat{\theta}_n(\underline{X})} \neq 0$. Then by the mean value theorem

$$\frac{1}{n}\frac{\partial^2}{\partial \theta^2} \log L_n(\theta|\underline{X})|_{\hat{\theta}_n(\underline{X})} - \frac{1}{n}\frac{\partial^2}{\partial \theta^2} \log L_n(\theta|\underline{X})|_{\theta_0}$$

$$= \frac{1}{n}(\hat{\theta}_n(\underline{X}) - \theta_0)\frac{\partial^3}{\partial \theta^3} \log L_n(\theta|\underline{X})|_{\theta_n^*(\underline{X})},$$

where $\theta_n^*(\underline{X})$ is a convex combination of $\hat{\theta}_n(\underline{X})$ and θ_0 given by $\theta_n^*(\underline{X}) = \alpha\hat{\theta}_n(\underline{X}) + (1 - \alpha)\theta_0, 0 < \alpha < 1$. Now using the fifth regularity condition we have

$$\left|\frac{1}{n}\frac{\partial^3}{\partial \theta^3} \log L_n(\theta|\underline{X})|_{\theta_n^*(\underline{X})}\right| \leq \frac{1}{n}\sum_{i=1}^{n}\left|\frac{\partial^3}{\partial \theta^3} \log f(X_i, \theta)|_{\theta_n^*(\underline{X})}\right|$$

$$\overset{P_{\theta_0}}{\to} E_{\theta_0}\left(\left|\frac{\partial^3}{\partial \theta^3} \log f(X, \theta)|_{\theta_n^*(\underline{X})}\right|\right)$$

$$\leq E_{\theta_0}(M(X)) < \infty.$$

Thus, using the fact that $\left|\frac{1}{n}\frac{\partial^3}{\partial \theta^3} \log L_n(\theta|\underline{X})|_{\theta_n^*(\underline{X})}\right|$ converges in probability to a finite number a say, which may depend on θ_0, we have

$$\left| \frac{1}{n} \frac{\partial^2}{\partial \theta^2} \log L_n(\theta|\underline{X})|_{\hat{\theta}_n(\underline{X})} - \frac{1}{n} \frac{\partial^2}{\partial \theta^2} \log L_n(\theta|\underline{X})|_{\theta_0} \right|$$

$$= \left| (\hat{\theta}_n(\underline{X}) - \theta_0) \right| \left| \frac{1}{n} \frac{\partial^3}{\partial \theta^3} \log L_n(\theta|\underline{X})|_{\theta_n^*(\underline{X})} \right|$$

$$\xrightarrow{P_{\theta_0}} 0,$$

as $\hat{\theta}_n(\underline{X})$ is consistent for θ_0. Since $\left| \frac{1}{n} \frac{\partial^2}{\partial \theta^2} \log L_n(\theta|\underline{X})|_{\hat{\theta}_n(\underline{X})} - \frac{1}{n} \frac{\partial^2}{\partial \theta^2} \log L_n(\theta|\underline{X})|_{\theta_0} \right|$ converges to 0 in probability, both the terms have the same limit for convergence in probability. But as proved in (4.3.4),

$$\frac{1}{n} \frac{\partial^2}{\partial \theta^2} \log L_n(\theta|\underline{X})|_{\theta_0} \xrightarrow{P_{\theta_0}} -I(\theta_0) \quad \Rightarrow \quad \frac{1}{n} \frac{\partial^2}{\partial \theta^2} \log L_n(\theta|\underline{X})|_{\hat{\theta}_n(\underline{X})} \xrightarrow{P_{\theta_0}} -I(\theta_0) < 0.$$

By Theorem 2.2.2, we conclude that $P_{\theta_0} \left[\frac{\partial^2}{\partial \theta^2} \log L_n(\theta|\underline{X})|_{\hat{\theta}_n(\underline{X})} < 0 \right] \to 1$ as $n \to \infty$ and the Result 4.3.3 is proved. $\qquad\square$

We now prove a lemma, proof of which is similar to that of Theorem 4.3.2. It is needed in the proof of Result 4.3.4.

▶ **Lemma 4.3.2**
Suppose T_n is any consistent estimator of θ_0. Then

$$P_{\theta_0} \left[\frac{\partial^2}{\partial \theta^2} \log L_n(\theta|\underline{X})|_{T_n} < 0 \right] \to 1 \text{ as } n \to \infty.$$

Proof Using arguments similar to those in the proof of Theorem 4.3.2 and using the consistency of T_n we have

$$\left| \frac{1}{n} \frac{\partial^2}{\partial \theta^2} \log L_n(\theta|\underline{X})|_{T_n} - \frac{1}{n} \frac{\partial^2}{\partial \theta^2} \log L_n(\theta|\underline{X})|_{\theta_0} \right|$$

$$= |(T_n - \theta_0)| \left| \frac{1}{n} \frac{\partial^3}{\partial \theta^3} \log L_n(\theta|\underline{X})|_{\theta_n^*(\underline{X})} \right|$$

$$\xrightarrow{P_{\theta_0}} 0,$$

where $\theta_n^*(\underline{X})$ is a convex combination of T_n and θ_0. Observe that

$$\frac{1}{n} \frac{\partial^2}{\partial \theta^2} \log L_n(\theta|\underline{X})|_{\theta_0} \xrightarrow{P_{\theta_0}} -I(\theta_0) \quad \Rightarrow \quad \frac{1}{n} \frac{\partial^2}{\partial \theta^2} \log L_n(\theta|\underline{X})|_{T_n} \xrightarrow{P_{\theta_0}} -I(\theta_0) < 0.$$

Hence, we conclude that

$$P_{\theta_0} \left[\frac{\partial^2}{\partial \theta^2} \log L_n(\theta|\underline{X})|_{T_n} < 0 \right] \to 1 \text{ as } n \to \infty.$$

$\qquad\square$

Theorem 4.3.3

With probability approaching 1 as $n \to \infty$, a consistent solution of the likelihood equation is unique.

Proof The proof is by contradiction. Suppose if possible, $\hat{\theta}_{1n}(\underline{X})$ and $\hat{\theta}_{2n}(\underline{X})$ are two consistent solutions of the likelihood equation in $(\theta_0 - \delta, \theta_0 + \delta)$, where $\underline{X} \in E_n \cap F_n$. Hence, with probability approaching 1 as $n \to \infty$,

$$\frac{1}{n}\frac{\partial}{\partial\theta}\log L_n(\theta|\underline{X})|_{\hat{\theta}_{1n}(\underline{X})} = 0 \quad \text{and} \quad \frac{1}{n}\frac{\partial}{\partial\theta}\log L_n(\theta|\underline{X})|_{\hat{\theta}_{2n}(\underline{X})} = 0.$$

Hence, by Rolle's theorem, with probability approaching 1 as $n \to \infty$,

$$\frac{1}{n}\frac{\partial^2}{\partial\theta^2}\log L_n(\theta|\underline{X})|_{\tilde{\theta}_n(\underline{X})} = 0 \quad \text{where}$$

$$\tilde{\theta}_n(\underline{X}) = \alpha\hat{\theta}_{1n}(\underline{X}) + (1-\alpha)\hat{\theta}_{2n}(\underline{X}), \quad 0 < \alpha < 1.$$

It is to be noted that $\hat{\theta}_{1n}(\underline{X})$ and $\hat{\theta}_{2n}(\underline{X})$ are consistent estimators of θ_0 and hence, being a convex combination, $\tilde{\theta}_n(\underline{X})$ is also a consistent estimator of θ_0. Hence by the Lemma 4.3.2,

$$\frac{1}{n}\frac{\partial^2}{\partial\theta^2}\log L_n(\theta|\underline{X})|_{\tilde{\theta}_n(\underline{X})} \overset{P_{\theta_0}}{\to} -I(\theta_0) < 0$$

$$\Rightarrow \quad P\left[\frac{\partial^2}{\partial\theta^2}\log L_n(\theta|\underline{X})|_{\tilde{\theta}_n(\underline{X})} < 0\right] \to 1, \text{ as } n \to \infty.$$

It is a contradiction to the statement that with probability approaching 1 as $n \to \infty$, $\frac{1}{n}\frac{\partial^2}{\partial\theta^2}\log L_n(\theta|\underline{X})|_{\tilde{\theta}_n(\underline{X})} = 0$. Hence, it is proved that with probability approaching 1 as $n \to \infty$, a consistent solution of the likelihood equation is unique. $\qquad\square$

We now proceed to prove the Result 4.3.2, which states that the large sample distribution of a consistent solution of the likelihood equation is normal.

Theorem 4.3.4

For large n, the distribution of $\hat{\theta}_n(\underline{X})$ is approximately normal $N(\theta_0, 1/nI(\theta_0))$, that is,

$$\sqrt{n}(\hat{\theta}_n(\underline{X}) - \theta_0) \overset{L}{\to} Z_1 \sim N(0, 1/I(\theta_0)).$$

Proof From the regularity conditions, likelihood is thrice differentiable, hence by the Taylor series expansion we have

$$0 = \frac{\partial}{\partial\theta}\log L_n(\theta|\underline{X})|_{\hat{\theta}_n(\underline{X})} = \frac{\partial}{\partial\theta}\log L_n(\theta|\underline{X})|_{\theta_0} + (\hat{\theta}_n(\underline{X}) - \theta_0)\frac{\partial^2}{\partial\theta^2}\log L_n(\theta|\underline{X})|_{\theta_0}$$

$$+ \frac{1}{2}(\hat{\theta}_n(\underline{X}) - \theta_0)^2\frac{\partial^3}{\partial\theta^3}\log L_n(\theta|\underline{X})|_{\theta^*(\underline{X})},$$

where $\theta^*(\underline{X}) = \alpha\,\hat\theta_n(\underline{X}) + (1-\alpha)\theta_0$. By rearranging the terms, we get

$$
(\hat\theta_n(\underline{X}) - \theta_0) = \frac{-\frac{\partial}{\partial\theta}\log L_n(\theta|\underline{X})|_{\theta_0}}{\frac{\partial^2}{\partial\theta^2}\log L_n(\theta|\underline{X})|_{\theta_0} + \frac{1}{2}(\hat\theta_n(\underline{X}) - \theta_0)\frac{\partial^3}{\partial\theta^3}\log L_n(\theta|\underline{X})|_{\theta^*}}
$$

$$
\Rightarrow \quad \sqrt{n}(\hat\theta_n(\underline{X}) - \theta_0)
$$

$$
= \frac{\frac{1}{\sqrt{n}}\frac{\partial}{\partial\theta}\log L_n(\theta|\underline{X})|_{\theta_0}}{-\frac{1}{n}\frac{\partial^2}{\partial\theta^2}\log L_n(\theta|\underline{X})|_{\theta_0} - \frac{1}{2}(\hat\theta_n(\underline{X}) - \theta_0)\frac{1}{n}\frac{\partial^3}{\partial\theta^3}\log L_n(\theta|\underline{X})|_{\theta^*(\underline{X})}}
$$

$$
= \frac{U_n}{V_n},
$$

where

$$
U_n = \frac{1}{\sqrt{n}}\frac{\partial}{\partial\theta}\log L_n(\theta|\underline{X})|_{\theta_0} \quad \&
$$

$$
V_n = -\frac{1}{n}\frac{\partial^2}{\partial\theta^2}\log L_n(\theta|\underline{X})|_{\theta_0} - \frac{1}{2}(\hat\theta_n(\underline{X}) - \theta_0)\frac{1}{n}\frac{\partial^3}{\partial\theta^3}\log L_n(\theta|\underline{X})|_{\theta^*(\underline{X})} .
$$

The denominator V_n is different from 0 for large n as is clear from the following:

$$
-\frac{1}{n}\frac{\partial^2}{\partial\theta^2}\log L_n(\theta|\underline{X})|_{\theta_0} = \frac{1}{n}\sum_{i=1}^{n}\left(-\frac{\partial^2}{\partial\theta^2}\log f(X_i,\theta)|_{\theta_0}\right) \overset{P_{\theta_0}}{\to} -I(\theta_0) ,
$$

by WLLN. Further, using the arguments similar to those in the proof of Theorem 4.3.2, we have

$$
\left|\frac{1}{n}\frac{\partial^3}{\partial\theta^3}\log L_n(\theta|\underline{X})|_{\theta_n^*(\underline{X})}\right| \le \frac{1}{n}\sum_{i=1}^{n}\left|\frac{\partial^3}{\partial\theta^3}\log f(X_i,\theta)|_{\theta_n^*(\underline{X})}\right|
$$

$$
\overset{P_{\theta_0}}{\to} E_{\theta_0}\left(\left|\frac{\partial^3}{\partial\theta^3}\log f(X,\theta)|_{\theta^*(\underline{X})}\right|\right)
$$

$$
\le E_{\theta_0}(M(X)) < \infty .
$$

Moreover, $\hat\theta_n(\underline{X})$ is consistent for θ_0, and hence
$\frac{1}{2}(\hat\theta_n(\underline{X}) - \theta_0)\frac{1}{n}\frac{\partial^3}{\partial\theta^3}\log L_n(\theta|\underline{X})|_{\theta^*(\underline{X})} \overset{P_{\theta_0}}{\to} 0.$
Consequently, the denominator

$$
V_n = -\frac{1}{n}\frac{\partial^2}{\partial\theta^2}\log L_n(\theta|\underline{X})|_{\theta_0} - \frac{1}{2}(\hat\theta_n(\underline{X}) - \theta_0)\frac{1}{n}\frac{\partial^3}{\partial\theta^3}\log L_n(\theta|\underline{X})|_{\theta^*(\underline{X})} \overset{P_{\theta_0}}{\to} I(\theta_0).
$$

The numerator U_n can be expressed as

$$
U_n = \frac{1}{\sqrt{n}}\frac{\partial}{\partial\theta}\log L_n(\theta|\underline{X})|_{\theta_0} = \frac{1}{\sqrt{n}}\sum_{i=1}^{n}\frac{\partial}{\partial\theta}\log f(X_i,\theta)|_{\theta_0}.
$$

It is given that $\{X_1, X_2, \ldots, X_n\}$ are independent and identically distributed random variables and being Borel functions, $\frac{\partial}{\partial \theta} \log f(X_i, \theta)|_{\theta_0}$ for $i = 1, 2, \ldots, n$ are also independent and identically distributed random variables with

$$E_{\theta_0}\left(\frac{\partial}{\partial \theta} \log f(X_i, \theta)|_{\theta_0}\right) = 0 \ \& \ Var_{\theta_0}\left(\frac{\partial}{\partial \theta} \log f(X_i, \theta)|_{\theta_0}\right) = I(\theta_0),$$
$$i = 1, 2, \ldots, n.$$

From the regularity conditions, we have $0 < I(\theta_0) < \infty$. Hence by the CLT for the independent and identically distributed random variables with positive finite variance,

$$\frac{\sum_{i=1}^{n} \frac{\partial}{\partial \theta} \log f(X_i, \theta)|_{\theta_0} - n * 0}{\sqrt{n} I(\theta_0)} \xrightarrow{L} Z \sim N(0, 1)$$

$$\Leftrightarrow \ U_n = \frac{1}{\sqrt{n}} \frac{\partial}{\partial \theta} \log L_n(\theta|\underline{X})|_{\theta_0} \xrightarrow{L} U \sim N(0, I(\theta_0)).$$

Using Slutsky's theorem,

$$\sqrt{n}(\hat{\theta}_n(\underline{X}) - \theta_0) = \frac{U_n}{V_n} \xrightarrow{L} Z_1 \sim N(0, 1/I(\theta_0)).$$

Thus, it is proved that $\hat{\theta}_n(\underline{X})$ is CAN for θ_0 with approximate variance $1/nI(\theta_0)$. $\qquad \square$

✍ Remark 4.3.2

Results 4.3.1 to 4.3.4 collectively convey that under the Cramér regularity conditions, for large n, the likelihood equation has a unique consistent solution $\hat{\theta}_n(\underline{X})$ and a relative maximum is attained at this solution. Further Result 4.3.2 asserts that it is CAN for θ_0 with approximate variance $1/nI(\theta_0)$. It is to be noted that for finite n, $1/nI(\theta_0)$ is the Cramér-Rao lower bound for the variance of an unbiased estimator of θ_0. For large n, $\hat{\theta}_n(\underline{X})$ is unbiased for θ_0, thus for large n, variance of $\hat{\theta}_n(\underline{X})$ attains the Cramér-Rao lower bound and hence it is an asymptotically efficient estimator. Hence, $\hat{\theta}_n(\underline{X})$ is referred to as the best asymptotically normal (BAN) estimator of θ_0, (Rohatgi and Saleh [7]). $1/nI(\theta_0)$ is also referred to as the Fisher lower bound for the variance, (Kale and Muralidharan [6]). Using similar arguments as in Results 4.3.1 to 4.3.4, it can be proved that all the four results are true for any interior point $\theta \in \Theta$.

It is already noted that all the standard distributions, such as $N(\theta, 1)$ when $\theta \in \mathbb{R}$, exponential distribution with mean $\theta > 0$, gamma distribution with scale 1 and shape $\theta > 0$, Binomial $B(n, \theta)$ where n is known and $\theta \in (0, 1)$, Poisson $Poi(\theta)$, $\theta > 0$, geometric distribution with success probability $\theta \in (0, 1)$ and the truncated versions of these, constitute a one-parameter exponential family of distributions. Hence, all these distributions satisfy the Cramér regularity conditions and Results 4.3.1 to 4.3.4 are valid. It is noted that a Cauchy distribution with location parameter θ and scale parameter 1 does not belong to a one-parameter exponential family. In the following example, we show that it belongs to a Cramér family and hence the maximum likelihood estimator of θ is BAN for θ. For a Cauchy distribution, it is not possible to solve the likelihood equation explicitly and hence it becomes necessary to apply suitable iterative procedures to get the numerical solution of the likelihood equation. Most commonly used iterative procedures are Newton-Raphson procedure obtained from the Taylor series expansion and the method of scoring as proposed by Fisher. These are discussed in Sect. 4.4.

✔ Example 4.3.1

Suppose a random variable X follows a Cauchy $C(\theta, 1)$ distribution with location parameter θ and scale parameter 1, then its probability density function $f(x, \theta)$ is given by

$$f(x, \theta) = \frac{1}{\pi} \frac{1}{1 + (x - \theta)^2}, \quad x \in \mathbb{R}, \quad \theta \in \mathbb{R}.$$

The support of X is a real line and it is free from θ. Further, the parameter space is also a real line, which an open set. Now,

$$\frac{\partial}{\partial \theta} \log f(x, \theta) = \frac{2(x - \theta)}{1 + (x - \theta)^2}, \quad \frac{\partial^2}{\partial \theta^2} \log f(x, \theta) = \frac{-2 + 2(x - \theta)^2}{[1 + (x - \theta)^2]^2}$$

$$\frac{\partial^3}{\partial \theta^3} \log f(x, \theta) = \frac{4(x - \theta)^3 - 12(x - \theta)}{[1 + (x - \theta)^2]^3} = \frac{4(x - \theta)((x - \theta)^2 - 12)}{[1 + (x - \theta)^2]^3}.$$

Thus, the partial derivatives $\frac{\partial}{\partial \theta} \log f(x, \theta)$, $\frac{\partial^2}{\partial \theta^2} \log f(x, \theta)$ and $\frac{\partial^3}{\partial \theta^3} \log f(x, \theta)$ exist for almost all values of $x \in S_f$. It is in view of the fact that for fixed x, $\log f(x, \theta)$ is a logarithm of a polynomial of degree 2 in θ and hence it is an analytic function of θ. Further,

$$\left| \frac{\partial^3}{\partial \theta^3} \log f(x, \theta) \right| \leq \frac{4|x - \theta|((x - \theta)^2 + 12)}{[1 + (x - \theta)^2]^3} = M(x), \quad \text{say}.$$

Now we examine whether $E(M(X)) < \infty$. We have

$$E(M(X)) = \frac{1}{\pi} \int_{-\infty}^{\infty} \frac{4|x - \theta|((x - \theta)^2 + 12)}{[1 + (x - \theta)^2]^3} \frac{1}{1 + (x - \theta)^2} \, dx$$

$$= \frac{1}{\pi} \int_{-\infty}^{\infty} \frac{4|y|(y^2 + 12)}{(1 + y^2)^3} \frac{1}{1 + y^2} \, dy$$

$$= \frac{1}{\pi} \int_{0}^{\infty} \frac{8y(y^2 + 12)}{(1 + y^2)^4} \, dy$$

$$= \frac{1}{\pi} \int_{0}^{\infty} \frac{P_3(y)}{P_8(y)} \, dy,$$

where $P_3(y)$ and $P_8(y)$ are polynomial functions of y with degrees 3 and 8 respectively. The infinite integral $\frac{1}{\pi}\int_0^\infty \frac{P_3(y)}{P_8(y)}dy$ is convergent as the degree of polynomial in denominator is larger by 1, than the degree of polynomial in numerator, which implies that the third partial derivative of $\log f(x,\theta)$ is bounded by an integrable function. Now to examine whether the differentiation and integration in the identity $\int_{S_f} f(x,\theta) = 1$ can be interchanged, we note that $\frac{\partial}{\partial\theta}f(x,\theta) = \frac{1}{\pi}\frac{2(x-\theta)}{(1+(x-\theta)^2)^2}$ is a continuous function of θ and hence integrable over a finite interval (a,b). Further, this integral is uniformly convergent as $a \to -\infty$ and $b \to \infty$ as the integrand behaves like $\frac{1}{x^3}$ near $\pm\infty$. Using similar arguments for $\frac{\partial^2}{\partial\theta^2}f(x,\theta)$, differentiation and integration can be interchanged second time. It follows that

$$E_\theta\left(\frac{\partial}{\partial\theta}\log f(X,\theta)\right) = \frac{1}{\pi}\int_{S_f}\frac{2(x-\theta)}{(1+(x-\theta)^2)^2} = \frac{1}{\pi}\int_{-\infty}^\infty\frac{2t}{(1+t^2)^2} = 0\,,$$

integrand being an odd function. We now find the information function $I(\theta)$ as follows:

$$I(\theta) = E\left(-\frac{\partial^2}{\partial\theta^2}\log f(X,\theta)\right) = E\left(\frac{2-2(X-\theta)^2}{[1+(X-\theta)^2]^2}\right)$$

$$= \frac{2}{\pi}\int_{-\infty}^\infty\frac{1-(X-\theta)^2}{[1+(X-\theta)^2]^3}\,dx$$

$$= \frac{2}{\pi}\int_{-\infty}^\infty\frac{1-y^2}{[1+y^2]^3}\,dy = \frac{4}{\pi}\int_0^\infty\frac{1-y^2}{[1+y^2]^3}\,dy$$

$$= \frac{4}{\pi}\left[\frac{1}{2}\int_0^\infty\frac{u^{-\frac{1}{2}}}{[1+u]^3}\,du - \frac{1}{2}\int_0^\infty\frac{u^{\frac{1}{2}}}{[1+u]^3}\,du\right]\quad\text{with}\quad y^2 = u$$

$$= \frac{4}{\pi}\left[\frac{1}{2}\int_0^\infty\frac{u^{1/2-1}}{[1+u]^{1/2+5/2}}\,du - \frac{1}{2}\int_0^\infty\frac{u^{3/2-1}}{[1+u]^{3/2+3/2}}\,du\right]$$

$$= \frac{2}{\pi}\left[B\left(\frac{1}{2},\frac{5}{2}\right) - B\left(\frac{3}{2},\frac{3}{2}\right)\right]\quad\text{by definition of beta function}$$

$$= \frac{2}{\pi}\left[\frac{\Gamma(1/2)\Gamma(5/2)}{\Gamma(3)} - \frac{\Gamma(3/2)\Gamma(3/2)}{\Gamma(3)}\right]$$

$$= \frac{1}{\pi}\left[\frac{3\pi}{4} - \frac{\pi}{4}\right] = \frac{1}{2}.$$

Thus $I(\theta)$ is positive and finite. Thus, a Cauchy $C(\theta,1)$ distribution satisfies all the Cramér regularity conditions, hence it belongs to a Cramér family. By the Cramér-Huzurbazar theorem, for large n, the maximum likelihood estimator $\hat\theta_n$ of θ is a CAN estimator of θ with the approximate variance $1/nI(\theta) = 2/n$. For large n, the approximate variance attains the Cramér lower bound for the variance, hence it is a BAN estimator. However, the likelihood equation given by

$\frac{\partial}{\partial \theta} \log L_n(\theta|\underline{X}) = \sum_{i=1}^{n} \frac{2(X_i - \theta)}{1 + (X_i - \theta)^2} = 0$ cannot be solved explicitly and we need to use the numerical methods, discussed in Sect. 4.4, to obtain the value of the maximum likelihood estimator corresponding to the given random sample. It is to be noted that for each n, there are multiple roots to the likelihood equation out of which some root is consistent. Since a Cauchy distribution belongs to a Cramér family, for large n, with high probability, the consistent solution of the likelihood equation is unique. □

✍ Remark 4.3.3

In Example 3.2.6, it is shown that for a Cauchy $C(\theta, 1)$ distribution, the sample median is CAN for θ with the approximate variance $\pi^2/4n$. It is to be noted that $\pi^2/4n = 2.4694/n > 2/n$ as expected.

In Example 4.2.3, we have noted that a normal $N(\theta, \theta^2)$ distribution, when $\theta > 0$, does not belong to a one-parameter exponential family, but still the maximum likelihood estimator of θ is CAN with approximate variance $1/nI(\theta)$. These results lead to a conjecture that this distribution may be a member of a Cramér family, which is verified in the next example.

✎ Example 4.3.2

Suppose $X \sim N(\theta, \theta^2)$ distribution, $\theta \in \Theta = (0, \infty)$. Thus, the parameter space $(0, \infty)$ is an open set. The support of the distribution is a real line, which is free from θ. The partial derivatives of $\log f(x, \theta)$ up to order three exist and are as follows:

$$\log f(x, \theta) = -\frac{1}{2}\log 2\pi - \log \theta - \frac{x^2}{2\theta^2} + \frac{x}{\theta} - \frac{1}{2},$$

$$\frac{\partial}{\partial \theta} \log f(x, \theta) = -\frac{1}{\theta} + \frac{x^2}{\theta^3} - \frac{x}{\theta^2}$$

$$\frac{\partial^2}{\partial \theta^2} \log f(x, \theta) = \frac{1}{\theta^2} - \frac{3x^2}{\theta^4} + \frac{2x}{\theta^3},$$

$$\frac{\partial^3}{\partial \theta^3} \log f(x, \theta) = -\frac{2}{\theta^3} + \frac{12x^2}{\theta^5} - \frac{6x}{\theta^4}.$$

To examine whether the third derivative of $\log f(x, \theta)$ is bounded by an integrable function, observe that for

$\theta \in N_\delta(\theta_0), \theta_0 - \delta < \theta < \theta_0 + \delta \iff \frac{1}{\theta_0+\delta} < \frac{1}{\theta} < \frac{1}{\theta_0-\delta}$. Hence,

$$\left| \frac{\partial^3}{\partial \theta^3} \log f(x, \theta) \right| = \left| -\frac{2}{\theta^3} + \frac{12x^2}{\theta^5} - \frac{6x}{\theta^4} \right| \le \left| \frac{2}{\theta^3} \right| + \left| \frac{12x^2}{\theta^5} \right| + \left| \frac{6x}{\theta^4} \right|$$

$$\le \left| \frac{2}{(\theta_0 - \delta)^3} \right| + \left| \frac{12x^2}{(\theta_0 - \delta)^5} \right| + \left| \frac{6x}{(\theta_0 - \delta)^4} \right| = M(x),$$

where $M(x) = \left|\frac{2}{(\theta_0-\delta)^3}\right| + \left|\frac{12x^2}{(\theta_0-\delta)^5}\right| + \left|\frac{6x}{(\theta_0-\delta)^4}\right|$. Now $E(X^2) = 2\theta^2$. We want to find a bound on $E(M(X))$, thus it is enough to find a bound on $E(|X|)$. Observe that $E(|X|) = E(|X - \theta + \theta|) \le E(|X - \theta|) + |\theta|$. Now,

$$E(|X - \theta|) = \frac{1}{\sqrt{2\pi}\theta} \int_{-\infty}^{\infty} |x - \theta| \exp\left\{-\frac{1}{2\theta^2}(x-\theta)^2\right\} dx$$

$$= \frac{1}{\sqrt{2\pi}\theta} \int_{-\infty}^{\infty} |y| \exp\left\{-\frac{1}{2\theta^2}y^2\right\} dy \quad \text{with} \quad y = x - \theta$$

$$= \frac{1}{\sqrt{2\pi}\theta} 2 \int_{0}^{\infty} y \exp\left\{-\frac{1}{2\theta^2}y^2\right\} dy$$

as integrand is an even function

$$= \frac{1}{\sqrt{2\pi}\theta} \int_{0}^{\infty} \exp\left\{-\frac{1}{2\theta^2}t\right\} dt \quad \text{with} \quad y^2 = t$$

$$= \frac{2\theta^2}{\sqrt{2\pi}\theta} \left[-\exp\{\frac{-t}{2\theta^2}\}\right]_{0}^{\infty} = \theta\sqrt{\frac{2}{\pi}}.$$

Thus, $E(|X|) = E(|X - \theta + \theta|) \le E(|X - \theta|) + |\theta| = \theta\sqrt{\frac{2}{\pi}} + \theta$. As a consequence,

$$E(M(X)) \le \left|\frac{2}{(\theta_0 - \delta)^3}\right| + 2\theta^2 \left|\frac{12}{(\theta_0 - \delta)^5}\right| + \left|\frac{6}{(\theta_0 - \delta)^4}\right| \left(\theta\sqrt{\frac{2}{\pi}} + \theta\right)$$

$$< \infty \; \forall \; \theta \in (\theta_0 - \delta, \theta_0 + \delta).$$

Thus, $N(\theta, \theta^2)$ distribution when $\theta \in (0, \infty)$ satisfies all the Cramér regularity conditions and hence it belongs to a Cramér family. $\square$

🖎 Remark 4.3.4

It is to be noted that the function $\left|\frac{2}{\theta^3}\right| + \left|\frac{12x^2}{\theta^5}\right| + \left|\frac{6x}{\theta^4}\right|$ for $\theta \in \Theta$ is not bounded as $\frac{2}{\theta^3}, \frac{12x^2}{\theta^5}$ and $\frac{6x}{\theta^4}$ can be made arbitrarily large by selecting θ sufficiently small. However, we want the Cramér regularity conditions to be satisfied for all $\theta \in N_\rho(\theta_0)$ and the function $\left|\frac{2}{(\theta_0-\delta)^3}\right| + \left|\frac{12x^2}{(\theta_0-\delta)^5}\right| + \left|\frac{6x}{(\theta_0-\delta)^4}\right|$ remains bounded for all $\theta \in N_\rho(\theta_0)$ for $\delta < \rho$.

In Example 3.3.9, it is shown that if $(X, Y)'$ has a bivariate normal distribution with parameter $(0, 0, 1, 1, \rho)'$, $\rho \in (-1, 1)$, then the sample correlation coefficient R_n is a CAN estimator of ρ with approximate variance $(1 - \rho^2)^2/n$. In the next example, we show that the family of a bivariate normal distribution with parameter $(0, 0, 1, 1, \rho)'$, $\rho \in (-1, 1)$, belongs to a Cramér family. Hence, for large n, the maximum likelihood estimator of ρ exists and is a CAN estimator of ρ with approximate variance $1/nI(\rho) = (1 - \rho^2)^2/n(1 + \rho^2)$, which is less than the approximate variance $(1 - \rho^2)^2/n$ of R_n as expected.

✔ Example 4.3.3

Suppose $(X, Y)'$ has a bivariate normal distribution with zero mean vector and dispersion matrix Σ given by

$$\Sigma = \begin{pmatrix} 1 & \rho \\ \rho & 1 \end{pmatrix},$$

$\rho \in (-1, 1)$. Then its probability density function $f(x, \rho)$ is given by

$$\frac{1}{2\pi\sqrt{1 - \rho^2}} \exp\left\{-\frac{(x^2 + y^2 - 2\rho xy)}{2(1 - \rho^2)}\right\}, \quad (x, y) \in \mathbb{R}^2, \quad -1 < \rho < 1.$$

It is to be noted that the parameter space $(-1, 1)$ is an open set and the support of the distribution is $\mathbb{R}^2$ which does not depend on the parameter ρ. Observe that

$$\log f(x, \rho) = -\frac{1}{2}\log(1 - \rho^2) - \frac{x^2 + y^2}{2(1 - \rho^2)} + \frac{\rho xy}{(1 - \rho^2)} - \log 2\pi$$

and it cannot be expressed in the form required for a one-parameter exponential family. Thus the distribution of $(X, Y)'$ does not belong to a one-parameter exponential family. To examine if it belongs to a Cramér family, we have already noted that the first two conditions are satisfied. Now we verify whether the derivatives of $\log f(x, \rho)$ up to order 3 exist. From the expression of $\log f(x, \rho)$ we have

$$\frac{\partial}{\partial \rho} \log f(x, \rho) = \frac{\rho}{1 - \rho^2} - \frac{\rho(x^2 + y^2)}{(1 - \rho^2)^2} + \frac{xy(1 + \rho^2)}{(1 - \rho^2)^2}$$

$$\frac{\partial^2}{\partial \rho^2} \log f(x, \rho) = \frac{1 + \rho^2}{(1 - \rho^2)^2} - \frac{(x^2 + y^2)(1 + 3\rho^2)}{(1 - \rho^2)^3} + \frac{xy(6\rho + 2\rho^3)}{(1 - \rho^2)^3}$$

$$\frac{\partial^3}{\partial \rho^3} \log f(x, \rho) = \frac{1 + 4\rho + 2\rho^3}{(1 - \rho^2)^3} - \frac{(x^2 + y^2)(12\rho(1 + \rho^2))}{(1 - \rho^2)^4}$$
$$+ \frac{6xy(1 + 6\rho^2 + \rho^4)}{(1 - \rho^2)^4}.$$

Suppose $\rho \in N_\delta(\rho_0)$, that is,
$\rho_0 - \delta < \rho < \rho_0 + \delta \quad \Leftrightarrow \quad 1/(\rho_0 + \delta) < 1/\rho < 1/(\rho_0 - \delta)$. Then

$$\left|\frac{\partial^3}{\partial \rho^3} \log f(x, \rho)\right| \leq \left|\frac{1 + 4\rho + 2\rho^3}{(1 - \rho^2)^3}\right| + \left|\frac{(x^2 + y^2)(12\rho(1 + \rho^2))}{(1 - \rho^2)^4}\right|$$
$$+ \left|\frac{6xy(1 + 6\rho^2 + \rho^4)}{(1 - \rho^2)^4}\right|$$

$$\leq \left| \frac{1 + 4(\rho_0 + \delta) + 2(\rho_0 + \delta)^3}{(1 - (\rho_0 - \delta)^2)^3} \right|$$

$$+ \left| \frac{(x^2 + y^2)(12(\rho_0 + \delta)(1 + (\rho_0 + \delta)^2))}{(1 - (\rho_0 - \delta)^2)^4} \right|$$

$$+ \left| \frac{6xy(1 + 6(\rho_0 + \delta)^2 + (\rho_0 + \delta)^4)}{(1 - (\rho_0 - \delta)^2)^4} \right|$$

$$= M(x, y).$$

Using the fact that $E(X^2 + Y^2) = 2$ and $E|XY| \leq (E(X^2)E(Y^2))^{1/2} = 1$, we note that $M(X, Y)$ is an integrable random variable $\forall\ \rho \in (\rho_0 - \delta, \rho_0 + \delta)$. From the second derivative of $\log f(x, \rho)$ we find the information function as follows:

$$I(\rho) = E\left(-\frac{\partial^2}{\partial \rho^2} \log f(X, \rho)\right)$$

$$= -\frac{1 + \rho^2}{(1 - \rho^2)^2} + \frac{E(X^2 + Y^2)(1 + 3\rho^2)}{(1 - \rho^2)^3} - \frac{E(XY)(6\rho + 2\rho^3)}{(1 - \rho^2)^3}$$

$$= -\frac{1 + \rho^2}{(1 - \rho^2)^2} + \frac{2(1 + 3\rho^2)}{(1 - \rho^2)^3} - \frac{\rho(6\rho + 2\rho^3)}{(1 - \rho^2)^3}$$

$$= \frac{2(1 - \rho^4)}{(1 - \rho^2)^3} - \frac{1 + \rho^2}{(1 - \rho^2)^2} = \frac{1 + \rho^2}{(1 - \rho^2)^2}.$$

Thus, $0 < I(\rho) < \infty$. Thus, all the Cramér regularity conditions are satisfied and the distribution of $(X, Y)'$ belongs to a Cramér family. Hence for large n, the maximum likelihood estimator of ρ exists and it is a CAN estimator of ρ with approximate variance $1/nI(\rho) = (1 - \rho^2)^2/n(1 + \rho^2)$. Now to find the maximum likelihood estimator of ρ, the log-likelihood function is given by

$$\log L_n(\rho|(\underline{X}, \underline{Y})) = -\frac{n}{2} \log(1 - \rho^2) - \frac{\sum_{i=1}^{n}(X_i^2 + Y_i^2)}{2(1 - \rho^2)} + \frac{\rho \sum_{i=1}^{n} X_i Y_i}{(1 - \rho^2)}.$$

Hence the likelihood equation is as follows:

$$\frac{\partial}{\partial \rho} \log L_n(\rho|(\underline{X}, \underline{Y})) = \frac{n\rho}{1 - \rho^2} - \frac{\rho \sum_{i=1}^{n}(X_i^2 + Y_i^2)}{(1 - \rho^2)^2} + \frac{(1 + \rho^2) \sum_{i=1}^{n} X_i Y_i}{(1 - \rho^2)^2} = 0$$

$$\Rightarrow \rho(1 - \rho^2) - \frac{\rho \sum_{i=1}^{n}(X_i^2 + Y_i^2)}{n} + \frac{(1 + \rho^2) \sum_{i=1}^{n} X_i Y_i}{n} = 0$$

$$\Rightarrow \rho^3 - \rho^2 V_n - \rho(1 - U_n) - V_n = 0$$

$$\Rightarrow a\rho^3 + 3b\rho^2 + 3c\rho + d = 0, \text{ say}$$

$$\text{where } U_n = \frac{\sum\limits_{i=1}^{n}(X_i^2 + Y_i^2)}{n}, \quad V_n = \frac{\sum\limits_{i=1}^{n}X_i Y_i}{n} \quad \& \ a = 1, \ 3b = -V_n,$$

$$3c = -(1 - U_n) \ \& \ d = -V_n.$$

Thus, the likelihood equation is a cubic equation in ρ. The condition for a unique real root for the cubic equation is $G^2 + 4H^3 > 0$, where $G = a^2d - 3abc + 2b^3$ and $H = ac - b^2$. To examine whether this condition is satisfied, suppose random variables G_n and H_n are defined as follows:

$$G_n = \frac{-4V_n}{3} + \frac{U_n V_n}{3} - \frac{2V_n^3}{27} \ \& \ H_n = -\frac{1}{3}\left[1 - U_n + \frac{V_n^2}{3}\right].$$

It is to be noted that

$$U_n = \frac{\sum\limits_{i=1}^{n}(X_i^2 + Y_i^2)}{n} \xrightarrow{P_\rho} E(X^2 + Y^2) = 2 \ \&$$

$$V_n = \frac{\sum\limits_{i=1}^{n}X_i Y_i}{n} \xrightarrow{P_\rho} E(XY) = \rho .$$

As a consequence,

$$-H_n = \frac{1}{3}\left[1 - U_n + \frac{V_n^2}{3}\right] \xrightarrow{P_\rho} \frac{1}{3}\left[1 - 2 + \frac{\rho}{3}\right]$$

$$= \frac{1}{3}\left[-1 + \frac{\rho}{3}\right] = C \in (-4/3, -2/3) \text{ as } \rho \in (-1, 1) .$$

Now

$$-H_n \xrightarrow{P_\rho} C < 0 \Rightarrow P[-H_n \le 0] \to 1 \text{ by Theorem 2.2.2}$$

$$\Rightarrow P[H_n \ge 0] \to 1 \quad \Rightarrow \quad P[H_n > 0] \to 1 \text{ as } P[H_n=0]=0$$

$$\Rightarrow P_\rho[G_n^2 + 4H_n^3 > 0] \to 1 \text{ as } n \to \infty.$$

Hence, with probability approaching 1 the cubic equation $\rho^3 - \rho^2 V_n - \rho(1 - U_n) - V_n = 0$ has a unique real root. Suppose $g(\rho) = \rho^3 - \rho^2 V_n - \rho(1 - U_n) - V_n$, then

$$g(-1) = -2V_n - U_n = -\frac{1}{n}\sum_{i=1}^{n}(X_i^2 + Y_i^2 + 2X_i Y_i)$$

$$= -\frac{1}{n}\sum_{i=1}^{n}(X_i + Y_i)^2 < 0 \ a.s.$$

$$\& \;\; g(1) = -2V_n + U_n = \frac{1}{n}\sum_{i=1}^{n}(X_i^2 + Y_i^2 - 2X_iY_i)$$

$$= \frac{1}{n}\sum_{i=1}^{n}(X_i - Y_i)^2 > 0 \;\; a.s.$$

Further g is a continuous function, hence with probability 1, the unique root is in $(-1, 1)$. Suppose the root is denoted by $\hat{\rho}_n$. Then by the Cramér-Huzurbazar theory, it is the maximum likelihood estimator of ρ and it is a CAN estimator of ρ with approximate variance

$$1/nI(\rho) = (1 - \rho^2)^2/n(1 + \rho^2).$$

Suppose $V_n = \sum_{i=1}^{n} X_iY_i/n = \sum_{i=1}^{n} U_i/n$, where $U_i = X_iY_i$, $i = 1, 2, \ldots, n$. Corresponding to a random sample from the distribution of $(X, Y)'$, we have a random sample $\{U_1, U_2, \ldots, U_n\}$. Further,

$$E(U_i) = \rho \;\; \& \;\; Var(U_i) = E(X^2Y^2) - (E(XY))^2 = E(X^2Y^2) - \rho^2 \, .$$

To find $E(X^2Y^2)$, we use the result that the conditional distribution of X given Y is normal $N(\rho Y, 1 - \rho^2)$. Observe that

$$
\begin{aligned}
E(X^2Y^2) &= E(\{E(X^2Y^2)|Y\}) = E(Y^2\{E(X^2)|Y\}) \\
&= E(Y^2\{Var(X|Y) + (E(X|Y))^2\}) = E(Y^2\{(1 - \rho^2) + (\rho Y)^2\}) \\
&= (1 - \rho^2)E(Y^2) + \rho^2 E(Y^4) = (1 - \rho^2) + 3\rho^2 = 1 + 2\rho^2 \, .
\end{aligned}
$$

Thus, $Var(U_i) = 1 + \rho^2$, which is positive and finite. By the WLLN and by the CLT

$$V_n \overset{P_\rho}{\to} \rho \;\; \& \;\; \sqrt{n}(V_n - \rho) \overset{L}{\to} Z1 \sim N(0, 1 + \rho^2) \, .$$

Hence, V_n is CAN for ρ with approximate variance $(1 + \rho^2)/n$. It is to be noted that V_n is a moment estimator of ρ. Thus, both V_n and $\hat{\rho}_n$ are CAN for ρ. Further,

$$1 + \rho^2 - \frac{(1 - \rho^2)^2}{(1 + \rho^2)} = \frac{(1 + \rho^2)^2 - (1 - \rho^2)^2}{(1 + \rho^2)} = \frac{4\rho^2}{(1 + \rho^2)} > 0 \, ,$$

implying that $\hat{\rho}_n$ is more efficient than T_n, which is expected as $\hat{\rho}_n$ is a BAN estimator of ρ. $\square$

✍ Remark 4.3.5

From Example 3.3.9 and Example 4.3.3, we note that if $(X, Y)'$ has a bivariate normal distribution with parameter $(0, 0, 1, 1, \rho)'$, $\rho \in (-1, 1)$, then we have the following three CAN estimators of ρ: (i) the sample correlation coefficient R_n with approximate variance $(1 - \rho^2)^2/n$, (ii) the maximum likelihood estimator $\hat{\rho}_n$ of ρ with approximate variance $(1 - \rho^2)^2/n(1 + \rho^2)$ and (iii) the moment estimator of ρ given by $V_n = \sum_{i=1}^{n} X_i Y_i /n$ with approximate variance $(1 + \rho^2)/n$. Among these three $\hat{\rho}_n$ has the smallest variance. In Example 4.5.4, R code is given to derive the maximum likelihood estimator $\hat{\rho}_n$ and to verify that it is a CAN estimator of ρ. In the same example it is shown that for a given sample, values of $\hat{\rho}_n$ and R_n are close. In Example 5.3.2, we also note that for various values of ρ, the values of the approximate variances of $\hat{\rho}_n$ and R_n are close. Using routine procedure, it can be proved that if $(X, Y)'$ has a bivariate normal distribution with parameter $(\mu_1, \mu_2, \sigma_1^2, \sigma_2^2, \rho)'$, where $\mu_1, \mu_2 \in \mathbb{R}$, $\sigma_1^2, \sigma_2^2 > 0$ and $\rho \in (-1, 1)$, then the distribution belongs to a five-parameter exponential family. The sample correlation coefficient R_n is the maximum likelihood estimator of ρ as well as a moment estimator of ρ based on a sufficient statistic and is a CAN estimator of ρ with approximate variance $(1 - \rho^2)^2/n$.

We have noted that if the probability distribution belongs to either a one-parameter exponential family or a Cramér family with indexing parameter θ, then for large n, the maximum likelihood estimator of θ is BAN for θ. Hence, it is better to obtain an interval estimator based on the maximum likelihood estimator of θ or to define a test statistic for testing certain hypotheses using the maximum likelihood estimator of θ.

Hodges and Lecam have given one example in 1953 in which they have proposed an estimator which is better than the maximum likelihood estimator, in the sense of having smaller variance at least at one parametric point, (Kale and Muralidharan [6]). The estimator proposed by Hodges and Lecam is hence known as super efficient estimator. We discuss it below:

Super efficient estimator: Suppose $\{X_1, X_2, \ldots, X_n\}$ is a random sample from a normal $N(\theta, 1)$ distribution. Then the sample mean $\overline{X}_n \sim N(\theta, 1/n)$ distribution for any n. It is to be noted that the variance $1/n$ of $\overline{X}_n$ is same as the Cramér Rao lower bound for the variance of an unbiased estimator. Hence, $\overline{X}_n$ is an efficient estimator of θ. Hodges and Lecam proposed an estimator T_n of θ as follows. Suppose for $0 < \alpha < 1$,

$$T_n = \begin{cases} \overline{X}_n, & \text{if } |\overline{X}_n| > n^{-1/4} \\ \alpha \overline{X}_n, & \text{if } |\overline{X}_n| \leq n^{-1/4} . \end{cases}$$

A technique of defining T_n in this way is known as a shrinkage technique. If $\overline{X}_n$ is small, it is made further small by multiplying it by a fraction, to define T_n. The cutoff $n^{-1/4}$ can be replaced by any sequence $\{a_n, n \geq 1\}$ such that $a_n \to 0$ and $\sqrt{n}a_n \to \infty$. For example, $a_n = 1/\log n$ or $a_n = n^{\frac{-1}{2}+\delta}$ where $0 < \delta < 1/2$. We now show that T_n with any such a_n is CAN for θ and find its approximate variance.

Suppose $Y_n = \sqrt{n}(T_n - \theta) - \sqrt{n}(\overline{X}_n - \theta) = \sqrt{n}(T_n - \overline{X}_n)$. Suppose $\theta \neq 0$. Then for $\epsilon > 0$,

$$P[|Y_n| < \epsilon] = P[|\sqrt{n}(T_n - \overline{X}_n)| < \epsilon] \geq P[T_n = \overline{X}_n] = P[|\overline{X}_n| > a_n]$$
$$= 1 - P[-a_n < \overline{X}_n < a_n] = 1 - \Phi(\sqrt{n}(a_n - \theta)) + \Phi(-\sqrt{n}(a_n + \theta)).$$

Now $\theta \neq 0 \Rightarrow d(\theta, 0) > 0$, where d is a distance function. Suppose $d(\theta, 0) = \lambda$. The sequence $\{a_n, n \geq 1\}$ is such that $a_n \to 0$, that is, given $\epsilon_1 > 0$, there exists $n_0(\epsilon_1)$ such that $\forall\ n \geq n_0(\epsilon_1)$, $|a_n| < \epsilon_1$. Suppose $\epsilon_1 = \lambda$. Then $\forall\ n \geq n_0(\lambda)$, $|a_n| < \lambda$.

Case(i): Suppose $\theta < -a_n$ then $\theta + a_n < 0$. Further, $\theta < -a_n < a_n \Rightarrow \theta - a_n < 0$. Thus, in this case $P[|Y_n| < \epsilon] \to 1$.

Case(ii): Suppose $\theta > a_n$ then $\theta - a_n > 0$. Further, $\theta > a_n > -a_n \Rightarrow \theta + a_n > 0$ and in this case also $P[|Y_n| < \epsilon] \to 1$. Hence, when $\theta \neq 0$, $Y_n \xrightarrow{P} 0$. As a consequence, asymptotic distribution of $\sqrt{n}(T_n - \theta)$ and of $\sqrt{n}(\overline{X}_n - \theta)$ is the same. But $\sqrt{n}(\overline{X}_n - \theta) \xrightarrow{L} Z \sim N(0, 1)$ and hence $\sqrt{n}(T_n - \theta) \xrightarrow{L} Z \sim N(0, 1)$.

Suppose $\theta = 0$. Observe that

$$\sqrt{n}(T_n) = \sqrt{n}(T_n - \alpha \overline{X}_n + \alpha \overline{X}_n) = \sqrt{n}(T_n - \alpha \overline{X}_n) + \sqrt{n}\alpha \overline{X}_n = U_n + \sqrt{n}\alpha \overline{X}_n.$$

Then for $\epsilon > 0$,

$$P[|U_n| < \epsilon] = P[|\sqrt{n}(T_n - \alpha \overline{X}_n)| < \epsilon] \geq P[T_n = \alpha \overline{X}_n] = P[|\overline{X}_n| < a_n]$$
$$= P[-\sqrt{n}a_n < \sqrt{n}(\overline{X}_n) < \sqrt{n}a_n] = \Phi(\sqrt{n}a_n) - \Phi(-\sqrt{n}a_n)$$
$$\to 1 \text{ as } n \to \infty \text{ as } \sqrt{n}a_n \to \infty.$$

Hence, when $\theta = 0$, $U_n \xrightarrow{P} 0$. As a consequence, asymptotic distribution of $\sqrt{n}(T_n)$ and of $\alpha\sqrt{n}(\overline{X}_n)$ is the same. But $\alpha\sqrt{n}(\overline{X}_n) \xrightarrow{L} Z_1 \sim N(0, \alpha^2)$ and hence $\sqrt{n}(T_n) \xrightarrow{L} Z_1 \sim N(0, \alpha^2)$. Thus for all θ, asymptotic distribution of T_n is normal with approximate variance $v(\theta)/n$, where $v(\theta)$ is given by

$$v(\theta) = \begin{cases} 1, & \text{if } \theta \neq 0 \\ \alpha^2, & \text{if } \theta = 0 \end{cases}$$

Thus at $\theta = 0$, $Var(T_n) < Var(\overline{X}_n)$ and hence T_n is labeled as a super efficient estimator.

In general, one can use similar shrinkage technique to improve a CAN estimator that minimizes the variance at a fixed point θ_0. More specifically, suppose S_n is a CAN estimator of θ with approximate variance $\sigma(\theta)/n$. Suppose $\{a_n, n \geq 1\}$ is a sequence of real numbers such that $a_n \to 0$ and $\sqrt{n}a_n \to \infty$. For $0 < \alpha < 1$, an estimator T_n is defined as

$$T_n = \begin{cases} S_n, & \text{if } |S_n - \theta_0| > a_n \\ \alpha(S_n - \theta_0) + \theta_0, & \text{if } |S_n - \theta_0| \leq a_n \end{cases}$$

Then it can be shown that T_n has asymptotically normal distribution with mean θ and approximate variance $w(\theta)/n$, where $w(\theta)$ is given by

$$w(\theta) \;=\; \begin{cases} \sigma(\theta), & \text{if} \quad \theta \neq \theta_0 \\ \alpha^2 \sigma(\theta), & \text{if} \quad \theta = \theta_0 \end{cases}$$

The shrinkage technique can be extended to reduce the variance at a finite number of points. Thus, the set of parameters, at which approximate variance of the super efficient estimator can be made smaller than the approximate variance of the maximum likelihood estimator, has Lebesgue measure 0. In view of the fact that the reduction in variance can be achieved only at finitely many parametric points, the super efficient estimator is not practically useful. For more discussion on this, we refer to Kale and Muralidharan [6].

🖎 Remark 4.3.6

In view of the above example of a super efficient estimator given by Hodges and Lecam, many statisticians object to label the CAN estimator with approximate variance $1/nI(\theta)$ as a BAN estimator. More discussion on this issue can be found in Rao [8], p. 348.

We now discuss the concept of asymptotic relative efficiency (ARE). It is useful to compare two CAN estimators via their asymptotic variances. It is defined as follows:

▶ Definition 4.3.1

Asymptotic Relative Efficiency: Suppose T_{1n} and T_{2n} are two CAN estimators of θ with approximate variance σ_1^2/n and σ_2^2/n respectively. Then ARE of T_{1n} with respect to T_{2n} is

$$ARE(T_{1n}, T_{2n}) = \frac{\sigma_2^2}{\sigma_1^2} \; .$$

If the norming factors of T_{1n} and T_{2n} are a_n and b_n respectively, then ARE of T_{1n} with respect to T_{2n} is defined as

$$ARE(T_{1n}, T_{2n}) = \lim_{n \to \infty} \frac{\sigma_2^2/b_n^2}{\sigma_1^2/a_n^2} \; .$$

🖎 Remark 4.3.7

If the norming factors a_n and b_n tend to ∞ at the same rate then, $ARE(T_{1n}, T_{2n}) = \sigma_2^2/\sigma_1^2$.

If $ARE(T_{1n}, T_{2n}) > 1$, then T_{1n} is preferred to T_{2n} as an estimator of θ. For a Cauchy $C(\theta, 1)$ distribution, the maximum likelihood estimator T_{1n} of θ is CAN for

θ with the approximate variance $2/n$. Further, the sample median T_{2n} is CAN for θ with the approximate variance $\pi^2/4n$. Thus

$$ARE(T_{1n}, T_{2n}) = \frac{\pi^2}{8} = 1.2324 > 1 .$$

Thus, the maximum likelihood estimator T_{1n} is preferred to the sample median. However, the gain in efficiency is marginal and it would be better to use the sample median in view of its computational ease.

For a normal $N(\theta, 1)$ distribution, the maximum likelihood estimator T_{1n} of θ is CAN for θ with the approximate variance $1/n$ and the sample median T_{2n} is CAN for θ with the approximate variance $\pi/2n$. Thus, ARE of sample mean with respect to sample median is

$$ARE(T_{1n}, T_{2n}) = \frac{\pi}{2} = 1.5714 > 1 .$$

Thus, the maximum likelihood estimator T_{1n} is again preferred to the sample median. In other words, ARE of the sample median with respect to the sample mean is $2/\pi = 0.6364$. It is interpreted as follows. If we use the sample mean instead of the sample median to estimate θ, then we get the same accuracy with 64% of observations.

Suppose X follows a Laplace $(\theta, 1)$ distribution with probability density function given by

$$f(x, \theta) = (1/2) \exp\{-|x - \theta|\}, \quad x \in \mathbb{R}, \ \theta \in \mathbb{R}.$$

It is easy to verify that the sample median $T_{1n} = X_{([n/2]+1)}$ is CAN for θ with approximate variance $1/n$. If X follows Laplace distribution, then $E(X) = \theta$ and $Var(X) = 2 < \infty$. Thus, by the WLLN and by the CLT, $T_{2n} = \overline{X}_n$ is CAN for θ with approximate variance $2/n$. Hence,

$$ARE(T_{1n}, T_{2n}) = \sigma_2^2/\sigma_1^2 = 2,$$

which implies that the sample median is a better estimator of θ than the sample mean. It is to be noted that the sample median is the maximum likelihood estimator of θ.

If the distribution belongs to either a one-parameter exponential family or a Cramér family, then we know that the maximum likelihood estimator is asymptotically efficient and the other estimator cannot be better than that. However, other estimators may have some desirable properties, such as ease of computation, robustness to underlying assumptions, which make them desirable. For example, in gamma $G(\alpha, \lambda)$ distribution, finding the maximum likelihood estimator of $(\alpha, \lambda)'$ is difficult than the moment estimator. In such cases, efficiency of the maximum likelihood estimator becomes important in calibrating what we are giving up, if we use another estimator.

We now briefly discuss the extension of Cramér-Huzurbazar theory to a multiparameter setup.

Cramér Huzurbazar theory in a multiparameter setup: Suppose X is a random variable or a random vector with the probability law $f(x, \underline{\theta})$ which is indexed by a vector parameter $\underline{\theta} = (\theta_1, \theta_2, \ldots, \theta_k)' \in \Theta \subset \mathbb{R}^k$. Suppose $\{X_1, X_2, \ldots, X_n\}$ is a random sample of size n from the distribution of X. Suppose the probability law $f(x, \underline{\theta})$ satisfies the following conditions in a neighborhood $N_\rho(\underline{\theta}_0) \subset \Theta$, where $\underline{\theta}_0$ is a true parameter value.

C-1 The support S_f is free from the parameter $\underline{\theta}$.

C-2 There exists an open subset of a parameter space which contains the true parameter point $\underline{\theta}_0$.

C-3 The partial derivatives $\frac{\partial}{\partial \theta_i} \log f(x, \underline{\theta})$, $i = 1, 2, \ldots, k$, $\frac{\partial^2}{\partial \theta_i \partial \theta_j} \log f(x, \underline{\theta})$, $i, j = 1, 2, \ldots, k$ and $\frac{\partial^3}{\partial \theta_i \partial \theta_j \partial \theta_l} \log f(x, \underline{\theta}), i, j, l = 1, 2, \ldots, k$ exist for almost all values of $x \in S_f$.

C-4 The identity $\int_{S_f} f(x, \underline{\theta}) dx = 1$ can be differentiated with respect to $\theta_i' s$ under the integral sign at least twice. As a consequence, $E\left(\frac{\partial}{\partial \theta_i} \log f(X, \underline{\theta})\right) = 0$, $i = 1, 2, \ldots, k$ and the information matrix $I(\underline{\theta}) = \left[I_{ij}(\underline{\theta})\right]$ is given by

$$I_{ij}(\underline{\theta}) = E\left(\frac{\partial}{\partial \theta_i} \log f(X, \underline{\theta}) \frac{\partial}{\partial \theta_j} \log f(X, \underline{\theta})\right)$$

$$= E\left(-\frac{\partial^2}{\partial \theta_i \partial \theta_j} \log f(X, \underline{\theta})\right) i, j = 1, 2, \ldots, k.$$

Further $I(\underline{\theta})$ is a positive definite matrix.

C-5 There exist functions $M_{ijl}(x)$ such that $\left|\frac{\partial^3}{\partial \theta_i \partial \theta_j \partial \theta_l} \log f(x, \underline{\theta})\right| < M_{ijl}(x)$, where $M_{ijl}(x)$ may depend on $\underline{\theta}_0$ and ρ and $E(M_{ijl}(X)) < \infty$ for all $i, j, l = 1, 2, \ldots, k$. Thus, the third order partial derivatives of $\log f(x, \underline{\theta})$ are bounded by integrable functions.

If the probability law $f(x, \underline{\theta})$ satisfies these Cramér regularity conditions in a neighborhood $N_\rho(\underline{\theta}_0) \subset \Theta$, then the corresponding family of distributions is a multiparameter Cramér family. It can be verified that a multiparameter exponential family is a subclass of a multiparameter Cramér family. As in Example 4.3.1, we can show that a Cauchy $C(\theta, \sigma)$ distribution with location parameter θ and scale parameter σ belongs to a two-parameter Cramér family. However, the Laplace distribution with probability density function

$$f(x, \theta, \alpha) = \frac{1}{2\alpha} \exp\left\{-\frac{|x - \theta|}{\alpha}\right\}, \quad x \in \mathbb{R}, \quad \theta \in \mathbb{R}, \quad \alpha > 0$$

does not belong to a two-parameter Cramér family, as the third condition gets violated. A uniform $U(\theta - \alpha, \theta + \alpha)$ distribution is not a member of a two-parameter

Cramér family, as its support depends on the parameters. A negative binomial distribution with probability mass function,

$$P_\theta[X = x] = \binom{x + k - 1}{x} p^k (1 - p)^x, \quad x = 0, 1, \ldots, \quad 0 < p < 1, \quad k \in I^+$$

does not belong to a two-parameter Cramér family as the parameter space in not open. If $k \in (0, \infty)$, then it belongs to a two-parameter Cramér family.

As in the case of the distributions belonging to a Cramér family with real indexing parameter, following four results are true in a multiparameter setup when the distribution satisfies the five regularity conditions stated above. We state these below, for more details, we refer to Kale and Muralidharan [6] and references cited therein.

Cramér-Huzurbazar theorem in a k-parameter Cramér family: Suppose $\{X_1, X_2, \ldots, X_n\}$ is a random sample from the distribution of X having the probability law $f(x, \underline{\theta})$ and $\underline{\theta}_0$ is a true parameter.

▶ **Result 4.3.5**
With probability approaching 1 as $n \to \infty$, a system of likelihood equations given by $\frac{\partial}{\partial \theta_i} \log L_n(\theta | \underline{X}) = 0$, $i = 1, 2, \ldots, k$ admits a solution $\hat{\underline{\theta}}_n(\underline{X})$ which is consistent for $\underline{\theta}_0$.

▶ **Result 4.3.6**
For large n, the distribution of $\hat{\underline{\theta}}_n(\underline{X})$ can be approximated by the normal distribution $N_k(\underline{\theta}_0, I^{-1}(\underline{\theta}_0)/n)$, that is,

$$\sqrt{n}\left(\hat{\underline{\theta}}_n(\underline{X}) - \underline{\theta}_0\right) \xrightarrow{L} \underline{Z}_1 \sim N_k\left(0, I^{-1}(\underline{\theta}_0)\right).$$

▶ **Result 4.3.7**
With probability approaching 1 as $n \to \infty$, there is a relative maximum of the likelihood function at $\hat{\underline{\theta}}_n(\underline{X})$, that is, the matrix $D = \left[\frac{\partial^2}{\partial \theta_i \partial \theta_j} \log L_n(\theta | \underline{X})|_{\hat{\underline{\theta}}_n(\underline{X})}\right]$ of second order partial derivatives evaluated at $\hat{\underline{\theta}}_n(\underline{X})$ is almost surely negative definite.

▶ **Result 4.3.8**
With probability approaching 1 as $n \to \infty$, a consistent solution of the system of likelihood equations is unique.

✍ **Remark 4.3.8**

If $X \sim N(\mu, \sigma^2)$, then we have seen in Example 4.2.6 that it belongs to a two-parameter exponential family, hence it also belongs to a Cramér family. If $X_1, X_2, \ldots, X_n$ is a random sample from normal $N(\mu, \sigma^2)$, then it is shown in Example 3.3.2, that the maximum likelihood estimator of (μ, σ^2) is CAN with approximate variance-covariance matrix $I^{-1}(\mu, \sigma^2)$. Similarly, it can be shown that the distribution of $(X, Y)'$ with the probability mass function as

$$P[X = x, Y = y] = \binom{x}{y} e^{-\lambda} \lambda^x p^y (1-p)^{x-y}/x!,$$
$$y = 0, 1, \ldots, x; \ \ x = 0, 1, 2, \ldots,$$

where $\lambda > 0$ and $0 < p < 1$, belongs to a two-parameter exponential family and hence it also belongs to a two-parameter Cramér family. In Example 3.3.4, it is shown that the maximum likelihood estimator of (λ, p) is CAN with approximate dispersion matrix $I^{-1}(\lambda, p)$. As expected these results are consistent with the Cramér-Huzurbazar theorem in a multiparameter setup.

We now discuss two examples in which the distributions belong to a multiparameter Cramér family.

Example 4.3.4

Suppose a random vector $(X, Y)'$ follows a bivariate normal $N_2(0, 0, \sigma_1^2, \sigma_2^2, \rho)$ distribution where $\rho \neq 0$ is a known correlation coefficient. The probability density function of $(X, Y)'$ is given by

$$f(x, y, \sigma_1^2, \sigma_2^2) = \frac{1}{2\pi\sigma_1\sigma_2\sqrt{1-\rho^2}} \exp\left\{-\frac{1}{2(1-\rho^2)}\left(\frac{x^2}{\sigma_1^2} - \frac{2\rho x y}{\sigma_1\sigma_2} + \frac{y^2}{\sigma_2^2}\right)\right\},$$
$$(x, y)' \in \mathbb{R}^2,$$

$\sigma_1^2, \sigma_2^2 > 0$. Thus, the support of the distribution is free from the parameters, parameter space Θ is $\Theta = \{(\sigma_1^2, \sigma_2^2)' | \sigma_1^2, \sigma_2^2 > 0\}$ and it is an open set. However, $f(x, y, \sigma_1^2, \sigma_2^2)$ cannot be expressed in a form required for a two-parameter family. Hence, the distribution does not belong to a two-parameter exponential family. We examine whether it belongs to a two-parameter Cramér family. We have noted that the first two conditions are satisfied. The partial derivatives up to order three, of $\log f(x, y, \sigma_1^2, \sigma_2^2)$ are as follows. Suppose c is a constant free from parameters. Then

$$\log f(x, y, \sigma_1^2, \sigma_2^2) = c - \frac{1}{2}\log\sigma_1^2 - \frac{1}{2}\log\sigma_2^2 - \frac{x^2}{2(1-\rho^2)\sigma_1^2} + \frac{\rho x y}{(1-\rho^2)(\sigma_1^2)^{1/2}(\sigma_2^2)^{1/2}} - \frac{y^2}{2(1-\rho^2)\sigma_2^2}$$

$$\frac{\partial}{\partial\sigma_1^2}\log f(x, y, \sigma_1^2, \sigma_2^2) = -\frac{1}{2\sigma_1^2} + \frac{x^2}{2(1-\rho^2)\sigma_1^4} - \frac{\rho x y}{2(1-\rho^2)(\sigma_1^2)^{3/2}(\sigma_2^2)^{1/2}}$$

$$\frac{\partial}{\partial\sigma_2^2} \log f(x, y, \sigma_1^2, \sigma_2^2) = -\frac{1}{2\sigma_2^2} + \frac{y^2}{2(1-\rho^2)\sigma_2^4}$$
$$- \frac{\rho x y}{2(1-\rho^2)(\sigma_1^2)^{1/2}(\sigma_2^2)^{3/2}}$$

$$\frac{\partial^2}{\partial(\sigma_1^2)^2} \log f(x, y, \sigma_1^2, \sigma_2^2) = \frac{1}{2\sigma_1^4} - \frac{x^2}{(1-\rho^2)\sigma_1^6}$$
$$+ \frac{3\rho x y}{4(1-\rho^2)(\sigma_1^2)^{5/2}(\sigma_2^2)^{1/2}}$$

$$\frac{\partial^2}{\partial(\sigma_2^2)^2} \log f(x, y, \sigma_1^2, \sigma_2^2) = \frac{1}{2\sigma_2^4} - \frac{y^2}{(1-\rho^2)\sigma_2^6}$$
$$+ \frac{3\rho x y}{4(1-\rho^2)(\sigma_2^2)^{5/2}(\sigma_1^2)^{1/2}}$$

$$\frac{\partial^2}{\partial(\sigma_2^2)\partial(\sigma_1^2)} \log f(x, y, \sigma_1^2, \sigma_2^2) = \frac{\rho x y}{4(1-\rho^2)(\sigma_2^2)^{3/2}(\sigma_1^2)^{3/2}}$$

$$\frac{\partial^3}{\partial(\sigma_1^2)^3} \log f(x, y, \sigma_1^2, \sigma_2^2) = -\frac{1}{\sigma_1^6} + \frac{3x^2}{(1-\rho^2)\sigma_1^8}$$
$$- \frac{15\rho x y}{8(1-\rho^2)(\sigma_1^2)^{7/2}(\sigma_2^2)^{1/2}}$$

$$\frac{\partial^3}{\partial(\sigma_2^2)^3} \log f(x, y, \sigma_1^2, \sigma_2^2) = -\frac{1}{\sigma_2^6} + \frac{3y^2}{(1-\rho^2)\sigma_2^8}$$
$$- \frac{15\rho x y}{8(1-\rho^2)(\sigma_1^2)^{1/2}(\sigma_2^2)^{7/2}}$$

$$\frac{\partial^3}{\partial(\sigma_2^2)\partial(\sigma_1^2)^2} \log f(x, y, \sigma_1^2, \sigma_2^2) = -\frac{3\rho x y}{8(1-\rho^2)(\sigma_1^2)^{5/2}(\sigma_2^2)^{3/2}}$$

$$\frac{\partial^3}{\partial(\sigma_1^2)\partial(\sigma_2^2)^2} \log f(x, y, \sigma_1^2, \sigma_2^2) = -\frac{3\rho x y}{8(1-\rho^2)(\sigma_1^2)^{3/2}(\sigma_2^2)^{5/2}}.$$

Thus, partial derivatives of $\log f(x, y, \sigma_1^2, \sigma_2^2)$ up to order three exist. We further examine whether the third order partial derivatives are bounded by integrable functions. Observe that for $\delta > 0$, $\sigma_1^2 \in (\sigma_{01}^2 - \delta, \sigma_{01}^2 + \delta)$ and $\sigma_2^2 \in (\sigma_{02}^2 - \delta, \sigma_{02}^2 + \delta)$, where σ_{01}^2 and σ_{02}^2 are true parameter values, we have

$$\left| \frac{\partial^3}{\partial(\sigma_1^2)^3} \log f(x, y, \sigma_1^2, \sigma_2^2) \right| \leq \left| \frac{1}{(\sigma_{01} - \delta)^6} \right| + \left| \frac{3x^2}{(1-\rho^2)(\sigma_{01} - \delta)^8} \right|$$
$$+ \left| \frac{15\rho x y}{8(1-\rho^2)(\sigma_{01} - \delta)^7(\sigma_{02} - \delta)} \right|$$
$$= M_{111}(x, y) \quad \text{say}$$

$$\&\ \ E(M_{111}(X,Y)) \le \left| \frac{1}{(\sigma_{01} - \delta)^6} \right| + \left| \frac{3\sigma_1^2}{(1 - \rho^2)(\sigma_{01} - \delta)^8} \right|$$

$$+ \left| \frac{15\rho\sigma_1\sigma_2}{8(1 - \rho^2)(\sigma_{01} - \delta)^7(\sigma_{02} - \delta)} \right| < \infty.$$

Hence, the third order partial derivative $\frac{\partial^3}{\partial(\sigma_1^2)^3} \log f(x, y, \sigma_1^2, \sigma_2^2)$ is bounded by an integrable function. On similar lines, we can show that the remaining third order partial derivatives are bounded by integrable functions. We now find the information matrix $I(\sigma_1^2, \sigma_2^2) = [I_{i,j}(\sigma_1^2, \sigma_2^2)]$. We have $E(X^2) = \sigma_1^2$, $E(Y^2) = \sigma_2^2$ & $E(XY) = \rho\sigma_1\sigma_2$. Hence,

$$I_{1,1}(\sigma_1^2, \sigma_2^2) = E\left(-\frac{\partial^2}{\partial(\sigma_1^2)^2} \log f(X, Y, \sigma_1^2, \sigma_2^2) \right)$$

$$= E\left(-\frac{1}{2\sigma_1^4} + \frac{X^2}{(1 - \rho^2)\sigma_1^6} - \frac{3\rho XY}{4(1 - \rho^2)(\sigma_1^2)^{5/2}(\sigma_2^2)^{1/2}} \right)$$

$$= -\frac{1}{2\sigma_1^4} + \frac{1}{(1 - \rho^2)\sigma_1^4} - \frac{3\rho^2}{4(1 - \rho^2)\sigma_1^4} = \frac{2 - \rho^2}{4(1 - \rho^2)\sigma_1^4}.$$

Similarly, $\ I_{2,2}(\sigma_1^2, \sigma_2^2) = \dfrac{2 - \rho^2}{4(1 - \rho^2)\sigma_2^4}$

Now, $\ I_{1,2}(\sigma_1^2, \sigma_2^2) = I_{2,1}(\sigma_1^2, \sigma_2^2) = E\left(-\dfrac{\partial^2}{\partial(\sigma_2^2)\partial(\sigma_1^2)} \log f(X, Y, \sigma_1^2, \sigma_2^2) \right)$

$$= E\left(-\frac{\rho XY}{4(1 - \rho^2)(\sigma_2^2)^{3/2}(\sigma_1^2)^{3/2}} \right) = -\frac{\rho^2}{4(1 - \rho^2)\sigma_1^2\sigma_2^2}.$$

Thus, the information matrix $I(\sigma_1^2, \sigma_2^2)$ is given by

$$I(\sigma_1^2, \sigma_2^2) = \begin{pmatrix} \dfrac{2 - \rho^2}{4(1 - \rho^2)\sigma_1^4} & -\dfrac{\rho^2}{4(1 - \rho^2)\sigma_1^2\sigma_2^2} \\[3mm] -\dfrac{\rho^2}{4(1 - \rho^2)\sigma_1^2\sigma_2^2} & \dfrac{2 - \rho^2}{4(1 - \rho^2)\sigma_2^4} \end{pmatrix}.$$

It is a positive definite matrix as its first principle minor is positive and $|I(\sigma_1^2, \sigma_2^2)| = 1/4(1 - \rho^2)\sigma_1^4\sigma_2^4 > 0$. Thus, all the Cramér regularity conditions are satisfied and hence the bivariate normal $N_2(0, 0, \sigma_1^2, \sigma_2^2, \rho)$ distribution belongs to a two-parameter Cramér family. Hence, based on the sample of size n from the distribution of $(X, Y)'$, the maximum likelihood estimator $(\hat{\sigma}_{1n}^2, \hat{\sigma}_{2n}^2)'$ of $(\sigma_1^2, \sigma_2^2)'$ is a CAN estimator with approximate dispersion matrix $I^{-1}(\sigma_1^2, \sigma_2^2)/n$. Thus,

$$\sqrt{n}(\hat{\sigma}_{1n}^2 - \sigma_1^2,\ \hat{\sigma}_{2n}^2 - \sigma_2^2)' \xrightarrow{L} \underline{Z} \sim N_2(\underline{0}, I^{-1}(\sigma_1^2, \sigma_2^2)).$$

The inverse of the information matrix $I(\sigma_1^2, \sigma_2^2)$ is given by

$$I^{-1}(\sigma_1^2, \sigma_2^2) = \begin{pmatrix} (2 - \rho^2)\sigma_1^4 & \rho^2\sigma_1^2\sigma_2^2 \\ \rho^2\sigma_1^2\sigma_2^2 & (2 - \rho^2)\sigma_2^4 \end{pmatrix}.$$

To find the maximum likelihood estimators, the system of likelihood equations is as given below.

$$-\frac{n}{2\sigma_1^2} + \frac{\sum\limits_{i=1}^{n} X_i^2}{2(1-\rho^2)\sigma_1^4} - \frac{\rho \sum\limits_{i=1}^{n} X_i Y_i}{2(1-\rho^2)(\sigma_1^3)(\sigma_2)} = 0 \qquad (4.3.5)$$

$$-\frac{n}{2\sigma_2^2} + \frac{\sum\limits_{i=1}^{n} Y_i^2}{2(1-\rho^2)\sigma_2^4} - \frac{\rho \sum\limits_{i=1}^{n} X_i Y_i}{2(1-\rho^2)(\sigma_1)(\sigma_2^3)} = 0. \qquad (4.3.6)$$

Multiplying Eq. (4.3.5) by $1/\sigma_2^2$ and subtracting from it $1/\sigma_1^2 \times$ Eq. (4.3.6), we have

$$\frac{1}{2(1-\rho^2)\sigma_1\sigma_2}\left(\sum_{i=1}^{n} X_i^2/\sigma_1^2 - \sum_{i=1}^{n} Y_i^2/\sigma_2^2\right) = 0.$$

Hence, $\displaystyle\sum_{i=1}^{n} X_i^2/\sigma_1^2 = \sum_{i=1}^{n} Y_i^2/\sigma_2^2 \Rightarrow 1/\sigma_2 = \left(\sum_{i=1}^{n} X_i^2/\sigma_1^2 \sum_{i=1}^{n} Y_i^2\right)^{1/2}$

$$\Rightarrow \sigma_1^2 = \frac{1}{1-\rho^2}\left(\frac{\sum\limits_{i=1}^{n} X_i^2}{n} - \rho\frac{\sum\limits_{i=1}^{n} X_i Y_i}{n}\left(\frac{\sum\limits_{i=1}^{n} X_i^2}{\sum\limits_{i=1}^{n} Y_i^2}\right)^{1/2}\right)$$

from Eq. (4.3.5)

$$\& \; \sigma_2^2 = \sigma_1^2\left(\sum_{i=1}^{n} Y_i^2 / \sum_{i=1}^{n} X_i^2\right)$$

$$= \frac{1}{1-\rho^2}\left(\frac{\sum\limits_{i=1}^{n} Y_i^2}{n} - \rho\frac{\sum\limits_{i=1}^{n} X_i Y_i}{n}\left(\frac{\sum\limits_{i=1}^{n} Y_i^2}{\sum\limits_{i=1}^{n} X_i^2}\right)^{1/2}\right).$$

Thus, we have a solution of a system of likelihood equations. It can be shown that the matrix of second order partial derivatives at the solution is almost surely negative definite matrix. Hence, the maximum likelihood estimator $(\hat{\sigma}_{1n}^2, \hat{\sigma}_{2n}^2)'$ of $(\sigma_1^2, \sigma_2^2)'$ is given by

$$\hat{\sigma}_{1n}^2 = \frac{1}{1-\rho^2}\left(\frac{\sum\limits_{i=1}^{n} X_i^2}{n} - \rho\frac{\sum\limits_{i=1}^{n} X_i Y_i}{n}\left(\frac{\sum\limits_{i=1}^{n} X_i^2}{\sum\limits_{i=1}^{n} Y_i^2}\right)^{1/2}\right)$$

$$\& \ \hat{\sigma}_{2n}^2 = \frac{1}{1-\rho^2} \left(\frac{\sum_{i=1}^{n} Y_i^2}{n} - \rho \frac{\sum_{i=1}^{n} X_i Y_i}{n} \left(\frac{\sum_{i=1}^{n} Y_i^2}{\sum_{i=1}^{n} X_i^2} \right)^{1/2} \right).$$

□

✍ Remark 4.3.9

In Example 4.3.4, it is shown that when $\rho \neq 0$, then

$$\sqrt{n}(\hat{\sigma}_{1n}^2 - \sigma_1^2, \ \hat{\sigma}_{2n}^2 - \sigma_2^2)' \overset{L}{\to} \underline{Z} \sim N_2(\underline{0}, \ I^{-1}(\sigma_1^2, \sigma_2^2)).$$

Hence,

$$\sqrt{n}(\hat{\sigma}_{1n}^2 - \sigma_1^2) \overset{L}{\to} Z_1 \sim N(0, (2-\rho^2)\sigma_1^4) \ \&$$
$$\sqrt{n}(\hat{\sigma}_{2n}^2 - \sigma_2^2) \overset{L}{\to} Z_2 \sim N(0, (2-\rho^2)\sigma_2^4). \tag{4.3.7}$$

If $(X, Y)'$ has bivariate normal distribution, then $X \sim N(0, \sigma_1^2)$ and $Y \sim N(0, \sigma_2^2)$. Suppose $\tilde{\sigma}_{1n}^2$ and $\tilde{\sigma}_{2n}^2$ denote the maximum likelihood estimator of σ_1^2 and σ_2^2 respectively, based on the random sample of size n from the distributions of X and Y. Then as shown in Example 3.3.2,

$$\sqrt{n}(\tilde{\sigma}_{1n}^2 - \sigma_1^2) \overset{L}{\to} Z_1 \sim N(0, 2\sigma_1^4) \ \& \ \sqrt{n}(\tilde{\sigma}_{2n}^2 - \sigma_2^2) \overset{L}{\to} Z_2 \sim N(0, 2\sigma_2^4). \tag{4.3.8}$$

Apparently results in Eq. (4.3.7) and Eq. (4.3.8) seem to be inconsistent, as the approximate variances are different. However, it is to be noted that the estimators in Eq. (4.3.7) and in Eq. (4.3.8) are obtained under different models. In Eq. (4.3.7), the estimators are derived under bivariate normal model, while in Eq. (4.3.8), these are derived under univariate normal model. Estimators of parameters involved in the marginal distribution in a bivariate model will have a different behavior than the estimators of the same parameters of the same distribution, but treated as a univariate model. In this example, $\hat{\sigma}_{1n}^2$ is the maximum likelihood estimator of σ_1^2 of a marginal distribution in a bivariate model, while $\tilde{\sigma}_{1n}^2$ is the maximum likelihood estimator of σ_1^2 in a univariate model. Observe that $\hat{\sigma}_{1n}^2$ involves data on Y variables and $\hat{\sigma}_{2n}^2$ involves data on X variables also. On the other hand, $\tilde{\sigma}_{1n}^2 = \sum_{i=1}^{n} X_i^2/n$ and $\tilde{\sigma}_{2n}^2 = \sum_{i=1}^{n} Y_i^2/n$. If $\rho = 0$, then X and Y are independent random variables, with $X \sim N(0, \sigma_1^2)$ and $Y \sim N(0, \sigma_2^2)$. In this case, the results based on bivariate and univariate models match as expected, in view of the fact that X and Y are uncorrelated and the bivariate probability density function is just the product of marginal probability density functions. One more important point to be noted is as follows:

$$\text{Approximate variance of } \hat{\sigma}_{1n}^2 = (2-\rho^2)\sigma_1^4 \ \leq \ 2\sigma_1^4$$
$$= \text{Approximate variance of } \tilde{\sigma}_{1n}^2.$$

In a bivariate model, information on σ_1^2 is available not only via X variable but also from Y variable through the correlation coefficient ρ and hence approximate variance of $\hat{\sigma}_{1n}^2$ is smaller than that of $\tilde{\sigma}_{1n}^2$. For the similar reason,

$$\text{Approximate variance of } \hat{\sigma}_{2n}^2 = (2 - \rho^2)\sigma_2^4 \leq 2\sigma_2^4$$
$$= \text{Approximate variance of } \tilde{\sigma}_{2n}^2.$$

Such a feature is reflected in the information function also. The $(1, 1)$-th element from the information matrix in a bivariate normal model is larger than the information function from the univariate normal model where $X \sim N(0, \sigma_1^2)$ as

$$I_{1,1}(\sigma_1^2, \sigma_2^2) - I(\sigma_1^2) = \frac{2 - \rho^2}{4(1 - \rho^2)\sigma_1^4} - \frac{1}{2\sigma_1^4} = \frac{\rho^2}{4(1 - \rho^2)\sigma_1^4} > 0.$$

Similarly,

$$I_{2,2}(\sigma_1^2, \sigma_2^2) - I(\sigma_2^2) = \frac{2 - \rho^2}{4(1 - \rho^2)\sigma_2^4} - \frac{1}{2\sigma_2^4} = \frac{\rho^2}{4(1 - \rho^2)\sigma_2^4} > 0.$$

In the following example of a bivariate Cauchy distribution, we observe similar features of a bivariate and corresponding univariate models.

In the next example, we verify whether a bivariate Cauchy distribution (Kotz et al. [9]) belongs to a two-parameter Cramér family.

✒ Example 4.3.5

Suppose a random vector $(X, Y)'$ follows a bivariate Cauchy $C_2(\theta_1, \theta_2)$ distribution with probability density function given by

$$f(x, y, \theta_1, \theta_2) = \frac{1}{2\pi} \left\{ 1 + (x - \theta_1)^2 + (y - \theta_2)^2 \right\}^{-3/2} \quad (x, y)' \in \mathbb{R}^2, \theta_1, \theta_2 \in \mathbb{R}.$$

It has been shown that the Cauchy $C(\theta, 1)$ distribution with location parameter θ does not belong to a one-parameter exponential family. On similar lines, we can show that the probability law of bivariate Cauchy $C_2(\theta_1, \theta_2)$ distribution cannot be expressed in a form required for a two-parameter exponential family and hence it does not belong to a two-parameter exponential family. In Example 4.3.1, we have shown that Cauchy $C(\theta, 1)$ distribution belongs to a Cramér family. On similar lines, we examine whether $C_2(\theta_1, \theta_2)$ distribution belongs to a two-parameter Cramér family. Observe that the parameter space is $\mathbb{R}^2$ and it is an open set. The support of the distribution is also $\mathbb{R}^2$ and it does not depend on the parameters. We now examine whether $\log f(x, y, \theta_1, \theta_2)$ has partial derivatives up to order three. We have

$$\log f(x, y, \theta_1, \theta_2) = -\log 2\pi - \frac{3}{2} \log\{1 + (x - \theta_1)^2 + (y - \theta_2)^2\}$$
$$\frac{\partial}{\partial \theta_1} \log f(x, y, \theta_1, \theta_2) = \frac{3(x - \theta_1)}{\{1 + (x - \theta_1)^2 + (y - \theta_2)^2\}}$$

$$\frac{\partial}{\partial \theta_2} \log f(x, y, \theta_1, \theta_2) = \frac{3(y - \theta_2)}{\{1 + (x - \theta_1)^2 + (y - \theta_2)^2\}}$$

$$\frac{\partial^2}{\partial \theta_1^2} \log f(x, y, \theta_1, \theta_2) = \frac{-3\{1 - (x - \theta_1)^2 + (y - \theta_2)^2\}}{\{1 + (x - \theta_1)^2 + (y - \theta_2)^2\}^2}$$

$$\frac{\partial^2}{\partial \theta_2^2} \log f(x, y, \theta_1, \theta_2) = \frac{-3\{1 + (x - \theta_1)^2 - (y - \theta_2)^2\}}{\{1 + (x - \theta_1)^2 + (y - \theta_2)^2\}^2}$$

$$\frac{\partial^2}{\partial \theta_2 \partial \theta_1} \log f(x, y, \theta_1, \theta_2) = \frac{6(x - \theta_1)(y - \theta_2)}{\{1 + (x - \theta_1)^2 + (y - \theta_2)^2\}^2}$$

$$\frac{\partial^3}{\partial \theta_1^3} \log f(x, y, \theta_1, \theta_2) = \frac{-6(x - \theta_1)\{3 - (x - \theta_1)^2 + 3(y - \theta_2)^2\}}{\{1 + (x - \theta_1)^2 + (y - \theta_2)^2\}^3}$$

$$\frac{\partial^3}{\partial \theta_2^3} \log f(x, y, \theta_1, \theta_2) = \frac{-6(y - \theta_2)\{3 + 3(x - \theta_1)^2 - (y - \theta_2)^2\}}{\{1 + (x - \theta_1)^2 + (y - \theta_2)^2\}^3}$$

$$\frac{\partial^3}{\partial \theta_2 \partial \theta_1^2} \log f(x, y, \theta_1, \theta_2) = \frac{-6(y - \theta_2)\{1 - 3(x - \theta_1)^2 + (y - \theta_2)^2\}}{\{1 + (x - \theta_1)^2 + (y - \theta_2)^2\}^3}$$

$$\frac{\partial^3}{\partial \theta_1 \partial \theta_2^2} \log f(x, y, \theta_1, \theta_2) = \frac{-6(x - \theta_1)\{1 + (x - \theta_1)^2 - 3(y - \theta_2)^2\}}{\{1 + (x - \theta_1)^2 + (y - \theta_2)^2\}^3} .$$

Thus, partial derivatives of $\log f(x, y, \theta_1, \theta_2)$ up to order three exist. We further examine whether the third order derivatives are bounded by integrable functions. Observe that

$$\left| \frac{\partial^3}{\partial \theta_1^3} \log f(x, y, \theta_1, \theta_2) \right| \leq \frac{6|x - \theta_1|\{3 + (x - \theta_1)^2 + 3(y - \theta_2)^2\}}{\{1 + (x - \theta_1)^2 + (y - \theta_2)^2\}^3}$$

$$= M_{111}(x, y), \quad \text{say} .$$

Suppose $u_1 = x - \theta_1$ and $u_2 = y - \theta_2$. Then using the fact that the integrand is an even function, we have

$$E(M_{111}(X, Y)) = \frac{1}{2\pi} \int_{-\infty}^{\infty} \int_{-\infty}^{\infty} \frac{6|u_1|\{3 + u_1^2 + 3u_2^2\}}{\{1 + u_1^2 + u_2^2\}^{9/2}} du_1 du_2$$

$$= \frac{12}{\pi} \int_0^{\infty} \int_0^{\infty} \frac{u_1\{3 + u_1^2 + 3u_2^2\}}{\{1 + u_1^2 + u_2^2\}^{9/2}} du_1 du_2$$

$$= \frac{12}{\pi} \int_0^{\infty} \frac{1}{(1 + u_2^2)^{7/2}} \left[\int_0^{\infty} \frac{u_1\{3 + u_1^2/(1 + u_2^2)\}}{\{1 + u_1^2/(1 + u_2^2)\}^{9/2}} du_1 \right] du_2$$

$$= \frac{6}{\pi} \int_0^{\infty} \frac{1}{(1 + u_2^2)^{5/2}} \left[\int_0^{\infty} \frac{3 + t}{(1 + t)^{9/2}} dt \right] du_2$$

$$\text{with } \frac{u_1^2}{1 + u_2^2} = t$$

$$= \frac{6}{\pi} \int_0^\infty \frac{1}{(1+u_2^2)^{5/2}} \left[3B(1, 7/2) + B(2, 5/2) \right] du_2$$

$$= \frac{132}{35\pi} \int_0^\infty \frac{t^{-1/2}}{(1+t)^{5/2}} dt$$

$$= \frac{132}{35\pi} B(1/2, 2) < \infty,$$

with $u_2^2 = t$. Hence, the third order partial derivative $\frac{\partial^3}{\partial \theta_1^3} \log f(x, y, \theta_1, \theta_2)$ is bounded by an integrable function. On similar lines, we can show that the remaining third order partial derivatives of $\log f(x, y, \theta_1, \theta_2)$ are bounded by integrable functions. We now find the information matrix $I(\theta_1, \theta_2) = [I_{i,j}(\theta_1, \theta_2)]$. Observe that

$$I_{1,2}(\theta_1, \theta_2) = E\left(-\frac{\partial^2}{\partial \theta_2 \partial \theta_1} \log f(x, y, \theta_1, \theta_2) \right)$$

$$= E\left(\frac{-6(X - \theta_1)(Y - \theta_2)}{\{1 + (X - \theta_1)^2 + (Y - \theta_2)^2\}^2} \right)$$

$$= -\frac{6}{2\pi} \int_{-\infty}^\infty \int_{-\infty}^\infty \frac{u_1 u_2}{\{1 + u_1^2 + u_2^2\}^{7/2}} du_1 du_2$$

$$= -\frac{6}{2\pi} \int_{-\infty}^\infty \frac{u_2}{\{1 + u_2^2\}^{7/2}} \left[\int_{-\infty}^\infty \frac{u_1}{\{1 + \frac{u_1^2}{1+u_2^2}\}^{7/2}} du_1 \right] du_2$$

$$= -\frac{6}{2\pi} \int_{-\infty}^\infty \frac{u_2}{\{1 + u_2^2\}^{7/2}} \times 0, \qquad \frac{u_1}{\{1 + \frac{u_1^2}{1+u_2^2}\}^{7/2}}$$

being an odd function

$$= 0.$$

$$I_{1,1}(\theta_1, \theta_2) = E\left(-\frac{\partial^2}{\partial \theta_1^2} \log f(x, y, \theta_1, \theta_2) \right)$$

$$= E\left(\frac{3\{1 - (X - \theta_1)^2 + (Y - \theta_2)^2\}}{\{1 + (X - \theta_1)^2 + (Y - \theta_2)^2\}^2} \right)$$

$$= \frac{3}{2\pi} \int_{-\infty}^\infty \int_{-\infty}^\infty \frac{1 - u_1^2 + u_2^2}{\{1 + u_1^2 + u_2^2\}^{7/2}} du_1 du_2$$

$$= \frac{3}{\pi} \int_{-\infty}^\infty \left[\int_0^\infty \frac{1 - u_1^2 + u_2^2}{\{1 + u_1^2 + u_2^2\}^{7/2}} du_1 \right] du_2$$

$$= \frac{3}{\pi} \int_{-\infty}^\infty \frac{1}{(1+u_2^2)^{5/2}} \left[\int_0^\infty \frac{1 - \frac{u_1^2}{1+u_2^2}}{\{1 + \frac{u_1^2}{1+u_2^2}\}^{7/2}} du_1 \right] du_2$$

$$= \frac{3}{2\pi} \int_{-\infty}^{\infty} \frac{1}{(1+u_2^2)^2} \left[\int_0^{\infty} \frac{t^{-1/2}(1-t)}{\{1+t\}^{7/2}} \, dt \right] du_2$$

$$\text{with} \quad \frac{u_1^2}{1+u_2^2} = t$$

$$= \frac{3}{2\pi} \int_{-\infty}^{\infty} \frac{1}{(1+u_2^2)^2}$$

$$\left[\int_0^{\infty} \frac{t^{1/2-1}}{\{1+t\}^{1/2+3}} \, dt - \int_0^{\infty} \frac{t^{3/2-1}}{\{1+t\}^{3/2+2}} \, dt \right] du_2$$

$$= \frac{3}{2\pi} \int_{-\infty}^{\infty} \frac{1}{(1+u_2^2)^2} \left[B(1/2,3) - B(3/2,2) \right] du_2$$

$$= \frac{3}{2\pi} \int_{-\infty}^{\infty} \frac{1}{(1+u_2^2)^2} \left[\frac{\Gamma(1/2)\Gamma(3)}{\Gamma(7/2)} - \frac{\Gamma(3/2)\Gamma(2)}{\Gamma(7/2)} \right] du_2$$

$$= \frac{6}{5\pi} \int_{-\infty}^{\infty} \frac{1}{(1+u_2^2)^2} \, du_2 = \frac{6}{5\pi} \int_0^{\infty} \frac{t^{1/2-1}}{\{1+t\}^{1/2+3/2}} \, dt$$

$$\text{with} \quad u_2^2 = t$$

$$= \frac{6}{5\pi} B(1/2,3/2) = \frac{6}{5\pi} \frac{\Gamma(1/2)\Gamma(3/2)}{\Gamma(2)} = \frac{3}{5}.$$

Similarly,

$$I_{2,2}(\theta_1,\theta_2) = E\left(-\frac{\partial^2}{\partial\theta_2^2} \log f(x,y,\theta_1,\theta_2) \right)$$

$$= E\left(\frac{3\{1+(X-\theta_1)^2-(Y-\theta_2)^2\}}{\{1+(X-\theta_1)^2+(Y-\theta_2)^2\}^2} \right)$$

$$= \frac{3}{2\pi} \int_{-\infty}^{\infty} \int_{-\infty}^{\infty} \frac{1+u_1^2-u_2^2}{\{1+u_1^2+u_2^2\}^{7/2}} \, du_1 du_2$$

$$= \frac{3}{\pi} \int_{-\infty}^{\infty} \int_0^{\infty} \frac{1+u_1^2-u_2^2}{\{1+u_1^2+u_2^2\}^{7/2}} \, du_1 du_2$$

$$= \frac{3}{\pi} \int_{-\infty}^{\infty} \frac{1}{(1+u_1^2)^{5/2}} \left[\int_0^{\infty} \frac{1-\frac{u_2^2}{1+u_1^2}}{\{1+\frac{u_2^2}{1+u_1^2}\}^{7/2}} \, du_2 \right] du_1$$

$$= \frac{3}{2\pi} \int_{-\infty}^{\infty} \frac{1}{(1+u_1^2)^2} \left[\int_0^{\infty} \frac{t^{-1/2}(1-t)}{\{1+t\}^{7/2}} \, dt \right] du_1$$

$$\text{with} \quad \frac{u_2^2}{1+u_1^2} = t$$

$$= \frac{6}{5\pi} \int_{-\infty}^{\infty} \frac{1}{(1+u_1^2)^2} \, du_1 = \frac{6}{5\pi} \int_0^{\infty} \frac{t^{1/2-1}}{\{1+t\}^{1/2+3/2}} \, dt$$

$$\text{with} \quad u_1^2 = t$$

$$= \frac{6}{5\pi} B(1/2, 3/2) = \frac{6}{5\pi} \frac{\Gamma(1/2)\Gamma(3/2)}{\Gamma(2)} = \frac{3}{5}.$$

Thus, the information matrix $I(\theta_1, \theta_2)$ is a diagonal matrix with diagonal elements $3/5$ each. It is a positive definite matrix. It then follows that the bivariate Cauchy distribution satisfies all the Cramér regularity conditions, hence it belongs to a two-parameter Cramér family. By the Cramér-Huzurbazar theorem, for large n the maximum likelihood estimator $(\hat{\theta}_{1n}, \hat{\theta}_{2n})'$ of $(\theta_1, \theta_2)'$ is a CAN estimator with the approximate dispersion matrix $I^{-1}(\theta_1, \theta_2)/n = \text{diag}[5/3, 5/3]/n$. The system of likelihood equations cannot be solved explicitly and we need to use the numerical methods, discussed in Sect. 4.4, to obtain the value of the maximum likelihood estimator corresponding to the given random sample.   □

🖉 Remark 4.3.10

As in the bivariate normal model in the Example 4.3.4, we note that the approximate variance $5/3n = 1.6667/n$ of $\hat{\theta}_{1n}$ in a bivariate Cauchy model is smaller than the approximate variance $2/n$ of the maximum likelihood estimator of θ_1 in a corresponding univariate Cauchy $C(\theta_1, 1)$ distribution based on the random sample from X. We have similar scenario for $\hat{\theta}_{2n}$. Similarly,

$$I_{1,1}(\theta_1, \theta_2) = 0.6 \ > \ 0.5 = I(\theta_1) \ \& \ I_{2,2}(\theta_1, \theta_2) = 0.6 \ > \ 0.5 = I(\theta_2).$$

We get such a relation in view of the dependence between X and Y which we cannot quantify as the correlation coefficient does not exist. Further, we cannot derive the explicit expressions for the maximum likelihood estimators, either for a bivariate Cauchy distribution or a univariate Cauchy distribution, so it is not visible whether data on Y is involved in the estimator of θ_1. However, the information on θ_1 is also derived from the component Y as X and Y are associated.

In Sect. 4.5, we discuss how to draw a random sample from a bivariate Cauchy distribution and based on it, how to obtain the maximum likelihood estimator of $(\theta_1, \theta_2)'$. We also obtain the maximum likelihood estimator of θ_1, treating it as a location parameter of a marginal distribution of X using data generated under bivariate Cauchy model and note that the two estimates are different. Similar scenario is observed for θ_2. We obtain Spearman's rank correlation coefficient as a measure of association between X and Y on the basis of generated sample.

4.4   Iterative Procedures

In Example 4.2.1, we have noted that for a Poisson $Poi(\theta)$ distribution, truncated at 0, likelihood equation or the moment equation to get the moment estimator based on a sufficient statistic, cannot be solved explicitly. A Cauchy distribution with location parameter θ and scale parameter 1 belongs to a Cramér family and we have to face the same problem to solve the likelihood equation to get the maximum likelihood estimator. In the above example of bivariate Cauchy distribution, we come across the same problem. In such situations, we adopt the numerical methods. We describe two such procedures, for a real parameter setup and for a vector parameter setup. Examples illustrating these procedures are discussed in Sect. 4.5, using R software.

Iterative procedures in a real parameter setup: Suppose the distribution of X is indexed by a real parameter θ and $L_n(\theta|\underline{X})$ is a likelihood function of θ corresponding to a random sample from the distribution of X.

Newton-Raphson procedure: This procedure has been derived from the Taylor's series expansion of $L_n(\theta|\underline{X})$ up to second order. In this procedure, the consecutive iterative values are obtained by the following formula:

$$\theta_{(i+1)} = \theta_{(i)} - \frac{\frac{\partial}{\partial\theta}\log L_n(\theta|\underline{X})|_{\theta_{(i)}}}{\frac{\partial^2}{\partial\theta^2}\log L_n(\theta|\underline{X})|_{\theta_{(i)}}},$$

where $\theta_{(i)}$ denotes the i th iterative value.

Method of scoring: This procedure is proposed by Fisher in 1925. The consecutive iterative values in it are obtained by the following formula:

$$\theta_{(i+1)} = \theta_{(i)} + \frac{\frac{\partial}{\partial\theta}\log L_n(\theta|\underline{X})|_{\theta_{(i)}}}{nI(\theta_{(i)})},$$

where $\theta_{(i)}$ denotes the i th iterative value. The method of scoring is essentially derived from the Newton-Raphson procedure by replacing the denominator of the second term by its expectation. Hence it is also known Fisher-Newton-Raphson procedure. It is well justified by the WLLN as $-\frac{1}{n}\frac{\partial^2}{\partial\theta^2}\log L_n(\theta|\underline{X})|_\theta$ converges in probability to the information function $I(\theta)$.

Iterative procedures in a multiparameter setup: Suppose the distribution of X is indexed by a vector parameter $\underline{\theta} = (\theta_1, \theta_2, \ldots, \theta_k)'$ and $L_n(\underline{\theta}|\underline{X})$ is a likelihood function of $\underline{\theta}$ corresponding to a random sample from the distribution of X.

Newton-Raphson procedure: In this procedure, the consecutive iterative values are obtained by the following formula:

$$\underline{\theta}_{(i+1)} = \underline{\theta}_{(i)} - \left[\frac{\partial^2}{\partial\theta_i\partial\theta_j}\log L_n(\underline{\theta}|\underline{X})\right]^{-1}_{|\underline{\theta}_{(i)}} \times$$

$$\left(\frac{\partial}{\partial\theta_1}\log L_n, \frac{\partial}{\partial\theta_2}\log L_n, \ldots, \frac{\partial}{\partial\theta_k}\log L_n\right)'_{|\underline{\theta}_{(i)}},$$

where $\underline{\theta}_{(i)}$ denotes the i th iterative value.

Method of scoring: In this procedure, the consecutive iterative values are obtained by the following formula:

$$\underline{\theta}_{(i+1)} = \underline{\theta}_{(i)} + \frac{1}{n} I^{-1}(\theta)|_{\underline{\theta}_{(i)}} \left(\frac{\partial}{\partial \theta_1} \log L_n, \frac{\partial}{\partial \theta_2} \log L_n, \dots, \frac{\partial}{\partial \theta_k} \log L_n \right)'_{|_{\underline{\theta}_{(i)}}},$$

where $\underline{\theta}_{(i)}$ denotes the i th iterative value.

In both the cases and for both the methods, the iterative procedure terminates when the consecutive iterative values are approximately close to each other. It has been proved in Kale [10] and Kale [11] that if the initial iterative value $\theta_{(0)}$ is taken as a value of any consistent estimator of θ, then under Cramér regularity conditions, the iterative procedure terminates with probability approaching 1 as $n \to \infty$. The method of scoring is preferred to the Newton-Raphson procedure, if the information function is free from θ as there is considerable simplification in the numerical procedure. For example, for Cauchy $C(\theta, 1)$, $I(\theta) = 1/2$ and it is better to use method of scoring than the Newton-Raphson procedure by taking value of the sample median as an initial iterative value.

4.5   Maximum Likelihood Estimation Using R

This section presents some examples illustrating the use of R software to find out the maximum likelihood estimator graphically, by numerical methods discussed in previous section and some built-in functions in R.

✏ Example 4.5.1

In Example 4.2.1, we discussed the maximum likelihood estimation of the parameter θ on the basis of a random sample from a truncated Poisson distribution, truncated at 0. It is noted that the likelihood equation given by $\overline{X}_n = \theta/(1 - e^{-\theta})$ cannot be solved explicitly. We now discuss how to approximately decide the root from the graph of log-likelihood and the graph of the first derivative of log-likelihood. We then use Newton-Raphson procedure and method of scoring to solve the equation using R software. We have

$$\log L_n(\theta|\underline{X}) = -n\theta - n\log(1 - e^{-\theta}) + \log\theta \sum_{i=1}^{n} X_i - \sum_{i=1}^{n} \log X_i!$$

$$\frac{\partial}{\partial\theta} \log L_n(\theta|\underline{X}) = -n - \frac{ne^{-\theta}}{1 - e^{-\theta}} + \frac{\sum_{i=1}^{n} X_i}{\theta} \quad \&$$

$$\frac{\partial^2}{\partial\theta^2} \log L_n(\theta|\underline{X}) = \frac{ne^{-\theta}}{(1 - e^{-\theta})^2} - \frac{\sum_{i=1}^{n} X_i}{\theta^2}.$$

Further the information function is given by
$I(\theta) = (1 - e^{-\theta} - \theta e^{-\theta})/\theta(1 - e^{-\theta})^2$. As a first step we generate a random
sample of size $n = 300$ from the truncated Poisson distribution with $\theta = 3$. Using
graphs as well as both the iterative procedures, we find the maximum likelihood
estimator of θ.

```
n=300; e=exp(1); a=3 ## parameter of  Poisson distribution
                                   truncated at 0
y=1:100; p=e^(-a)*a^y/((1-e^(-a))*factorial(y)); sum(p);
                                             set.seed(123)
x=sample(y,n,replace=TRUE, prob=p)  ## random sample
                # from truncated Poisson distribution
x1=sum(x); x2=sum(log(factorial(x))); th=seq(.5,5,.005)
logl= -n*th - n*log(1-e^(-th)) + log(th)*x1 - x2
par(mfrow=c(1,2))
plot(th,logl,"l",pch=20,xlab="Theta",ylab="Log-likelihood")
b=which.max(logl); mle = th[b]; mle; abline(v=mle)
dlogl=-n-n*e^(-th)/(1-e^(-th)) + x1/th; summary(dlogl)
plot(th,dlogl,"l",pch=20,xlab="Theta",
     ylab="First Derivative of Log-likelihood")
abline(h=0)
p=th[which.min(abs(dlogl))]; p; abline(v=p)
## Newton-Raphson procedure
dlogl=function(a)
{
t1=-n-n*e^(-a)/(1-e^(-a)) + x1/a
return(t1)
}
d2logl=function(a)
{
t2=-n*e^(-a)/(1-e^(-a))^2 - x1/a^2
return(t2)
}
thini=2.93         # initial value of th
r=c();r[1]=thini;k=1;diff=1
while(diff > 10^(-4))
{
r[k+1]=r[k]-dlogl(r[k])/d2logl(r[k])
diff=abs(r[k+1]-r[k])
k=k+1
}
thmle=r[k]; thmle
### Method of scoring
inf=function(a) {
t3=(1-e^(-a)-a*e^(-a))/(a*(1-e^(-a))^2) return(t3) } s=c();
s[1]=thini; k=1; diff=1
while(diff > 10^(-4)) {
s[k+1]=s[k]+dlogl(s[k])/(n*inf(s[k])) diff=abs(s[k+1]-s[k]) k=k+1 }
thmles=s[k]; thmles
```

From both the graphs, displayed in Fig. 4.1, the approximate value of the maximum likelihood estimator $\hat{\theta}_n$ is 2.93. By definition of the maximum likelihood estimator, it is a solution of $\frac{\partial}{\partial \theta} \log L_n(\theta|\underline{X}) = 0$. However, it may not be exactly 0 for the given realization. Hence we try to find that θ, for which it is minimum using the function p=th[which.min(abs(dlog1))]. By Newton-Raphson procedure, $\hat{\theta}_n = 2.9278$ and by the method of scoring, $\hat{\theta}_n = 2.9278$. In both these iterative procedures, we have taken the value of initial iterate as 2.93, as it is a solution of the likelihood equation obtained graphically. All the four procedures produce approximately the same value of $\hat{\theta}_n$, note that it will change for each realization of the truncated Poisson distribution.

One can find out the root of the likelihood equation directly in R using the uniroot function. We have to provide an interval such that at the lower and at the upper limit of the interval, $\frac{\partial}{\partial \theta} \log L_n(\theta|\underline{X})$ is of opposite sign. In the following, we adopt this approach to find $\hat{\theta}_n$ and examine whether the large sample distribution of $T_n = \sqrt{nI(\theta)}(\hat{\theta}_n - \theta)$ is standard normal.

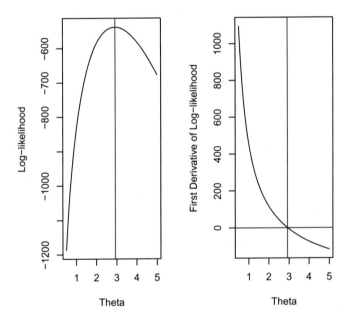

Fig. 4.1 Truncated Poisson distribution: MLE

```
e=exp(1); y=1:100; a=3; p=e^(-a)*a^y/((1-e^(-a))*factorial(y))
sum(p); n=600; nsim=1500; x1 = x2 = c()
for(m in 1:nsim)
{
set.seed(m)
x=sample(y,n,replace=TRUE, prob=p)
x1[m]=sum(x)
x2[m]=mean(x)
}
dlogl=function(par)
{
term=-n-n*e^(-par)/(1-e^(-par)) + x1[i]/par
return(term)
}
mle=c()
for(i in 1:nsim)
{
mle[i]=uniroot(dlogl,c(x2[i]-1,5))$root
}
summary(mle); var(mle)*(nsim-1)/nsim
v=a*(1-e^(-a))^2/(n*(1-e^(-a)-a*e^(-a))); v; v1=sqrt(v)
t=(mle-a)/v1; summary(t); shapiro.test(t)
r=seq(-4,4,.1); y=dnorm(r)
par(mfrow= c(2,2))
hist(t,freq=FALSE,main="Histogram", ylim=range(0,max(y)),
 xlab=expression(paste("T"[n])), col="light blue")
lines(r,y,"o",pch=20,col="dark blue")
boxplot(t,main="Box Plot", xlab=expression(paste("T"[n])))
qqnorm(t); qqline(t)
plot(ecdf(t),main="Empirical Distribution Function",
     ylab=expression(paste("F"[n](t))))
lines(r,pnorm(r),col="red")
```

We have generated 1500 random samples, each of size 600, from the truncated Poisson distribution, truncated at 0, with $\theta = 3$. The output of the function `summary(mle)` gives median 2.999 and mean to be 3 of 1500 values of the maximum likelihood estimator. Further, the variance comes out to be 0.0058, which is very close to the value 0.0056 of the approximate variance $1/nI(\theta)$ of $\hat{\theta}_n$ with $\theta = 3$. Approximate normality of the distribution of $T_n = \sqrt{nI(\theta)}(\hat{\theta}_n - \theta)$ is supported by the p-value 0.3914 of Shapiro-Wilk test and the four graphs in Fig. 4.2.   □

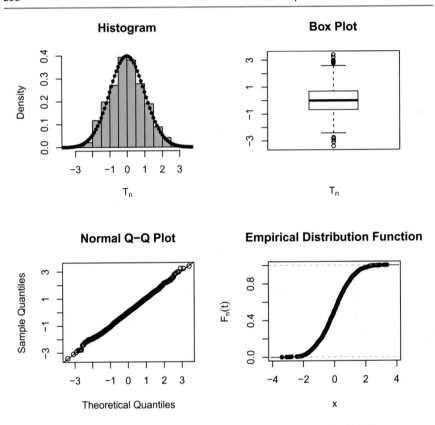

Fig. 4.2 Truncated Poisson distribution: approximate normality of normalized MLE

📝 **Example 4.5.2**

Suppose $\{X_1, X_2, \ldots, X_n\}$ is a random sample from a normal $N(\theta, \theta^2)$ distribution, $\theta > 0$. In Example 4.2.3, it is shown that $\hat{\theta}_n = (-m_1' + \sqrt{m_1'^2 + 4m_2'})/2$ is CAN for θ having the approximate variance $1/nI(\theta) = \theta^2/3n$. We verify these results by simulation. We first find the maximum likelihood estimate from the graph of log-likelihood function. Following is the R code for plotting the graph and locating the maximum.

```
th = 1.25; n = 250; set.seed(14); x = rnorm(n,th,th)
m1 = mean(x); m1 ;m2 = mean(x^2); m2; v = m2-m1^2; v
tn = (-m1+sqrt(m1^2+4*m2))/2; tn
loglik = function(a)
{
LL = 0
for(i in 1:n)
{
LL = LL + log(dnorm(x[i],a,a))
}
return(LL)
}
theta = seq(0.1,4,0.02);length(theta)
Ltheta = loglik(theta)
plot(theta,Ltheta,"1",xlab="Theta",ylab="Log-likelihood",
                                    ylim=range(-2500,-100))
b = which.max(Ltheta); mle = theta[b]; mle
abline(v=mle,col="blue")
```

On the basis of a random sample of size 250 generated with $\theta = 1.25$, we note that the maximum likelihood estimate is 1.22, both from the formula and from the graph of log-likelihood displayed in Fig. 4.3. Following R code verifies that the distribution of $T_n = \sqrt{nI(\theta)}(\hat{\theta}_n - \theta)$ can be approximated by the standard normal distribution.

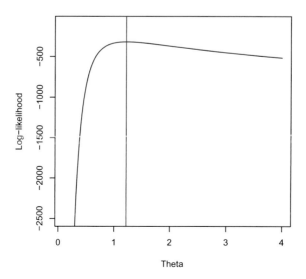

Fig. 4.3 Normal $N(\theta, \theta^2)$ distribution: log-likelihood

```
th=1.25; n=250; nsim=1500; m1 = m2 = c()
for(m in 1:nsim)
{
set.seed(m)
x=rnorm(n,th,th)
m1[m]=mean(x)
m2[m]=mean(x^2)
}
T1=(-m1+sqrt(m1^2+4*m2))/2
summary(T1); v=var(T1)*(nsim-1)/nsim; v
sigma=th^2/(3*n); sigma;
s1=(sqrt(3*n)/th)*(T1-th)
r=seq(-4,4,.08); length(r); y=dnorm(r)
par(mfrow= c(2,2))
hist(s1,freq=FALSE,main="Histogram", ylim=range(0,max(y)),
xlab=expression(paste("T"[n])), col="light blue")
lines(r,y,"o",pch=20,col="dark blue")
boxplot(s1,main="Box Plot",xlab=expression(paste("T"[n])))
qqnorm(s1); qqline(s1)
plot(ecdf(s1),main="Empirical Distribution Function",
        ylab=expression(paste("F"[n](t))))
lines(r,pnorm(r),col="blue")
shapiro.test(s1)
```

Summary statistic of T_n shows close agreement with the true parameter value $\theta = 1.25$ as the sample median and the sample mean based on 1500 simulations come out to be 1.25. Further, the variance 0.002053 of the simulated values is close to the approximate variance $\theta^2/3n = 0.002083$. Figure 4.4 and the p-value 0.8469 of Shapiro-Wilk test support the theoretical claim that the maximum likelihood estimator is CAN.   □

✐ Example 4.5.3

Suppose $\underline{X} \equiv \{X_1, X_2, \ldots, X_n\}$ is a random sample from the logistic distribution with probability density function

$$f(x, \theta) = \frac{\exp\{-(x-\theta)\}}{(1 + \exp\{-(x-\theta)\})^2}, \quad x \in \mathbb{R}, \ \theta \in \mathbb{R}.$$

The log-likelihood of θ corresponding to given random sample $\underline{X}$ is

$$\log L_n(\theta|\underline{X}) = -\sum_{i=1}^n (X_i - \theta) - 2\sum_{i=1}^n \log(1 + \exp\{-(x-\theta)\}).$$

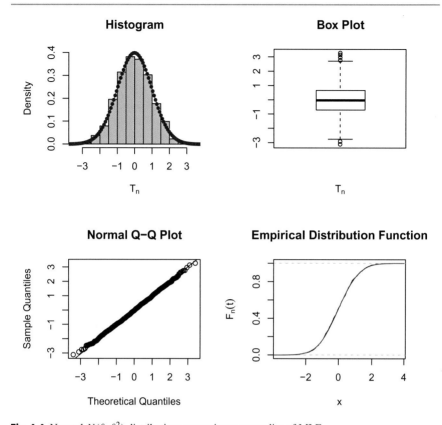

Fig. 4.4 Normal $N(\theta, \theta^2)$ distribution: approximate normality of MLE

Further, the first and the second order partial derivatives of the log likelihood are given by

$$\frac{\partial}{\partial \theta} \log L_n(\theta|\underline{X}) = n - 2 \sum_{i=1}^{n} \frac{\exp\{-(X_i - \theta)\}}{1 + \exp\{-(X_i - \theta)\}}$$

$$\frac{\partial^2}{\partial \theta^2} \log L_n(\theta|\underline{X}) = -2 \sum_{i=1}^{n} \frac{\exp\{-(X_i - \theta)\}}{(1 + \exp\{-(X_i - \theta)\})^2} .$$

Thus, the likelihood equation is given by $2 \sum_{i=1}^{n} \frac{\exp\{-(X_i-\theta)\}}{1+\exp\{-(X_i-\theta)\}} - n = 0$. We can find its solution either by the Newton-Raphson procedure or the method of scoring. Method of scoring is easier as the information function $I(\theta) = 1/3$ is free from θ. The information function $I(\theta)$ can be easily computed from the second derivative of the log-likelihood function as given above and the beta function. As an initial iterative value, one may take sample median or the sample mean. The logistic distribution is symmetric around the location parameter θ, hence the population mean and population median both are θ, further population variance is $\pi^2/3$ (Kotz et al. [12], p. 117). By WLLN and by the CLT, the sample mean $\overline{X}_n$ is CAN estimator of θ with approximate variance $\pi^2/3n = 3.2899/n$. By Theorem 3.2.3, the sample median $X_{[n/2]+1}$ is CAN estimator of θ with approximate variance $4/n$, which is larger than the approximate variance of $\overline{X}_n$. Thus, sample mean is a better CAN estimator of θ. It can be shown that the logistic distribution belongs to a Cramér family and hence the maximum likelihood estimator $\hat{\theta}_n$ of θ is a BAN estimator of θ with approximate variance $3/n$. Following is the R code to find the maximum likelihood estimate corresponding to the given data, by using uniroot function from R and to examine whether it is CAN for θ with approximate variance $3/n$, that is, to examine the asymptotic normality of $T_n = \sqrt{n/3}(\hat{\theta}_n - \theta)$.

```
e=exp(1); a=3; n=120; nsim=1000; x=matrix(nrow=n,ncol=nsim)
for(j in 1:nsim)
{
set.seed(j)
x[,j]=rlogis(n,a,1)
}
dlogl=function(par)
{
term2=0
for(i in 1:n)
{
term2=term2 -2*e^(-x[i,j]+par)/(1+e^(-x[i,j]+par))
}
dlogl= n + term2
return(dlogl)
}
mle=c()
for(j in 1:nsim)
{
mle[j]=uniroot(dlogl,c(10,.01))$root
}
summary(mle); var(mle)*(nsim-1)/nsim; v=3/n; v; v1=sqrt(v)
t=(mle-a)/v1; summary(t); shapiro.test(t)
r=seq(-4,4,.1); y=dnorm(r)
par(mfrow= c(2,2))
hist(t,freq=FALSE,main="Histogram", ylim=range(0,max(y)),
xlab=expression(paste("T"[n])), col="light blue")
lines(r,y,"o",pch=20,col="dark blue")
boxplot(t,main="Box Plot",xlab=expression(paste("T"[n])))
qqnorm(t); qqline(t)
plot(ecdf(t),main="Empirical Distribution Function",
        ylab=expression(paste("F"[n](t))))
lines(r,pnorm(r),col="blue")
```

From the output, the summary statistic of mle shows that from 1000 simulated values, the median is 3, mean is 2.997 with variance 0.02489 which is very close to $3/n = 0.025$. The Shapiro-Wilk test procedure with p-value 0.5313 supports the claim that the large sample distribution of the maximum likelihood estimator of θ is normal with approximate variance $3/n$. The four plots, which are not shown here, also support the claim. All these results are for the sample size $n = 120$ and for $\theta = 3$.   □

In Example 3.3.9, it is shown that if $(X, Y)'$ has a bivariate normal $N_2(0, 0, 1, 1, \rho)$ distribution, $\rho \in (-1, 1)$, then the sample correlation coefficient R_n is a CAN estimator of ρ with approximate variance $(1 - \rho^2)^2/n$ and in Example 4.3.3, it is shown that the maximum likelihood estimator $\hat{\rho}_n$ of ρ, which is a unique real root of a cubic equation, is a CAN estimator of ρ with approximate variance $(1 - \rho^2)^2/n(1 + \rho^2)$. In the following example, R code is given to find the maximum likelihood estimator of ρ and to verify that it is a CAN estimator.

✒ Example 4.5.4

Suppose $(X, Y)' \sim N_2(0, 0, 1, 1, \rho)$, $\rho \in (-1, 1)$. In Example 4.3.3, we have shown that the maximum likelihood estimator of ρ is a unique real root of the cubic equation given by

$$\rho^3 - \rho^2 V_n - \rho(1 - U_n) - V_n = 0 \text{ where}$$
$$U_n = \sum_{i=1}^{n}(X_i^2 + Y_i^2)/n \ \& \ V_n = \sum_{i=1}^{n} X_i Y_i/n.$$

We generate a random sample from the distribution of $(X, Y)'$ and use Newton-Raphson procedure to solve this equation. As an initial value in the iterative procedure, we take the value of $V_n = \sum_{i=1}^{n} X_i Y_i/n$, which is a moment estimator of ρ and is a consistent estimator of ρ. On the basis of multiple simulations, we examine whether $\hat{\rho}_n$ and R_n are CAN for ρ. We also examine the normality of corresponding Fisher's Z transformation.

```
rho = 0.3;th=0.5*log((1+rho)/(1-rho)); th
mu = c(0,0); sig=matrix(c(1,rho,rho,1),nrow=2)
n = 270; nsim = 1200
library(mvtnorm)
u = v = Z = T = s = R = mle = c()
g=function(a)
{
  term=a^3-a^2*v1-a*(1-u1)-v1
  return(term)
}
```

```
dg=function(a)
{
   term=3*a^2-a*2*v1 + u1-1
   return(term)
}
for(i in 1:nsim)
{
set.seed(i)
x = rmvnorm(n,mu,sig)
R[i] = cor(x)[1,2]
s[i] = 0.5*(log((1+R[i])/(1-R[i])))
Z[i] = sqrt(n)*(s-th)
T[i] = sqrt(n)*(R-rho)/(1-rho^2)
u[i] = sum((x[,1]^2+x[,2]^2))/n
v[i] =  sum((x[,1]*x[,2]))/n
}
m = 5; e = matrix(nrow=m,ncol=nsim)
for(i in 1:nsim)

{
e[1,i] = v[i]; v1 = v[i]; u1 = u[i]; j = 1; diff = 1
while(diff>10^(-4))
{
e[j+1,i] = e[j,i]-g(e[j,i])/dg(e[j,i])
diff = abs(e[j+1,i]-e[j,i]); j = j+1
}
mle[i] = e[j,i]
}
d = round(data.frame(mle,R),4); View(d); head(d); tail(d)
summary(mle); summary(s); summary(R)
vmle = (n-1)*var(mle)/n; avmle = (1-rho^2)^2/n*(1+rho^2); vmle; avmle
vZ = (n-1)*var(s)/n; avZ = 1/n;vZ;avZ
vR = (n-1)*var(R)/n; avR = (1-rho^2)^2/n;vR;avR
shapiro.test(mle); shapiro.test(R); shapiro.test(s)
```

We have generated 1200 random samples each of size $n = 270$ with $\rho = 0.3$. The function View(d) displays the values of $\hat{\rho}_n$ and R_n for all the simulations. The first six values are reported in Table 4.1. From summary(mle), the median and the mean of 1200 maximum likelihood estimators are 0.3007 and 0.3001 respectively, very close to $\rho = 0.3$. The variance of these 1200 maximum likelihood estimators is 0.0027 which is close to the approximate variance of $\hat{\rho}_n$ given by $(1 - \rho^2)^2/n(1 + \rho^2) = 0.0033$. For comparison, we have also computed the

Table 4.1 $N_2(0, 0, 1, 1, \rho)$ Distribution: values of $\hat{\rho}_n$ and R_n

$\hat{\rho}_n$	0.2672	0.2684	0.2573	0.3389	0.3008	0.3096
R_n	0.2672	0.2788	0.2627	0.3204	0.3078	0.2925

sample correlation coefficient R_n and corresponding Z transformation for each of 1200 random samples. From summary(s), we note that the median and the mean of 1200 Z values are 0.3111 and 0.3101 respectively and these are close to $\theta = (1/2)\log(1 + \rho)/(1 - \rho) = 0.3095$. From summary(R), we observe that the median and the mean of 1200 R_n values are 0.3014 and 0.2996 respectively, again very close to $\rho = 0.3$. The variance of these 1200 Z values is 0.0037, and it is the same as the approximate variance of Z which is $1/n = 0.0037$. Similarly, the variance of these 1200 R_n values is 0.0030, and it is also close to the approximate variance of Z given by $(1 - \rho^2)^2/n = 0.0031$. The asymptotic normality of $\hat{\rho}_n$, R_n and Z is supported by the p-values of the Shapiro-Wilk test for normality. The p-values corresponding to $\hat{\rho}_n$, R_n and Z are 0.3548, 0.2355 and 0.726 respectively. It is to be noted that the p-value corresponding to Z is larger than the other two, supporting Remark 3.3.3 that convergence to normality of Z is faster than that of R_n.   □

In the following examples, we discuss how to find a maximum likelihood estimator for a vector parameter using the Newton-Raphson iterative procedure and the method of scoring.

✐ Example 4.5.5

Suppose a random variable X follows a gamma distribution with scale parameter α and shape parameter λ. Its probability density function is given by

$$f(x, \alpha, \lambda) = \frac{\alpha^{\lambda}}{\Gamma(\lambda)} e^{-\alpha x} x^{\lambda - 1}, \quad x > 0, \alpha > 0, \lambda > 0.$$

It belongs to a two-parameter exponential family. In Example 2.7.3, we have verified that on the basis of a random sample $\underline{X} \equiv \{X_1, X_2, \ldots, X_n\}$ from the distribution of X, $(\hat{\alpha}_n, \hat{\lambda}_n)' = (m'_1/m_2, m'^2_1/m_2)'$ is a consistent estimator of $(\alpha, \lambda)'$. Using these as initial estimates in the Newton-Raphson procedure, we now find the maximum likelihood estimator of $(\alpha, \lambda)'$. We also find it by finding the point at which log-likelihood attains the maximum. The log-likelihood of $(\alpha, \lambda)'$ corresponding to the given random sample and the system of likelihood equations are given by

$$\log L_n(\alpha, \lambda | \underline{X}) = n\lambda \log \alpha - n \log \Gamma(\lambda) - \alpha \sum_{i=1}^{n} X_i + (\lambda - 1) \sum_{i=1}^{n} \log X_i$$

$$\frac{\partial}{\partial \alpha} \log L_n(\alpha, \lambda | \underline{X}) = \frac{n\lambda}{\alpha} - \sum_{i=1}^{n} X_i = 0 \quad \Leftrightarrow \quad \alpha = \lambda/\overline{X}_n$$

$$\frac{\partial}{\partial \lambda} \log L_n(\alpha, \lambda | \underline{X}) = n \log \alpha - n \text{digamma}(\lambda) + \sum_{i=1}^{n} \log X_i = 0 .$$

The two likelihood equations lead to an equation in λ given by

$$g(\lambda) = n \log \left(\frac{\lambda}{\overline{X}_n} \right) - n\text{digamma}(\lambda) + \sum_{i=1}^{n} \log X_i = 0 \,.$$

Here digamma(λ) is defined as $\text{digamma}(\lambda) = \frac{d}{d\lambda} \log \Gamma(\lambda)$. We solve the equation by Newton-Raphson procedure. Now,

$$g'(\lambda) = \frac{n}{\lambda} - n\text{trigamma}(\lambda), \quad \text{where} \quad \text{trigamma}(\lambda) = \frac{d^2}{d\lambda^2} \log \Gamma(\lambda) \,.$$

Thus, the iterative formula by Newton-Raphson procedure is
$\lambda_{(i+1)} = \lambda_{(i)} - g(\lambda_{(i)})/g'(\lambda_{(i)})$. As an initial iterate value we take
$\lambda(0) = m_1'^2/m_2$. Following R code gives the maximum likelihood estimator of
$(\alpha, \lambda)'$.

```
alpha=3; lambda=4; n=250; set.seed(120)
z=rgamma(n,shape = lambda,scale = 1/alpha ); summary(z)
v=(n-1)*var(z)/n;v; mu=lambda/alpha;mu
si=lambda/alpha^2;si
m1=mean(z); m2=mean(z^2); m3=m2-m1^2; alest=m1/m3; laest=m1^2/m3;
                                                alest; laest
# log-likelihood function
loglike=function(a,b)
{
n*b*log(a)  - n*log(gamma(b))-a*sum(z) + (b-1)*sum(log(z))
}
a=seq(2,4,by=0.01); length(a); b=seq(3,5,by=0.01); length(b)
L=matrix(nrow=length(a),ncol=length(b))
for(i in 1:length(a))
for(j in 1:length(b))
{
L[i,j]=loglike(a[i],b[j])
}
index=max(L);index
indices=which(L==max(L),arr.ind=TRUE);indices
alphamle=a[indices[1]];alphamle; lambdamle=b[indices[2]];lambdamle
### Newton-Raphson procedure
g=function(b)
{
term= n*log(b)  - n*log(m1)-n*digamma(b)+sum(log(z))
return(term)
}
dg=function(b)
{
term= n/b - n*trigamma(b)
return(term)
}
```

```
u=c(); u[1]=laest;i=1;diff=1
while(diff>10^(-4))
{
u[i+1]=u[i]- g(u[i])/dg(u[i])
diff=abs(u[1+i]-u[i])
i=i+1
}
mlelambda=u[i];mlealpha=mlelambda/m1;mlealpha;mlelambda
```

Summary statistics show that the sample mean 1.3555 is close to the population mean 1.3333 and the sample variance 0.4711 is close to population variance 0.4444. The moment estimator of α is 2.8775 and that of λ is 3.9004. The log-likelihood attains the maximum when α is 2.95 and λ is 4.00. By Newton-Raphson iterative procedure the maximum likelihood estimators of α and λ are 2.9515 and 4.0007 respectively.   □

In the above example, a gamma distribution involves two parameters, but the system of likelihood equations can be reduced to a single equation in λ by expressing α in terms of λ from the first equation. Thus, iterative procedures for a real parameter can be used to find the maximum likelihood estimator of λ and then of α. In the next example, we illustrate the application of both Newton-Raphson procedure and the method of scoring in the multiparameter setup.

✒ Example 4.5.6

Suppose a random variable X follows a Cauchy $C(\theta, \lambda)$ distribution with location parameter θ and shape parameter λ. Its probability density function is given by

$$f(x, \theta, \lambda) = \frac{\lambda}{\pi} \frac{1}{\lambda^2 + (x - \theta)^2} , \quad x \in \mathbb{R}, \ \theta \in \mathbb{R}, \ \lambda > 0 .$$

Our aim is to find the maximum likelihood estimator $(\theta, \lambda)'$ on the basis of a random sample $\underline{X} \equiv \{X_1, X_2, \ldots, X_n\}$ from the distribution of X. The log-likelihood function of $(\theta, \lambda)'$ is given by

$$\log L_n(\theta, \lambda | \underline{X}) = n \log \lambda - n \log \pi - \sum_{i=1}^{n} \log \left(\lambda^2 + (X_i - \theta)^2 \right) .$$

Hence, the system of likelihood equations is given by

$$\frac{\partial}{\partial \theta} \log L_n(\theta, \lambda | \underline{X}) = 2 \sum_{i=1}^{n} \frac{X_i - \theta}{\lambda^2 + (X_i - \theta)^2} = 0$$

$$\frac{\partial}{\partial \lambda} \log L_n(\theta, \lambda | \underline{X}) = \frac{n}{\lambda} - 2\lambda \sum_{i=1}^{n} \frac{1}{\lambda^2 + (X_i - \theta)^2} = 0 .$$

Using following R code, we solve these by (i) maximizing the log-likelihood function, (ii) by Newton-Raphson procedure and (iii) by the method of scoring. For Newton-Raphson procedure, we require the second order partial derivatives. These are as follows:

$$\frac{\partial^2}{\partial \theta^2} \log L_n(\theta, \lambda | \underline{X}) = 2 \sum_{i=1}^{n} \frac{(X_i - \theta)^2 - \lambda^2}{(\lambda^2 + (X_i - \theta)^2)^2}$$

$$\frac{\partial^2}{\partial \lambda \partial \theta} \log L_n(\theta, \lambda | \underline{X}) = -4\lambda \sum_{i=1}^{n} \frac{X_i - \theta}{(\lambda^2 + (X_i - \theta)^2)^2}$$

$$\frac{\partial^2}{\partial \lambda^2} \log L_n(\theta, \lambda | \underline{X}) = -\frac{n}{\lambda^2} - 2 \sum_{i=1}^{n} \frac{1}{\lambda^2 + (X_i - \theta)^2}$$

$$+ 4\lambda^2 \sum_{i=1}^{n} \frac{1}{(\lambda^2 + (X_i - \theta)^2)^2} \cdot$$

For the method of scoring, we first find the information matrix $I(\theta, \lambda) = [I_{i,j}(\theta, \lambda)], i, j = 1, 2$. Observe that

$$I_{1,2}(\theta, \lambda) = E\left(-\frac{\partial^2}{\partial \lambda \partial \theta} \log f(X, \theta, \lambda)\right) = E\left(\frac{4\lambda(X - \theta)}{(\lambda^2 + (X - \theta)^2)^2}\right)$$

$$= \frac{4}{\pi \lambda} \int_{-\infty}^{\infty} \frac{u}{(1 + u^2)^3} du, \quad \text{with} \quad \frac{x - \theta}{\lambda} = u$$

$$= 0, \text{ as } \int_{-\infty}^{\infty} \frac{u}{(1 + u^2)^3} du \text{ exists and integrand is an odd function}.$$

Now using beta function, we have

$$I_{1,1}(\theta, \lambda) = E\left(-\frac{\partial^2}{\partial \theta^2} \log f(X, \theta, \lambda)\right) = -2E\left(\frac{((X - \theta)^2 - \lambda^2)}{(\lambda^2 + (X - \theta)^2)^2}\right)$$

$$= -\frac{2}{\pi \lambda^2} \int_{-\infty}^{\infty} \frac{-1 + u^2}{(1 + u^2)^3} du, \quad \text{where} \quad \frac{x - \theta}{\lambda} = u$$

$$= \frac{2}{\pi \lambda^2} \int_{-\infty}^{\infty} \frac{1}{(1 + u^2)^3} du - \frac{2}{\pi \lambda^2} \int_{-\infty}^{\infty} \frac{u^2}{(1 + u^2)^3} du$$

$$= \frac{4}{\pi \lambda^2} \int_{0}^{\infty} \frac{1}{(1 + u^2)^3} du - \frac{4}{\pi \lambda^2} \int_{0}^{\infty} \frac{u^2}{(1 + u^2)^3} du,$$
$$\text{integrand being even function}$$

$$= \frac{2}{\pi \lambda^2} \int_{0}^{\infty} \frac{t^{\frac{1}{2} - 1}}{(1 + t)^3} dt - \frac{2}{\pi \lambda^2} \int_{0}^{\infty} \frac{t^{\frac{3}{2} - 1}}{(1 + t)^3} dt, \quad \text{with } u^2 = t$$

$$= \frac{2}{\pi \lambda^2} [B(1/2, 5/2) - B(3/2, 3/2)] = \frac{1}{2\lambda^2} \cdot$$

Proceeding with similar substitutions as in $I_{1,1}(\theta, \lambda)$, we get

$$
\begin{aligned}
I_{2,2}(\theta, \lambda) &= E\left(-\frac{\partial^2}{\partial \lambda^2} \log f(X, \theta, \lambda)\right) \\
&= E\left(\frac{1}{\lambda^2} + 2\frac{1}{(\lambda^2 + (X - \theta)^2)} - 4\lambda^2 \frac{1}{(\lambda^2 + (X - \theta)^2)^2}\right) \\
&= \frac{1}{2\lambda^2}.
\end{aligned}
$$

Thus, the information matrix is a diagonal matrix with each diagonal element $1/2\lambda^2$. Following R code gives the maximum likelihood estimates of θ and λ using the three approaches.

```
th=3; lambda=2; n=200; set.seed(50);
                z=rcauchy(n,location=th,scale=lambda )
summary(z); m1=median(z); m2=quantile(z,.75)-m1; m2
thest=m1; thest; laest=m2; laest
#### log-likelihood function
loglike=function(a,b)
{
n*log(b) - n*log(pi)-sum(log(b^2+(z-a)^2))
}
a=seq(2,4,by=0.01); b=seq(1,3,by=0.01); length(a); length(b)
L=matrix(nrow=length(a),ncol=length(b))
for(i in 1:length(a))
for(j in 1:length(b))
{
L[i,j]=loglike(a[i],b[j])
}
max=max(L); max; indices=which(L==max(L),arr.ind=TRUE);indices
thmle=a[indices[1]];thmle; lambdamle=b[indices[2]];lambdamle
###  Newton-Raphson procedure
dlog1th=function(a,b)
{
term=0
for(i in 1:n)
{
term= term + (z[i]-a)/(b^2 + (z[i]-a)^2)
}
term1=2*term
return(term1)
}
dlog1la=function(a,b)
```

```
{
term=0
for(i in 1:n)
{
term= term + 1/(b^2 + (z[i]-a)^2)
}
term2= n/b -2*b*term
return(term2)
}
L11=function(a,b)
{
term=0
for(i in 1:n)
{
term= term + ((z[i]-a)^2 - b^2)/((b^2 + (z[i]-a)^2)^2)
}
term1=2*term
return(term1)
}
L12=function(a,b)
{
term=0
for(i in 1:n)
{
term= term + (z[i]-a)/((b^2 + (z[i]-a)^2)^2)
}
term2=-4*b*term
return(term2)
}
L22=function(a,b)
{
term=0
for(i in 1:n)
{
term= term - 2/(b^2 + (z[i]-a)^2) + (4*b^2)/((b^2 + (z[i]-a)^2)^2)
}
term3=-n/b^2 +term
return(term3)
}
L=function(a,b)
{
f=matrix(c(L11(a,b),L12(a,b),L12(a,b),L22(a,b)), byrow=TRUE,ncol=2)
return(f)
}
v=function(a,b)
{
f=matrix(c(dloglth(a,b), dloglla(a,b)),byrow=TRUE,ncol=2)
return(f)
}
```

```
m=5; EstMat=matrix(nrow=m,ncol=2); EstMat[1,]=c(thest,laest);
                                  diff=2; i=1
while(diff>0.00001)
{
EstMat[i+1,]=EstMat[i,]
     - v(EstMat[i,1],EstMat[i,2])%*%solve(L(EstMat[i,1],EstMat[i,2]))
diff=sum((EstMat[i+1,]-EstMat[i,])^2)
i=i+1
}
EstMat
####   Method of scoring
I=matrix(nrow=2,ncol=2)
I=function(a,b)
{
f=matrix(c(1/(2*b^2),0,0,1/(2*b^2)), byrow=TRUE,ncol=2)
return(f)
}
Mat=matrix(nrow=m,ncol=2); Mat[1,]=c(thest,laest); diff=2; i=1
while(diff>0.00001)
{
Mat[i+1,]=Mat[i,] + (1/n)* v(Mat[i,1],Mat[i,2])
                   %*%solve(I(Mat[i,1],Mat[i,2]))
diff=sum((Mat[i+1,]-Mat[i,])^2)
i=i+1
}
Mat
```

A random sample of size 200 is generated from the Cauchy distribution with $\theta = 3$ and $\lambda = 2$. By maximizing the log-likelihood, the maximum likelihood estimate of θ is 3.15 and that of λ is 2.12. Using initial estimates as $\theta_0 = 3.14$, which is the sample median, and $\lambda_0 = 2.08$, which is the difference between the third sample quartile and the sample median, both the iterative procedures produce the maximum likelihood estimate of θ as 3.15 and that of λ as 2.12. It can be verified that the distribution belongs to a two-parameter Cramér family and hence the maximum likelihood estimator is CAN with approximate dispersion matrix $I^{-1}(\theta, \lambda)/n$. □

✍ Remark 4.5.1

In the above example, we have used three approaches to find the maximum likelihood estimators, but we note that, once we have an information matrix, the method of scoring is better than the Newton-Raphson method. In the next example of a bivariate Cauchy distribution, the information matrix is free from the parameters also. Hence, it is better to use the method of scoring for finding the maximum likelihood estimates.

In the following example, we verify the results established in Example 4.3.5. There is no built-in function to generate a random sample from a bivariate Cauchy distribution. Hence, we first discuss how to draw a random sample from a bivariate Cauchy distribution. If a random vector $(X, Y)'$ follows a bivariate Cauchy $C_2(\theta_1, \theta_2)$ distribution, then the marginal probability density function of Y is derived as follows:

$$f(y, \theta_2) = \int_{-\infty}^{\infty} \frac{1}{2\pi} \left\{1 + (x - \theta_1)^2 + (y - \theta_2)^2\right\}^{-3/2} dx$$

$$= \frac{1}{2\pi} \frac{1}{(1 + (y - \theta_2)^2)^{3/2}} \int_{-\infty}^{\infty} \frac{1}{(1 + (x - \theta_1)^2/(1 + (y - \theta_2)^2))^{3/2}} dx$$

$$= \frac{1}{\pi} \frac{1}{(1 + (y - \theta_2)^2)^{3/2}} \int_{0}^{\infty} \frac{1}{(1 + u^2/(1 + (y - \theta_2)^2))^{3/2}} du \quad u = (x - \theta_1)$$

$$= \frac{1}{\pi} \frac{1}{1 + (y - \theta_2)^2} \int_{0}^{\infty} \frac{t^{-1/2}}{2(1 + t)^{3/2}} dt \quad \text{with } u^2/(1 + (y - \theta_2)^2) = t$$

$$= \frac{1}{\pi} \frac{1}{1 + (y - \theta_2)^2} \int_{0}^{\infty} \frac{t^{1/2 - 1}}{B(1/2, 1)(1 + t)^{1/2 + 1}} dt$$

$$= \frac{1}{\pi} \frac{1}{1 + (y - \theta_2)^2}, \quad y \in \mathbb{R}.$$

Thus, the marginal distribution of Y is Cauchy with location parameter θ_2. On similar lines, it follows that the marginal distribution of X is also Cauchy with location parameter θ_1. Hence, the conditional distribution of X given $Y = y$ has the probability density function as

$$f(x, \theta_1, \theta_2|y) = \frac{1 + (y - \theta_2)^2}{2\{1 + (x - \theta_1)^2 + (y - \theta_2)^2\}^{3/2}} \quad x \in \mathbb{R}, \quad \theta_1, \theta_2 \in \mathbb{R}.$$

To generate a random sample from a bivariate Cauchy distribution, we first generate a random observation from Y and corresponding to realized value y, generate x from the conditional probability density function. Hence, we find the distribution function of a conditional distribution of X given $Y = y$ as follows. Observe that for each fixed y, the conditional distribution of X given $Y = y$ is symmetric around θ_1. Suppose the conditional distribution function $F_{X|Y=y}(x)$ of X given $Y = y$ is denoted by $F_y(x)$. With $1 + (y - \theta_2)^2 = a$ we have

$$F_y(x) = \frac{1}{2} \int_{-\infty}^{x} \frac{(1 + (y - \theta_2)^2)}{(1 + (u - \theta_1)^2 + (y - \theta_2)^2)^{3/2}} du$$

$$= \frac{1}{2\sqrt{a}} \int_{-\infty}^{x} \frac{1}{(1 + (u - \theta_1)^2/a)^{3/2}} \, du.$$

Suppose $x = \theta_1$. Then

$$F_y(\theta_1) = \frac{1}{2\sqrt{a}} \int_{-\infty}^{\theta_1} \frac{1}{(1 + (u - \theta_1)^2/a)^{3/2}} \, du$$

$$= \frac{1}{2\sqrt{a}} \int_{-\infty}^{0} \frac{1}{(1 + w^2/a)^{3/2}} \, dw \text{ with } (u - \theta_1) = w$$

$$= \frac{1}{2\sqrt{a}} \int_{0}^{\infty} \frac{1}{(1 + w^2/a)^{3/2}} \, dw = \frac{1}{2} \int_{0}^{\infty} \frac{t^{-1/2}}{2(1 + t)^{3/2}} \, dt \text{ with } w^2/a = t$$

$$= \frac{1}{2} \int_{0}^{\infty} \frac{1}{B(1/2, 1)} \frac{t^{1/2-1}}{(1 + t)^{1/2+1}} \, du = \frac{1}{2}.$$

Thus, θ_1 is the median of the conditional distribution of X given $Y = y$ and the distribution is symmetric around θ_1. Suppose $x > \theta_1$ and $((x - \theta_1)^2/a)/(1 + (x - \theta_1)^2/a) = b(x)$, say. Then

$$F_y(x) = \frac{1}{2\sqrt{a}} \int_{-\infty}^{x} \frac{1}{(1 + (u - \theta_1)^2/a)^{3/2}} \, du$$

$$= \frac{1}{2} + \frac{1}{2\sqrt{a}} \int_{\theta_1}^{x} \frac{1}{(1 + (u - \theta_1)^2/a)^{3/2}} \, du$$

$$= \frac{1}{2} + \frac{1}{2\sqrt{a}} \int_{0}^{x-\theta_1} \frac{1}{(1 + w^2/a)^{3/2}} \, du \text{ with } (u - \theta_1) = w$$

$$= \frac{1}{2} + \frac{1}{2} \int_{0}^{(x-\theta_1)^2/a} \frac{t^{-1/2}}{2(1 + t)^{3/2}} \, dt \text{ by } w^2/a = t$$

$$= \frac{1}{2} + \frac{1}{2} \int_{0}^{b(x)} \frac{1}{2} \frac{(v/(1 - v))^{-1/2}(1 - v)^{3/2}}{(1 - v)^2} \, dv \text{ by } t = v/(1 - v)$$

$$= \frac{1}{2} + \frac{1}{2} \int_{0}^{b(x)} \frac{1}{B(1/2, 1)} v^{1/2-1}(1 - v)^{1-1} \, dv = \frac{1}{2} + \frac{1}{2} G(b(x)),$$

where G is a distribution function of a beta distribution of first kind with shape 1 parameter $1/2$ and shape 2 parameter 1. By symmetry, for $x < \theta_1$

$$F_y(x) = 1 - \left(\frac{1}{2} + \frac{1}{2}G(b(x)) \right) = \frac{1}{2} - \frac{1}{2}G(b(x)).$$

Thus,

$$F_y(x) = \begin{cases} 1/2 - (1/2)G(b(x)), & \text{if } x < \theta_1 \\ 1/2 + (1/2)G(b(x)) & \text{if } x \geq \theta_1. \end{cases}$$

By the probability integral transformation, to obtain a random sample from the conditional distribution of X given $Y = y$, we solve $F_y(x) = u$, where u is a random observation from a $U(0, 1)$ distribution. Suppose $qbeta(1 - 2u, 1/2, 1) = q$, say. If $u < 1/2$ then

$$\frac{1}{2} - \frac{1}{2}G(b(x)) = u \implies b(x) = qbeta(1 - 2u, 1/2, 1) = q$$

$$\implies (x - \theta_1)^2 = aq/(1 - q) \implies x = \theta_1 - \sqrt{aq/(1 - q)},$$

we take negative root as $u < 1/2$ implies $x < \theta_1$. Suppose $qbeta(2u - 1, 1/2, 1) = p$, say. If $u \geq 1/2$ then

$$\frac{1}{2} + \frac{1}{2}G(b(x)) = u \implies b(x) = qbeta(2u - 1, 1/2, 1) = p$$

$$\implies (x - \theta_1)^2 = ap/(1 - p) \implies x = \theta_1 + \sqrt{ap/(1 - p)},$$

we take positive root as $u \geq 1/2$ implies $x \geq \theta_1$. Thus, we adopt the following stepwise procedure to generate a random sample from a bivariate Cauchy distribution.

1. Generate y from $C(\theta_2, 1)$ and hence find $a = 1 + (y - \theta_2)^2$.
2. Generate u from $U(0, 1)$.
3. Depending on the value of u find $d = qbeta((\cdot), 1/2, 1)$.
4. Find $x = \theta_1 - \sqrt{ad/(1 - d)}$ or $x = \theta_1 + \sqrt{ad/(1 - d)}$ corresponding to $u < 1/2$ and $u \geq 1/2$ respectively.

☑ Example 4.5.7

Suppose a random vector $(X, Y)'$ follows a bivariate Cauchy $C_2(\theta_1, \theta_2)$ distribution with probability density function given by

$$f(x, y, \theta_1, \theta_2) = \frac{1}{2\pi} \left\{ 1 + (x - \theta_1)^2 + (y - \theta_2)^2 \right\}^{-3/2} \quad (x, y)' \in \mathbb{R}^2,$$

$$\theta_1, \theta_2 \in \mathbb{R}.$$

In Example 4.3.5, we have obtained the information matrix of the distribution and it is free from the parameters. Hence, we use the method of scoring to obtain the maximum likelihood estimator of $(\theta_1, \theta_2)'$. We also obtain the maximum likelihood estimator of θ_1, which is a location parameter of the marginal distribution of X, but use data generated under the bivariate model. Similarly, we obtain the maximum likelihood estimator of θ_2, which is a location parameter of the marginal distribution of Y, but use data generated under the bivariate model. We compare these estimates with the estimates in the bivariate model. Following is a R code for these computations.

```
n=300; th1=1; th2=2; x=b=c(); set.seed(40)
y=rcauchy(n,location=th2,scale=1); a= 1+(y-th2)^2
set.seed(12); u=runif(n,0,1)
for(i in 1:n)
{
if(u[i]<1/2)
{
b[i]=qbeta(1-2*u[i],shape1=1/2,shape2=1)
x[i]=th1-(a[i]*b[i]/(1-b[i]))^(1/2)
}
else
{
b[i]=qbeta(2*u[i]-1,shape1=1/2,shape2=1)
x[i]=th1+(a[i]*b[i]/(1-b[i]))^(1/2)
}
}
r=cor(x,y,method="spearman");r
d=data.frame(x,y); summary(d)
s1=3*mean((x-th1)/(1+(x-th1)^2+(y-th2)^2))
s2=3*mean((y-th2)/(1+(x-th1)^2+(y-th2)^2))
i11=3*mean((1-(x-th1)^2+(y-th2)^2)/((1+(x-th1)^2+(y-th2)^2))^2)
i22=3*mean((1+(x-th1)^2-(y-th2)^2)/((1+(x-th1)^2+(y-th2)^2))^2)
i12=6*mean(((x-th1)*(y-th2))/((1+(x-th1)^2+(y-th2)^2))^2)
s1;s2;i11;i22;i12
#### mle by method of scoring
m1=median(x);m2=median(y);th1est=m1;th2est=m2;m1;m2
dlog1th1=function(a,b)
{
term=0
for(i in 1:n)
{
term=term + (x[i]-a)/(1 + (x[i]-a)^2+(y[i]-b)^2)
}
term1=3*term
return(term1)
}
```

```
dloglth2=function(a,b)
{
term=0
for(i in 1:n)
{
term=term + (y[i]-b)/(1 + (x[i]-a)^2+(y[i]-b)^2)
}
term1=3*term
return(term1)
}
v=function(a,b)
{
f=matrix(c(dloglth1(a,b), dloglth2(a,b)),byrow=TRUE,ncol=2)
return(f)
}
I=matrix(c(3/5,0,0,3/5),nrow=2,ncol=2)
m=5; Mat=matrix(nrow=m,ncol=2);Mat[1,]=c(th1est,th2est)
k=1; diff=1
while(diff>10^(-4))
{
Mat[k+1,]=Mat[k,] + (1/n)* v(Mat[k,1],Mat[k,2])%*%solve(I)
diff=sqrt(sum(Mat[k+1,]-Mat[k,])^2)
k=k+1
}
mle=Mat[k,] ;mle
### Marginal of X
dth1=function(a)
{
term=0
for(i in 1:n)
{
term=term + (x[i]-a)/(1 + (x[i]-a)^2)
term1=2*term
return(term1)
}
s=c(); s[1]=th1est; k=1; diff=1
while(diff > 10^(-4))
{
s[k+1]=s[k] + (2/n)* dth1(s[k])
diff=abs(s[k+1]-s[k])
k=k+1
}
mleth1=s[k]; mleth1
### Marginal of Y
dth2=function(a)
{
term=0
for(i in 1:n)
{
term=term + (y[i]-a)/(1 + (y[i]-a)^2)
}
```

```
term1=2*term
return(term1)
}
u=c(); u[1]=th2est; k=1; diff=1
while(diff > 10^(-4))
{
u[k+1]=u[k] + (2/n)* dth2(u[k])
diff=abs(u[k+1]-u[k])
k=k+1
}
mleth2=u[k]; mleth2
```

On the basis of a generated sample, we have obtained Spearman's rank correlation coefficient and it is -0.0355. It indicates the association between X and Y. The estimate of the information matrix on the basis of generated data is given by

$$\hat{I}_n(\theta_1, \theta_2) = \begin{pmatrix} 0.6417 & 0.0263 \\ 0.0263 & 0.6584 \end{pmatrix}.$$

The $(1, 1)$-th element of $\hat{I}_n(\theta_1, \theta_2)$ is obtained as the mean of the values of second derivative of the logarithm of joint density function, multiplied by -1. The other elements are obtained on similar lines. The diagonal elements are close to 0.6, but off-diagonal elements are not close to 0. If we increase the sample size, then these will approach to 0. Using the same method, estimates of expected values of score functions are obtained and these are -0.0039 and -0.0085, which are close to 0 as expected. In the method of scoring, we take the initial estimate of θ_1 as 0.9863, which is the sample median of X and the initial estimate of θ_2 as 1.9250, which is the sample median of Y. The maximum likelihood estimate of $(\theta_1, \theta_2)'$ comes out to be $(0.9934, 1.9868)'$, which is close to the true parameter value $(1, 2)'$ of $(\theta_1, \theta_2)'$. The maximum likelihood estimate of θ_1, treating it as a location parameter of the marginal distribution of X, but using the same data, is 0.9969. It is different from 0.9934, as expected. Similarly, the maximum likelihood estimate of θ_2, treating it as a location parameter of the marginal distribution of Y, but using the same data, is 1.9679. It is also different from 1.9868. In view of the association between X and Y, the maximum likelihood estimates from the bivariate model and the corresponding univariate models are different, when these are based on the bivariate data.   □

In the next example, we use the same model used in Example 4.5.7 and generate multiple random samples from a bivariate Cauchy distribution to obtain the estimate of approximate dispersion matrix of $(\hat{\theta}_{1n}, \hat{\theta}_{2n})'$. We also obtain the estimates of approximate variances of the estimators of parameters of the marginal distributions and compare with those of joint distribution.

✐ Example 4.5.8

Suppose a random vector $(X, Y)'$ follows a bivariate Cauchy $C_2(\theta_1, \theta_2)$ distribution as specified in Example 4.5.7. Following is a R code to generate multiple random samples from the bivariate Cauchy distribution and to obtain the estimate of approximate dispersion matrix of $(\hat{\theta}_{1n}, \hat{\theta}_{2n})'$ and approximate variances of the estimators of parameters of the marginal distributions.

```
n=300; nsim=1500; th1=1; th2=2
y=u=x=b=a=matrix(nrow=n,ncol=nsim);r=c()
for(j in 1:nsim)
{
set.seed(j)
y[,j]=rcauchy(n,location=th2,scale=1)
u[,j]=runif(n,0,1)
}
for(j in 1:nsim)
{
for(i in 1:n)
{
a[i,j]=1+(y[i,j]-th2)^2
if(u[i,j]<1/2)
{
b[i,j]=qbeta(1-2*u[i,j],shape1=1/2,shape2=1)
x[i,j]=th1-(a[i,j]*b[i,j]/(1-b[i,j]))^(1/2)
}
else
{
b[i,j]=qbeta(2*u[i,j]-1,shape1=1/2,shape2=1)
x[i,j]=th1+(a[i,j]*b[i,j]/(1-b[i,j]))^(1/2)
}
}
}
for(j in 1:nsim)
{
r[j]=cor(x[,j],y[,j],method="spearman")
}
mean(r);var(r); m1=m2=c()
dloglth1=function(a,b)
{
term=sum((xsamp-a)/(1 + (xsamp-a)^2+(ysamp-b)^2))
term1=3*term
return(term1)
}
dloglth2=function(a,b)
{
term=sum((ysamp-b)/(1 + (xsamp-a)^2+(ysamp-b)^2))
term1=3*term
return(term1)
}
```

```
v=function(a,b)
{
f=matrix(c(dloglth1(a,b), dloglth2(a,b)),byrow=TRUE,ncol=2)
return(f)
}
I=matrix(c(3/5,0,0,3/5),nrow=2,ncol=2); mle=matrix(nrow=nsim,ncol=2)
for(j in 1:nsim)
{
m1[j]=median(x[,j]); m2[j]=median(y[,j]); th1est=m1[j]; th2est=m2[j]
xsamp=x[,j]; ysamp=y[,j]
m=5; Mat=matrix(nrow=m,ncol=2); Mat[1,]=c(th1est,th2est)
k=1; diff=1
while(k < m && diff>10^(-4))
{
Mat[k+1,]=Mat[k,] + (1/n)* v(Mat[k,1],Mat[k,2])%*%solve(I)
diff=sqrt(sum(Mat[k+1,]-Mat[k,])^2)
k=k+1
}
mle[j,]=Mat[k,]
}
summary(mle); D1=round(cov(mle),5); D2=round(solve(I)/n,4); D1;D2
### Marginal
dth=function(a)
{
term=0
for(i in 1:n)
{
term=term + (samp[i]-a)/(1 + (samp[i]-a)^2)
}
term1=2*term
return(term1)
}
marmle=matrix(nrow=nsim,ncol=2)
for(j in 1:nsim)
{
m1[j]=median(x[,j]);m2[j]=median(y[,j]);th1est=m1[j];th2est=m2[j]
samp=x[,j]; s=c();s[1]=th1est; k=1; diff=1
while(diff > 10^(-4))
{
s[k+1]=s[k] + (2/n)* dth(s[k])
diff=abs(s[k+1]-s[k])
k=k+1
}
marmle[j,1]=s[k]
samp=y[,j]; s=c();s[1]=th2est; k=1; diff=1
while(diff > 10^(-4))
{
s[k+1]=s[k] + (2/n)* dth(s[k])
diff=abs(s[k+1]-s[k])
k=k+1
}
```

```
marmle[j,2]=s[k]
}
summary(marmle); apply(marmle,2,var); v=2/n; v
### Scatter plots
par(mfrow=c(1,2))
plot(mle[,1],marmle[,1],xlab="MLE of Th1: Joint Distribution",
     ylab="MLE of Th1: Marginal Distribution",col="blue")
abline(0,1,col="maroon")
plot(mle[,2],marmle[,2],xlab="MLE of Th2: Joint Distribution",
     ylab="MLE of Th2: Marginal Distribution",col="blue")
abline(0,1,col="maroon")
### Density plots
par(mfrow=c(1,2))
plot(density(mle[,1]),main="MLE of Theta_1")
lines(density(marmle[,1]),col=2,lty=2)
legend("bottomleft",legend=c("Joint","Marginal"),
                        col=1:2,lty=1:2,bty="n")
plot(density(mle[,2]),main="MLE of Theta_2")
lines(density(marmle[,2]),col=2,lty=2)
legend("bottomleft",legend=c("Joint","Marginal"),
                        col=1:2,lty=1:2,bty="n")
```

On the basis of 1500 random samples, each of size $n = 300$, the mean and variance of Spearman's rank correlation coefficients is -0.0032 and 0.0049 respectively. It indicates the association between X and Y, which is rather weak.

In the method of scoring, we take the initial estimate of θ_1 and of θ_2, as the sample medians of X and Y respectively. The summary of the maximum likelihood estimates of $(\theta_1, \theta_2)'$ gives mean to be $(1.0021, 2.000)'$ and median to be $(1.0035, 1.999)'$. Both are close to the true parameter value $(1, 2)'$ of $(\theta_1, \theta_2)'$. The estimate of approximate dispersion matrix D of $(\hat{\theta}_{1n}, \hat{\theta}_{2n})'$ and $I^{-1}(\theta_1, \theta_2)/n$ are as follows:

$$D = \begin{pmatrix} 0.00549 & -0.00008 \\ -0.00008 & 0.00556 \end{pmatrix} \quad \& \quad I^{-1}(\theta_1, \theta_2)/n = \begin{pmatrix} 0.0056 & 0 \\ 0 & 0.0056 \end{pmatrix}.$$

We note that the two are very close to each other. We also obtained the maximum likelihood estimates of $(\theta_1, \theta_2)'$, treating θ_1, θ_2 as the location parameters of the marginal distributions, but using the data generated under joint distribution. The summary of these gives mean to be $(1.0007, 1.998)'$ and median to be $(1.0004, 1.998)'$, slightly different than those obtained for the joint distribution. The estimates of approximate variances of $\tilde{\theta}_{1n}$ and $\tilde{\theta}_{2n}$, which are the maximum likelihood estimates of θ_1 and θ_2 respectively, treating these as the location parameters of the marginal distributions, are 0.0064 and 0.0069 respectively. These are close to $2/n = 0.0067$ but are different from those obtained under joint distribution. Under joint distribution, the approximate variance of $\hat{\theta}_{1n}$ is 0.00549 and it

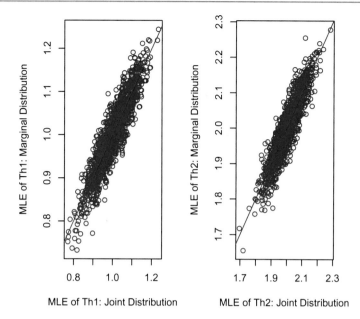

Fig. 4.5 Scatter plots: MLE in joint and marginal models

is smaller than the approximate variance of $\tilde{\theta}_{1n}$ which is 0.0064. Similarly, the approximate variance of $\hat{\theta}_{2n}$ is 0.00556 and it is smaller than the approximate variance of $\tilde{\theta}_{1n}$ which is 0.0069. Thus, simulation results do support the theoretical results as derived in Example 4.3.5. Figure 4.5 displays the scatter plots of the maximum likelihood estimators of θ_1 and of θ_2 under joint and marginal setup. We observe that the estimates under two setups are different, as expected. Figure 4.6 displays the density plots of the maximum likelihood estimators of θ_1 and of θ_2 under joint and marginal setup. From this figure also we note that the estimates under the two setups are different.   □

4.6   Conceptual Exercises

4.6.1 Suppose $\{X_1, X_2, \ldots, X_n\}$ is a random sample from a distribution of X with probability density function $f(x, \theta) = \theta/x^{\theta+1}$ $x > 1$, $\theta > 0$. (i) Examine whether the distribution belongs to a one-parameter exponential family. (ii) On the basis of a random sample of size n from the distribution of X, find the moment estimator of θ based on a sufficient statistic and the maximum likelihood estimator of θ. (iii) Examine whether these are CAN estimators of θ. (iv) Obtain the CAN estimator for $P[X \geq 2]$.

4.6.2 Suppose $\{X_1, X_2, \ldots, X_n\}$ is a random sample from a binomial $B(m, \theta)$ distribution, truncated at 0, $0 < \theta < 1$ and m is a known positive integer. Examine whether the distribution belongs to a one-parameter exponential

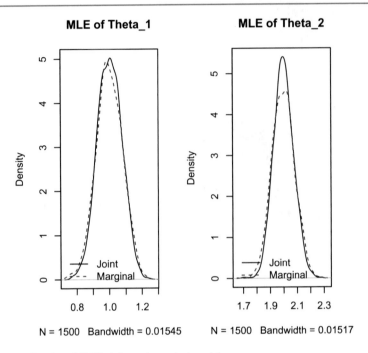

Fig. 4.6 Density plots: MLE in joint and marginal models

family. Find the moment estimator of θ based on a sufficient statistic and the maximum likelihood estimator of θ. Examine whether the two are the same and whether these are CAN. Find their approximate variances.

4.6.3 Suppose $\{X_1, X_2, \ldots, X_n\}$ is a random sample from a distribution of X with probability density function

(i) $f(x, \theta) = (x/\theta) \exp\{-x^2/2\theta\}, \quad x > 0, \quad \theta > 0$ and

(ii) $f(x, \theta) = (3x^2/\theta^3) \exp\{-x^3/\theta^3\} \quad x > 0, \quad \theta > 0.$

Examine whether the distribution belongs to a one-parameter exponential family. On the basis of a random sample of size n from these distributions, find the moment estimator based on a sufficient statistic and the maximum likelihood estimator of θ. Examine whether the two are the same and whether these are CAN. Find their approximate variances.

4.6.4 Suppose X has a logarithmic series distribution with probability mass function given by

$$p(x, \theta) = \frac{-1}{\log(1 - \theta)} \frac{\theta^x}{x} \quad x = 1, 2, \ldots, 0 < \theta < 1 \; .$$

Show that the logarithmic series distribution is a power series distribution. On the basis of a random sample from the logarithmic series distribution, find the moment estimator of θ based on a sufficient statistic and the maximum likelihood estimator of θ. Examine whether the two are the same and whether these are CAN. Find their approximate variances.

4.6.5 Suppose $(X, Y)'$ has a bivariate normal distribution with zero mean vector and dispersion matrix Σ given by

$$\Sigma = \sigma^2 \begin{pmatrix} 1 & \rho \\ \rho & 1 \end{pmatrix},$$

$\sigma^2 > 0$, $-1 < \rho < 1$. On the basis of a random sample of size n from the distribution of $(X, Y)'$ find the maximum likelihood estimator of $(\sigma^2, \rho)'$ and examine if it is CAN. Find the approximate dispersion matrix.

4.6.6 Suppose $(X, Y)'$ is random vector with a joint probability mass function as

$$P[X = x, Y = y] = \binom{x}{y} e^{-\lambda} \lambda^x p^y (1 - p)^{x-y} / x!,$$

$$y = 0, 1, \ldots, x; \quad x = 0, 1, 2, \ldots,$$

where $\lambda > 0$ and $0 < p < 1$. Examine if the distribution belongs to a two-parameter exponential family. Hence, find a CAN estimator for $(\lambda, p)'$ and its approximate dispersion matrix.

4.6.7 Suppose $(X, Y)'$ has a bivariate normal distribution with mean vector $(\mu_1, \mu_2)'$ and dispersion matrix Σ given by

$$\Sigma = \begin{pmatrix} 1 & \rho \\ \rho & 1 \end{pmatrix},$$

where $\rho \neq 0$ and is known and $\mu_1, \mu_2 \in \mathbb{R}$. Show that the distribution belongs to a two-parameter exponential family. Hence, find a CAN estimator $(\mu_1, \mu_2)'$ and its approximate dispersion matrix.

4.6.8 Suppose a random variable X has a negative binomial distribution with parameters (k, p) and with the following probability mass function.

$$P[X = x] = \binom{x + k - 1}{k - 1} p^k (1 - p)^x \quad x = 0, 1, 2, \ldots.$$

(i) Show that the distribution belongs to a one-parameter exponential family, if k is known and $p \in (0, 1)$ is unknown. Hence obtain a CAN estimator of p. (ii) Examine whether the distribution belongs to a one-parameter exponential family, if p is known and k is unknown positive integer. (iii) Examine whether the distribution belongs to a two-parameter exponential family, if both $p \in (0, 1)$ and k are unknown, where k is a positive integer.

4.6.9 Examine whether a logistic distribution with probability density function

$$f(x, \theta) = \frac{\exp\{-(x - \theta)\}}{(1 + \exp\{-(x - \theta)\})^2}, \quad x \in \mathbb{R}, \ \theta \in \mathbb{R}$$

belongs to a one-parameter exponential family. If not, examine if it belongs to a one-parameter Cramér family. If yes, find a CAN estimator of θ and its approximate variance.

4.6.10 Suppose a random variable X follows a Cauchy $C(\theta, \lambda)$ distribution with location parameter θ and shape parameter λ. Examine whether the distribution belongs to a two-parameter Cramér family.

4.6.11 Suppose $\{X_1, X_2, \ldots, X_n\}$ is a random sample from a Poisson distribution with parameter $\theta > 0$. An estimator T_n is defined as

$$T_n = \begin{cases} \overline{X}_n, & \text{if} \quad \overline{X}_n > 0 \\ 0.01, & \text{if} \quad \overline{X}_n = 0 \end{cases}$$

Show that $T_{1n} = e^{-T_n}$ is a CAN estimator of $P[X_1 = 0] = e^{-\theta}$. Find its approximate variance. Suppose random variables Y_i, $i = 1, 2, \ldots, n$ are defined as follows:

$$Y_i = \begin{cases} 1, & \text{if} \quad X_i = 0 \\ 0, & \text{otherwise} . \end{cases}$$

Obtain a CAN estimator T_{2n} for $e^{-\theta}$ based on $\{Y_1, Y_2, \ldots, Y_n\}$ and find its approximate variance. Find $ARE(T_{1n}, T_{2n})$.

4.7   Computational Exercises

Verify the results by simulation using R.

4.7.1 Suppose a random variable X follows a Cauchy $C(\theta, 1)$ distribution with location parameter θ and shape parameter 1. Using R draw a random sample from the distribution of X, plot the likelihood function and find the maximum likelihood estimator θ. Using simulation examine whether it is CAN for θ. (Hint: Use the code similar to that Example 4.5.3.)

4.7.2 In Exercise 4.6.2, you have obtained the moment estimator of θ based on the sufficient statistic and the maximum likelihood estimator of θ corresponding to a random sample $\{X_1, X_2, \ldots, X_n\}$ from the binomial $B(m, \theta)$ distribution, truncated at $0, 0 < \theta < 1$ and m is a known positive integer. Using R draw a random sample from the binomial $B(m, \theta)$ distribution, truncated at 0. Plot the likelihood function and approximately identify the solution of the likelihood equation. Use Newton-Raphson iterative procedure and method of scoring to solve the likelihood equation and find the maximum likelihood estimator of θ on the basis of a random sample generated from $B(m, \theta)$ distribution, truncated at 0 assuming some values for m and θ. Verify that the maximum likelihood estimator of θ is CAN. (Hint: Use the code similar to that for truncated Poisson distribution.)

4.7.3 Suppose $(X, Y)'$ has bivariate normal distribution with zero mean vector and dispersion matrix Σ given by

$$\Sigma = \begin{pmatrix} 1 & \rho \\ \rho & 1 \end{pmatrix},$$

$-1 < \rho < 1$. We have discussed the maximum likelihood estimation of ρ in Example 4.3.3. It is noted that we cannot find the explicit solution but a unique root of the likelihood equation exists. Hence use the simulation approach to get the solution of the likelihood equation. Draw a random sample from the above distribution using rmvnorm command from the MASS library to draw a sample from bivariate normal distribution. Plot the likelihood function and approximately identify the solution of the likelihood equation. Solve the likelihood equation by the Newton-Raphson procedure. Take the initial value of ρ as $\rho_0 = \sum_{i=1}^{n} X_i Y_i / n$ (Hint: Use the code similar to that for truncated Poisson distribution.)

4.7.4 Suppose $(X, Y)'$ has a bivariate normal distribution with zero mean vector and dispersion matrix Σ given by,

$$\Sigma = \sigma^2 \begin{pmatrix} 1 & \rho \\ \rho & 1 \end{pmatrix},$$

$\sigma^2 > 0$, $-1 < \rho < 1$. In Exercise 4.6.5, you have obtained the maximum likelihood estimators of $(\sigma^2, \rho)'$ and shown that it is CAN . Verify the results by simulation using R code. Solve the system of likelihood equations using Newton-Raphson procedure and also by the method of scoring.

4.7.5 Suppose a random vector $(X, Y)'$ follows a bivariate normal $N_2(0, 0, \sigma_1^2, \sigma_2^2, \rho)$ distribution where $\rho \neq 0$ is a known correlation coefficient. In Example 4.3.4, it is shown that the maximum likelihood estimator of $(\sigma_1^2, \sigma_2^2)'$ is a CAN estimator of $(\sigma_1^2, \sigma_2^2)'$. Verify the result by simulation. Based on the bivariate data, obtain the maximum likelihood estimator of σ_1^2 and of σ_1^2, treating these as parameters of marginal distributions. Comment on the results.

4.7.6 Suppose a random vector $(X, Y)'$ follows a bivariate normal $N_2(\mu_1, \mu_2, \sigma_1^2, \sigma_2^2, \rho)$ distribution where where $\mu_1, \mu_2 \in \mathbb{R}, \sigma_1^2, \sigma_2^2 > 0$ and $\rho \in (-1, 1)$. On the basis of simulated data, obtain the maximum likelihood estimator of $(\mu_1, \mu_2, \sigma_1^2, \sigma_2^2, \rho)'$. (Hint: The distribution belongs to five parameter exponential family, hence the maximum likelihood estimator of $(\mu_1, \mu_2, \sigma_1^2, \sigma_2^2, \rho)'$ is same as the moment estimator of $(\mu_1, \mu_2, \sigma_1^2, \sigma_2^2, \rho)'$ based on the sufficient statistic.)

4.7.7 Suppose a random variable X follows a Cauchy $C(\theta, \lambda)$ distribution with location parameter θ and shape parameter λ. Using R and Cramér-Wold device verify that the maximum likelihood estimator of $(\theta, \lambda)'$ is CAN with approximate dispersion matrix $I^{-1}(\theta, \lambda)/n$.

(Hint: $I(\theta, \lambda)$ is obtained in Example 4.5.5.)

4.7.8 For the multinomial distribution as specified in Example 4.2.7, simulate the
sample of size n and based on that find the maximum likelihood estimators
of the parameters θ and ϕ using Newton-Raphson procedure and using the
method of scoring. Comment on the result. It has been proved in Example
4.2.7 that the the maximum likelihood estimator of $(\theta, \phi)'$ is CAN. Verify it
by simulation.

4.7.9 Suppose a random variable X follows a gamma distribution with scale parameter α and shape parameter λ, with probability density function given by

$$f(x, \alpha, \lambda) = \frac{\alpha^\lambda}{\Gamma(\lambda)}e^{-\alpha x}x^{\lambda-1} \ , \quad x > 0, \ \alpha > 0, \ \lambda > 0 \ .$$

Find the maximum likelihood estimator of $(\alpha, \lambda)'$ using method of scoring.
(Hint: Use procedure as in Example 4.5.5)

4.7.10 Suppose a random variable X follows a logistic distribution with probability
density function

$$f(x, \theta) = \frac{\exp\{-(x - \theta)\}}{(1 + \exp\{-(x - \theta)\})^2}, \ x \in \mathbb{R}, \ \theta \in \mathbb{R}.$$

Find the maximum likelihood estimator of θ using the Newton-Raphson procedure and the method of scoring. As an initial iterative value one may take
the sample median or the sample mean as both are consistent for θ. (Hint:
Some part of the code of Example 4.5.3 will be useful.)

References

1. Lehmann, E. L., & Romano, J. P. (2005). *Testing of statistical hypothesis* (3rd ed.). New York: Springer.
2. van der Vaart, A. (1998). *Asymptotic statistics*. Cambridge: Cambridge University Press.
3. Apostol, T. (1967). *Calculus* (2nd ed., Vol. I). New York: Wiley.
4. Cramér, H. (1946). *Mathematical methods of statistics*. Princeton: Princeton University Press.
5. Huzurbazar, V. S. (1948). The likelihood equation, consistency and maxima of the likelihood function. *The Annals of Eugenics, 14,* 185–200.
6. Kale, B. K., & Muralidharan, K. (2016). *Parametric inference: An introduction*. Delhi: Narosa.
7. Rohatgi, V. K., & Saleh, A. K. Md. E. (2001). *Introduction to probability and statistics*. New York: Wiley.
8. Rao, C. R. (1978). *Linear statistical inference and its applications*. New York: Wiley.
9. Kotz, S., Balakrishnan, N., & Johnson, N. L. (2000). *Continuous multivariate distributions: Models and applications* (2nd ed., Vol. I). New York: Wiley.
10. Kale, B. K. (1961). On the solution of the likelihood equation by iteration processes. *Biometrika, 48,* 452–456.
11. Kale, B. K. (1962). On the solution of the likelihood equation by iteration processes, the multiparameter case. *Biometrika, 49,* 479–486.
12. Kotz, S., Balakrishnan, N., & Johnson, N. L. (1995). *Continuous univariate distributions* (2nd ed., Vol. II). New York: Wiley.

Large Sample Test Procedures

5

Contents

> **Learning Objectives** After going through this chapter, the readers should
> be able
> - to perform the large sample test procedures using the test statistic based
>   on the CAN estimator and judge the performance of a test procedure
>   using the power function
> - to carry out the likelihood ratio test procedure and decide
>   the asymptotic null distribution of the likelihood ratio test statistic
> - to use R software in the large sample test procedures and the likelihood
>   ratio test procedure

5.1   Introduction

In Chaps. 2, 3 and 4, we discussed point estimation of a parameter and studied the
large sample optimality properties of the estimators. We also discussed interval estimation for large n. The present and the next chapters are devoted to the large sample
test procedures. All the results about the estimators established in Chaps. 2, 3, and

4 are heavily used in both the chapters. Most of the theory of testing of hypotheses has revolved around the Neyman-Pearson lemma, which leads to the most powerful test for simple null against simple alternative hypothesis. It also leads to the uniformly most powerful tests in certain models, in particular for exponential families. A likelihood ratio test procedure, which we discuss in the second section, is also an extension of Neyman-Pearson lemma in some sense. Wald's test procedure, the score test procedure, which are frequently used in statistical modeling and analysis, are related to the likelihood ratio test procedure. All these test procedures, when the underlying probability model is a multinomial distribution, play a significant role in tests for validity of a proposed model, goodness of fit tests and tests for contingency tables.

In the present chapter, we discuss the likelihood ratio test procedure and its asymptotic theory. The next chapter is devoted to likelihood ratio test procedures associated with a multinomial distribution, in particular, the tests for goodness of fit and the tests for contingency tables.

Suppose X is a random variable or a random vector whose distribution is indexed by a real parameter θ. On the basis of a random sample of size n from the distribution of X, it is of interest to test the following hypotheses.

(i) $H_0 : \theta = \theta_0$ against the alternative $H_1 : \theta > \theta_0$.
(ii) $H_0 : \theta = \theta_0$ against the alternative $H_1 : \theta < \theta_0$.
(iii) $H_0 : \theta = \theta_0$ against the alternative $H_1 : \theta \neq \theta_0$.

θ_0 is a specified value of the parameter. Optimal test procedures, such as uniformly most powerful tests, uniformly most powerful unbiased tests, have been developed to test such hypotheses. If the distribution satisfies the monotone likelihood ratio property or if it belongs to an exponential family, then uniformly most powerful tests exist for the one-sided hypotheses. However, these may not exist for all the distributions. Thus, in a general setup, test procedures are based on a suitable test statistic, which is a function of the deviation of θ_0 from an appropriate estimator of θ. To account for the variation in the deviation, it is divided by the standard error of the estimator. Thus, in most of the cases the test statistic is given by

$$T_n = \frac{\hat{\theta}_n - \theta_0}{se(\hat{\theta}_n)},$$

where $\hat{\theta}_n$ is a suitable estimator of θ and $se(\hat{\theta}_n)$ is the standard error of the estimator, that is, the estimator of the standard deviation of $\hat{\theta}_n$. The test procedure is to reject H_0, if $T_n > c_1$ in case (i), if $T_n < c_2$, in case (ii) and in case (iii) if $|T_n| > c_3$. The constants c_1, c_2 and c_3 are determined so that the probability of type I error is fixed at a specified level of significance α. To determine the constants c_1, c_2 and c_3, it is essential to know the null distribution of T_n. In some situations, it is difficult to find out the null distribution of T_n for finite n. However, in most of the cases, it is possible to find the asymptotic null distribution of T_n. Using the asymptotic null distribution of T_n, we can find the approximate values of constants c_1, c_2 and c_3. Using the results studied in previous chapters, we now discuss how to obtain the asymptotic null distribution of T_n.

Suppose $\tilde{\theta}_n$ is a CAN estimator of θ with approximate variance $v(\theta)/n$. The test statistic T_n or S_n to test the null hypothesis H_0 can be defined as follows:

$$T_n = \sqrt{n/v(\theta_0)}\,(\tilde{\theta}_n - \theta_0) \quad \text{or} \quad S_n = \sqrt{n/v(\tilde{\theta}_n)}\,(\tilde{\theta}_n - \theta_0)\,.$$

Under H_0, the asymptotic distribution of T_n is standard normal, or equivalently, the asymptotic null distribution of T_n^2 is χ_1^2. If $v(\theta)$ is a continuous function of θ, then $v(\tilde{\theta}_n)$ is a consistent estimator of $v(\theta)$ and hence by Slutsky's theorem, the asymptotic null distribution of S_n is also standard normal, which implies that the asymptotic null distribution of S_n^2 is χ_1^2. Using these asymptotic null distributions, one can determine the approximate values of the constants c_1, c_2 and c_3.

If the distribution of X belongs to a one-parameter exponential family or a Cramér family, then it is proved in Chap. 4, that for large n, the maximum likelihood estimator $\hat{\theta}_n$ of θ is a CAN estimator of θ with approximate variance $1/nI(\theta)$. In such cases the test statistic is defined as

$$T_n = \sqrt{nI(\theta_0)}(\hat{\theta}_n - \theta_0) \quad \text{or} \quad S_n = \sqrt{nI(\hat{\theta}_n)}(\hat{\theta}_n - \theta_0)\,.$$

Under H_0, the asymptotic distribution of T_n is standard normal. If $I(\theta)$ is a continuous function of θ then the asymptotic null distribution of S_n is also standard normal or equivalently, the asymptotic null distributions of T_n^2 and S_n^2 are χ_1^2.

On similar lines in a k-parameter setup, If $\underline{\tilde{\theta}}_n$ is CAN for $\underline{\theta}$ with approximate dispersion matrix $\Sigma(\underline{\theta})/n$, then for testing $H_0 : \underline{\theta} = \underline{\theta}_0$ against the alternative $H_1 : \underline{\theta} \neq \underline{\theta}_0$, the test statistic is defined as

$$W_n = n(\underline{\tilde{\theta}}_n - \underline{\theta}_0)'\Sigma^{-1}(\underline{\theta}_0)(\underline{\tilde{\theta}}_n - \underline{\theta}_0) \quad \text{or} \quad U_n = n(\underline{\tilde{\theta}}_n - \underline{\theta}_0)'\Sigma^{-1}(\underline{\tilde{\theta}}_n)(\underline{\tilde{\theta}}_n - \underline{\theta}_0)\,.$$

The null hypothesis is rejected if $W_n > c$ or $U_n > c$. The asymptotic null distribution of W_n is χ_k^2. If each element of $\Sigma(\underline{\theta})$ is a continuous function of $\underline{\theta}$, then $\Sigma(\underline{\tilde{\theta}}_n)$ is a consistent estimator of $\Sigma(\underline{\theta})$. Consequently, $W_n - U_n \xrightarrow{P} 0$, hence the asymptotic null distribution of U_n is also χ_k^2. Thus, the constant c can be determined or corresponding p-value can be computed. If the distribution belongs to a k-parameter exponential family or a k-parameter Cramér family, the large sample distribution of the maximum likelihood estimator $\underline{\hat{\theta}}_n$ is $N_k(\underline{\theta}, I^{-1}(\underline{\theta})/n)$, where $I(\underline{\theta})$ is the information matrix. Hence, for testing $H_0 : \underline{\theta} = \underline{\theta}_0$ against the alternative $H_1 : \underline{\theta} \neq \underline{\theta}_0$, the test statistic is defined as

$$W_n = n(\underline{\hat{\theta}}_n - \underline{\theta}_0)'I(\underline{\theta}_0)(\underline{\hat{\theta}}_n - \underline{\theta}_0) \quad \text{or} \quad U_n = n(\underline{\hat{\theta}}_n - \underline{\theta}_0)'I(\underline{\hat{\theta}}_n)(\underline{\hat{\theta}}_n - \underline{\theta}_0)\,.$$

The null hypothesis is rejected if $W_n > c$ or $U_n > c$. The asymptotic null distributions of W_n and U_n are χ_k^2 and hence the constant c can be determined or corresponding p-value can be computed.

For some distributions more than one CAN estimators exist for the indexing parameter. In such situations, we select the estimator which has the smallest approximate variance, to propose the test statistic. Following examples illustrate these test procedures.

✒ Example 5.1.1

Suppose $\{X_1, X_2, \ldots, X_n\}$ is a random sample from a Laplace distribution with location parameter θ and scale parameter 1. Then the sample median is the maximum likelihood estimator of θ and it is CAN with approximate variance $1/n$. The sample mean is also CAN for θ with approximate variance $2/n$. Thus, the sample median is a better estimator of θ. Hence, the test procedure to test $H_0 : \theta = 1$ against $H_1 : \theta > 1$, is based on the sample median $X_{([n/2]+1)}$. We propose the test statistic T_n as $T_n = \sqrt{n}\left(X_{([n/2]+1)} - 1\right)$. For large n, under H_0, $T_n \sim N(0, 1)$ distribution. The null hypothesis $H_0 : \theta = 1$ is rejected against $H_1 : \theta > 1$, if $T_n > c$ where c is determined corresponding to the given level of significance α and the asymptotic null distribution of T_n. Thus, $c = a_{1-\alpha}$.   □

📝 Remark 5.1.1

It is to be noted that a Laplace distribution with location parameter θ and scale parameter 1 is not a member of Cramér family, but the maximum likelihood estimator of θ exists and it is CAN.

✒ Example 5.1.2

Suppose X follows a Laplace distribution with probability density function

$$f(x, \mu, \alpha) = \frac{1}{2\alpha} \exp\left(-\frac{|x - \mu|}{\alpha}\right), \quad x, \mu \in \mathbb{R}, \ \alpha > 0.$$

In Example 3.3.7, we have obtained CAN estimators of $(\mu, \alpha)'$ based on (i) the sample quantiles and (ii) the sample moments. On the basis of generalized variance, we have noted that the CAN estimator based on the sample moments is better than that based on the sample quantiles. Hence, the test statistic to test $H_0 : \mu = \mu_0, \alpha = \alpha_0$ against $H_1 : \mu \neq \mu_0, \alpha \neq \alpha_0$, is based on the sample moments. In Example 3.3.7, we have shown that $\left(m_1', \sqrt{m_2/2}\right)'$ is CAN for $(\mu, \alpha)'$ with the approximate dispersion matrix $\Sigma/n = \mathrm{diag}\left(2\alpha^2, 1.25\alpha^2\right)/n$. Hence, we propose the test statistic as

$$W_n = n\left(m_1' - \mu_0, \sqrt{m_2/2} - \alpha_0\right)' \Sigma^{-1}(\mu_0, \alpha_0)\left(m_1' - \mu_0, \sqrt{m_2/2} - \alpha_0\right)$$
$$= n(m_1' - \mu_0)^2/2\alpha_0^2 + n(\sqrt{m_2/2} - \alpha_0)^2/1.25\alpha_0^2.$$

For large n under H_0, $W_n \sim \chi_2^2$ distribution. The null hypothesis H_0 is rejected against H_1 if $W_n > c$, where c is determined corresponding to the given level of significance α and the asymptotic null distribution of W_n. Thus, $c = \chi^2_{1-\alpha,2}$. Another test statistic is defined as

$$U_n = n \left(m'_1 - \mu_0, \ \sqrt{m_2/2} - \alpha_0\right)' \Sigma^{-1} \left(m'_1, \ \sqrt{m_2/2}\right) \times$$
$$\left(m'_1 - \mu_0, \ \sqrt{m_2/2} - \alpha_0\right)$$
$$= n \left(m'_1 - \mu_0\right)^2 / m_2 + n \left(\sqrt{m_2/2} - \alpha_0\right)^2 / 0.625 m_2.$$

For large n, under H_0, $U_n \sim \chi^2_2$ distribution. The null hypothesis H_0 is rejected against H_1 if $U_n > c$, where $c = \chi^2_{1-\alpha,2}$. □

✒ Example 5.1.3

Suppose $\{X_1, X_2, \ldots, X_n\}$ is a random sample from a Poisson $Poi(\theta)$ distribution. Then the sample mean $\overline{X}_n$ is CAN for θ with approximate variance θ/n. To derive the large sample test procedure to test $H_0 : P[X > 0] = 2/3$ against $H_1 : P[X > 0] < 2/3$, note that $P[X > 0] = 1 - e^{-\theta} = g(\theta)$, say. Then g is a differentiable function with $g'(\theta) = e^{-\theta} \neq 0$ for all $\theta > 0$. Hence by the delta method, $1 - e^{-\overline{X}_n}$ is CAN for $1 - e^{-\theta}$ with approximate variance $\theta e^{-2\theta}/n$. By Slutsky's theorem,

$$\sqrt{\frac{n}{\overline{X}_n}} e^{\overline{X}_n} \left((1 - e^{-\overline{X}_n}) - (1 - e^{-\theta})\right) = \sqrt{\frac{n}{\overline{X}_n}} e^{\overline{X}_n} (e^{-\theta} - e^{\overline{X}_n}) \xrightarrow{L} Z \sim N(0, 1).$$

Hence, we propose the test statistic S_n as $S_n = \sqrt{n/\overline{X}_n} e^{\overline{X}_n} (1/3 - e^{-\overline{X}_n})$. For large n under H_0, $S_n \sim N(0, 1)$ distribution. The null hypothesis H_0 is rejected against H_1 if $S_n < c$, where c is determined corresponding to the given level of significance α and the asymptotic null distribution of S_n. Thus, $c = -a_{1-\alpha}$. □

✍ Remark 5.1.2

In Example 5.1.3, $P[X > 0] = 1 - e^{-\theta} = 2/3 \ \Rightarrow \ \theta = \log 3$, hence the null hypothesis $H_0 : P[X > 0] = 2/3$ can be expressed as $H_0 : \theta = \log 3 = \theta_0$, say. For a Poisson $Poi(\theta)$ distribution, it is known that the sample mean $\overline{X}_n$ is CAN for θ with approximate variance θ/n and hence the test statistic S_n is given by $S_n = \sqrt{n/\overline{X}_n}(\overline{X}_n - \theta_0)$ and the test statistic T_n is given by $T_n = \sqrt{n/\theta_0}(\overline{X}_n - \theta_0)$. For large n under H_0, $S_n \sim N(0, 1)$ distribution. The null hypothesis H_0 is rejected against H_1 if $S_n < c$, where $c = -a_{1-\alpha}$. The test procedure based on T_n will be similar to that based on S_n. Such a conversion may not be possible for all the distributions. For example, in Exercise 5.4.4 we cannot have such a conversion.

✒ Example 5.1.4

Suppose $\{X_1, X_2, \ldots, X_n\}$ is a random sample from a normal $N(\mu, \sigma^2)$ distribution and we want to derive a large sample test procedure to test $H_0 : \mu = \mu_0, \sigma^2 = \sigma_0^2$ against $H_1 : \mu \neq \mu_0, \sigma^2 \neq \sigma_0^2$. In Example 3.3.2 we have shown that $\hat{\theta}_n = (\overline{X}_n, S_n^2)'$ is CAN for $\underline{\theta} = (\mu, \sigma^2)'$ with approximate dispersion matrix Σ/n, where

$$\Sigma = \begin{pmatrix} \sigma^2 & 0 \\ 0 & 2\sigma^4 \end{pmatrix}.$$

As a consequence, for large n, $n(\hat{\theta}_n - \underline{\theta})'\Sigma^{-1}(\hat{\theta}_n - \underline{\theta}) \sim \chi_2^2$ and by Slutsky's theorem, for large n, $n(\hat{\theta}_n - \underline{\theta})'\hat{\Sigma}_n^{-1}(\hat{\theta}_n - \underline{\theta}) \sim \chi_2^2$, where $\hat{\Sigma}_n$ is diag$(S_n^2, 2S_n^4)$. Suppose $\underline{\theta}_0 = (\mu_0, \sigma_0^2)'$. We propose the test statistic as $T_n = n(\hat{\theta}_n - \underline{\theta}_0)'\hat{\Sigma}_n^{-1}(\hat{\theta}_n - \underline{\theta}_0)$. For large n under H_0, $T_n \sim \chi_2^2$ distribution. The null hypothesis H_0 is rejected against H_1, if $T_n > c$ where c is determined corresponding to the given level of significance α and the asymptotic null distribution of T_n. Thus, $c = \chi_{2,(1-\alpha)}^2$.   □

✒ Example 5.1.5

Suppose X and Y are independent random variables having Bernoulli $B(1, p_1)$ and $B(1, p_2)$ distributions respectively, $0 < p_1, p_2 < 1$. Suppose $\underline{X} = \{X_1, X_2, \ldots, X_{n_1}\}$ is a random sample from the distribution of X and $\underline{Y} = \{Y_1, Y_2, \ldots, Y_{n_2}\}$ is a random sample from the distribution of Y. On the basis of these samples we want to test $H_0 : p_1 = p_2$ against the alternative $p_1 \neq p_2$. Suppose

$$P_{1n_1} = \overline{X}_{n_1} = \sum_{i=1}^{n_1} X_i/n_1 \quad \& \quad P_{2n_2} = \overline{Y}_{n_2} = \sum_{i=1}^{n_1} Y_i/n_2$$

denote the proportion of successes in $\underline{X}$ and the proportion of successes in $\underline{Y}$ respectively. The maximum likelihood estimator of $(p_1, p_2)'$ when $0 < p_1, p_2 < 1$ is $(\overline{X}_{n_1}, \overline{Y}_{n_2})' \equiv (P_{1n_1}, P_{2n_2})'$. Further, by the WLLN and the CLT as $n_1 \to \infty$ & $n_2 \to \infty$,

$$P_{1n_1} \xrightarrow{P} p_1, \quad P_{2n_2} \xrightarrow{L} p_2, \quad \sqrt{n_1}(P_{1n_1} - p_1) \xrightarrow{L} Z_1 \quad \& \quad \sqrt{n_2}(P_{2n_2} - p_2) \xrightarrow{P} Z_2$$

where $Z_1 \sim N(0, p_1(1 - p_1))$ & $Z_2 \sim N(0, p_2(1 - p_2))$ distribution. Since X and Y are independent,

$$\sqrt{n_1}(P_{1n_1} - p_1) - \sqrt{n_2}(P_{2n_2} - p_2) \xrightarrow{L} Z_3,$$
$$\text{where } Z_3 \sim N(0, p_1(1 - p_1) + p_2(1 - p_2)).$$

Suppose $n_1 \to \infty$ & $n_2 \to \infty$ such that $(n_1 + n_2)/n_1 \to a$ & $(n_1 + n_2)/n_2 \to b$ where a and b are constants. Then

$$\sqrt{\frac{n_1 + n_2}{n_1}} \sqrt{n_1}(P_{1n_1} - p_1) - \sqrt{\frac{n_1 + n_2}{n_2}} \sqrt{n_2}(P_{2n_2} - p_2) \xrightarrow{L} Z_4,$$
where $Z_4 \sim N(0, v)$,

where $v = ap_1(1 - p_1) + bp_2(1 - p_2)$. Suppose under H_0 the common value of p_1 and p_2 is denoted by p. Then a test statistic W_n to test $H_0 : p_1 = p_2$ is defined as follows:

$$
\begin{aligned}
W_n &= \frac{(P_{1n_1} - P_{2n_2})}{\sqrt{P_{1n_1}(1 - P_{1n_1})/n_1 + P_{2n_2}(1 - P_{2n_2})/n_2}} \\
&= \frac{\sqrt{\frac{n_1+n_2}{n_1}} \sqrt{n_1}(P_{1n_1} - p) - \sqrt{\frac{n_1+n_2}{n_2}} \sqrt{n_2}(P_{2n_2} - p)}{\sqrt{((n_1 + n_2)/n_1)P_{1n_1}(1 - P_{1n_1}) + ((n_1 + n_2)/n_2)P_{2n_2}(1 - P_{2n_2})}} \\
&= \frac{U_n}{\sqrt{V_n}},
\end{aligned}
$$

where $U_n \xrightarrow{L} Z_4$ and $V_n \xrightarrow{P} ap_1(1 - p_1) + bp_2(1 - p_2)$. Hence, by Slutsky's theorem, the asymptotic null distribution of W_n is standard normal. The null hypothesis H_0 is rejected if $|W_n| > a_{1-\alpha/2}$. We define one more test statistic as follows. In the null set up $p_1 = p_2 = p$. Then the log-likelihood of p given random samples $\underline{X}$ and $\underline{Y}$, using independence of X and Y is given by

$$
\begin{aligned}
\log L_{n_1+n_2}(p|\underline{X}, \underline{Y}) = &\left(\sum_{i=1}^{n_1} X_i + \sum_{i=1}^{n_2} Y_i\right) \log p \\
&+ \left(n_1 + n_2 - \sum_{i=1}^{n_1} X_i - \sum_{i=1}^{n_2} Y_i\right) \log(1 - p).
\end{aligned}
$$

It follows that the maximum likelihood estimator of p is

$$
\hat{p}_n = \frac{\left(\sum_{i=1}^{n_1} X_i + \sum_{i=1}^{n_2} Y_i\right)}{(n_1 + n_2)} = \frac{(n_1 P_{1n_1} + n_2 P_{2n_2})}{(n_1 + n_2)} = P_n, \quad \text{say.}
$$

Further under H_0, $\hat{p}_n \xrightarrow{P} p$. Another test statistic S_n is defined as follows:

$$
\begin{aligned}
S_n &= \frac{(P_{1n_1} - P_{2n_2})}{\sqrt{P_n(1 - P_n)(1/n_1 + 1/n_2)}} \\
&= \frac{\sqrt{\frac{n_1+n_2}{n_1}} \sqrt{n_1}(P_{1n_1} - p) - \sqrt{\frac{n_1+n_2}{n_2}} \sqrt{n_2}(P_{2n_2} - p)}{\sqrt{P_n(1 - P_n)((n_1 + n_2)/n_1 + (n_1 + n_2)/n_2)}} \\
&= \frac{N_n}{\sqrt{D_n}},
\end{aligned}
$$

where $N_n \xrightarrow{L} Z_4 \sim N(0, v)$, where
$v = ap_1(1 - p_1) + bp_2(1 - p_2) = p(1 - p)(a + b)$ under H_0. Note that
$D_n \xrightarrow{P} p(1 - p)(a + b)$. Hence, by Slutsky's theorem, under H_0 the asymptotic distribution of S_n is standard normal. The null hypothesis H_0 is rejected if
$|S_n| > a_{1-\alpha/2}$.   □

We will discuss both these test procedures again in Sect. 6.4, where we prove that W_n is Wald's test statistic and S_n is a score test statistic for testing $H_0 : p_1 = p_2$ against the alternative $p_1 \neq p_2$ based on samples from two independent Bernoulli distributions.

In the next section, we discuss the most frequently used large sample test procedure, known as likelihood ratio test procedure. All the tests for contingency table and goodness of fit tests are likelihood ratio test procedures, when the underlying probability model is a multinomial distribution. These tests are discussed in Chap. 6.

5.2   Likelihood Ratio Test Procedure

Likelihood ratio test procedure is the most general test procedure when the parameter space is either a subset of $\mathbb{R}$ or $\mathbb{R}^k$. Whenever an optimal test exists, such as the most powerful test, uniformly most powerful test or uniformly most powerful unbiased test, the likelihood ratio test procedure leads to the optimal test procedure.

Suppose X is a random variable or a random vector whose probability law $f(x, \theta)$ is indexed by a parameter θ, which may be a real parameter or a vector parameter. Suppose Θ, Θ_0 and Θ_1 denote the parameter space, the parameter space corresponding to a null hypothesis and the parameter space corresponding to an alternative hypothesis respectively, where $\Theta_0 \cap \Theta_1 = \emptyset$ and $\Theta_0 \cup \Theta_1 = \Theta$. On the basis of a random sample $\underline{X} = \{X_1, X_2, \ldots, X_n\}$ of size n from the distribution of X, suppose we are interested in testing $H_0 : \theta \in \Theta_0$ against the alternative $H_1 : \theta \in \Theta_1$. Likelihood of θ corresponding to the data $\underline{X} = \{X_1, X_2, \ldots, X_n\}$ is given by $L_n(\theta|\underline{X}) = \prod_{i=1}^{n} f(X_i, \theta)$. If both the null and alternative hypotheses are simple, then by the Neyman-Pearson lemma, the most powerful test is based on the likelihood ratio $L_n(\theta_1|\underline{X})/L_n(\theta_0|\underline{X})$. For certain special models and certain composite hypotheses, the most powerful test turns out to be independent of $\theta_1 \in \Theta_1$. Thus, we get an uniformly most powerful test for testing H_0 against H_1. When both H_0 and H_1 are composite, a sensible extension of the idea behind the Neyman-Pearson lemma is to base a test on $T_n = \sup_{\Theta_0} L_n(\theta|\underline{X}) / \sup_{\Theta_1} L_n(\theta|\underline{X})$. Thus, the single points $\{\theta_0\}$ and $\{\theta_1\}$ are replaced by $\sup_{\Theta_0}$ and $\sup_{\Theta_1}$ respectively. For mathematical simplicity in the denominator of T_n, supremum over Θ_1 is replaced by the supremum over Θ and the likelihood ratio test statistic $\lambda(\underline{X})$ is defined as

$$\lambda(\underline{X}) = \frac{\sup_{\Theta_0} L_n(\theta|\underline{X})}{\sup_{\Theta} L_n(\theta|\underline{X})} .$$

The likelihood ratio test procedure is also proposed by Neyman and Pearson in 1928. It is to be noted that the likelihood ratio test statistic is a function of the minimal sufficient statistic and thus has the desirable property of achieving the reduction of data by sufficiency. If X is a discrete random variable then $\sup_{\Theta_0} L_n(\theta|\underline{X})$ denotes the maximum possible probability of obtaining the data $\underline{X}$ if $\theta \in \Theta_0$. This is compared with the maximum possible probability of obtaining the data $\underline{X}$ if $\theta \in \Theta$. It is to be noted that

$$\sup_{\Theta_0} L_n(\theta|\underline{X}) \geq 0 \ \& \ \sup_{\Theta} L_n(\theta|\underline{X}) > 0 \Rightarrow \lambda(\underline{X}) \geq 0$$

$$\Theta_0 \subset \Theta \ \Rightarrow \ \sup_{\Theta_0} L_n(\theta|\underline{X}) \leq \sup_{\Theta} L_n(\theta|\underline{X}) \Rightarrow \lambda(\underline{X}) \leq 1$$

$$\Rightarrow 0 \leq \lambda(\underline{X}) \leq 1.$$

If $\lambda(\underline{X})$ is near 1, then the numerator and denominator are close to each other and it indicates that the support of the data is to a null setup. On the other hand, if $\lambda(\underline{X})$ is small, then the data support an alternative setup. Hence, the likelihood ratio test procedure rejects H_0 if $\lambda(\underline{X}) < c, 0 < c < 1$, where c is determined so that size of the test is α, that is, $\sup_{\Theta_0} P_\theta[\lambda(\underline{X}) < c] = \alpha$.

In carrying out the test, one encounters two difficulties-one is finding the null distribution of $\lambda(\underline{X})$ and the second is finding supremum of the likelihood in the null setup as well as in the entire parameter space. The problem of finding supremum is essentially that of finding the maximum likelihood estimators of the parameters in Θ and in Θ_0. For some distributions, such as normal or exponential, the critical region $[\lambda(\underline{X}) < c]$ is equivalent to $[T_n(\underline{X}) > k]$ or $[T_n(\underline{X}) < k]$ or $[|T_n(\underline{X})| > k]$ where k is a constant so that size of the test is α and is determined using the null distribution of T_n, which is easy to obtain. In many cases, it is difficult to find the exact null distribution of $\lambda(\underline{X})$. But this problem is resolved by considering the null distribution for large n, to find the approximate value of the cut-off point. It has been proved by Wilks [1] that under certain conditions for large n, $-2\log \lambda(\underline{X}) \sim \chi_r^2$ where r is the difference between the number of parameters estimated in Θ and in Θ_0.

In spite of these apparent difficulties of a likelihood ratio test procedure, it does provide a unified approach for developing test procedures. Besides testing of hypotheses, a likelihood ratio test procedure is also used to construct a confidence interval for the desired parametric function. It is defined as usual by the acceptance region. The likelihood ratio test procedure is closely related to the score test and Wald's test. We will elaborate on this in the next chapter. In addition to the intuitive interpretation of the likelihood ratio test, in many cases of interest it is equivalent to optimal tests. We now show that if the most powerful test exists for testing a simple null hypothesis against a simple alternative, then the likelihood ratio test and the most powerful test are equivalent. Suppose $\Theta_0 = \{\theta_0\}$ and $\Theta_1 = \{\theta_1\}$. Then according to the Neyman-Pearson lemma the most powerful test is as given below.

$$\phi(\underline{X}) = \begin{cases} 1, & \text{if} & \frac{L_n(\theta_1|\underline{X})}{L_n(\theta_0|\underline{X})} > k \\ \\ 0, & \text{if} & \frac{L_n(\theta_1|\underline{X})}{L_n(\theta_0|\underline{X})} \leq k. \end{cases}$$

H_0 is rejected if

$$L_n(\theta_1|\underline{X}) > kL_n(\theta_0|\underline{X}). \qquad (5.2.1)$$

Now the likelihood ratio test statistic $\lambda(\underline{X})$ is given by

$$\lambda(\underline{X}) = \sup_{\Theta_0} L_n(\theta|\underline{X}) / \sup_{\Theta} L_n(\theta|\underline{X}) = L_n(\theta_0|\underline{X}) / \sup_{\Theta} L_n(\theta|\underline{X}).$$

H_0 is rejected if $\lambda(\underline{X}) < c \Leftrightarrow L_n(\theta_0|\underline{X}) < c \sup_{\Theta} L_n(\theta|\underline{X})$. If $L_n(\theta_0|\underline{X}) > L_n(\theta_1|\underline{X})$ then $\lambda(\underline{X}) = 1$ and H_0 is accepted. If $L_n(\theta_1|\underline{X}) > L_n(\theta_0|\underline{X})$ then H_0 is rejected if

$$\frac{L_n(\theta_0|\underline{X})}{L_n(\theta_1|\underline{X})} < c \Leftrightarrow L_n(\theta_1|\underline{X}) > \frac{1}{c}L_n(\theta_0|\underline{X}). \qquad (5.2.2)$$

From (5.2.1) and (5.2.2), it is clear that the likelihood ratio test procedure and the most powerful test procedure are equivalent. On similar lines, it can be shown that the likelihood ratio test procedure and the uniformly most powerful test procedure are equivalent.

Following examples illustrate the likelihood ratio test procedure and how the rejection region $\lambda(\underline{X}) < c$ is reduced to $T_n > k$ or $T_n < k$, where T_n is a test statistic whose null distribution can be obtained easily.

✍ Example 5.2.1

Suppose $\{X_1, X_2, \ldots, X_n\}$ is a random sample from a normal $N(\theta, 1)$ distribution. To derive the likelihood ratio test procedure for testing $H_0 : \theta = \theta_0$ against the alternative $H_1 : \theta \neq \theta_0$, the first step is to obtain the maximum likelihood estimator of θ in the entire parameter space Θ and in the null space Θ_0. Corresponding to a random sample $\underline{X} \equiv \{X_1, X_2, \ldots, X_n\}$ of size n from a normal $N(\theta, 1)$ distribution, the likelihood of θ is given by

$$L_n(\theta|\underline{X}) = \prod_{i=1}^{n} \frac{1}{\sqrt{2\pi}} \exp\left\{-\frac{1}{2}(X_i - \theta)^2\right\}$$

$$= \left(\sqrt{2\pi}\right)^{-n} \exp\left\{-\frac{1}{2}\sum_{i=1}^{n}(X_i - \theta)^2\right\}.$$

Here $\Theta = \mathbb{R}$ and it is well-known that the sample mean $\overline{X}_n$ is the maximum likelihood estimator of θ and $\overline{X}_n$ has normal $N(\theta, 1/n)$ distribution, which implies that $Z = \sqrt{n}(\overline{X}_n - \theta)$ has standard normal distribution. The null space is $\{\theta_0\}$, thus the supremum of the likelihood in null space is attained at $\theta = \theta_0$. Hence, the likelihood ratio test statistic $\lambda(\underline{X})$ is given by

$$\lambda(\underline{X}) = \frac{\sup_{\Theta_0} L_n(\theta|\underline{X})}{\sup_{\Theta} L_n(\theta|\underline{X})} = \exp\left\{-\frac{1}{2}\left\{\sum_{i=1}^{n}(X_i - \theta_0)^2 - \sum_{i=1}^{n}(X_i - \overline{X}_n)^2\right\}\right\}$$

$$= \exp\left\{-\frac{n}{2}(\overline{X}_n - \theta_0)^2\right\}. \qquad (5.2.3)$$

The likelihood ratio test procedure rejects H_0 if

$\lambda(\underline{X}) < c \iff \sqrt{n}|\overline{X}_n - \theta_0| > k$, where k is determined so that size of the test is α, that is, $P_{\theta_0}[\sqrt{n}|\overline{X}_n - \theta_0| > k] = \alpha$. Under H_0, $Z = \sqrt{n}(\overline{X}_n - \theta_0) \sim N(0, 1)$ and hence $k = a_{1-\alpha/2}$.

Suppose now $H_0 : \theta \geq \theta_0$ and the alternative is $H_1 : \theta < \theta_0$. In this case $\Theta_0 = [\theta_0, \infty)$ and $\Theta = \mathbb{R}$. As discussed in Example 2.2.3, if the parameter space is $\Theta_0 = [\theta_0, \infty)$, then the maximum likelihood estimator $\hat{\theta}_{0n}$ of θ is given by

$$\hat{\theta}_{0n} = \begin{cases} \theta_0, & \text{if} \quad \overline{X}_n < \theta_0 \\[2mm] \overline{X}_n, & \text{if} \quad \overline{X}_n \geq \theta_0. \end{cases}$$

Suppose $\overline{X}_n \geq \theta_0$, then the maximum likelihood estimator of θ in Θ and in Θ_0 is the same and hence $\lambda(\underline{X}) = 1$ and data support the null setup, hence $H_0 : \theta \geq \theta_0$ is not rejected, which is quite reasonable. If $\overline{X}_n < \theta_0$, the likelihood ratio test statistic $\lambda(\underline{X})$ is as in Eq. (5.2.3). The likelihood ratio test procedure rejects H_0 if

$$\lambda(\underline{X}) < c \iff n(\overline{X}_n - \theta_0)^2 > c_1 \iff \sqrt{n}(\overline{X}_n - \theta_0) < k, \quad \text{as } \overline{X}_n < \theta_0.$$

The constant k is determined so that size of the test is α, that is, $\sup_{\Theta_0} P_{\theta}[\sqrt{n}(\overline{X}_n - \theta_0) < k] = \alpha$. At $\theta = \theta_0$, $P_{\theta_0}[\sqrt{n}(\overline{X}_n - \theta_0) < k] = \alpha \implies k = a_{\alpha}$. Now for $\theta > \theta_0$ consider,

$$\begin{aligned} P_{\theta}[\sqrt{n}(\overline{X}_n - \theta_0) < k] &= P_{\theta}[\sqrt{n}(\overline{X}_n - \theta + \theta - \theta_0) < k] \\ &= P_{\theta}[\sqrt{n}(\overline{X}_n - \theta) < k_1], \end{aligned}$$

where $k_1 = k - \sqrt{n}(\theta - \theta_0) < k = a_{\alpha}$ as $(\theta - \theta_0) > 0$. Thus, $P_{\theta}[\sqrt{n}(\overline{X}_n - \theta_0) < k] < \alpha$ and hence $\sup_{\Theta_0} P_{\theta}[\sqrt{n}(\overline{X}_n - \theta_0) < k] = P_{\theta_0}[\sqrt{n}(\overline{X}_n - \theta_0) < k] = \alpha$. Thus, at $k = a_{\alpha}$, size of the test is α. Thus, if $\overline{X}_n < \theta_0$, then H_0 is rejected if $\sqrt{n}(\overline{X}_n - \theta_0) < k = a_{\alpha}$. Suppose now we want to test $H_0 : \theta \leq \theta_0$ against the alternative $H_1 : \theta > \theta_0$. In this case $\Theta_0 = (-\infty, \theta_0]$. As discussed in Example 2.2.3, the maximum likelihood estimator $\hat{\theta}_{0n}$ of θ is given by

$$\hat{\theta}_{0n} = \begin{cases} \overline{X}_n, & \text{if} \quad \overline{X}_n \leq \theta_0 \\[2mm] \theta_0, & \text{if} \quad \overline{X}_n > \theta_0. \end{cases}$$

Suppose $\overline{X}_n \leq \theta_0$, then the maximum likelihood estimator of θ in Θ and in Θ_0 is the same and hence $\lambda(\underline{X}) = 1$ and data support the null setup. Hence, $H_0 : \theta \leq \theta_0$ is not rejected. Now suppose $\overline{X}_n > \theta_0$. In this case the likelihood ratio test statistic $\lambda(\underline{X})$ is as in Eq. (5.2.3). The likelihood ratio test procedure rejects H_0 if

$$\lambda(\underline{X}) < c \iff n(\overline{X}_n - \theta_0)^2 > c_1 \iff \sqrt{n}(\overline{X}_n - \theta_0) > k, \quad \text{as } \overline{X}_n > \theta_0.$$

The constant k is determined so that size of the test is α, that is, $\sup_{\Theta_0} P_\theta[\sqrt{n}(\overline{X}_n - \theta_0) > k] = \alpha$. At $\theta = \theta_0$, $P_{\theta_0}[\sqrt{n}(\overline{X}_n - \theta_0) > k] = \alpha \Rightarrow k = a_{1-\alpha}$. Now for $\theta < \theta_0$ consider,

$$P_\theta[\sqrt{n}(\overline{X}_n - \theta_0) > k] = P_\theta[\sqrt{n}(\overline{X}_n - \theta + \theta - \theta_0) > k]$$
$$= P_\theta[\sqrt{n}(\overline{X}_n - \theta) > k_1],$$

where $k_1 = k - \sqrt{n}(\theta - \theta_0) > k = a_{1-\alpha}$ as $(\theta - \theta_0) < 0$. Thus, $P_\theta[\sqrt{n}(\overline{X}_n - \theta_0) > k] < \alpha$ and hence $\sup_{\Theta_0} P_\theta[\sqrt{n}(\overline{X}_n - \theta_0) > k] = P_{\theta_0}[\sqrt{n}(\overline{X}_n - \theta_0) > k] = \alpha$. Thus, at $k = a_{1-\alpha}$, size of the test is α. Hence, if $\overline{X}_n > \theta_0$, then H_0 is rejected if $\sqrt{n}(\overline{X}_n - \theta_0) > k = a_{1-\alpha}$. □

✒ Example 5.2.2

Suppose $X \sim N(\theta, \sigma^2)$. Suppose we want to derive the likelihood ratio test procedure for testing $H_0 : \theta = \theta_0$ against the alternative $H_1 : \theta \neq \theta_0$ when σ^2 unknown. Corresponding to a random sample $\underline{X} \equiv \{X_1, X_2, \ldots, X_n\}$ of size n from normal $N(\theta, \sigma^2)$ distribution, the likelihood of (θ, σ^2) is given by

$$L_n(\theta, \sigma^2|\underline{X}) = \prod_{i=1}^{n} \frac{1}{\sqrt{2\pi}\sigma} \exp\left\{-\frac{1}{2\sigma^2}(X_i - \theta)^2\right\}$$
$$= \left(\sqrt{2\pi}\sigma\right)^{-n} \exp\left\{-\frac{1}{2\sigma^2}\sum_{i=1}^{n}(X_i - \theta)^2\right\}.$$

If $\Theta = \{(\theta, \sigma^2)|\theta \in \mathbb{R}, \sigma^2 > 0\}$, then the sample mean $\overline{X}_n$ is the maximum likelihood estimator of θ and sample variance $S_n^2 = \sum_{i=1}^{n}(X_i - \overline{X}_n)^2/n$ is the maximum likelihood estimator of σ^2. Further, $\overline{X}_n$ has normal $N(\theta, \sigma^2/n)$ distribution which implies that $Z = \sqrt{n}(\overline{X}_n - \theta)/\sigma$ has standard normal distribution and $nS_n^2/\sigma^2 = \sum_{i=1}^{n}(X_i - \overline{X}_n)^2/\sigma^2 \sim \chi_{n-1}^2$ distribution. It is also known that $\overline{X}_n$ and nS_n^2/σ^2 are independent random variables. To derive the likelihood ratio test procedure for testing $H_0 : \theta = \theta_0$ against the alternative $H_1 : \theta \neq \theta_0$ when σ^2 unknown, the null space is $\Theta_0 = \{(\theta, \sigma^2)|\theta = \theta_0, \sigma^2 > 0\}$ and the maximum likelihood estimator of σ^2 is $S_{0n}^2 = \sum_{i=1}^{n}(X_i - \theta_0)^2/n$. Hence, the likelihood ratio test statistic $\lambda(\underline{X})$ is given by

$$\lambda(\underline{X}) = \frac{\sup_{\Theta_0} L_n(\theta|\underline{X})}{\sup_{\Theta} L_n(\theta|\underline{X})}$$

$$= \frac{\left(\sum_{i=1}^{n}(X_i - \theta_0)^2\right)^{-n/2} \exp\left\{-\frac{1}{2}\frac{\sum_{i=1}^{n}(X_i-\theta_0)^2}{\sum_{i=1}^{n}(X_i-\theta_0)^2/n}\right\}}{\left(\sum_{i=1}^{n}(X_i - \overline{X}_n)^2\right)^{-n/2} \exp\left\{-\frac{1}{2}\frac{\sum_{i=1}^{n}(X_i-\overline{X}_n)^2}{\sum_{i=1}^{n}(X_i-\overline{X}_n)^2/n}\right\}}$$

$$= \left(\frac{\sum\limits_{i=1}^{n}(X_i - \overline{X}_n)^2}{\sum\limits_{i=1}^{n}(X_i - \theta_0)^2} \right)^{n/2} = \left(\frac{\sum\limits_{i=1}^{n}(X_i - \overline{X}_n)^2}{\sum\limits_{i=1}^{n}(X_i - \overline{X}_n + \overline{X}_n - \theta_0)^2} \right)^{n/2}$$

$$= \left(\frac{\sum\limits_{i=1}^{n}(X_i - \overline{X}_n)^2}{\sum\limits_{i=1}^{n}(X_i - \overline{X}_n)^2 + n(\overline{X}_n - \theta_0)^2} \right)^{n/2} = \left(1 + \frac{n(\overline{X}_n - \theta_0)^2}{\sum\limits_{i=1}^{n}(X_i - \overline{X}_n)^2} \right)^{-n/2}$$

$$\tag{5.2.4}$$

The likelihood ratio test procedure rejects H_0 if

$$\lambda(\underline{X}) < c \quad \Leftrightarrow \quad \frac{n(\overline{X}_n - \theta_0)^2}{\sum\limits_{i=1}^{n}(X_i - \overline{X}_n)^2} > c_1 \quad \Leftrightarrow \quad |T_n| > k$$

$$\text{where } T_n = \frac{\sqrt{n}(\overline{X}_n - \theta_0)}{\sqrt{\sum\limits_{i=1}^{n}(X_i - \overline{X}_n)^2/(n - 1)}}$$

and k is determined so that size of the test is α, that is, $P_{\theta_0}[|T_n| > k] = \alpha$. Under H_0, $T_n \sim t_{n-1}$ distribution and hence $k = t_{n-1,1-\alpha/2}$.

Suppose now $H_0 : \theta \geq \theta_0$ and the alternative is $H_1 : \theta < \theta_0$. Further, σ^2 is unknown. In this case, null space is $\Theta_0 = \{(\theta, \sigma^2) | \theta \geq \theta_0, \sigma^2 > 0\}$ and the maximum likelihood estimators of θ and of σ^2 are given by

$$\hat{\theta}_{0n} = \begin{cases} \overline{X}_n, & \text{if} \quad \overline{X}_n \geq \theta_0 \\ \theta_0, & \text{if} \quad \overline{X}_n < \theta_0. \end{cases}$$

$$\hat{\sigma}_{0n}^2 = \begin{cases} \dfrac{1}{n}\sum\limits_{i=1}^{n}(X_i - \overline{X}_n)^2, & \text{if} \quad \overline{X}_n \geq \theta_0 \\ \dfrac{1}{n}\sum\limits_{i=1}^{n}(X_i - \theta_0)^2, & \text{if} \quad \overline{X}_n < \theta_0. \end{cases}$$

It is to be noted that if $\overline{X}_n \geq \theta_0$, then the maximum likelihood estimators of θ and of σ^2 are the same in Θ and in Θ_0. Hence, the likelihood ratio test statistic $\lambda(\underline{X}) = 1$ when $\overline{X}_n \geq \theta_0$ and $H_0 : \theta \geq \theta_0$ is not rejected. If $\overline{X}_n < \theta_0$ then the likelihood ratio test statistic $\lambda(\underline{X})$ is as given in Eq. (5.2.4). Thus, if $\overline{X}_n < \theta_0$ then the likelihood ratio test procedure rejects H_0 if

$$\lambda(\underline{X}) < c \iff \frac{n(\overline{X}_n - \theta_0)^2}{\sum_{i=1}^{n}(X_i - \overline{X}_n)^2} > c_1 \iff T_n < k$$

$$\text{where } T_n = \frac{\sqrt{n}(\overline{X}_n - \theta_0)}{\sqrt{\sum_{i=1}^{n}(X_i - \overline{X}_n)^2/(n-1)}},$$

as $\overline{X}_n < \theta_0$. The constant k is determined so that size of the test is α, that is, $\sup_{\Theta_0} P_\theta[T_n < k] = \alpha$. When $\theta = \theta_0$, $T_n \sim t_{n-1}$ distribution and as discussed in Example 5.2.1, supremum is attained at θ_0 and hence $k = t_{n-1,\alpha}$. □

✐ Example 5.2.3

Suppose $X \sim N(\theta, 1)$ and we want to derive the likelihood ratio test procedure for testing $H_0 : |\theta| \leq a$ against the alternative $H_1 : |\theta| > a$, where a is a positive real number. Corresponding to a random sample $\underline{X} \equiv \{X_1, X_2, \ldots, X_n\}$ of size n from a normal $N(\theta, 1)$ distribution, the likelihood of θ is given by

$$L_n(\theta|\underline{X}) = \prod_{i=1}^{n} \frac{1}{\sqrt{2\pi}} \exp\left\{-\frac{1}{2}(X_i - \theta)^2\right\}$$

$$= \left(\sqrt{2\pi}\right)^{-n} \exp\left\{-\frac{1}{2}\sum_{i=1}^{n}(X_i - \theta)^2\right\}.$$

Further, $\Theta = \mathbb{R}$ and the sample mean $\overline{X}_n$ is the maximum likelihood estimator of θ and it has normal $N(\theta, 1/n)$ distribution. In null space $\theta_0 \in [-a, a]$ hence as discussed in Example 2.2.3, the maximum likelihood estimator of θ is given by

$$\hat{\theta}_{0n} = \begin{cases} -a, & \text{if } \overline{X}_n < -a \\ \overline{X}_n, & \text{if } \overline{X}_n \in [-a, a] \\ a, & \text{if } \overline{X}_n > a. \end{cases}$$

As in the previous example, if $\overline{X}_n \in [-a, a]$, then the maximum likelihood estimator of θ is same in Θ and in Θ_0. Hence, the likelihood ratio test statistic $\lambda(\underline{X}) = 1$ when $\overline{X}_n \in [-a, a]$ and $H_0 : |\theta| \leq a$ is not rejected, which seems to be reasonable. If $\overline{X}_n < -a$ or $\overline{X}_n > a$ then the likelihood ratio test statistic $\lambda(\underline{X})$ is given by

$$\lambda(\underline{X}) = \begin{cases} \exp\{\frac{-n}{2}(\overline{X}_n + a)^2\}, & \text{if } \overline{X}_n < -a \\ \exp\{\frac{-n}{2}(\overline{X}_n - a)^2\}, & \text{if } \overline{X}_n > a \end{cases}$$

The likelihood ratio test procedure rejects H_0 if $\lambda(\underline{X}) < k$, that is, if

$$\exp\left\{\frac{-n}{2}(\overline{X}_n + a)^2\right\} < c_1 \ \& \ \overline{X}_n < -a \ \text{ or } \ \exp\left\{\frac{-n}{2}(\overline{X}_n - a)^2\right\} < c_2$$

$$\& \ \overline{X}_n > a$$
$$\Leftrightarrow (\overline{X}_n + a)^2 > c_3 \ \& \ \overline{X}_n < -a \ \text{ or } \ (\overline{X}_n - a)^2 > c_4$$
$$\& \ \overline{X}_n > a$$
$$\Leftrightarrow (\overline{X}_n + a) < -c_5 \ \text{ or } \ (\overline{X}_n - a) > c_6$$
$$\Leftrightarrow \overline{X}_n < -c_5 - a \ \text{ or } \ \overline{X}_n > c_6 + a$$
$$\Leftrightarrow \overline{X}_n < -c - a \ \text{ or } \ \overline{X}_n > c + a, \ \text{ if } c_5 = c_6 = c$$
$$\Leftrightarrow |\overline{X}_n| > a + c.$$

The constant c is determined so that size of the test is α, that is,

$$\alpha = \sup_{\theta_0} P_\theta[|\overline{X}_n| > a + c]$$

$$\Leftrightarrow \sup_{\theta_0}\{1 - P_\theta[-a - c < \overline{X}_n < a + c]\} = \alpha$$

$$\Leftrightarrow \sup_{\theta_0}\{1 - \Phi(\sqrt{n}(a + c - \theta)) + \Phi(\sqrt{n}(-a - c - \theta))\} = \alpha$$

$$\Leftrightarrow \inf_{-a \le \theta \le a}\{\Phi(\sqrt{n}(a + c - \theta)) - \Phi(\sqrt{n}(-a - c - \theta))\} = 1 - \alpha$$

$$\Leftrightarrow \Phi(\sqrt{n}c) - \Phi(-\sqrt{n}c) = 1 - \alpha$$

$$\Leftrightarrow \Phi(\sqrt{n}c) = 1 - \alpha/2$$

$$\Leftrightarrow c = \frac{1}{\sqrt{n}}\Phi^{-1}(1 - \alpha/2) = \frac{1}{\sqrt{n}}a_{1-\alpha/2}.$$

Thus, $H_0 : |\theta| \le a$ against the alternative $H_1 : |\theta| > a$ is rejected if $|\overline{X}_n| > a + c$, where $c = \frac{1}{\sqrt{n}}a_{1-\alpha/2}$. $\qquad\qquad\square$

In the three examples discussed above, it is possible to convert the critical region $[\lambda(\underline{X}) < c]$ to a critical region in terms of a test statistic T_n, whose null distribution can be obtained for each n. However, such a conversion is not possible for many distributions and hence we have to use the asymptotic distribution of $\lambda(\underline{X})$, which is discussed in the following theorems. It is assumed that the probability law $f(x, \theta)$, indexed by a parameter $\theta \in \Theta \subset \mathbb{R}$, belongs to a Cramér family. As a consequence, for large n, the maximum likelihood estimator of θ exists and is CAN with approximate variance $1/nI(\theta)$. This result is heavily used in deriving the asymptotic null distribution of $\lambda(\underline{X})$.

Theorem 5.2.1

Suppose X is a random variable or a random vector with probability law $f(x, \theta)$ indexed by a parameter $\theta \in \Theta \subset \mathbb{R}$. Suppose $f(x, \theta)$ belongs to a Cramér family. If $\lambda(\underline{X})$ is a likelihood ratio test statistic based on a random sample $\underline{X} = \{X_1, X_2, \ldots, X_n\}$ for testing $H_0 : \theta = \theta_0$ against the alternative $H_1 : \theta \neq \theta_0$, where θ_0 is a specified value of the parameter, then under H_0,

$$-2 \log \lambda(\underline{X}) \xrightarrow{L} U \sim \chi_1^2 \text{ as } n \to \infty.$$

Proof Suppose $\hat{\theta}_n$ is a maximum likelihood estimator of θ based on a random sample $\underline{X}$. Since the distribution of X belongs to a Cramér family, $\hat{\theta}_n$ is CAN with approximate variance $1/nI(\theta)$. The likelihood ratio test procedure rejects $H_0 : \theta = \theta_0$ against the alternative $H_1 : \theta \neq \theta_0$, if $\lambda(\underline{X}) < c \Leftrightarrow -2 \log \lambda(\underline{X}) > c_1$ where c and c_1 are constants and $-2 \log \lambda(\underline{X})$ is given by

$$-2 \log \lambda(\underline{X}) = 2[\log L_n(\hat{\theta}_n | \underline{X}) - \log L_n(\theta_0 | \underline{X})]. \tag{5.2.5}$$

Expanding $\log L_n(\theta_0 | \underline{X})$ around $\hat{\theta}_n$ using Taylor series expansion, we have

$$\log L_n(\theta_0 | \underline{X}) = \log L_n(\hat{\theta}_n | \underline{X}) + (\theta_0 - \hat{\theta}_n) \frac{\partial}{\partial \theta_0} \log L_n(\theta_0 | \underline{X})|_{\hat{\theta}_n}$$
$$+ \frac{(\hat{\theta}_n - \theta_0)^2}{2} \frac{\partial^2}{\partial \theta_0^2} \log L_n(\theta_0 | \underline{X})|_{\hat{\theta}_n} + R_n ,$$

where R_n is the remainder term given by $R_n = \frac{(\hat{\theta}_n - \theta_0)^3}{3!} \frac{\partial^3}{\partial \theta_0^3} \log L_n(\theta_0 | \underline{X})|_{\theta_n^*}$, where $\theta_n^* = \alpha\theta_0 + (1 - \alpha)\hat{\theta}_n$, $0 < \alpha < 1$. It is to be noted that under H_0, $\theta_n^* \xrightarrow{P} \theta_0$ as $n \to \infty$. Further, $\frac{\partial}{\partial \theta_0} \log L_n(\theta_0 | \underline{X})|_{\hat{\theta}_n} = 0$, as $\hat{\theta}_n$ is a solution of the likelihood equation. Substituting the expansion in (5.2.5) we have

$$-2 \log \lambda(\underline{X}) = 2[\log L_n(\hat{\theta}_n | \underline{X}) - \log L_n(\theta_0 | \underline{X})]$$
$$= (\hat{\theta}_n - \theta_0)^2 \left(-\frac{\partial^2}{\partial \theta_0^2} \log L_n(\theta_0 | \underline{X})|_{\hat{\theta}_n} \right) - 2R_n$$
$$= (\sqrt{n}(\hat{\theta}_n - \theta_0))^2 \left(-\frac{1}{n} \frac{\partial^2}{\partial \theta_0^2} \log L_n(\theta_0 | \underline{X})|_{\hat{\theta}_n} \right) - 2R_n .$$

Observe that

$$|R_n| = \left| \frac{(\hat{\theta}_n - \theta_0)^3}{3!} \frac{\partial^3}{\partial \theta_0^3} \log L_n(\theta_0 | \underline{X})|_{\theta_n^*} \right|$$
$$\leq \frac{1}{6\sqrt{n}} \left| (\sqrt{n}(\hat{\theta}_n - \theta_0))^3 \right| \left| \frac{1}{n} \frac{\partial^3}{\partial \theta_0^3} \log L_n(\theta_0 | \underline{X})|_{\theta_n^*} \right| .$$

As shown in the proof of Result 4.3.3 as given by Huzurbazar, $\left|\frac{1}{n}\frac{\partial^3}{\partial\theta_0^3}\log L_n(\theta_0|\underline{X})|_{\theta_n^*}\right|$
$\overset{P_{\theta_0}}{\to} K < \infty$, where K is a constant. Further, $\sqrt{n}(\hat{\theta}_n - \theta_0) \overset{L}{\to} Z_1$, hence by the
continuous mapping theorem $|(\sqrt{n}(\hat{\theta}_n - \theta_0))^3| \overset{L}{\to} |Z_1^3|$ and hence is bounded in
probability. Thus, $R_n \overset{P_{\theta_0}}{\to} 0$. By Cramér-Huzurbazar theory,

$$\sqrt{n}(\hat{\theta}_n - \theta_0) \overset{L}{\to} Z_1 \sim N(0, I^{-1}(\theta_0)) \quad \Rightarrow \quad nI(\theta_0)(\hat{\theta}_n - \theta_0)^2 \overset{L}{\to} U \sim \chi_1^2.$$

Further, $-\frac{1}{n}\frac{\partial^2}{\partial\theta_0^2}\log L_n(\theta_0|\underline{X})|_{\hat{\theta}_n} \overset{P_{\theta_0}}{\to} I(\theta_0)$. Hence by Slutsky's theorem under H_0,

$$-2\log\lambda(\underline{X}) = (\sqrt{n}(\hat{\theta}_n - \theta_0))^2 \frac{I(\theta_0)}{I(\theta_0)}\left(-\frac{1}{n}\frac{\partial^2}{\partial\theta_0^2}\log L_n(\theta_0|\underline{X})|_{\hat{\theta}_n}\right)$$

$$- 2R_n \overset{L}{\to} U \sim \chi_1^2.$$

$\square$

📖 Remark 5.2.1

From Theorem 5.2.1, we note that if $\lambda(\underline{X})$ is a likelihood ratio test statistic based
on a random sample $\underline{X} = \{X_1, X_2, \ldots, X_n\}$ for testing $H_0 : \theta = \theta_0$ against the
alternative $H_1 : \theta \neq \theta_0$, where θ_0 is a specified real number, then under H_0,
$-2\log\lambda(\underline{X})$ and $(\sqrt{n}(\hat{\theta}_n - \theta_0))^2\left(-\frac{1}{n}\frac{\partial^2}{\partial\theta_0^2}\log L_n(\theta_0|\underline{X})|_{\hat{\theta}_n}\right)$ have the same limiting distribution.

📝 Example 5.2.4

Suppose a coin with probability θ of getting heads is tossed 100 times and 60 heads
are observed. We derive a likelihood test procedure to test $H_0 : \theta = 0.5$ against
the alternative $H_1 : \theta \neq 0.5$. Suppose X denotes the outcome of a toss of a coin
with probability θ of getting heads. Then X has Bernoulli $B(1, \theta)$ distribution. On
the basis of a random sample of size 100 from the distribution of X, we want to test
the hypothesis $H_0 : \theta = 0.5$ against the alternative $H_1 : \theta \neq 0.5$. The likelihood
of θ given the data $\underline{X}$ is given by

$$L_n(\theta|\underline{X}) = \theta^{n\overline{X}_n}(1 - \theta)^{n - n\overline{X}_n}.$$

To derive a likelihood ratio test procedure, we note that the sample mean $\overline{X}_n$ is
the maximum likelihood estimator of θ in the entire parameter space $\Theta = (0, 1)$
and the null space Θ_0 is $\Theta_0 = \{0.5\}$ and hence the supremum of the likelihood

in the null space is attained at $\theta = 0.5$. The likelihood ratio test statistic $\lambda(\underline{X})$ is given by

$$\lambda(\underline{X}) = \frac{\sup_{\Theta_0} L_n(\theta|\underline{X})}{\sup_{\Theta} L_n(\theta|\underline{X})} = \frac{0.5^{n\overline{X}_n} 0.5^{n-n\overline{X}_n}}{(\overline{X}_n)^{n\overline{X}_n}(1-\overline{X}_n)^{n-n\overline{X}_n}}.$$

It is difficult to find the null distribution of $\lambda(\underline{X})$ and also difficult to convert the rejection region $\lambda(\underline{X}) < c$ in terms of some statistic T_n. Hence, we use Theorem 5.2.1 to get the asymptotic null distribution of $\lambda(\underline{X})$. Under H_0, $-2\log\lambda(\underline{X}) \sim \chi_1^2$ distribution. Thus, H_0 is rejected if $-2\log\lambda(\underline{X}) > c$ where $c = \chi_{1,1-\alpha}^2 = 3.8415$, if $\alpha = 0.05$. For the given data

$$-2\log\lambda(\underline{X}) = 2[60\log(0.6) + 40\log(0.4) - 100\log(0.5)]$$
$$= 4.0271 > c = 3.8415$$

and hence H_0 is rejected.   □

The next theorem is an extension of Theorem 5.2.1 to a multiparameter setup.

Theorem 5.2.2

Suppose X is a random variable or a random vector with probability law $f(x, \underline{\theta})$ indexed by a parameter $\underline{\theta} \in \Theta \subset \mathbb{R}^k$. Suppose $f(x, \underline{\theta})$ belongs to a Cramér family. If $\lambda(\underline{X})$ is a likelihood ratio test statistic based on a random sample $\underline{X} = \{X_1, X_2, \ldots, X_n\}$ for testing $H_0 : \underline{\theta} = \underline{\theta}_0$ against the alternative $H_1 : \underline{\theta} \neq \underline{\theta}_0$, where $\underline{\theta}_0$ is a specified vector, then under H_0, $-2\log\lambda(\underline{X}) \xrightarrow{L} U \sim \chi_k^2$ as $n \to \infty$.

Proof Suppose $\hat{\underline{\theta}}_n$ is a maximum likelihood estimator of $\underline{\theta}$ based on a random sample $\underline{X}$. Since the distribution of X belongs to a Cramér family, $\hat{\underline{\theta}}_n$ is CAN with approximate dispersion matrix $I^{-1}(\underline{\theta})/n$. According to the likelihood ratio test procedure, $H_0 : \underline{\theta} = \underline{\theta}_0$ is rejected against the alternative $H_1 : \underline{\theta} \neq \underline{\theta}_0$ if $\lambda(\underline{X}) < c \Leftrightarrow -2\log\lambda(\underline{X}) > c_1$ where c and c_1 are constants and $-2\log\lambda(\underline{X})$ is given by

$$-2\log\lambda(\underline{X}) = 2[\log L_n(\hat{\underline{\theta}}_n|\underline{X}) - \log L_n(\underline{\theta}_0|\underline{X})].$$

As in Theorem 5.2.1, we expand $\log L_n(\underline{\theta}_0|\underline{X})$ around $\hat{\underline{\theta}}_n$ using Taylor series expansion. Thus we have

$$\log L_n(\underline{\theta}_0|\underline{X}) = \log L_n(\hat{\underline{\theta}}_n|\underline{X}) + \sum_{i=1}^{k}(\theta_{i0} - \hat{\theta}_{in})\frac{\partial}{\partial\theta_{i0}}\log L_n(\underline{\theta}_0|\underline{X})|_{\hat{\underline{\theta}}_n}$$

$$+ \frac{1}{2}\sum_{i=1}^{k}\sum_{j=1}^{k}(\theta_{i0} - \hat{\theta}_{in})(\theta_{j0} - \hat{\theta}_{jn})\frac{\partial^2}{\partial\theta_{i0}\partial\theta_{j0}}\log L_n(\underline{\theta}_0|\underline{X})|_{\hat{\underline{\theta}}_n} + R_n,$$

where R_n is the remainder term. Further, $\frac{\partial}{\partial\theta_{i0}}\log L_n(\underline{\theta}_0|\underline{X})|_{\hat{\underline{\theta}}_n} = 0$, $\forall\ i = 1, 2, \ldots, k$, as $\hat{\underline{\theta}}_n$ is a solution of the system of likelihood equations. Hence we get

$$-2\log\lambda(\underline{X}) = 2[\log L_n(\hat{\underline{\theta}}_n|\underline{X}) - \log L_n(\underline{\theta}_0|\underline{X})]$$

$$= n\sum_{i=1}^{k}\sum_{j=1}^{k}(\theta_{i0} - \hat{\theta}_{in})(\theta_{j0} - \hat{\theta}_{jn}) \times$$

$$\left(-\frac{1}{n}\frac{\partial^2}{\partial\theta_{i0}\partial\theta_{j0}}\log L_n(\underline{\theta}_0|\underline{X})|_{\hat{\underline{\theta}}_n}\right) - 2R_n$$

$$= n(\hat{\underline{\theta}}_n - \underline{\theta}_0)'M_n(\hat{\underline{\theta}}_n - \underline{\theta}_0) - 2R_n \;,$$

where M_n is a matrix with (i, j)-th element given by
$M_n(i, j) = -\frac{1}{n}\frac{\partial^2}{\partial\theta_{0i}\partial\theta_{0j}}\log L_n(\underline{\theta}_0|\underline{X})|_{\hat{\underline{\theta}}_n}$. By the Cramér-Huzurbazar theorem,

$$\sqrt{n}(\hat{\underline{\theta}}_n - \underline{\theta}_0) \overset{L}{\to} Z_1 \sim N_k(\underline{0}, I^{-1}(\underline{\theta}_0))$$

$$\Rightarrow \quad n(\hat{\underline{\theta}}_n - \underline{\theta}_0)'I(\underline{\theta}_0)(\hat{\underline{\theta}}_n - \underline{\theta}_0) \overset{L}{\to} U \sim \chi_k^2 \;.$$

Further,

$$M_n(i, j) = -\frac{1}{n}\frac{\partial^2}{\partial\theta_{0i}\partial\theta_{0j}}\log L_n(\underline{\theta}_0|\underline{X})|_{\hat{\underline{\theta}}_n} \overset{P_{\theta_0}}{\to} I_{ij}(\underline{\theta}_0) \;,$$

for all $i, j = 1, 2, \ldots, k$. Hence, $M_n \overset{P_{\theta_0}}{\to} I(\underline{\theta}_0)$. As a consequence,

$$n(\hat{\underline{\theta}}_n - \underline{\theta}_0)'M_n(\hat{\underline{\theta}}_n - \underline{\theta}_0) - n(\hat{\underline{\theta}}_n - \underline{\theta}_0)'I(\underline{\theta}_0)(\hat{\underline{\theta}}_n - \underline{\theta}_0)$$

$$= n(\hat{\underline{\theta}}_n - \underline{\theta}_0)'(M_n - I(\underline{\theta}_0))(\hat{\underline{\theta}}_n - \underline{\theta}_0) \overset{P_{\theta_0}}{\to} 0 \;,$$

as $n \to \infty$. Hence, $n(\hat{\underline{\theta}}_n - \underline{\theta}_0)'M_n(\hat{\underline{\theta}}_n - \underline{\theta}_0) \overset{L}{\to} U \sim \chi_k^2$. Now the remainder term R_n is given by

$$R_n = \frac{1}{3!}\sum_{i=1}^{k}\sum_{j=1}^{k}\sum_{l=1}^{k}(\theta_{i0} - \hat{\theta}_{in})(\theta_{j0} - \hat{\theta}_{jn})(\theta_{l0} - \hat{\theta}_{ln})\left(\frac{\partial^3\log L_n(\underline{\theta}_0|\underline{X})}{\partial\theta_{i0}\partial\theta_{j0}\partial\theta_{l0}}|_{\underline{\theta}_n^*}\right) \;,$$

where, $\underline{\theta}_n^* = \alpha\underline{\theta}_0 + (1 - \alpha)\hat{\underline{\theta}}_n$, $0 < \alpha < 1$ and under H_0, $\underline{\theta}_n^* \overset{P_{\theta}}{\to} \underline{\theta}_0$. To show that $R_n \overset{P_{\theta}}{\to} 0$, consider

$$|R_n| = \left|\frac{1}{3!}\sum_{i=1}^{k}\sum_{j=1}^{k}\sum_{l=1}^{k}(\theta_{i0} - \hat{\theta}_{in})(\theta_{j0} - \hat{\theta}_{jn})(\theta_{l0} - \hat{\theta}_{ln})\left(\frac{\partial^3\log L_n(\underline{\theta}_0|\underline{X})}{\partial\theta_{i0}\partial\theta_{j0}\partial\theta_{l0}}|_{\underline{\theta}_n^*}\right)\right|$$

$$\leq \frac{1}{6\sqrt{n}}\sum_{i=1}^{k}\sum_{j=1}^{k}\sum_{l=1}^{k}|\sqrt{n}(\theta_{i0} - \hat{\theta}_{in})||\sqrt{n}(\theta_{j0} - \hat{\theta}_{jn})||\sqrt{n}(\theta_{l0} - \hat{\theta}_{ln})| \times$$

$$\left|\frac{1}{n}\frac{\partial^3\log L_n(\underline{\theta}_0|\underline{X})}{\partial\theta_{i0}\partial\theta_{j0}\partial\theta_{l0}}|_{\underline{\theta}_n^*}\right| \;.$$

Now we use the condition that the third order partial derivatives of $\log f(x, \theta)$ are bounded by integrable functions and proceed on the similar lines as in the proof of Result 4.3.3, as given by Huzurbazar. It then follows that $\left| \frac{1}{n} \frac{\partial^3 \log L_n(\underline{\theta}_0 | \underline{X})}{\partial \theta_{i0} \partial \theta_{j0} \partial \theta_{l0}} |_{\underline{\theta}_n^*} \right| \overset{P_{\theta_0}}{\to} K < \infty$, where K is a constant. Further,

$$\sqrt{n}(\underline{\hat{\theta}}_n - \underline{\theta}_0) = \sqrt{n}\left((\hat{\theta}_{1n} - \theta_{10}), (\hat{\theta}_{2n} - \theta_{20}), \ldots, (\hat{\theta}_{kn} - \theta_{k0})\right) \overset{L}{\to} \underline{Z}_1 .$$

Suppose $g : \mathbb{R}^k \to \mathbb{R}$ is defined as $g(x_1, x_2, \ldots, x_k) = \sum_{i=1}^{k} \sum_{j=1}^{k} \sum_{l=1}^{k} |x_i||x_j||x_l|$. Then g is a continuous function and hence by the continuous mapping theorem, $g\left(\sqrt{n}\left((\hat{\theta}_{1n} - \theta_{10}), (\hat{\theta}_{2n} - \theta_{20}), \ldots, (\hat{\theta}_{kn} - \theta_{k0})\right)\right) \overset{L}{\to} g(\underline{Z}_1)$ which implies that $\sum_{i=1}^{k} \sum_{j=1}^{k} \sum_{l=1}^{k} |\sqrt{n}(\theta_{i0} - \hat{\theta}_{in})||\sqrt{n}(\theta_{j0} - \hat{\theta}_{jn})||\sqrt{n}(\theta_{l0} - \hat{\theta}_{ln})|$ is bounded in probability and hence $R_n \overset{P_{\theta_0}}{\to} 0$. Thus, by Slutsky's theorem under H_0,

$$-2 \log \lambda(\underline{X}) = n(\underline{\hat{\theta}}_n - \underline{\theta}_0)' M_n (\underline{\hat{\theta}}_n - \underline{\theta}_0) - 2R_n \overset{L}{\to} U \sim \chi_k^2 . \qquad \square$$

✏ Remark 5.2.2

As in Theorem 5.2.1, in multiparameter set up, if $\lambda(\underline{X})$ is a likelihood ratio test statistic based on a random sample $\underline{X} = \{X_1, X_2, \ldots, X_n\}$ for testing $H_0 : \theta = \theta_0$ against the alternative $H_1 : \theta \neq \theta_0$, where θ_0 is a specified real number, then under H_0, $-2 \log \lambda(\underline{X})$ and $n(\hat{\theta}_n - \theta_0)' M_n(\hat{\theta}_n - \theta_0)$ have the same limiting distribution. In the next chapter we will discuss how these two statistics are related to a score test statistic and Wald's test statistic.

✒ Example 5.2.5

Suppose $\{X_1, X_2, \ldots, X_n\}$ are independent and identically distributed random variables with following probability mass function.

$$P[X_1 = 1] = (1 - \theta_1)/2, \quad P[X_1 = 2] = \theta_2/2, \quad P[X_1 = 3] = \theta_1/2$$
$$\& \quad P[X_1 = 4] = (1 - \theta_2)/2 .$$

We derive a likelihood ratio test procedure to test $H_0 : \theta_1 = 1/2, \ \theta_2 = 1/4$ against the alternative $H_1 : \theta \neq 1/2, \ \theta_2 \neq 1/4$. The likelihood of θ corresponding to the data $\underline{X} \equiv \{X_1, X_2, \ldots, X_n\}$ is given by

$$L_n(\theta_1, \theta_2 | \underline{X}) = \left(\frac{1 - \theta_1}{2}\right)^{n_1} \left(\frac{\theta_2}{2}\right)^{n_2} \left(\frac{\theta_1}{2}\right)^{n_3} \left(\frac{1 - \theta_2}{2}\right)^{n_4}$$
$$= 2^{-n}(1 - \theta_1)^{n_1} \theta_2^{n_2} \theta_1^{n_3} (1 - \theta_2)^{n_4} ,$$

as $n_1 + n_2 + n_3 + n_4 = n$, where n_i denotes the frequency of i, $i = 1, 2, 3, 4$ in the given sample of size n. To develop a likelihood ratio test procedure, we first obtain the maximum likelihood estimator of $(\theta_1, \theta_2)'$. Likelihood is a differentiable function of θ_1 and θ_2. Hence, the system of likelihood equations is given by

$$\frac{\partial}{\partial \theta_1} \log L_n(\theta_1, \theta_2 | \underline{X}) = -\frac{n_1}{1 - \theta_1} + \frac{n_3}{\theta_1} = 0$$

$$\& \quad \frac{\partial}{\partial \theta_2} \log L_n(\theta_1, \theta_2 | \underline{X}) = \frac{n_2}{\theta_2} - \frac{n_4}{1 - \theta_2} = 0$$

and its solution $\underline{\hat{\theta}}_n = (\hat{\theta}_{1n}, \hat{\theta}_{2n})'$ is given by

$$\hat{\theta}_{1n} = \frac{n_3}{n_1 + n_3} \quad \& \quad \hat{\theta}_{2n} = \frac{n_4}{n_2 + n_4}.$$

The matrix of second partial derivatives is $\text{diag}\left(-\frac{n_1}{(1-\theta_1)^2} - \frac{n_3}{\theta_1^2}, -\frac{n_2}{\theta_2^2} - \frac{n_4}{(1-\theta_2)^2}\right)$ and it is negative definite for all $(\theta_1, \theta_2)'$. Hence, $\underline{\hat{\theta}}_n = (\hat{\theta}_{1n}, \hat{\theta}_{2n})'$ is the maximum likelihood estimator of $\underline{\theta} = (\theta_1, \theta_2)'$. The entire parameter space Θ is $\Theta = \{(\theta_1, \theta_2)' | \theta_1, \theta_2 \in (0, 1)\}$ and the null space Θ_0 is $\Theta_0 = \{(\theta_1, \theta_2)' | \theta_1 = 1/2, \theta_2 = 1/4\}$. The likelihood ratio test statistic $\lambda(\underline{X})$ for testing the null hypothesis $H_0 : \theta_1 = 1/2, \ \theta_2 = 1/4$ against the alternative $H_1 : \theta \neq 1/2, \ \theta_2 \neq 1/4$ is given by

$$\lambda(\underline{X}) = \frac{\sup_{\Theta_0} L_n(\underline{\theta} | \underline{X})}{\sup_{\Theta} L_n(\underline{\theta} | \underline{X})} = \frac{(\frac{1}{4})^{n_1 + n_3} (\frac{1}{8})^{n_2} (\frac{3}{8})^{n_4}}{2^{-n} (1 - \hat{\theta}_{1n})^{n_1} \hat{\theta}_{2n}^{n_2} \hat{\theta}_{1n}^{n_3} (1 - \hat{\theta}_{2n})^{n_4}}.$$

The null hypothesis is rejected if $\lambda(\underline{X}) < c < 1 \ \Leftrightarrow \ -2 \log \lambda(\underline{X}) > c_1$. For the large sample size, $-2 \log \lambda(\underline{X}) \sim \chi_2^2$ distribution. H_0 is rejected if $-2 \log \lambda(\underline{X}) > \chi_{2,1-\alpha}^2$, where $\chi_{2,1-\alpha}^2$ is $(1 - \alpha)$-th quantile of χ_2^2 distribution.

$\square$

The next theorem is a further extension of Theorem 5.2.2 in a multiparameter setup. In many cases, $\underline{\theta}_0$ is not completely specified. In null setup, few parameters out of $\{\theta_1, \theta_2, \ldots, \theta_k\}$ are specified and remaining need to be estimated on the basis of the given data. More precisely, suppose $\underline{\theta}$ is partitioned as $\underline{\theta} = (\underline{\theta}^{(1)}, \underline{\theta}^{(2)})'$, where $\underline{\theta}^{(1)} = (\theta_1, \theta_2, \ldots, \theta_m)'$ and $\underline{\theta}^{(2)} = (\theta_{m+1}, \theta_{m+2}, \ldots, \theta_k)'$. In null setup $\underline{\theta}^{(1)}$ is completely specified and one needs to estimate the parameters involved in $\underline{\theta}^{(2)}$. $\underline{\theta}^{(2)}$ is known as a nuisance parameter. We come across with such a setup in many practical situations. For example, in regression analysis, the global F-test and the partial F-test for testing the significance of regression coefficients are the likelihood ratio test procedures where few parameters are specified under null setup and the remaining are estimated from the given data. Goodness of fit tests, test for validity of the model and the tests for contingency table are the likelihood ratio tests when underlying probability model is a multinomial distribution in k cells. In null setup, cell probabilities are either completely specified or indexed by an unknown parameter $\underline{\theta}$ or in some cases there is a certain relation amongst the cell probabilities. Following two theorems are related to the asymptotic null distribution of $-2\log\lambda(\underline{X})$, when the null hypothesis is a composite hypothesis. We outline the proof of this theorem, for details one may refer to Ferguson [2], Theorem 22.

Theorem 5.2.3

Suppose X is a random variable or a random vector with probability law $f(x,\underline{\theta})$ indexed by a parameter $\underline{\theta} \in \Theta \subset \mathbb{R}^k$ and the distribution of X belongs to a Cramér family. Suppose $\lambda(\underline{X})$ is a likelihood ratio test statistic based on a random sample $\underline{X} = \{X_1, X_2, \ldots, X_n\}$ for testing $H_0 : \underline{\theta}^{(1)} = \underline{\theta}_0^{(1)}$ against the alternative $H_1 : \underline{\theta}^{(1)} \neq \underline{\theta}_0^{(1)}$, where $\underline{\theta}^{(1)} = (\theta_1, \theta_2, \ldots, \theta_m)'$ and $\underline{\theta}^{(2)} = (\theta_{m+1}, \theta_{m+2}, \ldots, \theta_k)'$ is partition of $\underline{\theta}$ with $m < k$ and $\underline{\theta}_0^{(1)}$ is a specified vector. Then under H_0,

$$-2\log\lambda(\underline{X}) \xrightarrow{L} U \sim \chi_r^2 \quad \text{as } n \to \infty, \quad \text{where } r = k - (k-m) = m.$$

Proof Suppose $\hat{\underline{\theta}}_n = (\hat{\underline{\theta}}_n^{(1)}, \hat{\underline{\theta}}_n^{(2)})'$ is a maximum likelihood estimator of $\underline{\theta}=(\underline{\theta}^{(1)}, \underline{\theta}^{(2)})'$ in the entire parameter space and $\tilde{\underline{\theta}}_n^{(2)}$ is a maximum likelihood estimator of $\underline{\theta}^{(2)}$ when $\underline{\theta}^{(1)} = \underline{\theta}_0^{(1)}$, based on a random sample $\underline{X}$. Since the distribution of X belongs to a Cramér family, $\hat{\underline{\theta}}_n$ is CAN for $\underline{\theta}$ with approximate dispersion matrix $I^{-1}(\underline{\theta})/n$. Similarly, $\tilde{\underline{\theta}}_n^{(2)}$ is CAN for $\underline{\theta}^{(2)}$ with approximate dispersion matrix $\Sigma_{22}(\underline{\theta})/n$, say. The likelihood ratio test statistic $\lambda(\underline{X})$ is then given by

$$\lambda(\underline{X}) = \sup_{\Theta_0} L_n(\underline{\theta}|\underline{X}) / \sup_{\Theta} L_n(\underline{\theta}|\underline{X}) = L_n((\underline{\theta}_0^{(1)}, \tilde{\underline{\theta}}_n^{(2)})|\underline{X}) / L_n((\hat{\underline{\theta}}_n^{(1)}, \hat{\underline{\theta}}_n^{(2)})|\underline{X}).$$

Suppose $\underline{\theta}_0 = (\underline{\theta}_0^{(1)}, \underline{\theta}_0^{(2)})'$, then $-2\log\lambda(\underline{X})$ can be expressed as

$$
\begin{aligned}
-2\log\lambda(\underline{X}) &= 2[\log L_n((\hat{\underline{\theta}}_n^{(1)}, \hat{\underline{\theta}}_n^{(2)})|\underline{X}) - \log L_n((\underline{\theta}_0^{(1)}, \tilde{\underline{\theta}}_n^{(2)})|\underline{X})] \\
&= 2[\log L_n((\hat{\underline{\theta}}_n^{(1)}, \hat{\underline{\theta}}_n^{(2)})|\underline{X})) - \log L_n((\underline{\theta}_0^{(1)}, \underline{\theta}_0^{(2)})|\underline{X})] \\
&\quad - 2[\log L_n((\underline{\theta}_0^{(1)}, \tilde{\underline{\theta}}_n^{(2)})|\underline{X}) - \log L_n((\underline{\theta}_0^{(1)}, \underline{\theta}_0^{(2)})|\underline{X})] \\
&= 2[\log L_n(\hat{\underline{\theta}}_n|\underline{X}) - \log L_n(\underline{\theta}_0|\underline{X})] \\
&\quad - 2[\log L_n((\underline{\theta}_0^{(1)}, \tilde{\underline{\theta}}_n^{(2)})|\underline{X}) - \log L_n((\underline{\theta}_0^{(1)}, \underline{\theta}_0^{(2)})|\underline{X})].
\end{aligned}
$$

Thus, $-2\log\lambda(\underline{X}) = U_n - W_n$, where

$$U_n = 2[\log L_n(\hat{\underline{\theta}}_n|\underline{X}) - \log L_n(\underline{\theta}_0|\underline{X})]$$

$$\&\ W_n = 2[\log L_n((\underline{\theta}_0^{(1)}, \tilde{\underline{\theta}}_n^{(2)})|\underline{X}) - \log L_n((\underline{\theta}_0^{(1)}, \underline{\theta}_0^{(2)})|\underline{X})].$$

Since $\underline{\theta}_0$ is a known vector, by Theorem 5.2.2, $U_n \xrightarrow{L} U \sim \chi^2_k$. Expanding $\log L_n$ $((\underline{\theta}_0^{(1)}, \underline{\theta}_0^{(2)})|\underline{X})$ around $(\underline{\theta}_0^{(1)}, \tilde{\underline{\theta}}_n^{(2)})$ by Taylor series expansion and proceeding on similar lines as in the proof of Theorem 5.2.2, one can show that $W_n \xrightarrow{L} W \sim \chi^2_{k-m}$. Thus, for large n, $-2 \log \lambda(\underline{X})$ is distributed as $U - W$. It has been proved by Wilks [1] that Cochran's theorem on quadratic forms in normal variates holds for the asymptotic distributions and hence $U - W$ is distributed as $\chi^2_{k-(k-m)}$. Thus, under H_0 for large n, $-2 \log \lambda(\underline{X})$ has χ^2_m distribution. It is to be noted that the degrees of freedom m is the difference between the number of parameters estimated in the entire parameter space and the number of parameters estimated in the null parameter space.

$\square$

Theorem 5.2.3 is useful in a variety of tests in regression analysis. Following example illustrates its application.

✒ Example 5.2.6

Suppose $\{X_1, X_2, \ldots, X_n\}$ is a random sample from normal $N(\mu, \sigma^2)$ distribution. We illustrate Theorem 5.2.3 to derive the likelihood ratio test procedure for testing $H_0 : \sigma = \sigma_0$ against the alternative $H_0 : \sigma \neq \sigma_0$ when μ is unknown. Suppose $X \sim N(\mu, \sigma^2)$ distribution. We first derive a likelihood ratio test procedure for testing H_0, when μ is known. Then the null hypothesis $H_0 : \sigma = \sigma_0$ is a simple null hypothesis. The null parameter space is $\Theta_0 = \{\sigma_0\}$ and the entire parameter space is $\Theta = (0, \infty)$. Corresponding to a random sample $\underline{X} \equiv \{X_1, X_2, \ldots, X_n\}$ of size n from normal $N(\mu, \sigma^2)$ distribution, the likelihood of σ^2 is given by

$$L_n(\sigma^2|\underline{X}) = \prod_{i=1}^{n} \frac{1}{\sqrt{2\pi}\sigma} \exp\{-\frac{1}{2\sigma^2}(X_i - \mu)^2\} .$$

It then follows that the maximum likelihood estimator of σ^2 in the entire parameter space is given by $S_{0n}^2 = \sum_{i=1}^{n}(X_i - \mu)^2/n$. The likelihood ratio test statistic $\lambda(\underline{X})$ is then given by

$$\lambda(\underline{X}) = \frac{\sup_{\Theta_0} L_n(\underline{\theta}|\underline{X})}{\sup_{\Theta} L_n(\underline{\theta}|\underline{X})} = \left(\frac{S_{0n}^2}{\sigma_0^2}\right)^{n/2} \exp\left\{\frac{\sum_{i=1}^{n}(X_i - \mu)^2}{2}\left\{\frac{1}{S_{0n}^2} - \frac{1}{\sigma_0^2}\right\}\right\} .$$

The null hypothesis is rejected if $\lambda(\underline{X}) < c < 1 \iff -2 \log \lambda(\underline{X}) > c_1$. If sample size is large, then by Theorem 5.2.1, $-2 \log \lambda(\underline{X}) \sim \chi^2_1$ distribution and H_0 is rejected if $-2 \log \lambda(\underline{X}) > \chi^2_{1,1-\alpha}$ where $\chi^2_{1,1-\alpha}$ is $(1 - \alpha)$-th quantile of χ^2_1 distribution. Suppose μ is not known. Then the null hypothesis is also a composite hypothesis and μ is a nuisance parameter. The entire parameter space Θ is $\Theta = \{(\mu, \sigma^2)|\mu \in \mathbb{R}, \sigma^2 > 0\}$ and the null space Θ_0 is

$\Theta_0 = \{(\mu, \sigma^2)|\mu \in \mathbb{R}, \sigma^2 = \sigma_0^2\}$. In the null space the maximum likelihood estimators of μ is $\overline{X}_n$. In Example 3.3.2 we have obtained the maximum likelihood estimators of μ and σ^2 which are given by $\overline{X}_n$ and $S_n^2 = \sum_{i=1}^n (X_i - \overline{X}_n)^2/n$ respectively. Hence, the likelihood ratio test statistic $\lambda(\underline{X})$ is given by

$$\lambda(\underline{X}) = \frac{\sup_{\Theta_0} L_n(\underline{\theta}|\underline{X})}{\sup_{\Theta} L_n(\underline{\theta}|\underline{X})} = \left(\frac{S_n^2}{\sigma_0^2}\right)^{n/2} \exp\left\{\frac{\sum_{i=1}^n (X_i - \overline{X}_n)^2}{2}\left\{\frac{1}{S_n^2} - \frac{1}{\sigma_0^2}\right\}\right\}.$$

The null hypothesis is rejected if $\lambda(\underline{X}) < c < 1 \Leftrightarrow -2\log\lambda(\underline{X}) > c_1$. If sample size is large, then by Theorem 5.2.3, $-2\log\lambda(\underline{X}) \sim \chi_1^2$ distribution, since we estimate two parameters in the entire parameter space and only one parameter in the null space. H_0 is rejected if $-2\log\lambda(\underline{X}) > \chi_{1,1-\alpha}^2$ where $\chi_{1,1-\alpha}^2$ is $(1-\alpha)$-th quantile of χ_1^2 distribution. $\qquad\Box$

The next theorem is an extension of Theorem 5.2.3 where in null space the parameters have some functional relationship among themselves. More precisely, in the null setup, $\theta_i = g_i(\beta_1, \beta_2, \ldots, \beta_m)'$, $i = 1, 2, \ldots, k$, where $m \le k$ and $g_1, g_2, \ldots, g_k$ are Borel measurable functions from $\mathbb{R}^m$ to $\mathbb{R}$. Thus, k parameters are expressed in terms of m parameters. Such a scenario is very common in the tests for contingency tables and the goodness of fit tests. In Theorem 5.2.4, we derive the large sample distribution of $-2\log\lambda(\underline{X})$ in this case. The theorem is heavily used in all tests related to contingency tables and in all the goodness of fit tests, which are discussed in the next chapter.

Theorem 5.2.4

Suppose X is a random variable or a random vector with probability law $f(x, \underline{\theta})$ indexed by a parameter $\underline{\theta} \in \Theta \subset \mathbb{R}^k$ and the distribution of X belongs to a Cramér family. Suppose $\lambda(\underline{X})$ is a likelihood ratio test statistic based on a random sample $\underline{X} = \{X_1, X_2, \ldots, X_n\}$ for testing $H_0 : \underline{\theta} \in \Theta_0$ against the alternative $H_1 : \underline{\theta} \in \Theta_1$, where in Θ_0, $\theta_i = g_i(\beta_1, \beta_2, \ldots, \beta_m)'$, $i = 1, 2, \ldots, k$, where $m \le k$ and $g_1, g_2, \ldots, g_k$ are Borel measurable functions from $\mathbb{R}^m$ to $\mathbb{R}$, having continuous partial derivatives of first order. Then under

H_0, $-2\log\lambda(\underline{X}) \xrightarrow{L} U \sim \chi_r^2$ as $n \to \infty$, where $r = k - m$, k the number of parameters estimated in the entire parameter space and m the number of parameters estimated in the null parameter space.

Proof Suppose X is a random variable or a random vector with probability law $f(x, \underline{\theta})$ indexed by a parameter $\underline{\theta} \in \Theta \subset \mathbb{R}^k$, $k \ge 1$. Suppose $\hat{\underline{\theta}}_n$ is the maximum likelihood estimator of $\underline{\theta}$ in the entire parameter space, based on the random sample $\{X_1, X_2, \ldots, X_n\}$ from the distribution of X. Further, $\hat{\underline{\theta}}_n$ is a CAN estimator of $\underline{\theta}$ with approximate dispersion matrix $I^{-1}(\underline{\theta})/n$ of order $k \times k$. In the null setup $\theta_i = g_i(\beta_1, \beta_2, \ldots, \beta_m)'$, $i = 1, 2, \ldots, k$, where $m \leq k$ and $g_1, g_2, \ldots, g_k$ are Borel measurable functions from $\mathbb{R}^m$ to $\mathbb{R}$. Suppose $\tilde{\underline{\theta}}_n$ is the maximum likelihood estimator of $\underline{\theta}$ in the null setup, that is, $\tilde{\underline{\theta}}_n = \underline{g}(\tilde{\underline{\beta}}_n)$, where $\tilde{\underline{\beta}}_n$ is the maximum likelihood estimator of $\underline{\beta} = ((\beta_1, \beta_2, \ldots, \beta_m)'$ based on the random sample $\{X_1, X_2, \ldots, X_n\}$ from the distribution of X. Further, $\tilde{\underline{\beta}}_n$ is a CAN estimator of $\underline{\beta}$ with approximate dispersion matrix $I^{-1}(\underline{\beta})/n$ of order $m \times m$. The likelihood ratio test statistic $\lambda(\underline{X})$ is then given by

$$\lambda(\underline{X}) = \sup_{\Theta_0} L_n(\underline{\theta}|\underline{X}) / \sup_{\Theta} L_n(\underline{\theta}|\underline{X}) = L_n(\tilde{\underline{\theta}}_n|\underline{X})/L_n(\hat{\underline{\theta}}_n|\underline{X}) .$$

Hence,

$$
\begin{aligned}
-2\log \lambda(\underline{X}) &= 2[\log L_n(\hat{\underline{\theta}}_n|\underline{X}) - \log L_n(\tilde{\underline{\theta}}_n|\underline{X})] \\
&= 2[\log L_n(\hat{\underline{\theta}}_n|\underline{X}) - \log L_n(\underline{\theta}_0|\underline{X})] \\
&\quad - 2[\log L_n(\tilde{\underline{\beta}}_n|\underline{X}) - \log L_n(\underline{\beta}_0|\underline{X})] \\
&= U_n - W_n \\
\text{where } U_n &= 2[\log L_n(\hat{\underline{\theta}}_n|\underline{X}) - \log L_n(\underline{\theta}_0|\underline{X})] \\
W_n &= 2[\log L_n(\tilde{\underline{\beta}}_n|\underline{X}) - \log L_n(\underline{\beta}_0|\underline{X})],
\end{aligned}
$$

and $\underline{\beta}_0$ is a known vector such that $\underline{\theta}_0 = \underline{g}(\underline{\beta}_0)$ and $\log L_n(\underline{\beta}_0|\underline{X}) = \log L_n(\underline{\theta}_0|\underline{X})$. From Theorem 5.2.2, $U_n \xrightarrow{L} U \sim \chi_k^2$ and $W_n \xrightarrow{L} W \sim \chi_m^2$. Thus, for large n, $-2\log \lambda(\underline{X})$ is distributed as $U - W$. Again using the result by Wilks [1], we claim that $U - W$ is distributed as χ_{k-m}^2. Thus for large n under H_0, $-2\log \lambda(\underline{X})$ has χ_{k-m}^2 distribution.   □

✐ Example 5.2.7

Suppose $X \sim Poi(\theta_1)$ and $Y \sim Poi(\theta_2)$. Suppose $\underline{X} = \{X_1, X_2, \ldots, X_{n_1}\}$ is a random sample from a Poisson distribution with parameter θ_1 and $\underline{Y} = \{Y_1, Y_2, \ldots, Y_{n_2}\}$ is a random sample from a Poisson distribution with parameter θ_2. Suppose X and Y are independent. We derive the likelihood ratio test procedure to test $H_0 : \theta_1 = \theta_2$ against $H_1 : \theta_1 \neq \theta_2$. In the entire parameter space, the maximum likelihood estimator of θ_1 is $\hat{\theta}_{1n_1} = \overline{X}_{n_1}$ and the maximum likelihood estimator of θ_2 is $\hat{\theta}_{2n_2} = \overline{Y}_{n_2}$. In the null setup $\theta_1 = \theta_2 = \theta$, say. Then using independence of X and Y, the likelihood of θ given random samples $\underline{X}$ and $\underline{Y}$ is

$$L_n(\theta|\underline{X}, \underline{Y}) = e^{-(n_1+n_2)\theta} \, \theta^{(n_1\overline{X}_{n_1}+n_2\overline{Y}_{n_2})} \left(\prod_{i=1}^{n_1} X_i!\right)^{-1} \left(\prod_{i=1}^{n_2} Y_i!\right)^{-1} .$$

It then follows that the maximum likelihood estimator
$\hat{\theta}_{n_1+n_2}$ of θ is $\hat{\theta}_{n_1+n_2} = \left(n_1\overline{X}_{n_1} + n_2\overline{Y}_{n_2}\right)/(n_1 + n_2)$. The likelihood ratio test
statistic $\lambda(\underline{X})$ is then given by

$$\lambda(\underline{X}) = \frac{\sup_{\Theta_0} L_n(\theta|\underline{X})}{\sup_{\Theta} L_n(\theta|\underline{X})} = \frac{e^{-(n_1+n_2)\hat{\theta}_{n_1+n_2}} \hat{\theta}_{n_1+n_2}^{(n_1\overline{X}_{n_1}+n_2\overline{Y}_{n_2})}}{e^{-(n_1\hat{\theta}_{1n_1}+n_2\hat{\theta}_{2n_2})} \hat{\theta}_{1n_1}^{n_1\overline{X}_{n_1}} \hat{\theta}_{2n_2}^{n_2\overline{Y}_{n_2}}}.$$

The null hypothesis is rejected if $\lambda(\underline{X}) < c < 1 \Leftrightarrow -2\log\lambda(\underline{X}) > c_1$. If sample sizes are large, then by Theorem 5.2.4, $-2\log\lambda(\underline{X}) \sim \chi_1^2$ distribution, as in the entire parameter space we estimate two parameters and in null space we estimate one parameter. H_0 is rejected if $-2\log\lambda(\underline{X}) > \chi_{1,1-\alpha}^2$ where $\chi_{1,1-\alpha}^2$ is $(1-\alpha)$-th quantile of χ_1^2 distribution.   □

5.3 Large Sample Tests Using R

In this section we illustrate how the R software is useful in large sample test procedures and in likelihood ratio test procedures.

✒ Example 5.3.1

Suppose $\{X_1, X_2, \ldots, X_n\}$ is a random sample of size n from a Cauchy $C(\theta, 1)$ distribution. We derive a large sample test procedure for testing $H_0 : \theta = 0$ against the alternative $H_1 : \theta > 0$, based on the maximum likelihood estimator $\hat{\theta}_n$ of θ as it is the BAN estimator of θ with approximate variance $2/n$. Thus, the test statistic T_n is given by $T_n = \sqrt{n/2}\,\hat{\theta}_n$ and H_0 is rejected if $T_n > c$. Under H_0 for large n, $T_n \sim N(0, 1)$ distribution. Hence, $P_0[T_n > c] = \alpha \Rightarrow c = \Phi^{-1}(1-\alpha)$. The power function $\beta(\theta)$ for $\theta \geq 0$ is given by

$$\beta(\theta) = P_\theta[T_n > c] = P_\theta\left[\hat{\theta}_n > \sqrt{2/n}\,c\right]$$

$$= P_\theta\left[\sqrt{n/2}(\hat{\theta}_n - \theta) > c - \sqrt{n/2}\,\theta\right] = 1 - \Phi\left(c - \sqrt{n/2}\,\theta\right).$$

Following R code is used to find the maximum likelihood estimator $\hat{\theta}_n$ on the basis of a random sample generated when $\theta = 0.5$, it computes T_n and the power function with $\alpha = 0.05$.

```
th=.5; n=120; set.seed(123); x=rcauchy(n,th,1); summary(x)
dlogl=function(par)
{
term=0
for(i in 1:n)
{
term=term + (x[i] - par)/(1+(x[i]-par)^2)
}
dlogl= 2*term
return(dlogl)
}
dlogl(1); dlogl(0.1); mle=uniroot(dlogl,c(.1,1))$root; mle
T = sqrt(n/2)*mle; T; b = qnorm(.95); b ;
                        p = 1-pnorm(T); p ### p-value
th1=seq(0,1.5,.03); beta = 1 - pnorm(b-sqrt(n/2)*th1); beta[1]
plot(th1, beta, "o", pch = 20, main="Power Function", xlab="Theta",
    ylab = "Power",col="blue")
abline(h=beta[1], col="dark blue")
```

For the random sample of size $n = 120$ generated with $\theta = 0.5$, $\hat{\theta}_n = 0.5253$, $T_n = 4.0689$ which is larger than 1.65, 95% quantile of the standard normal distribution, p-value is almost 0. Hence, the data do not have sufficient support to null setup and we reject H_0. From Fig. 5.1, we observe that the power function $\beta(\theta)$ is an increasing function of θ and is almost 1 for all $\theta > 0.5$. It is 0.05 at $\theta = 0$, since we have decided the cut-off c so that power function at $\theta = 0$ is 0.05. It can be verified that if we generate a random sample with $\theta = 0$, then the

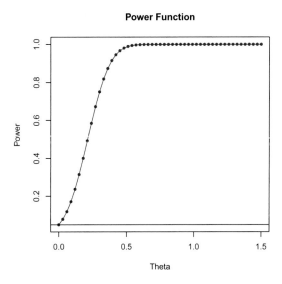

Fig. 5.1 Cauchy $C(\theta, 1)$ distribution: power function

data support to null setup. In fact generating random samples with $\theta = 0$, we can find the false positive rate, that is, estimate of probability of type I error, also known as empirical level of significance. Following R code computes the false positive rate, when random samples of size 120 are generated $m = 1000$ times with $\theta = 0$.

```
th=0; n=120; m=1000; x=matrix(nrow=n,ncol=m)
for(j in 1:m)
{
set.seed(j)
x[,j]=rcauchy(n,th,1)
}
dlogL=function(par)
(T > b))/m; FPR
{
term=0
for(i in 1:n)
{
term=term + (x[i,j] - par)/(1+(x[i,j]-par)^2)
}
dlogL= 2*term
return(dlogL)
}
mle=med=c()
for(j in 1:m)
{
med[j]=median(x[,j])
mle[j]=uniroot(dlogL,c(med[j]-3,med[j]+3))$root
}
summary(mle);T = sqrt(n/2)*mle; b = qnorm(.95); b
FPR=length(which(T > b))/m; FPR
```

From the 1000 simulations, estimate of probability of type I error comes out to be 0.051. Summary statistic of 1000 values of $\hat{\theta}_n$ shows that the sample median is 0.0004, which is close to 0.   □

In Example 3.3.9, we have shown that if $(X, Y)' \sim N_2(0, 0, 1, 1, \rho)$ distribution, $\rho \in (-1, 1)$, then the sample correlation coefficient R_n is a CAN estimator of ρ with approximate variance $(1 - \rho^2)^2/n$. In Example 4.3.3, it is shown that the maximum likelihood estimator $\hat{\rho}_n$ of ρ is also a CAN estimator of ρ with approximate variance $(1 - \rho^2)^2/n(1 + \rho^2)$ and it is smaller than that of R_n. In the following example, we derive a test procedure to test $H_0 : \rho = \rho_0$ against the alternative $H_1 : \rho \neq \rho_0$ based on R_n and $\hat{\rho}_n$ and compare their performance on the basis of the power function.

✒ Example 5.3.2

Suppose $(X, Y)'$ has a bivariate normal distribution with zero mean vector and dispersion matrix Σ given by

$$\Sigma = \begin{pmatrix} 1 & \rho \\ \rho & 1 \end{pmatrix},$$

$\rho \in (-1, 1)$. Suppose we want to test $H_0 : \rho = \rho_0$ against the alternative $H_1 : \rho \neq \rho_0$. In Example 3.3.9 and Example 4.3.3, we have shown that

$$\frac{\sqrt{n}}{(1 - \rho^2)}(R_n - \rho) \overset{L}{\to} N(0, 1) \ \& \ \frac{\sqrt{n(1 + \rho^2)}}{(1 - \rho^2)}(\hat{\rho}_n - \rho) \overset{L}{\to} N(0, 1).$$

Thus, to test $H_0 : \rho = \rho_0$ against the alternative $H_1 : \rho \neq \rho_0$, we have the following two test statistics.

$$T_n = \frac{\sqrt{n(1 + \rho_0^2)}}{(1 - \rho_0^2)}(\hat{\rho}_n - \rho_0) \ \& \ S_n = \frac{\sqrt{n}}{(1 - \rho_0^2)}(R_n - \rho_0).$$

The critical region is $[|T_n| > c]$ and $[|S_n| > c]$. The cut-off c is decided using the given level of significance α and the large sample null distribution of the test statistic, which is standard normal for both the test statistics. Hence, $c = a_{1-\alpha/2}$. We compare the performance of these test statistics, based on the empirical level of significance and the power function. The power function $\beta_{T_n}(\rho)$ of T_n is given by

$$\beta_{T_n}(\rho) = P_\rho[|T_n| > c] = 1 - P_\rho[|T_n| \leq c]$$

$$= 1 - P_\rho\left[-\frac{c(1 - \rho_0^2)}{\sqrt{n(1 + \rho_0^2)}} + (\rho_0 - \rho) < (\hat{\rho}_n - \rho) \leq \frac{c(1 - \rho_0^2)}{\sqrt{n(1 + \rho_0^2)}} + (\rho_0 - \rho)\right]$$

$$= 1 - \Phi\left(a(\rho) + c\,b(\rho)\right) + \Phi\left(a(\rho) - c\,b(\rho)\right),$$

where

$$a(\rho) = \frac{\sqrt{n(1 + \rho^2)}}{1 - \rho^2}(\rho_0 - \rho) \ \& \ b(\rho) = \frac{\sqrt{1 + \rho^2}}{\sqrt{1 + \rho_0^2}}\frac{(1 - \rho_0^2)}{(1 - \rho^2)}.$$

The power function $\beta_{S_n}(\rho)$ of S_n is given by

$$\beta_{S_n}(\rho) = P_\rho[|S_n| > c] = 1 - P_\rho[|S_n| \leq c]$$

$$= 1 - P_\rho[-c(1 - \rho_0^2)/\sqrt{n} + \rho_0 - \rho < (R_n - \rho) \leq c(1 - \rho_0^2)/\sqrt{n} + \rho_0 - \rho]$$

$$= 1 - P_\rho\left[\frac{\sqrt{n}(\rho_0 - \rho)}{(1 - \rho^2)} - \frac{c(1 - \rho_0^2)}{(1 - \rho^2)} < \frac{\sqrt{n}}{(1 - \rho^2)}(R_n - \rho) \leq \frac{\sqrt{n}(\rho_0 - \rho)}{(1 - \rho^2)} + \frac{c(1 - \rho_0^2)}{(1 - \rho^2)}\right]$$

$$= 1 - \Phi\left(\frac{\sqrt{n}(\rho_0 - \rho)}{(1 - \rho^2)} + \frac{c(1 - \rho_0^2)}{(1 - \rho^2)}\right) + \Phi\left(\frac{\sqrt{n}(\rho_0 - \rho)}{(1 - \rho^2)} - \frac{c(1 - \rho_0^2)}{(1 - \rho^2)}\right).$$

Using the following R code, we compute the empirical levels of significance and the power functions.

```
rho_0=0.3;mu = c(0,0); sig=matrix(c(1,rho_0,rho_0,1),nrow=2)
n = 80; nsim = 1500; R = u = v = mle = c(); library(mvtnorm)
g=function(a)
{
term=a^3-a^2*v1-a*(1-u1)-v1
return(term)
}
dg=function(a)
{
term=3*a^2-a*2*v1 + u1-1
return(term)
}
for(i in 1:nsim)
{
set.seed(i)
x = rmvnorm(n,mu,sig)
R[i] = cor(x)[1,2]
u[i] = sum((x[,1]^2+x[,2]^2))/n
v[i] = sum((x[,1]*x[,2]))/n
}
m=5;e=matrix(nrow=m,ncol=nsim)
for(i in 1:nsim)
{
 e[1,i]=v[i]
v1 = v[i]
u1 = u[i]
j = 1; diff = 1
while(diff > 10^(-4))
{
e[j+1,i]= e[j,i]-g(e[j,i])/dg(e[j,i])
diff=abs(e[j+1,i]-e[j,i])
j=j+1
}
mle[i]=e[j,i]
}
summary(mle); tn = sn  = pft = pfs = a = d = e = f = c()
alpha = 0.05; b = qnorm(1-alpha/2); b
for(i in 1:nsim)
{
 tn[i] = sqrt(n*(1+rho_0^2))*(mle[i]-rho_0)/(1-rho_0^2)
 sn[i] = sqrt(n)*(R[i]-rho_0)/(1-rho_0^2)
}
d0=data.frame(tn, sn); d1=round(d0,4); View(d1);head(d1); tail(d1)
elost = length(which(abs(tn)>b))/nsim; elost
eloss = length(which(abs(sn)>b))/nsim; eloss
### power function
```

```
rho = seq(-0.36,0.86,0.03); lrho = length(rho); lrho;
for(i in 1:lrho)
{
a[i] = sqrt(n*(1+rho[i]^2))*(rho_0-rho[i])/(1-rho[i]^2)
d[i] = sqrt((1+rho[i]^2))*(1-rho_0^2)/((1-rho[i]^2)*sqrt(1+rho_0^2))
e[i] = sqrt(n)*(rho_0-rho[i])/(1-rho[i]^2)
f[i] = (1-rho_0^2)/(1-rho[i]^2)
pft[i] = 1 - pnorm(a[i]+b*d[i]) + pnorm(a[i]- b*d[i])
pfs[i] = 1 - pnorm(e[i]+b*f[i]) + pnorm(e[i]-b*f[i])
}
s=seq(-.4,.9,.1);u=c(0,0.05,seq(.1,.9,.1))
plot(rho,pft,"o",lty=1,main="Power Function",
                    xlab = expression(paste("rho")),
ylab=expression(paste("Power")),col="dark blue",
xlim=c(-.4,1.3),xaxt="n",ylim=c(0,1),yaxt="n")
lines(rho,pfs,"o",lty=2,col="green")
abline(h=alpha,col="purple")
axis(1,at = s,las = 1,cex.axis = 0.7)
axis(2,at = u,las = 1,cex.axis = 0.7)
legend("topleft",legend=c("Tn","Sn"),lty=c(1,2),
        col=c("dark blue","green"),
        title=expression(paste("Test Statistic")))
umle=a+b*d; uR=e+b*f; lmle=a-b*d; lR=e-b*f
d2=data.frame(umle,uR,lmle,lR);d3=round(d2,4);head(d3)
avmle=(1-rho^2)^2/(n*(1+rho^2));avR=(1-rho^2)^2/n
d4=data.frame(rho,avmle,avR);d5=round(d4,4);View(d5);
                    head(d5);tail(d5)
```

The values of both the test statistics are displayed by the `View(d1)` function, the first 6 and the last 6 can be obtained from the `head(d1)` and `tail(d1)` function respectively. The empirical level of significance is 0.054 and 0.053 corresponding to T_n and S_n respectively, which is very close to the given level of significance 0.05. From Fig. 5.2, we note that power functions of the tests based on both the test statistics are almost the same. At $\rho = 0.3$, the power functions of both the test statistics have value 0.05, as the cut-off is determined corresponding to $\alpha = 0.05$. As the values of ρ shift away from $\rho = 0.3$ in either direction, the power functions increase as expected. From data frame $d3$, we note that 95% confidence intervals for ρ based on T_n and S_n are almost the same. Further, the behavior of the power functions of both the test statistics is the same. Such a similarity is in view of the fact that the large sample distributions of both $\hat{\rho}_n$ and of R_n are normal, with the same mean and their approximate variances are almost the same, for large n.

These are displayed in Table 5.1, for some values of ρ and for $n = 80$. These can be observed for all values of ρ with `View(d5)` function.   □

Power Function

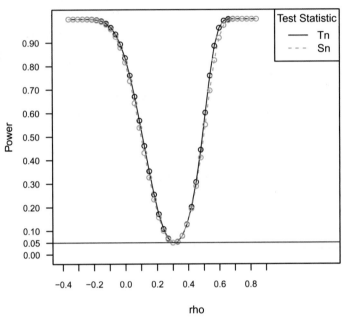

Fig. 5.2 Bivariate normal $N_2(0, 0, 1, 1, \rho)$ distribution: power function

Table 5.1 $N_2(0, 0, 1, 1, \rho)$ distribution: approximate variances of $\hat{\rho}_n$ and R_n

ρ	A. Variance of MLE	A. Variance of R
-0.36	0.0084	0.0095
-0.33	0.0090	0.0099
-0.30	0.0095	0.0104
-0.27	0.0100	0.0107
-0.24	0.0105	0.0111
-0.21	0.0109	0.0114

In the following examples we discuss how to carry out likelihood ratio test procedures using R. As discussed in Sect. 5.2, the first step in this procedure is to find the maximum likelihood estimator in the null and the entire parameter space. Thus, all the tools presented in Sect. 4.5 can be utilized to find the maximum likelihood estimator.

✐ Example 5.3.3

Suppose X has a truncated exponential distribution, truncated above at 5. Its probability density function is given by

$$f(x, \theta) = \frac{\theta e^{-x\theta}}{1 - e^{-5\theta}}, \qquad 0 < x < 5, \quad \theta > 0.$$

Suppose we want to test $H_0 : \theta = \theta_0 = 2$ against the alternative $H_1 : \theta \neq \theta_0$ on the basis of a random sample $\underline{X} \equiv \{X_1, X_2, \ldots X_n\}$ generated from this distribution. To generate a random sample, we use a probability integral transformation. Thus, invert the distribution function $F_X(x)$ of X. It is given by

$$F_X(x) = \begin{cases} 0, & \text{if} & x < 0 \\ \frac{(1-e^{-x\theta})}{(1-e^{-5\theta})}, & \text{if} & 0 \le x < 5 \\ 1 & \text{if} & x \ge 5 . \end{cases}$$

Thus for $0 \le x < 5$,

$$\frac{1 - e^{-x\theta}}{1 - e^{-5\theta}} = u \quad \Rightarrow \quad x = -\frac{1}{\theta} \log\{1 - u(1 - e^{-5\theta})\} .$$

The log-likelihood function of θ corresponding to a random sample $\underline{X}$ is given by

$$\log L_n(\theta | \underline{X}) = n \log \theta - \theta \sum_{i=1}^{n} X_i - n \log(1 - e^{-5\theta}) .$$

Hence, the likelihood equation is

$$-\frac{5ne^{-5\theta}}{1 - e^{-5\theta}} + \frac{n}{\theta} - \sum x_i = 0.$$

Following is the R code to carry out the likelihood ratio test procedure.

```
th0 = 2; n = 300; theta = 1.5 ##
    value of the parameter to generate sample
set.seed(12); u = runif(n,0,1)
x = -(1/theta)*log(1-u*(1-exp(-5*theta))); x1 = sum(x)
logL = function(par)
{
term = n*log(par)- x1*par - n*log(1-exp(-5*par))
return(term)
}
dLogL = function(par)
{
term = -5*n*exp(-5*par)/(1-exp(-5*par)) + n/par - x1
return(term)
}
t = 1/mean(x)
mle = uniroot(dLogL,c(t-1,5))$root; mle
LRTS = 2*(logL(mle) - logL(th0)); LRTS
b = qchisq(.95, 1); b
L1 = 2*(logL(mle) - logL(1.9)); L1
L2 = 2*(logL(mle) - logL(1.8)); L2
L3 = 2*(logL(mle) - logL(1.5)); L3
```

The maximum likelihood estimator of θ is 1.5496, when the sample is generated under $\theta = 1.5$ For testing $H_0 : \theta = 2$ against the alternative $H_1 : \theta \neq 2$, $-2\log(\lambda(\underline{X})) = 20.9513$. It is larger than the 95% quantile 3.8415 of χ_1^2 distribution. Thus, H_0 cannot be accepted. It is to be noted that H_0 rejected even if $\theta_0 = 1.9$ (L1 = 13.1195) and if $\theta_0 = 1.8$ (L2 = 6.9406) but not rejected if $\theta_0 = 1.5$ (L3 = 0.3054). The value of $-2\log(\lambda(\underline{X}))$ is small if $\theta_0 = 1.5$, which is expected as the random sample is drawn under $\theta = 1.5$. □

Following examples present the R code for the likelihood ratio test procedure in a multiparameter setup and to verify that the asymptotic null distribution of $-2\log \lambda(\underline{X})$ is χ^2.

✍ Example 5.3.4

Suppose $X \sim Poi(\theta_1)$ & $Y \sim Poi(\theta_2)$. In Example 5.2.7 we have developed a likelihood ratio test procedure to test $H_0 : \theta_1 = \theta_2 = \theta$, say, against $H_1 : \theta_1 \neq \theta_2$. The maximum likelihood estimator of θ is $\hat{\theta}_{n_1+n_2} = \left(\sum_{i=1}^{n_1} X_i + \sum_{i=1}^{n_2} Y_i\right) / (n_1 + n_2)$. In the entire parameter space, the maximum likelihood estimator of θ_1 is $\hat{\theta}_{1n_1} = \overline{X}_{n_1}$ and the maximum likelihood estimator of θ_2 is $\hat{\theta}_{2n_2} = \overline{Y}_{n_2}$. Following is the R code to test the hypothesis H_0 against H_1, using likelihood ratio test procedure, on the basis of random samples generated from two Poisson distributions. We adopt two approaches, in the first approach we use the results of Example 5.2.7, where we have obtained the maximum likelihood estimators of the parameters in the null and entire space and also the formula for the likelihood ratio test statistic. In the second approach, we do not use these formulae but use the R code to obtain the log-likelihood, and hence the maximum likelihood estimates of the parameters in the null and entire space. This approach is quite general and can be used for any other distribution.

```
#### Approach 1
m = 300; n = 275 # Sample size for X and Y respectively
th1 = 2; th2 = 2.5; set.seed(30)
x = rpois(m,th1); y = rpois(n,th2); mx = mean(x);
                                    my = mean(y); mx; my
m1 = (m*mx + n*my)/(m+n); m1 ## Estimate in null space
LRTS = -2*(-m1*(m+n)+(m*mx+n*my)*log(m1)+ m*mx-m*mx*log(mx)
                                        + n*my-n*my*log(my))
LRTS; b = qchisq(.95,1); b; p = 1-pchisq(LRTS,1); p
### Approach 2 without using formulae
logLx = function(a)
{
LL = 0
for(i in 1:m)
{
LL = LL + log(dpois(x[i],a))
}
```

```
return(LL)
}
logLy = function(a)
{
LL = 0
for(i in 1:n)
{
LL = LL + log(dpois(y[i],a))
}
return(LL)
}
th = seq(0.1,3,0.01); length(th)
Lx = logLx(th); bx = which.max(Lx); mlex = th[bx]; mlex
Ly = logLy(th); by = which.max(Ly); mley = th[by]; mley
Lnull = logLx(th)+logLy(th); b = which.max(Lnull);
                        mlenull = th[b]; mlenull
logL=logLx(mlex) + logLy(mley)
logLnull=logLx(mlenull) + logLy(mlenull)
Tn=-2*(logLnull-logL); Tn ### LRTS
b = qchisq(.95,1); b; p = 1-pchisq(Tn,1); p
```

From the output, we note that maximum likelihood estimates of the parameters in the null and entire space are the same in both the approaches. These are $\hat{\theta}_{1n_1} = 1.93$, $\hat{\theta}_{2n_2} = 2.37$ and $\hat{\theta}_{n_1+n_2} = 2.14$. Corresponding to the generated samples, $-2 \log \lambda(\underline{X}, \underline{Y}) = 12.60$, by both the approaches. It is larger than the 95% quantile 3.8415 of the χ_1^2 distribution and hence the null hypothesis cannot be accepted. The p-value comes out to be 0.00038, which is very small.   □

In the following example we use the same setup of Example 5.3.4 and verify the result that under the null setup $-2 \log(\lambda(\underline{X}, \underline{Y})) \sim \chi_1^2$ distribution. For simplicity, we adopt the first approach to derive the likelihood ratio test statistic.

✎ Example 5.3.5

Suppose $X \sim Poi(\theta_1)$ and $Y \sim Poi(\theta_2)$. In Examples 5.2.7 and 5.3.4 we have developed a likelihood ratio test procedure to test $H_0 : \theta_1 = \theta_2$ against $H_1 : \theta_1 \neq \theta_2$. In this example, we verify the result that under the null setup $-2 \log(\lambda(\underline{X}, \underline{Y})) \sim \chi_1^2$ distribution, graphically and by the chi-square test.

```
m = 550; n = 520; th1 = 2; th2 = 2; nsim = 1200
LRTS = c() ## Vector to store values of -2log
                    (likelihood ratio test statistic)
for(i in 1:nsim)
{
set.seed(i)
x=rpois(m,th1)
y=rpois(n,th2)
mx=mean(x); my = mean(y)
m1=(m*mx + n*my)/(m+n) ## Estimate under null space
LRTS[i]=-2*(-m1*(m+n)+(m*mx+n*my)*log(m1)
                    + m*mx-m*mx*log(mx)+ n*my-n*my*log(my))
}
summary(LRTS)
O=hist(LRTS,plot=FALSE)$counts; sum(O)
bk=hist(LRTS,plot=FALSE)$breaks; bk
M=max(bk); ep=c()
for(i in 1:(length(bk)-1))
{
ep[i] = pchisq(bk[i+1],1)-pchisq(bk[i],1)
}
a=1-sum(ep); a; ep1=c(ep, a); ef=sum(O)*ep1; O1=c(O,0)
d=data.frame(O1,round(ef,2)); d
ts=sum((O1-ef)^2/ef); ts
df=length(ef)-1; df; b=qchisq(.95,df); b
p1=1-pchisq(ts,df); p1; chisq.test(O1,p=ep1)
t = seq(0,M,.2); lt = length(t); lt
LRT1 = sort(LRTS); cdf = c()
for(j in 1:lt)
{
cdf[j] = length(which(LRT1<=t[j]))/nsim
}
dft = pchisq(t,1); pdft = dchisq(t,1); max(abs(cdf-dft))
par(mfrow = c(2,1))
hist(LRTS,freq=FALSE,main="Histogram of LRTS",
                ylim=c(0,1),xlim=c(0,M),
                xlab = expression(paste("LRTS")),col="light blue")
lines(t,pdft,"o",pch=20,lwd=1,col="dark blue")
plot(t,cdf,"l",main="Empirical Distribution Function of LRTS",
ylim=c(0,1), xlim=c(0,M),xlab = expression(paste("LRTS")),
                col="dark blue",lwd=2)
lines(t,dft,"o",pch=20,lwd=1,col="light blue")
```

The p-value 0.3672 of the chi-square test, which is the same as that of the `chisq.test` function, supports the theoretical result that the asymptotic null distribution of the likelihood ratio test statistic is χ_1^2. From Fig. 5.3, we observe that the graph of the histogram of the values of $-2 \log(\lambda(\underline{X}, \underline{Y}))$ and the probability density function of χ_1^2 distribution imposed on it, supports the result that $-2 \log(\lambda(\underline{X}, \underline{Y}))$ has asymptotically chi-square distribution. In the second plot,

Histogram of LRTS

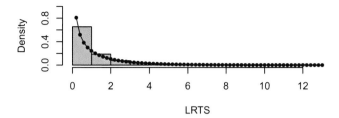

Empirical Distribution Function of LRTS

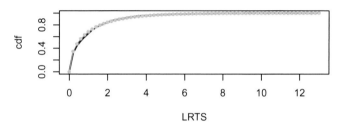

Fig. 5.3 Asymptotic null distribution of likelihood ratio test statistic

we have imposed the graph of distribution function of χ_1^2 distribution on the empirical distribution function of the values of $-2\log(\lambda(\underline{X}, \underline{Y}))$. It also supports that the asymptotic null distribution of $-2\log(\lambda(\underline{X}, \underline{Y}))$ is χ_1^2.   □

✏ Example 5.3.6

Suppose X and Y are independent random variables having negative binomial distribution with parameters (k, p_1) and (k, p_2) respectively. The probability mass functions of X and Y are given by

$$P[X = x] = \binom{x + k - 1}{k - 1} p_1^k (1 - p_1)^x \;\&\; P[Y = y]$$

$$= \binom{y + k - 1}{k - 1} p_2^k (1 - p_2)^y, \quad x, y = 0, 1, 2 \ldots.$$

We assume k is known and $0 < p_1, p_2 < 1$. In this example we derive a likelihood ratio test procedure to test $H_0 : p_1 = 2p_2$ against $H_1 : p_1 \neq 2p_2$ when $k = 10$, on the basis of random samples of sizes m and n from the distributions of X and Y respectively. We adopt the second approach of Example 5.3.4, so that we need not derive any formulae for the maximum likelihood estimators and the likelihood ratio test statistic. Following is the R code for the same.

```
m = 100;   n = 110 # Sample size for X and Y respectively
p1 = .4; p2 = .21; set.seed(30)
x = rnbinom(m, 10, p1); y = rnbinom(n, 10, p2);
logLx = function(a)
{
LL = 0
for(i in 1:m)
{
LL = LL + log(dnbinom(x[i],10,a))
}
return(LL)
}
logLy = function(a)
{
LL = 0
for(i in 1:n)
{
LL = LL + log(dnbinom(y[i],10,a))
}
return(LL)
}
p = seq(0.001,.8,0.01); length(p)
Lx = logLx(p); bx = which.max(Lx); mlex = p[bx]; mlex
Ly = logLy(p); by = which.max(Ly); mley = p[by]; mley
Lnull = logLx(2*p)+logLy(p); b = which.max(Lnull);
                         mlenull = p[b]; mlenull
logL = logLx(mlex) + logLy(mley)
logLnull = logLx(2*mlenull) + logLy(mlenull)
Tn = -2*(logLnull-logL);Tn
b = qchisq(.95,1); b; pv = 1-pchisq(Tn,1); pv
```

For the simulated data with $p_1 = 0.4$ and $p_2 = 0.21$, the value of the likelihood ratio test statistic is 0.9117 and corresponding p-value is 0.3397. Thus, data do support the null setup. If $p_1 = 0.4$ and $p_2 = 0.25$, the value of the likelihood ratio test statistic is 23.1605 and corresponding p-value is $1.4902e^{-6}$, which is almost 0. Hence, data do not support the null setup, which seems to be reasonable. □

5.4   Conceptual Exercises

5.4.1 Suppose $\{X_1, X_2, \ldots, X_n\}$ is a random sample from a Laplace distribution with probability density function $f(x, \theta)$ given by

$$f(x, \theta) = \frac{1}{2\theta} \exp\left\{-\frac{|x|}{\theta}\right\}, \quad x \in \mathbb{R}, \quad \theta > 0.$$

Derive a large sample test procedure for testing $H_0 : \theta = \theta_0$ against the alternative $H_1 : \theta < \theta_0$, when the test statistic is a function of $\sum_{i=1}^{n} |X_i|/n$. Find the power function.

5.4.2 Suppose $\{X_1, X_2, \ldots, X_n\}$ is a random sample from a Cauchy $C(\theta, 1)$ distribution. Derive a large sample test procedure for testing $H_0 : \theta = 0$ against the alternative $H_1 : \theta \neq 0$, when the test statistic is a function of sample median and when it is a function of the maximum likelihood estimator of θ. Obtain the power function in each case.

5.4.3 Suppose $\{X_1, X_2, \ldots, X_n\}$ is a random sample from an exponential distribution with scale parameter 1 and location parameter θ. Develop a test procedure to test $H_0 : \theta = \theta_0$ against the alternative $H_1 : \theta \neq \theta_0$, based on sample mean and based on the maximum likelihood estimator of θ. Find the power function in both the cases.

5.4.4 Suppose $\{X_1, X_2, \ldots, X_n\}$ is a random sample from a normal distribution with mean μ and variance σ^2. Develop a test procedure to test the hypothesis $H_0 : P[X < a] = p_0$ against $H_1 : P[X < a] \neq p_0$.

5.4.5 Suppose $\underline{X} \equiv \{X_1, X_2, \ldots, X_n\}$ is a random sample from a Poisson distribution with parameter θ. Obtain the likelihood ratio test to test $H_0 : P[X = 0] = 1/3$ against the alternative $H_1 : P[X = 0] \neq 1/3$.

5.4.6 Suppose $\{X_1, X_2, \ldots, X_n\}$ is a random sample from a lognormal distribution with parameters μ and σ^2. Derive a large sample test procedure to test $H_0 : \mu = \mu_0, \ \sigma^2 = \sigma_0^2$ against $H_1 : \mu \neq \mu_0, \ \sigma^2 \neq \sigma_0^2$.

5.4.7 Suppose $\{X_1, X_2, \ldots, X_n\}$ is a random sample from a Gamma $G(\alpha, \lambda)$ distribution. Derive a large sample test procedure to test $H_0 : \alpha = \alpha_0$ against $H_1 : \alpha \neq \alpha_0$ when λ is (i) known and (ii) unknown.

5.4.8 Suppose $\{X_1, X_2, \ldots, X_n\}$ is a random sample from a geometric distribution with probability mass function $p_x = P[X = x] = \theta(1 - \theta)^x$, $x = 0, 1, 2, \ldots$. Obtain the likelihood ratio test to test $H_0 : P[X = 0] = 0.3$ against the alternative $H_1 : P[X = 0] \neq 0.3$.

5.4.9 Suppose $\{X_1, X_2, \ldots, X_n\}$ are independent and identically distributed random variables with following probability mass function.

$$P[X_1 = 1] = (1 - \theta)/2, \quad P[X_1 = 2] = 1/2, \quad P[X_1 = 3] = \theta/2, \quad 0 < \theta < 1.$$

Derive a likelihood ratio test procedure to test $H_0 : \theta = 1/2$ against the alternative $H_1 : \theta \neq 1/2$. Explain how the critical region will be decided if the sample size is large.

5.4.10 Suppose X has a discrete distribution with possible values $1, 2, 3, 4$ with probabilities $(2 - \theta_1)/4, \theta_1/4, \theta_2/4, (2 - \theta_2)/4$ respectively. On the basis of a random sample from the distribution of X, derive a likelihood ratio test procedure to test $H_0 : \theta_1 = 1/3, \ \theta_2 = 2/3$ against the alternative $H_1 : \theta \neq 1/3, \theta_2 \neq 2/3$.

5.4.11 Suppose $\{X_1, X_2, \ldots, X_{n_1}\}$ is a random sample from a Bernoulli $B(1, p_1)$ distribution and $\{Y_1, Y_2, \ldots, Y_{n_2}\}$ is a random sample from a Bernoulli $B(1, p_2)$ distribution. Suppose X and Y are independent random variables.

Derive a likelihood ratio test procedure for testing $H_0 : p_1 = p_2$ against the alternative $H_1 : p_1 \neq p_2$.

5.5   Computational Exercises

Solve the following exercises by simulation using R

5.5.1 For Exercise 5.4.2 and Exercise 5.4.3, find the false positive rate and plot the power function for both the test procedures. Comment on which test procedure is better. Why? (Hint: Use code similar to that of Example 5.3.1.)

5.5.2 Suppose $\{X_1, X_2, \ldots, X_n\}$ is a random sample from a normal $N(\mu_1, \sigma_1^2)$ distribution and $\{Y_1, Y_2, \ldots, Y_n\}$ is an independent random sample from a normal $N(\mu_2, \sigma_2^2)$ distribution. Derive the likelihood ratio test procedure to test $H_0 : \mu_1 = \mu_2$ & $\sigma_1^2 = \sigma_2^2$ against the alternative H_1 where at least one of the statements $\mu_1 = \mu_2$ & $\sigma_1^2 = \sigma_2^2$ is not true. (Hint: Use code similar to that of Example 5.3.4.)

5.5.3 Suppose $\{X_1, X_2, \ldots, X_n\}$ is a random sample from an exponential distribution with mean $1/\theta$ and $\{Y_1, Y_2, \ldots, Y_n\}$ is an independent random sample from an exponential distribution with mean $1/\mu$. Derive the likelihood ratio test procedure to test $H_0 : \mu = 2\theta$ against the alternative $H_1 : \mu \neq 2\theta$. (Hint: Use code similar to that of Example 5.3.6.)

5.5.4 Verify the results of Exercise 5.4.7

5.5.5 Verify the results of Exercise 5.4.8

5.5.6 Verify the results of Exercise 5.4.9

5.5.7 Verify the results of Exercise 5.4.10

References

1. Wilks, S. S. (1938). The large sample distribution of the likelihood ratio test for testing composite hypotheses. *Annals of Mathematical Statistics, 9*, 60–62.
2. Ferguson, T. S. (1996). *A course in large sample theory*. London: Chapman and Hall.

Goodness of Fit Test and Tests for Contingency Tables

6

Contents

— **Learning Objectives** After going through this chapter, the readers should be able

- to understand the asymptotic optimality properties of the estimators of the parameters of a multinomial distribution
- to perform large sample test procedures associated with the multinomial distribution, such as test for the validity of the model
- to establish the association among the likelihood ratio test, the goodness of fit test and Karl Pearson's chi-square test for the multinomial distribution
- to explore the concepts of Wald's test, the score test and their relation with the likelihood ratio test
- to figure out the link between the score test and Karl Pearson's chi-square test, when the underlying probability model is multinomial

- to realize the relation among the likelihood ratio test, the score test and Karl Pearson's chi-square test in the context of tests for contingency tables
- to verify the consistency of a test procedure
- to validate the results associated with the large sample tests using R
- to execute the tests for the goodness of fit and tests for contingency tables using R

6.1   Introduction

The goodness of fit test and tests related to the contingency tables are the likelihood ratio tests, when the underlying probability model is a multinomial distribution. To elaborate on the tests for the contingency tables, suppose n objects are classified according to two criteria or two attributes A and B. Suppose the criterion A has r levels $\{A_1, A_2, \ldots, A_r\}$ and the criterion B has s levels $\{B_1, B_2, \ldots, B_s\}$. In other words, n objects are classified according to two categorical response variables X and Y, X with r categories and Y with s categories. Such a classification gives rise to a two-way table with r rows and s columns, displaying the values of n_{ij}, n_{ij} being the number of objects having i-th level of attribute A and j-th level of attribute B, $n_{ij} \geq 0 \ \forall \ i = 1, 2, \ldots, r, j = 1, 2, \ldots, s$ and $\sum_{i=1}^{r} \sum_{j=1}^{s} n_{ij} = n$. The count n_{ij} is known as a frequency of (i, j)-th cell. A two-way table of this type is known as an $r \times s$ contingency table, a term introduced by Karl Pearson in 1904. It is also known as a cross-classification table. The rs distinct classifications are known as the cells of the contingency table.

As an illustration of a two-way contingency table, suppose in a public opinion survey, n individuals are classified according to their political opinion, represented by attribute A say, with three levels such as (i) for the proposal, (ii) against the proposal and (iii) indifferent to the proposal. Another attribute B represents their gender, with two levels, male and female. When n individuals are classified according to these two criteria, we get a 3×2 contingency table. If individuals are further classified according to a third attribute C say, such as education or annual income, we get a three-way contingency table. In a variety of research fields, one is interested in the analysis of a contingency table. For example, a medical research worker might suspect that a certain environmental condition is responsible for a spread of some disease. He will then classify a group of individuals, according to whether they had the disease or not and whether they were exposed to the environmental conditions or not. Such a classification would lead to a 2×2 contingency table. An industrial engineer would use a contingency table to discover, whether two kinds of defects in a manufactured product were due to the same or different underlying causes.

On the basis of a contingency table, it is of interest to investigate the possible relationships among the criteria for classification. For example, one may suspect that males and females will react to a political proposal in different ways. Such conjectures can be tested using likelihood ratio tests. The underlying probability model is a multinomial distribution in $r \times s$ cells with cell probabilities as p_{ij} for the (i, j)-th cell, $p_{ij} > 0 \; \forall \; i = 1, 2, \ldots, r, \; j = 1, 2, \ldots, s$ and $\sum_{i=1}^{r} \sum_{j=1}^{s} p_{ij} = 1$. Thus, p_{ij} denotes the probability that a randomly selected object has i-th level of attribute A and j-th level of attribute B. The joint probability mass function of n_{ij}, $i = 1, 2, \ldots, r, \; j = 1, 2, \ldots, s$ is then given by

$$
n! \prod_{i=1}^{r} \prod_{j=1}^{s} p_{ij}^{n_{ij}} \left(\prod_{i=1}^{r} \prod_{j=1}^{s} n_{ij}! \right)^{-1}, \quad n_{ij} \geq 0 \quad \& \quad \sum_{i=1}^{r} \sum_{j=1}^{s} n_{ij} = 1.
$$

A possible relationship can be expressed in the form of a null hypothesis and one can use a likelihood ratio test procedure, discussed in the previous chapter, to test the hypothesis of interest. To carry out the likelihood ratio test, we need to find the maximum likelihood estimators of p_{ij} corresponding to the observed data n_{ij}, $i = 1, 2, \ldots, r, \; j = 1, 2, \ldots, s$ in the entire parameter space and in the null space. We discuss the tests for contingency tables in Sect. 6.5.

Another most frequently used test is a test for goodness of fit. It is also a likelihood ratio test, when the underlying probability model is a multinomial distribution. Suppose the observed data on a characteristic X under study are arranged in the form of (y_i, f_i) or $([x_i, x_{i+1}), f_i)$, $i = 1, 2, \ldots, k$ and are usually presented in a table with k rows and 2 columns. Here y_i is the possible value of the discrete random variable X and f_i denotes the frequency of y_i in a random sample of size n from the distribution of X, $i = 1, 2, \ldots, k$ where $n = \sum_{i=1}^{k} f_i$. A random sample of size n from a continuous distribution is usually classified as $([x_i, x_{i+1}), f_i)$, $i = 1, 2, \ldots, k$, where f_i denotes the number of observations in the class interval $[x_i, x_{i+1})$. On the basis of these observed data, we wish to test whether the data are from a specific distribution. More precisely, suppose X is discrete random variable with probability mass function (x_i, p_{x_i}), $i = 1, 2, \ldots$. A random sample from this distribution is classified as (y_i, f_i), $i = 1, 2, \ldots, k$, frequency of the last class corresponds to all observations beyond y_k if the support of X is not finite. On the basis of the observed data, it is of interest to test whether the data are generated from a specific probability distribution (x_i, p_{x_i}), $i = 1, 2, \ldots$, for example binomial or Poisson, etc. In such a setup, the underlying model can be taken as a multinomial distribution in k cells and with cell probabilities as p_{x_i}, which may depend on a parameter θ, either real or vector valued.

If the observed data are of the form $([x_i, x_{i+1}), f_i) \, i = 1, 2, \ldots, k$, then also the underlying model can be taken as a multinomial distribution in k cells and with cell probabilities $p_i = P[X \in [x_i, x_{i+1})]$, $i = 1, 2, \ldots, k$, which may depend on a parameter θ, either real or vector valued, and are according to some specific distribution, such as normal and exponential. To test the hypothesis that the data are from a specific distribution, we use likelihood ratio test, where in the null setup cell probabilities p_{x_i} are according to a particular distribution and we have to find their maximum likelihood estimators. In the entire setup, we consider cell probabilities as p_i, $i = 1, 2, \ldots, k$ with the only condition that $\sum_{i=1}^{k} p_i = 1$ and we have to find their maximum likelihood estimators corresponding to the given data. Thus, the goodness of fit tests are basically tests for validation of the model. These are studied in Sect. 6.3. Tests associated with a multinomial distribution are also useful for validating certain models in genetic experiments, as described in Example 6.2.3 in the next section. In such a setup, under the proposed model, the cell probabilities depend on parameters and we can use the likelihood ratio test for validating the model. The model to be validated is such that $p_i = p_i(\theta)$, where θ is either a real parameter or a vector parameter, $i = 1, 2, \ldots, k$, as in the goodness of fit test. Thus in null setup, we have $p_i = p_i(\theta)$, $i = 1, 2, \ldots, k$. The parameter θ is such that $p_i(\theta) > 0$ for all i and $\sum_{i=1}^{k} p_i(\theta) = 1$. The likelihood ratio test is useful to validate such a model.

The likelihood ratio test is related to Wald's test and the score test. Sect. 6.4 is concerned with Wald's test procedure and the score test procedure and their relation with the likelihood ratio test procedure. An important finding of this section is the link between a score test statistic and Karl Pearson's chi-square test statistic. It is proved that while testing any hypothesis about the parameters of a multinomial distribution, these two statistics are identical. Section 6.6 presents a brief introduction to the concept of a consistency of a test procedure. Section 6.7 elaborates on the application of R software to validate various results proved in earlier sections, to perform the goodness of test procedures and tests for contingency tables.

Thus, in this chapter most of the tests are associated with a multinomial distribution. Hence, the next section is devoted to the in-depth study of a multinomial distribution. Illustrative examples are given to show that in some cases a multinomial distribution belongs to an exponential family or a Cramér family and we can use all the results established in Chap. 4. We focus on the maximum likelihood estimation of cell probabilities p and the asymptotic properties of the maximum likelihood estimator of p. Some tests associated with multinomial distribution are also developed.

6.2   Multinomial Distribution and Associated Tests

In an experiment E with only two outcomes, the random variable X is defined as $X = 1$, if outcome of interest occurs and $X = 0$ otherwise. For example, in Bernoulli trials there are only two outcomes, the outcome of interest is labeled as success and the other as failure. The distribution of X is Bernoulli $B(1, \theta)$ distribution where θ is interpreted as the probability of success. Multinomial distribution is an extension of the Bernoulli distribution and it is a suitable model for the experiment with k outcomes. Suppose an experiment E results in k outcomes $\{O_1, O_2, \ldots, O_k\}$ and p_i denotes the probability that E results in outcome O_i, $p_i > 0$, $i = 1, 2, \ldots, k$ and $\sum_{i=1}^{k} p_i = 1$. A random variable Y_i, for $i = 1, 2, \ldots, k$, is defined as

$$Y_i = \begin{cases} 1, & \text{if outcome is } O_i, \\ 0, & \text{otherwise}, \end{cases}$$

where $\sum_{i=1}^{k} Y_i = 1$, implying that Y_i will be 1 only for one i and will be 0 for all other i. The joint probability mass function $p(y_1, y_2, \ldots, y_k)$ of $\{Y_1, Y_2, \ldots, Y_k\}$ is

given by

$$p(y_1, y_2, \ldots, y_k) = \prod_{i=1}^{k} p_i^{y_i}, \quad y_i = 0, 1, \quad \sum_{i=1}^{k} y_i = 1 \quad \& \quad \sum_{i=1}^{k} p_i = 1 .$$

The joint distribution of $\{Y_1, Y_2, \ldots, Y_k\}$ is a multinomial distribution in k cells, specified by the cell probabilities $p_1, p_2, \ldots, p_k$ with $p_i > 0$ for all $i = 1, 2, \ldots, k$ and $\sum_{i=1}^{k} p_i = 1$. Thus, the parameter space of the multinomial distribution in k cells is given by

$$\Theta = \left\{ \underline{p} = (p_1, p_2, \ldots, p_k) | p_i > 0 \ \forall \ i = 1, 2, \ldots, k \ \& \ \sum_{i=1}^{k} p_i = 1 \right\} .$$

It is clear that $Y_i \sim B(1, p_i)$ with $E(Y_i) = p_i$ and $Var(Y_i) = p_i(1 - p_i)$, $i = 1, 2, \ldots, k$. If $Y_i = 1$ then $Y_j = 0$ for any $j \neq i$, implies that $Y_i Y_j = 0$ with probability 1. As a consequence,

$$E(Y_i Y_j) = 0 \ \& \ Cov(Y_i, Y_j) = -E(Y_i)E(Y_j) = -p_i p_j \ \forall \ i \neq j = 1, 2, \ldots, k .$$

Suppose a random vector $\underline{Y}$ is defined as $\underline{Y} = (Y_1, Y_2, \ldots, Y_k)'$, then $E(\underline{Y}) = (p_1, p_2, \ldots, p_k)'$ and the dispersion matrix $D = [\sigma_{ij}]$ of $\underline{Y}$ is of order $k \times k$, where $\sigma_{ii} = p_i(1 - p_i)$ and $\sigma_{ij} = -p_i p_j$. If we add all the columns of D, then for each row, the sum comes out to be 0 implying that D is a singular matrix. This is a consequence of the fact that $\{Y_1, Y_2, \ldots, Y_k\}$ are linearly related by the identity $\sum_{i=1}^{k} Y_i = 1$. Hence, we consider the random vector $\underline{Y}$ as $\underline{Y} = (Y_1, Y_2, \ldots, Y_{k-1})'$, then $E(\underline{Y}) = (p_1, p_2, \ldots, p_{k-1})'$ and dispersion matrix $D = [\sigma_{ij}]$ of $\underline{Y}$ is of order $(k-1) \times (k-1)$, where $\sigma_{ii} = p_i(1 - p_i)$ and $\sigma_{ij} = -p_i p_j$. D can be shown to be positive definite. For example for $k = 3$,

$$D = \begin{pmatrix} p_1(1 - p_1) & -p_1 p_2 \\ -p_1 p_2 & p_2(1 - p_2). \end{pmatrix}$$

The first principal minor is $p_1(1 - p_1) > 0$ and the second principal minor, which is the same as the determinant $|D|$ is

$$|D| = p_1(1 - p_1)p_2(1 - p_2) - p_1^2 p_2^2 = p_1 p_2(1 - p_1 - p_2) = p_1 p_2 p_3 > 0$$

implying that D is a positive definite matrix. In general, $|D| = p_1 p_2 \ldots p_k$.

The joint probability mass function $p(\underline{y}) = p(y_1, y_2, \ldots, y_{k-1})$ of $\{Y_1, Y_2, \ldots, Y_{k-1}\}$ is given by

$$p(\underline{y}) = \prod_{i=1}^{k} p_i^{y_i} \quad \Leftrightarrow \quad \log p(\underline{y}) = \sum_{i=1}^{k} y_i \log p_i,$$

$$\text{where} \ p_k = 1 - \sum_{i=1}^{k-1} p_i \ \& \ y_k = 1 - \sum_{i=1}^{k-1} y_i .$$

In further discussion we assert that the joint distribution of $\{Y_1, Y_2, \ldots, Y_{k-1}\}$ is a multinomial distribution in k cells with cell probabilities p_i, $i = 1, 2, \ldots, k$ with $\sum_{i=1}^{k} p_i = 1$. We now find the information matrix $I(\underline{p}) = [I_{ij}(\underline{p})]$ of order $(k-1) \times (k-1)$ where $\underline{p} = (p_1, p_2, \ldots, p_{k-1})$. From $\log p(\underline{y}) = \sum_{i=1}^{k} y_i \log p_i$, we have

$$\frac{\partial}{\partial p_i} \log p(\underline{y}) = \frac{y_i}{p_i} - \frac{y_k}{p_k}, \quad \frac{\partial^2}{\partial p_i^2} \log p(\underline{y}) = -\frac{y_i}{p_i^2} - \frac{y_k}{p_k^2}, \quad i = 1, 2, \ldots, k-1$$

$$\frac{\partial^2}{\partial p_j \partial p_i} \log p(\underline{y}) = -\frac{y_k}{p_k^2}, \qquad j \neq i = 1, 2, \ldots, k-1.$$

Hence,

$$I_{ii}(\underline{p}) = E\left(-\frac{\partial^2}{\partial p_i^2} \log p(\underline{Y})\right) = E\left(\frac{Y_i}{p_i^2} + \frac{Y_k}{p_k^2}\right) = \frac{1}{p_i} + \frac{1}{p_k} \quad \text{and}$$

$$I_{ij}(\underline{p}) = E\left(-\frac{\partial^2}{\partial p_j \partial p_i} \log p(\underline{Y})\right) = E\left(\frac{Y_k}{p_k^2}\right) = \frac{1}{p_k}.$$

It can be shown that $D = I^{-1}(\underline{p})$, which can be easily verified for $k = 3$ and $k = 4$. For $k = 3$,

$$D^{-1} = \frac{1}{p_1 p_2 p_3} \begin{pmatrix} p_2(1-p_2) & p_1 p_2 \\ p_1 p_2 & p_1(1-p_1) \end{pmatrix} = \begin{pmatrix} \frac{1-p_2}{p_1 p_3} & \frac{1}{p_3} \\ \frac{1}{p_3} & \frac{1-p_1}{p_2 p_3} \end{pmatrix}$$

$$= \begin{pmatrix} \frac{1}{p_1} + \frac{1}{p_3} & \frac{1}{p_3} \\ \frac{1}{p_3} & \frac{1}{p_2} + \frac{1}{p_3} \end{pmatrix},$$

which is the same as $I(p_1, p_2)$. For $k = 4$,

$$D^{-1} = \frac{1}{p_1 p_2 p_3 p_4}$$
$$\begin{pmatrix} p_2 p_3(1-p_2-p_3) & p_1 p_2 p_3 & p_1 p_2 p_3 \\ p_1 p_2 p_3 & p_1 p_3(1-p_1-p_3) & p_1 p_2 p_3 \\ p_1 p_2 p_3 & p_1 p_2 p_3 & p_1 p_2(1-p_1-p_2) \end{pmatrix}$$

which simplifies to

$$D^{-1} = \begin{pmatrix} \frac{1}{p_1} + \frac{1}{p_4} & \frac{1}{p_4} & \frac{1}{p_4} \\ \frac{1}{p_4} & \frac{1}{p_2} + \frac{1}{p_4} & \frac{1}{p_4} \\ \frac{1}{p_4} & \frac{1}{p_4} & \frac{1}{p_3} + \frac{1}{p_4} \end{pmatrix} = I(p_1, p_2, p_3).$$

Suppose the experiment E is repeated under identical conditions, so that the probabilities of k outcomes remain the same for each repetition. We further assume that the repetitions are independent of each other. Suppose X_i denotes the frequency of occurrence of O_i in n repetitions. Then $X_i = \sum_{r=1}^{n} Y_{ir}$, $i = 1, 2, \ldots, k$ with $X_k = n - \sum_{i=1}^{k-1} X_i$. If a random vector $\underline{X}$ is defined as $\underline{X} = (X_1, X_2, \ldots, X_{k-1})'$, then $\underline{X} = \sum_{r=1}^{n} \underline{Y}_r$, where $\underline{Y}_r$ is the r-th observation on $\underline{Y} = (Y_1, Y_2, \ldots, Y_{k-1})'$. The likelihood of $\underline{p} = (p_1, p_2, \ldots, p_{k-1})$ given the data $\underline{U} = \{\underline{Y}_1, \underline{Y}_2, \ldots, \underline{Y}_n\} \equiv (X_1, X_2, \ldots, X_k)'$ is given by

$$L_n(\underline{p}|\underline{U}) = \prod_{r=1}^{n} \prod_{i=1}^{k} p_i^{Y_{ir}} = \prod_{i=1}^{k} p_i^{X_i}$$

$$\Leftrightarrow \quad \log L_n(\underline{p}|\underline{U}) = \sum_{r=1}^{n} \sum_{i=1}^{k} Y_{ir} \log p_i = \sum_{i=1}^{k} X_i \log p_i \; .$$

From the likelihood, it is clear that $\{X_1, X_2, \ldots, X_{k-1}\}$ is a sufficient statistic for the family. To find the maximum likelihood estimator of $\underline{p} = (p_1, p_2, \ldots, p_{k-1})$, we need to maximize $\log L_n(\underline{p}|\underline{U})$ with respect to the variation in $\underline{p}$ subject to the condition that $\sum_{i=1}^{k} p_i = 1$. Hence, we use Lagrange's method of multipliers. Suppose a function g defined as

$$g(p_1, p_2, \ldots, p_k, \lambda) = \sum_{i=1}^{k} X_i \log p_i + \lambda \left(\sum_{i=1}^{k} p_i - 1 \right) .$$

Solving the system of equations given by

$$\frac{\partial}{\partial p_i} g(p_1, p_2, \ldots, p_k, \lambda) = \frac{X_i}{p_i} - \lambda = 0, \quad i = 1, 2, \ldots, k \quad \text{and}$$

$$\frac{\partial}{\partial \lambda} g(p_1, p_2, \ldots, p_k, \lambda) = \sum_{i=1}^{k} p_i - 1 = 0$$

and using the condition that $\sum_{i=1}^{k} X_i = n$, we get the maximum likelihood estimator $\hat{p}_{in}$ of p_i as

$$\hat{p}_{in} = \frac{X_i}{n} = \frac{1}{n} \sum_{r=1}^{n} Y_{ir}, \quad i = 1, 2, \ldots, k.$$

It is to be noted that $\sum_{i=1}^{k} \hat{p}_{in} = 1$, as expected.

We now discuss the properties of $\hat{p}_{in}$. Observe that

$$X_i \sim B(n, p_i) \quad \Rightarrow \quad E(X_i) = np_i, \; Var(X_i) = np_i(1 - p_i), i = 1, 2, \ldots, k$$

$$\& \; Cov(X_i, X_j) = Cov\left(\sum_{r=1}^{n} Y_{ir}, \sum_{s=1}^{n} Y_{js}\right) = \sum_{r=1}^{n}\sum_{s=1}^{n} Cov(Y_{ir}, Y_{js})$$

$$= \sum_{r=1}^{n} Cov(Y_{ir}, Y_{jr}) = -np_i p_j,$$

as $Cov(Y_{ir}, Y_{js}) = 0 \; \forall \; r \neq s$, $\{\underline{Y}_1, \underline{Y}_2, \ldots, \underline{Y}_n\}$ being independent random vectors. Thus,

$$E(\hat{p}_{in}) = E(X_i/n) = p_i \quad \& \quad Var(\hat{p}_{in}) = p_i(1 - p_i)/n \to 0 \;\; \text{as} \;\; n \to \infty .$$

Thus, $\hat{p}_{in}$ is an unbiased estimator of p_i and its variance converges to 0 and hence it is an MSE consistent estimator of p_i. The consistency also follows from the WLLN as

$$\hat{p}_{in} = \frac{X_i}{n} = \frac{1}{n}\sum_{r=1}^{n} Y_{ir} \xrightarrow{P} E(Y_{ir}) = p_i , \quad i = 1, 2, \ldots, k - 1.$$

Since joint consistency is equivalent to the marginal consistency, $\underline{\hat{p}}_n = (\hat{p}_{1n}, \hat{p}_{2n}, \ldots, \hat{p}_{(k-1)n})'$ is consistent for $\underline{p}$. To examine whether it is a CAN estimator of $\underline{p}$ we note that

$$\underline{\hat{p}}_n = \left(\frac{X_1}{n}, \frac{X_2}{n}, \ldots, \frac{X_{k-1}}{n}\right)'$$

$$= \left(\frac{1}{n}\sum_{r=1}^{n} Y_{1r}, \frac{1}{n}\sum_{r=1}^{n} Y_{2r}, \ldots, \frac{1}{n}\sum_{r=1}^{n} Y_{(k-1)r}\right)' = \underline{\overline{Y}}_n ,$$

where $\underline{Y} = (Y_1, Y_2, \ldots, Y_{k-1})'$. Repeating the experiment E, n times gives a random sample $\{\underline{Y}_1, \underline{Y}_2, \ldots, \underline{Y}_n\}$ of size n from the distribution of $\underline{Y}$. Further, $E(\underline{Y}) = \underline{p}$ and dispersion matrix of $\underline{Y}$ is $D = I^{-1}(\underline{p})$ as specified above. It is positive definite. Hence, by the multivariate CLT,

$$\sqrt{n}(\underline{\overline{Y}}_n - \underline{p}) \xrightarrow{L} \underline{Z}_1 \sim N_{k-1}(\underline{0}, D) \quad \Leftrightarrow \quad \sqrt{n}(\underline{\hat{p}}_n - \underline{p}) \xrightarrow{L} \underline{Z}_1 \sim N_{k-1}(\underline{0}, D).$$

Thus, $\underline{\hat{p}}_n$ is a CAN estimator of $\underline{p}$, with approximate dispersion matrix $D/n = I^{-1}(\underline{p})/n$.

✏ Remark 6.2.1

It is to be noted that for each $i = 1, 2, \ldots, k - 1$,

$$\sqrt{n}(\underline{\hat{p}}_n - \underline{p}) \xrightarrow{L} \underline{Z}_1 \sim N_{k-1}(\underline{0}, I^{-1}(\underline{p}))$$

$$\Rightarrow \sqrt{n}(\hat{p}_{in} - p_i) \xrightarrow{L} Z_1 \sim N(0, p_i(1 - p_i)),$$

where $\hat{p}_{in} = X_i/n$. Suppose $k = 3$. Thus, the maximum likelihood estimator of $(p_1, p_2)'$ derived from the trinomial model and the two marginal univariate models are the same. Further, the large sample distribution of the first component in $\sqrt{n}(\hat{\underline{p}}_n - \underline{p})$ is the same as the large sample distribution of $\sqrt{n}(\hat{p}_{1n} - p_1)$, derived from the univariate Bernoulli model. However,

$$I_{1,1}(p_1, p_2) = \frac{1}{p_1} + \frac{1}{1 - p_1 - p_2} > I(p_1) = \frac{1}{p_1} + \frac{1}{1 - p_1}$$

$$\& \ I_{2,2}(p_1, p_2) = \frac{1}{p_2} + \frac{1}{1 - p_1 - p_2} > I(p_2) = \frac{1}{p_2} + \frac{1}{1 - p_2}$$

as observed in Exercise 4.6.7 for bivariate normal $N_2(\mu_1, \mu_2, 1, 1, \rho)$ model, where $\rho \neq 0$ is known. In Example 4.3.4, in which we discussed bivariate normal $N_2(0, 0, \sigma_1^2, \sigma_2^2, \rho)$ model with known $\rho \neq 0$, and in Example 4.3.5, where we discussed a bivariate Cauchy model with location parameters θ_1 and θ_2, we observed similar feature in information functions for bivariate and univariate models. In both these examples, the maximum likelihood estimator of the parameters derived from the bivariate and univariate models are different and the approximate variances in the asymptotic distributions are also different. Both these models belong to a two-parameter Cramér family while both the trinomial distribution and bivariate normal $N_2(\mu_1, \mu_2, 1, 1, \rho)$ distribution, belong to a two-parameter exponential family.

From the asymptotic distribution of $\hat{\underline{p}}_n$, we observe that

$$\sqrt{n}(\hat{\underline{p}}_n - \underline{p}) \overset{L}{\to} \underline{Z}_1 \sim N_{k-1}(\underline{0}, I^{-1}(\underline{p}))$$

$$\Rightarrow Q_n = n(\hat{\underline{p}}_n - \underline{p})' I(\underline{p})(\hat{\underline{p}}_n - \underline{p}) \overset{L}{\to} \chi_{k-1}^2.$$

Further, $\hat{\underline{p}}_n$ is consistent for $\underline{p}$ and each element of matrix $I(\underline{p})$ is a continuous function of $\underline{p}$, hence $I(\hat{\underline{p}}_n) \overset{P}{\to} I(\underline{p})$. Suppose W_n is defined as $W_n = n(\hat{\underline{p}}_n - \underline{p})' I(\hat{\underline{p}}_n)(\hat{\underline{p}}_n - \underline{p})$. Then,

$$W_n - Q_n = n(\hat{\underline{p}}_n - \underline{p})'(I(\hat{\underline{p}}_n) - I(\underline{p}))(\hat{\underline{p}}_n - \underline{p}) \overset{P}{\to} 0 \ \Rightarrow \ W_n \overset{L}{\to} \chi_{k-1}^2.$$

All these results are useful to obtain the asymptotic null distribution of a test statistic for certain tests of interest.

Following examples illustrate the large sample test procedures associated with a multinomial distribution.

✒ Example 6.2.1

Suppose we have a trinomial distribution with cell probabilities $p_1(\theta) = (1 + \theta)/2$ and $p_2(\theta) = p_3(\theta)$, $0 < \theta < 1$. Hence,

$p_2(\theta) = p_3(\theta) = (1 - \theta)/4$. The likelihood of θ corresponding to the data $\underline{X} \equiv \{X_1, X_2, X_3\}$ is given by

$$L_n(\theta|\underline{X}) = \left(\frac{1+\theta}{2}\right)^{X_1} \left(\frac{1-\theta}{4}\right)^{X_2+X_3}$$

$$= \left(\frac{1+\theta}{2}\right)^{X_1} \left(\frac{1-\theta}{4}\right)^{n-X_1}, \quad \text{as } X_1 + X_2 + X_3 = n,$$

where X_i denotes the frequency of cell i, $i = 1, 2, 3$. The likelihood is a differentiable function of θ, hence the likelihood equation and its solution $\hat{\theta}_n$ are given by

$$\frac{X_1}{1+\theta} - \frac{n-X_1}{1-\theta} = 0 \quad \Rightarrow \quad \hat{\theta}_n = \frac{2X_1 - n}{n} = 2\frac{X_1}{n} - 1.$$

The second order derivative
$\frac{\partial^2}{\partial \theta^2} \log L_n(\theta|\underline{X}) = -X_1/(1+\theta)^2 - (n-X_1)/(1-\theta)^2 < 0$ a.s. which implies that the maximum likelihood estimator $\hat{\theta}_n$ of θ is given by
$\hat{\theta}_n = (2X_1 - n)/n = 2X_1/n - 1$. To develop a procedure to test the null hypothesis $H_0 : p_1(\theta) = 0.6$ against the alternative $H_1 : p_1(\theta) < 0.6$, we obtain the large sample distribution of $\hat{\theta}_n$ with a suitable normalization. It is well-known that for the trinomial distribution, the cell frequency X_i has a binomial distribution, $i = 1, 2$. Thus, $X_1 \sim B(n, p_1(\theta))$ distribution. Further, X_1/n can be expressed as $X_1/n = \sum_{j=1}^{n} Y_{1j}/n$, where Y_{1j}, $j = 1, 2, \ldots, n$ are independent and identically distributed random variables each having Bernoulli $B(1, p_1(\theta))$ distribution with mean $p_1(\theta)$ and variance $0 < p_1(\theta)(1 - p_1(\theta)) < \infty$. Hence by the WLLN,

$$\frac{X_1}{n} = \frac{1}{n} \sum_{j=1}^{n} Y_{1j} \overset{P_\theta}{\to} p_1(\theta) \quad \Rightarrow \quad \hat{\theta}_n = \frac{2X_1 - n}{n} = 2\frac{X_1}{n} - 1 \overset{P_\theta}{\to} \theta$$

and by the CLT

$$\sqrt{n}\left(\frac{1}{n} \sum_{j=1}^{n} Y_{1j} - p_1(\theta)\right) \overset{L}{\to} Z_1 \sim N\left(0, p_1(\theta)(1 - p_1(\theta))\right)$$

$$\Rightarrow \quad \sqrt{n}(\hat{\theta}_n - \theta) \overset{L}{\to} Z_1 \sim N(0, 1 - \theta^2).$$

Further, to test the null hypothesis
$H_0 : p_1(\theta) = 0.6 \Leftrightarrow (1+\theta)/2 = 0.6 \Leftrightarrow \theta = \theta_0 = 0.2$ against the alternative
$H_1 : \theta < 0.2$, we propose two test statistics S_n and W_n as

$$S_n = \sqrt{n}(\hat{\theta}_n - 0.2)/\sqrt{1 - \theta_0^2} \quad \& \quad W_n = \sqrt{n}(\hat{\theta}_n - 0.2)/\sqrt{1 - \hat{\theta}_n^2}.$$

Under H_0, $S_n \overset{L}{\to} Z \sim N(0, 1)$ and by Slutsky's theorem $W_n \overset{L}{\to} Z \sim N(0, 1)$. H_0 is rejected if $S_n < c$ or $W_n < c$ where c is such that $P_{\theta_0}[S_n < c] = \alpha$. If $\alpha = 0.05$ then $c = -1.65$. If $X_1 = 33$ and $n = 50$, $S_n = 0.8660$ & $W_n = 0.8956$. Both are larger than c and hence we conclude that data do not have sufficient evidence to reject H_0. $\qquad\square$

When the cell probabilities are indexed by the parameter θ, real or vector valued, then in some cases a multinomial distribution belongs to an exponential family and all the results established for an exponential family in Chap. 4 are applicable. In some other cases it belongs to a Cramér family and results valid for a Cramér family proved in Chap. 4 are applicable. Following examples illustrate these applications. In the next example, we show that a multinomial distribution, as a model for a genetic experiment, belongs to a one-parameter exponential family and how this fact is used to derive a large sample test procedure to test some hypothesis.

☞ Example 6.2.2

According to a certain genetic model, the probabilities for three outcomes are θ^2, $2\theta(1 - \theta)$ and $(1 - \theta)^2$, $0 < \theta < 1$. The appropriate probability distribution for this model is a multinomial distribution in three cells. Suppose the random vector $(Y_1, Y_2)'$ has trinomial distribution with cell probabilities θ^2, $2\theta(1 - \theta)$ and $(1 - \theta)^2$, then its joint probability mass function is given by

$$P_\theta[Y_1 = y_1, Y_2 = y_2] = (\theta^2)^{y_1} (2\theta(1 - \theta))^{y_2} ((1 - \theta)^2)^{1 - y_1 - y_2}$$
$$= 2^{y_2} \theta^{2y_1 + y_2} (1 - \theta)^{2 - 2y_1 - y_2}.$$

To examine whether it belong to a one-parameter exponential family, observe that the joint probability mass function can be expressed as

$$\log P_\theta[Y_1 = y_1, Y_2 = y_2] = y_2 \log 2 + (2y_1 + y_2) \log \theta$$
$$+ (2 - 2y_1 - y_2) \log(1 - \theta)$$
$$= y_2 \log 2 + (2y_1 + y_2)(\log \theta - \log(1 - \theta))$$
$$+ 2 \log(1 - \theta)$$
$$= U(\theta) K(y_1, y_2) + V(\theta) + W(y_1, y_2),$$

where $U(\theta) = \log \theta - \log(1 - \theta)$, $K(y_1, y_2) = 2y_1 + y_2$, $V(\theta) = 2 \log(1 - \theta)$ and $W(y_1, y_2) = y_2 \log 2$. Thus, the probability law of $(Y_1, Y_2)'$ is expressible in the form required for a one-parameter exponential family. The support of the probability mass function is $\{(0, 0), (0, 1), (1, 0)\}$ and it is free from θ, the parameter space is $(0, 1)$, which is an open set. Further, $U'(\theta) = \frac{1}{\theta} + \frac{1}{1-\theta} = \frac{1}{\theta(1-\theta)} \neq 0$. $K(y_1, y_2)$ and 1 are linearly independent because in the identity $a + b(2y_1 + y_2) = 0$, if $y_1 = y_2 = 0$, then $a = 0$, if further in the identity $b(2y_1 + y_2) = 0$, if either $y_1 = 0$ and $y_2 = 1$ or $y_1 = 1$ and $y_2 = 0$, then $b = 0$. Thus, all the requirements of a one-parameter exponential family are satisfied and hence the joint probability mass function of $(Y_1, Y_2)'$

belongs to a one-parameter exponential family. To find the maximum likelihood estimator of θ, the likelihood of θ corresponding to the data $\underline{X} \equiv \{X_1, X_2, X_3\}$ is given by,

$$L_n(\theta|\underline{X}) = 2^{X_2}(\theta)^{2X_1+X_2}(1-\theta)^{2n-2X_1-X_2}, \quad X_1 + X_2 + X_3 = n,$$

where X_i is the frequency of i-th cell in the sample, $i = 1, 2, 3$. The likelihood is a differentiable function of θ, hence the likelihood equation and its solution $\hat{\theta}_n$ are given by

$$\frac{2X_1 + X_2}{\theta} - \frac{2n - 2X_1 - X_2}{1-\theta} = 0 \quad \Rightarrow \quad \hat{\theta}_n = \frac{2X_1 + X_2}{2n}.$$

Further, the second order derivative

$$\frac{\partial^2}{\partial\theta^2} \log L_n(\theta|\underline{X}) = -(2X_1 + X_2)/\theta^2 - (2n - 2X_1 - X_2)/(1-\theta)^2 < 0,$$
$$\forall \; \theta \in (0, 1) \; \& \; \forall \; X_1 \; \& \; X_2,$$

since $2n - 2X_1 - X_2 = X_2 + 2X_3$. Hence, $\hat{\theta}_n$ is the maximum likelihood estimator of θ. Next we find a moment estimator $\tilde{\theta}_n$ of θ based on a sufficient statistic. From the likelihood we observe that $2X_1 + X_2$ is a sufficient statistic. Thus, $\tilde{\theta}_n$ is solution of the equation

$$\frac{2X_1 + X_2}{n} = \frac{1}{n}\sum_{r=1}^{n}(2Y_{1r} + Y_{2r}) = E(2Y_1 + Y_2)$$

$$= 2\theta^2 + 2\theta(1-\theta) = 2\theta \quad \Rightarrow \quad \tilde{\theta}_n = \frac{2X_1 + X_2}{2n}.$$

Since the distribution of $(Y_1, Y_2)'$ belongs to a one-parameter exponential family, by Theorem 4.2.1, $\hat{\theta}_n = \tilde{\theta}_n$ is CAN for θ with approximate variance $1/nI(\theta)$. Now,

$$nI(\theta) = E_\theta\left(-\frac{\partial^2}{\partial\theta^2}\log L_n(\theta|\underline{X})\right)$$

$$= \frac{n(2\theta^2 + 2\theta(1-\theta))}{\theta^2} + \frac{2n - 2n\theta^2 - 2n\theta(1-\theta)}{(1-\theta)^2} = \frac{2n}{\theta(1-\theta)}.$$

Thus, $\hat{\theta}_n$ is a CAN estimator of θ with approximate variance $\theta(1-\theta)/2n$. We use these results to develop a test procedure for testing $H_0 : \theta = \theta_0$ against the alternative $H_1 : \theta \neq \theta_0$. Suppose two test statistics S_n and W_n are defined as

$$S_n = \sqrt{2n/(\theta_0(1-\theta_0))}\,(\hat{\theta}_n - \theta_0) \quad \& \quad W_n = \sqrt{2n/(\hat{\theta}_n(1-\hat{\theta}_n))}\,(\hat{\theta}_n - \theta_0).$$

Then under H_0, for large n, $S_n \sim N(0, 1)$. By Slutsky's theorem, under H_0, for large n, $W_n \sim N(0, 1)$ and hence $H_0 : \theta = \theta_0$ is rejected against the alternative $H_1 : \theta \neq \theta_0$ at level of significance α if $|S_n| > c$ or if $|W_n| > c$ where $c = a_{1-\alpha/2}$. One more approach to test $H_0 : \theta = \theta_0$ against the alternative $H_1 : \theta \neq \theta_0$ is the likelihood ratio test. The likelihood ratio test statistic $\lambda(\underline{X})$ is given by

$$\lambda(\underline{X}) = \frac{\sup_{\Theta_0} L_n(\theta|\underline{X})}{\sup_{\Theta} L_n(\theta|\underline{X})} = \frac{(\theta_0)^{2X_1+X_2}(1-\theta_0)^{2n-2X_1-X_2}}{(\hat{\theta}_n)^{2X_1+X_2}(1-\hat{\theta}_n)^{2n-2X_1-X_2}}.$$

It is difficult to get the finite sample distribution of the likelihood ratio. Hence, we use its asymptotic distribution. From Theorem 5.2.1 for large n under H_0, $-2\log\lambda(\underline{X}) \sim \chi_1^2$ distribution. For large n, H_0 is rejected if $-2\log\lambda(\underline{X}) > c$, where c is such that the size of the test is α and is determined using the χ_1^2 distribution. Thus, $c = \chi_{1,1-\alpha}^2$. For this model, we can obtain variance of $\hat{\theta}_n$ for finite n. It is given by

$$Var(\hat{\theta}_n) = Var\left(\frac{2X_1 + X_2}{2n}\right)$$

$$= \frac{1}{4n^2}\left(4Var(X_1) + Var(X_2) + 4Cov(X_1, X_2)\right)$$

$$= \frac{1}{4n^2}\left(4n\theta^2(1-\theta^2) + n2\theta(1-\theta)(1-2\theta(1-\theta)) - 4n\theta^2 2\theta(1-\theta)\right)$$

$$= \frac{\theta(1-\theta)}{2n},$$

which is the same as $1/nI(\theta)$. Further,

$$E(\hat{\theta}_n) = E\left(\frac{2X_1 + X_2}{2n}\right) = \frac{1}{2n}(2n\theta^2 + 2n\theta(1-\theta)) = \theta.$$

Thus, $\hat{\theta}_n$ is an unbiased estimator of θ. Since its variance attains the Cramér-Rao lower bound for the variance, it is MVBUE of θ. It is to be noted that $\hat{\theta}_n$ is a function of sufficient statistic $2X_1 + X_2$. Further, the dimension of the sufficient statistic and the dimension of the parameter space is the same, hence $2X_1 + X_2$ is a complete statistic. Thus, $\hat{\theta}_n$ is a function of complete and sufficient statistic, it is unbiased estimator of θ and hence by the Rao-Blackwell theorem and the Lehmann-Scheffe theorem it is UMVUE of θ. □

📣 Remark 6.2.2

In Example 6.2.2, we have discussed three test procedures to test $H_0 : \theta = \theta_0$ against the alternative $H_1 : \theta \neq \theta_0$. Tests based on S_n and W_n reject H_0 when $|S_n| > c$ and $|W_n| > c$ respectively. These rejection regions are also equivalent to $S_n^2 > c$ and $W_n^2 > c$ and the asymptotic distribution of both S_n^2 and W_n^2 is χ_1^2. Thus, the likelihood ratio test procedure, test procedure based on S_n^2 and the one based on W_n^2 for testing $H_0 : \theta = \theta_0$ against the alternative $H_1 : \theta \neq \theta_0$ are equivalent. This result holds true in a general setup and we discuss it in Sect. 6.4. The test based on S_n^2 is a score test and the one based on W_n^2 is Wald's test.

The following example shows that a multinomial distribution, when cell probabilities are indexed by a parameter θ, does not belong to a one-parameter exponential family, but belongs to a Cramér family and hence all the results established for a Cramér family are useful to develop large sample test procedures.

✎ Example 6.2.3

Fisher has used a multinomial distribution in 4 cells to analyze Carver's data on two varieties of maize classified as starchy versus sugary and further cross classified with the color as green and white, (refer to Kale and Muralidharan [1]). The cell probabilities depend on a parameter $\theta \in (0, 1)$, which is known as a linkage factor. The cell probabilities are given in Table 6.1, along with the observed frequencies as obtained in the experiment. An appropriate probability model for the outcome of the given genetic experiment is a multinomial distribution in four cells with cell probabilities $(2 + \theta)/4$, $(1 - \theta)/4$, $(1 - \theta)/4$ and $\theta/4$. Suppose $(Y_1, Y_2, Y_3)'$ has multinomial distribution in four cells with these cell probabilities. Then its joint probability mass function $P_\theta[Y_1 = y_1, Y_2 = y_2, Y_3 = y_3] = p_\theta(y_1, y_2, y_3)$ is given by

$$p_\theta(y_1, y_2, y_3) = ((2 + \theta)/4)^{y_1} ((1 - \theta)/4)^{y_2 + y_3} (\theta/4)^{y_4}$$
$$= 4^{-1}(2 + \theta)^{y_1} (1 - \theta)^{y_2 + y_3} \theta^{y_4},$$

$y_i = 0, 1$ for $i = 1, 2, 3$ and $\sum_{i=1}^{4} y_i = 1$. With $y_4 = 1 - y_1 - y_2 - y_3$, the joint probability mass function can be expressed as

$$\log p_\theta(y_1, y_2, y_3) = -\log 4 + y_1 \log(2 + \theta) + (y_2 + y_3) \log(1 - \theta) + y_4 \log \theta$$
$$= -\log 4 + y_1 \log \frac{2 + \theta}{\theta} + (y_2 + y_3) \log \frac{1 - \theta}{\theta} + \log \theta,$$

However, it cannot be expressed as $U(\theta)K(y_1, y_2, y_3) + V(\theta) + W(y_1, y_2, y_3)$, in particular we cannot get the term $U(\theta)K(y_1, y_2, y_3)$. Thus, the probability law of $(Y_1, Y_2, Y_3)'$ does not belong to a one-parameter exponential family. We

Table 6.1 Carver's data: two varieties of maize

Category	Starchy green	Starchy white	Sugary green	Sugary white
Cell probabilities	$\frac{2+\theta}{4}$	$\frac{1-\theta}{4}$	$\frac{1-\theta}{4}$	$\frac{\theta}{4}$
Observed frequencies	1977	906	904	32

now examine whether it belongs to a Cramér family. It is to be noted that the
parameter space Θ is $(0, 1)$ and it is an open set. The set of possible values of
Y_i for $i = 1, 2, 3$ is $\{0, 1\}$, which is free from θ. Further, all the cell probabilities
as a function of θ are analytical functions of θ and hence are differentiable any
number of times. We have

$$\frac{\partial}{\partial \theta} \log p_\theta(y_1, y_2, y_3) = \frac{y_1}{2 + \theta} - \frac{y_2 + y_3}{1 - \theta} + \frac{y_4}{\theta}$$

$$\frac{\partial^2}{\partial \theta^2} \log p_\theta(y_1, y_2, y_3) = -\frac{y_1}{(2 + \theta)^2} - \frac{y_2 + y_3}{(1 - \theta)^2} - \frac{y_4}{\theta^2}$$

$$\frac{\partial^3}{\partial \theta^3} \log p_\theta(y_1, y_2, y_3) = \frac{2y_1}{(2 + \theta)^3} - \frac{2(y_2 + y_3)}{(1 - \theta)^3} + \frac{2y_4}{\theta^3}.$$

All these partial derivatives exist for θ in any open subset of Θ. Further, if θ_0 is
a true parameter value, then for $\theta \in N_\delta(\theta_0)$, we have
$\theta_0 - \delta < \theta < \theta_0 + \delta \iff \frac{1}{\theta_0 + \delta} < \frac{1}{\theta} < \frac{1}{\theta_0 - \delta}$. Hence,

$$\left| \frac{\partial^3}{\partial \theta^3} \log p_\theta(y_1, y_2, y_3) \right| \leq \left| \frac{2y_1}{(2 + \theta)^3} \right| + \left| \frac{2(y_2 + y_3)}{(1 - \theta)^3} \right| + \left| \frac{2y_4}{\theta^3} \right|$$

$$= \frac{2y_1}{(2 + \theta)^3} + \frac{2(y_2 + y_3)}{(1 - \theta)^3} + \frac{2y_4}{\theta^3}$$

$$\leq \frac{2y_1}{(2 + \theta_0 - \delta)^3} + \frac{2(y_2 + y_3)}{(1 - \theta_0 - \delta)^3} + \frac{2y_4}{(\theta_0 - \delta)^3}$$

$$= M(y_1, y_2, y_3) \quad \text{say}$$

and

$$E_\theta(M(Y_1, Y_2, Y_3)) = \frac{2 + \theta}{2(2 + \theta_0 - \delta)^3} + \frac{1 - \theta}{(1 - \theta_0 - \delta)^3} + \frac{\theta}{2(\theta_0 - \delta)^3} < \infty.$$

Further observe that

$$E\left(\frac{\partial}{\partial \theta} \log p_\theta(Y_1, Y_2, Y_3) \right) = E\left(\frac{Y_1}{2 + \theta} - \frac{Y_2 + Y_3}{1 - \theta} + \frac{Y_4}{\theta} \right)$$

$$= \frac{2 + \theta}{4(2 + \theta)} - \frac{2(1 - \theta)}{4(1 - \theta)} + \frac{\theta}{4\theta} = 0.$$

Using the fact that $E(Y_i Y_j) = 0$ for all $i \neq j = 1, 2, 3$ we get

$$E\left(\frac{\partial}{\partial \theta} \log p_\theta(Y_1, Y_2, Y_3) \right)^2 = E\left(\frac{Y_1}{2 + \theta} - \frac{Y_2 + Y_3}{1 - \theta} + \frac{Y_4}{\theta} \right)^2$$

$$= \frac{2 + \theta}{4(2 + \theta)^2} + \frac{2(1 - \theta)}{4(1 - \theta)^2} + \frac{\theta}{4\theta^2}$$

$$= \frac{(2\theta + 1)}{2\theta(1 - \theta)(2 + \theta)}.$$

$$\& \ E\left(-\frac{\partial^2}{\partial\theta^2}\log p_\theta(Y_1, Y_2, Y_3)\right) = E\left(\frac{Y_1}{(2+\theta)^2} + \frac{Y_2+Y_3}{(1-\theta)^2} + \frac{Y_4}{\theta^2}\right)$$

$$= \frac{2+\theta}{4(2+\theta)^2} + \frac{2(1-\theta)}{4(1-\theta)^2} + \frac{\theta}{4\theta^2}$$

$$= \frac{(2\theta+1)}{2\theta(1-\theta)(2+\theta)}$$

$$= E\left(\frac{\partial}{\partial\theta}\log p_\theta(Y_1, Y_2, Y_3)\right)^2 .$$

Hence, the information function $I(\theta)$ is given by

$$I(\theta) = E\left(\frac{\partial}{\partial\theta}\log p_\theta(Y_1, Y_2, Y_3)\right)^2 = E\left(-\frac{\partial^2}{\partial\theta^2}\log p_\theta(Y_1, Y_2, Y_3)\right)$$

$$= \frac{(2\theta+1)}{2\theta(1-\theta)(2+\theta)} .$$

Thus, the probability law $p_\theta(y_1, y_2, y_3)$ satisfies all the Cramér regularity conditions and hence it belongs to a Cramér family. As a consequence, when observed cell frequencies are X_1, X_2, X_3, X_4 with $X_1 + X_2 + X_3 + X_4 = n$, for large n, the maximum likelihood estimator $\hat{\theta}_n$ of θ exists and it is CAN with approximate variance $1/nI(\theta)$. To find the maximum likelihood estimator of θ, the likelihood of θ given data $\underline{X} = \{X_1, X_2, X_3, X_4\}$ is given by

$$L(\theta|\underline{X}) = \prod_{i=1}^{4}(p_i(\theta))^{X_i} = \left(\frac{2+\theta}{4}\right)^{X_1}\left(\frac{1-\theta}{4}\right)^{X_2+X_3}\left(\frac{\theta}{4}\right)^{X_4} .$$

It is a differentiable function of θ and hence the likelihood equation is given by

$$\frac{\partial}{\partial\theta}\log L(\theta|\underline{X}) = 0$$

$$\Rightarrow \frac{X_1}{2+\theta} - \frac{X_2+X_3}{1-\theta} + \frac{X_4}{\theta} = 0$$

$$\Rightarrow -\theta^2(X_1+X_2+X_3+X_4)$$

$$+ \theta(X_1 - 2X_2 - 2X_3 - X_4) + 2X_4 = 0$$

$$\Rightarrow n\theta^2 - (X_1 - 2X_2 - 2X_3 - X_4)\theta - 2X_4 = 0 .$$

Thus, the likelihood equation is a quadratic equation in θ and has two roots, which may be real or complex. Suppose
$g(\theta) = n\theta^2 - (X_1 - 2X_2 - 2X_3 - X_4)\theta - 2X_4 = a\theta^2 + b\theta + c$. Then
$b^2 - 4ac = (X_1 - 2X_2 - 2X_3 - X_4)^2 + 8nX_4 > 0$. Hence, both the roots of
$g(\theta) = 0$ are real. Further, product of two roots is $c/a = -2X_4/n < 0$ which

implies that one root is positive and the other is negative. To examine whether the positive root is in $(0, 1)$, observe that

$$g(0) = -2X_4 < 0 \quad \& \quad g(1) = n - (X_1 - 2X_2 - 2X_3 - X_4) - 2X_4$$
$$= 3(X_2 + X_3) > 0 .$$

Further, g is a continuous function, thus there exists $\theta \in (0, 1)$ such that $g(\theta) = 0$. Moreover,

$$\frac{\partial^2}{\partial \theta^2} \log L(\theta | \underline{X}) = -\frac{X_1}{(2 + \theta)^2} - \frac{X_2 + X_3}{(1 - \theta)^2} - \frac{X_4}{\theta^2} < 0$$

for any set of frequencies $\{X_1, X_2, X_3, X_4\}$ and hence the likelihood attains its maximum at both the roots. Hence, the positive root $\hat{\theta}_n$ of the quadratic equation $g(\theta) = 0$ is the maximum likelihood estimator of θ. Thus,

$$\hat{\theta}_n = \frac{(X_1 - 2X_2 - 2X_3 - X_4) + \sqrt{(X_1 - 2X_2 - 2X_3 - X_4)^2 + 8nX_4}}{2n} .$$

Since the distribution belongs to a Cramér family, $\hat{\theta}_n$ is a CAN estimator of θ with approximate variance $1/nI(\theta) = 2\theta(1 - \theta)(2 + \theta)/n(2\theta + 1)$. For the given data, the likelihood equation is $38190\theta^2 + 16750\theta - 64 = 0$ and $\hat{\theta}_n = 0.0354$. We use this result to develop a test procedure for testing $H_0 : \theta = 0.02$ against the alternative $H_1 : \theta \neq 0.02$. Suppose the test statistic S_n is given by

$$S_n = \sqrt{nI(\theta_0)}(\hat{\theta}_n - \theta_0) = \sqrt{nI(0.02)}(\hat{\theta}_n - \theta_0) = \sqrt{13.134n}(\hat{\theta}_n - \theta_0) ,$$

then under H_0, for large n, $S_n \sim N(0, 1)$ and hence $H_0 : \theta = 0.02$ is rejected against the alternative $H_1 : \theta \neq 0.02$ at level of significance α if $|S_n| > c$ where $c = a_{1-\alpha/2}$. For the given data, $S_n = 3.4486$ at $\alpha = 0.05$, $c = 1.95$, hence H_0 is rejected. As in Example 6.2.2, the test statistic W_n is defined as $W_n = \sqrt{nI(\hat{\theta}_n)}(\hat{\theta}_n - \theta_0)$, then by Slutsky's theorem, under H_0, for large n, $W_n \sim N(0, 1)$ and hence H_0 is rejected against the alternative H_1 at level of significance α if $|W_n| > c$ where $c = a_{1-\alpha/2}$. For the given data, $W_n = 2.6411 > c = 1.95$ and hence H_0 is rejected. One more approach to test H_0 against the alternative H_1 is the likelihood ratio test. For the given data $T_n = -2 \log \lambda(\underline{X}) = 9.2321$. H_0 is rejected if $T_n > c$ and c is decided using the asymptotic null distribution of T_n which is χ_1^2 and $c = 3.8414$. Thus, H_0 is rejected according to likelihood ratio test procedure.   □

The multinomial distribution in Example 6.2.3 belongs to a Cramér family. We now demonstrate that if $p_i(\theta)$ satisfy certain regularity conditions, then a multinomial distribution in k cells with cell probabilities $p_i(\theta)$, $i = 1, 2, \ldots k$ indexed by a parameter θ belongs to a Cramér family. We derive these conditions when θ is a real parameter. Suppose $\underline{Y} = (Y_1, Y_2, \ldots, Y_{k-1})'$ has multinomial distribution in k cells with cell probabilities $p_i(\theta)$, $i = 1, 2, \ldots k$ indexed by a parameter $\theta \in \Theta \subset \mathbb{R}$. The joint probability mass function $p_\theta(y_1, y_2, \ldots, y_{k-1}) = p_\theta(\underline{y})$ of $\underline{Y}$ is given by

$$p_\theta(\underline{y}) = P_\theta[Y_1 = y_1, Y_2 = y_2, \ldots, Y_{k-1} = y_{k-1}] = \prod_{i=1}^{k}(p_i(\theta))^{y_i},$$

$$y_i = 0, 1 \quad \& \quad \sum_{i=1}^{k} y_i = 1.$$

The parameter space Θ is such that $\sum_{i=1}^{k} p_i(\theta) = 1$. We assume it to be an open set. Further, the support of each Y_i is $\{0, 1\}$, which is free from θ. Suppose the partial derivatives of $p_i(\theta)$ exist up to order 3 for all $i = 1, 2, \ldots k$. From the joint probability mass function $p_\theta(\underline{y})$, we have

$$\frac{\partial}{\partial\theta} \log p_\theta(\underline{y}) = \sum_{i=1}^{k} y_i \frac{\partial}{\partial\theta} \log p_i(\theta), \quad \frac{\partial^2}{\partial\theta^2} \log p_\theta(\underline{y}) = \sum_{i=1}^{k} y_i \frac{\partial^2}{\partial\theta^2} \log p_i(\theta)$$

$$\& \quad \frac{\partial^3}{\partial\theta^3} \log p_\theta(\underline{y}) = \sum_{i=1}^{k} y_i \frac{\partial^3}{\partial\theta^3} \log p_i(\theta),$$

as being finite summation, derivatives can be taken inside the sum. Thus, if partial derivatives of $p_i(\theta)$, $i = 1, 2, \ldots k$ up to order 3 exist, then partial derivatives of $\log p_\theta(\underline{y})$ up to order 3 exist. Now,

$$E_\theta\left(\frac{\partial}{\partial\theta} \log p_\theta(\underline{Y})\right) = E_\theta\left(\sum_{i=1}^{k} Y_i \frac{\partial}{\partial\theta} \log p_i(\theta)\right)$$

$$= \sum_{i=1}^{k} p_i(\theta) \frac{\partial}{\partial\theta} \log p_i(\theta) = \sum_{i=1}^{k} \frac{\partial}{\partial\theta} p_i(\theta).$$

To find its value, we note that

$$\sum_{i=1}^{k} p_i(\theta) = 1 \implies \frac{\partial}{\partial\theta} \sum_{i=1}^{k} p_i(\theta) = 0 \implies \sum_{i=1}^{k} \frac{\partial}{\partial\theta} p_i(\theta) = 0.$$

Thus, $E_\theta\left(\frac{\partial}{\partial\theta} \log p_\theta(\underline{Y})\right) = 0$. Further, using the fact that $E(Y_i Y_j) = 0 \ \forall \ i \neq j = 1, 2, \ldots, k$, we have

$$E_\theta\left(\frac{\partial}{\partial\theta} \log p_\theta(\underline{Y})\right)^2 = E_\theta\left(\sum_{i=1}^{k} Y_i \frac{\partial}{\partial\theta} \log p_i(\theta)\right)^2$$

$$= E_\theta\left(\sum_{i=1}^{k} Y_i^2 \left(\frac{\partial}{\partial\theta} \log p_i(\theta)\right)^2\right)$$

$$+ E_\theta \left(\sum_{j=1}^{k} \sum_{i=1}^{k} Y_i Y_j \frac{\partial}{\partial \theta} \log p_i(\theta) \frac{\partial}{\partial \theta} \log p_j(\theta) \right)$$

$$= \sum_{i=1}^{k} p_i(\theta) \left(\frac{\partial}{\partial \theta} \log p_i(\theta) \right)^2 .$$

Now observe that

$$E_\theta \left(\frac{\partial^2}{\partial \theta^2} \log p_\theta(\underline{Y}) \right) = E_\theta \left(\sum_{i=1}^{k} Y_i \frac{\partial^2}{\partial \theta^2} \log p_i(\theta) \right) = \sum_{i=1}^{k} p_i(\theta) \frac{\partial^2}{\partial \theta^2} \log p_i(\theta) .$$

To find a relation between $E_\theta \left(\frac{\partial}{\partial \theta} \log p_\theta(\underline{Y}) \right)^2$ and $E_\theta \left(\frac{\partial^2}{\partial \theta^2} \log p_\theta(\underline{Y}) \right)$ we note that

$$E_\theta \left(\frac{\partial}{\partial \theta} \log p_\theta(\underline{Y}) \right) = 0$$

$$\Rightarrow \sum_{i=1}^{k} p_i(\theta) \frac{\partial}{\partial \theta} \log p_i(\theta) = 0$$

$$\Rightarrow \frac{\partial}{\partial \theta} \left(\sum_{i=1}^{k} p_i(\theta) \frac{\partial}{\partial \theta} \log p_i(\theta) \right) = 0$$

$$\Rightarrow \sum_{i=1}^{k} \frac{\partial}{\partial \theta} p_i(\theta) \frac{\partial}{\partial \theta} \log p_i(\theta) + \sum_{i=1}^{k} p_i(\theta) \frac{\partial^2}{\partial \theta^2} \log p_i(\theta) = 0$$

$$\Rightarrow \sum_{i=1}^{k} p_i(\theta) \left(\frac{\partial}{\partial \theta} \log p_i(\theta) \right)^2 + \sum_{i=1}^{k} p_i(\theta) \frac{\partial^2}{\partial \theta^2} \log p_i(\theta) = 0$$

$$\Rightarrow E_\theta \left(\frac{\partial}{\partial \theta} \log p_\theta(\underline{Y}) \right)^2 + E_\theta \left(\frac{\partial^2}{\partial \theta^2} \log p_\theta(\underline{Y}) \right) = 0$$

$$\Rightarrow E_\theta \left(\frac{\partial}{\partial \theta} \log p_\theta(\underline{Y}) \right)^2 = E_\theta \left(-\frac{\partial^2}{\partial \theta^2} \log p_\theta(\underline{Y}) \right) .$$

Thus, for a multinomial distribution in k cells with cell probabilities $p_i(\theta)$, $i = 1, 2, \ldots k$ indexed by a parameter θ, the information function $I(\theta)$ is given by

$$I(\theta) = E_\theta \left(\frac{\partial}{\partial \theta} \log p_\theta(\underline{Y}) \right)^2 = E_\theta \left(-\frac{\partial^2}{\partial \theta^2} \log p_\theta(\underline{Y}) \right)$$

$$= \sum_{i=1}^{k} p_i(\theta) \left(\frac{\partial}{\partial \theta} \log p_i(\theta) \right)^2 = -\sum_{i=1}^{k} p_i(\theta) \frac{\partial^2}{\partial \theta^2} \log p_i(\theta) .$$

We assume that $0 < I(\theta) < \infty$. As an illustration, in Example 6.2.3 we have discussed a multinomial distribution in four cells with cell probabilities $(2 + \theta)/4$, $(1 - \theta)/4$, $(1 - \theta)/4$ and $\theta/4$. Hence,

$$\frac{\partial}{\partial \theta} \log p_1(\theta) = \frac{1}{2 + \theta}, \qquad \frac{\partial}{\partial \theta} \log p_2(\theta) = \frac{-1}{1 - \theta}$$

$$\frac{\partial}{\partial \theta} \log p_3(\theta) = \frac{-1}{1 - \theta} \quad \& \quad \frac{\partial}{\partial \theta} \log p_4(\theta) = \frac{1}{\theta}$$

and

$$\frac{\partial^2}{\partial \theta^2} \log p_1(\theta) = \frac{-1}{(2 + \theta)^2}, \qquad \frac{\partial^2}{\partial \theta^2} \log p_2(\theta) = \frac{-1}{(1 - \theta)^2}$$

$$\frac{\partial^2}{\partial \theta^2} \log p_3(\theta) = \frac{-1}{(1 - \theta)^2} \quad \& \quad \frac{\partial^2}{\partial \theta^2} \log p_4(\theta) = \frac{-1}{\theta^2}.$$

Thus,

$$I(\theta) = \sum_{i=1}^{4} p_i(\theta) \left(\frac{\partial}{\partial \theta} \log p_i(\theta) \right)^2 = \frac{2 + \theta}{4(2 + \theta)^2} + \frac{2(1 - \theta)}{4(1 - \theta)^2} + \frac{\theta}{4\theta^2}$$

$$= \frac{(2\theta + 1)}{2\theta(1 - \theta)(2 + \theta)}$$

$$\& \quad I(\theta) = -\sum_{i=1}^{4} p_i(\theta) \frac{\partial^2}{\partial \theta^2} \log p_i(\theta) = \frac{2 + \theta}{4(2 + \theta)^2} + \frac{2(1 - \theta)}{4(1 - \theta)^2} + \frac{\theta}{4\theta^2}$$

$$= \frac{(2\theta + 1)}{2\theta(1 - \theta)(2 + \theta)}.$$

To examine the last condition in the Cramér regularity conditions, consider

$$\left| \frac{\partial^3}{\partial \theta^3} \log p_\theta(\underline{y}) \right| = \left| \sum_{i=1}^{k} y_i \frac{\partial^3}{\partial \theta^3} \log p_i(\theta) \right|$$

$$< \sum_{i=1}^{k} |y_i| \left| \frac{\partial^3}{\partial \theta^3} \log p_i(\theta) \right|$$

$$< \sum_{i=1}^{k} \left| \frac{\partial^3}{\partial \theta^3} \log p_i(\theta) \right| \quad \text{as} \quad |y_i| = 0 \text{ or } 1$$

$$= C(\theta), \quad \text{say}.$$

In order to show that $\left| \frac{\partial^3}{\partial \theta^3} \log p_\theta(\underline{y}) \right| < M(\underline{y})$ such that $E(M(\underline{Y})) < \infty$, we require that $C(\theta)$ for $\theta \in N_\delta(\theta_0)$, θ_0 being the true parameter value, must be bounded so that $\sup_{\theta \in N_\delta(\theta_0)} C(\theta)$ can be taken as $M(\underline{Y})$.

To summarize, if a multinomial distribution in k cells with cell probabilities indexed by a real parameter satisfies the following conditions, then it belongs to a Cramér family.

(i) The parameter space is an open set.
(ii) Derivatives of $p_i(\theta)$, $i = 1, 2, \ldots k$ exist up to order 3.
(iii) $\sup_{\theta \in N_\delta(\theta_0)} C(\theta)$ is finite, which is equivalent to requirement that the third order partial derivative of log-likelihood is bounded by an integrable function.

Consequently, all results from the Cramér-Huzurbazar theory, proved in Chap. 4, are applicable. Thus, for large n, the maximum likelihood estimator $\hat{\theta}_n$ is CAN for θ with approximate variance $1/nI(\theta)$. All these conditions can be extended to a multinomial distribution in k cells with cell probabilities indexed by a vector parameter $\underline{\theta}$. These are as follows:

(i) The parameter space is an open set.
(ii) All partial derivatives of $p_i(\underline{\theta})$, $i = 1, 2, \ldots k$ exist up to order 3.
(iii) The third order partial derivatives of log-likelihood are bounded by integrable functions.

If these conditions are satisfied, then a multinomial distribution belongs to a multiparameter Cramér family and for large n, the maximum likelihood estimator $\underline{\theta}$ exists and is CAN. We need these results for proving the following Theorem 6.2.1. The proof is similar to that of Theorem 5.2.4.

Theorem 6.2.1

In a multinomial distribution in k cells with cell probabilities $\underline{p} = (p_1, p_2, \ldots, p_k)'$ with $p_i > 0$ $\forall$ $i = 1, 2, \ldots, k$ and $\sum_{i=1}^{k} p_i = 1$, suppose we want to test $H_0 : \underline{p} = \underline{p}(\underline{\theta})$ against the alternative $H_1 : \underline{p} \neq \underline{p}(\underline{\theta})$, where $\underline{\theta}$ is an indexing parameter of dimension $l < k$. Suppose $\lambda(\underline{X})$ is a likelihood ratio test statistic based on a random sample of size n. If the multinomial distribution with cell probabilities indexed by $\underline{\theta}$ belongs to a Cramér family, then for large n under H_0, $-2 \log \lambda(\underline{X})$ has χ^2_{k-1-l} distribution.

Proof Suppose $\underline{Y} = (Y_1, Y_2, \ldots, Y_{k-1})'$ has multinomial distribution in k cells with cell probabilities $\underline{p}$. Suppose $\underline{X} = (X_1, Y_2, \ldots, X_k)'$ denotes the vector of cell frequencies corresponding to a random sample of size n from $\underline{Y}$ with $\sum_{i=1}^{k} X_i = n$. Then the maximum likelihood estimator $\hat{\underline{p}}_n = (\hat{p}_{1n}, \hat{p}_{2n}, \ldots, \hat{p}_{kn})'$ of $\underline{p}$ is given by, $\hat{\underline{p}}_n = (X_1/n, X_2/n, \ldots, X_k/n)'$. Suppose the cell probabilities are indexed by a parameter $\underline{\theta}$, which is a vector valued parameter of dimension $l < k$. To test $H_0 : \underline{p} = \underline{p}(\underline{\theta})$ against the alternative $H_1 : \underline{p} \neq \underline{p}(\underline{\theta})$ using the likelihood ratio test procedure, the entire parameter space is

$$\Theta = \{\underline{p} = (p_1, p_2, \ldots, p_k)' | p_i > 0, i = 1, 2, \ldots, k \ \& \ \sum_{i=1}^{k} p_i = 1\}$$

and the null space is $\Theta_0 = \{\underline{p} = (p_1, p_2, \ldots, p_k)' | \underline{p} = \underline{p}(\underline{\theta})\}$. Suppose $\hat{\underline{\theta}}_n$ denotes the maximum likelihood estimator of $\underline{\theta}$ based on the observed data $\underline{X}$. Since the distribution belongs to the Cramér family, $\hat{\underline{\theta}}_n$ is CAN for $\underline{\theta}$ with approximate dispersion matrix $I^{-1}(\underline{\theta})/n$. Now the likelihood ratio test statistic $\lambda(\underline{X})$ is given by

$$\lambda(\underline{X}) = \sup_{\Theta_0} L_n(\underline{p}|\underline{X}) / \sup_{\Theta} L_n(\underline{p}|\underline{X}) = L_n(\hat{\underline{\theta}}_n|\underline{X}) / L_n(\hat{\underline{p}}_n|\underline{X}) \ .$$

Hence as in Theorem 5.2.4,

$$-2\log \lambda(\underline{X}) = 2[\log L_n(\hat{\underline{p}}_n|\underline{X}) - \log L_n(\hat{\underline{\theta}}_n|\underline{X})]$$
$$= 2[\log L_n(\hat{\underline{p}}_n|\underline{X}) - \log L_n(\underline{p}_0|\underline{X})]$$
$$- 2[\log L_n(\hat{\underline{\theta}}_n|\underline{X}) - \log L_n(\underline{\theta}_0|\underline{X})]$$
$$= U_n - W_n \ ,$$

where $U_n = 2[\log L_n(\hat{\underline{p}}_n|\underline{X}) - \log L_n(\underline{p}_0|\underline{X})]$ and $W_n = 2[\log L_n(\hat{\underline{\theta}}_n|\underline{X}) - \log L_n(\underline{\theta}_0|\underline{X})]$, $\underline{\theta}_0$ is a known vector and $\underline{p}_0 = \underline{p}(\underline{\theta}_0)$ so that $\log L_n(\underline{p}_0|\underline{X}) = \log L_n(\underline{\theta}_0|\underline{X})$. From Theorem 5.2.2, $U_n \overset{L}{\to} U \sim \chi^2_{k-1}$ and $W_n \overset{L}{\to} W \sim \chi^2_l$. Thus, for large n, $-2\log \lambda(\underline{X})$ is distributed as $U - W$, which has χ^2_{k-1-l} distribution, using the result of Wilks [2].   □

📝 Remark 6.2.3

As in Theorem 5.2.4, the parameters p_i, $i = 1, 2, \ldots, k - 1$ are functions of l parameters $\theta_1, \theta_2, \ldots, \theta_l$. In all the goodness of fit tests, in the null hypothesis cell probabilities are indexed by parameters.

Following example illustrates the application of Theorem 6.2.1 to examine validity of the probability model as proposed in Example 6.2.3.

✍ Example 6.2.4

In Example 6.2.3, Carver's data on two varieties of maize classified as starchy versus sugary and further cross classified with the color as green and white are given in Table 6.1. We examine whether the proposed theoretical model is valid on the basis of given data, using likelihood ratio test procedure. An appropriate probability model for the outcome of the given experiment is a multinomial distribution in four cells with cell probabilities $\underline{p} = (p_1, p_2, p_3, p_4)'$. We want to test $H_0 : \underline{p} = \underline{p}(\theta)$ against the alternative that p_i's do not depend on θ, the only restriction is these are positive and add up to 1. Corresponding to given data the maximum likelihood estimator of $\underline{p}$ is $\hat{\underline{p}}_n = (X_1/n, X_2/n.X_3/n, X_4/n)'$, where X_i denotes the frequency of i-th cell and its total is $n = 3819$. In the null setup, the cell probabilities $(2 + \theta)/4$, $(1 - \theta)/4$, $(1 - \theta)/4$ and $\theta/4$ depend on θ. On the basis of the given data, we have obtained the maximum likelihood estimator $\hat{\theta}_n$ of θ in Example 6.2.3. Its value is 0.0354. Using it, we obtain the maximum likelihood estimate of $\underline{p}(\theta)$. Thus, we compute the likelihood ratio test statistic $\lambda(\underline{X})$. For the given data,

$$-2\log(\lambda(\underline{X})) = 2\left(\sum_{i=1}^{k} X_i \log(\hat{p}_{in}) - \sum_{i=1}^{k} X_i \log(p_i(\hat{\theta}_n))\right) = 1.2398.$$

From Theorem 6.2.1 for large n, $-2\log\lambda(\underline{X}) \sim \chi^2_m$ where $m = 3 - 1 = 2$. H_0 is rejected if $-2\log\lambda(\underline{X}) > c$ where c is such that the size of the test is α and is determined using the χ^2_2 distribution. Thus $c = \chi^2_{1,0.95} = 5.9914$. The corresponding p-value is 0.5380. Hence on the basis of the given data, we conclude that the proposed model is a valid model. □

In the goodness of fit tests and in tests for validity of the model, the cell probabilities are usually indexed by a parameter θ, which may be real or vector valued, as in the above example. In general suppose we have a multinomial distribution in k cells with cell probabilities $p_i(\theta)$, where $p_i(\theta) > 0$ and $\sum_{i=1}^{k} p_i(\theta) = 1$. θ is an indexing parameter, in the sense that $p_i(\theta_1) \neq p_i(\theta_2)$ for any $i = 1, 2, \ldots, k$ implies that $\theta_1 \neq \theta_2$. On the basis of the given data in terms of cell frequencies $\{X_1, X_2, \ldots, X_k\}$, with $\sum_{i=1}^{k} X_i = n$, we obtain maximum likelihood estimator $\hat{\theta}_n$ of θ and hence of cell probabilities as $p_i(\hat{\theta}_n)$. For the likelihood ratio test procedure, in the null setup, the cell probabilities are indexed by θ while in the entire parameter space there is no restriction on the cell probabilities, except the condition that they add up to 1. The likelihood ratio test statistics $\lambda(\underline{X})$ is then given by

$$\lambda(\underline{X}) = \sup_{\Theta_0} L_n(\underline{p}|\underline{X}) / \sup_{\Theta} L_n(\underline{p}|\underline{X}) = \prod_{i=1}^{k}(p_i(\hat{\theta}_n))^{X_i} / \prod_{i=1}^{k}(\hat{p}_{in})^{X_i} .$$

The likelihood ratio test procedure is used to test the goodness of fit of the proposed distribution or to test the validity of the proposed model. For large n, H_0 is rejected if $-2\log\lambda(\underline{X}) > c$ where c such that the size of the test is α and is determined using the large sample distribution of $-2\log\lambda(\underline{X})$, which is χ^2_{k-1-l}, as derived in Theorem 6.2.1.

In the next section, we discuss the role of a multinomial distribution in a test for goodness of fit, which is essentially a test for validity of the model. Theorem 6.2.1 is heavily used in a goodness of fit test.

6.3   Goodness of Fit Test

As discussed in Sect. 6.1, suppose the observed data on a characteristic X are in the form of (y_i, f_i) or $([x_i, x_{i+1}), f_i)$ $i = 1, 2, \ldots, k$. where y_i is the possible value of a discrete random variable X and f_i denotes the frequency of y_i in a random sample of size n from X, $i = 1, 2, \ldots, k$. A random sample of size n from a continuous distribution is grouped as $([x_i, x_{i+1}), f_i)$ $i = 1, 2, \ldots, k$, where f_i denotes the number of observations in the class interval $[x_i, x_{i+1})$. On the basis of these observed data, we wish test whether the data are from a specific distribution. In the following example, we illustrate how the likelihood ratio test for a multinomial distribution is useful to test the conjecture that the data are from a specified distribution.

✐ Example 6.3.1

A computer program is written to generate random numbers from a uniform $U(0, 4)$ distribution. 200 observations are generated and are grouped in 8 classes. Table 6.2 displays frequencies of these 8 classes. We examine whether these data offer any evidence of the accuracy of the program, that is, we examine whether the data correspond to a random sample from a uniform $U(0, 4)$ distribution. Hence, we set our null hypothesis as $H_0 : X \sim U(0, 4)$ distribution against the alternative that X has any other continuous distribution. Thus, in the entire setup, the appropriate probability model for these data is a multinomial distribution in 8 cells with cell probabilities $\underline{p} = (p_1, p_2, \ldots, p_8)'$ with $\sum_{i=1}^{8} p_i = 1$. The conjecture that $X \sim U(0, 4)$ is converted in terms of the null hypothesis $H_0 : \underline{p} = \underline{p}_0$ where cell probabilities $\underline{p}_0 = (p_{01}, p_{02}, \ldots, p_{08})'$ are completely specified as follows:

$$p_{0r} = P[x_{r-1} \leq X \leq x_r] = 0.5/4 = 0.125, \quad r = 1, 2, \ldots, 8$$

as each interval $[x_{r-1}, x_r]$ is of the same length 0.5. The alternative $H_1 : \underline{p} \neq \underline{p}_0$ is equivalent to stating that X does not have $U(0, 4)$ distribution. To test $H_0 : \underline{p} = \underline{p}_0$ against the alternative $H_1 : \underline{p} \neq \underline{p}_0$, we adopt a likelihood ratio test procedure when underlying probability model is a multinomial distribution with 8 cells. Thus, the entire parameter space is $\Theta = \{\underline{p} = (p_1, p_2, \ldots p_8)' | p_i > 0, i = 1, 2, \ldots, 8 \ \& \ \sum_{i=1}^{8} p_i = 1\}$ and the maximum likelihood estimator $\hat{p}_{in}$ of p_i is given by $\hat{p}_{in} = X_i/n$ $i = 1, 2, \ldots, 8$, where X_i denotes the frequency of the i-th class, $i = 1, 2, \ldots, 8$. The null space Θ_0 is $\Theta_0 = \{\underline{p} | \underline{p} = \underline{p}_0\}$. Hence, the likelihood ratio test statistics $\lambda(\underline{X})$ is

Table 6.2 Uniform $U(0, 4)$ distribution: grouped frequency distribution

Class interval	[0, 0.5)	[0.5, 1)	[1, 1.5)	[1.5, 2)	[2, 2.5)	[2.5, 3)	[3, 3.5)	[3.5, 4]
Frequency	29	17	25	23	29	31	21	25

given by

$$\lambda(\underline{X}) = \frac{\sup\limits_{\Theta_0} L_n(\theta|\underline{X})}{\sup\limits_{\Theta} L_n(\theta|\underline{X})} = \frac{\prod\limits_{i=1}^{8}(p_{0i})^{X_i}}{\prod\limits_{i=1}^{8}(\hat{p}_{in})^{X_i}}.$$

From Theorem 5.2.2, for large n under H_0, $-2\log\lambda(\underline{X}) \sim \chi_7^2$ distribution. H_0 is rejected if $-2\log\lambda(\underline{X}) > c$ where c is such that the size of the test is α and is determined using the χ_7^2 distribution. Thus $c = \chi_{7,1-\alpha}^2$. For the given data $-2\log\lambda(\underline{X}) = 6.2827$, with $\alpha = 0.5$, $c = 14.06$ and p-value is 0.5072. So we may conclude that data offer evidence that the program is written properly and the observed data are from $U(0,4)$ distribution. There is a built-in function chisq.test to carry out this test for goodness of fit. We discuss it in Sect. 6.7. □

In the goodness of fit test procedures, the most frequently used test statistics is Karl Pearson's chi-square test statistic, proposed by Karl Pearson in 1900. It is given by $T_n(P) = \sum_{i=1}^{k}(o_i - e_i)^2/e_i$, where o_i and e_i denote the observed and expected frequencies of i-th class, $i = 1, 2, \ldots, k$, e_i is labeled as expected frequency, since it denotes a frequency expected under the null setup. Thus, the test statistic $T_n(P)$ measures the deviation of the observed frequencies from the frequencies expected under the hypothesized distribution. The deviation may be due to sampling fluctuations or may be large enough which suggests that the data may not be from the assumed distribution. Hence, large values of $T_n(P)$ do not support the null setup that the data are generated under a specific distribution. Thus, the null hypothesis $H_0 : \underline{p} = \underline{p}_0$ is rejected if $T_n(P) > c$, where c is determined using the null distribution of $T_n(P)$. In the following Theorem 6.3.1, we prove that the likelihood ratio test statistic and Karl Pearson's chi-square test statistic for testing $H_0 : \underline{p} = \underline{p}_0$ against the alternative $H_1 : \underline{p} \neq \underline{p}_0$ in a multinomial distribution are equivalent, in the sense that their asymptotic null distributions are the same.

Theorem 6.3.1

Suppose a multinomial distribution in k cells with cell probabilities
$\underline{p} = (p_1, p_2, \ldots, p_k)'$ *where* $p_i > 0 \ \forall \ i = 1, 2, \ldots, k$ *and* $\sum_{i=1}^{k} p_i = 1$ *belongs to a Cramér family. For testing* $H_0 : \underline{p} = \underline{p}_0$ *against the alternative* $H_1 : \underline{p} \neq \underline{p}_0$, *where* $\underline{p}_0$ *is a completely specified vector*
(i) the likelihood ratio test statistic and Karl Pearson's chi-square statistic
$T_n(P) = \sum_{i=1}^{k}(o_i - e_i)^2/e_i$ *have the same asymptotic null distribution, which is*
χ_{k-1}^2, *where* o_i *and* e_i *denote the observed and expected frequencies respectively of the i-th class,* $i = 1, 2, \ldots, k$.
(ii) the test statistic $W_n = \sum_{i=1}^{k}(o_i - e_i)^2/o_i$ *also has the asymptotic null distribution to be* χ_{k-1}^2.

Proof Suppose $\underline{Y} = (Y_1, Y_2, \ldots, Y_{k-1})'$ has a multinomial distribution in k cells with cell probabilities $\underline{p}$. Suppose $\underline{X} = (X_1, X_2, \ldots, X_k)'$ denotes the vector of cell frequencies corresponding to a random sample of size n from $\underline{Y}$ with $\sum_{i=1}^{k} X_i = n$. The maximum likelihood estimator $\hat{\underline{p}}_n = (\hat{p}_{1n}, \hat{p}_{2n}, \ldots, \hat{p}_{kn})'$ of $\underline{p}$ is then given by $\hat{\underline{p}}_n = (X_1/n, X_2/n, \ldots, X_k/n)'$. To test $H_0 : \underline{p} = \underline{p}_0$ against the alternative $H_1 : \underline{p} \neq \underline{p}_0$ using the likelihood ratio test procedure, note that the entire parameter space is

$$\Theta = \left\{ \underline{p} = (p_1, p_2, \ldots, p_k)' | p_i > 0, i = 1, 2, \ldots, k \ \& \ \sum_{i=1}^{k} p_i = 1 \right\}$$

and the null space Θ_0 is $\Theta_0 = \{\underline{p} | \underline{p} = \underline{p}_0\}$. Hence, the likelihood ratio test statistic $\lambda(\underline{X})$ is given by

$$\lambda(\underline{X}) = \frac{\sup_{\Theta_0} L_n(\theta | \underline{X})}{\sup_{\Theta} L_n(\theta | \underline{X})} = \frac{\prod_{i=1}^{k} (p_{0i})^{X_i}}{\prod_{i=1}^{k} (\hat{p}_{in})^{X_i}} = \prod_{i=1}^{k} \left(\frac{p_{0i}}{\hat{p}_{in}} \right)^{X_i}.$$

From Theorem 5.2.2, for large n under H_0, $-2 \log \lambda(\underline{X}) \sim \chi_{k-1}^2$ distribution. To prove that Karl Pearson's test chi-square statistic $T_n(P)$ also has χ_{k-1}^2 distribution under H_0, we proceed as follows. It is to be noted that $o_i = X_i$ and $e_i = np_{0i}$ denote the observed and expected frequencies of i-th class, $i = 1, 2, \ldots, k$. Suppose

$$u_i = \sqrt{n}(\hat{p}_{in} - p_{0i}) \ \Leftrightarrow \ \hat{p}_{in} = p_{0i} + \frac{u_i}{\sqrt{n}}, \ i = 1, 2, \ldots, k.$$

Further, observe that $\sum_{i=1}^{k} \frac{u_i}{\sqrt{n}} = \sum_{i=1}^{k} \hat{p}_{in} - \sum_{i=1}^{k} p_{0i} = 0$. Now, $\lambda(\underline{X}) = \prod_{i=1}^{k} \left(\frac{p_{0i}}{\hat{p}_{in}} \right)^{X_i}$ implies

$$-2 \log \lambda(\underline{X}) = 2 \sum_{i=1}^{k} X_i (\log \hat{p}_{in} - \log p_{0i}) = 2n \sum_{i=1}^{k} \hat{p}_{in} (\log \hat{p}_{in} - \log p_{0i})$$

$$= 2n \sum_{i=1}^{k} \left(p_{0i} + \frac{u_i}{\sqrt{n}} \right) \left(\log \left(p_{0i} + \frac{u_i}{\sqrt{n}} \right) - \log p_{0i} \right)$$

$$= 2n \sum_{i=1}^{k} \left(p_{0i} + \frac{u_i}{\sqrt{n}} \right) \log \left(1 + \frac{u_i}{\sqrt{n} \, p_{0i}} \right)$$

$$= 2n \sum_{i=1}^{k} \left(p_{0i} + \frac{u_i}{\sqrt{n}} \right) \left(\frac{u_i}{\sqrt{n} \, p_{0i}} - \frac{u_i^2}{2n p_{0i}^2} + \frac{u_i^3}{3n^{3/2} p_{0i}^3} - \cdots \right)$$

$$= 2n \sum_{i=1}^{k} \left(\frac{u_i}{\sqrt{n}} - \frac{u_i^2}{2np_{0i}} + \frac{u_i^3}{3n^{3/2}p_{0i}^2} + \frac{u_i^2}{np_{0i}} - \frac{u_i^3}{2n^{3/2}p_{0i}^2} + \frac{u_i^4}{3n^2 p_{0i}^3} + \cdots \right)$$

$$= 2n \sum_{i=1}^{k} \left(\frac{u_i^2}{2np_{0i}} + \frac{u_i}{\sqrt{n}} \right) + V_n \,,$$

where $V_n = 2n \sum_{i=1}^{k} a_1 \frac{u_i^3}{n^{3/2}} + 2n \sum_{i=1}^{k} a_2 \frac{u_i^4}{n^2} + \cdots$, where a_1 and a_2 are constants. Thus, using the fact that $\sum_{i=1}^{k} \frac{u_i}{\sqrt{n}} = 0$, we have

$$-2 \log \lambda(\underline{X}) = \sum_{i=1}^{k} \frac{u_i^2}{p_{0i}} + V_n \quad = \quad \sum_{i=1}^{k} \frac{n(\hat{p}_{in} - p_{0i})^2}{p_{0i}} + V_n$$

$$= \sum_{i=1}^{k} \frac{(n\hat{p}_{in} - np_{0i})^2}{np_{0i}} + V_n \quad = \quad \sum_{i=1}^{k} \frac{(o_i - e_i)^2}{e_i} + V_n.$$

Thus, $-2 \log \lambda(\underline{X}) - T_n(P) = V_n$. If we show that $V_n \xrightarrow{P} 0$ then $-2 \log \lambda(\underline{X})$ and $T_n(P)$ have the same limit law. But $-2 \log \lambda(\underline{X}) \xrightarrow{L} U \sim \chi_{k-1}^2$ and hence $T_n(P) \xrightarrow{L} U \sim \chi_{k-1}^2$. To prove that $V_n \xrightarrow{P} 0$, we consider the first term of V_n given by

$$T_{1n} = 2n \sum_{i=1}^{k} a_1 \frac{u_i^3}{n^{3/2}} = \frac{2a_1}{\sqrt{n}} \sum_{i=1}^{k-1} (\sqrt{n}(\hat{p}_{in} - p_{0i}))^3 + \frac{2a_1}{\sqrt{n}} (\sqrt{n}(\hat{p}_{kn} - p_{0k}))^3 \,.$$

In Sect. 6.2, it is proved that for a multinomial distribution with k cells,

$$\underline{Z}_n = \left(\sqrt{n}(\hat{p}_{1n} - p_{01}), \sqrt{n}(\hat{p}_{2n} - p_{02}), \ldots, \sqrt{n}(\hat{p}_{(k-1)n} - p_{0(k-1)}) \right)'$$
$$\xrightarrow{L} \underline{Z} \sim N_{k-1}(\underline{0}, I^{-1}(\underline{p}_0)),$$

where $I(\underline{p}_0)$ is a positive definite matrix. Suppose a function $g : \mathbb{R}^{k-1} \to \mathbb{R}$ is defined as $g(\underline{x}) = \sum_{i=1}^{k-1} x_i^3$, then it is a continuous function and by the continuous mapping theorem

$$g(\underline{Z}_n) = \sum_{i=1}^{k-1} (\sqrt{n}(\hat{p}_{in} - p_{0i}))^3 \xrightarrow{L} g(\underline{Z}),$$

which implies that $g(\underline{Z}_n)$ is bounded in probability and hence $\frac{2a_1}{\sqrt{n}} \sum_{i=1}^{k-1} (\sqrt{n}(\hat{p}_{in} - p_{0i}))^3 \xrightarrow{P} 0$. Now using the fact that $p_k = 1 - p_1 - p_2 - \cdots p_{k-1}$, the second term in T_{1n} can be expressed as

$$\frac{2a_1}{\sqrt{n}} (\sqrt{n}(\hat{p}_{kn} - p_{0k}))^3 = \frac{2a_1}{\sqrt{n}} \left(\sum_{i=1}^{k-1} (\sqrt{n}(\hat{p}_{in} - p_{0i})) \right)^3 \,.$$

Further, we define a function $g : \mathbb{R}^{k-1} \to \mathbb{R}$ as $g(\underline{x}) = \left(\sum_{i=1}^{k-1} x_i\right)^3$, which is a continuous function. By the continuous mapping theorem

$$g(\underline{Z}_n) = \left(\sum_{i=1}^{k-1} (\sqrt{n}(\hat{p}_{in} - p_{0i}))\right)^3 \xrightarrow{L} g(\underline{Z}) \Rightarrow \frac{2a_1}{\sqrt{n}}\left(\sqrt{n}(\hat{p}_{kn} - p_{0k})\right)^3 \xrightarrow{P} 0.$$

Thus, $T_{1n} \xrightarrow{P} 0$. Using similar logic, we can prove that the remaining terms in V_n converge in probability to 0 and hence $V_n \xrightarrow{P} 0$. Thus under H_0, $-2\log\lambda(\underline{X})$ and $T_n(P)$ have the same limit law and it is χ^2_{k-1}.

(ii) The test statistic $T_n(P)$ can be expressed as

$$T_n(P) = \sum_{i=1}^{k} \frac{n(\hat{p}_{in} - p_{0i})^2}{p_{0i}} = n(\hat{\underline{p}}_n - \underline{p}_0)' A (\hat{\underline{p}}_n - \underline{p}_0)$$

$$\text{where } A = \text{diag}\left(\frac{1}{p_{01}}, \frac{1}{p_{02}}, \ldots, \frac{1}{p_{0k}}\right).$$

Similarly, W_n can be expressed as

$$W_n = \sum_{i=1}^{k} \frac{(o_i - e_i)^2}{o_i} = \sum_{i=1}^{k} \frac{(n\hat{p}_{in} - np_{0i})^2}{n\hat{p}_{in}}$$

$$= \sum_{i=1}^{k} \frac{n(\hat{p}_{in} - p_{0i})^2}{\hat{p}_{in}} = n(\hat{\underline{p}}_n - \underline{p}_0)' A_n (\hat{\underline{p}}_n - \underline{p}_0),$$

where $A_n = \text{diag}\left(\frac{1}{\hat{p}_{1n}}, \frac{1}{\hat{p}_{2n}}, \ldots, \frac{1}{\hat{p}_{kn}}\right)$. As in Theorem 5.2.2, the fact that $\hat{\underline{p}}_n \xrightarrow{P} \underline{p}_0$ under H_0, implies that $A_n \xrightarrow{P} A$ and hence $W_n - T_n(P) \xrightarrow{P} 0$. Thus, the limit law of W_n and $T_n(P)$ is the same and it is χ^2_{k-1}.   □

For the data in Example 6.3.1, $T_n(P) = 6.08$ & $W_n = 6.9652$. Further, $c = \chi^2_{7,0.95} = 14.06$ and values of both $T_n(P)$ and W_n is less than the cut-off c. The p-values corresponding to $T_n(P)$ and W_n are 0.5304 and 0.4325 respectively. Hence, data do not have sufficient evidence to reject H_0.

Next example illustrates a goodness of fit test procedure for a discrete distribution, it is similar to that in Example 6.3.1.

☑ Example 6.3.2

Table 6.3 displays possible values of a discrete random variable Y and the corresponding frequencies f_i in a random sample of size $n = 50$ from the distribution of Y. We test the claim that the data are from binomial $B(6, 0.6)$ distribution,

Table 6.3 Truncated binomial distribution: frequency distribution

i	1	2	3	4	5	6
f_i	3	8	12	15	9	3

truncated at 0, using likelihood ratio test statistic, Pearson's chi-square statistic and W_n.

We wish to test the conjecture on the basis of data conveying the frequency $\underline{X} = \{f_1, f_2, f_3, f_4, f_5, f_6\}$ of 6 possible values of Y in a random sample of size 50 from Y. The appropriate probability model for these data is a multinomial distribution in 6 cells with cell probabilities $\underline{p} = (p_1, p_2, p_3, p_4, p_5, p_6)'$ with $\sum_{i=1}^{6} p_i = 1$. The conjecture that Y follows binomial $B(6, .6)$ distribution, truncated at 0, can be converted in terms of the null hypothesis $H_0 : \underline{p} = \underline{p}_0$ where cell probabilities $\underline{p}_0 = (p_{01}, p_{02}, p_{03}, p_{04}, p_{05}, p_{06})'$ are completely specified by the $B(6, 0.6)$ distribution, truncated at 0. Thus,

$$p_{0i} = \frac{\binom{6}{i}(0.6)^i\,(0.4)^{6-i}}{1 - (0.4)^6}, \quad i = 1, 2, \ldots, 6.$$

In the entire parameter space, the maximum likelihood estimator of p is given by $\hat{p}_{in} = X_i/n, i = 1, 2, \ldots, 6$. Hence, the likelihood ratio test statistic $\lambda(\underline{X})$ is given by

$$\lambda(\underline{X}) = \frac{\sup_{\Theta_0} L_n(\underline{p}|\underline{X})}{\sup_{\Theta} L_n(\underline{p}|\underline{X})} = \frac{\prod\limits_{i=1}^{6}(p_{0i})^{X_i}}{\prod\limits_{i=1}^{6}(\hat{p}_{in})^{X_i}}.$$

Under H_0, $-2\log \lambda(\underline{X}) \sim \chi_5^2$ distribution and H_0 is rejected if $-2\log \lambda(\underline{X}) > c$, where $c = \chi^2_{1-\alpha, 5}$. For the given data, $-2\log \lambda(\underline{X}) = 1.2297$ and $c = 11.0705$ with $\alpha = 0.05$ and the p-value is 0.9420. Hence, data provides strong support to H_0. We now find the value of Karl Pearson's chi-square test statistic and of W_n. The expected frequencies e_i are then given by $e_i = np_{0i}$. The observed frequencies are $o_i = f_i$. Table 6.4 displays the values of observed and expected frequencies.

The value of Pearson's chi-square test statistic $T_n(P) = \sum_{i=1}^{k}(o_i - e_i)^2/e_i$ is 1.3537 and of $W_n = \sum_{i=1}^{k}(o_i - e_i)^2/o_i$ is 1.06 with corresponding p-values 0.9293 and 0.9576. Thus, on the basis of these two test procedures, we note that the data strongly support the null setup. From Table 6.4, we note that the observed and the expected frequencies of the first and the last class are less than 5. Hence, according to the convention, we may pool the observed and the expected frequency of the first two classes and the last two classes. For the pooled data, $T_n(P) = 0.8412$. Under H_0, $T_n \sim \chi_3^2$. For large n, H_0 is rejected if $T_n(P) > c$, with $\alpha = 0.05, c = \chi^2_{3,0.95} = 7.8147$, which is larger than value of $T_n(P)$. Hence, the conclusion remains the same that data do not show sufficient evidence to reject

Table 6.4 Truncated binomial distribution: observed and expected frequencies

i	o_i	p_{0i}	e_i
1	3	0.0370	1.8508
2	8	0.1388	6.9404
3	12	0.2776	13.8809
4	15	0.3123	15.6159
5	9	0.1874	9.3696
6	3	0.0468	2.3424

H_0. The claim that the data are from binomial $B(6, .6)$ distribution, truncated at 0 may be accepted. In R, there is a built-in function chisq.test to test the goodness of fit. It is based on the Karl Pearson's chi-square test statistic. We demonstrate it in Sect. 6.7.   □

In the goodness of fit test in Example 6.3.1 and in Example 6.3.2, the null hypothesis is simple. Hence, it is expressed as $H_0 : p = p_0$ in the setup of multinomial distribution, where p_0 is a completely specified vector. Suppose we have data on scores of students, classified in k classes with corresponding frequencies. The conjecture is that the scores have a normal $N(\mu, \sigma^2)$ distribution, then the hypothesis can be expressed as $H_0 : p = p(\theta)$, where $\theta = (\mu, \sigma^2)'$ and the cell probabilities are functions of an unknown parameter θ. Thus, the null hypothesis is a composite hypothesis as in Theorem 5.2.4 or Theorem 6.2.1. In such a setup, the first step is to estimate θ from the given data and then estimate cell probabilities. We use the likelihood ratio test, as we have used for testing the validity of the model. This is the most common scenario in the goodness of fit test. The next example is a typical example of fitting a continuous distribution, normal in this case, to the data presented in the form of grouped frequency distribution. The procedure is similar for any other continuous distribution. We use Theorem 6.2.1 to determine the critical region.

✏ Example 6.3.3

It is often assumed the IQ scores of human beings are normally distributed. We test this claim for the data given in Table 6.5, using the likelihood ratio test procedure. Suppose a random variable Y denotes the IQ score. Then the conjecture is $Y \sim N(\mu, \sigma^2)$ distribution, which we test on the basis of the data conveying the number $\underline{X} = \{X_1, X_2, X_3, X_4, X_5, X_6\}$ of human beings with IQ score within a specified class interval for 6 class intervals. Observe that $X_1 + X_2 + X_3 + X_4 + X_5 + X_6 = n = 100$. The parameters μ and σ^2 are unknown and we estimate these on the basis of the given data. The appropriate probability model for these data is again a multinomial distribution in 6 cells with cell probabilities depending on $\theta = (\mu, \sigma^2)'$ in the null setup. Thus, for $i = 1, 2, \ldots, 6$,

$$p_i(\theta) = P_{\underline{\theta}}[x_{i-1} \le X \le x_i] = \Phi\left(\frac{x_i - \mu}{\sigma}\right) - \Phi\left(\frac{x_{i-1} - \mu}{\sigma}\right)$$

$$= g_i(\mu, \sigma^2), \text{ say .}$$

Table 6.5 IQ scores: grouped frequency distribution

IQ score	≤ 90	(90, 100]	(100, 110]	(110, 120]	(120, 130]	> 130
Frequency	10	18	23	22	18	9

In the entire parameter space, the parameter is $p = (p_1, p_2, p_3, p_4, p_5, p_6)'$ with $\sum_{i=1}^{6} p_i = 1$. The conjecture that $Y \sim N(\mu, \sigma^2)$ distribution can be converted in terms of the null hypothesis $H_0 : p = p(\theta)$ against the alternative $H_1 : p \neq p(\theta)$, where $p(\theta) = (p_1(\theta), p_2(\theta), \ldots, p_6(\theta))'$. Suppose $\hat{\theta}_n$ denotes the maximum likelihood estimator of θ on the basis of given data. As the underlying probability model is a normal distribution, $\hat{\theta}_n$ is CAN for θ. Further, $\hat{p}_n(\hat{\theta}_n)$ is the maximum likelihood estimator of $p(\theta)$. In the entire parameter space, the maximum likelihood estimator of p is given by $\hat{p}_{in} = X_i/n$, $i = 1, 2, \ldots, 6$ and in the null space $\hat{p}_{in}(\hat{\theta}_n) = \Phi((x_i - \hat{\mu}_n)/\hat{\sigma}_n) - \Phi((x_{i-1} - \hat{\mu}_n)/\hat{\sigma}_n)$, $i = 1, 2, \ldots, k$. From Theorem 6.2.1, for large n under H_0, $-2 \log \lambda(\underline{X}) \sim \chi_m^2$ where $m = 5 - 2 = 3$ as in the entire parameter space we estimate 5 parameters and in the null space we estimate 2 parameters. For large n, H_0 is rejected if $-2 \log \lambda(\underline{X}) > c$ where c is such that the size of the test is α and is determined using the χ_3^2 distribution. Thus $c = \chi_{3, 1-\alpha}^2$. For the given data $\hat{\mu}_n = \overline{X}_n = 109.7$ and $\hat{\sigma}_n^2 = 210.91$, which are calculated by taking the frequency of class (<80) as 0, of $(80\text{--}90)$ as 10, of $(130\text{--}140)$ as 9 and of (>140) as 0. Further,

$$-2 \log \lambda(\underline{X}) = 2 \sum_{i=1}^{6} [X_i(\log \hat{p}_{in} - \log \hat{p}_{in}(\hat{\theta}_n))] = 5.694188$$

$$\&\ c = \chi_{3, 0.95}^2 = 7.8147 > -2 \log \lambda(\underline{X})$$

with $\alpha = 0.05$. Thus, data do not have sufficient evidence to reject H_0 and we may conclude that the normal distribution seems to be an appropriate model for IQ scores. $\square$

Example 6.3.3 is a typical example of a goodness of fit test, in which null hypothesis is $H_0 : p = p(\theta)$, where the cell probabilities are functions of an unknown parameter θ of dimension $l \times 1$ say. The most commonly used test statistic is Karl Pearson's chi-square statistic given by $T_n(P) = \sum_{i=1}^{k} (o_i - e_i)^2/e_i$. In this setup, the Pearson-Fisher theorem (Kale and Muralidharan [1]) states that under H_0, $T_n(P) \sim \chi_{k-1-l}^2$ distribution where l is the number of parameters estimated in the null setup. In the following Theorem 6.3.2, we prove that for a multinomial distribution when cell probabilities are indexed by θ, the likelihood ratio test statistic and Karl Pearson's chi-square test statistic for testing $H_0 : p = p(\theta)$ against the alternative $H_1 : p \neq p(\theta)$ have the same asymptotic null distribution. We further show that a test statistic $W_n = \sum_{i=1}^{k} (o_i - e_i)^2/o_i$ also has the same asymptotic null distribution, which is chi-square.

Theorem 6.3.2

In a multinomial distribution with k cells having cell probabilities
$p = (p_1, p_2, \ldots, p_k)'$ *where* $p_i > 0 \ \forall \ i = 1, 2, \ldots, k$ *and* $\sum_{i=1}^{k} p_i = 1$, *sup-*
pose we want to test $H_0 : p = p(\theta)$ *against the alternative* $H_1 : p \neq p(\theta)$, *where*
θ *is an indexing parameter of dimension* $l < k$. *It is assumed that a multinomial*
distribution, when cell probabilities are indexed by θ, *belongs to a Cramér family.*
Then
(i) the likelihood ratio test statistic $-2 \log \lambda(\underline{X})$ *and Karl Pearson's chi-square*
statistic $T_n(P) = \sum_{i=1}^{k} (o_i - e_i)^2 / e_i$ *have the same asymptotic null distribution*
as χ^2_{k-1-l},
(ii) the test statistic $W_n = \sum_{i=1}^{k} (o_i - e_i)^2 / o_i$ *also has the asymptotic null distri-*
bution as χ^2_{k-1-l}.

Proof Suppose $\underline{Y} = (Y_1, Y_2, \ldots, Y_{k-1})'$ has a multinomial distribution in k cells
with cell probabilities p. Suppose $\underline{X} = (X_1, X_2, \ldots, X_k)'$ denotes the vector of cell
frequencies corresponding to a random sample of size n from $\underline{Y}$ with $\sum_{i=1}^{k} X_i = n$.
The maximum likelihood estimator $\underline{\hat{p}}_n = (\hat{p}_{1n}, \hat{p}_{2n}, \ldots, \hat{p}_{kn})'$ of p is then given
by $\underline{\hat{p}}_n = (X_1/n, X_2/n, \ldots, X_k/n)'$. Suppose the cell probabilities are indexed by
a parameter θ, which is a vector valued parameter of dimension $l < k$. To test
$H_0 : p = p(\theta)$ against the alternative $H_1 : p \neq p(\theta)$ using the likelihood ratio test
procedure, note that the null space Θ_0 is given by
$\Theta_0 = \{p | p = p(\theta), \ \sum_{i=1}^{k} p_i(\theta) = 1\}$. Suppose $\hat{\theta}_n$ denotes the maximum likelihood estimator of θ based on the observed data $\underline{X}$. Since the distribution belongs to
a Cramér family, $\hat{\theta}_n$ is CAN for θ with approximate dispersion matrix $I^{-1}(\theta)/n$. In
Theorem 6.2.1, we have proved that $-2 \log \lambda(\underline{X})$ has χ^2_{k-l-1} distribution.

(i) To prove that $T_n(P) \sim \chi^2_{k-l-l}$ distribution under H_0, we proceed on similar lines
as in Theorem 6.3.1. Note that $o_i = X_i$ and $e_i = n p_i(\hat{\theta}_n)$ denote the observed and
expected   frequencies   of i-th   class, $i = 1, 2, \ldots, k$.   Suppose
$u_i = \sqrt{n}(\hat{p}_{in} - p_i(\hat{\theta}_n)) \ \Leftrightarrow \ \hat{p}_{in} = p_i(\hat{\theta}_n) + u_i/\sqrt{n}$, $i = 1, 2, \ldots, k$. Further,
it is to be noted that $\sum_{i=1}^{k} u_i / \sqrt{n} = \sum_{i=1}^{k} \hat{p}_{in} - \sum_{i=1}^{k} p_i(\hat{\theta}_n) = 0$. With these substitutions, $-2 \log \lambda(\underline{X})$ can be expressed as follows:

$$-2 \log \lambda(\underline{X}) = 2 \sum_{i=1}^{k} X_i (\log \hat{p}_{in} - \log p_i(\hat{\theta}_n)) = 2n \sum_{i=1}^{k} \hat{p}_{in} (\log \hat{p}_{in} - \log p_i(\hat{\theta}_n))$$

$$= 2n \sum_{i=1}^{k} \left(p_i(\hat{\theta}_n) + \frac{u_i}{\sqrt{n}} \right) \left(\log \left(p_i(\hat{\theta}_n) + \frac{u_i}{\sqrt{n}} \right) - \log p_i(\hat{\theta}_n) \right)$$

$$= 2n \sum_{i=1}^{k} \left(p_i(\hat{\theta}_n) + \frac{u_i}{\sqrt{n}} \right) \log \left(1 + \frac{u_i}{\sqrt{n} \, p_i(\hat{\theta}_n)} \right)$$

$$= 2n \sum_{i=1}^{k} \left(p_i(\hat{\underline{\theta}}_n) + \frac{u_i}{\sqrt{n}} \right) \times$$

$$\left(\frac{u_i}{\sqrt{n}\, p_i(\hat{\underline{\theta}}_n)} - \frac{u_i^2}{2n(p_i(\hat{\underline{\theta}}_n))^2} + \frac{u_i^3}{3n^{3/2}(p_i(\hat{\underline{\theta}}_n))^3} - \cdots \right)$$

$$= 2n \sum_{i=1}^{k} \left(\frac{u_i}{\sqrt{n}} - \frac{u_i^2}{2np_i(\hat{\underline{\theta}}_n)} + \frac{u_i^3}{3n^{3/2}(p_i(\hat{\underline{\theta}}_n))^2} + \cdots \right)$$

$$+ 2n \sum_{i=1}^{k} \left(\frac{u_i^2}{np_i(\hat{\underline{\theta}}_n)} - \frac{u_i^3}{2n^{3/2}(p_i(\hat{\underline{\theta}}_n))^2} + \frac{u_i^4}{3n^2(p_i(\hat{\underline{\theta}}_n))^3} + \cdots \right)$$

$$= 2n \sum_{i=1}^{k} \left(\frac{u_i^2}{2np_i(\hat{\underline{\theta}}_n)} + \frac{u_i}{\sqrt{n}} \right) + V_n \,,$$

where $V_n = 2n \sum_{i=1}^{k} a_1 \frac{u_i^3}{(p_i(\hat{\underline{\theta}}_n))^2 n^{3/2}} + 2n \sum_{i=1}^{k} a_2 \frac{u_i^4}{(p_i(\hat{\underline{\theta}}_n))^3 n^2} + \cdots$, where a_1 and a_2 are constants. Thus, using the fact that $\sum_{i=1}^{k} \frac{u_i}{\sqrt{n}} = 0$, we have

$$-2 \log \lambda(\underline{X}) = \sum_{i=1}^{k} \frac{u_i^2}{p_i(\hat{\underline{\theta}}_n)} + V_n \quad = \quad \sum_{i=1}^{k} \frac{n(\hat{p}_{in} - p_i(\hat{\underline{\theta}}_n))^2}{p_i(\hat{\underline{\theta}}_n)} + V_n$$

$$= \sum_{i=1}^{k} \frac{(n\hat{p}_{in} - np_i(\hat{\underline{\theta}}_n))^2}{np_i(\hat{\underline{\theta}}_n)} + V_n \quad = \quad \sum_{i=1}^{k} \frac{(o_i - e_i)^2}{e_i} + V_n.$$

Thus, $-2 \log \lambda(\underline{X}) - T_n(P) = V_n$. If we show that $V_n \xrightarrow{P} 0$ then $-2 \log \lambda(\underline{X})$ and $T_n(P)$ have the same limit law. To prove $V_n \xrightarrow{P} 0$, we consider the first term of V_n given by

$$T_{1n} = 2n \sum_{i=1}^{k} a_1 \frac{u_i^3}{(p_i(\hat{\underline{\theta}}_n))^2 n^{3/2}} = \frac{2a_1}{\sqrt{n}} \sum_{i=1}^{k} \frac{(\sqrt{n}(\hat{p}_{in} - p_i(\hat{\underline{\theta}}_n)))^3}{(p_i(\hat{\underline{\theta}}_n))^2}$$

$$= \frac{2a_1}{\sqrt{n}} \sum_{i=1}^{k} \left(\frac{(\sqrt{n}(\hat{p}_{in} - p_{0i}) - \sqrt{n}(p_i(\hat{\underline{\theta}}_n) - p_{0i}))^3}{(p_i(\hat{\underline{\theta}}_n))^2} \right)$$

$$= \frac{2a_1}{\sqrt{n}} \sum_{i=1}^{k} \left(\frac{(\sqrt{n}(\hat{p}_{in} - p_{0i}))^3 - (\sqrt{n}(p_i(\hat{\underline{\theta}}_n) - p_{0i}))^3}{(p_i(\hat{\underline{\theta}}_n))^2} \right)$$

$$+ \frac{2a_1}{\sqrt{n}} \sum_{i=1}^{k} \left(\frac{3(\sqrt{n}(\hat{p}_{in} - p_{0i}))^2(\sqrt{n}(p_i(\hat{\underline{\theta}}_n) - p_{0i})) - 3(\sqrt{n}(\hat{p}_{in} - p_{0i}))(\sqrt{n}(p_i(\hat{\underline{\theta}}_n) - p_{0i})^2)}{(p_i(\hat{\underline{\theta}}_n))^2} \right),$$

where $p_{0i} = p_i(\underline{\theta}_0)$. We first consider the term

$$\frac{2a_1}{\sqrt{n}} \sum_{i=1}^{k} \frac{(\sqrt{n}(\hat{p}_{in} - p_{0i}))^3}{(p_i(\hat{\underline{\theta}}_n))^2} = \frac{2a_1}{\sqrt{n}} \sum_{i=1}^{k-1} \frac{(\sqrt{n}(\hat{p}_{in} - p_{0i}))^3}{(p_i(\hat{\underline{\theta}}_n))^2}$$

$$+ \frac{2a_1}{\sqrt{n}} \frac{(\sqrt{n}(\hat{p}_{kn} - p_{0k}))^3}{(p_k(\hat{\underline{\theta}}_n))^2}.$$

Suppose $\underline{Y}_n = \left(\dfrac{\sqrt{n}(\hat{p}_{1n} - p_{01})}{(p_1(\hat{\underline{\theta}}_n))^2}, \dfrac{\sqrt{n}(\hat{p}_{2n} - p_{02})}{(p_2(\hat{\underline{\theta}}_n))^2}, \ldots, \dfrac{\sqrt{n}(\hat{p}_{(k-1)n} - p_{0(k-1)})}{(p_{k-1}(\hat{\underline{\theta}}_n))^2} \right)' = A_n \underline{Z}_n$,

where $A_n = \text{diag}\left(1/(p_1(\hat{\underline{\theta}}_n))^2, 1/(p_2(\hat{\underline{\theta}}_n))^2, \ldots, 1/(p_{k-1}(\hat{\underline{\theta}}_n))^2 \right)$. Suppose

$\hat{\underline{\theta}}_n \overset{P}{\to} \underline{\theta}_0$, then $A_n \overset{P}{\to} A = \text{diag}\left(1/(p_1(\underline{\theta}_0))^2, 1/(p_2(\underline{\theta}_0))^2, \ldots, 1/(p_{k-1}(\underline{\theta}_0))^2 \right)$.
Further,

$$\underline{Z}_n = \left(\sqrt{n}(\hat{p}_{1n} - p_{01}), \sqrt{n}(\hat{p}_{2n} - p_{02}), \ldots, \sqrt{n}(\hat{p}_{(k-1)n} - p_{0(k-1)}) \right)'$$
$$\overset{L}{\to} \underline{Z} \sim N_{k-1}(\underline{0}, I^{-1}(\underline{p}_0)),$$

where $I(\underline{p}_0)$ is a positive definite matrix. Hence by Slutsky's theorem,

$\underline{Y}_n = A_n \underline{Z}_n \overset{L}{\to} A\underline{Z}$. As in Theorem 6.3.1, a function $g : \mathbb{R}^{k-1} \to \mathbb{R}$ is defined
as $g(\underline{x}) = \sum_{i=1}^{k-1} x_i^3$, then it is a continuous function. By the continuous mapping
theorem

$$g(\underline{Y}_n) = \sum_{i=1}^{k-1} \left(\dfrac{(\sqrt{n}(\hat{p}_{in} - p_{0i}))^3}{(p_i(\hat{\underline{\theta}}_n))^2} \right) \overset{L}{\to} g(A\underline{Z}),$$

which implies that $g(\underline{Y}_n)$ is bounded in probability and hence
$\dfrac{2a_1}{\sqrt{n}} \sum_{i=1}^{k-1} \left(\dfrac{(\sqrt{n}(\hat{p}_{in} - p_{0i}))^3}{(p_i(\hat{\underline{\theta}}_n))^2} \right) \overset{P}{\to} 0$. Now using the fact that
$p_k = 1 - p_1 - p_2 - \cdots p_{k-1}$, the term $\dfrac{2a_1}{\sqrt{n}} \dfrac{(\sqrt{n}(\hat{p}_{kn} - p_{0k}))^3}{(p_k(\hat{\underline{\theta}}_n))^2}$ can be expressed as

$$\dfrac{2a_1}{\sqrt{n}} \dfrac{(\sqrt{n}(\hat{p}_{kn} - p_{0k}))^3}{(p_k(\hat{\underline{\theta}}_n))^2} = \dfrac{2a_1}{\sqrt{n}} \dfrac{1}{(p_k(\hat{\underline{\theta}}_n))^2} \left(\sum_{i=1}^{k-1} (\sqrt{n}(\hat{p}_{in} - p_{0i})) \right)^3.$$

Now defining a function $g : \mathbb{R}^{k-1} \to \mathbb{R}$ as $g(\underline{x}) = \left(\sum_{i=1}^{k-1} x_i \right)^3$, which is a continu-
ous function, by the continuous mapping theorem $g(\underline{Z}_n) = \left(\sum_{i=1}^{k-1} (\sqrt{n}(\hat{p}_{in} - p_{0i})) \right)^3$
$\overset{L}{\to} g(\underline{Z})$. Further, $(p_k(\hat{\underline{\theta}}_n))^2 \overset{P}{\to} (p_k(\underline{\theta}_0))^2$. Hence,
$(2a_1/\sqrt{n})(\sqrt{n}(\hat{p}_{kn} - p_{0k}))^3/(p_k(\hat{\underline{\theta}}_n))^2 \overset{P}{\to} 0$. To examine whether
$\sum_{i=1}^{k} \left((\sqrt{n}(p_i(\hat{\underline{\theta}}_n) - p_{0i}))^3/(p_i(\hat{\underline{\theta}}_n))^2 \right)$ converges in law, it is to be noted that by
the mean value theorem,

$$\sqrt{n}(p_i(\hat{\underline{\theta}}_n) - p_{0i}) = \delta_i'|_{\underline{\theta}_n^*} \times \sqrt{n}(\hat{\underline{\theta}}_n - \underline{\theta}_0)$$
$$\text{where } \delta_i' = \left(\dfrac{\partial}{\partial \theta_1} p_i(\hat{\underline{\theta}}_n), \dfrac{\partial}{\partial \theta_2} p_i(\hat{\underline{\theta}}_n), \ldots, \dfrac{\partial}{\partial \theta_l} p_i(\hat{\underline{\theta}}_n) \right)$$

and $\underline{\theta}_n^* = \alpha \underline{\theta}_0 + (1 - \alpha)\hat{\underline{\theta}}_n, 0 < \alpha < 1$. Since $\hat{\underline{\theta}}_n \overset{P}{\to} \underline{\theta}_0$, we have $\underline{\theta}_n^* \overset{P}{\to} \underline{\theta}_0$. Hence,

$$
\begin{aligned}
\underline{U}_n &= \left(\frac{\sqrt{n}(p_1(\hat{\underline{\theta}}_n) - p_{01})}{(p_1(\hat{\underline{\theta}}_n))^2}, \frac{\sqrt{n}(p_2(\hat{\underline{\theta}}_n) - p_{02})}{(p_2(\hat{\underline{\theta}}_n))^2}, \ldots, \frac{\sqrt{n}(p_k(\hat{\underline{\theta}}_n) - p_{0k})}{(p_k(\hat{\underline{\theta}}_n))^2} \right)' \\
&= B_n \left(\sqrt{n}(p_1(\hat{\underline{\theta}}_n) - p_{01}), \sqrt{n}(p_2(\hat{\underline{\theta}}_n) - p_{02}), \ldots, \sqrt{n}(p_k(\hat{\underline{\theta}}_n) - p_{0k}) \right)' \\
&= B_n \left(\delta_1'|_{\underline{\theta}_n^*} \sqrt{n}(\hat{\underline{\theta}}_n - \underline{\theta}_0), \delta_2'|_{\underline{\theta}_n^*} \sqrt{n}(\hat{\underline{\theta}}_n - \underline{\theta}_0), \ldots, \delta_k'|_{\underline{\theta}_n^*} \sqrt{n}(\hat{\underline{\theta}}_n - \underline{\theta}_0) \right)' \\
&= B_n M_n \sqrt{n}(\hat{\underline{\theta}}_n - \underline{\theta}_0),
\end{aligned}
$$

where B_n is a diagonal matrix of order $k \times k$ with diagonal elements $1/(p_i(\hat{\underline{\theta}}_n))^2$, $i = 1, 2, \ldots, k$ and M_n is a matrix of order $k \times l$ with i-th row as $\delta_i'|_{\underline{\theta}_n^*}$, $i = 1, 2, \ldots, k$. Since $\hat{\underline{\theta}}_n \xrightarrow{P} \underline{\theta}_0$, we have $B_n \xrightarrow{P} B$ where B is a matrix of order $k \times k$ with diagonal elements $1/(p_i(\underline{\theta}_0))^2$, $i = 1, 2, \ldots, k$. Since $\underline{\theta}_n^* \xrightarrow{P} \underline{\theta}_0$, we have $M_n \xrightarrow{P} M$ where M is a matrix of order $k \times l$ with i-th row as $\delta_i'|_{\underline{\theta}_0}$, $i = 1, 2, \ldots, k$. Further, $\sqrt{n}(\hat{\underline{\theta}}_n - \underline{\theta}_0) \xrightarrow{L} \underline{Z} \sim N_l(\underline{0}, \Sigma)$, where Σ is a positive definite matrix. Hence, $B_n M_n \sqrt{n}(\hat{\underline{\theta}}_n - \underline{\theta}_0) \xrightarrow{L} BM\underline{Z}$. Again defining a function $g : \mathbb{R}^k \to \mathbb{R}$ as $g(\underline{x}) = \sum_{i=1}^{k} x_i^3$, which is a continuous function and by using the continuous mapping theorem, we get that $(2a_1/\sqrt{n}) \sum_{i=1}^{k}((p_i(\hat{\underline{\theta}}_n) - p_{0i}))^3/(p_1(\hat{\underline{\theta}}_n))^2 \xrightarrow{P} 0$. Now using the similar arguments, it can be shown that the remaining terms in T_{1n} and V_n converge in probability to 0. Thus, $V_n \xrightarrow{P} 0$ and hence $-2\log \lambda(\underline{X})$ and $T_n(P) = \sum_{i=1}^{k}(o_i - e_i)^2/e_i$ both have the limit law as χ^2_{k-l-1} under H_0.

(ii) As in Theorem 6.2.1, $T_n(P)$ can be expressed as

$$
T_n(P) = \sum_{i=1}^{k} n(\hat{p}_{in} - p_i(\hat{\underline{\theta}}_n))^2/p_i(\hat{\underline{\theta}}_n) = n(\hat{\underline{p}}_n - \underline{p}(\hat{\underline{\theta}}_n))' A_n (\hat{\underline{p}}_n - \underline{p}(\hat{\underline{\theta}}_n)),
$$

where $A_n = \text{diag}\left(1/p_1(\hat{\underline{\theta}}_n), 1/p_2(\hat{\underline{\theta}}_n), \ldots, 1/p_k(\hat{\underline{\theta}}_n)\right)$. Similarly, W_n can be expressed as

$$
\begin{aligned}
W_n &= \sum_{i=1}^{k} \frac{(o_i - e_i)^2}{o_i} = \sum_{i=1}^{k} \frac{(n\hat{p}_{in} - np_i(\hat{\underline{\theta}}_n))^2}{n\hat{p}_{in}} \\
&= \sum_{i=1}^{k} \frac{n(\hat{p}_{in} - p_i(\hat{\underline{\theta}}_n))^2}{\hat{p}_{in}} = n(\hat{\underline{p}}_n - \underline{p}(\hat{\underline{\theta}}_n))' B_n (\hat{\underline{p}}_n - \underline{p}(\hat{\underline{\theta}}_n)),
\end{aligned}
$$

where $B_n = \text{diag}\left(1/\hat{p}_{1n}, 1/\hat{p}_{2n}, \ldots, 1/\hat{p}_{kn}\right)$. Suppose $A = \text{diag}(1/p_{01}, 1/p_{02}, \ldots, 1/p_{0k})$. Now under H_0,

$$
\hat{\underline{\theta}}_n \xrightarrow{P} \underline{\theta}_0 \Rightarrow A_n \xrightarrow{P} A \ \& \ \hat{\underline{p}}_n \xrightarrow{P} \underline{p}_0 \Rightarrow B_n \xrightarrow{P} A
$$

$$
\Rightarrow T_n(P) - W_n = n(\hat{\underline{p}}_n - \underline{p}(\hat{\underline{\theta}}_n))'(A_n - B_n)(\hat{\underline{p}}_n - \underline{p}(\hat{\underline{\theta}}_n)) \xrightarrow{P} 0
$$

$$
\Rightarrow W_n = \sum_{i=1}^{k}(o_i - e_i)^2/o_i \xrightarrow{L} U \sim \chi^2_{k-l-1} \text{ as } T_n(P) \xrightarrow{L} U \sim \chi^2_{k-l-1}.
$$

$\square$

Table 6.6 IQ scores: observed and expected frequencies

IQ score	$o_i = X_i$	$\hat{p}_{in}(\hat{\theta}_n)$	e_i
≤ 90	10	0.067046	6.7046
$(90, 100]$	18	0.164623	16.4622
$(100, 110]$	23	0.256148	25.6148
$(110, 120]$	22	0.252670	25.2670
$(120, 130]$	18	0.158005	15.8005
> 130	9	0.062613	6.2613

For the data given in Example 6.3.3, we find the values of T_n and W_n. We have $\hat{\mu}_n = \overline{X}_n = 109.7$ and $\hat{\sigma}_n^2 = 210.91$. Further, we note that $o_i = X_i$ and we find e_i as $e_i = n\hat{p}_{in}(\hat{\theta}_n)$, $i = 1, 2, \ldots, 6$. Table 6.6 displays the values of observed and expected frequencies.

For the given data, $T_n(P) = 3.9567$. For large n, H_0 is rejected if $T_n > c$ where c is such that the size of the test is α and the asymptotic null distribution of $T_n(P)$, which is χ_3^2. Thus $c = \chi_{3,1-\alpha}^2$. Further for the given data, the value of the test statistic W_n is 3.1019. For large n, H_0 is rejected if $W_n > c$, where c is such that the size of the test is α and is determined using the χ_3^2 distribution. Thus if $\alpha = 0.05$ then $c = \chi_{3,0.95}^2 = 7.8147$ which is larger than values of both $T_n(P)$ and W_n. Hence, data do not show the sufficient evidence to reject H_0 and normal distribution can be taken as a good model for IQ scores.

✐ Remark 6.3.1

In Example 6.3.2, the parameters of the binomial distribution are known. In most of the situations, the numerical parameter is known, but the probability of success θ is not known. We then estimate it from the given data. Thus, for binomial $B(m, \theta)$ distribution, truncated at 0, the maximum likelihood estimator which is same as the moment estimator based on the sufficient statistic is given by the solution of the equation $\overline{X}_n = m\theta/(1 - (1 - \theta)^m)$. In such a situation the degrees of freedom of the asymptotic null chi-square distribution is reduced by 1, as we estimate one parameter in the null space.

In the next section, we discuss a class of large sample tests and exhibit how likelihood ratio test, test based on Karl Pearson's chi-square statistic and a test based on $W_n = (o_i - e_i)^2/o_i$ are related to each other in a general setup.

6.4   Score Test and Wald's Test

In Theorem 6.3.1 and Theorem 6.3.2, we have noted that for a multinomial distribution in k cells, for testing $H_0 : \underline{p} = \underline{p}_0$ against the alternative $H_1 : \underline{p} \neq \underline{p}_0$ or for testing $H_0 : \underline{p} = \underline{p}(\theta)$ against the alternative $H_1 : \underline{p} \neq \underline{p}(\theta)$, the likelihood ratio test, test based on Karl Pearson's chi-square statistic and a test based on the test statistic $W_n = \sum_{i=1}^{k}(o_i - e_i)^2/o_i$ are asymptotically equivalent, in the sense that the asymptotic null distributions of all the three test statistics are the same. In this section, we study a general class of large sample tests where these three test statistics are asymptotically equivalent. This class includes a score test and Wald's test.

Suppose X is a random variable or a random vector with probability law $f(x, \underline{\theta})$ indexed by a parameter $\underline{\theta} \in \Theta \subset \mathbb{R}^k$, $k \geq 1$. Suppose $\hat{\underline{\theta}}_n$ is the maximum likelihood estimator of $\underline{\theta}$ based on a random sample $\{X_1, X_2, \ldots, X_n\}$ from the distribution of X. Further, we assume that $\hat{\underline{\theta}}_n$ is a CAN estimator of $\underline{\theta}$ with approximate dispersion matrix $I^{-1}(\underline{\theta})/n$, which is true if the probability law $f(x, \underline{\theta})$ belongs to a k-parameter exponential family or a k-parameter Cramér family. As a consequence

$$Q_n(\underline{X}, \underline{\theta}) = n(\hat{\underline{\theta}}_n - \underline{\theta})' I(\underline{\theta})(\hat{\underline{\theta}}_n - \underline{\theta}) \xrightarrow{L} U \sim \chi_k^2 \,.$$

Suppose $S_n(\underline{X}, \underline{\theta})$ is defined as

$$S_n(\underline{X}, \underline{\theta}) = n(\hat{\underline{\theta}}_n - \underline{\theta})' M_n(\hat{\underline{\theta}}_n - \underline{\theta}),$$

$$\text{where } M_n = [M_n(i, j)] \, \& \, M_n(i, j) = -\frac{1}{n} \frac{\partial^2 \log L_n(\underline{\theta}|X)}{\partial \theta_i \partial \theta_j}\Big|_{\hat{\underline{\theta}}_n}.$$

Since $M_n(i, j) \xrightarrow{P_\theta} I_{ij}(\underline{\theta})$ for all $i, j = 1, 2, \ldots, k$, we have $M_n \xrightarrow{P_\theta} I(\underline{\theta})$. Hence as $n \to \infty$,

$$S_n(\underline{X}, \underline{\theta}) - Q_n(\underline{X}, \underline{\theta}) = n(\hat{\underline{\theta}}_n - \underline{\theta})'(M_n - I(\underline{\theta}))(\hat{\underline{\theta}}_n - \underline{\theta})$$

$$\xrightarrow{P_\theta} 0 \quad \Rightarrow \quad S_n(\underline{X}, \underline{\theta}) \xrightarrow{L} U \sim \chi_k^2 \,.$$

Further, we define $W_n(\underline{X}, \underline{\theta})$ as

$$W_n(\underline{X}, \underline{\theta}) = n(\hat{\underline{\theta}}_n - \underline{\theta})' I(\hat{\underline{\theta}}_n)(\hat{\underline{\theta}}_n - \underline{\theta}) \,.$$

We have $\hat{\underline{\theta}}_n$ to be consistent for $\underline{\theta}$ and hence if each element of matrix $I(\underline{\theta})$ is a continuous function of $\underline{\theta}$, then $I(\hat{\underline{\theta}}_n) \xrightarrow{P} I(\underline{\theta})$. Consequently,

$$W_n(\underline{X}, \underline{\theta}) - Q_n(\underline{X}, \underline{\theta}) = n(\hat{\underline{\theta}}_n - \underline{\theta})'(I(\hat{\underline{\theta}}_n) - I(\underline{\theta}))(\hat{\underline{\theta}}_n - \underline{\theta})$$

$$\xrightarrow{P} 0 \quad \Rightarrow \quad W_n(\underline{X}, \underline{\theta}) \xrightarrow{L} U \sim \chi_k^2.$$

Suppose $U_i(\underline{X}, \underline{\theta})$ for $i = 1, 2, \ldots, k$ is defined as

$$U_i(\underline{X}, \underline{\theta}) = \frac{1}{\sqrt{n}} \frac{\partial \log L_n(\underline{\theta}|X)}{\partial \theta_i} = \frac{1}{\sqrt{n}} \sum_{r=1}^{n} \frac{\partial \log f(X_r, \underline{\theta})}{\partial \theta_i}$$

and a random vector $\underline{V}_n(\underline{X}, \underline{\theta})$ is defined as $\underline{V}_n(\underline{X}, \underline{\theta}) = (U_1(\underline{X}, \underline{\theta}), U_2(\underline{X}, \underline{\theta}), \ldots,$
$U_k(\underline{X}, \underline{\theta}))'$. $\underline{V}_n(\underline{X}, \underline{\theta})$ is known as a vector of score functions. Further it is known
that

$$E_{\underline{\theta}}\left(\frac{\partial \log f(X, \underline{\theta})}{\partial \theta_i}\right) = 0 \quad \& \quad Cov\left(\frac{\partial \log f(X, \underline{\theta})}{\partial \theta_i}, \frac{\partial \log f(X, \underline{\theta})}{\partial \theta_j}\right)$$

$$= I_{ij}(\underline{\theta}), \quad i, j = 1, 2, \ldots, k \ .$$

As a consequence, $E_{\underline{\theta}}(\underline{V}_n(\underline{X}, \underline{\theta})) = \underline{0}$ and dispersion matrix D of $\underline{V}_n(\underline{X}, \underline{\theta})$ is $I(\underline{\theta})$.
If $\frac{\partial \log f(X_r, \underline{\theta})}{\partial \theta_i}$ is denoted by Y_{ir}, and $(Y_{1r}, Y_{2r}, \ldots, Y_{kr})'$ by $\underline{Y}_r$, $r = 1, 2, \ldots, n$,
then $E(\underline{Y}_r) = \underline{0}$ and its dispersion matrix is given by $I(\underline{\theta})$, which is positive definite.
Further, $\{X_1, X_2, \ldots, X_n\}$ are independent and identically distributed random vari-
ables, implies $\{\underline{Y}_1, \underline{Y}_2, \ldots, \underline{Y}_n\}$ are independent and identically distributed random
vectors. Hence by the multivariate CLT,

$$\underline{V}_n(\underline{X}, \underline{\theta}) = (U_1(\underline{X}, \underline{\theta}), U_2(\underline{X}, \underline{\theta}), \ldots, U_k(\underline{X}, \underline{\theta})'$$

$$= \frac{1}{\sqrt{n}} \sum_{r=1}^{n} \underline{Y}_r \xrightarrow{L} \underline{Z}_2 \sim N_k(\underline{0}, I(\underline{\theta})) \ .$$

Consequently,

$$F_n(\underline{X}, \underline{\theta}) = \underline{V}_n'(\underline{X}, \underline{\theta}) I^{-1}(\underline{\theta}) \underline{V}_n(\underline{X}, \underline{\theta}) \xrightarrow{L} U \sim \chi_k^2 \ .$$

Thus, $Q_n(\underline{X}, \underline{\theta})$, $S_n(\underline{X}, \underline{\theta})$, $W_n(\underline{X}, \underline{\theta})$ and $F_n(\underline{X}, \underline{\theta})$ have the same limiting distribution and it is χ_k^2. Further, observe that

$$\underline{V}_n(\underline{X}, \underline{\theta}) \xrightarrow{L} \underline{Z}_2 \sim N_k(\underline{0}, I(\underline{\theta})) \implies I^{-1}(\underline{\theta}) \underline{V}_n(\underline{X}, \underline{\theta}) \xrightarrow{L} \underline{Z}_1 \sim N_k(\underline{0}, I^{-1}(\underline{\theta})) \ .$$

We know that $\sqrt{n}(\hat{\underline{\theta}}_n - \underline{\theta}) \xrightarrow{L} \underline{Z}_1 \sim N_k(\underline{0}, I^{-1}(\underline{\theta}))$. Thus, $I^{-1}(\underline{\theta}) \underline{V}_n(\underline{X}, \underline{\theta})$ and
$\sqrt{n}(\hat{\underline{\theta}}_n - \underline{\theta})$ have the same limit law.

In Table 6.7 we summarize all these results. These results are heavily used in
testing null hypothesis $H_0 : \underline{\theta} = \underline{\theta}_0$ against the alternative $H_1 : \underline{\theta} \neq \underline{\theta}_0$ based on a
random sample $\{X_1, X_2, \ldots, X_n\}$ from the distribution of X. We assume that $\underline{\theta}_0$ is
a completely specified vector, thus H_0 is a simple null hypothesis.

We now discuss three test procedures to test H_0 against H_1.

Table 6.7 Limit laws of quadratic forms

Quadratic form	Limit law
$Q_n(\underline{X}, \underline{\theta}) = n(\hat{\underline{\theta}}_n - \underline{\theta})' I(\underline{\theta})(\hat{\underline{\theta}}_n - \underline{\theta})$	χ_k^2
$S_n(\underline{X}, \underline{\theta}) = n(\hat{\underline{\theta}}_n - \underline{\theta})' M_n(\hat{\underline{\theta}}_n - \underline{\theta})$	χ_k^2
$W_n(\underline{X}, \underline{\theta}) = n(\hat{\underline{\theta}}_n - \underline{\theta})' I(\hat{\underline{\theta}}_n)(\hat{\underline{\theta}}_n - \underline{\theta})$	χ_k^2
$F_n(\underline{X}, \underline{\theta}) = \underline{V}_n'(\underline{X}, \underline{\theta}) I^{-1}(\underline{\theta}) \underline{V}_n(\underline{X}, \underline{\theta})$	χ_k^2

Likelihood ratio test: In Chap. 5 we have discussed this test procedure in detail. Suppose $\lambda(\underline{X})$ is a test statistic for testing $H_0 : \underline{\theta} = \underline{\theta}_0$ against the alternative $H_1 : \underline{\theta} \neq \underline{\theta}_0$. In Theorem 5.2.2, it is proved that

$$-2 \log \lambda(\underline{X}) - S_n(\underline{X}, \underline{\theta}) \xrightarrow{P} 0 \quad \Rightarrow \quad -2 \log \lambda(\underline{X}) \xrightarrow{L} U \sim \chi_k^2 .$$

H_0 is rejected if $-2 \log \lambda(\underline{X}) > c$ where the cut-off point c is determined corresponding to the given level of significance α and using the asymptotic null distribution of $-2 \log \lambda(\underline{X})$, which is χ_k^2, which implies that $c = \chi_{k,1-\alpha}^2$. Neyman and Pearson proposed this test procedure in 1928.

Wald's test: Wald's test for testing $H_0 : \underline{\theta} = \underline{\theta}_0$ against the alternative $H_1 : \underline{\theta} \neq \underline{\theta}_0$ is based on the test statistic $T_n(W)$ given by

$$T_n(W) = W_n(\underline{X}, \underline{\theta}_0) = n(\hat{\underline{\theta}}_n - \underline{\theta}_0)' I(\hat{\underline{\theta}}_n)(\hat{\underline{\theta}}_n - \underline{\theta}_0) .$$

H_0 is rejected if $T_n(W) > c$ where the cut-off point c is determined corresponding to the given level of significance α and using the null distribution of $T_n(W)$. We have proved above that $W_n(\underline{X}, \underline{\theta}) = n(\hat{\underline{\theta}}_n - \underline{\theta})' I(\hat{\underline{\theta}}_n)(\hat{\underline{\theta}}_n - \underline{\theta}) \xrightarrow{L} U \sim \chi_k^2$, hence the asymptotic null distribution of Wald's test statistic $T_n(W)$ is χ_k^2, which implies $c = \chi_{k,1-\alpha}^2$. Wald proposed this test procedure in 1943.

Score test: It is proposed by C.R.Rao in 1947. The test statistic for a score test for testing $H_0 : \underline{\theta} = \underline{\theta}_0$ against the alternative $H_1 : \underline{\theta} \neq \underline{\theta}_0$ is based on a score function and is given by

$$T_n(S) = F_n(\underline{X}, \underline{\theta}_0) = \underline{V}_n'(\underline{X}, \underline{\theta}_0) I^{-1}(\underline{\theta}_0) \underline{V}_n(\underline{X}, \underline{\theta}_0) .$$

H_0 is rejected if $T_n(S) > c$ where the cut-off point c is determined corresponding to the given level of significance α and using the asymptotic null distribution of $T_n(S)$. We have proved above that $F_n(\underline{X}, \underline{\theta}) \xrightarrow{L} U \sim \chi_k^2$, hence the asymptotic null distribution of the score test statistic $T_n(S)$ is χ_k^2, which implies that $c = \chi_{k,1-\alpha}^2$.

Thus, all the three test procedures described above are asymptotically equivalent, in the sense that, the asymptotic null distribution for all the three statistics is the same. For a score test, computation of $\hat{\underline{\theta}}_n$ is not necessary, which can be a major advantage for some probability models.

✍ Remark 6.4.1

It is to be noted that $s.e.(\hat{\theta}_n) = \sqrt{1/nI(\hat{\theta}_n)}$ is the standard error of $\hat{\theta}_n$. Hence, in real parameter setup, Wald's test statistic and a score test statistic can also be defined as

$$T_n^*(W) = \sqrt{T_n(W)} = \sqrt{nI(\hat{\theta}_n)}(\hat{\theta}_n - \theta_0) = (\hat{\theta}_n - \theta_0)/s.e.(\hat{\theta}_n)$$
$$T_n^*(S) = \sqrt{T_n(S)} = \sqrt{nI(\theta_0)}(\hat{\theta}_n - \theta_0) = (\hat{\theta}_n - \theta_0)/s.e.(\hat{\theta}_n)|_{\theta_0} ,$$

where $s.e.(\hat{\theta}_n)|_{\theta_0}$ is the standard error of $\hat{\theta}_n$ evaluated at θ_0. In both the procedures under H_0, the asymptotic null distributions of the test statistics $T_n^*(W)$ and $T_n^*(S)$ are standard normal. In Wald's test procedure H_0 is rejected at level of significance α if $|T_n^*(W)| > c$, similarly in a score test procedure H_0 is rejected if $|T_n^*(S)| > c$, where $c = a_{1-\alpha/2}$.

In simple or multiple liner regression or in logistic regression, for testing significance of the regression coefficients, the most frequently used test is Wald's test. For example, in a simple liner regression model $Y = \beta_0 + \beta_1 X + \epsilon$, the test statistic for testing $H_0 : \beta_1 = 0$ against the alternative $H_1 : \beta_1 \neq 0$ is $T_n = \hat{\beta}_{1n}/s.e.(\hat{\beta}_{1n})$, which is Wald's test statistic. For large n, its null distribution is standard normal and H_0 is rejected at level of significance α, if $|T_n| > a_{1-\alpha/2}$. Many statistical software use Wald's test.

In the following examples we illustrate these test procedures.

☑ Example 6.4.1

Suppose $\{X_1, X_2, \ldots, X_n\}$ is a random sample from Cauchy $C(\theta, 1)$ distribution. We derive Wald's test procedure and a score test procedure for testing $H_0 : \theta = \theta_0$ against the alternative $H_1 : \theta \neq \theta_0$. In Chap. 4, we have proved that Cauchy $C(\theta, 1)$ distribution belongs to a Cramér family. Hence for large n, the maximum likelihood estimator $\hat{\theta}_n$ is CAN for θ with approximate variance $1/nI(\theta) = 2/n$ as $I(\theta) = 1/2$ for Cauchy $C(\theta, 1)$ distribution. Thus $s.e.(\hat{\theta}_n) = \sqrt{2/n}$, which is free from θ. Hence, Wald's test statistic $T_n^*(W)$ and the score test statistic $T_n^*(S)$ are the same and are given by

$$T_n^*(W) = T_n^*(S) = (\hat{\theta}_n - \theta_0)/s.e.(\hat{\theta}_n) = \sqrt{n/2}(\hat{\theta}_n - \theta_0) .$$

For large n, its null distribution is standard normal and H_0 is rejected at level of significance α if $|T_n^*(W)| > a_{1-\alpha/2}$.   □

In the next example for a Bernoulli $B(1, \theta)$ distribution, we compute the limit laws of four quadratic forms as listed in Table 6.7 and illustrate how these are useful to test the null hypothesis $H_0 : \theta = \theta_0$.

☑ Example 6.4.2

Suppose X follows Bernoulli $B(1, \theta)$ distribution and $\underline{X} = \{X_1, X_2, \ldots, X_n\}$ is a random sample from the distribution of X. We derive likelihood ratio test procedure, Wald's test procedure and the score test procedure for testing $H_0 : \theta = \theta_0$ against the alternative $H_1 : \theta \neq \theta_0$, where θ_0 is a specified constant, assuming sample size n is large. Since $X \sim B(1, \theta)$, its probability mass function $p(x, \theta)$ is

$$p(x, \theta) = P_\theta[X = x] = \theta^x (1 - \theta)^{1-x}, \quad x = 0, 1; \quad 0 < \theta < 1.$$

Hence, the likelihood function of θ corresponding to the random sample $\underline{X}$ is

$$L_n(\theta|\underline{X}) = \theta^{\sum_{i=1}^{n} X_i} (1 - \theta)^{\sum_{i=1}^{n}(1-X_i)} \quad \Leftrightarrow \quad \log L_n(\theta|\underline{X})$$
$$= \log \theta \sum_{i=1}^{n} X_i + \log(1 - \theta)(n - \sum_{i=1}^{n} X_i).$$

$\log L_n(\theta|\underline{X})$ is a differentiable function of θ. Hence,

$$\frac{\partial \log L_n(\theta|\underline{X})}{\partial \theta} = \frac{\sum_{i=1}^{n} X_i}{\theta} - \frac{n - \sum_{i=1}^{n} X_i}{1 - \theta}$$

$$\& \quad \frac{\partial^2 \log L_n(\theta|\underline{X})}{\partial \theta^2} = -\frac{\sum_{i=1}^{n} X_i}{\theta^2} - \frac{n - \sum_{i=1}^{n} X_i}{(1 - \theta)^2}.$$

We assume that all $X_i's$ are not 0 or not 1. It is to be noted that $\frac{\partial^2 \log L_n(\theta|\underline{X})}{\partial \theta^2} < 0$ for any realization of the random sample $\{X_1, X_2, \ldots, X_n\}$ which implies that at the solution of the likelihood equation, the likelihood attains the maximum. Thus, the maximum likelihood estimator $\hat{\theta}_n$ of θ is given by $\hat{\theta}_n = \overline{X}_n$, which is the relative frequency of occurrence of outcome 1 in the random sample of size n. By the WLLN, $\hat{\theta}_n \overset{L}{\to} \theta$ and by the CLT

$$\sqrt{n}(\hat{\theta}_n - \theta) \overset{L}{\to} Z_1 \sim N(0, \theta(1 - \theta))$$
$$\Leftrightarrow \quad Z_n = \frac{\sqrt{n}}{\sqrt{\theta(1 - \theta)}}(\hat{\theta}_n - \theta) \overset{L}{\to} Z \sim N(0, 1).$$

Thus, $\hat{\theta}_n$ is CAN estimator of θ with approximate variance $\theta(1 - \theta)/n$. Now,

$$Z_n \overset{L}{\to} Z \sim N(0, 1), \quad \Rightarrow \quad Q_n(\underline{X}, \theta) = Z_n^2 = \frac{n}{\theta(1 - \theta)}(\hat{\theta}_n - \theta)^2 \overset{L}{\to} U \sim \chi_1^2.$$

Further,

$$M_n = -\frac{1}{n} \frac{\partial^2 \log L_n(\theta|\underline{X})}{\partial \theta^2}|_{\hat{\theta}_n} = \frac{1}{\hat{\theta}_n(1 - \hat{\theta}_n)}$$
$$\Rightarrow \quad S_n(\underline{X}, \theta) = \frac{n}{\overline{X}_n(1 - \overline{X}_n)}(\hat{\theta}_n - \theta)^2 \overset{L}{\to} U \sim \chi_1^2.$$

The information function $I(\theta)$ is given by
$I(\theta) = E_\theta \left(-\frac{\partial^2 \log p(X, \theta)}{\partial \theta^2} \right) = 1/\theta(1 - \theta)$. Hence,

$$W_n(\underline{X}, \theta) = nI(\hat{\theta}_n)(\hat{\theta}_n - \theta)^2 = \frac{n}{\hat{\theta}_n(1 - \hat{\theta}_n)}(\hat{\theta}_n - \theta)^2 \xrightarrow{L} U \sim \chi_1^2.$$

The score function $V_n(\underline{X}, \underline{\theta})$ is given by

$$V_n(\underline{X}, \theta) = \frac{1}{\sqrt{n}} \frac{\partial \log L_n(\theta | \underline{X})}{\partial \theta}$$

$$= \frac{1}{\sqrt{n}} \left(\frac{\sum_{i=1}^{n} X_i}{\theta} - \frac{n - \sum_{i=1}^{n} X_i}{1 - \theta} \right) = \frac{\sqrt{n}(\hat{\theta}_n - \theta)}{\theta(1 - \theta)}.$$

$$\text{Now,} \quad V_n(\underline{X}, \theta) \xrightarrow{L} Z_2 \sim N\left(0, \frac{1}{\theta(1 - \theta)} \right)$$

$$\Rightarrow \quad F_n(\underline{X}, \theta) = \frac{n}{\theta(1 - \theta)}(\hat{\theta}_n - \theta)^2 \xrightarrow{L} U \sim \chi_1^2.$$

Thus, corresponding to a random sample from $B(1, \theta)$, $S_n(\underline{X}, \theta) \xrightarrow{L} U$, $W_n(\underline{X}, \theta) \xrightarrow{L} U$ and $F_n(\underline{X}, \theta) \xrightarrow{L} U$ & $U \sim \chi_1^2$. Now to test $H_0 : \theta = \theta_0$ against the alternative $H_1 : \theta \neq \theta_0$, the likelihood ratio test statistic $-2 \log \lambda(\underline{X})$, Wald's test statistic $T_n(W)$ and the score test statistic is $T_n(S)$ are given by

$$-2 \log \lambda(\underline{X}) = 2n \left[\overline{X}_n \log \frac{\overline{X}_n}{\theta_0} + (1 - \overline{X}_n) \log \frac{1 - \overline{X}_n}{1 - \theta_0} \right]$$

$$T_n(W) = W_n(\underline{X}, \theta_0) = \frac{n}{\hat{\theta}_n(1 - \hat{\theta}_n)}(\hat{\theta}_n - \theta_0)^2$$

$$\& \ T_n(S) = F_n(\underline{X}, \theta_0) = \frac{n}{\theta_0(1 - \theta_0)}(\hat{\theta}_n - \theta_0)^2.$$

For large n under H_0, $-2 \log \lambda(\underline{X})$ is distributed as $S_n(\underline{X}, \theta_0)$ and under H_0,

$$S_n(\underline{X}, \theta_0) \xrightarrow{L} U, \ W_n(\underline{X}, \theta_0) \xrightarrow{L} U, \ \& \ F_n(\underline{X}, \theta_0) \xrightarrow{L} U \ \& \ U \sim \chi_1^2.$$

H_0 is rejected if the value of the test statistic is larger than c, where the cutoff point c is determined corresponding to the given level of significance α and using the null distribution of the test statistic. For all the three test procedures, the asymptotic null distribution is the same as χ_1^2 which implies that $c = \chi_{1, 1-\alpha}^2$.

We now express $T_n(W)$ and $T_n(S)$ as follows. Note that the sample mean $\overline{X}_n$ is nothing but the proportion of successes in the sample of size n and is given by $\hat{\theta}_n = \overline{X}_n = o_1/n$ where o_1 denotes the number of 1's, that is, number of successes in n trials. Suppose o_0 denotes the number of 0's, that is, number of failures in n trials. Suppose $e_1 = n\theta_0$ and $e_0 = n(1 - \theta_0)$ denote the expected number of successes and failures in n trials under the null setup. Further, $o_1 + o_0 = n$ and $e_1 + e_0 = n$. With this notation, $T_n(W)$ can be rewritten as follows:

$$T_n(W) = \frac{n}{\hat{\theta}_n(1-\hat{\theta}_n)}(\hat{\theta}_n - \theta_0)^2 = \frac{n}{o_1/n(1 - o_1/n)}(o_1/n - \theta_0)^2$$

$$= \frac{(o_1 - e_1)^2}{o_1}\frac{n}{o_0} = \frac{(o_1 - e_1)^2}{o_1}\frac{(o_1 + o_0)}{o_0}$$

$$= \frac{(o_1 - e_1)^2}{o_1}(1 + \frac{o_1}{o_0}) = \frac{(o_1 - e_1)^2}{o_1} + \frac{(o_1 - e_1)^2}{o_1}\frac{o_1}{o_0}$$

$$= \frac{(o_1 - e_1)^2}{o_1} + \frac{(n - o_0 - n + e_0)^2}{o_1}\frac{o_1}{o_0}$$

$$= \frac{(o_1 - e_1)^2}{o_1} + \frac{(o_0 - e_0)^2}{o_0}$$

similarly, $$T_n(S) = \frac{n}{\theta_0(1-\theta_0)}(\hat{\theta}_n - \theta_0)^2 = \frac{n^3}{n\theta_0 n(1 - \theta_0)}(o_1/n - \theta_0)^2$$

$$= \frac{(o_1 - e_1)^2}{e_1}\frac{n}{e_0} = \frac{(o_1 - e_1)^2}{e_1}\frac{(e_0 + e_1)}{e_0}$$

$$= \frac{(o_1 - e_1)^2}{e_1} + \frac{(o_1 - e_1)^2}{e_1}\frac{e_1}{e_0} = \frac{(o_1 - e_1)^2}{e_1} + \frac{(o_0 - e_0)^2}{e_0}.$$

Thus, $T_n(S)$ is exactly same as Karl Pearson's chi-square test statistic $T_n(P)$. $\square$

📖 Remark 6.4.2

In the above example, it is to be noted that $s.e.(\hat{\theta}_n) = \sqrt{\hat{\theta}_n(1 - \hat{\theta}_n)}/\sqrt{n}$ is the standard error of $\hat{\theta}_n$. Hence, for testing $H_0 : \theta = \theta_0$ against the alternative $H_1 : \theta \neq \theta_0$ for $B(1, \theta)$ distribution, Wald's test statistic $T_n^*(W)$ and the score test statistic $T_n^*(S)$ are as follows:

$$T_n^*(W) = \sqrt{T_n(W)} = \frac{\sqrt{n}}{\sqrt{\hat{\theta}_n(1-\hat{\theta}_n)}}(\hat{\theta}_n - \theta_0) = \frac{\hat{\theta}_n - \theta_0}{s.e.(\hat{\theta}_n)}$$

$$T_n^*(S) = \sqrt{T_n(S)} = \frac{\sqrt{n}}{\sqrt{\theta_0(1-\theta_0)}}(\hat{\theta}_n - \theta_0) = \frac{\hat{\theta}_n - \theta_0}{s.e.(\hat{\theta}_n)|_{\theta_0}}.$$

In both the procedures under H_0, the asymptotic null distribution of the test statistics is standard normal. In Wald's test procedure H_0 is rejected if

$|T_n^*(W)| > c$ and in the score test procedure H_0 is rejected if $|T_n^*(S)| > c$, where $c = a_{1-\alpha/2}$. We rewrite $T_n^*(S)$ as follows:

$$T_n^*(S) = \frac{\sqrt{n}}{\sqrt{\theta_0(1-\theta_0)}}(\hat{\theta}_n - \theta_0) = \frac{\sqrt{n}(U/n - p_0)}{\sqrt{p_0(1-p_0)}}$$

$$= \frac{\sqrt{n}(P_n - p_0)}{\sqrt{p_0(1-p_0)}}, \quad \text{where} \ p_0 = \theta_0,$$

U denotes the total number of successes in a random sample of size n and P_n is the proportion of successes. The function prop.test from R uses the score test statistic for testing the null hypothesis $H_0 : p = p_0$ based on the sample proportion P_n, where p denotes the population proportion. We illustrate it in Sect. 6.7.

In Theorem 6.3.1, it is proved that for a multinomial distribution with k cells, the likelihood ratio test statistic and Karl Pearson's chi-square statistic $T_n(P) = \frac{(o_i - e_i)^2}{e_i}$ and a statistic $W_n = \frac{(o_i - e_i)^2}{o_i}$ for testing $H_0 : \underline{p} = \underline{p}_0$ against the alternative $H_1 : \underline{p} \neq \underline{p}_0$, have the same limiting null distribution as χ^2_{k-1}. In the following theorem we prove an additional feature of these test statistics. In Example 6.4.2 for a Bernoulli distribution, we have shown that Wald's test statistic $T_n(W)$ can be expressed as $\sum_{i=1}^{2}(o_i - e_i)^2/o_i$ and the score test statistic $T_n(S)$ reduces to $\sum_{i=1}^{2}(o_i - e_i)^2/e_i$. In the following theorem, we extend the results of this example and prove that in a multinomial distribution with k cells, for testing $H_0 : \underline{p} = \underline{p}_0$ against the alternative $H_1 : \underline{p} \neq \underline{p}_0$, Wald's test statistic $T_n(W)$ simplifies to $\sum_{i=1}^{k}(o_i - e_i)^2/o_i$ while the score test statistic $T_n(S)$ simplifies to Karl Pearson's chi-square test statistic $T_n(P) = \sum_{i=1}^{k}(o_i - e_i)^2/e_i$.

Theorem 6.4.1

Suppose $\underline{Y} = \{Y_1, Y_2, \ldots, Y_{k-1}\}$ has a multinomial distribution in k cells with cell probabilities $\underline{p} = \{p_1, p_2, \ldots, p_{k-1}\}$, where $p_i > 0$, $i = 1, 2, \ldots, k$ with $p_k = 1 - \sum_{i=1}^{k-1} p_i$. On the basis of a random sample of size n from the distribution of $\underline{Y}$, suppose we want to test $H_0 : \underline{p} = \underline{p}_0$ against the alternative $H_1 : \underline{p} \neq \underline{p}_0$, where $\underline{p}_0$ is a completely specified vector. Suppose o_i and e_i denote that observed and expected cell frequencies of the i-th cell, $i = 1, 2, \ldots, k$. Then (i) Wald's test statistic $T_n(W)$ is the same as $\sum_{i=1}^{k}(o_i - e_i)^2/o_i$ and (ii) The score test statistic $T_n(S)$ is the same as $T_n(P) = \sum_{i=1}^{k}(o_i - e_i)^2/e_i$.

Proof Suppose $\underline{X} = (X_1, X_2, \ldots, X_k)'$ denotes the vector of cell frequencies corresponding to a random sample of size n from $\underline{Y}$ with $\sum_{i=1}^{k} X_i = n$. The maximum likelihood estimator $\hat{\underline{p}}_n = (\hat{p}_{1n}, \hat{p}_{2n}, \ldots, \hat{p}_{(k-1)n})'$ of $\underline{p}$ is then given by $\hat{\underline{p}}_n = (X_1/n, X_2/n, \ldots, X_{k-1}/n)'$.
(i) Suppose $I(\underline{p})_{(k-1)\times(k-1)} = [I_{ij}(\underline{p})]$ is the information matrix for the multinomial distribution in k cells, which is obtained in Sect. 6.2. Wald's test for testing

$H_0 : \underline{p} = \underline{p}_0$ against the alternative $H_1 : \underline{p} \neq \underline{p}_0$ is based on the test statistic

$$T_n(W) = n(\hat{\underline{p}}_n - \underline{p}_0)' I(\hat{\underline{p}}_n)(\hat{\underline{p}}_n - \underline{p}_0) \text{ with } I_{ii}(\underline{p}) = \frac{1}{p_i} + \frac{1}{p_k} \quad \& \quad I_{ij}(\underline{p}) = \frac{1}{p_k}.$$

Suppose $\underline{o} = (o_1, o_2, \ldots, o_{k-1})' = n\hat{\underline{p}}_n$ and $\underline{e} = (e_1, e_2, \ldots, e_{k-1})' = n\underline{p}_0$ denote the vector of observed and expected cell frequencies. Then $T_n(W)$ can be expressed as

$$T_n(W) = n(\hat{\underline{p}}_n - \underline{p}_0)' I(\hat{\underline{p}}_n)(\hat{\underline{p}}_n - \underline{p}_0) = \frac{1}{n}(n\hat{\underline{p}}_n - n\underline{p}_0)' I(\hat{\underline{p}}_n)(n\hat{\underline{p}}_n - n\underline{p}_0)$$

$$= \frac{1}{n}(\underline{o} - \underline{e})' I(\hat{\underline{p}}_n)(\underline{o} - \underline{e}),$$

which after using the form of $I(\hat{\underline{p}}_n)$, simplifies as follows:

$$T_n(W) = \sum_{i=1}^{k-1} (o_i - e_i)^2 \left(\frac{1}{n\hat{p}_{in}} + \frac{1}{n\hat{p}_{kn}} \right) + \sum_{i=1}^{k-1} \sum_{j \neq i=1}^{k-1} (o_i - e_i)(o_j - e_j) \frac{1}{n\hat{p}_{kn}}$$

$$= \sum_{i=1}^{k-1} \frac{(o_i - e_i)^2}{o_i} + \frac{1}{o_k} \left\{ \sum_{i=1}^{k-1} (o_i - e_i)^2 + \sum_{i=1}^{k-1} \sum_{j \neq i=1}^{k-1} (o_i - e_i)(o_j - e_j) \right\}$$

$$= \sum_{i=1}^{k-1} \frac{(o_i - e_i)^2}{o_i} + \frac{1}{o_k} \left\{ \sum_{i=1}^{k-1} (o_i - e_i) \right\}^2$$

$$= \sum_{i=1}^{k-1} \frac{(o_i - e_i)^2}{o_i} + \frac{1}{o_k}(o_k - e_k)^2 \quad \text{as} \quad \sum_{i=1}^{k}(o_i - e_i) = 0$$

$$= \sum_{i=1}^{k} \frac{(o_i - e_i)^2}{o_i}.$$

(ii) Score test for testing $H_0 : \underline{p} = \underline{p}_0$ against the alternative $H_1 : \underline{p} \neq \underline{p}_0$ is based on the test statistic $T_n(S) = \underline{V}_n'(\underline{X}, \underline{p}_0) I^{-1}(\underline{p}_0) \underline{V}_n(\underline{X}, \underline{p}_0)$, where $\underline{V}_n(\underline{X}, \underline{p})$ is a vector of score functions. It is defined as

$$\underline{V}_n(\underline{X}, \underline{p}) = (U_1(\underline{X}, \underline{p}), U_2(\underline{X}, \underline{p}), \ldots, U_{k-1}(\underline{X}, \underline{p}))'$$

$$\text{where } U_i(\underline{X}, \underline{p}) = \frac{1}{\sqrt{n}} \frac{\partial \log L_n(\underline{p}|\underline{X})}{\partial p_i}.$$

It is proved that $\underline{V}_n(\underline{X}, \underline{p}) \xrightarrow{L} \underline{Z}_1 \sim N_{k-1}(\underline{0}, I(\underline{p}))$. Consequently,

$\underline{V}_n'(\underline{X}, \underline{p}) I^{-1}(\underline{p}) \underline{V}_n(\underline{X}, \underline{p}) \xrightarrow{L} U$ which has χ^2_{k-1} distribution. In Sect. 6.2, we have verified that the information matrix $I(\underline{p})$ is the inverse of the dispersion matrix

$D = [\sigma_{ij}]$ where $\sigma_{ii} = p_i(1 - p_i)$ and $\sigma_{ij} = -p_i p_j$ of $\underline{Y} = (Y_1, Y_2, \ldots, Y_{k-1})'$ having multinomial distribution in k cells. Hence, with $\underline{V}_n \equiv \underline{V}_n(\underline{X}, \underline{p}_0)$, $T_n(S)$ can be written as $T_n(S) = \underline{V}'_n D(\underline{p}_0)\underline{V}_n$. To show that $T_n(S) = T_n(P)$, note that $X_i = o_i$ and $e_i = np_{0i}$. The log-likelihood of p corresponding to the observed cell frequencies $\underline{X}$ is given by $\log L_n(\underline{p}|\underline{X}) = \sum_{i=1}^{k} X_i \log p_i$. Hence,

$$U_i(\underline{X}, \underline{p}) = \frac{1}{\sqrt{n}} \frac{\partial \log L_n(\underline{p}|\underline{X})}{\partial p_i} = \frac{1}{\sqrt{n}} \left(\frac{X_i}{p_i} - \frac{X_k}{p_k} \right)$$

$$\Rightarrow \underline{V}'_n = \frac{1}{\sqrt{n}} \left(\frac{X_1}{p_{01}} - \frac{X_k}{p_{0k}}, \frac{X_2}{p_{02}} - \frac{X_k}{p_{0k}}, \ldots, \frac{X_{k-1}}{p_{0(k-1)}} - \frac{X_k}{p_{0k}} \right)$$

$$= \frac{1}{\sqrt{n}} n \left(\frac{o_1}{e_1} - \frac{o_k}{e_k}, \frac{o_2}{e_2} - \frac{o_k}{e_k}, \ldots, \frac{o_{k-1}}{e_{k-1}} - \frac{o_k}{e_k} \right)$$

$$= \sqrt{n} \left(\frac{o_1 - e_1}{e_1} - \frac{o_k - e_k}{e_k}, \frac{o_2 - e_2}{e_2} - \frac{o_k - e_k}{e_k}, \ldots, \frac{o_{k-1} - e_{k-1}}{e_{k-1}} - \frac{o_k - e_k}{e_k} \right).$$

Further, the dispersion matrix $D = [\sigma_{ij}]$ under H_0 can be rewritten as

$$\sigma_{ii} = p_{0i}(1 - p_{0i}) = \frac{e_i}{n} \left(1 - \frac{e_i}{n} \right) = \frac{e_i(n - e_i)}{n^2} \quad \& \quad \sigma_{ij} = -p_{0i}\,p_{0j} = -\frac{e_i e_j}{n^2}.$$

With these substitutions, $T_n(S)$, which is a quadratic form, can be expressed as follows.

$$nT_n(S) = \sum_{i=1}^{k-1} \left(\frac{o_i - e_i}{e_i} - \frac{o_k - e_k}{e_k} \right)^2 e_i(n - e_i)$$

$$- 2 \sum_{i=1}^{k-1} \sum_{j=i+1}^{k-1} \left(\frac{o_i - e_i}{e_i} - \frac{o_k - e_k}{e_k} \right) \left(\frac{o_j - e_j}{e_j} - \frac{o_k - e_k}{e_k} \right) e_i e_j$$

$$= \sum_{i=1}^{k-1} \left(\frac{o_i - e_i}{e_i} \right)^2 e_i(n - e_i) + \left(\frac{o_k - e_k}{e_k} \right)^2 \sum_{i=1}^{k-1} e_i(n - e_i)$$

$$- 2 \left(\frac{o_k - e_k}{e_k} \right) \sum_{i=1}^{k-1} \left(\frac{o_i - e_i}{e_i} \right) e_i(n - e_i)$$

$$- 2 \sum_{i=1}^{k-1} \sum_{j=i+1}^{k-1} \left(\frac{o_i - e_i}{e_i} \right) \left(\frac{o_j - e_j}{e_j} \right) e_i e_j - 2 \left(\frac{o_k - e_k}{e_k} \right)^2 \sum_{i=1}^{k-1} \sum_{j=i+1}^{k-1} e_i e_j$$

$$+ 2 \left(\frac{o_k - e_k}{e_k} \right) \sum_{i=1}^{k-1} \sum_{j=i+1}^{k-1} \left[\frac{o_i - e_i}{e_i} + \frac{o_j - e_j}{e_j} \right] e_i e_j$$

$$= n \sum_{i=1}^{k-1} \frac{(o_i - e_i)^2}{e_i} - \sum_{i=1}^{k-1}(o_i - e_i)^2 - 2 \sum_{i=1}^{k-1} \sum_{j=i+1}^{k-1} (o_i - e_i)(o_j - e_j)$$

$$+ \left(\frac{o_k - e_k}{e_k} \right)^2 \left[\sum_{i=1}^{k-1} e_i(n - e_i) - 2 \sum_{i=1}^{k-1} \sum_{j=i+1}^{k-1} e_i e_j \right]$$

$$+ 2 \left(\frac{o_k - e_k}{e_k} \right)$$

$$\left[\sum_{i=1}^{k-1} \sum_{j=i+1}^{k-1} \left[\frac{o_i - e_i}{e_i} + \frac{o_j - e_j}{e_j} \right] e_i e_j - \sum_{i=1}^{k-1} \left(\frac{o_i - e_i}{e_i} \right) e_i(n - e_i) \right] \qquad .$$

$$(6.4.1)$$

From the definition of expected frequencies we have $\sum_{i=1}^{k} e_i = n$. As a consequence,

$$\sum_{i=1}^{k-1} e_i(n - e_i) - 2 \sum_{i=1}^{k-1} \sum_{j=i+1}^{k-1} e_i e_j = n(n - e_k) - \sum_{i=1}^{k-1} e_i^2 - 2 \sum_{i=1}^{k-1} \sum_{j=i+1}^{k-1} e_i e_j$$

$$= n(n - e_k) - \left(\sum_{i=1}^{k-1} e_i \right)^2$$

$$= n(n - e_k) - (n - e_k)^2 = n e_k - e_k^2.$$

Hence the term

$$\left(\frac{o_k - e_k}{e_k} \right)^2 \left[\sum_{i=1}^{k-1} e_i(n - e_i) - 2 \sum_{i=1}^{k-1} \sum_{j=i+1}^{k-1} e_i e_j \right] = \frac{n(o_k - e_k)^2}{e_k} - (o_k - e_k)^2.$$

Now, we simplify the sum $U_n = \sum_{i=1}^{k-1} \sum_{j=i+1}^{k-1} \left[\frac{o_i - e_i}{e_i} + \frac{o_j - e_j}{e_j} \right] e_i e_j$ as follows.

$$U_n = \sum_{i=1}^{k-1} \sum_{j=i+1}^{k-1} [(o_i - e_i)e_j + (o_j - e_j)e_i]$$

$$= \sum_{i=1}^{k-1} (o_i - e_i) \sum_{j=i+1}^{k-1} e_j + \sum_{i=1}^{k-1} e_i \sum_{j=i+1}^{k-1} (o_j - e_j)$$

$$= \sum_{i=1}^{k-1} (o_i - e_i) \sum_{j=i+1}^{k-1} e_j + \sum_{j=1}^{k-1} (o_j - e_j) \sum_{i=1}^{j-1} e_i \quad \text{by interchanging the sums}$$

$$= \sum_{i=1}^{k-1} (o_i - e_i) \sum_{j=i+1}^{k-1} e_j$$

$$+ \sum_{i=1}^{k-1} (o_i - e_i) \sum_{j=1}^{i-1} e_j \quad \text{by interchanging i and j in second term}$$

$$= \sum_{i=1}^{k-1} (o_i - e_i) \left[\sum_{j=1}^{i-1} e_j + \sum_{j=i+1}^{k-1} e_j \right] = \sum_{i=1}^{k-1} (o_i - e_i)(n - e_i - e_k).$$

Further,

$$U_n - \sum_{i=1}^{k-1} \left(\frac{o_i - e_i}{e_i} \right) e_i (n - e_i) = \sum_{i=1}^{k-1} (o_i - e_i)[(n - e_i - e_k) - (n - e_i)]$$

$$= - \sum_{i=1}^{k-1} (o_i - e_i)e_k .$$

Substituting this expression in $2 \left(\frac{o_k - e_k}{e_k} \right) \left[\sum_{i=1}^{k-1} \sum_{j=i+1}^{k-1} \left[\frac{o_i - e_i}{e_i} + \frac{o_j - e_j}{e_j} \right] \right.$
$\left. e_i e_j - \sum_{i=1}^{k-1} \left(\frac{o_i - e_i}{e_i} \right) e_i (n - e_i) \right]$, it reduces to $-2(o_k - e_k) \sum_{i=1}^{k-1} (o_i - e_i)$. With these simplifications, expression for $nT_n(S)$ in (6.4.1) can be written as,

$$nT_n(S) = n \sum_{i=1}^{k-1} \frac{(o_i - e_i)^2}{e_i} - \sum_{i=1}^{k-1} (o_i - e_i)^2 - 2 \sum_{i=1}^{k-1} \sum_{j=i+1}^{k-1} (o_i - e_i)(o_j - e_j)$$

$$+ \frac{n(o_k - e_k)^2}{e_k} - (o_k - e_k)^2 - 2(o_k - e_k) \sum_{i=1}^{k-1} (o_i - e_i)$$

$$= n \sum_{i=1}^{k} \frac{(o_i - e_i)^2}{e_i} - \sum_{i=1}^{k} (o_i - e_i)^2 - 2 \sum_{i=1}^{k} \sum_{j=i+1}^{k-1} (o_i - e_i)(o_j - e_j)$$

$$= n \sum_{i=1}^{k} \frac{(o_i - e_i)^2}{e_i} - \left(\sum_{i=1}^{k} (o_i - e_i) \right)^2$$

$$= n \sum_{i=1}^{k} \frac{(o_i - e_i)^2}{e_i} \quad \text{as} \quad \sum_{i=1}^{k} (o_i - e_i) = 0 .$$

Thus, the score test statistic $T_n(S)$ simplifies to $T_n(P) = \sum_{i=1}^{k} (o_i - e_i)^2 / e_i$, the most frequently used Karl Pearson's chi-square statistic. In Theorem 6.3.1, it is proved that for large n, $T_n(W)$ and $T_n(S)$ have χ^2_{k-1} distribution under H_0.   □

In Sect. 6.7 we verify the results of Theorem 6.4.1 by simulation using R. For a given sample from a multinomial distribution, it is always simple to compute $\sum_{i=1}^{k} (o_i - e_i)^2 / o_i$, instead of $T_n(W)$ and $T_n(P)$ instead of $T_n(S)$. It will be clear from Example 6.7.6 in Sect. 6.7.

📖 Remark 6.4.3

At the beginning of this section we have shown that for a class of distributions with probability law $f(x, \underline{\theta})$ indexed by a parameter $\underline{\theta} \in \Theta \subset \mathbb{R}^k$, $k \geq 1$, if the maximum likelihood estimator $\hat{\underline{\theta}}_n$ of $\underline{\theta}$ is a CAN estimator of $\underline{\theta}$ with approximate dispersion matrix $I^{-1}(\underline{\theta})/n$, then the asymptotic null distributions of Wald's test statistic $T_n(W)$ and the score test statistic $T_n(S)$ is χ^2_{k-1}. In Theorem 6.3.1, it is proved that for a class of multinomial distributions in k cells, $T_n(W)$ and $\sum_{i=1}^{k} (o_i - e_i)^2/o_i$, also $T_n(S)$ and $\sum_{i=1}^{k} (o_i - e_i)^2/e_i$ are identically distributed for large n in null setup. Theorem 6.4.1 proves that $T_n(W)$ and $\sum_{i=1}^{k} (o_i - e_i)^2/o_i$ are identical random variables, similarly $T_n(S)$ and $T_n(P) = \sum_{i=1}^{k} (o_i - e_i)^2/e_i$ are identical random variables, for any n.

📖 Remark 6.4.4

Suppose $\underline{Y} = \{Y_1, Y_2, \ldots, Y_{k-1}\}$ has a multinomial distribution in k cells with cell probabilities being function of θ, which may be real or vector valued parameter. Thus, $\underline{p}(\theta) = \{p_1(\theta), p_2(\theta), \ldots, p_k(\theta)\}$, where $p_i(\theta) > 0$, $i = 1, 2, \ldots, k$ and $\sum_{i=1}^{k} p_i(\theta) = 1$. On the basis of a random sample of size n from the distribution of $\underline{Y}$, suppose we want to test $H_0 : \theta = \theta_0$ which is equivalent to $H_0 : \underline{p} = \underline{p}_0 = \underline{p}(\theta_0)$ against the alternative $H_1 : \underline{p} \neq \underline{p}_0$. Thus, $\underline{p}_0$ is a again completely specified vector. However, in this setup Wald's test statistic is $T_n(W)$ is in general not equal to $\sum_{i=1}^{k} (o_i - e_i)^2/o_i$ but score test statistic is $T_n(S) = \sum_{i=1}^{k} (o_i - e_i)^2/e_i$. It is illustrated in Example 6.7.7.

📖 Remark 6.4.5

In Theorem 6.4.1, the null hypothesis is simple. In many test procedures the null hypothesis is composite, for example, goodness of fit test procedures as in Example 6.3.3. Suppose $\underline{Y} = \{Y_1, Y_2, \ldots, Y_{k-1}\}$ has a multinomial distribution in k cells with cell probabilities $\underline{p}$ being a function of $\underline{\theta}$, a vector valued parameter of dimension $l \times 1, l < k$. On the basis of a random sample of size n from the distribution of $\underline{Y}$, suppose we want to test $H_0 : \underline{p} = \underline{p}(\underline{\theta})$ against the alternative $H_1 : \underline{p} \neq \underline{p}(\underline{\theta})$. Observe that the null hypothesis is a composite hypothesis. In null space we first obtain the maximum likelihood estimator $\hat{\underline{\theta}}_n$ of $\underline{\theta}$ and hence of $\underline{p}(\underline{\theta})$. Thus, the expected frequency of i-th cell is $e_i = np_i(\hat{\underline{\theta}}_n)$. It is to be noted that in the proof of Theorem 6.4.1, when the null hypothesis is simple, the derivation to show that $T_n(S) = T_n(P)$ depends on the vector of score functions, expressed in terms of o_i and e_i and the inverse of information matrix, again expressed in terms of e_i's. The derivation remains valid when the null hypothesis is $H_0 : \underline{p} = \underline{p}(\underline{\theta})$. Thus, in this setup also the score test statistic and Karl Pearson's chi-square test statistic are the same. However, Wald's test statistic does not reduce to $\sum_{i=1}^{k} \frac{(o_i - e_i)^2}{o_i}$. In the next section, we prove that the score test statistic and Karl Pearson's test statistic are the same in more general setup as well. For example, in a $r \times s$ contingency table, underlying model is again a multinomial distribution. Suppose we wish to test that the two criteria A and B are not associated with each other, then the null hypothesis is composite and in the null setup the cell probabilities have some relations among them. In particular, the $rs - 1$ cell probabilities are expressed in terms of $r + s - 2$ parameters. In this case also, the score test statistic and Karl Pearson's test statistic are the same. We will elaborate on this in Sect. 6.5.

We now briefly discuss the score test and Wald's test for testing a composite null hypothesis, when in the null setup the parameters have some functional relations among themselves and when underlying model need not be multinomial.

Suppose X is a random variable or a random vector with probability law $f(x, \underline{\theta})$ indexed by a parameter $\underline{\theta} \in \Theta \subset \mathbb{R}^k$, $k \geq 1$. Suppose $\hat{\underline{\theta}}_n$ is the maximum likelihood estimator of $\underline{\theta}$ in the entire parameter space, based on a random sample $\{X_1, X_2, \ldots, X_n\}$ from the distribution of X. Further, we assume that $\hat{\underline{\theta}}_n$ is a CAN estimator of $\underline{\theta}$ with approximate dispersion matrix $I^{-1}(\underline{\theta})/n$ of order $k \times k$. Suppose the null hypothesis is

$$H_0 : \theta_i = g_i(\beta_1, \beta_2, \ldots, \beta_m)', \ i = 1, 2, \ldots, k, \ m \leq k$$

and $g_1, g_2, \ldots, g_k$ are Borel measurable functions from $\mathbb{R}^m$ to $\mathbb{R}$, having continuous partial derivatives of first order. Thus k parameters are expressed in terms of m parameters. Suppose $\tilde{\underline{\theta}}_n$ is the maximum likelihood estimator of $\underline{\theta}$ in a null setup, that is, $\tilde{\underline{\theta}}_n = \underline{g}(\tilde{\underline{\beta}}_n)$, where $\tilde{\underline{\beta}}_n$ is the maximum likelihood estimator of $\underline{\beta} = (\beta_1, \beta_2, \ldots, \beta_m)'$ based on the random sample $\{X_1, X_2, \ldots, X_n\}$ from the distribution of X. Further, we assume that $\tilde{\underline{\beta}}_n$ is a CAN estimator of $\underline{\beta}$ with approximate dispersion matrix $I^{-1}(\underline{\beta})/n$ of order $m \times m$. The likelihood ratio test statistic $\lambda(\underline{X})$ is then given by

$$\lambda(\underline{X}) = \frac{\sup_{\Theta_0} L_n(\underline{\theta}|\underline{X})}{\sup_{\Theta} L_n(\underline{\theta}|\underline{X})} = \frac{L_n(\tilde{\underline{\theta}}_n|\underline{X})}{L_n(\hat{\underline{\theta}}_n|\underline{X})}.$$

In Theorem 5.2.4, we have proved that $-2 \log \lambda(\underline{X}) \overset{L}{\to} U \sim \chi^2_{k-m}$ in the null setup. The score test statistic in such a setup is given by

$$T_n^{(c)}(S) = \underline{V}_n'(\underline{X}, \tilde{\underline{\theta}}_n) I^{-1}(\tilde{\underline{\theta}}_n) \underline{V}_n(\underline{X}, \tilde{\underline{\theta}}_n),$$

where $\underline{V}_n(\underline{X}, \underline{\theta})$ is a vector of score functions (Rao [3], p. 418). It is to be noted that the score test statistic is obtained by replacing $\underline{\theta}$ in the score function by its maximum likelihood estimator in the null setup.

To define Wald's test statistic when the null hypothesis is $H_0 : \theta_i = g_i(\beta_1, \beta_2, \ldots, \beta_m)'$, $i = 1, 2, \ldots, k$, we express the conditions imposed by null hypothesis on $\underline{\theta}$ as $R_i(\underline{\theta}) = 0$, $i = 1, 2, \ldots, k - m$. It is assumed that R_i's admit continuous partial derivative of first order. Wald's test statistic is then given by

$$T_n^{(c)}(W) = \sum_{i=1}^{k-m} \sum_{j=1}^{k-m} \lambda^{ij}(\hat{\theta}_n) R_i(\hat{\theta}_n) R_j(\hat{\theta}_n),$$

where $\hat{\theta}_n$ is the maximum likelihood estimator of θ in the entire parameter space and $\lambda^{ij}(\hat{\theta}_n)$ is the (i, j)-th element of inverse of the approximate dispersion matrix of $(R_1(\hat{\theta}_n), R_2(\hat{\theta}_n), \ldots, R_{k-m}(\hat{\theta}_n))'$, evaluated at $\hat{\theta}_n$ (Rao [3], p. 419).

For both the procedures, the null hypothesis is rejected if the value of the test statistic is larger than c, where c is determined using the given level of significance and the asymptotic null distribution. The asymptotic null distribution of both $T_n^{(c)}(S)$ and $T_n^{(c)}(W)$ is χ_{k-m}^2 (Rao [3], p. 419).

We illustrate these two test procedures in the following example.

✒ Example 6.4.3

Suppose X and Y are independent random variables having Bernoulli $B(1, p_1)$ and $B(1, p_2)$ distributions respectively, $0 < p_1, p_2 < 1$. Suppose $\underline{X} = \{X_1, X_2, \ldots, X_{n_1}\}$ is a random sample from the distribution of X and $\underline{Y} = \{Y_1, Y_2, \ldots, Y_{n_2}\}$ is a random sample from the distribution of Y. On the basis of these samples we want to test $H_0 : p_1 = p_2$ against the alternative $H_1 : p_1 \neq p_2$. Note that the null hypothesis is a composite hypothesis. In this example, we derive a score test procedure and Wald's test procedure for testing H_0 against the alternative H_1. Suppose

$$P_{1n_1} = \overline{X}_{n_1} = \sum_{i=1}^{n_1} X_i/n_1 \;\&\; P_{2n_2} = \overline{Y}_{n_2} = \sum_{i=1}^{n_1} Y_i/n_2$$

denote the proportion of successes in $\underline{X}$ and the proportion of successes in $\underline{Y}$ respectively. In Example 5.1.5, we have derived two large sample test procedures for testing H_0 against the alternative H_1, based on the following two test statistics.

$$W_n = \frac{(P_{1n_1} - P_{2n_2})}{\sqrt{P_{1n_1}(1 - P_{1n_1})/n_1 + P_{2n_2}(1 - P_{2n_2})/n_2}}$$

$$\&\; S_n = \frac{(P_{1n_1} - P_{2n_2})}{\sqrt{P_n(1 - P_n)(1/n_1 + 1/n_2)}},$$

where $P_n = (n_1 P_{1n_1} + n_2 P_{2n_2})/(n_1 + n_2)$ and $n = n_1 + n_2$. In this example we show that W_n is a square root of Wald's test statistic, while S_n is a square root of the score test statistic. The log-likelihood of $(p_1, p_2)'$ in the entire parameter space, corresponding to random samples $\underline{X}$ and $\underline{Y}$, using independence of X and Y is given by

$$\log L_n(p_1, p_2 | \underline{X}, \underline{Y}) = \sum_{i=1}^{n_1} X_i \log p_1 + \left(n_1 - \sum_{i=1}^{n_1} X_i\right) \log(1 - p_1)$$

$$+ \sum_{i=1}^{n_2} Y_i \log p_2 + \left(n_2 - \sum_{i=1}^{n_2} Y_i\right) \log(1 - p_2)$$

$$
= n_1 P_{1n_1} \log p_1 + n_1(1 - P_{1n_1}) \log(1 - p_1)
$$
$$
+ n_2 P_{2n_2} \log p_2 + n_2(1 - P_{2n_2}) \log(1 - p_2).
$$

From the log-likelihood it is easy to show that the maximum likelihood estimator of $(p_1, p_2)'$ is $(P_{1n_1}, P_{2n_2})'$. In the null setup $p_1 = p_2 = p$, say. Then the log-likelihood of p corresponding to given random samples $\underline{X}$ and $\underline{Y}$, using independence of X and Y, is given by

$$
\log L_n(p | \underline{X}, \underline{Y}) = \left(\sum_{i=1}^{n_1} X_i + \sum_{i=1}^{n_2} Y_i \right) \log p
$$
$$
+ \left(n_1 + n_2 - \sum_{i=1}^{n_1} X_i - \sum_{i=1}^{n_2} Y_i \right) \log(1 - p).
$$

It then follows that the maximum likelihood estimator of p is

$$
\hat{p}_n = \frac{\left(\sum_{i=1}^{n_1} X_i + \sum_{i=1}^{n_2} Y_i \right)}{(n_1 + n_2)} = \frac{(n_1 P_{1n_1} + n_2 P_{2n_2})}{(n_1 + n_2)} = P_n.
$$

To derive the score test statistic, we obtain a vector of score functions as follows. From the log-likelihood of $(p_1, p_2)'$ in the entire parameter space, we have

$$
\frac{\partial \log L_n(p_1, p_2 | \underline{X}, \underline{Y})}{\partial p_1} = \frac{n_1 P_{1n_1}}{p_1} - \frac{n_1(1 - P_{1n_1})}{1 - p_1} = \frac{n_1(P_{1n_1} - p_1)}{p_1(1 - p_1)}
$$
$$
\frac{\partial \log L_n(p_1, p_2 | \underline{X}, \underline{Y})}{\partial p_2} = \frac{n_2 P_{2n_2}}{p_2} - \frac{n_2(1 - P_{2n_2})}{1 - p_2} = \frac{n_2(P_{2n_2} - p_2)}{p_2(1 - p_2)}
$$
$$
\Rightarrow V_n(p_1, p_2) = \left(\frac{1}{\sqrt{n_1}} \frac{n_1(P_{1n_1} - p_1)}{p_1(1 - p_1)}, \frac{1}{\sqrt{n_2}} \frac{n_2(P_{2n_2} - p_2)}{p_2(1 - p_2)} \right)'
$$
$$
\Rightarrow V_n(\hat{p}_n, \hat{p}_n) = \left(\frac{1}{\sqrt{n_1}} \frac{n_1(P_{1n_1} - \hat{p}_n)}{\hat{p}_n(1 - \hat{p}_n)}, \frac{1}{\sqrt{n_2}} \frac{n_2(P_{2n_2} - \hat{p}_n)}{\hat{p}_n(1 - \hat{p}_n)} \right)',
$$

as in the null setup $p_1 = p_2 = p$ and $\hat{p}_n$ is its maximum likelihood estimator. It is to be noted that in the first component of $V_n(p_1, p_2)$, the first factor is $1/\sqrt{n_1}$ and not $1/\sqrt{n_1 + n_2}$ as P_{1n_1} is based on n_1 observations. Similarly, in the second component of $V_n(p_1, p_2)$, the first factor is $1/\sqrt{n_2}$. From the log-likelihood, it is easy to find the information matrix. Its inverse at $\hat{p}_n$ is given by, $I^{-1}(\hat{p}_n, \hat{p}_n) = diag(\hat{p}_n(1 - \hat{p}_n), \hat{p}_n(1 - \hat{p}_n)) = \hat{p}_n(1 - \hat{p}_n)I_2$, where I_2 is an identity matrix of order 2. Observe that

$$
P_{1n_1} - \hat{p}_n = P_{1n_1} - (n_1 P_{1n_1} + n_2 P_{2n_2})/(n_1 + n_2)
$$
$$
= n_2(P_{1n_1} - P_{2n_2})/(n_1 + n_2)
$$
$$
P_{2n_2} - \hat{p}_n = -n_1(P_{1n_1} - P_{2n_2})/(n_1 + n_2)
$$

$$\Rightarrow V_n(\hat{p}_n, \hat{p}_n) = \frac{\sqrt{n_1 n_2}(P_{1n_1} - P_{2n_2})}{(n_1 + n_2)\hat{p}_n(1 - \hat{p}_n)}(\sqrt{n_2}, -\sqrt{n_1})'$$

$$\Rightarrow T_n^{(c)}(S) = \underline{V}_n'(\underline{X}, \hat{p}_n)I^{-1}(\hat{p}_n, \hat{p}_n)\underline{V}_n(\underline{X}, \hat{p}_n)$$

$$= \frac{n_1 n_2 (P_{1n_1} - P_{2n_2})^2}{(n_1 + n_2)^2(\hat{p}_n(1 - \hat{p}_n))^2}\hat{p}_n$$

$$(1 - \hat{p}_n(\sqrt{n_2}, -\sqrt{n_1})'I_2(\sqrt{n_2}, -\sqrt{n_1})$$

$$= \frac{n_1 n_2 (P_{1n_1} - P_{2n_2})^2}{(n_1 + n_2)^2(\hat{p}_n(1 - \hat{p}_n))}(n_1 + n_2)$$

$$= \frac{n_1 n_2 (P_{1n_1} - P_{2n_2})^2}{(n_1 + n_2)(\hat{p}_n(1 - \hat{p}_n))} = \left(\frac{(P_{1n_1} - P_{2n_2})}{\sqrt{P_n(1 - P_n)(\frac{1}{n_1} + \frac{1}{n_2})}}\right)^2$$

$$\Rightarrow \sqrt{T_n^{(c)}(S)} = \frac{(P_{1n_1} - P_{2n_2})}{\sqrt{P_n(1 - P_n)(\frac{1}{n_1} + \frac{1}{n_2})}} = S_n.$$

The asymptotic null distribution of $T_n^{(c)}(S)$ is χ_1^2, as $k = 2$ parameters are estimated in the entire space and in the null setup both are expressed in term of $m = 1$ parameter. The null hypothesis is rejected if $T_n^{(c)}(S) > c$, where $c = \chi_{1-\alpha,1}^2$, where α is the given level of significance. Equivalently the test statistic S_n can also be used to test the null hypothesis. Its asymptotic null distribution is standard normal and H_0 is rejected if $|S_n| > a_{1-\alpha/2}$.

We now find the Wald's test statistic $T_n^{(c)}(W)$. In the null setup, the condition $p_1 = p_2$ can be expressed as

$$R(p_1, p_2) = p_1 - p_2 = 0,$$

$$\Rightarrow R(\hat{p}_{1n}, \hat{p}_{2n}) = \hat{p}_{1n} - \hat{p}_{2n} = P_{1n_1} - P_{2n_2}$$

$$Var(R(\hat{p}_{1n}, \hat{p}_{2n})) = \frac{p_1(1 - p_1)}{n_1} + \frac{p_2(1 - p_2)}{n_2}$$

$$\Rightarrow T_n^{(c)}(W) = \frac{(P_{1n_1} - P_{2n_2})^2}{\hat{p}_{1n}(1 - \hat{p}_{1n})/n_1 + \hat{p}_{2n}(1 - \hat{p}_{2n})/n_2}$$

$$\Rightarrow \sqrt{T_n^{(c)}(W)} = \frac{(P_{1n_1} - P_{2n_2})}{\sqrt{P_{1n_1}(1 - P_{1n_1})/n_1 + P_{2n_2}(1 - P_{2n_2})/n_2}} = W_n.$$

The asymptotic null distribution of $T_n^{(c)}(W)$ is χ_1^2. The null hypothesis is rejected if $T_n^{(c)}(W) > c$, where $c = \chi_{1-\alpha,1}^2$. Equivalently the test statistic W_n can also be used to test the null hypothesis. Its asymptotic null distribution is standard normal and H_0 is rejected if $|W_n| > a_{1-\alpha/2}$.   □

📖 Remark 6.4.6

The function `prop.test` from R uses the score test statistic for testing the hypothesis of equality of population proportions based on the sample proportions, with the sample of size n_1 from population 1 and the sample of size n_2 from population 2, when two populations are independent. We illustrate it in Sect. 6.7.

Following example illustrates the score test and Wald's test for a composite hypothesis in a multinomial distribution.

✒ Example 6.4.4

Suppose $\underline{Y} = (Y_1, Y_2, Y_3)'$ has a multinomial distribution in 4 cells with cell probabilities $p_i > 0$, $i = 1, 2, 3, 4$ and $\sum_{i=1}^{4} p_i = 1$. Suppose $\underline{p} = (p_1, p_2, p_3)$. On the basis of a random sample of size n from the distribution of $\underline{Y}$, we want to test the null hypothesis $H_0 : p_1 = p_3$ & $p_2 = p_4$ against the alternative that at least one of two equalities in the null setup are not valid. Suppose in the null setup $p_1 = p_3 = \alpha$ & $p_2 = p_4 = \beta$ then $2\alpha + 2\beta = 1$, thus in null setup there is only one unknown parameter. We obtain its maximum likelihood estimator as follows. The log-likelihood of $(\alpha, \beta)'$ given observed cell frequencies $\underline{X} = (X_1, X_2, X_3, X_4)$ with $X_1 + X_2 + X_3 + X_4 = n$, is given by.

$$\log L_n(\alpha, \beta | X_1, X_2, X_3, X_4) = (X_1 + X_3) \log \alpha + (X_2 + X_4) \log \beta.$$

Maximizing it subject to the condition $\alpha + \beta = 1/2$, the maximum likelihood estimator of α is $\hat{\alpha}_n = (X_1 + X_3)/2n$ and of β is $\hat{\beta}_n = (X_2 + X_4)/2n$. Note that $\hat{\alpha}_n + \hat{\beta}_n = 1/2$. Suppose $\hat{\underline{p}}_{0n}$ denote the maximum likelihood estimator of $\underline{p}$ in the null setup. To derive the score test statistic, we first find the vector of score functions from the log-likelihood $\sum_{i=1}^{4} X_i \log p_i$. Thus, the vector $\underline{V}_n$ of score functions is given by

$$\underline{V}_n(\underline{X}, \underline{p}) = \frac{1}{\sqrt{n}} \left(\left(\frac{X_1}{p_1} - \frac{X_4}{p_4} \right), \left(\frac{X_2}{p_2} - \frac{X_4}{p_4} \right), \left(\frac{X_3}{p_3} - \frac{X_4}{p_4} \right) \right)$$

$$\Rightarrow \underline{V}_n(\underline{X}, \hat{\underline{p}}_{0n}) = \sqrt{n} \left(\left(\frac{o_1}{e_1} - \frac{o_4}{e_4} \right), \left(\frac{o_2}{e_2} - \frac{o_4}{e_4} \right), \left(\frac{o_3}{e_3} - \frac{o_4}{e_4} \right) \right),$$

where $o_i = X_i$ and $e_1 = e_3 = (X_1 + X_3)/2$ & $e_2 = e_4 = (X_2 + X_4)/2$. Further, Information matrix $I(p)$ can also be expressed in terms of e_i and proceeding as in Theorem 6.4.1, it follows that the score test statistic $T_n^{(c)}(S) = T_n(P) = \sum_{i=1}^{4} (o_i - e_i)^2 / e_i$.
To obtain Wald's test statistic, observe that in the null setup, the condition

$$p_1 = p_3 \ \& \ p_2 = p_4 \ \Leftrightarrow \ R_1(\underline{p}) = p_1 - p_3 = 0 \ \& \ R_2(\underline{p}) = p_2 - p_4 = 0.$$

We now find the approximate dispersion matrix of $D/n = [\lambda_{ij}]$ of $(R_1(\hat{\underline{p}}_n), R_2(\hat{\underline{p}}_n))$, where $\hat{\underline{p}}_n$ is the maximum likelihood estimator of $\underline{p}$ in the entire parameter space and is given by $\hat{\underline{p}}_n = (X_1/n, X_2/n, X_3/n)'$. Suppose $\hat{p}_{4n} = X_4/n$. Observe that

$$
\begin{aligned}
\lambda_{11} &= Var(R_1(\hat{\underline{p}}_n)) = Var(\hat{p}_{1n} - \hat{p}_{3n}) \\
&= Var(\hat{p}_{1n}) + Var(\hat{p}_{3n}) - 2cov(\hat{p}_{1n}, \hat{p}_{3n}) \\
&= \frac{p_1(1 - p_1)}{n} + \frac{p_3(1 - p_3)}{n} + \frac{2p_1 p_3}{n} \\
\lambda_{22} &= Var(R_2(\hat{\underline{p}}_n)) = Var(\hat{p}_{2n} - \hat{p}_{4n}) \\
&= Var(\hat{p}_{2n}) + Var(\hat{p}_{4n}) - 2cov(\hat{p}_{2n}, \hat{p}_{4n}) \\
&= \frac{p_2(1 - p_2)}{n} + \frac{p_4(1 - p_4)}{n} + \frac{2p_2 p_4}{n} \\
\lambda_{12} &= \lambda_{21} = Cov(R_1(\hat{\underline{p}}_n), R_2(\hat{\underline{p}}_n)) = Cov(\hat{p}_{1n} - \hat{p}_{3n}, \hat{p}_{2n} - \hat{p}_{4n}) \\
&= Cov(\hat{p}_{1n}, \hat{p}_{2n}) - Cov(\hat{p}_{1n}, \hat{p}_{4n}) - Cov(\hat{p}_{3n}, \hat{p}_{2n}) + Cov(\hat{p}_{3n}, \hat{p}_{4n}) \\
&= -\frac{p_1 p_2}{n} + \frac{p_1 p_4}{n} + \frac{p_3 p_2}{n} - \frac{p_3 p_4}{n}.
\end{aligned}
$$

Suppose λ^{ij} is the (i, j)-th element of $(D/n)^{-1}$. Then by definition, Wald's test statistic $T_n^{(c)}(W)$ is given by

$$
\begin{aligned}
T_n^{(c)}(W) &= \sum_{i=1}^{2} \sum_{j=1}^{2} \lambda^{ij}(\hat{\underline{p}}_n) R_i(\hat{\underline{p}}_n) R_j(\hat{\underline{p}}_n) \\
&= \lambda^{11}(\hat{\underline{p}}_n)(\hat{p}_{1n} - \hat{p}_{3n})^2 + \lambda^{22}(\hat{\underline{p}}_n)(\hat{p}_{2n} - \hat{p}_{4n})^2 \\
&\quad + 2\lambda^{12}(\hat{\underline{p}}_n)(\hat{p}_{1n} - \hat{p}_{3n})(\hat{p}_{2n} - \hat{p}_{4n}).
\end{aligned}
$$

The asymptotic null distribution of both $T_n^{(c)}(S)$ and $T_n^{(c)}(W)$ is χ_2^2. The null hypothesis is rejected if value of the test statistic is $> c$, where $c = \chi_{1-\alpha,2}^2$. □

We illustrate the computation of the test statistics in Example 6.7.9 in Sect. 6.7.

In the next section we find $T_n^{(c)}(S)$ for testing hypothesis of independence of two attributes in a $r \times s$ contingency table and show that it is the same as $T_n(P)$.

6.5   Tests for Contingency Tables

As discussed in Sect. 6.1, when n objects are classified according to two criteria A and B, with r levels of A and s levels of B, then the count data are presented as an $r \times s$ contingency table. Suppose n_{ij} is a frequency of (i, j)-th cell, n_{ij} being the number of objects having i-th level of attribute A and j-th level of attribute B, $n_{ij} \geq 0 \ \forall \ i = 1, 2, \ldots, r, \ j = 1, 2, \ldots, s$ and $\sum_{i=1}^{r} \sum_{j=1}^{s} n_{ij} = n$. The probability model underlying the $r \times s$ contingency table is a multinomial distribution in $r \times s$ cells with cell probabilities as p_{ij} for the (i, j)th cell. We assume $p_{ij} > 0$ $\forall \ i = 1, 2, \ldots, r, \ j = 1, 2, \ldots, s$ and $\sum_{i=1}^{r} \sum_{j=1}^{s} p_{ij} = 1$. On the basis of the data in a contingency table, we can investigate relationship between the two criteria. In tests related to a contingency table, we set up a null hypothesis which reflects some relation among the cell probabilities, depending on the possible relationship. We investigate such relations using the likelihood ratio tests for the multinomial distribution, a score test which comes out to be the same as Karl Pearson's test and also a test based on a statistic similar to W_n as defined in Sect. 6.3. We obtain the maximum likelihood estimator of the cell probabilities, which are governed by the assumed relationship among cell probabilities, to carry out various test procedures. In the entire setup the parameter p is a vector of $rs - 1$ dimension with components as p_{ij}, which are positive $\forall$ $i = 1, 2, \ldots, r$, $j = 1, 2, \ldots, s - 1$ and $\sum_{i=1}^{r} \sum_{j=1}^{s-1} p_{ij} \leq 1$. As shown in Sect. 6.2, using Lagrange's method of multipliers the maximum likelihood estimator of p corresponding to the observed cell frequencies n_{ij} is given by $\hat{p}_{ij} = n_{ij}/n$, $i = 1, 2, \ldots, r$, $j = 1, 2, \ldots, s$, where $\hat{p}_{rsn} = 1 - \sum_{i=1}^{r} \sum_{j=1}^{s-1} \hat{p}_{ijn}$. Further, $\sqrt{n}(\hat{\underline{p}}_n - \underline{p}) \overset{L}{\to} \underline{Z}_1 \sim N_{rs-1}(\underline{0}, I^{-1}(\underline{p}))$.

We begin with the most frequently used test procedure in a two-way contingency table. It is about investigating the conjecture that A and B are two independent criteria which is equivalent to the statement that A and B are not associated with each other.

Test for independence of two attributes in a two-way contingency table: A conjecture of no association between the two attributes A and B in a two-way contingency table can be expressed in terms of the parameters indexing the underlying probability model. The statement that A and B are two independent criteria is equivalent to the statement that $p_{ij} = p_{i.} p_{.j}$ $\forall$ $i = 1, 2, \ldots, r$, $j = 1, 2, \ldots, s$, where $p_{i.} = \sum_{j=1}^{s} p_{ij}$ is the probability that the object possesses the i-th level of A and $p_{.j} = \sum_{i=1}^{r} p_{ij}$ is the probability that the object possesses the j-th level of B. To elaborate on this relation, suppose two categorical random variables X_1 and X_2 are defined as $X_1 = i$ & $X_2 = j$, if the given object possesses i-th level of A and j-th level of B. Hence, independence of A and B can be expressed as

$$p_{ij} = P[X_1 = i, X_2 = j] = P[X_1 = i]P[X_2 = j]$$
$$= p_{i.} p_{.j} \ \forall \ i = 1, 2, \ldots, r, \ j = 1, 2, \ldots, s.$$

Thus, if we are interested in testing the hypothesis of independence of two attributes then the null hypothesis is $H_0 : p_{ij} = p_{i.} p_{.j}$ $\forall$ i & j against the alternative $H_1 : p_{ij} \neq p_{i.} p_{.j}$ for at least one pair (i, j) $i = 1, 2, \ldots, r$ & $j = 1, 2, \ldots, s$. The null parameter space Θ_0 in this setup is

$$\Theta_0 = \left\{ (p_{11}, p_{12}, \ldots, p_{rs})' \mid p_{ij} = p_{i.} p_{.j} \ \forall \ i \ \& j, \ \sum_{j=1}^{s} p_{.j} = 1 \ \& \ \sum_{i=1}^{r} p_{i.} = 1 \right\}.$$

To find the maximum likelihood estimators of the parameters in the null space, observe that the likelihood of p under H_0 is

$$L_n(\underline{p}|n_{11}, n_{12}, \ldots, n_{rs}) = \prod_{i=1}^{r} \prod_{j=1}^{s} (p_{i.} p_{.j})^{n_{ij}} = \prod_{i=1}^{r} p_{i.}^{n_{i.}} \prod_{j=1}^{s} p_{.j}^{n_{.j}},$$

where $n_{i.} = \sum_{j=1}^{s} n_{ij}$ is the marginal frequency of the i-th level of A and $n_{.j} = \sum_{i=1}^{r} n_{ij}$ is the marginal frequency of j-th level of B. We maximize the likelihood with respect to variations in $p_{i.}$ and $p_{.j}$ under the condition that $\sum_{i=1}^{r} p_{i.} = 1$ and $\sum_{j=1}^{s} p_{.j} = 1$. Again using Lagrange's method of multipliers and proceeding on similar lines as in Sect. 6.2, we get the maximum likelihood estimator $\hat{p}_{i.n}$ of $p_{i.}$ and $\hat{p}_{.jn}$ of $p_{.j}$ as

$$\hat{p}_{i.n} = n_{i.}/n, \ i = 1, 2, \ldots, r \ \& \ \hat{p}_{.jn} = n_{.j}/n, \ j = 1, 2, \ldots, s .$$

It is to be noted that the joint distribution of $(n_{1.}, n_{2.}, \ldots, n_{r.})'$ is multinomial in r cells with cell probabilities $(p_{1.}, p_{2.}, \ldots, p_{r.})'$ and joint distribution of $(n_{.1}, n_{.2}, \ldots, n_{.s})'$ is also multinomial in s cells with cell probabilities $(p_{.1}, p_{.2}, \ldots, p_{.s})'$. As a consequence, the maximum likelihood estimator $(\hat{p}_{1.n}, \hat{p}_{2.n}, \ldots, \hat{p}_{(r-1).n})$ is a CAN estimator of $(p_{1.}, p_{2.}, \ldots, p_{(r-1).})'$. Similarly, the maximum likelihood estimator $(\hat{p}_{.1n}, \hat{p}_{.2n}, \ldots, \hat{p}_{.(s-1)n})$ is a CAN estimator of $(p_{.1}, p_{.2}, \ldots, p_{.(s-1)})'$. The likelihood ratio test statistic $\lambda(\underline{X})$ is given by

$$\lambda(\underline{X}) = \sup_{\Theta_0} L_n(\underline{p}|\underline{X}) / \sup_{\Theta} L_n(\underline{p}|\underline{X})$$

$$= \left(\prod_{i=1}^{r} \left(\frac{n_{i.}}{n} \right)^{n_{i.}} \prod_{j=1}^{s} \left(\frac{n_{.j}}{n} \right)^{n_{.j}} \right) \left(\prod_{i=1}^{r} \prod_{j=1}^{s} \left(\frac{n_{ij}}{n} \right)^{n_{ij}} \right)^{-1} . \quad (6.5.1)$$

From Theorem 5.2.4, for large n under H_0, $-2 \log \lambda(\underline{X})$ has χ_l^2 distribution, where $l = rs - 1 - (r - 1 + s - 1) = (r - 1)(s - 1)$. H_0 is rejected if $-2 \log \lambda(\underline{X}) > c$, where c is determined by the size of the test and the null distribution of $-2 \log \lambda(\underline{X})$. Thus, $c = \chi^2_{(r-1)(s-1), 1-\alpha}$.

Proceeding as in Theorem 6.4.1, in the following theorem we now prove that, for testing the hypothesis of independence of two attributes in a $r \times s$ contingency table, the likelihood ratio test statistic and Karl Pearson's chi-square test statistic $T_n(P)$ are asymptotically equivalent, under H_0. In practice, it is always simpler to compute $T_n(P)$ than $-2 \log \lambda(\underline{X})$.

Theorem 6.5.1

In a $r \times s$ contingency table for testing H_0 : Two attributes A and B are independent against the alternative H_1 : A and B are not independent, the asymptotic null distribution of the likelihood ratio test statistic $-2 \log \lambda(\underline{X})$, Karl Pearson's chi-square test statistic $T_n(P)$ and W_n is $\chi^2_{(r-1)(s-1)}$, where

$$T_n(P) = \sum_{i=1}^{r}\sum_{j=1}^{s}\frac{(o_{ij}-e_{ij})^2}{e_{ij}} \quad \& \quad W_n = \sum_{i=1}^{r}\sum_{j=1}^{s}\frac{(o_{ij}-e_{ij})^2}{o_{ij}},$$

where o_{ij} and e_{ij} are the observed and the expected frequencies of (i, j)-th cell respectively.

Proof For testing H_0 : Two attributes A and B are independent against the alternative H_1 : A and B are not independent, Karl Pearson's chi-square test statistic $T_n(P)$ is given by

$$T_n(P) = \sum_{i=1}^{r}\sum_{j=1}^{s}\frac{(o_{ij}-e_{ij})^2}{e_{ij}} \quad ,$$

where $o_{ij} = n\hat{p}_{ijn} = n_{ij}$ & $e_{ij} = n\tilde{p}_{ijn} = n\hat{p}_{i.n}\hat{p}_{.jn} = \dfrac{n_{i.}n_{.j}}{n}$

are the observed and the expected frequencies of (i, j)-th cell respectively, $\hat{p}_{ijn}$ is the maximum likelihood estimator of p_{ij} in the entire parameter space and $\tilde{p}_{ijn} = \hat{p}_{i.n}\hat{p}_{.jn}$ is the maximum likelihood estimator of p_{ij} in the null space. Suppose

$$u_{ij} = \sqrt{n}(\hat{p}_{ijn} - \hat{p}_{i.n}\hat{p}_{.jn}) \quad \Leftrightarrow \quad \hat{p}_{ijn} = \hat{p}_{i.n}\hat{p}_{.jn} + \frac{u_{ij}}{\sqrt{n}} ,$$

$$i = 1, 2, \ldots, r, \ j = 1, 2, \ldots, s.$$

Further it is to be noted that
$\sum_{i=1}^{r}\sum_{j=1}^{s}u_{ij}/\sqrt{n} = \sum_{i=1}^{r}\sum_{j=1}^{s}\hat{p}_{ijn} - \sum_{i=1}^{r}\hat{p}_{i.n}\sum_{j=1}^{s}\hat{p}_{.jn} = 0$. Now, from the expression of $\lambda(\underline{X})$ as given in Eq. (6.5.1), we have

$$-2\log\lambda(\underline{X}) = 2\left(\sum_{i=1}^{r}\sum_{j=1}^{s}n_{ij}\log\hat{p}_{ijn} - \sum_{i=1}^{r}n_{i.}\log\hat{p}_{i.n} - \sum_{j=1}^{s}n_{.j}\log\hat{p}_{.jn}\right)$$

$$= 2n\sum_{i=1}^{r}\sum_{j=1}^{s}\frac{n_{ij}}{n}\left(\log\hat{p}_{ijn} - \log\hat{p}_{i.n}\hat{p}_{.jn}\right)$$

$$= 2n\sum_{i=1}^{r}\sum_{j=1}^{s}\left(\hat{p}_{i.n}\hat{p}_{.jn} + \frac{u_{ij}}{\sqrt{n}}\right)\left(\log\left(\hat{p}_{i.n}\hat{p}_{.jn} + \frac{u_{ij}}{\sqrt{n}}\right) - \log\hat{p}_{i.n}\hat{p}_{.jn}\right)$$

$$= 2n\sum_{i=1}^{r}\sum_{j=1}^{s}\left(\hat{p}_{i.n}\hat{p}_{.jn} + \frac{u_{ij}}{\sqrt{n}}\right)\log\left(1 + \frac{u_{ij}}{\sqrt{n}\hat{p}_{i.n}\hat{p}_{.jn}}\right)$$

$$= 2n\sum_{i=1}^{r}\sum_{j=1}^{s}\left(\hat{p}_{i.n}\hat{p}_{.jn} + \frac{u_{ij}}{\sqrt{n}}\right)\left(\frac{u_{ij}}{\sqrt{n}\hat{p}_{i.n}\hat{p}_{.jn}} - \frac{u_{ij}^2}{2n\hat{p}_{i.n}^2\hat{p}_{.jn}^2} + \frac{u_{ij}^3}{3n^{3/2}\hat{p}_{i.n}^3\hat{p}_{.jn}^3} - \cdots\right)$$

$$= 2n\sum_{i=1}^{r}\sum_{j=1}^{s}\left(\frac{u_{ij}}{\sqrt{n}} - \frac{u_{ij}^2}{2n\hat{p}_{i.n}\hat{p}_{.jn}} + \frac{u_{ij}^3}{3n^{3/2}\hat{p}_{i.n}^2\hat{p}_{.jn}^2} + \frac{u_{ij}^2}{n\hat{p}_{i.n}\hat{p}_{.jn}} - \frac{u_{ij}^3}{2n^{3/2}\hat{p}_{i.n}^2\hat{p}_{.jn}^2} + \cdots\right)$$

$$= 2n\sum_{i=1}^{r}\sum_{j=1}^{s}\left(\frac{u_{ij}^2}{2n\hat{p}_{i.n}\hat{p}_{.jn}} + \frac{u_{ij}}{\sqrt{n}}\right) + V_n ,$$

where V_n consists of the terms with powers of n in the denominator. As in Theorem 6.4.1, we can show that the numerator of V_n is bounded in probability and hence $V_n \xrightarrow{P} 0$ as $n \to \infty$. Observe that $\sum_{i=1}^{r} \sum_{j=1}^{s} u_{ij}/\sqrt{n} = 0$ and

$$\sum_{i=1}^{r} \sum_{j=1}^{s} \frac{u_{ij}^2}{\hat{p}_{i.n}\hat{p}_{.jn}} = \sum_{i=1}^{r} \sum_{j=1}^{s} \frac{n^2(\hat{p}_{ijn} - \hat{p}_{i.n}\hat{p}_{.jn})^2}{n\hat{p}_{i.n}\hat{p}_{.jn}} = \sum_{i=1}^{r} \sum_{j=1}^{s} \frac{(o_{ij} - e_{ij})^2}{e_{ij}} = T_n(P).$$

Hence, $-2\log \lambda(\underline{X}) - T_n(P) \xrightarrow{P} 0$. Hence, for large n under H_0, $-2\log \lambda(\underline{X})$ and $T_n(P)$ have the same distribution and it is $\chi^2_{(r-1)(s-1)}$. The null hypothesis H_0 is rejected if $T_n(P) > c$, where the cut-off c is determined corresponding to the size of the test and the asymptotic null distribution of $T_n(P)$.

To prove that W_n also has the same asymptotic null distribution, suppose elements of a matrix with (i, j)-th element $\sqrt{n}(\hat{p}_{ijn} - \hat{p}_{i.n}\hat{p}_{.jn})$ are presented in a vector $\underline{Y}_n$ of dimension rs. Then $T_n(P)$ can be expressed as $\underline{Y}_n' A_n \underline{Y}_n$, where A_n is a diagonal matrix with typical element $1/\hat{p}_{i.n}\hat{p}_{.jn}$. It is known that $\hat{p}_{i.n} \xrightarrow{P} p_{i.}$ and $\hat{p}_{.jn} \xrightarrow{P} p_{.j}$. Hence, $A_n \xrightarrow{P} A$ where A is a diagonal matrix with typical element $1/p_{i.}p_{.j} = 1/p_{ij}$ under H_0. Observe that a test statistic W_n defined as

$$W_n = \sum_{i=1}^{r} \sum_{j=1}^{s} (o_{ij} - e_{ij})^2/o_{ij} = \sum_{i=1}^{r} \sum_{j=1}^{s} n(\hat{p}_{ijn} - \hat{p}_{i.n}\hat{p}_{.jn})^2/\hat{p}_{ijn}$$

can be expressed as $\underline{Y}_n' B_n \underline{Y}_n$, where B_n is a diagonal matrix with typical element $1/\hat{p}_{ijn}$. It is known that $\hat{p}_{ijn} \xrightarrow{P} p_{ij}$ and hence, $B_n \xrightarrow{P} B$ where B is a diagonal matrix with typical element $1/p_{ij}$. Observe that under H_0, $A_n - B_n \xrightarrow{P} 0$, a null matrix, which implies that $\underline{Y}_n' B_n \underline{Y}_n - \underline{Y}_n' A_n \underline{Y}_n \xrightarrow{P} 0$. Thus under H_0, the asymptotic distribution of W_n is the same as that of $T_n(P)$, which is $\chi^2_{(r-1)(s-1)}$. □

✍ Remark 6.5.1

It is to be noted that W_n is similar to the Wald's statistic for testing $H_0 : \underline{p} = \underline{p}_0$ against the alternative $H_0 : \underline{p} \neq \underline{p}_0$ in a multinomial distribution.

In Sect. 6.4, we have defined a score test statistic for a composite null hypothesis. A null hypothesis of independence of two attributes in a $r \times s$ contingency table is a composite null hypothesis. In the following theorem we prove that a score test statistic is the same as Karl Pearson's chi-square test statistic for testing a null hypothesis of independence of two attributes in a $r \times s$ contingency table.

Theorem 6.5.2

In a $r \times s$ contingency table for testing the null hypothesis of independence of two attributes, a score test statistic is the same as Karl Pearson's chi-square test statistic.

Proof A score test statistic $T_n^{(c)}(S)$ is given by

$$T_n^{(c)}(S) = \underline{V}_n'(\underline{X}, \tilde{\underline{\theta}}_n)I^{-1}(\tilde{\underline{\theta}}_n)\underline{V}_n(\underline{X}, \tilde{\underline{\theta}}_n),$$

where $\tilde{\underline{\theta}}_n$ is the maximum likelihood estimator of $\underline{\theta}$ in the null setup. For a $r \times s$ contingency table $\underline{\theta} = \underline{p}$, a vector of p_{ij} of dimension $rs - 1$ and in the null setup $p_{ij} = p_{i.}p_{.j} = g_{ij}(p_{1.}, p_{2.}, \dots, p_{r.}, p_{.1}, p_{.2}, \dots, p_{.s})$, with the condition $\sum_{i=1}^{r}\sum_{j=1}^{s} p_{ij} = 1$, $\sum_{i=1}^{r} p_{i.} = 1$ and $\sum_{j=1}^{s} p_{.j} = 1$. Thus, $rs - 1$ parameters are expressed in terms of $r + s - 2$ parameters. The maximum likelihood estimators of these in the null setup are given by $\tilde{p}_{ijn} = \hat{p}_{i.n}\hat{p}_{.jn}$, $i = 1, 2, \dots, r$ and $j = 1, 2, \dots, s$. To find the vector $\underline{V}_n$ of score functions, note that $o_{ij} = n_{ij}$ is the observed frequency and $e_{ij} = n\hat{p}_{i.n}\hat{p}_{.jn}$ is the expected frequency of (i, j)-th cell. Thus,

$$\log L_n(\underline{p}|\underline{X}) = \sum_{i=1}^{r}\sum_{j=1}^{s} n_{ij}\log p_{ij}$$

$$\Rightarrow U_{ij}(\underline{X}, \underline{p}) = \frac{1}{\sqrt{n}}\frac{\partial \log L_n(\underline{p}|\underline{X})}{\partial p_{ij}} = \frac{1}{\sqrt{n}}\left(\frac{n_{ij}}{p_{ij}} - \frac{n_{rs}}{p_{rs}}\right)$$

$$\Rightarrow U_{ij}(\underline{X}, \hat{p}_{i.n}\hat{p}_{.jn}) = \frac{1}{\sqrt{n}}\left(\frac{o_{ij}}{\hat{p}_{i.n}\hat{p}_{.jn}} - \frac{o_{rs}}{\hat{p}_{r.n}\hat{p}_{.sn}}\right) = \sqrt{n}\left(\frac{o_{ij}}{e_{ij}} - \frac{o_{rs}}{e_{rs}}\right)$$

$$= \sqrt{n}\left(\frac{o_{ij} - e_{ij}}{e_{ij}} - \frac{o_{rs} - e_{rs}}{e_{rs}}\right).$$

Suppose the elements in a two-way table are organized in a vector, then the diagonal element in $I^{-1}(\underline{p})$ corresponding to (i, j)-th cell is $p_{ij}(1 - p_{ij})$. In the null setup, the maximum likelihood estimator of p_{ij} is $\hat{p}_{i.n}\hat{p}_{.jn}$. Thus, the diagonal element of $I^{-1}(\underline{p})$ at $\hat{p}_{i.n}\hat{p}_{.jn}$ is given by $\hat{p}_{i.n}\hat{p}_{.jn}(1 - \hat{p}_{i.n}\hat{p}_{.jn}) = (e_{ij}/n)(1 - e_{ij}/n)$. The off-diagonal element is $-p_{kl}p_{uv}$. At the maximum likelihood estimator in the null setup, it is given by $-\hat{p}_{k.n}\hat{p}_{.ln}\hat{p}_{u.n}\hat{p}_{.vn} = -(e_{kl}/n)(e_{uv}/n)$. The expressions of the score function U_{ij} and of the elements in the inverse of information matrix are similar to those as in Theorem 6.4.1. Hence, proceeding exactly on the same lines as in Theorem 6.4.1, we can show that $T_n^{(c)}(S) = \sum_{i=1}^{r}\sum_{j=1}^{s}(o_{ij} - e_{ij})^2/e_{ij}$, which is Karl Pearson's chi-square test statistic. $\square$

In the following example, we illustrate the derivation of Theorem 6.5.2 for a 2×2 contingency table.

✒ Example 6.5.1

Suppose we want to test the hypothesis of independence of two attributes in a 2×2 contingency table. Suppose $o_{ij} = n_{ij}$ and $e_{ij} = n \hat{p}_{i.n} \hat{p}_{.jn} = n_{i.} n_{.j}/n$ are the observed and expected frequency of (i, j)-th cell respectively. The vector $\underline{V}'_n \equiv \underline{V}'_n(\underline{X}, \tilde{p}_n)$ of score functions is given by

$$\underline{V}'_n = \sqrt{n} \left(\frac{o_{11} - e_{11}}{e_{11}} - \frac{o_{22} - e_{22}}{e_{22}}, \ \frac{o_{12} - e_{12}}{e_{12}} - \frac{o_{22} - e_{22}}{e_{22}}, \ \frac{o_{21} - e_{21}}{e_{21}} - \frac{o_{22} - e_{22}}{e_{22}} \right)$$

$$= \sqrt{n} \underline{U}'_n, \text{ say.}$$

The inverse of information matrix $I^{-1}(\tilde{p}_n)$ is given by $\frac{1}{n^2} M_n$, where

$$M_n = \begin{pmatrix} e_{11}(n - e_{11}) & -e_{11}e_{12} & -e_{11}e_{21} \\ -e_{11}e_{12} & e_{12}(n - e_{12}) & -e_{12}e_{21} \\ -e_{11}e_{21} & -e_{12}e_{21} & e_{21}(n - e_{21}) \end{pmatrix}.$$

Suppose the column of M_n are denoted by C_1, C_2, C_3. Then

$$\begin{aligned}
\underline{U}'_n \times C_1 &= (o_{11} - e_{11})(n - e_{11}) - (o_{12} - e_{12})e_{11} - (o_{21} - e_{21})e_{11} \\
&\quad + \frac{(o_{22} - e_{22})}{e_{22}} (e_{11}e_{12} + e_{11}e_{21} - e_{11}(n - e_{11})) \\
&= n(o_{11} - e_{11}) - e_{11}((o_{11} - e_{11}) + (o_{12} - e_{12}) + (o_{21} - e_{21})) \\
&\quad + \frac{(o_{22} - e_{22})}{e_{22}} (e_{11}(e_{12} + e_{21} - n + e_{11})) \\
&= n(o_{11} - e_{11}) + e_{11}(o_{22} - e_{22}) - \frac{(o_{22} - e_{22})}{e_{22}} e_{11}e_{22} \\
&= n(o_{11} - e_{11}).
\end{aligned}$$

Simplifying in a similar way, $\underline{U}'_n M_n$ reduces to $n((o_{11} - e_{11}), (o_{12} - e_{12}), (o_{21} - e_{21}))$. Hence, observe that

$$\begin{aligned}
T_n^{(c)}(S) &= \underline{V}'_n(\underline{X}, \tilde{p}_n) I^{-1}(\tilde{p}_n) \underline{V}_n(\underline{X}, \tilde{p}_n) \\
&= (\sqrt{n}\underline{U}'_n)(\frac{1}{n^2} M_n)(\sqrt{n}\underline{U}_n) \\
&= ((o_{11} - e_{11}), (o_{12} - e_{12}), (o_{21} - e_{21})) \times \underline{U}_n \\
&= \frac{(o_{11} - e_{11})^2}{e_{11}} + \frac{(o_{12} - e_{12})^2}{e_{12}} + \frac{(o_{21} - e_{21})^2}{e_{21}} \\
&\quad - \frac{(o_{22} - e_{22})}{e_{22}} ((o_{11} - e_{11}) + (o_{12} - e_{12}) + (o_{21} - e_{21})) \\
&= \frac{(o_{11} - e_{11})^2}{e_{11}} + \frac{(o_{12} - e_{12})^2}{e_{12}} + \frac{(o_{21} - e_{21})^2}{e_{21}} \\
&\quad - \frac{(o_{22} - e_{22})}{e_{22}} (n - o_{22} - n + e_{22}) \\
&= \sum_{i=1}^{2} \sum_{j=1}^{2} (o_{ij} - e_{ij})^2 / e_{ij}.
\end{aligned}$$

Thus, Karl Pearson's chi-square test statistic and the score test statistic are the same.   □

In Example 6.7.9, we verify that for testing independence of two attributes for data in a 2×3 contingency table, values of Karl Pearson's chi-square test statistic and of the score test statistic are the same.

For a two-way contingency table, other hypotheses of interest are as follows:

(i) Irrelevance of criterion B: In this setup the p_{ij}'s do not change as levels of criterion B change, for all i, that is, $p_{ij} = a_i$, say $\forall\ j$ and for $i = 1, 2, \ldots, r$. Then

$$p_{i.} = \sum_{j=1}^{s} p_{ij} = \sum_{j=1}^{s} a_i = s a_i \;\Rightarrow\; a_i = p_{i.}\frac{1}{s} \;\Rightarrow\; p_{ij} = p_{i.}\frac{1}{s}.$$

Hence, the null hypothesis is expressible as $H_0 : p_{ij} = p_{i.}/s$. The maximum likelihood estimator of p_{ij} in null setup is $\hat{p}_{ijn} = (n_{i.}/n)(1/s) = n_{i.}/ns$.

(ii) Irrelevance of criterion A: As discussed in (i) above, in this setup p_{ij}'s do not change as levels of criterion A change, for all j. Hence the null hypothesis is given by, $H_0 : p_{ij} = p_{.j}/r$ and the maximum likelihood estimator of p_{ij} in null setup is $\hat{p}_{ijn} = n_{.j}/nr$.

(iii) Complete irrelevance: In this setup p_{ij}'s do not change as levels of either criterion A or B change. Hence, the null hypothesis is $H_0 : p_{ij} = 1/rs$. Hence, the maximum likelihood estimator of p_{ij} in null setup is $\hat{p}_{ijn} = 1/rs$.

All these hypotheses can be tested using either the likelihood ratio test or the score test or Karl Pearson's chi-square test. The asymptotic null distribution of all these statistics is χ^2_{rs-1-l}, where l is the number of parameters estimated in the null setup. Thus values of l are $r - 1, s - 1$ and 0 respectively in above three cases. If the value of the test statistic is larger than $\chi^2_{rs-1-l,1-\alpha}$, then H_0 is rejected.

✐ Remark 6.5.2

From the proof of Theorem 6.5.2, we note that the vector of score functions and the inverse of information matrix are expressed in terms of o_{ij} and e_{ij}. Hence the proof remains valid for any other null hypotheses, which are listed above. The formula for e_{ij} will change according to the null hypothesis of interest. Thus, for a two-way contingency table, to test any hypothesis the score test statistic and Karl Pearson's chi-square test statistic are identical.

We may come across a situation, when p_{ij}'s do not change as levels of criterion B change, for some i. In such a case we write the hypothesis accordingly and adopt the same procedure as outlined above. For example, suppose in a 3×3 contingency table, we want to test that p_{ij}'s do not change as levels of criterion B change, for $i = 1, 2$, but change for $i = 3$. Then the null hypothesis is expressed as $H_0 : p_{1j} = a_1, p_{2j} =$

a_2, $j = 1, 2, 3$. In the null setup, we have 5 parameters a_1, a_2, p_{31}, p_{32} and p_{33} such that $3a_1 + 3a_2 + p_{31} + p_{32} + p_{33} = 1$. Subject to this condition, we maximize the likelihood to obtain the maximum likelihood estimators of these parameters and then use either the likelihood ratio test or the score test to test the given hypothesis.

Test procedures for a two-way contingency table can be extended to a three-way contingency table. Suppose n_{ijk} denotes the frequency of (i, j, k)-th cell, $i = 1, 2, \ldots, r$, $j = 1, 2, \ldots, s$, $k = 1, 2, \ldots, m$, when n objects are classified according to three criteria, A with r levels, B with s levels and C with m levels. As in a two-way contingency table, joint distribution of n_{ijk} is a multinomial distribution in $rsm - 1$ cells, with cell probabilities p_{ijk}, $i = 1, 2, \ldots, r$, $j = 1, 2, \ldots, s$, $k = 1, 2, \ldots, m$. In the entire parameter space, the maximum likelihood estimator of p_{ijk} is $\hat{p}_{ijkn} = n_{ijk}/n$. Once we have the maximum likelihood estimators of the parameters in null and in the entire parameter space, we can find likelihood ratio test statistic $-2 \log \lambda(\underline{X})$. Similarly, we can find the expected frequencies and hence Karl Pearson's chi-square test statistic. It can be proved that for a three-way contingency table also, to test any hypothesis, the score test statistic and Karl Pearson's chi-square test statistic are the same, as underlying probability model is again a multinomial distribution.

✍ Remark 6.5.3

In summary, if underlying probability model is a multinomial distribution, then the score test statistic and Karl Pearson's chi-square test statistic are the same for the following three types of null hypotheses.

  (i) $H_0 : \underline{p} = \underline{p}_0$, where $\underline{p}_0$ is a completely specified vector,
 (ii) $H_0 : \underline{p} = \underline{p}(\underline{\theta})$, where $\underline{\theta}$ is an unknown vector and
(iii) H_0 specifies some functional relations among the cell probabilities, as in two-way or three-way contingency table.

We get the two statistics to be the same in view of the fact that
$T_n(S) = \underline{V}'_n(\underline{X}, \tilde{\underline{p}}_n) I^{-1}(\tilde{\underline{p}}_n) \underline{V}_n(\underline{X}, \tilde{\underline{p}}_n)$ is a quadratic form and Karl Pearson's chi-square test statistic $T_n(P) = \sum_{i=1}^{r} \sum_{j=1}^{s} (o_{ij} - e_{ij})^2 / e_{ij}$ can also be expressed as a quadratic form $\underline{Y}'_n A_n \underline{Y}_n$. With a peculiar form of the probability mass function of the multinomial distribution and its dispersion matrix, which is nothing but the inverse of the information matrix, the vector of sore functions and the inverse of the information matrix evaluated at the maximum likelihood estimator of $\underline{p}$ in null setup, result in $\underline{V}_n(\underline{X}, \tilde{\underline{p}}_n) = \underline{Y}_n$ and $I^{-1}(\tilde{\underline{p}}_n) = A_n$ and hence the two test statistics are exactly the same.

For a three-way contingency table, Karl Pearson's chi-square test statistic is given by

$$T_n(P) = \sum_{i=1}^{r} \sum_{j=1}^{s} \sum_{k=1}^{m} \frac{(o_{ijk} - e_{ijk})^2}{e_{ijk}}.$$

The null hypothesis is rejected when the value of the test statistic is larger than a constant c, where c is determined corresponding to the given size of the test and the asymptotic null distribution of the test statistic, which is $\chi^2_{rsm-1-l}$, where l is the number of parameters estimated in the null space.

The most frequently encountered hypotheses in a three-way contingency table are listed below.

(i) Complete or mutual independence among A, B and C: In this case the null hypothesis is $H_0 : p_{ijk} = p_{i..}p_{.j.}p_{..k} \; \forall \; i, j, k$, where

$$p_{i..} = \sum_{j=1}^{s} \sum_{k=1}^{m} p_{ijk}, \quad p_{.j.} = \sum_{i=1}^{r} \sum_{k=1}^{m} p_{ijk} \; \& \; p_{..k} = \sum_{i=1}^{r} \sum_{j=1}^{s} p_{ijk}.$$

To clarify such a relation under H_0, as in a two-way contingency table, we define three categorical random variables X_1, X_2 and X_3 as $X_1 = i$, $X_2 = j$ & $X_3 = k$, if the given object possesses i-th level of A, j-th level of B and k-th level of C. Hence, the mutual independence can be expressed as

$$p_{ijk} = P[X_1 = i, X_2 = j, X_3 = k] = P[X_1 = i]P[X_2 = j]P[X_3 = k]$$
$$= p_{i..}p_{.j.}p_{..k} \; \forall \; i, j, k.$$

The number l of parameters one need to estimate in null setup is $l = (r - 1) + (s - 1) + (m - 1)$. The alternative in this case is $H_1 : p_{ijk} \neq p_{i..}p_{.j.}p_{..k}$ for at least one triplet (i, j, k). Proceeding on similar lines as in the case of a two-way contingency table, the maximum likelihood estimator of p_{ijk} in the null setup is $\hat{p}_{ijkn} = (n_{i..}/n)(n_{.j.}/n)(n_{..k}/n)$.

(ii) Conditional independence: Suppose it is of interest to test whether two attributes are independent given the levels of the third attribute. In particular, suppose we want to test whether A and C are conditionally independent given B. In terms of probability distribution of random variables it can expressed as follows:

$$P[X_1 = i, X_3 = k|X_2 = j] = P[X_1 = i|X_2 = j]P[X_3 = k|X_2 = j]$$
$$\Leftrightarrow \frac{p_{ijk}}{p_{.j.}} = \frac{p_{ij.}}{p_{.j.}} \frac{p_{.jk}}{p_{.j.}}$$
$$\Leftrightarrow p_{ijk} = \frac{p_{ij.}p_{.jk}}{p_{.j.}}.$$

Thus, the null hypothesis that A and C are conditionally independent given B is expressed as $H_0 : p_{ijk} = p_{ij.}p_{.jk}/p_{.j.} \; \forall \; i, j, k$. Hence, the maximum likelihood estimator of p_{ijk} in the null setup is $\hat{p}_{ijkn} = (n_{ij.}/n)(n_{.jk}/n)(n/n_{.j.}) = (n_{ij.}n_{.jk})/(nn_{.j.})$. In the null setup, we estimate $p_{ij.}$, these are $rs - 1$ parameters, $p_{.jk}$, these are $sm - 1$ parameters and $p_{.j.}$, which are $s - 1$ parameters. However, $p_{.j.}$ can be obtained from $p_{ij.}$ by taking sum over i or from $p_{.jk}$ by taking sum over k. Hence, the number l of parameters one estimates in null setup is $l = rs - 1 + sm - 1 + s - 1 - 2(s - 1) = rs + sm - s - 1$.

(iii) Independence between A and (B, C): In this case the null hypothesis is $H_0 : p_{ijk} = p_{i..} p_{.jk} \ \forall \ i, j, k$, which again follows from

$$p_{ijk} = P[X_1 = i, X_2 = j, X_3 = k]$$
$$= P[X_1 = i]P[X_2 = j, X_3 = k] = p_{i..} p_{.jk}.$$

Thus the maximum likelihood estimator of p_{ijk} in null setup is $\hat{p}_{ijkn} = (n_{i..}/n)(n_{.jk}/n)$. The number l of parameters we have to estimate in null set up is $l = (r - 1) + (sm - 1)$.

(iv) Suppose we want to test the hypothesis that the probabilities of classification according to criterion A are known, given by π_i, say. Then $H_0 : p_{ijk} = \pi_i p_{.jk} \ \forall \ i, j, k$. The maximum likelihood estimator of p_{ijk} in the null setup is $\hat{p}_{ijkn} = \pi_i (n_{.jk}/n)$. The number l of parameters one need to estimate in null set up is $l = sm - 1$.

In Sect. 6.7, we discuss some examples in which we carry out these tests for contingency table using R.

In the next section, we briefly discuss the concept of a consistency of a test procedure.

6.6 Consistency of a Test Procedure

Consistency of a test procedure is an optimality criterion for a test procedure. It is defined as follows:

▶ **Definition 6.6.1**

Consistency of a Test Procedure: Suppose $\underline{X} = \{X_1, X_2, \ldots, X_n\}$ is a random sample from a distribution of X, whose probability law is indexed by a parameter $\theta \in \Theta$, which may be real or vector valued. Suppose $\{\phi_n(\underline{X}), n \geq 1\}$ is a sequence of test functions based on $\underline{X}$ for testing $H_0 : \theta \in \Theta_0$ against the alternative $H_1 : \theta \in \Theta_1$ where $\Theta_0 \cap \Theta_1 = \emptyset$ and $\Theta_0 \cup \Theta_1 = \Theta$. The test procedure governed by a test function ϕ_n is said to be consistent if

$$(i) \ \sup_{\theta \in \Theta_0} E_\theta(\phi_n(\underline{X})) \to \alpha \in (0, 1) \quad \& \quad (ii) \ E_\theta(\phi_n(\underline{X})) \to 1 \ \forall \ \theta \in \Theta_1,$$

where α is a size of the test.

Most of the test procedures discussed in Chap. 5 and in this chapter are consistent. In view of this fact, the consistency of a test procedure is a too weak property to be really useful. If a given test procedure is not consistent, then it conveys that something must be fundamentally wrong with the test. If a test procedure is not consistent against a large class of alternatives, then it is considered as an undesirable test. For example, suppose we want to test $H_0 : \theta = 0$ against the alternative $H_1 : \theta > 0$ based on a random sample of size n from a Cauchy $C(\theta, 1)$ distribution. Suppose the test function ϕ_n is given by

$$\phi_n = \begin{cases} 1, & \text{if} \quad \overline{X}_n > k \\ 0, & \text{otherwise.} \end{cases}$$

The cut-off point c is determined so that $P_{\theta=0}[\overline{X}_n > k] = \alpha$, the given level of significance. It is known that if $X \sim C(\theta, 1)$ distribution then $\overline{X}_n \sim C(\theta, 1)$ distribution. Hence, c is the $(1 - \alpha)$-th quantile of the $C(\theta, 1)$ distribution. Thus, $E_{\theta=0}(\phi_n(X)) = \alpha$ and the first requirement of the consistency of a test procedure is satisfied. Suppose $\beta(\theta)$ denotes the power function, then for

$$\theta > 0, \quad \beta(\theta) = P_\theta[\overline{X}_n > k] = P_\theta[\overline{X}_n - \theta > k - \theta]$$
$$= P[U > k - \theta], \quad \text{where} \quad U \sim C(0, 1).$$

Thus, $P_\theta[\overline{X}_n > k]$ does not depend on n at all, so will not converge to 1 as $n \to \infty$. Hence, the test procedure based on $\overline{X}_n$ is not consistent. We have noted in Exercise 2.8.15 of Chap. 2, that for a $C(\theta, 1)$ distribution, $\overline{X}_n$ is not consistent for θ. However, the sample median is consistent for θ. In Chap. 4, we have proved that $C(\theta, 1)$ distribution belongs to a Cramér family and hence the maximum likelihood estimator of θ is CAN for θ. Thus, $\overline{X}_n$ is not at all a desirable estimator for θ which is reflected in the test procedure based on $\overline{X}_n$. In the following example we show that for a Cauchy $C(\theta, 1)$ distribution, the test procedure based on the maximum likelihood estimator of θ is consistent.

✎ Example 6.6.1

Suppose $\underline{X} = \{X_1, X_2, \ldots, X_n\}$ is a random sample from a Cauchy $C(\theta, 1)$ distribution. Suppose a test procedure for testing $H_0 : \theta = 0$ against the alternative $H_1 : \theta > 0$ is given by

$$\phi_n(\underline{X}) = \begin{cases} 1, & \text{if} \quad T_n > k \\ 0, & \text{otherwise,} \end{cases}$$

where T_n is the maximum likelihood estimator of θ. We examine if the test procedure with level of significance α is consistent. It is known that $C(\theta, 1)$ distribution belongs to a Cramér family and hence the maximum likelihood estimator T_n of θ is CAN for θ with approximate variance $1/nI(\theta) = 2/n$. Thus $\sqrt{n/2}(T_n - \theta) \xrightarrow{L} Z \sim N(0, 1)$. For large n, the cut-off point k is decided so that

$$P_0[T_n > k] = \alpha \quad \Leftrightarrow \quad P_0\left[\sqrt{n/2}\, T_n > \sqrt{n/2}\, k\right] = \alpha \quad \Rightarrow \quad k = a_{1-\alpha}\sqrt{2/n},$$

where $a_{1-\alpha}$ is $(1 - \alpha)$-th quantile of standard normal distribution. Thus, $E_{\theta=0}(\phi_n(\underline{X})) = \alpha$. Suppose $\beta(\theta)$ denotes the power function, then

$$\beta(\theta) = P_\theta\left[T_n > a_{1-\alpha}\sqrt{2/n}\right] = P_\theta\left[\sqrt{n/2}\,(T_n - \theta) > a_{1-\alpha} - \sqrt{n/2}\,\theta\right]$$

$$= 1 - \Phi\left(a_{1-\alpha} - \sqrt{n/2}\,\theta\right) \to 1, \;\; \forall\; \theta > 0.$$

Thus, the test procedure is consistent.   □

The next section is devoted to illustrations of various test procedures discussed in this chapter using R software.

6.7   Large Sample Tests Using R

In the present chapter, we discussed tests for validity of the model, test for goodness of fit and tests for contingency tables. All these tests are the likelihood ratio test, when the underlying probability model is a multinomial distribution. Further, it is noted that the score test, Wald's test and the likelihood ratio test are asymptotically equivalent. In addition, we have also proved that in a multinomial distribution for testing simple and the composite null hypotheses about the probability vector, the score test statistic and Karl Pearson's chi-square test statistic are identical. In this section, we verify all these results and illustrate how to carry out these tests using R software.

✐ Example 6.7.1

In Example 6.2.2, we have discussed a genetic model in which the probabilities for three outcomes are θ^2, $2\theta(1 - \theta)$ and $(1 - \theta)^2$, $0 < \theta < 1$. The appropriate probability distribution for this model is a multinomial distribution in three cells. We have shown that the multinomial distribution, with these cell probabilities, belongs to a one-parameter exponential family and the maximum likelihood estimator of θ is $\hat{\theta}_n = (2X_1 + X_2)/2n$. Further, it shown that it is a CAN estimator of θ with approximate variance $\theta(1 - \theta)/2n$. We have defined two test statistics $S_n = \sqrt{2n/\theta_0(1 - \theta_0)}\,(\hat{\theta}_n - \theta_0)$ and $W_n = \sqrt{2n/\hat{\theta}_n(1 - \hat{\theta}_n)}\,(\hat{\theta}_n - \theta_0)$ for testing $H_0 : \theta = \theta_0$ against the alternative $H_1 : \theta \neq \theta_0$. In both the cases, H_0 is rejected if the absolute values of the test statistics are larger than $c = a_{1-\alpha/2}$. We verify these results by simulation using the following code. It is to be noted that when a random sample of size n is drawn from a multinomial distribution with k cells, the cell frequencies $(X_1, X_2, \ldots, X_k)'$ with $X_1 + X_2 + \cdots + X_k = n$ is a sufficient statistic and the joint distribution of $(X_1, X_2, \ldots, X_k)'$ is also a multinomial distribution with parameter n and with the same cell probabilities. In all the test procedures related to a multinomial distribution, the observed data are $(X_1, X_2, \ldots, X_k)'$. Hence to generate such data, we draw a random sample of size 1 from a multinomial distribution with parameters n and cell probabilities $\underline{p} = (p_1, p_2, \ldots, p_k)'$.

```
th =.4;   th0 =.3; b = qnorm(.95); b;
          p = c(th^2,2*th*(1-th),(1-th)^2); p
n = 150; set.seed(21); x = rmultinom(1,n,p); x;
dlogl=function(par)
{
dlogl=(2*x[1]+x[2])/par -(2*n-2*x[1]-x[2])/(1-par)
return(dlogl)
}
dlogl(.5); dlogl(0.3)
mle=uniroot(dlogl,c(.5,.3))$root; mle
a = (2*x[1]+x[2])/(2*n);a ## mle by formula
Sn = sqrt(2*n/(th0*(1-th0)))*(mle-th0);Sn
Wn = sqrt(2*n/(mle*(1-mle)))*(mle-th0);Wn
pv1 = 1-pnorm(Sn)+pnorm(-Sn); pv1
pv2 = 1-pnorm(Wn)+pnorm(-Wn); pv2
### Verification of CAN property
nsim = 1000; n = 150; x = matrix(nrow=length(p),ncol=nsim);
mle = c()
for(j in 1:nsim)
{
set.seed(j)
x[,j] = rmultinom(1,n,p)
}
dlogl = function(par)
{
dlogl = (2*x[1,j]+x[2,j])/par -(2*n-2*x[1,j]-x[2,j])/(1-par)
return(dlogl)
}
for(j in 1:nsim)
{
mle[j] = uniroot(dlogl,c(.6,.2))$root
}
summary(mle); Tn = sqrt(2*n/(th*(1-th)))*(mle-th)
Sn = sqrt(2*n/(th0*(1-th0)))*(mle-th0);
Wn = sqrt(2*n/(mle*(1-mle)))*(mle-th0)
shapiro.test(Tn); shapiro.test(Sn); shapiro.test(Wn)
```

For $\theta = 0.4$, the vector of cell probabilities is $(0.16, 0.48, 0.36)$ and the observed cell frequencies are $(28, 71, 51)$ which add up to 150. The maximum likelihood estimator of θ using uniroot function is 0.4233. From the formula derived in Example 6.2.2, the estimate is the same as 0.4233. To test $H_0 : \theta = \theta_0 = 0.3$ against the alternative $H_1 : \theta \neq \theta_0$, the value of test statistic S_n corresponding to observed data, is 3.4380 and that of W_n is 3.2719, the corresponding p-values are 0.00058 and 0.00107 respectively. Thus, according to both the test procedures, data do not support the null setup. In Example 6.2.2, we have shown that the maximum likelihood estimator of θ is a CAN estimator. Further, large sample distribution of both S_n and W_n is standard normal. It is verified on the basis of 1000 simulations, each of sample size 150. The p-value of the Shapiro-Wilk test comes out to be 0.216, 0.216 and 0.225 for T_n, S_n and W_n respectively, supporting the claim that the maximum likelihood estimator of θ is a CAN estimator and large sample distribution of both S_n and W_n are standard normal.   □

The following example is concerned with the test for validity of a model.

✎ Example 6.7.2

According to genetic linkage theory, observed frequencies of four phenotypes resulting from crossing tomato plants are in the ratio $9/16 + \theta : 3/16 - \theta : 3/16 - \theta : 1/16 + \theta$. A researcher reported the frequencies of four phenotypes as displayed in Table 6.8.

Our aim is to check whether genetic linkage theory seems plausible on the basis of the given data. In the entire parameter space, the data in 4 cells are modeled by a multinomial distribution in 4 cells with cell probabilities p_1, p_2, p_3, p_4, which are positive and add up to 1. The maximum likelihood estimator $\hat{p}_{in}$ of p_i is given by $\hat{p}_{in} = n_i/n$, where n_i denotes the observed frequency of the i-th cell, $i = 1, 2, 3, 4$. To test the validity of the proposed model, we use Karl Pearson's chi-square test, Wald's test and the likelihood ratio test procedure. Under H_0, Karl Pearson's chi-square test statistic, Wald's test statistic and $-2\log(\lambda(\underline{X}))$ follow χ_r^2 distribution, where $r = 3 - 1 = 2$. In the null space the cell probabilities are given by

$$p_1(\theta) = \frac{9}{16} + \theta, \quad p_2(\theta) = p_3(\theta) = \frac{3}{16} - \theta$$
$$\& \ p_4(\theta) = \frac{1}{16} + \theta, \ 0 < \theta < \frac{3}{16}.$$

In all the test procedures, the first step is to find the maximum likelihood estimator of θ corresponding to above data $\underline{X}$, say. The log-likelihood function of θ is given by

$$\log L(\theta|\underline{X}) = n_1 \log\left(\frac{9}{16} + \theta\right) + (n_2 + n_3)\log\left(\frac{3}{16} - \theta\right) + n_4 \log\left(\frac{1}{16} + \theta\right).$$

Hence the log-likelihood equation is

$$\frac{n_1}{\frac{9}{16} + \theta} - \frac{n_2 + n_3}{\frac{3}{16} - \theta} + \frac{n_4}{\frac{1}{16} + \theta} = 0.$$

Using the following R code, we find the maximum likelihood estimate of θ and the values of Karl Pearson's chi-square test statistic, Wald's statistic and of $-2\log(\lambda(\underline{X}))$.

Table 6.8 Frequencies of phenotypes

Phenotype	Frequency
Tall, cut-leaf	926
Tall, potato-leaf	288
Dwarf, cut-leaf	293
Dwarf, potato-leaf	104
Total	1611

```
n = c(926,288,293,104)    ## Vector of observed frequencies
phat=n/sum(n);   phat      ## Vector of mle of cell probabilities
                              in entire space
dlogL = function(pr)
{
term = n[1]*(9/16+pr)^(-1)-(n[2]+n[3])*(3/16-pr)^(-1)
                          +n[4]*(1/16+pr)^(-1)
return(term)
}
dlogL(.002); dlogL(.125)
mle = uniroot(dlogL,c(0.002,.125))$root; mle    ## mle of parameter
ep = c(9/16+mle,3/16-mle,3/16-mle,1/16+mle); ep
ef = sum(n)*ep      ## Vector of expected frequencies
d = data.frame(n,round(ef,4)); d
KP = sum((n-ef)^2/ef); KP    ## Karl Pearson's test statistic
W = sum((n-ef)^2/n);  W      ## Wald type statistic
e1 = sum(n*log(phat)); e2 = sum(n*log(ep))
LRT = 2*(e1-e2); LRTS        ## -2log(likelihood ratio test statistic)
b = qchisq(.95,2); b         ## cut-off point
p1 = 1-pchisq(KP,2); p2 = 1-pchisq(W,2); p3 = 1-pchisq(LRT,2)
p1; p2; p3 ## p-values
```

On the basis of the given data, the maximum likelihood estimate of θ is 0.004727. Frequencies expected under null setup for cells 1 to 4 are given by 913.80, 294.45, 294.45, 108.30 respectively. It is to be noted that these are quite close to the corresponding observed frequencies. The value of Karl Pearson's test statistic is 0.4820, Wald type statistic is 0.4902 and of $-2\log(\lambda(\underline{X}))$ is 0.4847, these are close to each other and are less than the cut-off 5.99. Further, the p-values are 0.7858, 0.7826 and 0.7848 for Karl Pearson's test, Wald type test and the likelihood ratio test respectively. Hence, the data do support the null setup and the proposed genetic model seems to be plausible.   □

Following examples illustrate how to carry out the tests for goodness of fit.

✐ Example 6.7.3

In Example 6.3.1, we discussed a likelihood test procedure to test whether a uniform $U(0, 4)$ distribution is a good model when 200 observations are grouped in 8 classes. It is noted that we can also use test procedures based on $T_n(P) = \sum_{i=1}^{k}(o_i - e_i)^2/e_i$ and $W_n = \sum_{i=1}^{k}(o_i - e_i)^2/o_i$. Further, from Theorem 6.4.1, we have noted that $T_n(P)$ is same as the score test statistic. We have already used a built-in function chisq.test(o,p=pr) for testing goodness of fit of data in Chap. 3, with o denoting the vector of observed frequencies and pr denoting the corresponding vector of expected probabilities. In this example we show that this built-in function is based on the score test statistic. We generate a random sample from a uniform $U(0, 3)$ distribution and compute the likelihood ratio test statistic, Wald's test statistic, score test statistic, which is same as Karl Pearson's chi-square test statistic and use the built-in function chisq.test to examine that the data are indeed from $U(0, 3)$ distribution.

```
n = 200; set.seed(20); x = runif(n,0,3)
O = hist(x,plot=FALSE)$counts; O; sum(O)   ## observed frequencies
bk = hist(x,plot=FALSE)$breaks; bk          ## class boundaries
pn = O/sum(O); pn   ## mle of cell probabilities in entire space
pr = c()
for(i in 1:(length(bk)-1))
{
pr[i]=punif(bk[i+1],0,3)-punif(bk[i],0,3)
}
pr ## vector of expected probabilities
T = 2*(sum(O*log(pn))-sum(O*log(pr))); T ## LRTS
e = pr*n   ## vector of expected frequencies
S = sum((O-e)^2/e); S ## Karl Pearson's test
                      statistic/score test statistic
W = sum((O-e)^2/O); W ## Wald's test statistic
b = qchisq(.95,length(bk)-2); b; PT = 1-pchisq(T,length(bk)-2)
PS = 1-pchisq(S,length(bk)-2); PW = 1-pchisq(W,length(bk)-2)
PT; PS; PW; chisq.test(O,p=pr)
```

Extracting counts from the function O=hist(x,plot=FALSE) gives a vector of observed frequencies to be $(38, 27, 37, 30, 25, 43)'$ corresponding to six classes $[0, 0.5], (0.5, 1.0], (1.0, 1.5], (1.5, 2.0], (2.0, 2.5], (2.5, 3.0]$ respectively. The results of the test procedures are summarized in Table 6.9.

From the table we note that values of Karl Pearson's test statistic and the statistic used in the built-in function chisq.test are the same. Further, $\chi^2_{.95,5} = 11.0705$. Thus, values of all the test statistics are less than the cut-off point. Hence, we conclude that data are generated by $U(0, 3)$ distribution. p-values also support the conclusion.   □

Table 6.9 Test for goodness of fit: summary of test procedures

Test procedure	Value of test statistic	p-value
Likelihood ratio test	$T_n = 7.4954$	0.1863
Wald's test	$W_n = 7.7433$	0.1710
Score test	$S_n = 7.4800$	0.1873
Karl Pearson's test		
chisq.test	7.4800	0.1873

Observe that for a uniform $U(0, 3)$ distribution, there is an upper bound for the support of the distribution. If the support of the distribution is $(0, \infty)$ or $(-\infty, \infty)$ as for the normal distribution, we need to augment one/two class intervals at the beginning and at the end. We have done it for the data on IQ scores while testing whether normal distribution is a good fit. We illustrate it in the following code. It is similar to the code for the above example, with few modifications. We generate data from a gamma distribution. Extracting breaks from the function bk=hist(x,plot=FALSE) gives the class boundaries to group the data in class intervals. We augment one more class at the end with its probability under null setup calculated by subtracting from 1 the sum of the probabilities of previous classes. Note that observed frequency of the last class is 0, so we cannot compute Wald's test statistic. There is no contribution to the likelihood from the last class with frequency 0 as probability of that class raised to 0 is 1 and in the log-likelihood, the contribution is 0. Hence, while computing likelihood ratio test statistic, we do not consider the last class. To compute score test statistic which is same as Karl Pearson's test statistic and for the built-in function, we consider the last class also. Observe the differences in the following code and the code for the uniform $U(0, 3)$ distribution. Run the code and observe the output. Results are similar to those for the uniform $U(0, 3)$ distribution in the above example.

```
### Gamma distribution G(al,la)
al = 2; la = 3; n = 220; set.seed(20)
x = rgamma(n,shape=la,scale=1/al)
O = hist(x,plot=FALSE)$counts; O; sum(O) ## observed frequencies
pn = O/sum(O); pn ## mle of cell probabilities in entire space
bk = hist(x,plot=FALSE)$breaks; bk ## class boundaries
p = c()
for(i in 1:(length(bk)-1))
{
p[i] = pgamma(bk[i+1],shape=la,scale=1/al)-pgamma
            (bk[i],shape=la,scale=1/al)
}
T = 2*(sum(O*log(pn))-sum(O*log(p)));T ## LRTS
PT = 1-pchisq(T,length(bk)-2); PT
sum(p)
pr = c(p,1-sum(p)); O = c(O,0)
```

```
e = pr*n ## vector of expected frquencies
l = length(bk)
S = sum((O-e)^2/e); S
b = qchisq(.95,1-1);b; PS = 1-pchisq(S,1-1); PS
chisq.test(O,p=pr)
```

✒ Example 6.7.4

Frequencies of cars passing during a one-minute period have been observed for 360 consecutive minutes and the resulting observations are given in Table 6.10. It is of interest to examine whether a Poisson distribution is a good model for the given data. In the entire parameter space, the data are modeled by a multinomial distribution in 5 cells with cell probabilities $p_1, p_2, \ldots, p_5$, which are positive and add up to 1. The maximum likelihood estimator $\hat{p}_{in}$ of p_i is given by $\hat{p}_{in} = n_i/n$, where n_i denotes the observed frequency of the i-th cell, $i = 1, 2, \ldots, 5$. To test whether a Poisson distribution is a good model, we use Karl Pearson's chi-square test statistic and the likelihood ratio test statistic. In the null space, the cell probabilities are given by $p_1(\theta), p_2(\theta), \ldots, p_5(\theta)$ where $p_{i+1}(\theta) = P[X = i], i = 0, 1, 2, 3$ and $p_5(\theta) = P[X \geq 4]$, where $X \sim Poi(\theta)$. Thus, to obtain the maximum likelihood estimator of θ, we write the likelihood function corresponding to above data $\underline{X}$, say. It is given by

$$L(\theta|\underline{X}) = \prod_{i=1}^{5} [p_i(\theta)]^{n_i} .$$

The log-likelihood function can be simplified as

$$\log L(\theta|\underline{X}) = -\theta \sum_{i=1}^{4} n_i + \log(\theta) \sum_{i=1}^{4}(i - 1)n_i$$
$$- \sum_{i=1}^{4} n_i \log((i - 1)!) + n_5 \log\left(1 - \sum_{i=0}^{3} \frac{e^{-\theta}\theta^i}{i!}\right).$$

Hence the likelihood equation is

$$-\sum_{i=1}^{4} n_i + \frac{1}{\theta} \sum_{i=1}^{4}(i - 1)n_i - \frac{n_5}{1 - \sum_{i=0}^{3} \frac{e^{-\theta}\theta^i}{i!}} \sum_{i=0}^{3} \frac{e^{-\theta}\theta^{i-1}(i - \theta)}{i!} = 0 .$$

Table 6.10 Number of cars passing during a unit interval

Number of cars	0	1	2	3	4 or more
Number of occurrences	145	128	55	25	7

With the following R code, we compute the maximum likelihood estimate of θ and the values of Karl Pearson's test statistic and of $-2\log(\lambda(\underline{X}))$. Under H_0, Karl Pearson's test statistic and $-2\log(\lambda(\underline{X}))$ follow χ_3^2 distribution.

```
n = c(145,128,55,25,7)    ## Vector of observed frequencies
phat = n/sum(n);   phat   ## Vector of mle of cell probabilities
                             in entire space
a = c(0,1,2,3); n1 = n[1:4]
dLogL = function(pr)
{
prob = dpois(a,pr)
term1 = -sum(n1)+sum(a*n1)/pr
term2 = 0
for(i in 0:3)
{
term2 = term2 + exp(-pr)*pr^(i-1)*(i-pr)/factorial(i)
}
term3 = n[5]*term2/(1-sum(prob))
dlog = term1 - term3
return(dlog)
}
mle = uniroot(dLogL,c(0.05,3))$root; mle   ## mle of theta
ep = dpois(a,mle); ep5 = 1-sum(ep); e = c(ep,ep5)
ef = sum(n)*e     ## Vector of expected frequencies
d = data.frame(n,round(ef,4)); d
KP = sum((n-ef)^2/ef); KP    ## Karl Pearson's test statistic
e1 = sum(n*log(phat)); e2 = sum(n*log(e))
LRT=2*(e1-e2); LRTS         ## -2log(likelihood ratio test statistic)
b=qchisq(.95,3);    b       ## cut-off point
p1 = 1-pchisq(KP,3); p2 = 1-pchisq(LRT,3);  p1; p2  ## p-values
```

On the basis of the given data, the maximum likelihood estimate of θ is 0.9514. The expected frequencies of cells 1 to 5 are 139.03, 132.28, 62.93, 19.96, 5.82. For the given data, value of Karl Pearson's test statistic, which is same as the score test statistic, is 2.9089 and $-2\log(\lambda(\underline{X})) = 2.8412$. These values are less than the cut-off 7.8147, the p-values are 0.4059 and 0.4168 for Karl Pearson's test and likelihood ratio test respectively. Hence, we may accept the claim that the data are from a Poisson distribution.   □

✍ **Remark 6.7.1**

In the above example, one can use the built-in function `chisq.test(n,p=e)` where n is a vector of observed frequencies and $p = e$ specifies the vector of expected probabilities. The value of the test statistic comes out to be the same as 2.9089, as the function uses the score test statistic. However, the p-value is calculated assuming χ_4^2 distribution, which is not correct since while computing expected frequencies, we have estimated θ and hence the degrees of freedom has to be 3. In the built-in function `chisq.test`, degrees of freedom are 4 as we specify the vector e, but to find e, we have to estimate one parameter and we need to adjust for it.

✒ Example 6.7.5

In Example 6.4.2, we have derived the likelihood ratio test procedure, Wald's test procedure and the score test procedure for testing $H_0 : p = p_0$ against the alternative $H_1 : p \neq p_0$, on the basis of a random sample from the distribution of $X \sim B(1, p)$. It is assumed that the sample size n is large and p_0 is a specified constant. It is also shown that Wald's test statistic can be expressed as $\sum_{i=1}^{2}(o_i - e_i)^2/o_i$ and the score test statistic reduces to $\sum_{i=1}^{2}(o_i - e_i)^2/e_i$, which is Karl Pearson's chi-square test statistic. In this example we verify all these results. We further show that the built-in function `prop.test` is based on the score test statistic. We also discuss how Wald's test, score test and the built-in function `prop.test` can be used for one sided alternatives.

```
n = 250; p = .4; p0 = .3; set.seed(30)
x = rbinom(n,1,p); mx = mean(x); mx
logLx = function(a)
{
LL = 0
for(i in 1:n)
{
LL = LL + log(dbinom(x[i],1,a))
}
return(LL)
}
logL = logLx(mx); logLnull = logLx(p0)
Tn = -2*(logLnull-logL); Tn ## LRTS
pTn = 1-pchisq(Tn,1); pTn
Wn = (mx-p0)/(mx*(1-mx)/n)^(0.5); Wn; Wn^2 ## Wald's test statistic
pWn = 1-pnorm(abs(Wn))+pnorm(-abs(Wn)); pWn
Sn = (mx-p0)/(p0*(1-p0)/n)^(0.5); Sn; Sn^2 ## score test statistic
pSn = 1-pnorm(abs(Sn))+pnorm(-abs(Sn)); pSn
o = c(sum(x),n-sum(x)); e = c(n*p0,n*(1-p0))
                        ## observed and expected frequencies
Un = sum((o-e)^2/o); Un ## Un = Wn^2, Wald's test statistic
Vn = sum((o-e)^2/e); Vn## Vn = Sn^2, Karl Pearson's test statistic
prop.test(sum(x),n,p=p0,alt="two.sided",correct=FALSE)
                        ## built-in function
pWn2 = 1-pchisq(Wn^2,1); pWn2
pSn2 = 1-pchisq(Sn^2,1); pSn2
### Null:p < p0, H_0 is rejected if value of Wn or Sn > constant
```

```
pWnl = 1-pnorm(Wn); pWnl
pSnl = 1-pnorm(Sn); pSnl
prop.test(sum(x),n,p=p0,alt="greater",correct=FALSE)
### Null:p > p0, H_0 is rejected if value of Wn or Sn < constant
pWng = pnorm(Wn); pWng
pSng = pnorm(Sn);pSng
prop.test(sum(x),n,p=p0,alt="less",correct=FALSE)
```

The output for the alternative $H_1 : p \neq p_0$ is summarized in Table 6.11. From the p-values reported in Table 6.11, we note that on the basis of the simulated data, the null hypothesis $H_0 : p = p_0$ against the alternative $H_1 : p \neq p_0$ gets rejected. It is to be expected as the sample is generated with $p = 0.4$ and p_0 is taken as 0.3. It is to be noted that the p-value for Wald's test or the score test can be computed using the asymptotic null distribution of W_n and S_n which is standard normal or using the asymptotic null distribution of W_n^2 and S_n^2 which is χ_1^2. These come out to be the same. We note that the value of Karl Pearson's test statistic $T_n(P)$ is the same as that of the score test statistic S_n^2 and the value of the test statistic U_n is the same as that of W_n^2. Further, observe that the value of the test statistic and the p-value given by the built-in function prop.test are the same as that of the score test statistic S_n^2. If the alternative hypothesis is $H_1 : p > p_0$, then H_0 is rejected if $W_n > c$ or $S_n > c$. The cut-off c and the corresponding p-values are obtained using the asymptotic null distribution of W_n and S_n which is standard normal. Similarly, if the alternative hypothesis is $H_1 : p < p_0$, then H_0 is rejected if $W_n < c$ or $S_n < c$. The cut-off c and the corresponding p-values are obtained using the asymptotic null distribution of W_n and S_n. We note that when the alternative hypothesis is $H_1 : p > p_0$, the p-values are 0.0046 and 0.0029 for Wald's test procedure and the score test procedure respectively. Thus, H_0 is rejected on the basis of simulated data, again it is as per the expectations as $p = 0.4$ and $p_0 = 0.3$. When the alternative hypothesis is $H_1 : p < p_0$, the p-values are 0.9954 and 0.9971 for Wald's test procedure and the score test procedure respectively, giving the strong support to null setup, as $p = 0.4$ and $p_0 = 0.3$. Again observe that the value of the test statistic and the p-value given by the built-in function prop.test with the respective options "greater" and

Table 6.11 Test for proportion: summary of test procedures

Test procedure	Value of test statistic	p-value
Likelihood ratio test	$T_n = 7.2920$	0.0069
Wald's test	$W_n = 2.6060$	0.0092
	$U_n = W_n^2 = 6.7912$	
Score test	$S_n = 2.7603$	0.0058
	$S_n^2 = 7.6190$	
Karl Pearson's test	$T_n(P) = 7.6190$	0.0058
prop.test	7.6190	0.0058

"less" for the alternative hypothesis, are the same as that of the score test statistic S_n. To compute p-values the asymptotic null distribution of S_n is used which is standard normal. □

The next example verifies the result proved in Theorem 6.4.1.

☞ Example 6.7.6

In Theorem 6.4.1, it is proved that for a multinomial distribution in k cells, for testing $H_0 : \underline{p} = \underline{p}_0$ against the alternative $H_1 : \underline{p} \neq \underline{p}_0$, Wald's test statistic $T_n(W)$ simplifies to $\sum_{i=1}^{k}(o_i - e_i)^2/o_i$ while the score test statistic $T_n(S)$ simplifies to Karl Pearson's chi-square test statistic. In this example we verify these results by simulation using R. We find the value of Wald's test statistic using following two formulae.

$$(i) \ T_n(W) = n(\hat{\underline{p}}_n - \underline{p}_0)' I(\hat{\underline{p}}_n)(\hat{\underline{p}}_n - \underline{p}_0) \quad \& \quad (ii) \ T_n(W) = \sum_{i=1}^{k} \frac{(o_i - e_i)^2}{o_i}.$$

We find the value of the score test statistic using following two formulae .

$$(i) \ T_n(S) = \underline{V}_n' D(\underline{p}_0) \underline{V}_n,$$

$$\text{where } \underline{V}_n' = \frac{1}{\sqrt{n}} \left(\frac{X_1}{p_{01}} - \frac{X_k}{p_{0k}}, \quad \frac{X_2}{p_{02}} - \frac{X_k}{p_{0k}}, \quad \ldots, \quad \frac{X_{k-1}}{p_{0(k-1)}} - \frac{X_k}{p_{0k}} \right)$$

and (ii) $T_n(S) = \sum_{i=1}^{k} \frac{(o_i - e_i)^2}{e_i}$, where o_i and e_i denote the observed and expected cell frequencies of the i-th cell.

```
p = c(.3,.2,.3,.1,.1); p0 = c(.35,.15,.2,.15,.15);
                  n = 100; set.seed(20)
x = rmultinom(1,n,p); mle = x[1:4]/n; mle
D = function(u)  ### Information matrix
{
D = matrix(nrow = 4,ncol=4)
for(i in 1:4)
{
for(j in 1:4)
{
if(i==j)
{
D[i,j] = 1/u[i]+1/u[5]
}
else
{
```

```
D[i,j] = 1/u[5]
}
}
}
return(D)
}
u1 = x/n; p1 = p0[1:4]
TWn = n*t((mle-p1))%*%D(u1)%*%(mle-p1); TWn
    ### Wald's test statistic
u2 = x[1:4]/p0[1:4]; u3 = rep(x[5]/p0[5],4)
vn = 1/sqrt(n)*(u2-u3); M = solve(D(p0))
TSn = t(vn)%*%M%*%vn; TSn   ### score test statistic
o = x; e = n*p0           ### vectors of observed
                              and expected frequencies
Wn = sum((o-e)^2/o); Wn
Sn = sum((o-e)^2/e); Sn
```

We note that $T_n(W) = 11.1127 = \sum_{i=1}^{k}(o_i - e_i)^2/o_i$ and $T_n(S) = 11.6405 = \sum_{i=1}^{k}(o_i - e_i)^2/e_i$.   □

☑ Example 6.7.7

Suppose we have a multinomial distribution as discussed in Example 6.2.2 with cell probabilities θ^2, $2\theta(1 - \theta)$ and $(1 - \theta)^2$, $0 < \theta < 1$. Thus the cell probabilities depend on θ. Suppose we want to test $H_0 : \theta = 0.3$ against the alternative $H_1 : \theta \neq 0.3$. As stated in Remark 6.4.4, we examine whether in this setup Wald's test statistic $T_n(W)$ is equal to $\sum_{i=1}^{k}(o_i - e_i)^2/o_i$ and the score test statistic $T_n(S)$ is equal to $\sum_{i=1}^{k}(o_i - e_i)^2/e_i$.

```
th = .4; n = 100; set.seed(20); p = c(th^2,2*th*(1-th),(1-th)^2); p
x = rmultinom(1,n,p); x;
dlogl = function(par)
{
dlogl = (2*x[1]+x[2])/par -(2*n-2*x[1]-x[2])/(1-par)
return(dlogl)
}
dlogl(.5); dlogl(0.3)
mle = uniroot(dlogl,c(.5,.3))$root; mle
th0 =.3; o = x; e = n*c(th0^2,2*th0*(1-th0),(1-th0)^2); e; o
Wn = sum((o-e)^2/o); Wn;   Sn = sum((o-e)^2/e); Sn
D=function(u)
{
D = matrix(nrow = 2,ncol=2)
for(i in 1:2)
{
```

```
for(j in 1:2)
{
if(i==j)
{
D[i,j] = 1/u[i]+1/u[3]
}
else
{
D[i,j] = 1/u[3]
}
}
}
return(D)
}
u1 = c(mle^2,2*mle*(1-mle),(1-mle)^2);
p0 = c(th0^2,2*th0*(1-th0),(1-th0)^2)
p1 = p0[1:2]; u2 = u1[1:2]
TWn = n*t((u2-p1))%*%D(u1)%*%(u2-p1); TWn
u3 = x[1:2]/p0[1:2]; u4 = rep(x[3]/p0[3],2)
vn = 1/sqrt(n)*(u3-u4); M = solve(D(p0))
TSn = t(vn)%*%M%*%vn; TSn
```

From the output, we note that for the simulated sample

$$\sum_{i=1}^{k}(o_i - e_i)^2/o_i = 8.6385 \neq T_n(W) = 9.3597$$

$$\& \sum_{i=1}^{k}(o_i - e_i)^2/e_i = 15.5090 = T_n(S).$$

$\square$

📝 Example 6.7.8

In Example 6.4.3 we have derived test procedures for testing $H_0 : p_1 = p_2$ against the alternative $H_1 : p_1 \neq p_2$ on the basis of random samples drawn from $X \sim B(1, p_1)$ and $Y \sim B(1, p_2)$ where X and Y are independent random variables. In this example we simulate samples from two Bernoulli distributions carry out these test procedures. Further, we show that a built-in function prop.test is based on a score test statistic.

```
m = 150;  n = 170 # Sample size for X and Y respectively
p1 = .4; p2 = .5; set.seed(20)
x = rbinom(m,1,p1); y = rbinom(n,1,p2); mx = mean(x);
                                        my = mean(y); mx; my
mp = (sum(x)+sum(y))/(m+n); mp
logLx = function(a)
{
LL = 0
for(i in 1:m)
{
LL = LL + log(dbinom(x[i],1,a))
}
return(LL)
}
logLy = function(a)
{
LL = 0
for(i in 1:n)
{
LL = LL + log(dbinom(y[i],1,a))
}
return(LL)
}
p = seq(0.1,.9,0.01); length(p)
Lx = logLx(p); bx = which.max(Lx); mlex = p[bx]; mlex
Ly = logLy(p); by = which.max(Ly); mley = p[by]; mley
Lnull = logLx(p) + logLy(p); b = which.max(Lnull);
                        mlenull = p[b]; mlenull
logL = logLx(mlex) + logLy(mley)
logLnull = logLx(mlenull) + logLy(mlenull)
Tn = -2*(logLnull-logL); Tn ## LRTS
b = qchisq(.95,1); b; p=1-pchisq(Tn,1); p
Wn = (mx-my)/(mx*(1-mx)/m +my*(1-my)/n)^(0.5); Wn;
                        Wn^2 ## Wald's test
pWn = 1-pnorm(abs(Wn)) + pnorm(-abs(Wn)); pWn
Sn = (mx-my)/(mp*(1-mp)*(1/m +1/n))^(0.5); Sn; Sn^2 ## score test
pSn = 1-pnorm(abs(Sn)) + pnorm(-abs(Sn)); pSn
a = c(sum(x),sum(y))
d = c(m,n)
prop.test(a,d,alt="two.sided",correct=FALSE) ## built-in function
pWn2 = 1-pchisq(Wn^2,1);pWn2
pSn2 = 1-pchisq(Sn^2,1);pSn2
### Null:p1 < p2
pWn1 = 1-pnorm(Wn); pWn1
pSn1 = 1-pnorm(Sn);pSn1
prop.test(a,d,alt="greater",correct=FALSE)
### Null:p1 > p2
pWng = pnorm(Wn); pWng
pSng = pnorm(Sn);pSng
prop.test(a,d,alt="less",correct=FALSE)
```

Table 6.12 Test for equality of proportions: summary of test procedures

Test procedure	Value of test statistic	p-value
Likelihood ratio test	$T_n = 8.3891$	0.0038
Wald's test	$W_n = -2.9274$	0.0035
	$W_n^2 = 8.5699$	
Score test	$S_n = -2.8844$	0.0039
	$S_n^2 = 8.3198$	
prop.test	8.3198	0.0039

On the basis of simulated samples, the maximum likelihood estimate of p_1 is 0.38 and that of p_2 is 0.54. Under the null setup $p_1 = p_2 = p$ and the maximum likelihood estimate of p is 0.4656. The output for the alternative $H_1 : p_1 \neq p_2$ is summarized in Table 6.12. From the p-values reported in Table 6.12, we note that on the basis of simulated data, the null hypothesis $H_0 : p_1 = p_2$ against the alternative $H_1 : p_1 \neq p_2$ gets rejected. It is to be expected as the samples are generated with $p_1 = 0.4$ and $p_2 = 0.5$. Note that the p-values for Wald's test and the score test, computed using the asymptotic null distribution of W_n and S_n, which is standard normal, are same as those computed using the asymptotic null distribution of W_n^2 and S_n^2 which is χ_1^2. Further, observe that the value of the test statistic and the p-value given by the built-in function prop.test are the same as that of the score test statistic S_n^2. If the alternative hypothesis is $H_1 : p_1 > p_2$, then H_0 is rejected if $W_n > c$ or $S_n > c$. Similarly, if the alternative hypothesis is $H_1 : p_1 < p_2$, then H_0 is rejected if $W_n < c$ or $S_n < c$. The cut-off c and the corresponding p-values are obtained using the asymptotic null distribution of W_n and S_n. We note that when the alternative hypothesis is $H_1 : p_1 > p_2$, the p-values are 0.9982 and 0.9980 for Wald's test procedure and the score test procedure respectively, giving the strong support to the null setup, as $p_1 = 0.4$ and $p_2 = 0.5$. When the alternative hypothesis is $H_1 : p_1 < p_2$, the p-values are 0.0017 and 0.0019 for Wald's test procedure and the score test procedure respectively. Thus, H_0 is rejected giving the strong support to the alternative setup, as $p = 0.4$ and $p_2 = 0.5$. Again observe that the value of the test statistic and the p-value given by the built-in function prop.test with the respective options "greater" and "less" for the alternative hypothesis are the same as that of the score test statistic S_n. To compute p-values, the asymptotic null distribution of S_n is used which is standard normal.   □

In the next example, we illustrate the computation of Wald's test statistic in Example 6.4.4.

✏ Example 6.7.9

Suppose $\underline{Y} = (Y_1, Y_2, Y_3)'$ has a multinomial distribution in 4 cells with cell probabilities $p_i > 0$, $i = 1, 2, 3, 4$ and $\sum_{i=1}^{4} p_i = 1$. Suppose $\underline{p} = (p_1, p_2, p_3)$. On the basis of a random sample of size n from the distribution of $\underline{Y}$, we want to test the null hypothesis $H_0 : p_1 = p_3$ & $p_2 = p_4$ against the alternative that at least one of the two equalities in the null setup are not valid. Suppose in the null setup $p_1 = p_3 = \alpha$ & $p_2 = p_4 = \beta$ then $2\alpha + 2\beta = 1$. The maximum likelihood estimator of α is $\hat{\alpha}_n = (X_1 + X_3)/2n$ and of β is $\hat{\beta}_n = (X_2 + X_4)/2n$. The score test statistic is $T_n^{(c)}(S) = T_n(P) = \sum_{i=1}^4 (o_i - e_i)^2/e_i$ and Wald's test statistic is given by

$$T_n^{(c)}(W) = \lambda^{11}(\underline{\hat{p}}_n)(\hat{p}_{1n} - \hat{p}_{3n})^2 + \lambda^{22}(\underline{\hat{p}}_n)(\hat{p}_{2n} - \hat{p}_{4n})^2$$
$$+ 2\lambda^{12}(\underline{\hat{p}}_n)(\hat{p}_{1n} - \hat{p}_{3n})(\hat{p}_{2n} - \hat{p}_{4n})$$

where λ^{ij}, $i, j = 1, 2$ are as given in Example 6.4.4 and $\hat{p}_{in} = X_i/n, i = 1, 2, 3, 4$. We generate a random sample of size $n = 100$ from a multinomial distribution in 4 cells to test the null hypothesis.

```
p = c(.35,.23,.25,.17);  n = 100; set.seed(110)
x = rmultinom(1,n,p); mle = x/n; mle
e1 = e3 = (x[1]+x[3])/2; e2 = e4 = (x[2]+x[4])/2; e1/n; e4/n
o = x; e = c(e1,e2,e3,e4)
Wn = sum((o-e)^2/o); Wn; Sn = sum((o-e)^2/e); Sn
pWn = 1-pchisq(Wn,2); pSn = 1-pchisq(Sn,2); pWn; pSn
yn = c((mle[1]-mle[3]), (mle[2]-mle[4])); yn = as.vector(yn)
a11 = (mle[1]*(1-mle[1]) + mle[3]*(1-mle[3])+2*mle[1]*mle[3])/n
a22 = (mle[2]*(1-mle[2]) + mle[4]*(1-mle[4])+2*mle[2]*mle[4])/n
a12 = (-mle[1]*mle[2] + mle[1]*mle[4]
                     + mle[2]*mle[3] - mle[3]*mle[4])/n
M=matrix(c(a11,a12,a12,a22),nrow=2,byrow=TRUE)
Tn=t(yn)%*%solve(M)%*%yn; Tn; pTn=1-pchisq(Tn,2); pTn
```

On the basis of the generated data, the maximum likelihood estimate of p in the entire parameter space is $(0.44, 0.22, 0.23, 0.11)$ while in the null space it is $(0.335, 0.165, 0.335, 0.165)$. The value of the score test statistic, which is same as Karl Pearson's chi-square test statistic, is 10.2488, while value of the Wald's test statistic is 11.4191. The respective p-values are 0.0060 and 0.0033. Hence, the data do not support null setup. It seems reasonable as we have generated data under $p = (0.35, 0.23, 0.25, 0.17)'$, in which p_1 and p_3 are not close to each other, similarly p_2 and p_4 are also not close. We have also computed the test statistic $W_n = \sum_{i=1}^4 (o_i - e_i)^2/o_i$, its value is 11.4242, with p-value 0.0033. Hence, the null hypothesis is rejected on the basis of W_n also. It is to be noted that the value of W_n is close to the value of Wald's test statistic, but these are not the same. Observe that the null hypothesis $H_0 : p_1 = p_3$ & $p_2 = p_4$ can be expressed as $H_0 : \underline{p} = \underline{p}(\alpha)$ where $\underline{p}(\alpha) = (\alpha, 1/2 - \alpha, \alpha, 1/2 - \alpha)'$, where $0 < \alpha < 1/2$. Thus, the cell probabilities are indexed by a real parameter α. As stated in

Remark 6.4.4 and as noted in Example 6.7.7, if the null hypothesis is composite, in general Wald's test statistic does not simplify to $W_n = \sum_{i=1}^{4} (o_i - e_i)^2/o_i$.

In the following example, we verify Theorem 6.5.1, which states that values of Karl Pearson's chi-square test statistic and the score test statistic are the same for testing hypothesis of independence of attributes in a two-way contingency table.

✒ Example 6.7.10

Table 6.13 presents cross-classification of two attributes gender (A) with two levels as female and male, and political party identification (B) with three levels as democratic, republican party and independents. Data are from the book Agresti [4], p. 38. We want to test whether there is any association between gender and political party identification. In this example we verify Theorem 6.5.1. Thus we find Karl Pearson's chi-square test statistic and the score test statistic and show that the two are same.

```
A = matrix(c(762,327,468,484,239,477), byrow=TRUE,ncol=3); A;
r = 2; s = 3
E = matrix(nrow=2,ncol=3)
for(i in 1:r)
{
for(j in 1:s)
{
E[i,j] = (sum(A[i,])*sum(A[,j]))/sum(A)
}
}
T = sum((A-E)^2/E); T; ### Karl Pearson's test statistic
df = (r-1)*(s-1); df; b = qchisq(.95,df); b; p = 1-pchisq(T,df); p
chisq.test(A)
A1 = as.vector(A); A1; E1 = as.vector(E); E1; n = sum(A1); n
A2 = A1[-length(A1)]; A2; E2 = E1[-length(A1)]; E2
vn = n^(.5)*(A2/E2-A1[length(A1)]/E1[length(A1)]); vn; b = E2/n; b
D=function(u)   ### inverse of Information matrix
{
D = matrix(nrow = 5,ncol=5)
for(i in 1:5)
{
for(j in 1:5)
{
if(i==j)
{
D[i,j] = u[i]*(1-u[i])
}
else
{
```

Table 6.13 Cross-classification by gender and political party identification

	Political party identification		
	Democratic	Republican	Independents
Female	762	327	468
Male	484	239	477

```
D[i,j] = -u[i]*u[j]
   }
  }
 }
return(D)
}
kp = t(vn)%*%D(b)%*%vn;kp ## score test statistic
```

From the output we note that the value of Karl Pearson's chi-square test statistic is 30.0702 and of score test statistic is also 30.0702. The built-in function `chisq.test(A)` also gives the same value. On the basis of the given data, the null hypothesis of association between gender and part identification is rejected, p-value being almost 0.   ☐

In the next example, we test the hypothesis of independence of attributes, again in a two-way contingency table, by using three different approaches.

☑ Example 6.7.11

Table 6.14 displays a report on relationship between aspirin use and heart attacks by the Physicians' Health Study Research Group at Harvard Medical School. Data are from the book Agresti [5] p. 37. The attribute A "Myocardial Infraction" has three levels as "Fatal Attack", "Nonfatal Attack" and "No Attack". The attribute B has two levels as "Placebo" and "Use of Aspirin". On the basis of these data, it is of interest to examine whether the use of aspirin and incidence of heart attack are associated. It is examined by applying the test of independence of two attributes on these data. Following R code performs the test in three different ways. In the first approach, we find the expected frequencies and use Karl Pearson's chi-square test statistic. In the second approach, a built-in function `chisq.test(data)` gives the result of Karl Pearson's chi-square test. Thirdly, we use a function `xtab` from R to prepare a 2 × 3 contingency table and a built-in function `summary()` gives the result.

```
A = matrix(c(18,171,10845,5,99,10933), byrow=TRUE,ncol=3);
A ## Given data
r = 2; s = 3; E = matrix(nrow=r,ncol=s)
for(i in 1:r)
{
for(j in 1:s)
{
E[i,j] = (sum(A[i,])*sum(A[,j]))/sum(A)
}
}
E       ## Matrix of expected frequencies
T = sum((A-E)^2/E); T        # Karl Pearson's chi-square
                                  test statistic
df = (r-1)*(s-1); df; b = qchisq(.95,2); b;
p = 1-pchisq(T,2) ; p  # p-value
chisq.test(A)          # built-in function
#### To construct a two-way contingency table
A = c("P","A","P","A","P","A");
B = c("FA","FA","NFA","NFA","NA","NA")
D = data.frame(A,B); D; Dt = c(18,5,171,99,10845,10933)
U = xtabs(Dt~.,D); U ##  contingency table
summary(U)
```

The value of Karl Pearson's chi-square test statistic is 26.9030, which is larger than the cut-off 5.9915 with p-value $1.439e - 06$ and hence on the basis of the given data we can say that the use of aspirin and incidence of heart attack is associated. The output of a built-in function chisq.test(A) is given below. The results are same as stated above.

```
> chisq.test(A)
          Pearson's Chi-squared test
data:   A
X-squared = 26.903, df = 2, p-value = 1.439e-06
```

Output of the cross tabulation is given below.

Table 6.14 Cross-classification of aspirin use and myocardial infraction

B	A		
	Fatal attack	Nonfatal attack	No attack
Placebo	18	171	10845
Aspirin	5	99	10933

```
  D
    A   B
1 P   FA
2 A   FA
3 P  NFA
4 A  NFA
5 P   NA
6 A   NA
> Dt=c(18,5,171,99,10845,10933)
> U
    B
A       FA    NA    NFA
    A    5 10933    99
    P   18 10845   171
> summary(U)
Call: xtabs(formula = Dt~., data = D)
Number of cases in table: 22071
Number of factors: 2
Test for independence of all factors:
        Chisq = 26.903, df = 2, p-value = 1.439e-06
```

The data frame D specifies levels of 6 cells and the vector Dt gives the counts of the 6 cells according to the levels as specified in D. The function U=xtabs(Dt ~ ., D) provides the 2 × 3 contingency table and summary(U) gives the value of Karl Pearson's chi-square test statistic and corresponding p-value. The results are the same as above.   □

The next example gives R code to carry out tests procedures in a three-way contingency table as discussed in Sect. 6.5. We extend the techniques used in the previous example.

✒ Example 6.7.12

Table 6.15 displays data on presence and absence of a coronary artery disease, serum cholesterol level and blood pressure.
It is of interest to examine various relations among these three attributes based on the given data. We list these below.

Table 6.15 Count data in a three-way contingency table

Serum cholesterol	Disease			
	Present	Present	Absent	Absent
	Low BP	High BP	Low BP	High BP
Low	10	38	421	494
High	11	34	432	322

1. One would like to test, whether the presence of disease depends on the blood pressure and serum cholesterol, that is, in terms of tests for contingency tables, whether the three attributes are associated with each other.
2. Another hypothesis of interest is, whether blood pressure levels and presence or absence of the disease are independent given the serum cholesterol levels
3. Similarly, one may like to see, whether serum cholesterol levels and presence or absence of the disease are independent given the blood pressure levels.
4. One more test of interest is, whether the attribute "disease" is independent of blood pressure and serum cholesterol, while blood pressure and serum cholesterol may not be independent.

We have discussed these four types of hypothesis in Sect. 6.5. The first conjecture is about the mutual independence of three attributes, while second and third are related to conditional independence. The fourth is independence of A with $B \& C$, where A denotes the attribute "disease", B denotes "BP" and C denotes "serum cholesterol". We carry out the test procedures to examine these claims, under the assumption that the joint distribution of cell counts is multinomial. As a first step, we construct a three-way contingency table for the above data using `xtab` function in R. Depending on the hypothesis, we find the expected frequency for each cell and use Karl Pearson's chi-square test statistic with appropriate degrees of freedom. In Table 6.16 we list the null hypothesis, formulae for expected frequencies (E) and for degrees of freedom.

There is one more approach to analyze the count data in a contingency table and it is via Poisson regression, with three factors as presence and absence of a coronary artery disease, serum cholesterol level and blood pressure. Using `loglm`, function in library `MASS` of R, (Venables and Ripley [6]) we can analyze these count data by Poisson regression approach. Both the approaches are illustrated below and they yield the same results.

Table 6.16 Three-way contingency table: formulae for expected frequencies and degrees of freedom

Hypothesis	Null hypothesis	E	Degrees of freedom
1. Mutual independence	$H_0 : p_{ijk} = p_{i..}p_{.j.}p_{..k}$	$\frac{n_{i..}n_{.j.}n_{..k}}{n^2}$	$rsm - r - s - m + 2$
2. Conditional independence given cholesterol	$H_0 : p_{ijk} = \frac{p_{i.k}p_{.jk}}{p_{..k}}$	$\frac{n_{i.k}n_{.jk}}{n_{..k}}$	$(r-1)(s-1)m$
3. Conditional independence given BP	$H_0 : p_{ijk} = \frac{p_{ij.}p_{.jk}}{p_{.j.}}$	$\frac{n_{ij.}n_{.jk}}{n_{.j.}}$	$(r-1)s(m-1)$
4. Independence of disease with BP and cholesterol	$H_0 : p_{ijk} = p_{i..}p_{.jk}$	$\frac{n_{i..}n_{.jk}}{n}$	$(r-1)(sm-1)$

```
### To  construct 2x2x2 contingency table
Disease = c("Yes","Yes","Yes","Yes","No","No","No","No")
BP = c("Low","Low","High","High","Low","Low","High","High")
Cholesterol = c("Low","High","Low","High","Low","High","Low","High")
D = data.frame(Disease,BP,Cholesterol); D
Dt = c(10,11,38,34,421,432,494,322)
T = xtabs(Dt~.,D); T ## contingency table
summary(T)### Results for mutual independence
r = 2; s = 2; m = 2; n = sum(T); n; E1 = E2 = E3 = E4 = O = c()
t=1
for(i in 1:r)
for(j in 1:s)
for(k in 1:m)
{
O[t] = T[i,j,k]
E1[t] = sum(T[i,,])*sum(T[,j,])*sum(T[,,k])/(n^2)
E2[t] = sum(T[i,,k])*sum(T[,j,k])/sum(T[,,k])
E3[t] = sum(T[i,j,])*sum(T[,j,k])/sum(T[,j,])
E4[t] = sum(T[i,,])*sum(T[,j,k])/n
t = t+1
}
d = round(data.frame(O,E1,E2,E3,E4),4); d
d = as.matrix(d)
df = c(r*s*m-r-s-m+2,m*(r-1)*(s-1),s*(r-1)*(m-1),(s*m-1)*(r-1)); df
TS = p = b = c()
for(i in 1:4)
{
TS[i] = sum(((d[,1]-d[,i+1])^2)/d[,i+1])
p[i] = 1-pchisq(TS[i],df[i])
b[i] = qchisq(0.95,df[i])
}
d1 = round(data.frame(TS,df,p,b),4); d1
#### Poison regression approach
library("MASS")
data = data.frame(Disease,BP,Cholesterol,Dt)
loglm(Dt~Disease+BP+Cholesterol,data=data)
loglm(Dt~Disease*Cholesterol+BP*Cholesterol,data=data)
loglm(Dt~Disease*BP+Cholesterol*BP,data=data)
loglm(Dt~Disease+(BP*Cholesterol),data=data)
```

A partial output corresponding to function T=xtabs(Dt~ ., D) is given below, to specify how the three-way contingency table is constructed.

```
### Output
D
   Disease    BP Cholesterol
1      Yes   Low          Low
2      Yes   Low         High
3      Yes  High          Low
4      Yes  High         High
5       No   Low          Low
6       No   Low         High
7       No  High          Low
8       No  High         High
  Dt=c(10,11,38,34,421,432,494,322)
  T
, , Cholesterol = High
        BP
Disease High Low
     No   322 432
    Yes    34  11

, , Cholesterol = Low
        BP
Disease High Low
     No   494 421
    Yes    38  10
```

Data frame D specifies the levels of three attributes according to which the data in the vector Dt of counts is entered, for example, there are 10 patients suffering from the disease, when the levels of BP and cholesterol are at low level and there are 322 patients not suffering from the disease, when the levels of BP and cholesterol are at high level. Object T displays the cross tabulation of data in vector Dt according to levels specified in D. It prepares two tables corresponding to two levels of third attribute "cholestorel". Thus, when cholesterol level is high, we get a 2×2 table where rows correspond to presence of disease and columns correspond to levels of BP. Table 6.17 (data frame d) displays the observed frequencies O and the expected frequencies E_i, where E_i denotes the frequency expected under the hypotheses $i, i = 1, 2, 3, 4$. The vector O of observed frequencies is formed as follows. Index i corresponds to attribute "Disease" with level 1 for presence and level 2 for absence, index j corresponds to attribute "BP" with level 1 for high and level 2 for low and index k corresponds to attribute "cholesterol" with level 1 for high and level 2 for low. As indices i, j and k run from 1 to $r = 2$, 1 to $s = 2$ and 1 to $m = 2$ respectively, vector O corresponds to counts according to the combinations of levels $(1, 1, 1)$, $(1, 1, 2)$, $(1, 2, 1)$, $(1, 2, 2)$, $(2, 1, 1)$, $(2, 1, 2)$, $(2, 2, 1)$, $(2, 2, 2)$.

The vectors of the expected frequencies follow the same pattern.

The results of the test procedures corresponding to the hypotheses 1 to 4 are displayed in Table 6.18. From the first row of Table 6.18, we note that the data do not have sufficient evidence to accept the hypothesis of mutual independence.

Table 6.17 Three-way contingency table: observed and expected frequencies

O	322	494	432	421	34	38	11	10
E_1	381.42	459.71	375.41	452.46	21.25	25.62	20.92	25.21
E_2	335.95	505.48	418.05	409.52	20.05	26.52	24.95	21.48
E_3	327.14	488.86	432.35	420.64	28.86	43.14	10.64	10.36
E_4	337.21	503.92	419.62	408.251	18.79	28.08	23.38	22.75

Table 6.18 Three-way contingency table: values of test statistic and p-values

Hypothesis	Test statistic	df	p-value	Cut-off point
Mutual independence	50.0468	4	0.0000	9.4877
Conditional independence given Cholesterol	30.2432	2	0.0000	5.9915
Conditional independence given BP	1.6841	2	0.4308	5.9915
Independence of disease with BP and cholesterol	31.1632	3	0.0000	7.8147

Thus, the presence of disease and levels of BP and cholesterol are associated with each other. The built-in function summary(T) gives the results only for the hypothesis of mutual independence. These are the same as displayed in Table 6.18. Further, we note that given the attribute "cholesterol", disease and BP are associated with each other, but the two attributes "Disease" and "cholesterol" are not associated with each other given the attribute "BP". It is to be noted from Table 6.17 that under this hypothesis, the observed frequencies and the frequencies expected under null setup are in close agreement with each other. The fourth hypothesis of independence of "Disease" with "BP" and "cholesterol" is rejected.

Results for the four tests using Poisson regression approach are displayed in Table 6.19. In addition to Karl Pearson's test, this approach also uses likelihood ratio test procedure. The values of the test statistic for Karl Pearson's test and likelihood ratio test are close to each other. The results obtained by the Poisson regression approach are exactly the same as in Table 6.18.   □

Table 6.19 Three-way contingency table: analysis by Poisson regression

Hypothesis	Test procedure	Test statistic	df	p-value
Mutual independence	Likelihood ratio	51.93	4	$1.42e^{-10}$
	Karl Pearson	50.04	4	$3.53e^{-10}$
Conditional independence given cholesterol	Likelihood ratio	31.58	2	$1.39e^{-7}$
	Karl Pearson	30.24	2	$2.71e^{-7}$
Conditional independence given BP	Likelihood ratio	1.66	2	0.4360
	Karl Pearson	1.68	2	0.4308
Independence of disease with BP and cholesterol	Likelihood ratio	31.94	3	$5.38e^{-7}$
	Karl Pearson	31.16	3	$7.85e^{-7}$

6.8   Conceptual Exercises

6.8.1 In a multinomial distribution with 3 cells, the cell probabilities are

$$p_1(\theta) = p_2(\theta) = (1 + \theta)/3 \quad \text{and} \quad p_3(\theta) = (1 - 2\theta)/3, \quad 0 < \theta < 1/2.$$

(i) Examine whether the distribution belongs to a one-parameter exponential family. On the basis of a random sample of size n from this distribution find the maximum likelihood estimator and the moment estimator based on the sufficient statistic for θ and examine if these are CAN.
(ii) Use the result to derive Wald's test and the score test procedure for testing $H_0 : \theta = \theta_0$ against the alternative $H_1 : \theta \neq \theta_0$.

6.8.2 In a multinomial distribution with four cells, the cell probabilities are

$$p_1(\theta) = p_4(\theta) = (2 - \theta)/4 \quad \text{and} \quad p_2(\theta) = p_3(\theta) = \theta/4, \quad 0 < \theta < 2.$$

Examine whether the distribution belongs to a one-parameter exponential family. On the basis of a random sample of size n from this distribution find the maximum likelihood estimator of θ and examine if it is CAN. Use the result to derive (i) a likelihood ratio test, (ii) Wald's test, (iii) a score test and (iv) Karl Pearson's chi-square test to test $H_0 : \theta = \theta_0$ against the alternative $H_1 : \theta \neq \theta_0$.

6.8.3 In a certain genetic experiment two different varieties of certain species are crossed. A specific characteristic of an offspring can occur at three levels A, B and C. According to the proposed model, probabilities for three levels A, B and C are $1/12$, $3/12$ and $8/12$ respectively. Out of fifty offspring 6, 8 and 36 have levels A, B and C respectively. Test the validity of the proposed model by a score test, Karl Pearson's test and by Wald's test.

6.8.4 On the basis of data in a 3×3 contingency table, derive a likelihood ratio test procedure and Karl Pearson's test procedure to test

$$H_0 : p_{ij} = p_{ji}, \ i \neq j = 1, 2, 3 \ \text{against the alternative}$$
$$H_1 : p_{ij} \neq p_{ji}, \ i \neq j = 1, 2, 3$$

for at least one pair.

6.8.5 On the basis of data in a 2×3 contingency table, derive a likelihood ratio test procedure and Karl Pearson's test procedure to test $H_0 : p_{11} = p_{12} = p_{13}$ against the alternative that there is no restriction as specified in H_0.

6.8.6 Suppose $\{X_1, X_2, \ldots, X_n\}$ is a random sample from a Laplace distribution with location parameter θ and scale parameter 1. Derive a large sample test procedure to test $H_0 : \theta = \theta_0$ against the alternative $H_0 : \theta > \theta_0$ and examine whether it is a consistent test procedure.

6.9   Computational Exercises

Verify the results by simulation using R.

6.9.1 For the multinomial distribution of Example 6.2.1, obtain the maximum likelihood estimator of θ on the basis of a simulated sample. Find the value of the test statistic T_n and write the conclusion about the test for H_0 against H_1 as specified in Example 6.2.1. Verify that the maximum likelihood estimator of θ is a CAN estimator of θ. (Hint: Use code similar to Example 6.7.1.)

6.9.2 On the basis of data given in Example 6.2.3, examine whether the proposed model is valid using a test based on a score test and $W_n = \sum_{i=1}^{4}(o_i - e_i)^2/o_i$. Use the formula for score test statistic in terms of a quadratic form and find its value. Examine whether it is the same as the value of Karl Pearson's test statistic. (Hint: Use code similar to Example 6.7.6.)

6.9.3 For the multinomial distribution in Exercise 6.8.1, on the basis of simulated sample, test $H_0 : \theta = 1/4$ against the alternative $H_1 : \theta \neq 1/4$ using Wald's test. Find the p-value. Plot the power function and hence examine whether the test is consistent. (Hint: Use code similar to Example 6.7.2.)

6.9.4 For the multinomial distribution in Exercise 6.8.2, on the basis of simulated sample, test $H_0 : \theta = 1$ against the alternative $H_1 : \theta \neq 1$ using the score test. Find the p-value. Plot the power function and hence examine whether the test is consistent.(Hint: Use code similar to Example 6.7.2.)

6.9.5 Suppose $(Y_1, Y_2)'$ has a multinomial distribution in three cells with cell probabilities $(\theta + \phi)/2$, $(1 - \theta)/2$ and $(1 - \phi)/2$, $0 < \theta$, $\phi < 1$. On the basis of simulated sample, test the hypothesis $H_0 : \theta = \theta_0, \phi = \phi_0$ against the alternative $H_1 : \theta \neq \theta_0, \phi \neq \phi_0$ using (i) a likelihood ratio test, (ii) Wald's test and (iii) a score test. (Hint: Use code similar to Example 6.7.2.)

6.9.6 A gene in a particular organism is either dominant (A) or recessive (a). Under the assumption that the members of the population of this organism choose their mating partner in a manner that is completely unrelated to the type of the gene, there are three possible genotypes which can be observed namely, AA, Aa and aa respectively. Table 6.20 provides the number of organisms possessing these genotypes when a sample of 600 organisms is selected. Test the claim that the proportions of the organisms in a population corresponding to these genotypes are as provided in Table 6.20. (Hint: Use code similar to Example 6.7.2.)

6.9.7 Table 6.21 shows the number (x) of a particular organism found in 100 samples of water from a pond. Test the hypothesis that these data are from a binomial $B(6, p)$ distribution. (Hint: Use code similar to Example 6.7.4.)

6.9.8 Table 6.22 presents the distribution of heights collected on 300 8-year old girls. Examine whether the data are from normal distribution. (Hint: Use code similar to Example 6.7.4.)

6.9.9 Table 6.23 displays data on number of individuals classified according to party and race. Test the hypothesis of independence between party identification and race, using three approaches discussed in Example 6.7.10.

6.9.10 A sample from 200 married couples was taken from a certain population. Husbands and wives were interviewed separately to determine whether their main source of news was from the newspapers, radio or television. The results are displayed in Table 6.24.
(i) Test the hypothesis of symmetry specified by $H_0 : p_{ij} = p_{ji}$. (ii) Test the hypothesis of symmetry and independence specified by $H_0 : p_{ij} = \pi_i \pi_j$ where $\pi_1 + \pi_2 + \pi_3 = 1$.

6.9.11 When a new TV serial is launched, the producer wants to get a feedback from the viewers. Random samples of 250, 200 and 350 consumers from three cities are selected and the following data is obtained from them. Suppose three categories A, B and C are defined as A: Never heard about the serial, B: Heard about the serial but did not watch and C: saw it at least once. Can we claim on the basis of the data in Table 6.25 that the viewers preferences differ in the three cities?

Table 6.20 Number of organisms with specific genotype

Genotype	Proportion	Number of organisms
AA	θ^2	200
Aa	$2\theta(1 - \theta)$	300
aa	$(1 - \theta)^2$	100

Table 6.21 Number of organisms

x	0	1	2	3	4	5	6
Frequency	15	30	25	20	5	4	1

Table 6.22 Heights of eight year old girls

Height (in cms)	[114, 120)	[120, 126)	[126, 132)	[132, 138)	[138, 144)
Observed frequency	29	91	130	46	4

Table 6.23 Cross-classification by race and party identification

Race	Party identification		
	Democrat	Independent	Republican
Black	103	15	11
White	341	105	405

Table 6.24 Classification according to source of news

		Husband		
		Papers	Radio	TV
Wife	Papers	15	6	10
	Radio	11	10	20
	TV	23	15	90

Table 6.25 Data on feedback of viewers of TV serial

	A	B	C	Total
City 1	51	70	129	250
City 2	60	71	69	200
City 3	69	95	188	350
Total	180	234	386	800

6.9.12 The resident data set "HairEyeColor" in R gives the distribution of hair and eye color and sex for 592 statistics students. It is a three-dimensional array resulting from cross-tabulating 592 observations on 3 attributes. The attributes and their levels are as displayed in Table 6.26. The data can be obtained by giving commands data(HairEyeColor). The commands HairEyeColor and help(HairEyeColor) give the description of the data set. On the basis of these data, test whether (i) the three attributes are associated with each other, (ii) hair color and eye color are independent attributes given the sex of an individual and (iii) hair color and sex are independent attributes given the eye color of an individual. (Hint: Use code similar to Example 6.7.12.)

Table 6.26 Levels of three variables: hair color, eye color and sex

Number	Attribute	Levels
1	Hair Color	Black, Brown, Red, Blond
2	Eye Color	Brown, Blue, Hazel, Green
3	Sex	Male, Female

References

1. Kale, B. K., & Muralidharan, K. (2016). *Parametric inference: An introduction.* Delhi: Narosa.
2. Wilks, S. S. (1938). The large sample distribution of the likelihood ratio test for testing composite hypotheses. *Annals of Mathematical Statistics, 9,* 60–62.
3. Rao, C. R. (1978). *Linear statistical inference and its applications.* New York: Wiley.
4. Agresti, A. (2007). *An introduction to categorical data analysis* (2nd ed.). New York: Wiley.
5. Agresti, A. (2002). *Categorical data analysis* (2nd ed.). New York: Wiley.
6. Venables, W. N., & Ripley, B. D. (2002). *Modern applied statistics with S* (4th ed.). New York: Springer. http://www.stats.ox.ac.uk/pub/MASS4.

Solutions to Conceptual Exercises 7

Contents

7.1   Chapter 2

2.8.1 Suppose T_n is a consistent estimator of θ. Obtain conditions on the sequence $\{a_n, n \geq 1\}$ such that the following are also consistent estimators of θ. (i) $a_n T_n$, (ii) $a_n + T_n$ and (iii) $(a_n + nT_n)/(n+1)$.

Solution: It is given that T_n is a consistent estimator of θ, that is $T_n \overset{P_\theta}{\to} \theta$, $\forall \; \theta \in \Theta$. In (i) if $a_n \to 1$ as $n \to \infty$ then $a_n T_n \overset{P_\theta}{\to} \theta$, $\forall \; \theta \in \Theta$. In (ii) if $a_n \to 0$ as $n \to \infty$ then $a_n + T_n \overset{P_\theta}{\to} \theta$, $\forall \; \theta \in \Theta$. In (iii) $(a_n + nT_n)/(n+1) = a_n/(n+1) + n/(n+1)T_n$. Further, $n/(n+1) \to 1$, hence if the sequence $\{a_n, n \geq 1\}$ is such that $a_n/(n+1) \to 0$ as $n \to \infty$, then $(a_n + nT_n)/(n+1) \overset{P_\theta}{\to} \theta$, $\forall \; \theta \in \Theta$. For example, $a_n = n^\delta$, $\delta < 1$ or

$a_n = \exp(-n)$. It is to be noted that if $a_n = n^{\delta}, 0 < \delta < 1$, then $\{a_n, n \geq 1\}$ is not a convergent sequence.

2.8.2 Suppose $\{X_1, X_2, \ldots, X_n\}$ is a random sample of size n from a distribution of X, where $E(X) = \theta$ and $E(X^2) = V < \infty$. Show that $T_n = \frac{2}{n(n+1)} \sum_{i=1}^{n} i X_i$ is a consistent estimator of θ.

Solution: It is given that $E(X) = \theta$, hence

$$E(T_n) = E \left(\frac{2}{n(n+1)} \sum_{i=1}^{n} i X_i \right) = \frac{2}{n(n+1)} \sum_{i=1}^{n} i E(X_i)$$

$$= \frac{2\theta}{n(n+1)} \frac{n(n+1)}{2} = \theta .$$

Thus, T_n is an unbiased estimator of θ. Since $\{X_1, X_2, \ldots, X_n\}$ is a random sample, $\{i X_i, i = 1, 2, \ldots, n\}$ are independent random variables. Further, $Var(X) = V - \theta^2 = \sigma^2$, say and it is finite. Hence,

$$Var(T_n) = \left(\frac{2}{n(n+1)} \right)^2 Var \left(\sum_{i=1}^{n} i X_i \right) = \frac{4}{n^2(n+1)^2} \sum_{i=1}^{n} i^2 \sigma^2$$

$$= \frac{4\sigma^2}{n^2(n+1)^2} \frac{n(n+1)(2n+1)}{6}$$

$$= \frac{2\sigma^2(2n+1)}{3n(n+1)} \to 0 \text{ as } n \to \infty .$$

Thus, T_n is an unbiased estimator of θ and its variance $\frac{2\sigma^2(2n+1)}{3n(n+1)}$ converges to 0 and hence it is a MSE consistent estimator of θ.

2.8.3 Suppose $\{X_1, X_2, \ldots, X_n\}$ is a random sample from a distribution with mean θ and variance σ^2. Suppose $T_n = \sum_{i=1}^{n} a_i X_i$, where $\sum_{i=1}^{n} a_i \to 1$ and $\sum_{i=1}^{n} a_i^2 \to 0$. Show that T_n is consistent for θ.

Solution: Since $E(X_i) = \theta$ and $Var(X_i) = \sigma^2$, we have

$$E(T_n) = E \left(\sum_{i=1}^{n} a_i X_i \right) = \theta \sum_{i=1}^{n} a_i \to \theta$$

$$\Rightarrow Bias_\theta(T_n) = (E(T_n) - \theta) \to 0 \text{ and}$$

$$Var(T_n) = Var \left(\sum_{i=1}^{n} a_i X_i \right) = \sigma^2 \sum_{i=1}^{n} a_i^2 \to 0 .$$

Hence, T_n is consistent for θ.

2.8.4 Suppose $g(x)$ is an even, non-decreasing and non-negative function on $[0, \infty)$. Then show that T_n is consistent for $\eta(\theta)$ if $E(g(T_n - \eta(\theta)) \to 0 \; \forall \; \theta$.

Solution: The basic inequality from probability theory states that if X is an arbitrary random variable and $g(\cdot)$ is even, non-decreasing and non-negative Borel function on $[0, \infty)$, then for every $a > 0$,

$$P[|X| \geq a] \leq \frac{E(g(X))}{g(a)}.$$

Using the basic inequality and the fact that g is an even function, for all $\epsilon > 0$, we have,

$$P_\theta[|T_n - \eta(\theta)| > \epsilon] \leq \frac{E(g(|T_n - \eta(\theta)|))}{g(\epsilon)} = \frac{E(g(T_n - \eta(\theta))}{g(\epsilon)} \to 0 \; \forall \; \theta .$$

Hence, T_n is consistent for $\eta(\theta)$.

2.8.5 Suppose $\{X_1, X_2, \ldots, X_n\}$ is a random sample of size n from the following distributions—(i) Bernoulli $B(1, \theta)$, (ii) Poisson $Poi(\theta)$, (iii) uniform $U(0, 2\theta)$ and exponential distribution with mean θ. Show that the sample mean $\overline{X}_n$ is consistent for θ using the four approaches discussed in Example 2.2.2.

Solution: For Bernoulli $B(1, \theta)$, Poisson $Poi(\theta)$, uniform $U(0, 2\theta)$ and exponential distribution, mean is θ and respective variances $v(\theta)$ are $\theta(1 - \theta)$, θ, $\theta^2/3$ and θ^2 and these are finite. Hence, by the CLT $\sqrt{n/v(\theta)}(\overline{X}_n - \theta) \xrightarrow{L} Z \sim N(0, 1)$. We use this result to find the limit of coverage probability. As discussed in Example 2.2.2, (i) the first approach is verification of consistency by the definition. Observe that, for given $\epsilon > 0$,

$$P_\theta[|\overline{X}_n - \theta| < \epsilon] = P_\theta\left[\sqrt{n/v(\theta)}\,|\overline{X}_n - \theta| < \sqrt{n/v(\theta)}\epsilon\right]$$
$$\approx \Phi\left(\sqrt{n/v(\theta)}\epsilon\right) - \Phi\left(-\sqrt{n/v(\theta)}\epsilon\right) \to 1,$$
$$\text{as } n \to \infty, \; \forall \; \theta \in \Theta.$$

Thus, the coverage probability converges to 1 as $n \to \infty$, $\forall \; \theta \in \Theta$ and hence the sample mean $\overline{X}_n$ is a consistent estimator of θ.

(ii) It is to be noted that

$$E(\overline{X}_n - \theta)^2 = Var(\overline{X}_n) = v(\theta)/n \to 0, \text{ as } n \to \infty, \; \forall \; \theta \in \Theta.$$

Thus, $\overline{X}_n$ converges in quadratic mean to θ and hence converges in probability to θ.

(iii) Suppose $F_n(x)$, $x \in \mathbb{R}$ denotes the distribution function of $\overline{X}_n - \theta$. Then

$$F_n(x) = P_\theta[\overline{X}_n - \theta \le x] = P_\theta \left[\sqrt{n/v(\theta)}(\overline{X}_n - \theta) \le \sqrt{n/v(\theta)}\, x \right]$$
$$\approx \Phi \left(\sqrt{n/v(\theta)}\, x \right), \quad x \in \mathbb{R}$$

and limiting behavior of $F_n(x)$ as $n \to \infty$ is as follows:

$$F_n(x) \quad \to \quad \begin{cases} 0, & \text{if } x < 0 \\ 1/2, & \text{if } x = 0 \\ 1, & \text{if } x > 0. \end{cases}$$

Thus, $F_n(x) \to F(x)$, $\forall\ x \in C_F(x) = \mathbb{R} - \{0\}$, where F is a distribution function of a random variable degenerate at 0 and $C_F(x)$ is a set of points of continuity of F. It implies that $(\overline{X}_n - \theta) \xrightarrow{L} 0$, where the limit law is degenerate and hence, $(\overline{X}_n - \theta) \xrightarrow{P_\theta} 0$, for all $\theta \in \Theta$, which proves that $\overline{X}_n$ is consistent for θ.

(iv) Further, by Khinchine's WLLN, $\overline{X}_n \xrightarrow{P_\theta} \theta$, for all $\theta \in \Theta$.

2.8.6 Suppose $\underline{X} \equiv \{X_1, X_2, \ldots, X_n\}$ is a random sample of size n from a Poisson $Poi(\theta)$ distribution. Find the maximum likelihood estimator of θ and examine whether it is consistent for θ when (i) $\theta \in [a, b] \subset (0, \infty)$ and (ii) $\theta \in \{1, 2\}$.

Solution: Corresponding to a random sample $\underline{X}$ from $Poi(\theta)$ distribution, the likelihood of θ is given by

$$L_n(\theta|\underline{X}) = \prod_{i=1}^{n}(1/X_i!)\exp(-\theta)\theta^{X_i} = \left(\prod_{i=1}^{n}(1/X_i!) \right) \exp(-n\theta)\, \theta^{\sum\limits_{i=1}^{n} X_i}.$$

(i) The log likelihood function $Q(\theta) = \log L_n(\theta|\underline{X})$ and its first and second derivatives are given by

$$Q(\theta) = c - n\theta + n\overline{X}_n \log\theta, \quad Q'(\theta) = -n + \frac{n\overline{X}_n}{\theta} = n(\overline{X}_n - \theta)/\theta$$

$$\text{and } Q''(\theta) = -\frac{n\overline{X}_n}{\theta^2},$$

where c is a constant free from θ. Thus, the solution of the likelihood equation is given by $\theta = \overline{X}_n$ and at this solution the second derivative is negative if $\overline{X}_n > 0$. Hence, the maximum likelihood estimator $\hat{\theta}_n$ of θ is given by $\hat{\theta}_n = \overline{X}_n$, provided $\overline{X}_n \in [a, b]$. However, for any $\theta \in [a, b]$, it is possible that $\overline{X}_n < a$ and $\overline{X}_n > b$ as shown below. Suppose $U_n = \sum_{i=1}^{n} X_i$ then $U_n \sim Poi(n\theta)$. Hence,

$$P_\theta[\overline{X}_n < a] = P_\theta[U_n < na] > 0 \ \& \ P_\theta[\overline{X}_n > b] = P_\theta[U_n > nb] > 0.$$

Now

$$\overline{X}_n < a \le \theta \ \Rightarrow \ \overline{X}_n - \theta < 0 \ \Rightarrow \ Q'(\theta) = n(\overline{X}_n - \theta)/\theta < 0$$

and hence $Q(\theta)$ is a decreasing function of θ. It attains maximum at the smallest possible value of θ which is a. Similarly,

$$\overline{X}_n > b \ge \theta \ \Rightarrow \ \overline{X}_n - \theta > 0 \ \Rightarrow \ Q'(\theta) > 0$$

and hence $Q(\theta)$ is an increasing function of θ. It attains maximum at the largest possible value of θ which is b. Thus, the maximum likelihood estimator $\hat{\theta}_n$ of θ is given by

$$\hat{\theta}_n \ = \ \begin{cases} a, & \text{if } \overline{X}_n < a \\ \overline{X}_n, & \text{if } \overline{X}_n \in [a, b] \\ b, & \text{if } \overline{X}_n > b. \end{cases}$$

To verify its consistency, observe that by WLLN $\overline{X}_n \overset{P_\theta}{\to} \theta$, for all $\theta \in [a, b]$. Now for $\epsilon > 0$ and $\theta \in (a, b)$, for large n

$$\begin{aligned} P_\theta[|\hat{\theta}_n - \overline{X}_n| < \epsilon] &\ge P_\theta[\hat{\theta}_n = \overline{X}_n] = P_\theta[a \le \overline{X}_n \le b] \\ &\approx \Phi(\sqrt{n}(b - \theta)/\sqrt{\theta}) - \Phi(\sqrt{n}(a - \theta)/\sqrt{\theta}) \end{aligned}$$

and it converges to 1 as $n \to \infty$. Hence, $\forall \ \theta \in (a, b)$, $\hat{\theta}_n \overset{P_\theta}{\to} \theta$. Now to examine convergence in probability at the boundary points a and b, note that for $\theta = a$,
$$P_a[|\hat{\theta}_n - a| > \epsilon] = P_a[\hat{\theta}_n - a > \epsilon] = P_a[\hat{\theta}_n > a + \epsilon] \text{ and}$$

$$P_a[\hat{\theta}_n > a + \epsilon] \ = \ \begin{cases} 0, & \text{if } \epsilon > b - a \\ P_a[\overline{X}_n > a + \epsilon] \approx 1 - \Phi(\sqrt{n}\epsilon/\sqrt{a}) \ \to 0, & \text{if } \epsilon \le b - a. \end{cases}$$

Thus, $\hat{\theta}_n \overset{P_a}{\to} a$. Further, for the boundary point b,
$$P_b[|\hat{\theta}_n - b| > \epsilon] = P_b[b - \hat{\theta}_n > \epsilon] = P_b[\hat{\theta}_n < b - \epsilon] \text{ and}$$

$$P_b[\hat{\theta}_n < b - \epsilon] \ = \ \begin{cases} 0, & \text{if } \epsilon > b - a \\ P_b[\overline{X}_n < b - \epsilon] \approx \Phi(-\sqrt{n}\epsilon/\sqrt{b}) \ \to 0, & \text{if } \epsilon \le b - a. \end{cases}$$

Thus, $\hat{\theta}_n \overset{P_b}{\to} b$ and hence $\hat{\theta}_n$ is a consistent estimator of θ.

(ii) It is to be noted that if $\theta \in \Theta = \{1, 2\}$, the likelihood is not even a continuous function of θ and hence to find the maximum likelihood estimator of θ, we compare $L_n(2|\underline{X})$ with $L_n(1|\underline{X})$. Observe that, $\frac{L_n(2|\underline{X})}{L_n(1|\underline{X})} = \exp(-n) \, 2^{n\overline{X}_n}$. Now

$$\exp(-n)2^{n\overline{X}_n} > 1 \quad \Leftrightarrow \quad -n + n\overline{X}_n \log 2 > 0 \quad \Leftrightarrow \quad \overline{X}_n > 1/\log 2 = 1.4427 \, .$$

Thus, $L_n(2|\underline{X}) > L_n(1|\underline{X})$ if $\overline{X}_n > 1.4427$. Hence, the maximum likelihood estimator $\hat{\theta}_n$ of θ is given by

$$\hat{\theta}_n = \begin{cases} 2, & \text{if } \overline{X}_n > 1.4427 \\ 1, & \text{if } \overline{X}_n \leq 1.4427. \end{cases}$$

To verify consistency of $\hat{\theta}_n$, we have to check whether $\hat{\theta}_n \overset{P_\theta}{\to} \theta$ for every θ, that is, whether $\hat{\theta}_n \overset{P_2}{\to} 2$ and $\hat{\theta}_n \overset{P_1}{\to} 1$. Observe that

$$P_2[|\hat{\theta}_n - 2| < \epsilon] = \begin{cases} 1, & \text{if } \epsilon > 1 \\ P_2[\hat{\theta}_n = 2], & \text{if } 0 < \epsilon \leq 1. \end{cases}$$

Further,

$$P_2[\hat{\theta}_n = 2] = P_2\left[\overline{X}_n > 1.4427\right]$$
$$\approx 1 - \Phi\left(\sqrt{n}(1.4427 - 2)/\sqrt{2}\right) \quad \to 1 \quad \text{as} \quad n \to \infty.$$

On similar lines,

$$P_1[|\hat{\theta}_n - 1| < \epsilon] = P_1 [1 - \epsilon < \hat{\theta}_n < 1 + \epsilon] = \begin{cases} 1, & \text{if } \epsilon > 1 \\ P_1[\hat{\theta}_n = 1], & \text{if } 0 < \epsilon \leq 1 \end{cases}$$

and,

$$P_1[\hat{\theta}_n = 1] = P_1\left[\overline{X}_n < 1.4427\right] \approx \Phi\left(\sqrt{n}(1.4427 - 1)\right) \quad \to 1 \quad \text{as} \quad n \to \infty.$$

Thus, $\hat{\theta}_n \overset{P_2}{\to} 2$ and $\hat{\theta}_n \overset{P_1}{\to} 1$ implying that $\hat{\theta}_n$ is consistent for θ.

2.8.7 Suppose $\{X_1, X_2, \ldots, X_n\}$ is a random sample from a uniform $U(\theta - 1, \theta + 1)$ distribution, $\theta \in \mathbb{R}$. (i) Examine whether $T_{1n} = \overline{X}_n$, $T_{2n} = X_{(1)} + 1$ and $T_{3n} = X_{(n)} - 1$ are consistent estimators for θ. Which one is better? Why? (ii) Find an uncountable family of consistent estimators of θ based on sample quantiles.

Solution: Suppose $X \sim U(\theta - 1, \theta + 1)$ distribution, then $E(X) = \theta$ and hence by the WLLN, $\overline{X}_n$ is consistent for θ. Distribution function $F_X(x, \theta)$ of X is given by

$$F_X(x, \theta) = \begin{cases} 0, & \text{if} \quad x < \theta - 1 \\ \frac{x - (\theta - 1)}{2}, & \text{if } \theta - 1 \leq x < \theta + 1 \\ 1, & \text{if} \quad x \geq \theta + 1. \end{cases}$$

Suppose a random variable Y is defined as $Y = F_X(X, \theta) = \frac{X - \theta + 1}{2}$. Then by the probability integral transformation, it is known that $Y \sim U(0, 1)$. In Example 2.5.2, we have shown that $Y_{(1)} \xrightarrow{P} 0$ and $Y_{(n)} \xrightarrow{P} 1$. Now, $Y_{(1)} \xrightarrow{P} 0 \Rightarrow X_{(1)} + 1 \xrightarrow{P_\theta} \theta \ \forall \ \theta \in \mathbb{R}$. Hence, $T_{2n} = X_{(1)} + 1$ is consistent for θ. On similar lines, $Y_{(n)} \xrightarrow{P} 1 \Rightarrow X_{(n)} - 1 \xrightarrow{P_\theta} \theta \ \forall \ \theta \in \mathbb{R}$. Hence, $T_{3n} = X_{(n)} - 1$ is consistent for θ. We compare these three consistent estimators by comparing their MSEs. $T_{1n} = \overline{X}_n$ is an unbiased estimator of θ. Hence,

$$MSE_\theta(T_{1n}) = Var(\overline{X}_n) = Var(X)/n = 1/3n.$$

To find MSE of T_{2n} and T_{3n}, note that

$$Y = (X - \theta + 1)/2 \sim U(0, 1) \Rightarrow E(Y_{(1)}) = 1/(n + 1),$$
$$E(Y_{(1)}^2) = 2/n + 1)(n + 2)$$
$$E(Y_{(n)}) = n/(n + 1), \ \& \ E(Y_{(n)}^2) = n/(n + 2).$$

Using these expressions we have,

$$MSE_\theta(X_{(1)} + 1) = E_\theta(X_{(1)} + 1 - \theta)^2 = 4E_\theta\left((X_{(1)} + 1 - \theta)/2\right)^2$$
$$= 4E_\theta(Y_{(1)})^2 = 8/(n + 1)(n + 2)$$
$$\& \ MSE_\theta(X_{(n)} - 1) = E_\theta(X_{(n)} - 1 - \theta)^2 = 4E_\theta\left((X_{(n)} + 1 - \theta - 2)/2\right)^2$$
$$= 4E_\theta(Y_{(n)} - 1)^2 = 4\left(E_\theta(Y_{(n)}^2) - 2E_\theta(Y_{(n)}) + 1\right)$$
$$= 8/(n + 1)(n + 2).$$

It is to be noted that MSE of T_{2n} and T_{3n} is the same and

$$MSE_\theta(T_{1n}) - MSE_\theta(T_{2n}) = \frac{1}{3n} - \frac{8}{(n + 1)(n + 2)}$$
$$= \frac{n^2 - 21n + 2}{n(n + 1)(n + 2)} = \frac{n(n - 21) + 2}{n(n + 1)(n + 2)} > 0,$$

if $n > 21$. If we compare the rate of convergence of MSE to 0, then also MSE of T_{2n}, which is same as that of T_{3n}, converges to 0 faster than that of T_{1n}. Thus, T_{2n} and T_{3n} are better than T_{1n}.

(ii) From the distribution function of X, the p-th population quantile $a_p(\theta)$ is given by the solution of the equation,

$$F_X(x, \theta) = p \Leftrightarrow (x - (\theta - 1))/2 = p \Rightarrow a_p(\theta) = 2p - 1 + \theta, \ 0 < p < 1.$$

Suppose $r_n = [np] + 1$. Then the p-th sample quantile $X_{(r_n)}$ is consistent for $a_p(\theta)$ and hence $X_{(r_n)} - 2p + 1$ is consistent for θ. Thus, the uncountable family of consistent estimators of θ based on the sample quantiles is given by $\{X_{(r_n)} - 2p + 1, \ 0 < p < 1\}$.

2.8.8 Suppose $\{X_1, X_2, \ldots, X_n\}$ are independent random variables where X_i follows a uniform $U(i(\theta - 1), i(\theta + 1))$ distribution, $\theta \in \mathbb{R}$. Find the maximum likelihood estimator of θ and examine whether it is a consistent estimator of θ.

Solution: Suppose a random variable Y_i is defined as $Y_i = X_i/i$, $i = 1, 2, \ldots, n$. Then $\{Y_1, Y_2, \ldots, Y_n\}$ are independent and identically distributed random variables where $Y_i \sim U(\theta - 1, \theta + 1)$ distribution. Corresponding to observations $\{X_1, X_2, \ldots, X_n\}$, we have a random sample $\underline{Y} \equiv \{Y_1, Y_2, \ldots, Y_n\}$ from the distribution of $Y \sim U(\theta - 1, \theta + 1)$. The likelihood of θ given $\underline{Y}$ is given by

$$L_n(\theta | \underline{Y}) = \prod_{i=1}^{n} \frac{1}{2} = \left(\frac{1}{2}\right)^n, \quad \theta - 1 \leq Y_i \leq \theta + 1,$$

$$\forall \ i \ \Leftrightarrow \ Y_{(1)} \geq \theta - 1, \ Y_{(n)} \leq 1 + \theta.$$

Thus, the likelihood is constant for $Y_{(n)} - 1 \leq \theta \leq Y_{(1)} + 1$. Thus, any value between $(Y_{(n)} - 1, Y_{(1)} + 1)$ can be taken as the maximum likelihood estimator of θ. Thus, we define the maximum likelihood estimator $\hat{\theta}_n$ of θ as $\hat{\theta}_n = \alpha(Y_{(n)} - 1) + (1 - \alpha)(Y_{(1)} + 1)$. Now as shown in Example 2.5.2,

$$Y_{(1)} \xrightarrow{P_\theta} \theta - 1 \ \& \ Y_{(n)} \xrightarrow{P_\theta} \theta + 1 \ \forall \ \theta.$$

Thus, $\hat{\theta}_n \xrightarrow{P_\theta} \theta \ \forall \ \theta$, which implies that $\hat{\theta}_n$ is consistent for θ.

2.8.9 Suppose $\{X_1, X_2, \ldots, X_n\}$ are independent random variables where $X_i \sim U(0, i\theta)$ distribution, $\theta \in \Theta = \mathbb{R}^+$. (i) Find the maximum likelihood estimator of θ and examine whether it is consistent for θ. (ii) Find the moment estimator of θ and examine whether it is consistent for θ.

Solution: Suppose $X_i \sim U(0, i\theta)$, then it is easy to verify that $Y_i = X_i/i \sim U(0, \theta)$. Thus, $\{Y_1, Y_2, \ldots, Y_n\}$ are independent and identically distributed random variables each having a uniform $U(0, \theta)$. Proceeding on similar lines as in Example 2.2.1, we get that the maximum likelihood estimator of θ is $Y_{(n)}$ and it is consistent for θ. The moment estimator of θ is $2\overline{Y}_n$ and it is consistent for θ.

2.8.10 Suppose $\{X_1, X_2, \ldots, X_n\}$ is a random sample from a uniform $U(-\theta, \theta)$ distribution. Examine whether $-X_{(1)}$ and $X_{(n)}$ are both consistent for θ. Is $(X_{(n)} - X_{(1)})/2$ consistent for θ? Justify your answer.

Solution: We define a random variable Y as $Y = (X + \theta)/2\theta$, then by the probability integral transformation, $Y \sim U(0, 1)$. In Example 2.5.2, we have shown that corresponding to a random sample of size n from the uniform $U(0, 1)$ distribution,

$$Y_{(1)} \stackrel{P}{\rightarrow} 0 \ \& \ Y_{(n)} \stackrel{P}{\rightarrow} 1 \ \Rightarrow \ X_{(1)} \stackrel{P_\theta}{\rightarrow} -\theta \ \& \ X_{(n)} \stackrel{P_\theta}{\rightarrow} \theta \,.$$

Thus, $-X_{(1)}$ and $X_{(n)}$ both are consistent for θ. Further, convergence in probability is closed under all arithmetic operations, thus $(X_{(n)} - X_{(1)})/2 \stackrel{P_\theta}{\rightarrow} \theta$ and hence is consistent for θ.

2.8.11 Suppose $\{X_1, X_2, \ldots, X_{2n+1}\}$ is a random sample from a uniform $U(\theta - 1, \theta + 1)$ distribution, $\theta \in \mathbb{R}$. Examine whether $X_{(n)} - 1$ and $X_{([n/4]+1)}$ is consistent for θ.

Solution: Suppose $X \sim U(\theta - 1, \theta + 1)$ distribution. Then its distribution function $F(x, \theta)$ is given by

$$F(x, \theta) \;=\; \begin{cases} 0, & \text{if} \quad x < \theta - 1 \\ (x - \theta + 1)/2, & \text{if } \theta - 1 \le x < \theta + 1 \\ 1, & \text{if} \quad x \ge \theta + 1. \end{cases}$$

It then follows that the median of X is θ. From Theorem 2.2.6, p-th sample quantile $X_{(r_n)} \stackrel{P_\theta}{\rightarrow} a_p(\theta)$, the p-th population quantile, where $r_n = [np] + 1$, $0 < p < 1$ and n is the sample size. Here, sample size is $2n + 1$. Hence, with $p = 1/2$, $r_{2n+1} = [(2n + 1)1/2] + 1 = n + 1 \Rightarrow X_{(n+1)} \stackrel{P_\theta}{\rightarrow} a_{1/2}(\theta) = \theta$. Now to examine whether $X_{(n)} - 1$ is consistent for θ, we decide p so that

$$r_{2n+1} \;=\; [(2n + 1)p] + 1 = n \ \Rightarrow \ [(2n + 1)p] = n - 1$$
$$\Rightarrow (2n + 1)p < n \ \Rightarrow \ p < n/(2n + 1) \ < \ 1/2$$
$$\Rightarrow X_{(n)} \stackrel{P_\theta}{\rightarrow} a_p(\theta) < \theta$$
$$\Rightarrow X_{(n)} - 1 \stackrel{P_\theta}{\rightarrow} a_p(\theta) - 1 < \theta - 1$$

and hence $X_{(n)} - 1$ is not consistent for θ. To examine whether $X_{([n/4]+1)}$ is consistent for θ we proceed on similar lines. Observe that

$$X_{(r_{2n+1})} \;=\; X_{([n/4]+1)} \ \Rightarrow \ r_{2n+1} = [(2n + 1)p] + 1 = [n/4] + 1$$
$$\Rightarrow [(2n + 1)p] = [n/4] \ \Rightarrow \ (2n + 1)p + x = n/4 + y,$$
$$0 < x, \ y < 1.$$

If $x = y$, then $p = n/4(2n + 1) < 1/8$. If $x > y$, then $(2n + 1)p = n/4 + y - x < n/4$ which implies that $p < n/4(2n + 1) < 1/8$. Thus, $X_{([n/4]+1)} \overset{P_\theta}{\to} a_p(\theta) < a_{1/8}(\theta) < \theta$. If $x < y$, then maximum possible value for $y - x$ can be 1. Hence,

$$(2n + 1)p = n/4 + y - x < n/4 + 1$$
$$\Rightarrow p < n/4(2n + 1) + 1/(2n + 1) < 1/2.$$

Thus, $X_{([n/4]+1)} \overset{P_\theta}{\to} a_p(\theta) < \theta$. Hence, $X_{([n/4]+1)}$ does not converge in probability to θ and hence $X_{([n/4]+1)}$ is not consistent for θ.

2.8.12 Suppose $\underline{X} \equiv \{X_1, X_2, \ldots, X_n\}$ is a random sample of size n from a binomial $B(1, \theta)$ distribution, $\theta \in \Theta = (0, 1)$. (i) Find the maximum likelihood estimator of θ and examine whether it is consistent for θ. (ii) Find the moment estimator of θ and examine whether it is consistent for θ. (iii) Find the maximum likelihood estimator of θ and examine whether it is consistent for θ, if $\Theta = (a, b) \subset (0, 1)$.

Solution: Suppose $X \sim B(1, \theta)$, then the likelihood of θ corresponding to the given random sample is given by

$$\log L_n(\theta | \underline{X}) = \log \theta \sum_{i=1}^n X_i - \log(1 - \theta)(n - \sum_{i=1}^n X_i).$$

To find the maximum likelihood estimator of θ, the first and the second derivative of $\log L_n(\theta | \underline{X})$ are given by

$$Q(\theta) = \frac{\partial}{\partial \theta} \log L_n(\theta | \underline{X}) = \frac{n \overline{X}_n}{\theta} - \frac{n - n \overline{X}_n}{1 - \theta} = \frac{n \overline{X}_n}{\theta(1 - \theta)} - \frac{n}{1 - \theta}$$

$$= \frac{n}{1 - \theta} \left(\frac{\overline{X}_n - \theta}{\theta} \right)$$

$$\& \quad \frac{\partial^2}{\partial \theta^2} \log L_n(\theta | \underline{X}) = n \overline{X}_n \left(-\frac{1}{\theta^2} - \frac{1}{(1 - \theta)^2} \right)$$

$$- \frac{n}{(1 - \theta)^2} < 0 \; \forall \; \theta \in (0, 1).$$

Solving the likelihood equation $Q(\theta) = 0$, we get the solution as $\theta = \overline{X}_n$ and at this solution, the second derivative is almost surely negative.
(i) If the parameter space is $\Theta = (0, 1)$, then $\overline{X}_n$ is an estimator provided $\overline{X}_n \in (0, 1)$. However, it is possible that $\overline{X}_n = 0 \Leftrightarrow X_i = 0 \; \forall \; i = 1, 2, \ldots, n$, the probability of which is $(1 - \theta)^n > 0$. In this case, the likelihood of θ is given by $(1 - \theta)^n$. It is a decreasing function of θ and attains

supremum at $\theta = 0$. Similarly, it is possible that $\overline{X}_n = 1 \Leftrightarrow X_i = 1 \; \forall \; i = 1, 2, \ldots, n$, the probability of which is $\theta^n > 0$. In this case, the likelihood of θ is given by θ^n. It is an increasing function of θ and attains supremum at $\theta = 1$. However, 0 and 1 are not included in the parameter space. Hence, the maximum likelihood estimator of θ does not exist if $\theta \in (0, 1)$. However, it is to be noted that both $P[\overline{X}_n = 0] = (1 - \theta)^n$ and $P[\overline{X}_n = 1] = \theta^n$ converge to 0 as n increases, that is with probability approaching 1, $0 < \overline{X}_n < 1$ and for large n, $\overline{X}_n$ as the maximum likelihood estimator of θ. By the WLLN it is consistent for θ.

(ii) The moment estimator of θ is $\overline{X}_n$ provided $\overline{X}_n \in (0, 1)$, however as discussed above it is possible that $\overline{X}_n = 0$ and 1. Thus, a moment estimator does not exist if $\overline{X}_n \notin (0, 1)$. If $0 < \overline{X}_n < 1$, by the WLLN it is consistent for θ.

(iii) In (i) we have seen that the solution of the likelihood equation is $\theta = \overline{X}_n$ and the second derivative is negative at this solution. Thus $\overline{X}_n$ is the maximum likelihood estimator if $\overline{X}_n \in [a, b]$. However,

$$P\left[\overline{X}_n < a\right] = P\left[\sum_{i=1}^n X_i < an\right] = \sum_{r=0}^{[an]-1} \binom{n}{r} \theta^r (1 - \theta)^{n-r} > 0.$$

Similarly, $P\left[\overline{X}_n > b\right] > 0$. Suppose $\overline{X}_n < a$. Further, $a \leq \theta \Rightarrow \overline{X}_n < \theta \Rightarrow Q(\theta) < 0$. In this case, the likelihood is a decreasing function of θ and attains supremum at smallest possible value of θ, which is a. Similarly, if $\overline{X}_n > b$ then $\theta \leq b \Rightarrow \overline{X}_n > \theta \Rightarrow Q(\theta) > 0$. In this case, the likelihood is an increasing function of θ and attains supremum at largest possible value of θ, which is b. Hence, the maximum likelihood estimator $\hat{\theta}_n$ of θ is given by

$$\hat{\theta}_n = \begin{cases} a, & \text{if } \overline{X}_n < a \\ \overline{X}_n, & \text{if } a \leq \overline{X}_n \leq b \\ b, & \text{if } \overline{X}_n > b. \end{cases}$$

Now to verify whether it is consistent, we proceed as follows. By WLLN, $\overline{X}_n \xrightarrow{P_\theta} \theta, \; \forall \; \theta \in [a, b]$ and by the CLT $\sqrt{n}(\overline{X}_n - \theta) \xrightarrow{L} Z_1 \sim N(0, \theta(1 - \theta))$. For $\epsilon > 0$

$$P_\theta[|\hat{\theta}_n - \overline{X}_n| < \epsilon] \geq P_\theta[\hat{\theta}_n = \overline{X}_n] = P_\theta[a \leq \overline{X}_n \leq b]$$

$$= P_\theta\left[\frac{\sqrt{n}(a - \theta)}{\sqrt{\theta(1 - \theta)}} \leq \frac{\sqrt{n}(\overline{X}_n - \theta)}{\sqrt{\theta(1 - \theta)}} \leq \frac{\sqrt{n}(b - \theta)}{\sqrt{\theta(1 - \theta)}}\right]$$

$$\approx \Phi\left(\frac{\sqrt{n}(b - \theta)}{\sqrt{\theta(1 - \theta)}}\right) - \Phi\left(\frac{\sqrt{n}(a - \theta)}{\sqrt{\theta(1 - \theta)}}\right) \to 1 \quad \text{as } n \to \infty,$$

$\forall \ \theta \in (a,b)$. Thus, $(\hat{\theta}_n - \overline{X}_n) \overset{P_\theta}{\to} 0$, $\forall \ \theta \in (a,b)$ and hence, $\hat{\theta}_n \overset{P_\theta}{\to} \theta$, $\forall \ \theta \in (a,b)$. Now for $\theta = a$, $P_a[|\hat{\theta}_n - a| > \epsilon] = P_a[\hat{\theta}_n - a > \epsilon] = P_a[\hat{\theta}_n > a + \epsilon] = 0$ if $a + \epsilon > b \Leftrightarrow \epsilon > b - a$. Suppose $0 < \epsilon \le b - a$, then

$$P_a\left[\hat{\theta}_n > a + \epsilon\right] = P_a\left[a + \epsilon < \overline{X}_n < b\right]$$

$$= P_a\left[\frac{\sqrt{n}(a + \epsilon - a)}{\sqrt{a(1-a)}} < \frac{\sqrt{n}(\overline{X}_n - a)}{\sqrt{a(1-a)}} < \frac{\sqrt{n}(b - a)}{\sqrt{a(1-a)}}\right]$$

$$\approx \Phi\left(\frac{\sqrt{n}(b-a)}{\sqrt{a(1-a)}}\right) - \Phi\left(\frac{\sqrt{n}\epsilon}{\sqrt{a(1-a)}}\right) \to 0 \ \text{ as } n \to \infty.$$

Thus, $\hat{\theta}_n \overset{P_a}{\to} a$. Suppose $\theta = b$, then $P_b[|b - \hat{\theta}_n| > \epsilon] = P_b[b - \hat{\theta}_n > \epsilon] = P_b[\hat{\theta}_n < b - \epsilon] = 0$ if $b - \epsilon < a \Leftrightarrow \epsilon > b - a$. Suppose $0 < \epsilon \le b - a$, then

$$P_b\left[\hat{\theta}_n < b - \epsilon\right] = P_b\left[a < \overline{X}_n < b - \epsilon\right]$$

$$= P_b\left[\frac{\sqrt{n}(a - b)}{\sqrt{b(1-b)}} < \frac{\sqrt{n}(\overline{X}_n - b)}{\sqrt{b(1-b)}} < \frac{\sqrt{n}(b - \epsilon - b)}{\sqrt{b(1-b)}}\right]$$

$$\approx \Phi\left(-\frac{\sqrt{n}\epsilon}{\sqrt{b(1-b)}}\right) - \Phi\left(\frac{\sqrt{n}(a-b)}{\sqrt{b(1-b)}}\right) \to 0 \ \text{ as } n \to \infty.$$

Hence, $\hat{\theta}_n \overset{P_b}{\to} b$. Thus, $\hat{\theta}_n \overset{P_\theta}{\to} \theta$ for all $\theta \in [a,b]$ and hence is consistent for θ.

2.8.13 Suppose $\{X_1, X_2, \ldots, X_n\}$ is a random sample of size n from a normal $N(\theta, 1)$ distribution, $\theta \in \Theta = \{0, 1\}$. An estimator $T_k(\overline{X}_n)$ is defined as follows. Prove that it is consistent for θ, if and only if $0 < k < 1$.

$$T_k(\overline{X}_n) = \begin{cases} 0, & \text{if } \overline{X}_n < k \\ 1, & \text{if } \overline{X}_n \ge k. \end{cases}$$

Solution: Suppose $0 < k < 1$. To verify consistency of $T_k(\overline{X}_n)$, we examine whether $T_k(\overline{X}_n) \overset{P_0}{\to} 0$ and $T_k(\overline{X}_n) \overset{P_1}{\to} 1$. Observe that

$$P_0[|T_k(\overline{X}_n) - 0| < \epsilon] = \begin{cases} 1, & \text{if } \epsilon > 1 \\ P_0[T_k(\overline{X}_n) = 0], & \text{if } 0 < \epsilon \le 1 \end{cases}$$

$$\& \ \ P_0[T_k(\overline{X}_n) = 0] = P_0\left[\overline{X}_n \le k\right] = \Phi\left(\sqrt{n}k\right) \to 1 \text{ as } n \to \infty, \text{ as } k > 0.$$

On similar lines,

$$P_1[|T_k(\overline{X}_n) - 1| < \epsilon] = P_1 [1 - \epsilon < T_k(\overline{X}_n) < 1 + \epsilon]$$
$$= \begin{cases} 1, & \text{if } \epsilon > 1 \\ P_1[T_k(\overline{X}_n) = 1], & \text{if } 0 < \epsilon \leq 1 \end{cases}$$

Further, $P_1[T_k(\overline{X}_n) = 1] = P_1 [\overline{X}_n > k] = 1 - \Phi\left(\sqrt{n}(k - 1)\right) \to 1$
as $n \to \infty$, as $k < 1$.

Thus, if $0 < k < 1$, $T_k(\overline{X}_n)$ is consistent for θ. Now suppose $T_k(\overline{X}_n)$ is consistent for θ, that is, $T_k(\overline{X}_n) \overset{P_0}{\to} 0$ and $T_k(\overline{X}_n) \overset{P_1}{\to} 1$, that is $\forall \; \epsilon > 0$, $P_0[|T_k(\overline{X}_n) - 0| < \epsilon] \to 1$ and $P_1[|T_k(\overline{X}_n) - 1| < \epsilon] \to 1$ as $n \to \infty$. Since $T_k(\overline{X}_n)$ is either 0 or 1, for $\epsilon > 1$, coverage probability is 1 for both $\theta = 0$ and $\theta = 1$. For $\epsilon \leq 1$, $\Phi\left(\sqrt{n}k\right) \to 1$ implies $k > 0$ and $1 - \Phi\left(\sqrt{n}(k - 1)\right) \to 1$ implies $k < 1$. For these two implications to be true, k must be in $(0, 1)$.

2.8.14 Suppose $\underline{X} \equiv \{X_1, X_2, \ldots, X_n\}$ is a random sample of size n from a normal $N(\theta, 1)$ distribution, $\theta \in \Theta = \{-1, 0, 1\}$. (i) Find the maximum likelihood estimator of θ and examine whether it is consistent for θ. (ii) Examine whether it is unbiased for θ. Examine whether it is asymptotically unbiased for θ.

Solution: Corresponding to a random sample $\underline{X}$ from a normal $N(\theta, 1)$ distribution, the likelihood of θ is given by

$$L_n(\theta|\underline{X}) = \prod_{i=1}^{n} \frac{1}{\sqrt{2\pi}} \exp\left\{-\frac{1}{2}(X_i - \theta)^2\right\}$$
$$= (\sqrt{2\pi})^{-n} \exp\left\{-\frac{1}{2} \sum_{i=1}^{n}(X_i - \theta)^2\right\}.$$

(i) It is to be noted for $\theta \in \Theta = \{-1, 0, 1\}$, the likelihood is not a continuous function of θ and hence to find the maximum likelihood estimator of θ, we compare $L_n(-1|\underline{X})$ with $L_n(0|\underline{X})$ and $L_n(0|\underline{X})$ with $L_n(1|\underline{X})$. Observe that, with $c = (\sqrt{2\pi})^{-n}$,

$$L_n(-1|\underline{X}) = c \exp\left\{-\frac{1}{2}\left(\sum_{i=1}^{n} X_i^2 + 2\sum_{i=1}^{n} X_i + n\right)\right\}$$

$$L_n(0|\underline{X}) = c \exp\left\{-\frac{1}{2} \sum_{i=1}^{n} X_i^2\right\}$$

$$L_n(1|\underline{X}) = c \exp\left\{-\frac{1}{2}\left(\sum_{i=1}^{n} X_i^2 - 2\sum_{i=1}^{n} X_i + n\right)\right\}.$$

Further,
$$\frac{L_n(-1|\underline{X})}{L_n(0|\underline{X})} = \exp\left\{-\sum_{i=1}^{n} X_i - n/2\right\} = \exp\left\{-n(\overline{X}_n - (-1/2))\right\}$$
$$> 1 \text{ if } \overline{X}_n < -1/2 \quad \& \quad \leq 1 \text{ if } \overline{X}_n \geq -1/2.$$

Similarly,
$$\frac{L_n(0|\underline{X})}{L_n(1|\underline{X})} = \exp\left\{-\frac{1}{2}(-2\sum_{i=1}^{n} X_i + n)\right\} = \exp\left\{-n(\overline{X}_n - 1/2)\right\}$$
$$> 1 \text{ if } \overline{X}_n < 1/2 \quad \& \quad \leq 1 \text{ if } \overline{X}_n \geq 1/2.$$

Hence, the maximum likelihood estimator $\hat{\theta}_n$ of θ is given by

$$\hat{\theta}_n \; = \; \begin{cases} -1, & \text{if} \quad \overline{X}_n < -1/2 \\ 0, & \text{if } -1/2 \leq \overline{X}_n < 1/2 \\ 1, & \text{if} \quad \overline{X}_n \geq 1/2 \, . \end{cases}$$

$\hat{\theta}_n$ will be a consistent estimator of θ if and only if $\hat{\theta}_n \overset{P_{-1}}{\to} -1$, $\hat{\theta}_n \overset{P_0}{\to} 0$ and $\hat{\theta}_n \overset{P_1}{\to} 1$. Observe that for $0 < \epsilon \leq 1$,
$$P_{-1}[|\hat{\theta}_n - (-1)| < \epsilon] = P_{-1}[-1-\epsilon < \hat{\theta}_n < -1+\epsilon] = P_{-1}[\hat{\theta}_n = -1]$$
$$= P_{-1}\left[\overline{X}_n < -1/2\right] = \Phi\left(\sqrt{n}/2\right) \quad \to \quad 1 \text{ as } n \to \infty.$$

Further, for $1 < \epsilon \leq 2$,

$$P_{-1}[|\hat{\theta}_n - (-1)| < \epsilon] = P_{-1}[-1-\epsilon < \hat{\theta}_n < -1+\epsilon]$$
$$= P_{-1}[\hat{\theta}_n = -1, \hat{\theta}_n = 0]$$
$$= P_{-1}\left[\overline{X}_n < 1/2\right] = \Phi\left(3\sqrt{n}/2\right) \quad \to \quad 1 \text{ as } n \to \infty.$$

Now suppose $\epsilon > 2$, then
$$P_{-1}[|\hat{\theta}_n - (-1)| < \epsilon] = P_{-1}[-1-\epsilon < \hat{\theta}_n < -1+\epsilon]$$
$$= P_{-1}[\hat{\theta}_n = -1, \hat{\theta}_n = 0, \hat{\theta}_n = 1] = 1.$$

Thus, $\hat{\theta}_n \overset{P_{-1}}{\to} -1$. For $\theta = 0$ and for $0 < \epsilon \leq 1$,

$$P_0[|\hat{\theta}_n| < \epsilon] = P_0[-\epsilon < \hat{\theta}_n < \epsilon] = P_0[\hat{\theta}_n = 0]$$
$$= P_0\left[-1/2 < \overline{X}_n < 1/2\right]$$
$$= \Phi\left(\sqrt{n}/2\right) - \Phi\left(-\sqrt{n}/2\right) \quad \to \quad 1 \text{ as } n \to \infty.$$

Further for $\epsilon > 1$,

$$P_0[|\hat{\theta}_n| < \epsilon] = P_0[-\epsilon < \hat{\theta}_n < \epsilon] = P_0[\hat{\theta}_n = -1, \hat{\theta}_n = 0, \hat{\theta}_n = 1] = 1.$$

Hence, $\hat{\theta}_n \xrightarrow{P_0} 0$. Now for $\theta = 1$ and for $0 < \epsilon \le 1$,

$$P_1[|\hat{\theta}_n - 1| < \epsilon] = P_1[1 - \epsilon < \hat{\theta}_n < 1 + \epsilon] = P_1[\hat{\theta}_n = 1]$$
$$= P_1[\overline{X}_n \ge 1/2] = 1 - \Phi(-\sqrt{n}/2) \quad \to \quad 1 \text{ as } n \to \infty.$$

For $1 < \epsilon \le 2$,

$$P_1[|\hat{\theta}_n - 1| < \epsilon] = P_1[1 - \epsilon < \hat{\theta}_n < 1 + \epsilon] = P_1[\hat{\theta}_n = 0, \hat{\theta}_n = 1]$$
$$= P_1[\overline{X}_n \ge -1/2] = 1 - \Phi(-3\sqrt{n}/2) \quad \to \quad 1 \text{ as } n \to \infty.$$

Further for $\epsilon \ge 2$,

$$P_1[|\hat{\theta}_n - 1| < \epsilon] = P_1[1 - \epsilon < \hat{\theta}_n < 1 + \epsilon] = 1.$$

Hence, $\hat{\theta}_n \xrightarrow{P_1} 1$. Thus, $\hat{\theta}_n \xrightarrow{P_\theta} \theta$, $\forall \ \theta \in \Theta = \{-1, 0, 1\}$ and hence $\hat{\theta}_n$ is consistent for θ.

(ii) The estimator $\hat{\theta}_n$ is unbiased for θ, if $E_\theta(\hat{\theta}_n) = \theta$, $\forall \ \theta \in \Theta$. The possible values of $\hat{\theta}_n$ are $\{-1, 0, 1\}$ with probabilities $P_\theta[\overline{X}_n < -1/2]$, $P_\theta[-1/2 \le \overline{X}_n < 1/2]$ and $P_\theta[\overline{X}_n \ge 1/2]$ for $\theta = -1, 0, 1$ respectively. Hence,

$$E_{-1}(\hat{\theta}_n) = -1P_{-1}[\overline{X}_n < -1/2] + 0P_{-1}[-1/2 \le \overline{X}_n < 1/2]$$
$$+ 1P_{-1}[\overline{X}_n \ge 1/2]$$
$$= -1P_{-1}[\sqrt{n}(\overline{X}_n - (-1)) < \sqrt{n}(-1/2 - (-1))]$$
$$+ 1P_{-1}[\sqrt{n}(\overline{X}_n - (-1)) \ge \sqrt{n}(1/2 - (-1))]$$
$$= -\Phi(\sqrt{n}/2) + 1 - \Phi(3\sqrt{n}/2).$$

For $n = 4$, $E_{-1}(\hat{\theta}_n) = -0.8386$, for $n = 49$, $E_{-1}(\hat{\theta}_n) = -0.99976$ and for $n = 100$, it is approximately -1. Hence, we conclude that $\hat{\theta}_n$ is not unbiased for θ. However, for $n = 100$, it is approximately -1, indicates that it may be asymptotically unbiased for θ, which is verified below. It is to be noted that

$$E_{-1}(\hat{\theta}_n) = -\Phi(\sqrt{n}/2) + 1 - \Phi(3\sqrt{n}/2) \to -1 \text{ as } n \to \infty,$$

$$E_0(\hat{\theta}_n) = -1P_0[\overline{X}_n < -1/2] + 0P_0[-1/2 \le \overline{X}_n < 1/2] + 1P_0[\overline{X}_n \ge 1/2]$$
$$= -1P_0[\sqrt{n}(\overline{X}_n - 0) < \sqrt{n}(-1/2 - 0)]$$
$$+ 1P_0[\sqrt{n}(\overline{X}_n - 0) \ge \sqrt{n}(1/2 - 0)]$$
$$= -\Phi(-\sqrt{n}/2) + 1 - \Phi(\sqrt{n}/2) \to 0 \text{ as } n \to \infty.$$

& $E_1(\hat{\theta}_n) = -1 P_1 \left[\overline{X}_n < -1/2 \right] + 0 P_1 \left[-1/2 \le \overline{X}_n < 1/2 \right]$
$+ 1 P_1 \left[\overline{X}_n \ge 1/2 \right]$
$= -1 P_0 \left[\sqrt{n}(\overline{X}_n - 1) < \sqrt{n}(-1/2 - 1) \right]$
$+ 1 P_0 \left[\sqrt{n}(\overline{X}_n - 1) \ge \sqrt{n}(1/2 - 1) \right]$
$= -\Phi\left(-3\sqrt{n}/2 \right) + 1 - \Phi\left(-\sqrt{n}/2 \right) \to 1 \text{ as } n \to \infty.$

Thus, $E_\theta(\hat{\theta}_n) \to \theta$ as $n \to \infty$ for all $\theta \in \Theta$. Hence, $\hat{\theta}_n$ is asymptotically unbiased for θ.

2.8.15 Suppose $\{X_1, X_2, \ldots, X_n\}$ is a random sample from a Cauchy $C(\theta, 1)$ distribution, where $\theta \in \mathbb{R}$. Examine whether the sample mean is consistent for θ.

Solution: If $\{X_1, X_2, \ldots, X_n\}$ is a random sample from a Cauchy $C(\theta, 1)$ distribution, then the sample mean $\overline{X}_n$ also follows a Cauchy $C(\theta, 1)$ distribution. Hence for $\epsilon > 0$,

$$P_\theta[|\overline{X}_n - \theta| < \epsilon] = \int_{\theta-\epsilon}^{\theta+\epsilon} \frac{1}{\pi} \frac{1}{1 + (x - \theta)^2} dx$$
$$= \frac{1}{\pi} \left[\tan^{-1}(x - \theta) \right]_{\theta-\epsilon}^{\theta+\epsilon} = \frac{2}{\pi} \tan^{-1}(\epsilon),$$

which is a constant free from n and hence does not converge to 1 as $n \to \infty$. Hence, the sample mean is not consistent for θ.

2.8.16 Suppose $\underline{X} \equiv \{X_1, X_2, \ldots, X_n\}$ is a random sample from X with probability density function $f(x, \theta) = \theta/x^2$, $x \ge \theta$, $\theta > 0$. (i) Find the maximum likelihood estimator of θ and examine its consistency for θ by computing the coverage probability and also the MSE. (ii) Examine if $X_{(n)}$ is consistent for θ.

Solution: (i) The probability density function of a random variable X is given by

$$f_X(x, \theta) = \theta/x^2, \quad x \ge \theta.$$

Corresponding to a random sample $\underline{X}$ from this distribution, the likelihood of θ is given by

$$L_n(\theta|\underline{X}) = \prod_{i=1}^n \theta/X_i^2 = \theta^n \prod_{i=1}^n X_i^{-2}, \quad X_i \ge \theta, \forall i \Leftrightarrow X_{(1)} \ge \theta \Leftrightarrow \theta \le X_{(1)}.$$

Thus, the likelihood is an increasing function of θ and attains maximum at the maximum possible value of θ given data $\{X_1, X_2, \ldots, X_n\}$. The maximum possible value of θ given data is $X_{(1)}$ and hence the maximum likelihood estimator $\hat{\theta}_n$ of θ is given by $X_{(1)}$. To verify the consistency of $X_{(1)}$ as an estimator of θ, we find the coverage probability using the distribution function of $X_{(1)}$. The distribution function $F_X(x)$ of X is given by

$$F_X(x) = \begin{cases} 0, & \text{if } x < \theta \\ 1 - \theta/x, & \text{if } x \geq \theta. \end{cases}$$

Hence, the distribution function of $X_{(1)}$ is given by

$$F_{X_{(1)}}(x) = 1 - [1 - F_X(x)]^n = \begin{cases} 0, & \text{if } x < \theta \\ 1 - \theta^n/x^n, & \text{if } x \geq \theta. \end{cases}$$

For $\epsilon > 0$, the coverage probability is given by

$$\begin{aligned} P_\theta[|X_{(1)} - \theta| < \epsilon] &= P_\theta[\theta - \epsilon < X_{(1)} < \theta + \epsilon] \\ &= P_\theta[\theta < X_{(1)} < \theta + \epsilon] \quad \text{as} \quad X_{(1)} \geq \theta \\ &= F_{X_{(1)}}(\theta + \epsilon) - F_{X_{(1)}}(\theta) = 1 - \theta^n/(\theta + \epsilon)^n \\ &\to 1 \;\; \forall \; \epsilon > 0 \;\; \text{and} \;\; \forall \; \theta \;\; \text{as} \;\; n \to \infty. \end{aligned}$$

Hence, $X_{(1)}$ is consistent for θ. To compute the MSE of $X_{(1)}$ as an estimator of θ, we find the probability density function $g(x, \theta)$ of $X_{(1)}$ from its distribution function. It is given by $g(x, \theta) = n\theta^n/x^{n+1}$, $x \geq \theta$. Hence,

$$E(X_{(1)}) = \frac{n\theta}{n - 1}, \quad E(X_{(1)}^2) = \frac{n\theta^2}{n - 2}$$

$$\Rightarrow \; MSE_\theta(X_{(1)}) = \frac{2\theta^2}{(n - 1)(n - 2)} \; \to 0, \;\; \text{as } n \to \infty.$$

Thus, $X_{(1)}$ is MSE consistent for θ.

(ii) To examine if $X_{(n)}$ is consistent for θ, consider for $\epsilon > 0$, the coverage probability,

$$\begin{aligned} P_\theta[|X_{(n)} - \theta| < \epsilon] &= P_\theta[\theta - \epsilon < X_{(n)} < \theta + \epsilon] \\ &= P_\theta[\theta < X_{(n)} < \theta + \epsilon] \quad \text{as} \quad X_{(n)} \geq \theta \\ &= F_{X_{(n)}}(\theta + \epsilon) - F_{X_{(n)}}(\theta) = (1 - \theta/(\theta + \epsilon))^n \\ &\to 0 \;\; \forall \; \epsilon > 0 \;\; \text{and} \;\; \forall \; \theta \;\; \text{as} \;\; n \to \infty. \end{aligned}$$

Hence, $X_{(n)}$ is not consistent for θ.

2.8.17 Suppose X follows a Laplace distribution with probability density function $f(x,\theta) = \exp\{-|x-\theta|\}/2$, $x \in \mathbb{R}$, $\theta \in \mathbb{R}$. A random sample of size n is drawn from the distribution of X. Examine whether $\overline{X}_n$ is consistent for θ. Examine if it is MSE consistent. Is sample median consistent for θ? Justify. Find the maximum likelihood estimator of θ and examine whether it is consistent for θ. Find a family of consistent estimators of θ based on the sample quantiles.

Solution: Since X follows a Laplace distribution, $E(X) = \theta < \infty$ and $Var(X) = 2$. Thus, by the WLLN $\overline{X}_n \xrightarrow{P_\theta} \theta$. Hence, $\overline{X}_n$ is a consistent estimator of θ. Further, it is unbiased and hence its MSE is given by $Var(\overline{X}_n) = 2/n \to 0$. Thus, $\overline{X}_n$ is MSE consistent for θ. A Laplace distribution is a symmetric distribution, symmetric around θ, thus the population median is also θ. Hence by Theorem 2.2.6, the sample median is consistent for θ. To find the maximum likelihood estimator of θ, note that the likelihood of θ given the data $\underline{X}$ is,

$$L_n(\theta|\underline{X}) = \prod_{i=1}^{n} \exp\{-|X_i - \theta|\}/2 = (1/2^n) \exp\left\{-\sum_{i=1}^{n} |X_i - \theta|\right\}.$$

It is maximum with respect to variations in θ when $\sum_{i=1}^{n} |X_i - \theta|$ is minimum and it is minimized when θ is the sample median. Thus, the maximum likelihood estimator of θ is the sample median and it is consistent for θ. To find a family of consistent estimators of θ based on the sample quantiles, we first find the distribution function $F(x,\theta)$ of X. If $x < \theta$ then $|x - \theta| = -(x - \theta)$. Hence for $x < \theta$,

$$F(x,\theta) = \int_{-\infty}^{x} \frac{1}{2} \exp\{-|u-\theta|\}\, du = \int_{-\infty}^{x} \frac{1}{2} \exp\{u - \theta\}\, du$$

$$= \frac{1}{2} \exp\{x - \theta\}$$

For $x \geq \theta$, $F(x,\theta) = \int_{-\infty}^{\theta} \frac{1}{2} \exp\{-|u-\theta|\}\, du + \int_{\theta}^{x} \frac{1}{2} \exp\{-|u-\theta|\}\, du$

$$= \frac{1}{2} + \int_{\theta}^{x} \frac{1}{2} \exp\{-(u-\theta)\}\, du$$

$$= \frac{1}{2} + \frac{1}{2}[1 - \exp\{-(x-\theta)\}]$$

$$= 1 - \frac{1}{2} \exp\{-(x-\theta)\}.$$

Now the p-th population quantile $a_p(\theta)$ is a solution of the equation $F(x, \theta) = p$.

For $p < 1/2$, $F(x, \theta) = p \Rightarrow (1/2)\exp\{x - \theta\} = p \Rightarrow x = a_p(\theta) = \theta + \log 2p$.

For $p \geq 1/2$, $F(x, \theta) = p \Rightarrow 1 - (1/2)\exp\{-(x - \theta)\} = p \Rightarrow x = a_p(\theta) = \theta - \log 2(1 - p)$.

By Theorem 2.2.6, the p-th sample quantile $X_{([np]+1)} \overset{P_\theta}{\to} a_p(\theta)$. Hence, for all $p \in (0, 1/2)$, $X_{([np]+1)} - \log 2p$ is consistent for θ and, for all $p \in [1/2, 1)$, $X_{([np]+1)} + \log 2(1 - p)$ is consistent for θ.

2.8.18 Suppose $\{X_1, X_2, \ldots, X_n\}$ is a random sample from an exponential distribution with mean θ. Show that $\overline{X}_n$ is MSE consistent for θ. Find a constant $c \in \mathbb{R}$ such that $T_n = n\overline{X}_n/(n + c)$ has MSE smaller than that of $\overline{X}_n$. Is T_n consistent for θ? Justify.

Solution: Since X follows an exponential distribution with mean θ, $Var(X) = \theta^2$. Further, $\overline{X}_n$ is an unbiased estimator of θ, hence $MSE_\theta(\overline{X}_n) = Var(\overline{X}_n) = \theta^2/n \to 0$ as $n \to \infty$. Thus, $\overline{X}_n$ is MSE consistent for θ. Now for

$$T_n = \frac{n\overline{X}_n}{n + c}, \quad E(T_n) = \frac{n\theta}{n + c}, \quad Bias_\theta(T_n) = \frac{-c\theta}{n + c} \quad \& \quad Var(T_n) = \frac{n\theta^2}{(n + c)^2}$$

$$\Rightarrow MSE_\theta(T_n) = Var(T_n) + (Bias_\theta(T_n))^2$$
$$= \frac{n\theta^2}{(n + c)^2} + \frac{c^2\theta^2}{(n + c)^2} = \frac{(n + c^2)\theta^2}{(n + c)^2}.$$

We find $c \in \mathbb{R}$ such that $MSE_\theta(T_n)$ is minimum. Suppose

$$g(c) = \frac{(n + c^2)}{(n + c)^2} \Rightarrow g'(c) = \frac{(n + c)^2 2c - 2(n + c^2)(n + c)}{(n + c)^2}$$
$$= \frac{2(n + c)n(c - 1)}{(n + c)^2}.$$

Now $g'(c) = 0 \Rightarrow c = 1$. c cannot be $-n$ as in that case T_n is not defined. Thus for $c = 1$, MSE of T_n is smaller than that of $\overline{X}_n$. Consistency of T_n follows from the consistency of $\overline{X}_n$ and the fact that $n/(n + c) \to 1$ as $n \to \infty$.

2.8.19 Suppose $\{X_1, X_2, \ldots, X_n\}$ is a random sample from a Laplace distribution with probability density function $f(x, \theta)$ given by

$$f(x, \theta) = (1/2\theta)\exp\{-|x|/\theta\}, \quad x \in \mathbb{R}, \quad \theta > 0.$$

Examine whether following estimators are consistent for θ. (i) Sample mean, (ii) sample median, (iii) $\sum_{i=1}^n |X_i|/n$ and (iv) $(\sum_{i=1}^n X_i^2/n)^{1/2}$.

Solution: Since X follows a Laplace distribution, it is distributed as $Y_1 - Y_2$, where Y_1 and Y_2 are independent random variables each having an exponential distribution with scale parameter $1/\theta$. Hence, $E(X) = 0$ and $Var(X) = 2\theta^2$. By the WLLN, the sample mean $\overline{X}_n \xrightarrow{P_\theta} E(X) = 0$, hence $\overline{X}_n$ cannot be consistent for θ as the limit random variable in convergence in probability is almost surely unique. Since the distribution of X is symmetric around 0, the median of X is the same as the mean of X which is zero. From Theorem 2.2.6, the sample median is consistent for population median which is 0 hence sample median cannot be consistent for θ. Now,

$$E(|X|) = (1/2\theta) \int_{-\infty}^{\infty} |x| \exp\{-|x|/\theta\}\, dx = (2/2\theta) \int_0^{\infty} x \exp\{-x/\theta\}\, dx = \theta.$$

Hence by the WLLN, $\sum_{i=1}^n |X_i|/n \xrightarrow{P_\theta} \theta$ and hence is consistent for θ. It is to be noted that $E(X^2) = Var(X) = 2\theta^2$ as $E(X) = 0$. Hence by the WLLN, $\sum_{i=1}^n X_i^2/n \xrightarrow{P_\theta} 2\theta^2$. By the invariance property of consistency under continuous transformations, $(\sum_{i=1}^n X_i^2/n)^{1/2} \xrightarrow{P_\theta} \sqrt{2}\theta$ and hence it is not consistent for θ.

2.8.20 Suppose $\{X_1, X_2, \ldots, X_n\}$ is a random sample of size n from a distribution of X with probability density function $\theta/x^{\theta+1}$, $x > 1, \theta > 0$. Examine whether $\overline{X}_n$ is consistent for θ for $\theta > 1$. What happens if $0 < \theta \le 1$? Obtain a consistent estimator of θ based on the transformations $g(x) = \log x$ and $g(x) = 1/x$.

Solution: For the given distribution $E(X) = \theta \int_1^{\infty} x/x^{\theta+1}\, dx$. The integral is convergent if the degree of polynomial in denominator is larger by 1 than the degree of polynomial in numerator, that is if $\theta + 1 - 1 > 1 \Leftrightarrow \theta > 1$. If $0 < \theta \le 1$, then the integral is divergent and $E(X)$ does not exist. For $\theta > 1$, $E(X) = \theta/(\theta - 1)$. By the WLLN, $\overline{X}_n \xrightarrow{P_\theta} \theta/(\theta - 1)$ and hence it cannot be consistent for θ, since the limit random variable in convergence in probability is almost surely unique. To find a consistent estimator of θ based on the transformation $\log x$, define $Y = \log X$, then using jacobian of transformation method, it follows that Y has an exponential distribution with scale parameter θ, that is, mean $1/\theta$. Corresponding to a random sample $\{X_1, X_2, \ldots, X_n\}$, we have a random sample $\{Y_1, Y_2, \ldots, Y_n\}$ and by the WLLN, $\overline{Y}_n = \sum_{i=1}^n \log X_i/n \xrightarrow{P_\theta} 1/\theta$, $\forall\, \theta > 0$. Thus, $n/\sum_{i=1}^n \log X_i$ is a consistent estimator of θ. Now to find a consistent estimator of θ based on the transformation $1/x$, observe that

$$E\left(\frac{1}{X}\right) = \theta \int_1^{\infty} \frac{1}{x^{\theta+2}}\, dx = \frac{\theta}{\theta + 1} < \infty \;\; \forall\, \theta > 0.$$

Hence by the WLLN,

$$T_n = \frac{1}{n}\sum_{i=1}^{n}\frac{1}{X_i} \xrightarrow{P_\theta} \frac{\theta}{\theta+1} \quad\Rightarrow\quad \frac{T_n}{1-T_n} \xrightarrow{P_\theta} \theta \ .$$

Thus, $T_n/(1 - T_n)$ is a consistent estimator of θ.

2.8.21 Suppose $\{X_1, X_2, \ldots, X_n\}$ is a random sample of size n from an exponential distribution with scale parameter 1 and location parameter θ. Examine whether $X_{(1)}$ is strongly consistent for θ.

Solution: If X follows an exponential distribution with scale parameter 1 and location parameter θ, then its probability density function $f_X(x, \theta)$ is given by $f_X(x, \theta) = \exp\{-(x - \theta)\}$, $x \geq \theta$. It is shown in Example 2.2.16 that the distribution of $X_{(1)}$ is again exponential with scale parameter n and location parameter θ. Hence, $E(X_{(1)}) = \theta + 1/n$ and $Var(X_{(1)}) = 1/n^2$, which implies that

$$MSE_\theta(X_{(1)}) = E(X_{(1)} - \theta)^2 = (Bias_\theta X_{(1)})^2 + Var(X_{(1)}) = 2/n^2.$$

To examine strong consistency we use the sufficient condition which states that if for some $r > 0$, $\sum_{n \geq 1} E(|X_n - X|^r) < \infty$, then $X_n \xrightarrow{a.s.} X$. Observe that

$$\sum_{n \geq 1} E(|X_{(1)} - \theta|^2) = \sum_{n \geq 1}\frac{2}{n^2} < \infty \quad\Rightarrow\quad X_{(1)} \xrightarrow{a.s.} \theta.$$

Hence, $X_{(1)}$ is strongly consistent for θ. We may use another sufficient condition of almost sure convergence which states that if $\forall \ \epsilon > 0$, $\sum_{n \geq 1} P[|X_n - X| > \epsilon] < \infty$, then $X_n \xrightarrow{a.s.} X$. Using the exponential distribution of $X_{(1)}$, we have for $\epsilon > 0$,

$$P[|X_{(1)} - \theta| > \epsilon] = 1 - P[\theta < X_{(1)} < \theta + \epsilon] = e^{-n\epsilon}$$
$$\Rightarrow \sum_{n \geq 1} P[|X_n - X| > \epsilon]$$
$$= \sum_{n \geq 1}(e^{-\epsilon})^n < \infty \quad \text{as } e^{-\epsilon} < 1 \text{ for } \epsilon > 0$$
$$\Rightarrow X_{(1)} \xrightarrow{a.s.} \theta.$$

2.8.22 Suppose $\underline{X} \equiv \{X_1, X_2, \ldots, X_n\}$ is a random sample from a normal $N(\theta, 1)$ distribution, where $\theta \in \{0, 1\}$. Examine whether the maximum likelihood estimator $\hat\theta_n$ of θ is strongly consistent for θ. Examine whether $\hat\theta_n = \theta$ almost surely for large n.

Solution: In Example 2.2.3, we have obtained the maximum likelihood estimator of θ corresponding to a random sample $\underline{X}$ from a normal $N(\theta, 1)$ distribution, when $\theta \in \{0, 1\}$. It is given by

$$\hat{\theta}_n = \begin{cases} 1, & \text{if } \overline{X}_n > \frac{1}{2} \\ 0, & \text{if } \overline{X}_n \leq \frac{1}{2}. \end{cases}$$

Further, we have shown that it is consistent for θ, since for $\epsilon > 1$,
$P_\theta[|\hat{\theta}_n - \theta| < \epsilon] = 1$ for $\theta = 0$ and $\theta = 1$. For $0 < \epsilon \leq 1$,
$P_\theta[|\hat{\theta}_n - \theta| < \epsilon] = \Phi\left(\sqrt{n}/2\right) \to 1$, as $n \to \infty$, for both $\theta = 0$ and $\theta = 1$.
Thus, for large n and $0 < \epsilon \leq 1$,

$$P_\theta[|\hat{\theta}_n - \theta| \geq \epsilon] = 1 - \Phi\left(\frac{\sqrt{n}}{2}\right)$$

$$= \Phi\left(-\frac{\sqrt{n}}{2}\right) = \phi\left(\frac{\sqrt{n}}{2}\right)\frac{2}{\sqrt{n}} = \frac{2}{\sqrt{2\pi}}\frac{e^{-n/8}}{\sqrt{n}},$$

where $\phi(\cdot)$ is the probability density function of the standard normal distribution. By the ratio test for convergence of series,

$$\frac{e^{-(n+1)/8}}{\sqrt{n+1}}\frac{\sqrt{n}}{e^{-n/8}} = \frac{\sqrt{n}}{\sqrt{n+1}}e^{-1/8} \to e^{-1/8} < 1$$

$$\Rightarrow \sum_{n\geq 1} P_\theta[|\hat{\theta}_n - \theta| > \epsilon] = \frac{2}{\sqrt{2\pi}}\sum_{n\geq 1}\frac{e^{-n/8}}{\sqrt{n}} < \infty$$

$$\Rightarrow \hat{\theta}_n \overset{a.s.}{\to} \theta,$$

by the sufficient condition of almost sure convergence. Further we have,

$$P_0[\hat{\theta}_n = 0] = P_1[\hat{\theta}_n = 1] = \Phi\left(\sqrt{n}/2\right)$$

$$\Leftrightarrow P_\theta[\hat{\theta}_n = \theta] = \Phi\left(\sqrt{n}/2\right) \,\forall\, \theta$$

$$\Rightarrow P_\theta[\hat{\theta}_n \neq \theta] = 1 - \Phi\left(\sqrt{n}/2\right) = \Phi\left(-\sqrt{n}/2\right)$$

$$= \frac{2}{\sqrt{2\pi}}\frac{e^{-n/8}}{\sqrt{n}} \;\&\; \frac{2}{\sqrt{2\pi}}\sum_{n\geq 1}\frac{e^{-n/8}}{\sqrt{n}} < \infty.$$

Suppose an event A_n is defined as $A_n = \{\omega | \hat{\theta}_n(\omega) \neq \theta\}$, then by the Borel-Cantelli lemma,

$$\sum_{n\geq 1} P_\theta[\hat{\theta}_n \neq \theta] < \infty \Rightarrow \sum_{n\geq 1} P(A_n) < \infty$$

$$\Rightarrow P(\limsup A_n) = 0 \Leftrightarrow P(\liminf A_n^c) = 1$$

$$\Rightarrow P_\theta\{\omega | \hat{\theta}_n(\omega) = \theta, \,\forall\, n \geq n_0(\omega)\} = 1.$$

Hence, we conclude that $\hat{\theta}_n = \theta$ almost surely for large n.

2.8.23 Suppose $\{X_1, X_2, \ldots, X_n\}$ is a random sample of size n from a normal $N(\theta, 1)$ distribution, $\theta \in \Theta = I$, the set of integers. Examine whether the maximum likelihood estimator $\hat{\theta}_n$ of θ is strongly consistent for θ. Examine whether $\hat{\theta}_n = \theta$ almost surely for large n.

Solution: Suppose $X \sim N(\theta, 1)$ distribution, where $\theta \in \Theta = I$. In Example 2.2.3 we have obtained the maximum likelihood estimator $\hat{\theta}_n$ of θ, which is given by

$$\hat{\theta}_n = k \quad \text{if} \quad \overline{X}_n \in [k - 1/2, k + 1/2), \quad \text{where} \quad k \in I.$$

In Example 2.2.3, we have proved that $\hat{\theta}_n$ is consistent for θ. To examine whether it is strongly consistent, we find the probability mass function of $\hat{\theta}_n$ as follows. For $k \in I$:

$$\begin{aligned}
P_\theta[\hat{\theta}_n = \theta] &= P_\theta[\theta - 1/2 \leq \overline{X}_n < \theta + 1/2] \\
&= \Phi(\sqrt{n}(\theta + 1/2 - \theta)) - \Phi(\sqrt{n}(\theta - 1/2 - \theta)) \\
&= \Phi(\sqrt{n}/2) - \Phi(-\sqrt{n}/2) \\
&= 1 - 2\Phi(-\sqrt{n}/2) \quad \forall \ \theta \in I.
\end{aligned}$$

Hence, $\quad P_\theta[\hat{\theta}_n \neq \theta] = 2\Phi(-\sqrt{n}/2) = 2\phi\left(\dfrac{\sqrt{n}}{2}\right)\dfrac{2}{\sqrt{n}} = \dfrac{4}{\sqrt{2\pi}}\dfrac{e^{-n/8}}{\sqrt{n}}$.

Now, as in the previous example,

$$\sum_{n \geq 1} P_\theta[\hat{\theta}_n \neq \theta] = \frac{4}{\sqrt{2\pi}} \sum_{n \geq 1} \frac{e^{-n/8}}{\sqrt{n}} < \infty$$

$$\Rightarrow \quad \sum_{n \geq 1} P_\theta[|\hat{\theta}_n - \theta| > \epsilon] \leq \sum_{n \geq 1} P_\theta[\hat{\theta}_n \neq \theta] < \infty$$

$\forall \ \epsilon > 0$. Hence, by the sufficient condition of almost sure convergence, $\hat{\theta}_n \overset{a.s.}{\to} \theta$. Suppose an event A_n is defined as $A_n = \{\omega|\hat{\theta}_n(\omega) \neq \theta\}$, then

$$\sum_{n \geq 1} P_\theta[\hat{\theta}_n \neq \theta] < \infty \Leftrightarrow \sum_{n \geq 1} P(A_n) < \infty$$

$$\Rightarrow P(\limsup A_n) = 0 \Leftrightarrow P(\liminf A_n^c) = 1$$

$$\Rightarrow P_\theta\{\omega|\hat{\theta}_n(\omega) = \theta, \ \forall \ n \geq n_0(\omega)\} = 1,$$

by the Borel-Cantelli lemma. Hence, we conclude that $\hat{\theta}_n = \theta$ almost surely for large n.

2.8.24 Suppose $\{X_1, X_2, \ldots, X_n\}$ is a random sample of size n from a Bernoulli $B(1, \theta)$ distribution. Examine whether $\overline{X}_n$ is uniformly consistent for $\theta \in \Theta = (0, 1)$.

Solution: Since $X \sim B(1, \theta)$ distribution, $E(\overline{X}_n) = \theta$ and $Var(\overline{X}_n) = \theta(1 - \theta)/n$. Consistency of $\overline{X}_n$ for θ follows from the WLLN. We use Chebyshev's inequality to find the minimum sample size $n_0(\epsilon, \delta, \theta)$. By Chebyshev's inequality,

$$P_\theta[|\overline{X}_n - \theta| < \epsilon] \geq 1 - \frac{E(\overline{X}_n - \theta)^2}{\epsilon^2}$$

$$= 1 - \frac{\theta(1 - \theta)}{n\epsilon^2} \geq 1 - \frac{1}{4n\epsilon^2} \quad \text{as } \theta(1 - \theta) \leq 1/4, \ \forall \ \theta \in \Theta.$$

We select $n_0(\epsilon, \delta, \theta)$ such that

$$1 - \frac{1}{4n\epsilon^2} \geq 1 - \delta \ \Rightarrow \ n \geq \frac{1}{4\epsilon^2\delta} \ \Rightarrow \ n_0(\epsilon, \delta, \theta) = \left[\frac{1}{4\epsilon^2\delta}\right] + 1,$$

thus, $n_0(\epsilon, \delta, \theta)$ does not depend on θ and hence $\overline{X}_n$ is uniformly consistent for θ.

2.8.25 Suppose X follows an exponential distribution with location parameter μ and scale parameter σ, with probability density function $f(x, \mu, \sigma)$ as

$$f(x, \mu, \sigma) = (1/\sigma) \exp\{-(x - \mu)/\sigma\} \quad x \geq \mu, \ \sigma > 0, \ \mu \in \mathbb{R}.$$

Suppose $\{X_1, X_2, \ldots, X_n\}$ is a random sample from the distribution of X. (i) Verify whether $\overline{X}_n$ is consistent for μ or σ. (ii) Find a consistent estimator for $\underline{\theta} = (\mu, \sigma)'$ based on the sample median and the sample mean.

Solution: (i) For a random variable X following an exponential distribution with location parameter μ and scale parameter σ, $E(X) = \mu + \sigma$. Hence, given a random sample $\{X_1, X_2, \ldots, X_n\}$, by the WLLN $\overline{X}_n \overset{P_\theta}{\to} E(X) = \mu + \sigma$. Thus, $\overline{X}_n$ is consistent for $\mu + \sigma$ and it cannot be consistent for any other parametric function, since the limit law in convergence in probability is almost surely unique. Hence, $\overline{X}_n$ is not consistent for μ or σ.
(ii) We find the median of X from its distribution function $F_X(x)$ given by

$$F_X(x) = \begin{cases} 0, & \text{if } x < \mu \\ 1 - \exp\{-(x - \mu)/\sigma\}, & \text{if } x \geq \mu. \end{cases}$$

Solution of the equation $F_X(x) = 1/2$ gives the median $a_{1/2}(\underline{\theta}) = \mu + \sigma \log_e 2$. Sample median $X_{([n/2]+1)}$ is consistent for the population median $a_{1/2}(\underline{\theta}) = \mu + \sigma \log_e 2$. Thus,

$$\overline{X}_n \overset{P_\theta}{\to} \mu + \sigma \ \ \& \ \ X_{([n/2]+1)} \overset{P_\theta}{\to} \mu + \sigma \log_e 2$$

$$\Rightarrow \ \overline{X}_n - \frac{X_{([n/2]+1)} - \overline{X}_n}{\log_e 2 - 1} \overset{P_\theta}{\to} \mu \ \ \& \ \ \frac{X_{([n/2]+1)} - \overline{X}_n}{\log_e 2 - 1} \overset{P_\theta}{\to} \sigma,$$

as convergence in probability is closed under arithmetic operations. Thus, $(\overline{X}_n + 3.2589(X_{([n/2]+1)} - \overline{X}_n), \ -3.2589(X_{([n/2]+1)} - \overline{X}_n))'$ is a consistent estimator of $(\mu, \sigma)'$.

2.8.26 Suppose $\{(X_1, Y_1)', (X_2, Y_2)', \ldots, (X_n, Y_n)'\}$ is a random sample from a bivariate Cauchy $C_2(\theta_1, \theta_2, \lambda)$ distribution, with probability density function,

$$f(x, y, \theta_1, \theta_2, \lambda) = (\lambda/2\pi)\{\lambda^2 + (x - \theta_1)^2 + (y - \theta_2)^2\}^{-3/2}$$
$$(x, y)' \in \mathbb{R}^2, \theta_1, \theta_2 \in \mathbb{R}, \lambda > 0.$$

Using the sample quartiles based on the samples from marginal distributions, obtain two distinct consistent estimators of $(\theta_1, \theta_2, \lambda)'$. Hence, obtain a family of consistent estimators of $(\theta_1, \theta_2, \lambda)'$.

Solution: Since $(X, Y)'$ has a bivariate Cauchy distribution, the marginal distribution of X is Cauchy $C(\theta_1, \lambda)$ and that of Y is Cauchy $C(\theta_2, \lambda)$. Hence, the quartiles of X and Y are given by

$$Q_1(X) = \theta_1 - \lambda, \ Q_2(X) = \theta_1, \ Q_3(X) = \theta_1 + \lambda$$
$$\& \ Q_1(Y) = \theta_2 - \lambda, \ Q_2(Y) = \theta_2, \ Q_3(Y) = \theta_2 + \lambda.$$

From Theorem 2.2.6, the sample quartiles are consistent for the corresponding population quartiles. Hence,

$$X_{([n/4]+1)} \overset{P_{(\theta_1,\lambda)}}{\to} \theta_1 - \lambda, \ \ X_{([n/2]+1)} \overset{P_{(\theta_1,\lambda)}}{\to} \theta_1 \ \ \& \ \ X_{([3n/4]+1)} \overset{P_{(\theta_1,\lambda)}}{\to} \theta_1 + \lambda$$

$$\& \ Y_{([n/4]+1)} \overset{P_{(\theta_2,\lambda)}}{\to} \theta_2 - \lambda, \ \ Y_{([n/2]+1)} \overset{P_{(\theta_2,\lambda)}}{\to} \theta_2 \ \ \& \ \ Y_{([3n/4]+1)} \overset{P_{(\theta_2,\lambda)}}{\to} \theta_2 + \lambda.$$

Thus, $T_{n1} = X_{([n/2]+1)}$ is consistent for θ_1. To get another consistent estimator for θ_1, we use the result that convergence in probability is closed under arithmetic operations. Hence, we have $T_{n2} = \left(X_{([n/4]+1)} + X_{([3n/4]+1)}\right)/2$ to be consistent for θ_1. On similar lines, $S_{n1} = Y_{([n/2]+1)}$ and $S_{n2} = \left(Y_{([n/4]+1)} + Y_{([3n/4]+1)}\right)/2$ are consistent for θ_2. Now to get consistent estimator of λ, observe that $U_{n1} = \left(X_{([n/2]+1)} - X_{([n/4]+1)}\right)$ and $U_{n2} = \left(Y_{([n/2]+1)} - Y_{([n/4]+1)}\right)$ both are consistent for λ. Thus, $\underline{V}_{n1} = (T_{n1}, S_{n1}, U_{n1})'$ and $\underline{V}_{n2} = (T_{n2}, S_{n2}, U_{n2})'$ are two distinct consistent estimators of $(\theta_1, \theta_2, \lambda)'$. Further, the convex combination of $\underline{V}_{n1}$ and $\underline{V}_{n2}$ given by $\alpha\underline{V}_{n1} + (1 - \alpha)\underline{V}_{n2}, \ 0 < \alpha < 1$ is a consistent estimator of $(\theta_1, \theta_2, \lambda)'$. Thus, a family of consistent estimators of $(\theta_1, \theta_2, \lambda)'$ is given by $\{\alpha\underline{V}_{n1} + (1 - \alpha)\underline{V}_{n2}, \ 0 < \alpha < 1\}$.

2.8.27 Suppose $\{X_1, X_2, \ldots, X_n\}$ is a random sample from an exponential distribution with probability density function $f(x, \theta, \alpha)$ given by

$$f(x, \theta, \alpha) = (1/\alpha) \exp\{-(x - \theta)/\alpha\}, \quad x \geq \theta, \ \theta \in \mathbb{R}, \ \alpha > 0 .$$

Show that $\left(X_{(1)}, \sum_{i=2}^{n}(X_{(i)} - X_{(1)})/(n - 1)\right)'$ is consistent for $(\theta, \alpha)'$. Obtain a consistent estimator of $(\theta, \alpha)'$ based on the sample moments.

Solution: To verify the consistency of $X_{(1)}$ as an estimator of θ, we find the distribution function of $X_{(1)}$, it is given by
$F_{X_{(1)}}(x) = 1 - [1 - F_X(x)]^n, \quad x \in \mathbb{R}$. The distribution function $F_X(x)$ is given by

$$F_X(x) = \begin{cases} 0, & \text{if } x < \theta \\ 1 - \exp\{-(x - \theta)/\alpha\}, & \text{if } x \geq \theta. \end{cases}$$

Hence, the distribution function of $X_{(1)}$ is given by

$$F_{X_{(1)}}(x) = \begin{cases} 0, & \text{if } x < \theta \\ 1 - \exp\{-n(x - \theta)/\alpha\}, & \text{if } x \geq \theta. \end{cases}$$

Thus, the distribution of $X_{(1)}$ is again exponential with location parameter θ and scale parameter n/α. Hence, $E(X_{(1)}) = \theta + \alpha/n$ which implies that the bias of $X_{(1)}$ as an estimator of θ is α/n and it converges to 0 as $n \to \infty$. Further, $Var(X_{(1)}) = \alpha^2/n^2 \to 0$ as $n \to \infty$. Hence, $X_{(1)}$ is consistent for θ. We have derived this result in Example 2.2.16 when $\alpha = 1$. Now to examine whether $T_n = \sum_{i=2}^{n}(X_{(i)} - X_{(1)})/(n - 1)$ is consistent for α, we define random variables Y_i as $Y_i = (n - i + 1)(X_{(i)} - X_{(i-1)}), \ i = 2, 3, \ldots, n$. Then

$$\sum_{i=2}^{n} Y_i = \sum_{i=2}^{n} X_{(i)} - (n - 1)X_{(1)} = \sum_{i=2}^{n}(X_{(i)} - X_{(1)}) = (n - 1)T_n .$$

It can be proved that $\{Y_2, Y_3, \ldots, Y_n\}$ are independent and identically distributed random variables each following an exponential distribution with location parameter 0 and scale parameter $1/\alpha$. Thus, $E(Y_2) = \alpha < \infty$. Hence by the WLLN, $\sum_{i=2}^{n} Y_i/(n - 1) \overset{P_{\theta,\alpha}}{\to} \alpha$ and hence T_n is consistent for α. Since joint consistency is equivalent to the marginal consistency, $\left(X_{(1)}, \sum_{i=2}^{n}(X_{(i)} - X_{(1)})/(n - 1)\right)'$ is consistent for $(\theta, \alpha)'$. To obtain a consistent estimator for $(\theta, \alpha)'$ based on the sample moments, observe that $E(X) = \theta + \alpha$ and $Var(X) = \alpha^2$. By the WLLN $\overline{X}_n \overset{P_\theta}{\to} \theta + \alpha$ and $m_2 \overset{P_\theta}{\to} \alpha^2$. Convergence in probability is closed under all arithmetic operations, hence $\sqrt{m_2} \overset{P_\theta}{\to} \alpha$ and $m_1' - \sqrt{m_2} \overset{P_\theta}{\to} \theta$. Thus, a consistent estimator of $(\theta, \alpha)'$ based on the sample moments is $(m_1' - \sqrt{m_2}, \sqrt{m_2})'$.

2.8.28 Suppose $\{X_{ij}, \ i = 1, 2, \ j = 1, 2, \ldots, n\}$ are independent random variables such that $X_{ij} \sim N(\mu_i, \sigma^2)$ distribution. Find the maximum likelihood estimators of $\mu_i \ i = 1, 2$ and σ^2. Examine whether these are consistent.

Solution: It is given that the random variables $X_{ij} \sim N(\mu_i, \sigma^2)$ distribution, hence the likelihood of $\mu_i, \ i = 1, 2$ and σ^2 given the data $\underline{X} = \{X_{ij}, \ i = 1, 2, n, \ j = 1, 2, \ldots, n\}$ is given by

$$L_n(\mu_1, \mu_2, \sigma^2 | \underline{X}) = \left(\sqrt{2\pi\sigma^2}\right)^{-2n} \exp\left\{-\frac{1}{2\sigma^2} \sum_{i=1}^{2} \sum_{j=1}^{n} (X_{ij} - \mu_i)^2\right\}$$

and the maximum likelihood estimators are given by

$$\hat{\mu}_{in} = \frac{1}{n} \sum_{j=1}^{n} X_{ij} = \overline{X}_{in}, \ i = 1, 2$$

$$\& \quad \hat{\sigma}_n^2 = \frac{1}{2n} \sum_{i=1}^{2} T_{in}, \ \text{where} \ T_{in} = \sum_{j=1}^{n} (X_{ij} - \overline{X}_{in})^2, \ i = 1, 2.$$

By the WLLN $\hat{\mu}_{in} \xrightarrow{P} \mu_i, \ i = 1, 2$ and hence it is consistent for $\mu_i, \ i = 1, 2$. To examine whether $\hat{\sigma}_n^2$ is consistent for σ^2, observe that T_{in}/n is the sample central moment of order 2 for $i = 1, 2$, hence it is consistent for the population central moment of order 2, which is σ^2. Thus,

$$\frac{1}{n} T_{in} \xrightarrow{P} \sigma^2 \ \Rightarrow \ \hat{\sigma}_n^2 = \frac{1}{2n} \sum_{i=1}^{2} T_{in} \xrightarrow{P} \sigma^2$$

and $\hat{\sigma}_n^2$ is consistent for σ^2.

2.8.29 Suppose $\{X_1, X_2, \ldots, X_n\}$ is a random sample from a normal $N(\theta, \sigma^2)$ distribution. Suppose $S_n^2 = \sum_{i=1}^{n} (X_i - \overline{X}_n)^2$. (i) Examine whether $T_{1n} = S_n^2/n$ and $T_{2n} = S_n^2/(n-1)$ are consistent for θ. (ii) Show that $MSE_\theta(T_{1n}) < MSE_\theta(T_{2n}), \ \forall \ n \geq 2$. (iii) Show that $T_{3n} = S_n^2/(n+k)$ is consistent for θ. Determine k such that $MSE_\theta(T_{3n})$ is minimum.

Solution: (i) Suppose $X \sim N(\theta, \sigma^2)$ distribution. Then $T_n = S_n^2/\sigma^2 \sim \chi_{n-1}^2$ distribution. Hence, $E(T_n) = n - 1$ and $Var(T_n) = 2(n - 1)$. As a consequence,

$$E(T_{1n}) = \frac{\sigma^2(n-1)}{n}, \ Var(T_{1n}) = \frac{2\sigma^4(n-1)}{n^2}$$

$$\& \ MSE_\theta(T_{1n}) = \frac{\sigma^4(2n-1)}{n^2} \to 0,$$

as $n \to \infty$. Similarly,

$$E(T_{2n}) = \sigma^2, \; Var(T_{2n}) = MSE_\theta(T_{2n}) = \frac{2\sigma^4}{n-1} \to 0, \; \text{as } n \to \infty.$$

Thus, both $T_{1n} = S_n^2/n$ and $T_{2n} = S_n^2/(n-1)$ are consistent for θ.
(ii) It is easy to verify that

$$MSE_\theta(T_{1n}) - MSE_\theta(T_{2n}) = \frac{\sigma^4(2n-1)}{n^2} - \frac{2\sigma^4}{n-1}$$

$$= \frac{\sigma^4(1-3n)}{n^2(n-1)} < 0, \; \forall \; n \geq 2.$$

(iii) As in (i) we have

$$E(T_{3n}) = \frac{\sigma^2(n-1)}{n+k}, \; Var(T_{3n}) = \frac{2\sigma^4(n-1)}{(n+k)^2}$$

$$\& \; MSE_\theta(T_{3n}) = \frac{\sigma^4(2n-2+(k+1)^2)}{(n+k)^2},$$

and it converges to 0 as $n \to \infty$. Thus, T_{3n} is consistent for θ. To determine k such that $MSE_\theta(T_{3n})$ is minimum, suppose $MSE_\theta(T_{3n}) = \sigma^4 g(k)$, where

$$g(k) = \frac{2n-2+(k+1)^2}{(n+k)^2}$$

$$\Rightarrow \; g'(k) = \frac{(n+k)^2 2(k+1) - (2n-2+(k+1)^2)2(n+k)}{(n+k)^4}$$

$$g'(k) = 0 \Rightarrow (n+k)(k+1) - (2n-2+(k+1)^2) = 0$$

$$\Rightarrow k(n-1) - n + 1 = 0 \Rightarrow k = 1.$$

Thus, for $k = 1$, $MSE_\theta(T_{3n})$ is minimum.

2.8.30 An electronic device is such that the probability of its instantaneous failure is θ, that is, if X denotes the life length random variable of the device, then $P[X = 0] = \theta$. Given that $X > 0$, the conditional distribution of life length is exponential with mean α. In a random sample of size n, it is observed that r items failed instantaneously and remaining $n - r$ items had life times $\{X_{i_1}, X_{i_2}, \ldots, X_{i_{n-r}}\}$. On the basis of these data, find a consistent estimator of θ and of α.

Solution: Suppose F denotes the distribution function of X, then $F(x, \theta, \alpha) = 0 \; \forall \; x < 0$ as $X \geq 0$ a. s. For $x = 0$, $F(x, \theta, \alpha) = P[X \leq 0] =$

$P[X = 0] = \theta$. For $x > 0$,

$$
\begin{aligned}
F(x, \theta, \alpha) = P[X \le x] &= P[X = 0] + P[0 < X \le x] \\
&= P[X = 0] + P[X \le x | X > 0] P[X > 0] \\
&= \theta + (1 - e^{-x/\alpha})(1 - \theta).
\end{aligned}
$$

Hence, the distribution function $F(x, \theta, \alpha)$ is given by

$$
F(x, \theta, \alpha) = \begin{cases} 0, & \text{if } x < 0 \\ \theta, & \text{if } x = 0 \\ \theta + (1 - e^{-x/\alpha})(1 - \theta), & \text{if } x > 0. \end{cases}
$$

It is neither continuous nor discrete. To find a consistent estimator of θ, we define a random variable Y_i, $i = 1, 2, \ldots, n$ as

$$
Y_i = \begin{cases} 1, & \text{if } X_i = 0 \\ 0, & \text{if } X_i > 0. \end{cases}
$$

Thus, Y_i is a Borel function of X_i, $i = 1, 2, \ldots, n$ and hence $\{Y_1, Y_2, \ldots, Y_n\}$ are independent and identically distributed random variables each having Bernoulli $B(1, \theta)$ distribution. Hence by the WLLN, $\overline{Y}_n = r/n$ is a consistent estimator of $E(Y_1) = \theta$. To find a consistent estimator for α, we define a random variable U_i, $i = 1, 2, \ldots, n$ as

$$
U_i = \begin{cases} 0, & \text{if } X_i = 0 \\ X_i, & \text{if } X_i > 0. \end{cases}
$$

$\{U_1, U_2, \ldots, U_n\}$ are also independent and identically distributed random variables. Observe that

$$
\begin{aligned}
E(U_1) = E(X_1 I_{[X_1 > 0]}) &= E\left(E(X_1 I_{[X_1 > 0]}) | I_{[X_1 > 0]}\right) \\
&= E\left(I_{[X_1 > 0]} E(X_1 | X_1 > 0)\right) = \alpha(1 - \theta).
\end{aligned}
$$

Hence by the WLLN,

$$
\overline{U}_n = \frac{X_{i_1} + X_{i_2} + \cdots + X_{i_{n-r}}}{n} = \frac{S_n}{n} \xrightarrow{P} \alpha(1 - \theta),
$$

where $S_n = X_{i_1} + X_{i_2} + \cdots + X_{i_{n-r}}$.

Hence,

$$
T_n = \frac{S_n}{n} \frac{n}{n-r} = \frac{S_n}{n-r} = \frac{\left(X_{i_1} + X_{i_2} + \cdots + X_{i_{n-r}}\right)}{n-r} \xrightarrow{P} \alpha.
$$

2.8.31 A linear regression model is given by $Y = a + bX + \epsilon$, where $E(\epsilon) = 0$ and $Var(\epsilon) = \sigma^2$. Suppose $\{(X_i, Y_i)', i = 1, 2, \ldots, n\}$ is a random sample from the distribution of $(X, Y)'$. Examine whether the least square estimators of a and b are consistent for a and b respectively.

Solution: For a linear regression model $Y = a + bX + \epsilon$, it is known that $a = E(Y) - bE(X)$ and $b = Cov(X, Y)/Var(X)$. Further, corresponding to a random sample $\{(X_i, Y_i)', i = 1, 2, \ldots, n\}$, the least square estimators of a and b are given by

$$\hat{b}_n = \frac{S_{XY}}{S_{XX}} \quad \& \quad \hat{a}_n = \overline{Y}_n - \hat{b}_n \overline{X}_n \,,$$

where S_{XY} is the sample covariance and S_{XX} is the sample variance of X. In Theorem 2.5.4, it is proved that the sample covariance is consistent for the population covariance. From Theorem 2.5.3, the sample variance is consistent for the population variance. Further, the sample mean is consistent for a population mean. Convergence in probability is closed under arithmetic operations, hence it follows that $\hat{a}_n$ and $\hat{b}_n$ are consistent for a and b respectively.

2.8.32 Suppose $\{X_1, X_2, \ldots, X_n\}$ is a random sample of size n from a normal $N(\mu, \sigma^2)$ distribution. Find a consistent estimator of $P[X_1 < a]$ where a is any real number.

Solution: Suppose $X \sim N(\mu, \sigma^2)$, then $(X - \mu)/\sigma \sim N(0, 1)$. Thus,

$$P[X_1 < a] = \Phi\left((a - \mu)/\sigma\right) = g(\mu, \sigma^2), \quad \text{say} \,.$$

Since $\Phi(\cdot)$ is a distribution function of a continuous random variable, g is a continuous function from $\mathbb{R}^2$ to $\mathbb{R}$. In Example 2.5.3, it shown that if $\{X_1, X_2, \ldots, X_n\}$ is a random sample of size n from a normal $N(\mu, \sigma^2)$ distribution then $(m_1', m_2)'$ is a consistent estimator of $(\mu, \sigma^2)'$. Hence, by the invariance property of consistency under continuous transformation, $\Phi\left((a - m_1')/\sqrt{m_2}\right)$ is a consistent estimator of $P[X_1 < a] = \Phi\left((a - \mu)/\sigma\right)$.

2.8.33 Suppose $\{\underline{Z}_1, \underline{Z}_2, \ldots, \underline{Z}_n\}$ is a random sample of size n from a multivariate normal $N_p(\underline{\mu}, \Sigma)$ distribution. Find a consistent estimator of $\underline{\theta} = (\underline{\mu}, \Sigma)$. Also find a consistent estimator of $\underline{l}'\underline{\mu}$ where $\underline{l}$ is a vector in $\mathbb{R}^p$.

Solution: Suppose $\underline{Z} = (X_1, X_2, \ldots, X_p)' \sim N_p(\underline{\mu}, \Sigma)$, where $\underline{\mu} = (\mu_1, \mu_2, \ldots, \mu_p)'$ and $\Sigma = [\sigma_{ij}]_{p \times p}$. Then $X_i \sim N(\mu_i, \sigma_{ii}), i = 1, 2, \ldots, p$. A random sample $\{\underline{Z}_1, \underline{Z}_2, \ldots, \underline{Z}_n\}$ gives a random sample of size n on each X_i. Hence as shown in Example 2.5.3, the sample mean $\overline{X}_{in} = \sum_{r=1}^{n} X_{ir}/n$ is consistent for $\mu_i, i = 1, 2, \ldots, p$. Since joint consistency and marginal consistency are equivalent,

$\overline{Z}_n = (\overline{X}_{1n}, \overline{X}_{2n}, \ldots, \overline{X}_{pn})'$ is consistent for $\mu = (\mu_1, \mu_2, \ldots, \mu_p)'$. Now $\sigma_{ii} = Var(X_i)$, hence as shown in Example 2.5.3, sample variance $S_{iin} = \sum_{r=1}^{n}(X_{ir} - \overline{X}_{in})^2/n$ is consistent for σ_{ii}. Further, $\sigma_{ij} = Cov(X_i, X_j)$. In Theorem 2.5.4, it is proved that the sample covariance is consistent for population covariance. Hence, the sample covariance $S_{ijn} = \sum_{r=1}^{n}(X_{ir} - \overline{X}_{in})(X_{jr} - \overline{X}_{jn})/n$ is consistent for population covariance σ_{ij}. Thus, if $S_n = [S_{ijn}]$ denotes the sample dispersion matrix then S_n is consistent for Σ. Hence, the consistent estimator for $\underline{\theta} = (\underline{\mu}, \Sigma)$ is $(\overline{Z}_n, S_n)$. To find a consistent estimator for $\underline{l}'\underline{\mu}$, we use invariance property of consistency under continuous transformation. We define a function $g : \mathbb{R}^p \to \mathbb{R}$ as $g(\underline{x}) = \underline{l}'\underline{x}$. It is a continuous function and hence the consistent estimator for $\underline{l}'\underline{\mu} = \sum_{i=1}^{p} l_i \mu_i$ is $\underline{l}'\overline{Z}_n = \sum_{i=1}^{p} l_i \overline{X}_{in}$.

2.8.34 On the basis of a random sample of size n from a multinomial distribution in k cells with cell probabilities $(p_1, p_2, \ldots, p_k)$, with $\sum_{i=1}^{k} p_i = 1$, find a consistent estimator for $\underline{p} = (p_1, p_2, \ldots, p_k)$.

Solution: Suppose $\underline{Y} = (Y_1, Y_2, \ldots, Y_k)$ has a multinomial distribution in k cells with cell probabilities $(p_1, p_2, \ldots, p_k)$, with $\sum_{i=1}^{k} p_i = 1$ and $\sum_{i=1}^{k} Y_i = 1$. Then for each i, Y_i has Bernoulli $B(1, p_i)$ distribution with $E(Y_i) = p_i$, $i = 1, 2, \ldots, k$. Suppose $\{\underline{Y}_r, r = 1, 2, \ldots, n\}$ is a random sample from the distribution of $\underline{Y}$. If X_i denotes the frequency of i-th cell in the sample, then $X_i = \sum_{r=1}^{n} Y_{ir}$, $i = 1, 2, \ldots, k$ and $\sum_{i=1}^{k} X_i = n$. By the WLLN,

$$\hat{p}_{in} = \frac{X_i}{n} = \frac{1}{n} \sum_{r=1}^{n} Y_{ir} \xrightarrow{P_{\underline{p}}} p_i, i = 1, 2, \ldots, k .$$

Thus, $(X_1/n, X_2/n, \ldots, X_k/n)'$ is a consistent estimator of $\underline{p} = (p_1, p_2, \ldots, p_k)$. It is to be noted that $(X_1/n, X_2/n, \ldots, X_k/n)'$ is a vector of relative frequencies of k cells in a sample of size n which add up to 1.

2.8.35 Suppose $\{X_1, X_2, \ldots, X_n\}$ is a random sample of size n from a uniform $U(\theta_1, \theta_2)$ distribution, $-\infty < \theta_1 < x < \theta_2 < \infty$. Examine whether $(X_{(1)}, X_{(n)})'$ is a consistent estimator of $\underline{\theta} = (\theta_1, \theta_2)'$. Obtain a consistent estimator of $(\theta_1 + \theta_2)/2$ and of $(\theta_2 - \theta_1)^2/12$ based on $(X_{(1)}, X_{(n)})'$ and also based on the sample moments.

Solution: Suppose we define a random variable Y as $Y = (X - \theta_1)/(\theta_2 - \theta_1)$, then by probability integral transformation, $Y \sim U(0, 1)$. In Example 2.5.2, we have shown that corresponding to a random sample of size n from the uniform $U(0, 1)$ distribution,

$$Y_{(1)} \xrightarrow{P} 0 \quad \& \quad Y_{(n)} \xrightarrow{P} 1 \quad \Rightarrow \quad X_{(1)} \xrightarrow{P_\theta} \theta_1 \quad \& \quad X_{(n)} \xrightarrow{P_\theta} \theta_2 .$$

Thus, $(X_{(1)}, X_{(n)})'$ is consistent for $(\theta_1, \theta_2)'$. Further, convergence in probability is closed under all arithmetic operations, hence

$$\frac{(X_{(n)} + X_{(1)})}{2} \xrightarrow{P_\theta} \frac{\theta_1 + \theta_2}{2} \quad \& \quad \frac{(X_{(n)} - X_{(1)})^2}{12} \xrightarrow{P_\theta} \frac{(\theta_2 - \theta_1)^2}{12}.$$

To obtain a consistent estimator of $(\theta_1 + \theta_2)/2$ and of $(\theta_2 - \theta_1)^2/12$ based on the sample moments, it is to be noted that $(\theta_1 + \theta_2)/2$ is the mean and $(\theta_2 - \theta_1)^2/12$ is the variance of the $U(\theta_1, \theta_2)$ distribution. Hence by the WLLN, the sample mean $\overline{X}_n$ is consistent for $(\theta_1 + \theta_2)/2$ and by Theorem 2.5.3, the sample variance m_2 is consistent for $(\theta_2 - \theta_1)^2/12$.

2.8.36 Suppose $\underline{X} \equiv \{X_1, X_2, \ldots, X_n\}$ is a random sample of size n from a Laplace distribution with probability density function given by

$$f(x, \theta, \lambda) = (1/2\lambda) \exp\{-|x - \theta|/\lambda\}, \quad x \in \mathbb{R}, \quad \theta \in \mathbb{R}, \quad \lambda > 0.$$

Using stepwise maximization procedure, find the maximum likelihood estimator of θ and of λ and examine if those are consistent for θ and λ respectively.

Solution: Corresponding to a random sample $\underline{X}$ from a Laplace distribution, the log-likelihood of (θ, λ) is given by

$$\log L_n(\theta, \lambda | \underline{X}) = -n \log 2\lambda - \frac{1}{\lambda} \sum_{i=1}^{n} |X_i - \theta|.$$

Suppose λ is fixed at λ_0, the log likelihood is maximized with respect to variations in θ when θ is the sample median $X_{([n/2]+1)}$, it does not depend on the fixed value of λ. Now we consider a function $h(\lambda) = -n \log 2\lambda - \sum_{i=1}^{n} |X_i - X_{([n/2]+1)}|/\lambda$. It is a differentiable function of λ and by the calculus method we get that $h(\lambda)$ is maximum when $\lambda = \sum_{i=1}^{n} |X_i - X_{([n/2]+1)}|/n$. Hence, the maximum likelihood estimators of θ and λ are given by $\hat{\theta}_n = X_{([n/2]+1)}$ and $\hat{\lambda}_n = \sum_{i=1}^{n} |X_i - X_{([n/2]+1)}|/n$. To examine whether these are consistent, we proceed as follows. For a Laplace distribution with location parameter θ and scale parameter λ, the population median is θ and hence by Theorem 2.2.6, the sample median $X_{([n/2]+1)} = \hat{\theta}_n$ is consistent for θ. Now to examine whether $\hat{\lambda}_n$ is consistent, we define $Y = (X - \theta)/\lambda$, then Y follows a Laplace distribution with location parameter 0 and scale parameter 1. Hence, $E(|Y|) = 1$. Since $\{X_1, X_2, \ldots, X_n\}$ are independent and identically distributed random variables, being Borel functions, $\{|Y_1|, |Y_2|, \ldots, |Y_n|\}$ are also independent and identically distributed random variables with mean 1. Hence by the WLLN,

$$\frac{1}{n} \sum_{i=1}^{n} |Y_i| \xrightarrow{P} 1 \quad \Rightarrow \quad T_n = \frac{1}{n} \sum_{i-1}^{n} |X_i - \theta| \xrightarrow{P} \lambda.$$

Now we use following two inequalities related to the absolute values, which are given by, $|a + b| \leq |a| + |b|$ and $||a| - |b|| \leq |a - b|$ to establish consistency of $\hat{\lambda}_n$. We have,

$$
\begin{aligned}
|T_n - \hat{\lambda}_n| &= \left| \frac{1}{n} \sum_{i-1}^{n} |X_i - \theta| - \frac{1}{n} \sum_{i-1}^{n} |X_i - \hat{\theta}_n| \right| \\
&\leq \frac{1}{n} \sum_{i-1}^{n} \left| |X_i - \theta| - |X_i - \hat{\theta}_n| \right| \\
&\leq \frac{1}{n} \sum_{i-1}^{n} \left| X_i - \theta - X_i + \hat{\theta}_n \right| = |\hat{\theta}_n - \theta|.
\end{aligned}
$$

Thus, we get $|T_n - \hat{\lambda}_n| \leq |\hat{\theta}_n - \theta|$, which implies that if $|\hat{\theta}_n - \theta| < \epsilon$, then $|T_n - \hat{\lambda}_n| < \epsilon$. Hence for every $\epsilon > 0$,

$$
P[|T_n - \hat{\lambda}_n| < \epsilon] \geq P[|\hat{\theta}_n - \theta| < \epsilon] \rightarrow 1 \text{ as } n \rightarrow \infty.
$$

As a consequence, $T_n - \hat{\lambda}_n \xrightarrow{P} 0$. We have proved that $T_n \xrightarrow{P} \lambda$ and hence $\hat{\lambda}_n \xrightarrow{P} \lambda$.

7.2 Chapter 3

3.5.1 Suppose $\{X_1, X_2, \ldots, X_n\}$ is a random sample from a uniform $U(0, \theta)$ distribution, $\theta > 0$. (i) Examine whether the maximum likelihood estimator of θ is a CAN estimator of θ. (ii) Examine whether the moment estimator of θ is a CAN estimator of θ. (iii) Solve (i) and (ii) if $\{X_1, X_2, \ldots, X_n\}$ are independent random variables where X_i follows a uniform $U(0, i\theta)$ distribution, $\theta > 0$.

Solution: (i) Corresponding to a random sample of size n from a uniform $U(0, \theta)$ distribution, it is shown in Example 2.2.1 that the maximum likelihood estimator $\hat{\theta}_n$ of θ is given by $X_{(n)}$ and is consistent for θ. To derive the asymptotic distribution of $X_{(n)}$ with suitable norming, we define $Y_n = n(\theta - X_{(n)})$ and derive its distribution function $G_{Y_n}(y)$ for $y \in \mathbb{R}$ from the distribution function of $X_{(n)}$. It is given by

$$
F_{X_{(n)}}(x) = [F_X(x)]^n =
\begin{cases}
0, & \text{if } x < 0 \\
(\frac{x}{\theta})^n, & \text{if } 0 \leq x < \theta \\
1, & \text{if } x \geq \theta.
\end{cases}
$$

Since $X_{(n)} \leq \theta$, $Y_n \geq 0$, hence for $y < 0$, $G_{Y_n}(y) = 0$. Suppose $y \geq 0$, then

$$G_{Y_n}(y) = P_\theta[n(\theta - X_{(n)}) \le y] = P_\theta[X_{(n)} \ge \theta - y/n]$$
$$= 1 - F_{X_{(n)}}(\theta - y/n), \text{ hence}$$

$$G_{Y_n}(y) = \begin{cases} 1 - 0 = 1, & \text{if } \theta - y/n < 0 \Leftrightarrow y \ge n\theta \\ 1 - (\frac{\theta - y/n}{\theta})^n = 1 - (1 - \frac{y}{n\theta})^n, & \text{if } \quad 0 < y \le n\theta \\ 1 - 1 = 0, & \text{if } \quad y \le 0. \end{cases}$$

As $n \to \infty$,

$$G_{Y_n}(y) \to \begin{cases} 0, & \text{if } y \le 0 \\ 1 - e^{-\frac{y}{\theta}}, & \text{if } y \ge 0. \end{cases}$$

Thus, $Y_n = n(\theta - X_{(n)})$ converges in distribution to an exponential distribution with location parameter 0 and scale parameter $1/\theta$. Thus, with norming factor n, the asymptotic distribution of $X_{(n)}$ is not normal. Proceeding on similar lines as in Example 3.2.1, there exists no sequence $\{a_n, n \ge 1\}$ of real numbers tending to ∞ as $n \to \infty$, such that the asymptotic distribution of $a_n(X_{(n)} - \theta)$ is normal. Hence, $X_{(n)}$ is not CAN for θ.

(ii) If a random variable X follows a uniform $U(0, \theta)$ distribution, then $E(X) = \theta/2 < \infty$. Hence by the WLLN, $\overline{X}_n \overset{P_\theta}{\to} E(X) = \theta/2$, $\forall \theta$. Hence, $\tilde{\theta}_n = 2\overline{X}_n$ is consistent for θ. Further, $Var(X) = \theta^2/12$, which is positive and finite and hence by the CLT,

$$\sqrt{n}\left(\overline{X}_n - \theta/2\right) \overset{L}{\to} Z_1 \sim N\left(0, \theta^2/12\right)$$
$$\Leftrightarrow \quad \sqrt{n}(2\overline{X}_n - \theta) \overset{L}{\to} Z_2 \sim N\left(0, \theta^2/3\right),$$

as $n \to \infty$. Hence, $\tilde{\theta}_n = 2\overline{X}_n$ is CAN for θ with approximate variance $\theta^2/3n$.

(iii) Suppose $X_i \sim U(0, i\theta)$, then it is easy to verify that $Y_i = X_i/i \sim U(0, \theta)$. Thus, $\{Y_1, Y_2, \dots, Y_n\}$ are independent and identically distributed random variables each having a uniform $U(0, \theta)$. Hence, proceeding on similar lines as in (i) and (ii), the maximum likelihood estimator of θ is $Y_{(n)}$ and it is consistent but not CAN for θ. The moment estimator of θ is $2\overline{Y}_n$ and it is CAN for θ with approximate variance $\theta^2/3n$.

3.5.2 Suppose $\underline{X} \equiv \{X_1, X_2, \dots, X_n\}$ is a random sample from a uniform $U(0, \theta)$ distribution. Obtain $100(1 - \alpha)\%$ asymptotic confidence interval for θ based on a sufficient statistic.

Solution: In Example 2.2.1 we have shown that corresponding to a random sample $\underline{X} = \{X_1, X_2, \dots, X_n\}$ from a uniform $U(0, \theta)$ distribution, $X_{(n)}$

is a sufficient statistic for the family of $U(0, \theta)$ distributions for $\theta > 0$. In the solution of Exercise 3.5.1 it is shown that $Y_n = n(\theta - X_{(n)})$ converges in distribution to an exponential distribution with location parameter 0 and scale parameter $1/\theta$. Hence, $Q_n = (\theta - X_{(n)})/\theta = n(1 - X_{(n)}/\theta)$ converges in distribution to the exponential distribution with location parameter 0 and scale parameter 1. Thus, Q_n is a pivotal quantity. Given a confidence coefficient $(1 - \alpha)$, we can find a and b so that $P[a < Q_n < b] = 1 - \alpha$. Inverting the inequality $a < Q_n < b$, we get

$$\frac{X_{(n)}}{1 - a/n} < \theta < \frac{X_{(n)}}{1 - b/n}.$$

Hence, $100(1 - \alpha)\%$ large sample confidence interval for θ is given by

$$\left(\frac{X_{(n)}}{1 - a/n}, \ \frac{X_{(n)}}{1 - b/n} \right).$$

3.5.3 Suppose $\underline{X} \equiv \{X_1, X_2, \ldots, X_n\}$ is a random sample from a uniform $U(\theta, 1)$ distribution, $0 < \theta < 1$. Find the maximum likelihood estimator of θ and the moment estimator of θ. Examine whether these are CAN estimators for θ.

Solution: If $X \sim U(\theta, 1)$, then its probability density function $f(x, \theta)$ and the distribution function $F(x, \theta)$ are as follows:

$$f(x, \theta) = \begin{cases} \frac{1}{1-\theta}, & \text{if } \theta < x < 1 \\ 0, & \text{otherwise.} \end{cases}$$

$$F(x, \theta) = \begin{cases} 0, & \text{if } x < \theta \\ \frac{x-\theta}{1-\theta}, & \text{if } \theta \leq x < 1 \\ 1, & \text{if } x \geq 1. \end{cases}$$

Corresponding to a random sample $\underline{X}$ from this distribution, the likelihood of θ is given by

$$L_n(\theta|\underline{X}) = \prod_{i=1}^{n} \frac{1}{1-\theta} = \frac{1}{(1-\theta)^n}, \quad X_i \geq \theta, \ \forall \ i \ \Leftrightarrow \ X_{(1)} \geq \theta.$$

Thus, the likelihood is an increasing function of θ and attains maximum at the maximum possible value of θ given data $\underline{X}$. The maximum possible value of θ given data is $X_{(1)}$ and hence the maximum likelihood estimator $\hat{\theta}_n$ of θ is given by $X_{(1)}$.

To verify the consistency of $X_{(1)}$ as an estimator of θ, we find the coverage probability using the distribution function of $X_{(1)}$. The distribution function of $X_{(1)}$ is given by

$$F_{X_{(1)}}(x) = 1 - [1 - F_X(x)]^n \;\; = \;\; \begin{cases} 0, & \text{if } \;\; x < \theta \\ 1 - (\frac{1-x}{1-\theta})^n, & \text{if } \theta \le x < 1 \\ 1, & \text{if } \;\; x \ge 1. \end{cases}$$

For $\epsilon > 0$, the coverage probability is given by

$$
\begin{aligned}
P_\theta[|X_{(1)} - \theta| < \epsilon] &= P_\theta[\theta - \epsilon < X_{(1)} < \theta + \epsilon] \\
&= P_\theta[\theta < X_{(1)} < \theta + \epsilon] \;\; \text{as} \;\; X_{(1)} \ge \theta \\
&= F_{X_{(1)}}(\theta + \epsilon) - F_{X_{(1)}}(\theta) \\
&= 1 - \left(\frac{1 - \theta - \epsilon}{1 - \theta}\right)^n - 0 \\
&\to 1 \;\; \forall \; \epsilon > 0 \;\; \text{and} \;\; \forall \; \theta \;\; \text{as} \;\; n \to \infty.
\end{aligned}
$$

Hence, $X_{(1)}$ is consistent for θ. To derive the asymptotic distribution of $X_{(1)}$ with suitable norming, we define $Y_n = n(X_{(1)} - \theta)$ and derive its distribution function $G_{Y_n}(y)$ for $y \in \mathbb{R}$. Since $X_{(1)} \ge \theta$, $Y_n \ge 0$, hence for $y < 0$, $G_{Y_n}(y) = 0$. Suppose $y \ge 0$, then

$$G_{Y_n}(y) = P_\theta[n(X_{(1)} - \theta) \le y] = P_\theta[X_{(1)} \le \theta + y/n] = F_{X_{(1)}}(\theta + y/n).$$

Thus,

$$
G_{Y_n}(y) \;\; = \;\; \begin{cases} 1 - (1 - 0)^n = 0, & \text{if} & \theta + y/n < \theta \;\; \Leftrightarrow \;\; y < 0 \\ 1 - \left(\frac{1 - \theta - y/n}{1 - \theta}\right)^n, & \text{if } \theta \le \theta + y/n < 1 & \Leftrightarrow \;\; 0 \le y < n(1 - \theta) \\ 1 - [1 - 1] = 1, & \text{if} & y \ge n(1 - \theta). \end{cases}
$$

Hence,

$$
G_{Y_n}(y) \;\; \to \;\; \begin{cases} 0, & \text{if } y < 0 \\ 1 - e^{-\frac{y}{1 - \theta}}, & \text{if } y \ge 0. \end{cases}
$$

Thus, the asymptotic distribution of $Y_n = n(X_{(1)} - \theta)$ is exponential with location parameter 0 and scale parameter $1/(1 - \theta)$ and with norming factor n, the asymptotic distribution of $X_{(1)}$ is not normal. Proceeding on similar lines as in Example 3.2.1, it follows that there exists no sequence $\{a_n, n \ge 1\}$ of real numbers tending to ∞ as $n \to \infty$ such that the asymptotic distribution of $a_n(X_{(1)} - \theta)$ is normal, hence we claim that $X_{(1)}$ is not CAN for θ. Another approach to claim that $X_{(1)}$ is not CAN is as follows. Suppose $Z_n = a_n(X_{(1)} - \theta)$. Then $F_n(x) = P[Z_n \le x] = 0$ if $x < 0$. Hence, $\lim_{n \to \infty} F_n(x) = 0$ if $x < 0$. However, $\Phi(x) > 0$ if $x < 0$, where $\Phi(\cdot)$ denotes the distribution function of the standard normal distribution. Hence, there exists no sequence $\{a_n, n \ge 1\}$ of real numbers tending to ∞ as $n \to \infty$ such that the asymptotic distribution of $a_n(X_{(1)} - \theta)$ is normal.

(ii) If a random variable $X \sim U(\theta, 1)$ distribution, then
$E(X) = (1 + \theta)/2 < \infty$. Hence, by the WLLN,

$\overline{X}_n \xrightarrow{P_\theta} E(X) = (1 + \theta)/2$, $\forall$ θ. Hence, $\tilde{\theta}_n = 2\overline{X}_n - 1$ is consistent for θ.
Further, $Var(X) = (1 - \theta)^2/12$, which is positive and finite and hence by
the CLT,

$$\sqrt{n} \left(\overline{X}_n - \frac{1 + \theta}{2} \right) \xrightarrow{L} Z_1 \sim N \left(0, \frac{(1 - \theta)^2}{12} \right)$$

$$\Leftrightarrow \quad \sqrt{n}(2\overline{X}_n - 1 - \theta) \xrightarrow{L} Z_2 \sim N \left(0, \frac{(1 - \theta)^2}{3} \right),$$

as $n \to \infty$. Hence, $\tilde{\theta}_n = 2\overline{X}_n - 1$ is CAN for θ with approximate variance
$(1 - \theta)^2/3n$.

3.5.4 Suppose $\{X_1, X_2, \ldots, X_n\}$ is a random sample from a Bernoulli $B(1, \theta)$
distribution, $\theta \in (0, 1)$. (i) Suppose an estimator $\hat{\theta}_n$ is defined as follows:

$$\hat{\theta}_n = \begin{cases} 0.01, \text{ if } & \overline{X}_n = 0 \\ \overline{X}_n, \text{ if } 0 < \overline{X}_n < 1 \\ 0.98, \text{ if } & \overline{X}_n = 1. \end{cases}$$

Examine whether it is a CAN estimator of θ. (ii) Examine whether the maxi-
mum likelihood estimator of θ is a CAN estimator of θ if $\theta \in [a, b] \subset (0, 1)$.

Solution: (i) Suppose the distribution of a random variable X is a Bernoulli
$B(1, \theta)$, then its probability mass function is given by,
$P[X = x] = \theta^x (1 - \theta)^{1-x}$, $x = 0, 1$, $\theta \in (0, 1)$. Given a random sample
$\{X_1, X_2, \ldots, X_n\}$ from Bernoulli $B(1, \theta)$ distribution, by the WLLN $\overline{X}_n$
is consistent for θ and by the CLT $\sqrt{n}(\overline{X}_n - \theta) \xrightarrow{L} Z_1 \sim N(0, \theta(1 - \theta))$.
Further for any $\epsilon > 0$,

$$P[|\overline{X}_n - \hat{\theta}_n| < \epsilon] \geq P[\hat{\theta}_n = \overline{X}_n] = P[0 < \overline{X}_n < 1]$$
$$= 1 - \theta^n - (1 - \theta)^n \quad \to \quad 1, \text{ if } \theta \in (0, 1).$$

Hence, $\overline{X}_n$ and $\hat{\theta}_n$ have the same limit in convergence in probability. Thus,
consistency of $\overline{X}_n$ implies consistency of $\hat{\theta}_n$. To find its asymptotic distribution, note that

$$\sqrt{n}(\overline{X}_n - \theta) - \sqrt{n}(\hat{\theta}_n - \theta) = \sqrt{n}(\overline{X}_n - \hat{\theta}_n) \text{ and}$$

$$P[\sqrt{n}|\overline{X}_n - \hat{\theta}_n| < \epsilon] \geq P[\hat{\theta}_n = \overline{X}_n] = P[0 < \overline{X}_n < 1]$$
$$= 1 - \theta^n - (1 - \theta)^n \quad \to \quad 1, \text{ if } \theta \in (0, 1).$$

Thus, for $\theta \in (0, 1)$, the limit law of $\sqrt{n}(\overline{X}_n - \theta)$ and of $\sqrt{n}(\hat{\theta}_n - \theta)$ is the same. But by the CLT $\sqrt{n}(\overline{X}_n - \theta) \xrightarrow{L} Z_1 \sim N(0, \theta(1 - \theta))$ and hence $\sqrt{n}(\hat{\theta}_n - \theta) \xrightarrow{L} Z_1 \sim N(0, \theta(1 - \theta)), \forall\ \theta \in (0, 1)$.

(ii) Here the parameter space is $[a, b]$. It is shown in the solution of Exercise 2.8.12 that the maximum likelihood estimator $\hat{\theta}_n$ of θ is given by

$$\hat{\theta}_n = \begin{cases} a, & \text{if } \overline{X}_n < a \\ \overline{X}_n, & \text{if } a \le \overline{X}_n \le b \\ b, & \text{if } \overline{X}_n > b. \end{cases}$$

It is shown to be consistent for θ. To find its asymptotic distribution, note that $\sqrt{n}(\overline{X}_n - \theta) - \sqrt{n}(\hat{\theta}_n - \theta) = \sqrt{n}(\overline{X}_n - \hat{\theta}_n)$ and

$$P[\sqrt{n}|\overline{X}_n - \hat{\theta}_n| < \epsilon] \ge P[\hat{\theta}_n = \overline{X}_n] = P[a \le \overline{X}_n \le b] \;\rightarrow\; 1, \text{ if } \theta \in (a, b),$$

as shown in the solution of Exercise 2.8.12. Thus, for $\theta \in (a, b)$, the limit law of $\sqrt{n}(\overline{X}_n - \theta)$ and of $\sqrt{n}(\hat{\theta}_n - \theta)$ is the same. But by the CLT $\sqrt{n}(\overline{X}_n - \theta) \xrightarrow{L} Z_1 \sim N(0, \theta(1 - \theta))$ and hence $\sqrt{n}(\hat{\theta}_n - \theta) \xrightarrow{L} Z_1 \sim N(0, \theta(1 - \theta)), \quad \forall\ \theta \in (a, b)$. We now investigate the asymptotic distribution of $\sqrt{n}(\hat{\theta}_n - \theta)$ at $\theta = a$ and at $\theta = b$. Suppose $\theta = a$, then $P_a[\sqrt{n}(\hat{\theta}_n - a) \le x] = 0$ if $x < 0$ as $\hat{\theta}_n \ge a$. If $x = 0$, then

$$P_a[\sqrt{n}(\hat{\theta}_n - a) \le 0] = P_a[\hat{\theta}_n = a] = P_a[\overline{X}_n < a]$$
$$= P_a\left[\frac{\sqrt{n}(\overline{X}_n - a)}{\sqrt{a(1 - a)}} < \frac{\sqrt{n}(a - a)}{\sqrt{a(1 - a)}} \right] \rightarrow \Phi(0),$$

which is $1/2$. Suppose $x > 0$ then,

$$P_a[\sqrt{n}(\hat{\theta}_n - a) \le x] = P_a[\sqrt{n}(\hat{\theta}_n - a) \le 0] + P_a[0 < \sqrt{n}(\hat{\theta}_n - a) \le x]$$
$$= P_a[\sqrt{n}(\hat{\theta}_n - a) \le 0]$$
$$+ P_a\left[0 < \frac{\sqrt{n}(\overline{X}_n - a)}{\sqrt{a(1 - a)}} \le \frac{x}{\sqrt{a(1 - a)}} \right]$$
$$\rightarrow \frac{1}{2} + \Phi\left(\frac{x}{\sqrt{a(1 - a)}} \right) - \Phi(0)$$
$$= \Phi\left(\frac{x}{\sqrt{a(1 - a)}} \right).$$

Thus,

$$P_a[\sqrt{n}(\hat{\theta}_n - a) \le x] \;\rightarrow\; \begin{cases} 0, & \text{if } x < 0 \\ \frac{1}{2}, & \text{if } x = 0 \\ \Phi\left(\frac{x}{\sqrt{a(1-a)}} \right), & \text{if } x > 0. \end{cases}$$

Suppose $\theta = b$, then $\hat{\theta}_n \leq b \Rightarrow P_b[\sqrt{n}(\hat{\theta}_n - b) \geq x] = 0$ if $x > 0$. If $x = 0$, then

$$P_b[\sqrt{n}(\hat{\theta}_n - b) \geq 0] = P_b[\hat{\theta}_n = b] = P_b[\overline{X}_n > b]$$
$$= P_b\left[\frac{\sqrt{n}(\overline{X}_n - b)}{\sqrt{b(1-b)}} > \frac{\sqrt{n}(b-b)}{\sqrt{b(1-b)}}\right] \to \frac{1}{2}.$$

Suppose $x < 0$ then,

$$P_b[\sqrt{n}(\hat{\theta}_n - b) \geq x] = P_b[x \leq \sqrt{n}(\hat{\theta}_n - b) < 0] + P_b[\sqrt{n}(\hat{\theta}_n - b) \geq 0]$$
$$= P_b\left[\frac{x}{\sqrt{b(1-b)}} \leq \frac{\sqrt{n}(\overline{X}_n - b)}{\sqrt{b(1-b)}} < 0\right]$$
$$+ P_b[\sqrt{n}(\hat{\theta}_n - b) \geq 0]$$
$$\to \Phi(0) - \Phi\left(\frac{x}{\sqrt{b(1-b)}}\right) + \frac{1}{2}$$
$$= 1 - \Phi\left(\frac{x}{\sqrt{b(1-b)}}\right).$$

Thus,

$$P_b[\sqrt{n}(\hat{\theta}_n - b) \geq x] \to \begin{cases} 0, & \text{if } x > 0 \\ \frac{1}{2}, & \text{if } x = 0 \\ 1 - \Phi\left(\frac{x}{\sqrt{b(1-b)}}\right), & \text{if } x < 0. \end{cases}$$

which is equivalent to

$$P_b[\sqrt{n}(\hat{\theta}_n - b) \leq x] \to \begin{cases} \Phi\left(\frac{x}{\sqrt{b(1-b)}}\right), & \text{if } x < 0 \\ \frac{1}{2}, & \text{if } x = 0 \\ 1, & \text{if } x > 0. \end{cases}$$

Thus, $\sqrt{n}(\hat{\theta}_n - \theta) \overset{L}{\to} Z_1 \sim N(0, \theta(1-\theta))$, $\forall\ \theta \in (a, b)$, but at $\theta = a$ and $\theta = b$, which are the boundary points of the parameter space, asymptotic distribution of $\sqrt{n}(\hat{\theta}_n - \theta)$ is not normal and hence we conclude that $\hat{\theta}_n$ is not CAN for $\theta \in [a, b]$. It is noted that for $\theta \in (a, b)$, $P_\theta[a \leq \overline{X}_n \leq b] \to 1$ as $n \to \infty$, thus for large n, $\hat{\theta}_n = \overline{X}_n$ and will have approximate normal distribution.

3.5.5 Suppose $\{X_1, X_2, \ldots, X_n\}$ is a random sample from a normal $N(\theta, 1)$ distribution. Find the maximum likelihood estimator of θ and examine if it is CAN for θ if $\theta \in [0, \infty)$. Identify the limiting distribution at $\theta = 0$.

Solution: In Example 2.2.3, we have obtained the maximum likelihood estimator $\hat{\theta}_n$ of θ as

$$\hat{\theta}_n = \begin{cases} \overline{X}_n, & \text{if } \overline{X}_n \geq 0 \\ 0, & \text{if } \overline{X}_n < 0, \end{cases}$$

and it is shown to be consistent. To examine whether it is CAN we proceed as follows. Since $X \sim N(\theta, 1)$, $\sqrt{n}(\overline{X}_n - \theta) \sim N(0, 1)$. Further, $\sqrt{n}(\overline{X}_n - \theta) - \sqrt{n}(\hat{\theta}_n - \theta) = \sqrt{n}(\overline{X}_n - \hat{\theta}_n)$. Observe that for $\theta > 0$ and for $\epsilon > 0$,

$$P_\theta[|\sqrt{n}(\overline{X}_n - \hat{\theta}_n)| < \epsilon] \geq P_\theta[\overline{X}_n = \hat{\theta}_n] = P_\theta[\overline{X}_n \geq 0]$$
$$= 1 - \Phi(-\sqrt{n}\theta) \quad \rightarrow \quad 1 \quad \text{if } \theta > 0.$$

Thus, for $\theta > 0$, $\sqrt{n}(\overline{X}_n - \theta) - \sqrt{n}(\hat{\theta}_n - \theta) \xrightarrow{P_\theta} 0$, hence $\sqrt{n}(\overline{X}_n - \theta)$ and $\sqrt{n}(\hat{\theta}_n - \theta)$ have the same limit law. But $\sqrt{n}(\overline{X}_n - \theta) \xrightarrow{L} Z \sim N(0, 1)$ and hence $\sqrt{n}(\hat{\theta}_n - \theta) \xrightarrow{L} Z \sim N(0, 1)$, for $\theta > 0$. Suppose now $\theta = 0$.

If $x < 0$, $\quad P_0[\sqrt{n}(\hat{\theta}_n - 0) \leq x] = 0 \quad$ as $\hat{\theta}_n \geq 0$.

For $x = 0$, $\quad P_0[\sqrt{n}\hat{\theta}_n \leq 0] = P_0[\sqrt{n}\hat{\theta}_n = 0] = P_0[\overline{X}_n < 0] = \Phi(0) = 1/2$.

For $x > 0$, $\quad P_0[\sqrt{n}\hat{\theta}_n \leq x] = P_0[\sqrt{n}\hat{\theta}_n \leq 0] + P_0[0 < \sqrt{n}\hat{\theta}_n \leq x]$
$$= 1/2 + P_0[0 < \sqrt{n}(\overline{X}_n) \leq x]$$
$$= 1/2 + \Phi(x) - \Phi(0) = \Phi(x).$$

Thus for $\theta = 0$,

$$P_0[\sqrt{n}\hat{\theta}_n \leq x] = \begin{cases} 0, & \text{if } x < 0 \\ 1/2, & \text{if } x = 0, \\ \Phi(x), & \text{if } x > 0, \end{cases}$$

which shows that at $\theta = 0$, the asymptotic distribution of $\sqrt{n}\hat{\theta}_n$ is not normal and 0 is a point of discontinuity. Suppose U_1 is a random variable with a distribution degenerate at 0. Then its distribution function is given by

$$F_{U_1}(x) = \begin{cases} 0, & \text{if } x < 0 \\ 1, & \text{if } x \geq 0 \end{cases}$$

Suppose a random variable U_2 is defined as $U_2 = |U|$ where $U \sim N(0, 1)$. Then $P[U_2 \leq x] = 0$ if $x < 0$. Suppose $x \geq 0$, then

$$P[U_2 \leq x] = P[|U| \leq x] = P[-x \leq U \leq x] = \Phi(x) - \Phi(-x) = 2\Phi(x) - 1.$$

Thus, the distribution function of U_2 is given by

$$F_{U_2}(x) = \begin{cases} 0, & \text{if } x < 0 \\ 2\Phi(x) - 1, & \text{if } x \geq 0 \end{cases}$$

It is easy to verify that $P_0[\sqrt{n}(\hat{\theta}_n - 0) \leq x] = (1/2)F_{U_1}(x) + (1/2)F_{U_2}(x)$.

3.5.6 Suppose $\underline{X} \equiv \{X_1, X_2, \ldots, X_n\}$ is a random sample from a distribution of a random variable X with probability density function $f(x, \theta) = k\theta^k / x^{k+1}$, $x \geq \theta$, $\theta > 0$ & $k \geq 3$ is a fixed positive integer.
(i) Find the maximum likelihood estimator of θ and examine whether it is CAN for θ.
(ii) Find the moment estimator of θ and examine whether it is CAN for θ.
(iii) Find 95% asymptotic confidence interval for θ based on the maximum likelihood estimator.

Solution: Corresponding to a random sample $\underline{X}$ from this distribution, the likelihood of θ is given by

$$L_n(\theta | \underline{X}) = \prod_{i=1}^{n} k\theta^k / X_i^{k+1} = k^n \theta^{kn} \left(\prod_{i=1}^{n} X_i^{k+1} \right)^{-1}, \quad X_i \geq \theta, \ \forall \ i \ \Leftrightarrow \ X_{(1)} \geq \theta.$$

Thus, the likelihood is an increasing function of θ and attains the maximum at the maximum possible value of θ given data $\underline{X}$. The maximum possible value of θ given data is $X_{(1)}$ and hence the maximum likelihood estimator $\hat{\theta}_n$ of θ is given by $X_{(1)}$. To verify the consistency of $X_{(1)}$ as an estimator of θ, we find the coverage probability using the distribution function of $X_{(1)}$. The distribution function $F_X(x)$ of X is given by

$$F_X(x) = \begin{cases} 0, & \text{if } x < \theta \\ 1 - \theta^k / x^k, & \text{if } x \geq \theta. \end{cases}$$

Hence, the distribution function of $X_{(1)}$ is given by

$$F_{X_{(1)}}(x) = 1 - [1 - F_X(x)]^n = \begin{cases} 0, & \text{if } x < \theta \\ 1 - \theta^{kn} / x^{kn}, & \text{if } x \geq \theta. \end{cases}$$

For $\epsilon > 0$, the coverage probability is given by

$$\begin{aligned} P_\theta[|X_{(1)} - \theta| < \epsilon] &= P_\theta[\theta - \epsilon < X_{(1)} < \theta + \epsilon] \\ &= P_\theta[\theta < X_{(1)} < \theta + \epsilon] \quad \text{as} \quad X_{(1)} \geq \theta \\ &= F_{X_{(1)}}(\theta + \epsilon) - F_{X_{(1)}}(\theta) \\ &= 1 - \frac{\theta^{kn}}{(\theta + \epsilon)^{kn}} - 0 \\ &\to 1 \ \forall \ \epsilon > 0 \ \text{and} \ \forall \ \theta \ \text{as} \ n \to \infty. \end{aligned}$$

Hence, $X_{(1)}$ is consistent for θ. To derive the asymptotic distribution of $X_{(1)}$ with suitable norming, we define $Y_n = n(X_{(1)} - \theta)$ and derive its distribution function $G_{Y_n}(y)$ for $y \in \mathbb{R}$. Since $X_{(1)} \geq \theta$, $Y_n \geq 0$, hence for $y < 0$, $G_{Y_n}(y) = 0$. Suppose $y \geq 0$, then

$$G_{Y_n}(y) = P_\theta[n(X_{(1)} - \theta) \leq y] = P_\theta[X_{(1)} \leq \theta + y/n] = F_{X_{(1)}}(\theta + y/n).$$

$$\text{Hence} \quad G_{Y_n}(y) = \begin{cases} 0, & \text{if } \theta + y/n < \theta \Leftrightarrow y < 0 \\ 1 - (\frac{\theta}{\theta + y/n})^{kn}, & \text{if } \theta + y/n \geq \theta \Leftrightarrow y \geq 0. \end{cases}$$

As $n \to \infty$,

$$G_{Y_n}(y) \to \begin{cases} 0, & \text{if } y \leq 0 \\ 1 - \exp(-ky/\theta), & \text{if } y \geq 0. \end{cases}$$

Thus, the asymptotic distribution of $Y_n = n(X_{(1)} - \theta)$ is exponential distribution with location parameter 0 and scale parameter k/θ. Thus, with norming factor n, the asymptotic distribution of $X_{(1)}$ is not normal. Proceeding on similar lines as in Example 3.2.1, it follows that there exists no sequence $\{a_n, n \geq 1\}$ of real numbers tending to ∞ as $n \to \infty$ such that the asymptotic distribution of $a_n(X_{(1)} - \theta)$ is normal. Hence we conclude that $X_{(1)}$ is not CAN for θ.

(ii) For a random variable X with probability density function
$f(x, \theta) = k\theta^k/x^{k+1}$, $x \geq \theta$, $E(X) = k\theta/(k-1) < \infty$ as $k \geq 3$. Hence, by the WLLN, $\overline{X}_n \overset{P_\theta}{\to} E(X) = k\theta/(k-1)$, $\forall$ θ. Hence, $(k-1)\overline{X}_n/k$ is consistent for θ. Further, $E(X^2) = k\theta^2/(k-2)$ and
$Var(X) = k\theta^2/(k-2)(k-1)^2$, which is positive and finite for $k \geq 3$ and hence by the CLT,

$$\sqrt{n}\left(\overline{X}_n - \frac{k\theta}{k-1}\right) \overset{L}{\to} Z_1 \sim N\left(0, \frac{k\theta^2}{(k-2)(k-1)^2}\right) \quad \text{as } n \to \infty.$$

Using delta method, $(k-1)\overline{X}_n/k$ is CAN for θ with approximate variance $\theta^2/nk(k-2)$.

(iii) The maximum likelihood estimator $\hat{\theta}_n$ of θ is $X_{(1)}$, it is consistent and $Y_n = n(X_{(1)} - \theta) \overset{L}{\to} Y \sim \exp(k/\theta)$. To find a pivotal quantity based on the maximum likelihood estimator, note that if $Y \sim \exp(k/\theta)$, then its moment generating function is $(1 - \theta t/k)^{-1}$. Thus, if a random variable U is defined as $U = kY/\theta$, then its moment generating function is $(1 - t)^{-1}$, implying that U has the exponential distribution with location parameter 0 and scale parameter 1. Suppose Q_n is defined as $Q_n = (k/X_{(1)})n(X_{(1)} - \theta)$ then by Slutsky's theorem,

$$Q_n = \frac{k}{X_{(1)}}n(X_{(1)} - \theta) = \frac{\theta}{X_{(1)}}\frac{k}{\theta}Y_n \overset{L}{\to} 1 \times U = U \sim \exp(1).$$

Thus for large n, Q_n is a pivotal quantity and we can find uncountably many pairs (a, b) such that $\int_a^b e^{-y}dy = 1 - \alpha$. Now

$$a < \frac{k}{X_{(1)}}n(X_{(1)} - \theta) < b \Leftrightarrow X_{(1)} - \frac{bX_{(1)}}{kn} < \theta < X_{(1)} - \frac{aX_{(1)}}{kn}.$$

Thus, $100(1 - \alpha)\%$ large sample confidence interval for θ is given by

$$\left(X_{(1)} - \frac{bX_{(1)}}{kn}, \ X_{(1)} - \frac{aX_{(1)}}{kn} \right).$$

3.5.7 Suppose $\{X_1, X_2, \ldots, X_n\}$ is a random sample from an exponential distribution with mean θ. (i) Find a CAN estimator for the mean residual life $E(X - t | X > t)$, $t > 0$. (ii) Show that for some constant $c(p)$, $\sqrt{n}(c(p)X_{([np]+1)} - \theta)$ converges in law to $N(0, \sigma^2(p))$. Find the constant $c(p)$ and $\sigma^2(p)$.

Solution: (i) If X has an exponential distribution with mean θ, then it is known that the distribution of residual life random variable $X - t | X > t$ is also exponential with mean θ. If $\{X_1, X_2, \ldots, X_n\}$ is a random sample from an exponential distribution with mean $\theta < \infty$, then by the WLLN $\overline{X}_n \xrightarrow{P_\theta} \theta$. Further, $Var(X) = \theta^2$ which is positive and finite and hence by the CLT, $\sqrt{n}(\overline{X}_n - \theta) \xrightarrow{L} Z_1 \sim N(0, \theta^2)$. Thus, $\overline{X}_n$ is CAN for θ with approximate variance θ^2/n and it is also CAN for $E(X - t | X > t)$ with approximate variance θ^2/n.

(ii) For an exponential distribution with mean θ, the p-th population quantile is given by $a_p(\theta) = -\theta \log(1 - p)$. Hence,

$\sqrt{n}(X_{([np]+1)} - (-\theta \log(1 - p))) \xrightarrow{L} Z_1 \sim N(0, v(\theta, p))$ where $v(\theta, p) = \theta^2 p/(1 - p)$. Thus,

$\sqrt{n}(c(p)X_{([np]+1)} - \theta) \xrightarrow{L} Z_2 \sim N(0, \sigma^2(p))$ where $\sigma^2(p) = \theta^2 p/(1 - p)(\log(1 - p))^2$ and $c(p) = -1/\log(1 - p)$.

3.5.8 Suppose $\{X_1, X_2, \ldots, X_n\}$ are independent and identically distributed random variables each having a Poisson distribution with mean θ, $\theta > 0$. Find a CAN estimator of $P[X_1 = 1]$. Is it necessary to impose any condition on the parameter space? Under this condition, using the CAN estimator of $P[X_1 = 1]$, obtain a large sample confidence interval for $P[X_1 = 1]$.

Solution: In Example 2.2.5, it is shown that if $\{X_1, X_2, \ldots, X_n\}$ is a random sample of size n from Poisson $Poi(\theta)$ distribution with $\theta > 0$, then an estimator T_n of θ defined as

$$T_n = \begin{cases} \overline{X}_n, & \text{if} \quad \overline{X}_n > 0 \\ 0.05, & \text{if} \quad \overline{X}_n = 0 \end{cases}$$

is consistent for θ. We now examine if its asymptotic distribution with suitable norming is normal. By the CLT it immediately follows that

$\sqrt{n}(\overline{X}_n - \theta) \xrightarrow{L} Z_1 \sim N(0, \theta)$. Now, $\sqrt{n}(T_n - \theta) - \sqrt{n}(\overline{X}_n - \theta) = \sqrt{n}(T_n - \overline{X}_n)$ and if it converges to 0 in probability, then normality of $\sqrt{n}(\overline{X}_n - \theta)$ implies normality of $\sqrt{n}(T_n - \theta)$. Observe that, for $\epsilon > 0$,

$$P[\sqrt{n}|T_n - \overline{X}_n| < \epsilon] \geq P[T_n = \overline{X}_n] = P[\overline{X}_n > 0]$$
$$= 1 - \exp(-n\theta) \rightarrow 1, \ \forall \ \theta > 0.$$

Thus, $\sqrt{n}(T_n - \overline{X}_n) \xrightarrow{P_\theta} 0, \ \forall \ \theta > 0$ and hence,

$$\sqrt{n}(\overline{X}_n - \theta) \xrightarrow{L} Z_1 \sim N(0, \theta) \ \Rightarrow \ \sqrt{n}(T_n - \theta) \xrightarrow{L} Z_1 \sim N(0, \theta), \ \forall \ \theta > 0$$

which proves that T_n is CAN for θ with approximate variance θ/n. Now $P[X_1 = 1] = \theta e^{-\theta} = g(\theta)$, say. It is clear that g is a differentiable function and $g'(\theta) = (1 - \theta)e^{-\theta} \neq 0, \ \forall \ \theta \neq 1$. Hence $\forall \ \theta \neq 1$, by the delta method $g(T_n) = T_n e^{-T_n}$ is CAN for $g(\theta) = \theta e^{-\theta}$ with approximate variance $e^{-2\theta}\theta(1 - \theta)^2/n$, that is,

$$\sqrt{n}\left(T_n e^{-T_n} - P[X_1 = 1]\right) \xrightarrow{L} Z_1 \sim N(0, e^{-2\theta}\theta(1 - \theta)^2).$$

Thus by Slutsky's theorem,

$$Q_n = \sqrt{\frac{n}{e^{-2T_n}T_n(1 - T_n)^2}}\left(T_n e^{-T_n} - P[X_1 = 1]\right) \xrightarrow{L} Z \sim N(0, 1).$$

Hence, Q_n is a pivotal quantity and is useful to find a large sample confidence interval for $P[X_1 = 1]$. Thus, given a confidence coefficient $(1 - \alpha)$, we can find the quantile $a_{1-\alpha/2}$ of the standard normal distribution so that $P[-a_{1-\alpha/2} < Q_n < a_{1-\alpha/2}] = 1 - \alpha$. Inverting the inequality $-a_{1-\alpha/2} < Q_n < a_{1-\alpha/2}$, we get

$$T_n e^{-T_n} - a_{1-\alpha/2}\, e^{-T_n}(1 - T_n)\sqrt{T_n/n} < P[X_1 = 1]$$
$$< T_n e^{-T_n} + a_{1-\alpha/2}\, e^{-T_n}(1 - T_n)\sqrt{T_n/n}.$$

Hence, for all $\theta \neq 1$, $100(1 - \alpha)\%$ large sample confidence interval for $P[X_1 = 1]$ is given by

$$\left(T_n e^{-T_n} - a_{1-\alpha/2}\, e^{-T_n}(1 - T_n)\sqrt{T_n/n},\ T_n e^{-T_n} + a_{1-\alpha/2}\, e^{-T_n}(1 - T_n)\sqrt{T_n/n}\right).$$

3.5.9 Suppose $\{X_1, X_2, \ldots, X_n\}$ is a random sample from $f(x, \theta) = \theta x^{\theta-1}$, $0 < x < 1$, $\theta > 0$. Find a CAN estimator of $e^{-\theta}$ based on the sample mean and also based on a sufficient statistic. Compare the two estimators.

Solution: In Example 3.2.2, it is shown that if $\{X_1, X_2, \ldots, X_n\}$ is a random sample from a distribution with probability density function
$f(x, \theta) = \theta x^{\theta-1}$, $0 < x < 1$, $\theta > 0$, then a CAN estimator of θ based on the sample mean is given by $\hat{\theta}_n = \overline{X}_n/(1 - \overline{X}_n)$ with approximate variance $\theta(1 + \theta)^2/n(\theta + 2)$. A CAN estimator for θ based on a sufficient statistic is

given by, $\tilde{\theta}_n = n/S_n = n/(-\sum_{i=1}^n \log X_i)$ with approximate variance θ^2/n. We now use delta method to find CAN estimator of $e^{-\theta}$ based on the sample mean and also based on the sufficient statistic. Suppose $g(\theta) = e^{-\theta}$, then g is a differentiable function with $g'(\theta) = -e^{-\theta} \neq 0$ for all $\theta > 0$. Hence, by the delta method, $g(\hat{\theta}_n) = \exp\{-\overline{X}_n/(1 - \overline{X}_n)\}$ is CAN for $g(\theta) = e^{-\theta}$ with approximate variance $e^{-2\theta}\theta(1+\theta)^2/n(\theta+2)$. Similarly, CAN estimator of $e^{-\theta}$ based on the sufficient statistic is given by, $g(\tilde{\theta}_n) = \exp\{-n/S_n\}$ with approximate variance $e^{-2\theta}\theta^2/n$. As discussed in Example 3.2.2, it can be shown that $g(\tilde{\theta}_n)$ is a better CAN estimator than $g(\hat{\theta}_n)$ as an estimator of $e^{-\theta}$.

3.5.10 Suppose $\{X_1, X_2, \ldots, X_n\}$ is a random sample from a Bernoulli $B(1, \theta)$. Find a CAN estimator of $\theta(1 - \theta)$ when $\theta \in (0, 1) - \{1/2\}$. What is the limiting distribution of the estimator when $\theta = 1/2$ and when the norming factor is $\sqrt{n}$ and n?

Solution: Suppose $X \sim B(1, \theta)$. Then $E(X) = \theta$ and $Var(X) = \theta(1 - \theta)$ which is positive and finite for all $\theta \in (0, 1)$. Then, by the WLLN and the CLT, it follows that $\overline{X}_n$ is CAN for θ with approximate variance $\theta(1 - \theta)/n$. Suppose $g(\theta) = \theta(1 - \theta)$, then g is a differentiable function and $g'(\theta) = (1 - 2\theta) \neq 0$ for all $\theta \in (0, 1) - \{1/2\}$. Hence by the delta method, $g(\overline{X}_n) = \overline{X}_n(1 - \overline{X}_n)$ is CAN for $g(\theta) = \theta(1 - \theta)$ with approximate variance $\theta(1 - \theta)(1 - 2\theta)^2/n$. Suppose $\theta = 1/2$. Then $\sqrt{n}(\overline{X}_n - 1/2) \xrightarrow{L} Z_1 \sim N(0, 1/4)$ and hence it is bounded in probability. Hence,

$$\sqrt{n}\left(\overline{X}_n(1 - \overline{X}_n) - 1/4\right) = \sqrt{n}\left(\overline{X}_n - \overline{X}_n^2 - 1/4\right) = -\sqrt{n}\left(\overline{X}_n - 1/2\right)^2$$
$$= \frac{-1}{\sqrt{n}}\left(\sqrt{n}(\overline{X}_n - 1/2)\right)^2 \xrightarrow{P} 0.$$

Thus, at $\theta = 1/2$, $\sqrt{n}(\overline{X}_n(1 - \overline{X}_n)) - 1/4$ does not have limiting distribution as a normal distribution. Proceeding on similar lines as in Example 3.2.1, we claim that there exists no sequence $\{a_n, n \geq 1\}$ of real numbers tending to ∞ as $n \to \infty$ such that the asymptotic distribution of $a_n\left(\overline{X}_n(1 - \overline{X}_n) - 1/4\right)$ is normal. If instead of $\sqrt{n}$ norming, suppose we take the norming factor as n, then

$$n\left(\overline{X}_n(1 - \overline{X}_n) - 1/4\right) = n\left(\overline{X}_n - \overline{X}_n^2 - 1/4\right) = -n\left(\overline{X}_n - 1/2\right)^2$$
$$= -\left(\sqrt{n}(\overline{X}_n - 1/2)\right)^2 \xrightarrow{L} -U, \text{ where } U \sim \chi_1^2.$$

Thus, with norming factor n, limiting distribution is non-degenerate but not normal.

3.5.11 Suppose $\{X_1, X_2, \ldots, X_n\}$ is a random sample from a geometric distribution, with probability mass function

$$p(x, \theta) = \theta(1 - \theta)^x, \quad x = 0, 1, \ldots.$$

However, $X_1, X_2, \ldots, X_n$ are not directly observable, but one can note whether $X_i \geq 2$ or not. (i) Find a CAN estimator for θ, based on the observed data. (ii) Find a CAN estimator for θ, if $X_i \geq 2$ is replaced by $X_i > 2$.

Solution: (i) Suppose X has geometric distribution, with probability mass function $p(x, \theta) = \theta(1 - \theta)^x$, $x = 0, 1, \ldots$. Then

$$h(\theta) = P_\theta[X \geq 2] = \theta \sum_{x \geq 2}(1 - \theta)^x = \theta(1 - \theta)^2 \sum_{x \geq 2}(1 - \theta)^{x-2}$$

$$= \theta(1 - \theta)^2 \theta^{-1} = (1 - \theta)^2.$$

We now define a random variable Y_i, $i = 1, 2, \ldots, n$ as

$$Y_i = \begin{cases} 1, & \text{if } X_i \geq 2 \\ 0, & \text{if } X_i < 2. \end{cases}$$

Since $\{X_1, X_2, \ldots, X_n\}$ are independent and identically distributed random variables, being Borel functions, $\{Y_1, Y_2, \ldots, Y_n\}$ are also independent and identically distributed random variables, each having Bernoulli $B(1, h(\theta))$ distribution. By the WLLN and the CLT, it follows that $\overline{Y}_n$ is CAN for $h(\theta) = \phi$ say, with approximate variance $h(\theta)(1 - h(\theta))/n$. To find a CAN estimator of θ, we find a transformation $g : \mathbb{R} \to \mathbb{R}$ such that $g(\phi) = \theta$. Now $(1 - \theta)^2 = \phi \Rightarrow \theta = 1 - \phi^{1/2}$. Thus, we have $g(\phi) = 1 - \phi^{1/2}$. It is a differentiable function of ϕ and $g'(\phi) = -1/2\sqrt{\phi} \neq 0$. Hence by the delta method, $g(\overline{Y}_n) = 1 - \overline{Y}_n^{1/2}$ is CAN for $g(\phi) = \theta$ with approximate variance $(h(\theta)(1 - h(\theta))/n) \times (1/4\phi) = (1 - (1 - \theta)^2)/4n$.

(ii) In this setup we define a random variable Y_i, $i = 1, 2, \ldots, n$ as

$$Y_i = \begin{cases} 1, & \text{if } X_i > 2 \\ 0, & \text{if } X_i \leq 2. \end{cases}$$

Observe that

$$h(\theta) = P_\theta[X > 2] = \theta \sum_{x \geq 3}(1 - \theta)^x = \theta(1 - \theta)^3 \sum_{x \geq 3}(1 - \theta)^{x-3}$$

$$= \theta(1 - \theta)^3 \theta^{-1} = (1 - \theta)^3.$$

Thus, $\{Y_1, Y_2, \ldots, Y_n\}$ are independent and identically distributed random variables, each having Bernoulli $B(1, h(\theta))$ distribution. By the WLLN and the CLT, $\overline{Y}_n$ is CAN for $h(\theta) = \phi$ say, with approximate variance $h(\theta)(1 - h(\theta))/n$. Proceeding on similar lines as in (i) we get that $1 - \overline{Y}_n^{1/3}$ is CAN for θ with approximate variance $(3\theta - 3\theta^2 + \theta^3)/(9n(1 - \theta))$.

3.5.12 Suppose $\{X_1, X_2, \ldots, X_n\}$ is a random sample from a normal $N(\theta, 1)$ distribution, $\theta \in \mathbb{R}$. However, $X_1, X_2, \ldots, X_n$ are not directly observable, but one can note whether $X_i > 2$ or not. Find a CAN estimator for θ, based on the observed data.

Solution: Suppose $\sim N(\theta, 1)$ distribution. Then $h(\theta) = P_\theta[X > 2] = 1 - \Phi(2 - \theta)$. As in the previous example, we define a random variable Y_i, $i = 1, 2, \ldots, n$ as

$$Y_i \;=\; \begin{cases} 1, \text{ if } X_i > 2 \\ 0, \text{ if } X_i \leq 2. \end{cases}$$

Proceeding on similar lines as in the previous example, it can be shown that $2 - \Phi^{-1}(1 - \overline{Y}_n)$ is CAN for θ with approximate variance

$$(1/n)(1 - \Phi(2 - \theta))\Phi(2 - \theta)\Phi^{-4}(\Phi(2 - \theta))\phi^2(\Phi(2 - \theta)).$$

3.5.13 Suppose $\{X_1, X_2, \ldots, X_n\}$ is a random sample from a distribution of X, with probability density function $f(x, \alpha, \theta)$ as given by

$$f(x, \alpha, \theta) \;=\; \begin{cases} \frac{2x}{\alpha\theta}, & \text{if } 0 < x \leq \theta \\ \frac{2(\alpha - x)}{\alpha(\alpha - \theta)}, & \text{if } \theta < x \leq \alpha. \end{cases}$$

Find a CAN estimator of θ when α is known.

Solution: From the probability density function $f(x, \alpha, \theta)$ of X we have $E(X) = (\alpha + \theta)/3$ and $Var(X) = (\alpha^2 - \alpha\theta + \theta^2)/18 < \infty$. Hence by the WLLN, $\overline{X}_n \xrightarrow{P_\theta} E(X) = (\alpha + \theta)/3$ and by the CLT, $\sqrt{n}(\overline{X}_n - E(X)) \xrightarrow{L} Z_1 \sim N(0, Var(X))$. Hence, $\hat{\theta}_n = 3\overline{X}_n - \alpha$ is consistent for θ and $\sqrt{n}(\hat{\theta}_n - \theta) \xrightarrow{L} Z_2 \sim N\left(0, (\alpha^2 - \alpha\theta + \theta^2)/2\right)$. Thus, $\hat{\theta}_n$ is CAN for θ with approximate variance $(\alpha^2 - \alpha\theta + \theta^2)/2n$.

3.5.14 Suppose $\{X_1, X_2, \ldots, X_n\}$ is a random sample from a negative binomial distribution with probability mass function given by

$$P_\theta[X = x] = \binom{x + k - 1}{x} p^k (1 - p)^x, \quad x = 0, 1, \ldots, \quad 0 < p < 1, \quad k > 0.$$

Obtain a CAN estimator for p assuming k to be known.

Solution: If X follows a negative binomial distribution then

$$E(X) = \frac{k(1-p)}{p} \quad \& \quad Var(X) = \frac{k(1-p)}{p^2} < \infty.$$

Hence by the WLLN and by the CLT,

$$\overline{X}_n \overset{L}{\to} \frac{k(1-p)}{p} \quad \& \quad \sqrt{n}\left(\overline{X}_n - \frac{k(1-p)}{p}\right) \overset{L}{\to} Z_1 \sim N\left(0, \frac{k(1-p)}{p^2}\right).$$

Thus, $\overline{X}_n$ is CAN for $k(1-p)/p = \phi$, say, with approximate variance $k(1-p)/np^2$. To get a CAN estimator for p, we find a transformation g such that $g(\phi) = p$. Suppose $g(y) = k/(k+y)$, $y > 0$, then g is a differentiable function with $g'(y) = -k/(k+y)^2 \neq 0$, $\forall\ y > 0$. Hence by the delta method, $g(\overline{X}_n) = k/(k + \overline{X}_n)$ is CAN for $g(\phi) = p$ with approximate variance $(k(1-p)/np^2) \times (p^4/k^2) = p^2(1-p)/nk$.

3.5.15 Suppose $\{X_1, X_2, \ldots, X_n\}$ is a random sample from an exponential distribution with mean θ. Suppose $T_{1n} = \sum_{i=1}^{n} X_i/n$ and $T_{2n} = \sum_{i=1}^{n} X_i/(n+1)$. (i) Examine whether T_{1n} and T_{2n} are consistent for θ. (ii) Prove that $\sqrt{n}(T_{2n} - T_{1n}) \overset{P_\theta}{\to} 0$ and hence both T_{1n} and T_{2n} are CAN for θ with the same approximate variance, but $MSE_\theta(T_{2n}) < MSE_\theta(T_{1n})$ $\forall\ n \geq 1$. (iii) Find a CAN estimator for $P[X_1 > t]$, where t is a positive real number.

Solution: (i) Suppose X has an exponential distribution with mean θ. Then $Var(X) = \theta^2$ which is positive and finite. Hence by the WLLN and by the CLT $\forall\ \theta \in \Theta$,

$$\overline{X}_n \overset{P_\theta}{\to} \theta \quad \& \quad \sqrt{n}(\overline{X}_n - \theta) = \sqrt{n}(T_{1n} - \theta) \overset{L}{\to} Z_1 \sim N(0, \theta^2).$$

Thus, $T_{1n} = \overline{X}_n$ is CAN for θ with approximate variance θ^2/n. Now,

$$T_{2n} = \frac{1}{n+1} \sum_{i=1}^{n} X_i = \frac{n}{n+1} T_{1n} \overset{P_\theta}{\to} \theta.$$

(ii) To examine whether T_{2n} is CAN observe that

$$\sqrt{n}(T_{1n} - \theta) - \sqrt{n}(T_{2n} - \theta) = \sqrt{n}(T_{1n} - T_{2n}) = \sqrt{n}\left(T_{1n} - \frac{n}{n+1}T_{1n}\right)$$

$$= \frac{\sqrt{n}}{\sqrt{n+1}}\frac{1}{\sqrt{n+1}}T_{1n}$$

$$\overset{P_\theta}{\to} 0 \text{ as } T_{1n} \overset{P_\theta}{\to} \theta$$

& hence is bounded in probability.

Thus, $\sqrt{n}(T_{1n} - \theta)$ and $\sqrt{n}(T_{2n} - \theta)$ have the same asymptotic distribution as $N(0, \theta^2)$. Hence, both T_{1n} and T_{2n} are CAN for θ with the same approximate variance θ^2/n. It is to be noted that T_{1n} is unbiased for θ and hence $MSE_\theta(T_{1n}) = Var(T_{1n}) = \theta^2/n$. To find $MSE_\theta(T_{2n})$, observe that

$$T_{2n} = \frac{n}{n+1} T_{1n} \implies E(T_{2n}) = \frac{n}{n+1}\theta$$

$$\&\ Var(T_{2n}) = \left(\frac{n}{n+1}\right)^2 \frac{\theta^2}{n} = \frac{n\theta^2}{(n+1)^2}.$$

Thus,

$$MSE_\theta(T_{2n}) = \frac{n\theta^2}{(n+1)^2} + \frac{\theta^2}{(n+1)^2} = \frac{\theta^2}{n+1} < \frac{\theta^2}{n} = MSE_\theta(T_{1n}) \ \forall \ n \geq 1.$$

(iii) If X follows an exponential distribution with mean θ, then $P[X > t] = \exp(-t/\theta) = g(\theta)$, say. It is clear that g is a differentiable function and hence continuous. Thus, the consistent estimator for $g(\theta)$ is given by $g(T_{1n})$. To examine if it is CAN, we use delta method. Note that $g'(\theta) = \exp(-t/\theta)(t/\theta^2) \neq 0, \ \forall \ \theta > 0$. Hence, $g(T_{1n}) = \exp(-t/T_{1n})$ is CAN for $g(\theta) = \exp(-t/\theta)$ with approximate variance $(\exp(-t/\theta)(t/\theta^2))^2(\theta^2/n) = t^2 \exp(-2t/\theta)/n\theta^2$.

3.5.16 Suppose $\{X_1, X_2, \ldots, X_n\}$ is a random sample from a Poisson $Poi(\theta)$ distribution. Examine whether the sample variance is CAN for θ.

Solution: Suppose $X \sim Poi(\theta)$, then $E(X) = Var(X) = \theta < \infty$. From Theorem 2.5.3, the sample variance $m_2 = S_n^2 = \frac{1}{n}\sum_{i=1}^{n}(X_i - \overline{X}_n)^2$ is consistent for $Var(X) = \theta$. To examine whether it is asymptotically normal, observe that

$$S_n^2 = \frac{1}{n}\sum_{i=1}^{n}(X_i - \overline{X}_n)^2 = \frac{1}{n}\sum_{i=1}^{n}((X_i - \theta) - (\overline{X}_n - \theta))^2$$

$$= \frac{1}{n}\sum_{i=1}^{n}(Y_i - \overline{Y}_n)^2 = \frac{1}{n}\sum_{i=1}^{n}Y_i^2 - \overline{Y}_n^2,$$

where $Y_i = X_i - \theta$. Thus, $E(Y_i) = 0$ and $E(Y_i^2) = Var(Y_i) = \theta$. Hence by the WLLN, $\overline{Y}_n \overset{P}{\to} E(Y_i) = 0$. By the CLT $\sqrt{n}(\overline{Y}_n) \overset{L}{\to} Z_1 \sim N(0, \theta)$. Hence, $\sqrt{n}(\overline{Y}_n)(\overline{Y}_n) \overset{P}{\to} 0$. Further, by the CLT applied to $\{Y_1^2, Y_2^2, \ldots, Y_n^2\}$, $\sqrt{n}(\frac{1}{n}\sum_{i=1}^{n}Y_i^2 - \theta) \overset{L}{\to} Z_2 \sim N(0, v(\theta))$, where $v(\theta) = Var(Y_i^2) = E(X_i - \theta)^4 - (E(X_i - \theta)^2)^2 = \mu_4(\theta) - \theta^2$. As a consequence,

$$\sqrt{n}(S_n^2 - \theta) = \sqrt{n}\left(\frac{1}{n}\sum_{i=1}^{n}Y_i^2 - \overline{Y}_n^2 - \theta\right)$$

$$= \sqrt{n}\left(\frac{1}{n}\sum_{i=1}^{n} Y_i^2 - \theta\right) - \sqrt{n}(\bar{Y}_n)(\bar{Y}_n) \xrightarrow{L} Z_2$$

where $Z_2 \sim N(0, v(\theta))$. Thus, the sample variance S_n^2 is CAN for θ with approximate variance $v(\theta)/n$. To find $v(\theta)$ we need to find the fourth central moment of $Poi(\theta)$ distribution. We obtain it from that the cumulant generating function $C_X(t)$ of X. The moment generating function $M_X(t)$ of X is $M_X(t) = \exp(\theta\{e^t - 1\})$ and hence the cumulant generating function $C_X(t)$ of X is $C_X(t) = \theta(e^t - 1) = \theta(t + t^2/2! + t^3/3! + t^4/4! + \cdots)$. Using the relation between cumulants k_i and central moments μ_i we have $k_i = \mu_i = \theta, i = 2, 3$ and $\mu_4 = k_4 + 3k_2^2 = \theta + 3\theta^2$. Thus, $v(\theta) = \mu_4(\theta) - \theta^2 = \theta + 2\theta^2$.

3.5.17 Show that the empirical distribution function $F_n(a)$ is CAN for $F(a)$, where a is a fixed real number. Hence obtain a large sample confidence interval for $F(a)$.

Solution: Suppose $\{X_1, X_2, \ldots, X_n\}$ is a random sample from the distribution of X with the distribution function $F(x)$, $x \in \mathbb{R}$. For each fixed $a \in \mathbb{R}$, the empirical distribution function $F_n(a)$ corresponding to the given random sample, is defined as

$$F_n(a) = \frac{\text{number of } X_i \leq a}{n} = \frac{1}{n}\sum_{i=1}^{n} Y_i = \bar{Y}_n,$$

where for $i = 1, 2, \ldots, n$,

$$Y_i = \begin{cases} 1, \text{ if } & X_i \leq a \\ 0, \text{ if } & X_i > a. \end{cases}$$

In Example 2.2.9, it is shown that $F_n(a)$ is consistent for $F(a)$. Now $\{X_1, X_2, \ldots, X_n\}$ are independent and identically distributed random variables, hence being Borel functions, $\{Y_1, Y_2, \ldots, Y_n\}$ are also independent and identically distributed random variables with $E(Y_i) = P_\theta[X_i \leq a] = F(a)$ and $Var(Y_i) = v = F(a)(1 - F(a)) < \infty$. Hence by the CLT,

$$\sqrt{n}(\bar{Y}_n - F(a)) \xrightarrow{L} Z_1 \sim N(0, v)$$
$$\Leftrightarrow \quad \sqrt{n}(F_n(a) - F(a)) \xrightarrow{L} Z_1 \sim N(0, v).$$

Thus, the empirical distribution function $F_n(a)$ is CAN for $F(a)$ with approximate variance v/n. To obtain the asymptotic confidence interval for $F(a)$, note that by Slutsky's theorem,

$$Q_n = \sqrt{\frac{n}{F_n(a)(1 - F_n(a))}}((F_n(a) - F(a)) \xrightarrow{L} Z \sim N(0, 1).$$

Hence, Q_n is a pivotal quantity and is useful to find the asymptotic confidence interval for $F(a)$. Thus, given the confidence coefficient $(1 - \alpha)$, we find the $(1 - \alpha/2)$-th quantile $a_{1-\alpha/2}$ of the standard normal distribution and invert the inequality $-a_{1-\alpha/2} < Q_n < a_{1-\alpha/2}$. Hence, $100(1 - \alpha)\%$ asymptotic confidence interval for $F(a)$ is given by

$$\left(F_n(a) - a_{1-\alpha/2}\sqrt{\frac{F_n(a)(1 - F_n(a))}{n}}, \quad F_n(a) + a_{1-\alpha/2}\sqrt{\frac{F_n(a)(1 - F_n(a))}{n}} \right).$$

3.5.18 Suppose $\{X_1, X_2, \ldots, X_n\}$ is a random sample from a normal $N(\theta, a\theta^2)$ distribution, $\theta > 0$ and a is a known positive real number. Find the maximum likelihood estimator of θ. Examine whether it is CAN for θ.

Solution: In Example 4.2.3, we have obtained the maximum likelihood estimator of θ and shown it to be CAN for θ when $X \sim N(\theta, \theta^2)$. Here $X \sim N(\theta, a\theta^2)$, hence we proceed on same lines to find the maximum likelihood estimator of θ. It is given by $\hat{\theta}_n = (-m_1' + \sqrt{m_1'^2 + 4am_2'})/2a$. Now,

$$\hat{\theta}_n = \frac{-m_1' + \sqrt{m_1'^2 + 4am_2'}}{2a} \xrightarrow{P_\theta} \frac{-\theta + \sqrt{\theta^2 + 4a(a\theta^2 + \theta^2}}{2a}$$
$$= \frac{-\theta + \theta(1 + 2a)}{2a} = \theta .$$

Hence, it is consistent for θ. To examine whether $\hat{\theta}_n$ is CAN, we use Theorem 3.3.2 and an appropriate transformation. From Theorem 3.3.2, $\underline{T}_n = (m_1', m_2')'$ is CAN for $\phi = (\mu_1', \mu_2')' = (\theta, \theta^2(1 + a))'$ with approximate dispersion matrix Σ/n where Σ is given by

$$\Sigma = \begin{pmatrix} a\theta^2 & 2a\theta^3 \\ 2a\theta^3 & 2a\theta^4(a + 2) \end{pmatrix}.$$

We further define a transformation $g : \mathbb{R}^2 \to \mathbb{R}$ such that $g(x_1, x_2) = (-x_1 + \sqrt{x_1^2 + 4ax_2})/2a$. Then with the routine procedure we get $\hat{\theta}_n$ is CAN for θ with approximate variance $a\theta^2/n(1 + 2a) > 0 \ \forall \ \theta > 0$. It is to be noted that

$$nI(\theta) = E_\theta \left(-\frac{\partial^2}{\partial\theta^2} \log L_n(\theta|\underline{X}) \right) = n \left(-\frac{1}{\theta^2} + \frac{3\theta^2(a + 1)}{a\theta^4} - \frac{2\theta}{a\theta^3} \right)$$
$$= \frac{(1 + 2a)n}{a\theta^2}.$$

Thus, $\hat{\theta}_n$ is CAN for θ with approximate variance $a\theta^2/n(1 + 2a) = 1/nI(\theta)$.

3.5.19 Suppose $\{X_1, X_2, \ldots, X_n\}$ is a random sample from a uniform $U(-\theta, \theta)$ distribution. Find a CAN estimator of θ based on $\sum_{i=1}^{n} |X_i|$. Are the sample mean and the sample median CAN for θ? Justify your answer. Find a consistent estimator for θ based on $X_{(1)}$ and find a consistent estimator for θ based on $X_{(n)}$. Examine if these are CAN for θ.

Solution: Suppose $X \sim U(-\theta, \theta)$. Then its distribution function $F_X(x, \theta)$ is given by

$$F_X(x, \theta) = \begin{cases} 0, & \text{if } x \leq -\theta \\ \frac{x+\theta}{2\theta}, & \text{if } -\theta \leq x < \theta \\ 1, & \text{if } x \geq \theta. \end{cases}$$

Suppose $Y = |X|$, then $Y \geq 0$ hence $P[Y \leq y] = 0$ if $y < 0$. Further, $Y \leq \theta$. Hence, $P[Y \leq y] = 1$ if $y > \theta$. Suppose $0 \leq y < \theta$, then

$$P[Y \leq y] = P[|X| \leq y] = F_X(y, \theta) - F_X(-y, \theta) = \frac{y+\theta}{2\theta} - \frac{-y+\theta}{2\theta} = \frac{y}{\theta}.$$

Thus, distribution function $F_Y(y, \theta)$ of $Y = |X|$ is given by

$$F_Y(y, \theta) = \begin{cases} 0, & \text{if } y \leq 0 \\ \frac{y}{\theta}, & \text{if } 0 \leq y < \theta \\ 1, & \text{if } y \geq \theta. \end{cases}$$

Thus, $Y = |X| \sim U(0, \theta)$. Hence, $E(|X|) = \theta/2$ and $Var(|X|) = \theta^2/12$ and by the WLLN and by the CLT,

$$T_n = \frac{1}{n} \sum_{i=1}^{n} |X_i| \xrightarrow{P_\theta} \frac{\theta}{2} \quad \& \quad \sqrt{n}\left(T_n - \frac{\theta}{2}\right) \xrightarrow{L} Z_1 \sim N\left(0, \frac{\theta^2}{12}\right).$$

As a consequence $2T_n$ is CAN for θ with approximate variance $\theta^2/3n$. For $U(-\theta, \theta)$ distribution, both the population mean and median are 0. Hence, the sample mean and the sample median will converge to 0 in probability. Thus, the sample mean and the sample median are not consistent for θ and hence not CAN for θ. From Example 2.5.2, $X_{(1)} \xrightarrow{P_\theta} -\theta$ and $X_{(n)} \xrightarrow{P_\theta} \theta$. Thus, $-X_{(1)}$ and $X_{(n)}$ are consistent for θ. However, using arguments similar to those in the solution of Exercise 3.5.1, we can show that these are not CAN.

3.5.20 Suppose $\{X_1, X_2, \ldots, X_n\}$ is a random sample from a uniform $U(0, \theta)$ distribution, $\theta > 0$. Examine whether $S_n = \left(\prod_{i=1}^{n} X_i\right)^{1/n}$ is CAN for θe^{-1}.

Solution: Suppose $T_n = -\log S_n = \frac{1}{n}\sum_{i=1}^{n}(-\log X_i) = \frac{1}{n}\sum_{i=1}^{n} Y_i$ where $Y_i = -\log X_i$, $i = 1, 2, \ldots, n$. Suppose $X \sim U(0, \theta)$ and $Y = -\log X$. Then the probability density function $f_Y(y, \theta)$ of Y is given by

$f_Y(y, \theta) = (1/\theta)e^{-y}, \quad -\log\theta < y < \infty$. Further using the method of integration by parts, we have

$$E(Y) = \frac{1}{\theta} \int_{-\log\theta}^{\infty} ye^{-y} = \frac{1}{\theta} \left[\int_{-\log\theta}^{\infty} y\frac{d}{dy}(-e^{-y}) \right] = 1 - \log\theta$$

$$E(Y^2) = \frac{1}{\theta} \int_{-\log\theta}^{\infty} y^2 e^{-y} = \frac{1}{\theta} \left[\int_{-\log\theta}^{\infty} y^2 \frac{d}{dy}(-e^{-y}) \right]$$

$$= \frac{1}{\theta} \left[(-y^2 e^{-y})_{-\log\theta}^{\infty} + 2 \int_{-\log\theta}^{\infty} ye^{-y} \right]$$

$$= (\log\theta)^2 + 2 - 2\log\theta = 1 + (1 - \log\theta)^2 .$$

Thus, $Var(Y) = 1$ and $\{Y_1, Y_2, \ldots, Y_n\}$ are independent and identically distributed random variables with finite mean $1 - \log\theta$ and variance 1. Hence by the WLLN and CLT,

$$T_n = \frac{1}{n} \sum_{i=1}^{n} Y_i \overset{P_\theta}{\to} 1 - \log\theta \quad \& \quad \sqrt{n}(T_n - (1 - \log\theta)) \overset{L}{\to} Z \sim N(0, 1).$$

Thus, T_n is CAN for $1 - \log\theta$ with approximate variance $1/n$. To examine whether S_n is CAN for θe^{-1}, we consider a transformation $g(x) = e^{-x}$, it is differentiable function with $g'(x) = -e^{-x} \neq 0$. Hence by the delta method, $g(T_n) = e^{-T_n} = S_n = \left(\prod_{i=1}^{n} X_i \right)^{1/n}$ is CAN for $g(1 - \log\theta) = e^{-1+\log\theta} = \theta e^{-1}$ with approximate variance $(1/n)(-e^{-1+\log\theta})^2$ which reduces to $\theta^2 e^{-2}/n$.

3.5.21   Suppose $\{X_1, X_2, \ldots, X_{2n+1}\}$ is a random sample from a uniform $U(\theta - 1, \theta + 1)$ distribution. (i) Show that $\overline{X}_{2n+1}$ and $X_{(n+1)}$ are both CAN for θ. Compare the two estimators.
(ii) Using the large sample distribution, obtain the minimum sample size n_0 required for both the estimators to attain a given level of accuracy specified by ϵ and δ, such that $P[|T_n - \theta| < \epsilon] \geq 1 - \delta$, $\forall n \geq n_0$, where T_n is either $\overline{X}_{2n+1}$ or $X_{(n+1)}$.

Solution: (i) Suppose $X \sim U(\theta - 1, \theta + 1)$ distribution. Then $E(X) = \theta$, $Var(X) = 1/3$ and population median is θ. By the WLLN and the CLT $\overline{X}_{2n+1}$ is CAN for θ with approximate variance $1/3(2n + 1)$. Further by Theorem 3.3.3, the sample median $X_{(n+1)}$ is CAN for θ with approximate variance $1/(2n + 1)$. Comparing the approximate variances of the two estimators we get that sample mean is better than the sample median. (ii) By the CLT $\sqrt{3(2n + 1)} \left(\overline{X}_{2n+1} - \theta \right) \overset{L}{\to} Z \sim N(0, 1)$. Hence,

$$P\left[|\overline{X}_{2n+1} - \theta| < \epsilon \right] \approx \Phi\left(\sqrt{3(2n + 1)}\,\epsilon \right) - \Phi\left(-\sqrt{3(2n + 1)}\,\epsilon \right)$$

$$= 2\Phi\left(\sqrt{3(2n + 1)}\,\epsilon \right) - 1 > 1 - \delta$$

$$\Rightarrow \Phi\left(\sqrt{3(2n+1)}\,\epsilon\right) \geq 1 - \frac{\delta}{2}$$

$$\Rightarrow \sqrt{3(2n+1)}\,\epsilon \geq \Phi^{-1}\left(1 - \frac{\delta}{2}\right)$$

$$\Rightarrow n \geq \frac{\left(\Phi^{-1}(1-\frac{\delta}{2})\right)^2}{6\epsilon^2} - \frac{1}{2}$$

$$\Rightarrow n_0 = \left[\frac{\left(\Phi^{-1}(1-\frac{\delta}{2})\right)^2}{6\epsilon^2} - \frac{1}{2}\right] + 1.$$

For large n, $X_{(n+1)} \sim N(\theta, 1/(2n+1))$. Hence,

$$P\left[|X_{(n+1)} - \theta| < \epsilon\right] \approx \Phi\left(\sqrt{(2n+1)}\,\epsilon\right) - \Phi\left(-\sqrt{(2n+1)}\,\epsilon\right)$$

$$= 2\Phi\left(\sqrt{(2n+1)}\,\epsilon\right) - 1 > 1 - \delta$$

$$\Rightarrow \Phi\left(\sqrt{(2n+1)}\,\epsilon\right) \geq 1 - \frac{\delta}{2}$$

$$\Rightarrow \sqrt{(2n+1)}\,\epsilon \geq \Phi^{-1}\left(1 - \frac{\delta}{2}\right)$$

$$\Rightarrow n \geq \frac{\left(\Phi^{-1}(1-\frac{\delta}{2})\right)^2}{2\epsilon^2} - \frac{1}{2}$$

$$\Rightarrow n_0 = \left[\frac{\left(\Phi^{-1}(1-\frac{\delta}{2})\right)^2}{2\epsilon^2} - \frac{1}{2}\right] + 1,$$

it is larger than that for $\overline{X}_{2n+1}$.

3.5.22 Suppose $\{X_1, \ldots, X_n\}$ is a random sample from X with probability density function $f(x, \theta) = \theta/x^2$, $x \geq \theta$, $\theta > 0$. (i) Find the maximum likelihood estimator of θ and examine if it is CAN for θ. (ii) Find a CAN estimator of θ based on the sample quantiles.

Solution: (i) The probability density function of a random variable X is given by

$$f_X(x, \theta) = \theta/x^2, \quad x \geq \theta.$$

In Exercise 2.8.16, we have obtained the maximum likelihood estimator $\hat{\theta}_n$ of θ which is $X_{(1)}$ and it is shown to be consistent for θ. To derive the asymptotic distribution of $X_{(1)}$ with suitable norming, we define $Y_n = n\left(X_{(1)} - \theta\right)$ and derive its distribution function $G_{Y_n}(y)$ for $y \in \mathbb{R}$. Since $X_{(1)} \geq \theta$, $Y_n \geq 0$, hence for $y < 0$, $G_{Y_n}(y) = 0$. Suppose $y \geq 0$, then

$$G_{Y_n}(y) = P_\theta[n(X_{(1)} - \theta) \leq y] = P_\theta[X_{(1)} \leq \theta + y/n] = F_{X_{(1)}}(\theta + y/n).$$

Now, the distribution function $F_X(x)$ of X is given by

$$F_X(x) = \begin{cases} 0, & \text{if } x < \theta \\ 1 - \theta/x, & \text{if } x \geq \theta. \end{cases}$$

Hence, the distribution function of $X_{(1)}$ is given by

$$F_{X_{(1)}}(x) = 1 - [1 - F_X(x)]^n = \begin{cases} 0, & \text{if } x < \theta \\ 1 - \theta^n/x^n, & \text{if } x \geq \theta. \end{cases}$$

Consequently, distribution function $G_{Y_n}(y)$ is given by

$$G_{Y_n}(y) = \begin{cases} 0, & \text{if } \theta + y/n < \theta \quad \Leftrightarrow y < 0 \\ 1 - \left(\frac{\theta}{\theta + y/n}\right)^n, & \text{if } \theta + y/n \geq \theta \quad \Leftrightarrow y \geq 0. \end{cases}$$

As $n \to \infty$,

$$G_{Y_n}(y) \to \begin{cases} 0, & \text{if } y \leq 0 \\ 1 - \exp(-y/\theta), & \text{if } y \geq 0. \end{cases}$$

Thus, the asymptotic distribution of $Y_n = n\left(X_{(1)} - \theta\right)$ is exponential with location parameter 0 and scale parameter $1/\theta$. Thus, with norming factor n, the asymptotic distribution of $X_{(1)}$ is not normal. Proceeding on similar lines as in Example 3.2.1, we conclude that there exists no sequence $\{a_n, n \geq 1\}$ of real numbers tending to ∞ as $n \to \infty$ such that the asymptotic distribution of $a_n\left(X_{(1)} - \theta\right)$ is normal, hence we claim that $X_{(1)}$ is not CAN for θ.

(ii) From the distribution function $F_X(x, \theta)$ of X, the p-th population quantile $a_p(\theta)$ is given by the solution of the equation $F_X(x, \theta) = p$. Thus, $a_p(\theta) = \theta/(1 - p)$. Hence by Theorem 3.3.3, it follows that the p-th sample quantile defined as $X_{([np]+1)}$ is CAN for $a_p(\theta) = \theta/(1 - p)$ with approximate variance $\theta^2 p/n(1 - p)^3$. Thus, $(1 - p)X_{([np]+1)}$ is CAN for θ with approximate variance $\theta^2 p/n(1 - p)$.

3.5.23 Suppose $\{X_1, X_2, \ldots, X_n\}$ is a random sample from a Weibull distribution with probability density function

$$f(x, \theta) = \theta \, x^{\theta-1} \exp\{-x^\theta\}, \quad x > 0, \quad \theta > 0.$$

Obtain an estimator of θ based on the sample quantiles. Is it CAN? Justify your answer.

Solution: The distribution function $F(x, \theta)$ of a Weibull distribution is given by

$$F(x, \theta) = \theta \int_0^x u^{\theta-1} \exp\{-u^\theta\} du, \quad x > 0$$

$$= \int_0^{x^\theta} \exp\{-y\}dy, \quad \text{by substituting } u^\theta = y$$

$$= 1 - \exp\{-x^\theta\}.$$

To find the p-th quantile $a_p(\theta)$, we solve

$$F(x,\theta) = p \quad \Leftrightarrow \quad -\log(1-p) = x^\theta$$
$$\Rightarrow \quad a_p(\theta) = (-\log(1-p))^{1/\theta}, \quad 0 < p < 1.$$

By Theorem 2.2.6, the p-th sample quantile $X_{([np]+1)}$ is consistent for p-th population quantile $a_p(\theta)$ and by Theorem 3.3.3, $X_{([np]+1)}$ is CAN for $a_p(\theta) = \phi$, say, with approximate variance
$p\left(n\theta^2(1-p)(\log(1-p))^{(2-2/\theta)}\right)^{-1}$. We now define a function
$g : \mathbb{R}^+ \to \mathbb{R}^+$ as $g(x) = (\log(-\log(1-p)))/\log x$ so that $g(\phi) = \theta$. It is clear that g is a differentiable function with
$g'(x) = (-\log(-\log(1-p)))/((\log x)^2 x) \neq 0$ for all $x > 0$, but is not defined for $x = 1 \Leftrightarrow p = 1 - e^{-1}$. We hence assume that $p \neq 1 - e^{-1}$.
By the delta method, $g(X_{([np]+1)}) = (\log(-\log(1-p)))/(\log X_{([np]+1)})$ is CAN for $g(\phi) = \theta$ with approximate variance
$p\theta^2 \left(n(1-p)(\log(1-p))^2(\log(-\log(1-p)))^2\right)^{-1}$.

3.5.24 Suppose $\{X_1, X_2, \ldots, X_n\}$ is a random sample from a Laplace $(\theta, 1)$ distribution. Find a family of CAN estimators of θ based on sample quantiles. Also find the CAN estimator of θ based on the sample mean and the sample median. Which one is better? Why?

Solution: Suppose X follows a Laplace $(\theta, 1)$ distribution. Then its probability density function is given by, $f(x, \theta) = (1/2)\exp\{-|x - \theta|\}$, $x \in \mathbb{R}$, $\theta \in \mathbb{R}$. In Exercise 2.8.17, we have obtained the distribution function $F(x, \theta)$ of X and it is given by

$$F(x,\theta) = \begin{cases} \frac{1}{2}\exp\{x - \theta\}, & \text{if } x < \theta \\ 1 - \frac{1}{2}\exp\{-(x - \theta)\}, & \text{if } x \geq \theta. \end{cases}$$

To find a family of CAN estimators of θ based on sample quantiles, we note that

$$\text{for } p < \frac{1}{2}, \ F(x,\theta) = p \ \Rightarrow \ \frac{1}{2}\exp\{x - \theta\} = p \ \Rightarrow \ x = a_p(\theta) = \theta + \log 2p$$

$$\& \ \text{for } p \geq \frac{1}{2}, \ F(x,\theta) = p \ \Rightarrow \ 1 - \frac{1}{2}\exp\{-(x - \theta)\} = p$$
$$\Rightarrow \ x = a_p(\theta) = \theta - \log 2(1 - p).$$

By Theorem 2.2.6, the p-th sample quantile $X_{([np]+1)} \xrightarrow{P_\theta} a_p(\theta)$. Hence for all $p \in (0, 1/2)$, $X_{([np]+1)} - \log 2p$ is consistent for θ and for all

$p \in [1/2, 1)$, $X_{([np]+1)} + \log 2(1 - p)$ is consistent for θ. By Theorem 3.3.3, $X_{([np]+1)}$ is CAN for $a_p(\theta)$ with approximate variance $p(1 - p)/n(f(a_p(\theta), \theta))^2$. Hence, for $p < 1/2$, $X_{([np]+1)} - \log 2p$ is CAN for θ with approximate variance $(1 - p)/np$ and for $p \geq 1/2$, $X_{([np]+1)} + \log 2(1 - p)$ is CAN for θ with approximate variance $p/n(1 - p)$. With $p = 1/2$, the sample median $X_{([n/2]+1)}$ is CAN for θ with approximate variance $1/n$. If X follows Laplace distribution then $E(X) = \theta$ and $Var(X) = 2 < \infty$. Thus, by the WLLN and by the CLT, $\overline{X}_n$ is CAN for θ with approximate variance $2/n$. Approximate variance of the sample median is smaller than that of sample mean, hence sample median is a better estimator of θ than the sample mean. It is to be noted that the sample median is the maximum likelihood estimator of θ.

3.5.25 Suppose $\{X_1, X_2, \ldots, X_n\}$ is a random sample from a gamma $G(\theta, \lambda)$ distribution. Find $100(1 - \alpha)\%$ asymptotic confidence interval for the mean of the distribution.

Solution: Suppose $X \sim G(\theta, \lambda)$ distribution. Then $E(X) = \lambda/\theta = \alpha$, say and $Var(X) = \lambda/\theta^2 = \sigma^2$, say and $\sigma^2 < \infty$. By the WLLN and the CLT, $\overline{X}_n$ is CAN for $E(X) = \alpha$ with approximate variance σ^2/n. Thus, $(\sqrt{n}/\sigma)(\overline{X}_n - \alpha) \overset{L}{\to} Z \sim N(0, 1)$. However, this pivotal quantity cannot be used to find the asymptotic confidence interval for the mean, since σ involves the parameters. It is known that the sample variance S_n^2 is a consistent estimator of σ^2. Hence by Slutsky's theorem,

$$Q_n = \frac{\sqrt{n}}{S_n}(\overline{X}_n - \alpha) \overset{L}{\to} Z \sim N(0, 1)$$

and we use the pivotal quantity Q_n to find asymptotic confidence interval for the mean. Thus, given a confidence coefficient $(1 - \alpha)$, we can find the quantile $a_{1-\alpha/2}$ of the standard normal distribution so that $P[-a_{1-\alpha/2} < Q_n < a_{1-\alpha/2}] = 1 - \alpha$. Inverting the inequality $-a_{1-\alpha/2} < Q_n < a_{1-\alpha/2}$, we get

$$\overline{X}_n - a_{1-\alpha/2}\sqrt{S_n/n} < E(X) < \overline{X}_n + a_{1-\alpha/2}\sqrt{S_n/n}.$$

Hence, $100(1 - \alpha)\%$ large sample confidence interval for $E(X)$ is given by

$$\left(\overline{X}_n - a_{1-\alpha/2}\sqrt{S_n/n}, \quad \overline{X}_n + a_{1-\alpha/2}\sqrt{S_n/n}\right).$$

3.5.26 Suppose $\{X_1, X_2, \ldots, X_n\}$ is a random sample from a Bernoulli distribution with mean θ. Find a asymptotic confidence interval for θ with confidence coefficient $(1 - \alpha)$, using both the studentization procedure and the variance stabilization technique.

Solution: Suppose $X \sim B(1, \theta)$, then by WLLN $\overline{X}_n$ is consistent for θ and by the CLT it is CAN for θ with approximate variance $\theta(1 - \theta)/n$, that is,

$$T_n = \frac{\sqrt{n}}{\sqrt{\theta(1 - \theta)}} (\overline{X}_n - \theta) \overset{L}{\to} Z \sim N(0, 1) \text{ distribution}$$

However, the pivotal quantity T_n based on $\overline{X}_n$ is not useful to find the asymptotic confidence interval for θ. Hence we adopt the studentization procedure and the variance stabilization technique to find the asymptotic confidence interval for θ.

(i) Studentization procedure: In this procedure we define a pivotal quantity $\tilde{Q}_n$ as

$$\tilde{Q}_n = \frac{\sqrt{n}}{\sqrt{\overline{X}_n(1 - \overline{X}_n)}} (\overline{X}_n - \theta) \sim N(0, 1) \text{ distribution}$$

by Slutsky's theorem for large n. Hence, given the confidence coefficient $(1 - \alpha)$, we find $a_{(1-\alpha/2)}$ such that $P[-a_{(1-\alpha/2)} < \tilde{Q}_n < a_{(1-\alpha/2)}] = 1 - \alpha$. Now, using routine method, the asymptotic confidence interval for θ with confidence coefficient $(1 - \alpha)$, is given by

$$\left(\overline{X}_n - \frac{\sqrt{\overline{X}_n(1 - \overline{X}_n)}}{\sqrt{n}} a_{(1-\alpha/2)}, \ \overline{X}_n + \frac{\sqrt{\overline{X}_n(1 - \overline{X}_n)}}{\sqrt{n}} a_{(1-\alpha/2)} \right).$$

(ii) Variance stabilization technique: In this approach we find a differentiable function g such that $(g'(\theta))^2 \neq 0$ and $Var(g(\overline{X}_n))$ is free from θ, that is

$$(g'(\theta))^2 \theta(1 - \theta) = c^2 \ \Rightarrow \ g(\theta) = \int \frac{c \, d\theta}{\sqrt{\theta(1 - \theta)}} \ \Rightarrow \ g(\theta) = 2c \sin^{-1}(\sqrt{\theta}).$$

Hence,

$$\tilde{Q}_n = \frac{\sqrt{n}}{c} \left(2c \sin^{-1}(\sqrt{\overline{X}_n}) - 2c \sin^{-1}(\sqrt{\theta}) \right)$$

$$= 2\sqrt{n} \left(\sin^{-1}(\sqrt{\overline{X}_n}) - \sin^{-1}(\sqrt{\theta}) \right) \sim N(0, 1),$$

for large n. Hence, taking $\tilde{Q}_n$ as a pivotal quantity, corresponding to the confidence coefficient $(1 - \alpha)$, the asymptotic confidence interval for $\sin^{-1} \sqrt{\theta}$ is given by

$$\left(\sin^{-1}(\sqrt{\overline{X}_n}) - \frac{1}{2\sqrt{n}} a_{(1-\alpha/2)}, \ \sin^{-1}(\sqrt{\overline{X}_n}) + \frac{1}{2\sqrt{n}} a_{(1-\alpha/2)} \right).$$

Consequently, the asymptotic confidence interval for θ is given by

$$\left(\left(\sin\left(\sin^{-1}(\sqrt{\overline{X}_n}) - \frac{1}{2\sqrt{n}}a_{(1-\alpha/2)} \right) \right)^2 , \left(\sin\left(\sin^{-1}(\sqrt{\overline{X}_n}) + \frac{1}{2\sqrt{n}}a_{(1-\alpha/2)} \right) \right)^2 \right) .$$

3.5.27 Suppose $\{X_1, X_2, \ldots, X_n\}$ is a random sample from a normal $N(\mu, \sigma^2)$ distribution. Find a CAN estimator of the p-th population quantile based on p-th sample quantile as well as based on a CAN estimator of $\underline{\theta} = (\mu, \sigma^2)'$. Which is better? Why? Obtain the asymptotic confidence interval for the p-th population quantile using the estimator which is better between the two.

Solution: Suppose $X \sim N(\mu, \sigma^2)$ distribution with probability density function $f(x, \mu, \sigma^2)$. Then p-th quantile $a_p(\underline{\theta})$ is a solution of the equation,

$$P[X \le a_p(\underline{\theta})] = p \quad \Leftrightarrow \quad P\left[(X - \mu)/\sigma \le (a_p(\underline{\theta}) - \mu)/\sigma \right] = p$$
$$\Leftrightarrow \quad a_p(\underline{\theta}) = \mu + \sigma \Phi^{-1}(p) .$$

From Theorem 3.3.3, the p-th sample quantile $X_{([np]+1)}$ is CAN for the p-th population quantile $a_p(\underline{\theta})$ with approximate variance $v(\underline{\theta})/n = p(1-p)(n(f(a_p(\underline{\theta}), \mu, \sigma^2))^2)^{-1}$. Now,

$$\begin{aligned} f(a_p(\underline{\theta}), \mu, \sigma^2) &= \frac{1}{\sqrt{2\pi}\sigma} \exp\left\{ -\frac{1}{2\sigma^2}(a_p(\underline{\theta}) - \mu)^2 \right\} \\ &= \frac{1}{\sqrt{2\pi}\sigma} \exp\left\{ -\frac{1}{2\sigma^2}(\mu + \sigma\Phi^{-1}(p) - \mu)^2 \right\} \\ &= \frac{1}{\sqrt{2\pi}\sigma} \exp\left\{ -\frac{1}{2}\Phi^{-2}(p) \right\} . \end{aligned}$$

Hence,

$$\frac{v(\underline{\theta})}{n} = \frac{p(1-p)}{n(f(a_p(\underline{\theta}), \mu, \sigma^2))^2} = \frac{p(1-p)2\pi\sigma^2 \exp\{\Phi^{-2}(p)\}}{n} .$$

It is to be noted that $a_p(\underline{\theta}) = \mu + \sigma\Phi^{-1}(p)$ is a function of μ and σ^2. Hence we obtain its CAN estimator by using the delta method. In Example 3.3.2, it is shown that $(m_1', m_2)'$ is the CAN estimator of $\underline{\theta} = (\mu, \sigma^2)'$ with approximate dispersion matrix D/n, where

$$D = \begin{pmatrix} \sigma^2 & 0 \\ 0 & 2\sigma^4 \end{pmatrix}.$$

we further define a transformation $g : \mathbb{R}^2 \to \mathbb{R}$ such that

$$g(x_1, x_2) = x_1 + \sqrt{x_2}\,\Phi^{-1}(p) \quad \Rightarrow \quad \frac{\partial}{\partial x_1} g(x_1, x_2) = 1$$

$$\&\quad \frac{\partial}{\partial x_2} g(x_1, x_2) = \frac{1}{2\sqrt{x_2}}\Phi^{-1}(p).$$

These partial derivatives are continuous and hence g is a totally differentiable function. The gradient vector Δ evaluated at $(\mu, \sigma^2)'$ is given by $\Delta = [1, \Phi^{-1}(p)/2\sigma]'$. Hence, by Theorem 3.3.4, $g(m'_1, m_2) = m'_1 + \sqrt{m_2}\,\Phi^{-1}(p)$ is CAN for $g(\mu, \sigma^2) = \mu + \sigma\Phi^{-1}(p) = a_p(\underline{\theta})$ with approximate variance $\Delta D\Delta'/n$, where $\Delta D\Delta' = \sigma^2 + \sigma^2\Phi^{-2}(p)/2$. To decide which one of the two is a better estimator we compare $v(\underline{\theta})$ with $\Delta D\Delta'$. Suppose $\Phi^{-2}(p) = u$ then as p varies over $(0, 1)$, $\Phi^{-1}(p)$ is approximately between -3 to 3 which implies that $0 < u < 9$, where upper limit 9 is approximate. Observe that

$$\begin{aligned}
\Delta D\Delta' - v(\underline{\theta}) &= \sigma^2\left(1 + u/2 - 6.28p(1-p)e^u\right) \\
&= \sigma^2\left(1 + u/2 - 6.28p(1-p)(1 + u + u^2/2 + \cdots)\right) \\
&= \sigma^2(1 - 6.28p(1-p)) \\
&\quad + \sigma^2\left((1/2 - 6.28p(1-p))u - 6.28p(1-p)(u^2/2 + \cdots)\right).
\end{aligned}$$

The first two terms in the above expression are negative as $p(1-p) \in (0, 1/4]$ and the third term is also negative as $(u^2/2 + \cdots) > 0$. Hence, $\Delta D\Delta' - v(\underline{\theta}) < 0$. Thus, $T_n = m'_1 + \sqrt{m_2}\,\Phi^{-1}(p)$ is a better CAN estimator of $a_p(\underline{\theta})$ with approximate variance $\sigma^2 + \sigma^2\Phi^{-2}(p)/2$. To find the asymptotic confidence interval for $a_p(\underline{\theta})$ based on T_n we define a pivotal quantity Q_n as $Q_n = \left((m_2 + m_2\Phi^{-2}(p)/2)/n\right)^{-1/2}(T_n - a_p(\underline{\theta}))$. Given a confidence coefficient $(1-\alpha)$, we can find the quantile $a_{1-\alpha/2}$ of the standard normal distribution so that $P[-a_{1-\alpha/2} < Q_n < a_{1-\alpha/2}] = 1 - \alpha$. Inverting the inequality $-a_{1-\alpha/2} < Q_n < a_{1-\alpha/2}$, we get $100(1-\alpha)\%$ large sample confidence interval for $a_p(\underline{\theta})$ as

$$\left(T_n - a_{1-\alpha/2}\sqrt{\frac{m_2 + m_2\frac{\Phi^{-2}(p)}{2}}{n}}, \quad T_n + a_{1-\alpha/2}\sqrt{\frac{m_2 + m_2\frac{\Phi^{-2}(p)}{2}}{n}}\right).$$

3.5.28 Suppose $\{X_1, X_2, \ldots, X_n\}$ is a random sample from an exponential distribution with location parameter μ and scale parameter $1/\sigma$. (i) Obtain an asymptotic confidence interval for μ when σ is known and when it is unknown. (ii) Obtain an asymptotic confidence interval for σ when μ is known and when it is unknown. (iii) Obtain an asymptotic confidence interval for the p-th quantile when both μ and σ are unknown.

Solution: X follows an exponential distribution with location parameter μ and scale parameter $1/\sigma$, hence its probability density function is $f(x, \mu, \sigma) = (1/\sigma) \exp\{-(x - \mu)/\sigma\}, x \geq \mu$. (i) Suppose σ is known. Proceeding as in Example 2.2.16, the maximum likelihood estimator of μ is $X_{(1)}$ and it is consistent for μ. Further, the distribution function $F_{X_{(1)}}(x, \mu, \sigma)$ of $X_{(1)}$ is given by

$$F_{X_{(1)}}(x, \mu, \sigma) = \begin{cases} 0, & \text{if } x < \mu \\ 1 - \exp\{-\frac{n}{\sigma}(x - \mu)\}, & \text{if } x \geq \mu. \end{cases}$$

It thus follows that for each n, $X_{(1)}$ has an exponential distribution with location parameter μ and scale parameter n/σ which further implies that $Y_n = (n/\sigma)(X_{(1)} - \mu)$ has the exponential distribution with location parameter 0 and scale parameter 1 for each n and Y_n is a pivotal quantity. The asymptotic confidence interval for μ when σ is known is based on Y_n. Given a confidence coefficient $(1 - \alpha)$, we can find a and b so that $P[a < Y_n < b] = 1 - \alpha$. Inverting the inequality $a < Y_n < b$, we get

$$X_{(1)} - \frac{b\sigma}{n} < \mu < X_{(1)} - \frac{a\sigma}{n}.$$

Hence, $100(1 - \alpha)\%$ large sample confidence interval for μ is given by

$$\left(X_{(1)} - \frac{b\sigma}{n}, \quad X_{(1)} - \frac{a\sigma}{n} \right).$$

When σ is unknown, pivotal quantity Y_n is of no use to find the confidence interval. We then use the studentization procedure and replace σ by its consistent estimator. Now $Var(X) = \sigma^2$, hence the sample variance S_n^2 is a consistent estimator of σ^2. Hence,

$$Q_n = \frac{n}{S_n}(X_{(1)} - \mu) = \left\{ \frac{\sigma}{S_n} \right\} \left\{ \frac{n}{\sigma}(X_{(1)} - \mu) \right\} \xrightarrow{L} U,$$

where U follows the exponential distribution with location parameter 0 and scale parameter 1. Given a confidence coefficient $(1 - \alpha)$, we can find a and b so that $P[a < Q_n < b] = 1 - \alpha$. Inverting the inequality $a < Q_n < b$, we get $100(1 - \alpha)\%$ large sample confidence interval for μ as

$$\left(X_{(1)} - \frac{bS_n}{n}, \quad X_{(1)} - \frac{aS_n}{n} \right).$$

(ii) Suppose μ is known, then $Y = X - \mu$ follows an exponential distribution with location parameter 0 and scale parameter $1/\sigma$. A random sample $\{X_1, X_2, \ldots, X_n\}$ gives a random sample $\{Y_1, Y_2, \ldots, Y_n\}$ from the distribution of Y. Hence, by the WLLN and the CLT, $\overline{Y}_n = \overline{X}_n - \mu$ is CAN for σ with approximate variance σ^2/n. Thus,

$$\sqrt{n/\sigma}\,(\overline{Y}_n - \sigma) \overset{L}{\to} Z \sim N(0, 1)$$

$$\Rightarrow \quad Q_n = \sqrt{n/\overline{Y}_n}\,(\overline{Y}_n - \sigma) \overset{L}{\to} Z \sim N(0, 1)\,,$$

by Slutsky's theorem. Thus, Q_n is a pivotal quantity. Given a confidence coefficient $(1 - \alpha)$, we can find the quantile $a_{1-\alpha/2}$ of the standard normal distribution so that $P[-a_{1-\alpha/2} < Q_n < a_{1-\alpha/2}] = 1 - \alpha$. Inverting the inequality $-a_{1-\alpha/2} < Q_n < a_{1-\alpha/2}$, we get

$$\overline{Y}_n - a_{1-\alpha/2}\sqrt{\overline{Y}_n/n} \; < \; \sigma \; < \; \overline{Y}_n + a_{1-\alpha/2}\sqrt{\overline{Y}_n/n}\,.$$

Hence using the studentization technique, $100(1 - \alpha)\%$ large sample confidence interval for σ is given by

$$\left(\overline{Y}_n - a_{1-\alpha/2}\sqrt{\overline{Y}_n/n}, \;\; \overline{Y}_n + a_{1-\alpha/2}\sqrt{\overline{Y}_n/n}\right)\,.$$

Suppose now that μ is unknown. In Example 3.3.5, we have shown that $(m_1' - \sqrt{m_2}, \sqrt{m_2})'$ is CAN for $(\mu, \sigma)'$ with the approximate dispersion matrix D/n, where

$$D = \begin{pmatrix} \sigma^2 & -\sigma^2 \\ -\sigma^2 & 2\sigma^2 \end{pmatrix}.$$

Thus when μ is unknown, $\sqrt{m_2}$ is CAN for σ with approximate variance $2\sigma^2/n$. Suppose Q_n is defined as $Q_n = \sqrt{n/2m_2}\,(\sqrt{m_2} - \sigma)$. Then

$$Q_n = \sqrt{\frac{n}{2m_2}}\,(\sqrt{m_2} - \sigma) = \left\{\sqrt{\frac{2\sigma^2}{2m_2}}\right\}\left\{\sqrt{\frac{n}{2\sigma^2}}\,(\sqrt{m_2} - \sigma)\right\}$$

$$\overset{L}{\to} Z \sim N(0, 1),$$

by Slutsky's theorem. Thus, Q_n is a pivotal quantity. Given a confidence coefficient $(1 - \alpha)$, we can find the quantile $a_{1-\alpha/2}$ of the standard normal distribution so that $P[-a_{1-\alpha/2} < Q_n < a_{1-\alpha/2}] = 1 - \alpha$. Inverting the inequality $-a_{1-\alpha/2} < Q_n < a_{1-\alpha/2}$, we get

$$\sqrt{m_2} - a_{1-\alpha/2}\sqrt{2m_2/n} \; < \; \sigma \; < \; \sqrt{m_2} + a_{1-\alpha/2}\sqrt{2m_2/n}\,.$$

Hence using the studentization technique, $100(1 - \alpha)\%$ large sample confidence interval for σ is given by

$$\left(\sqrt{m_2} - a_{1-\alpha/2}\sqrt{2m_2/n}, \;\; \sqrt{m_2} + a_{1-\alpha/2}\sqrt{2m_2/n}\right)\,.$$

(iii) Suppose $\underline{\theta} = (\mu, \sigma)'$. The p-th quantile $a_p(\underline{\theta})$ is a solution of the equation

$$F_X(a_p(\underline{\theta}), \mu, \sigma) = p \quad \Leftrightarrow \quad 1 - \exp\{-\frac{1}{\sigma}(a_p(\underline{\theta}) - \mu)\} = p$$
$$\Rightarrow \quad a_p(\underline{\theta}) = \mu - \sigma \log(1 - p).$$

Thus, $a_p(\underline{\theta})$ is a function of μ and σ. Now $(m_1' - \sqrt{m_2}, \sqrt{m_2})'$ is CAN for $(\mu, \sigma)'$ with the approximate dispersion matrix D/n. Suppose a function $g : \mathbb{R}^2 \to \mathbb{R}$ is defined as $g(x_1, x_2) = x_1 - x_2 \log(1 - p)$. Further, $\frac{\partial}{\partial x_1} g(x_1, x_2) = 1$ and $\frac{\partial}{\partial x_2} g(x_1, x_2) = -\log(1 - p)$. These partial derivatives are continuous and hence g is a totally differentiable function. The gradient vector Δ evaluated at $(\mu, \sigma)'$ is given by $\Delta = [1, -\log(1 - p)]'$. Hence by the delta method,

$$g(m_1' - \sqrt{m_2}, \sqrt{m_2}) = m_1' - \sqrt{m_2} - \log(1 - p)\sqrt{m_2}$$
$$= m_1' - \sqrt{m_2}(1 + \log(1 - p)) = T_n,$$

say, is CAN for $g(\mu, \sigma) = \mu - \sigma \log(1 - p) = a_p(\underline{\theta})$ with approximate variance $\Delta' D \Delta / n$, where
$\Delta' D \Delta = \sigma^2[1 + 2\log(1 - p) + 2(\log(1 - p))^2] = \sigma^2 h(p)$, say. Suppose Q_n is defined as $Q_n = (\sqrt{n}/\sqrt{m_2 h(p)}) (T_n - a_p(\underline{\theta}))$ Then

$$Q_n = \left\{\frac{\sigma\sqrt{h(p)}}{\sqrt{m_2}\sqrt{h(p)}}\right\}\left\{\frac{n}{\sigma\sqrt{h(p)}}\right\}(T_n - a_p(\underline{\theta})) \xrightarrow{L} Z \sim N(0, 1),$$

by Slutsky's theorem. Thus, Q_n is a pivotal quantity. Given a confidence coefficient $(1 - \alpha)$, we can find the quantile $a_{1-\alpha/2}$ of the standard normal distribution so that $P[-a_{1-\alpha/2} < Q_n < a_{1-\alpha/2}] = 1 - \alpha$. Inverting the inequality $-a_{1-\alpha/2} < Q_n < a_{1-\alpha/2}$, we get $100(1 - \alpha)\%$ large sample confidence interval for $a_p(\underline{\theta})$ as

$$\left(T_n - a_{1-\alpha/2} \sqrt{m_2 h(p)/n}, \quad T_n + a_{1-\alpha/2} \sqrt{m_2 h(p)/n}\right).$$

3.5.29 Suppose $\{X_1, X_2, \ldots, X_n\}$ is a random sample of size n from a uniform $U(\theta_1, \theta_2)$ distribution, $-\infty < \theta_1 < \theta_2 < \infty$. Obtain a CAN estimator of p-th population quantile and hence based on it obtain an asymptotic confidence interval for p-th population quantile.

Solution: Suppose $X \sim U(\theta_1, \theta_2)$ distribution. Then its distribution function $F(x, \theta_1, \theta_2)$ is given by

$$F(x, \theta_1, \theta_2) = \begin{cases} 0, & \text{if} \quad x < \theta_1 \\ \frac{x - \theta_1}{\theta_2 - \theta_1}, & \text{if } \theta_1 \le x < \theta_2 \\ 1, & \text{if} \quad x \ge \theta_2. \end{cases}$$

Hence the p-th population quantile $a_p(\underline{\theta}) = \theta_1 + p(\theta_2 - \theta_1)$, where $\underline{\theta} = (\theta_1, \theta_2)'$. From Theorem 3.3.3, the p-th sample quantile $X_{([np]+1)}$ is CAN for p-th population quantile $a_p(\underline{\theta})$ with approximate variance $p(1-p)(\theta_2 - \theta_1)^2/n$. In the solution of Exercise 2.8.35, we have shown that $(X_{(1)}, X_{(n)})'$ is consistent for $(\theta_1, \theta_2)'$. Thus, the consistent estimator for the approximate variance $p(1-p)(\theta_2 - \theta_1)^2/n$ is $p(1-p)(X_{(n)} - X_{(1)})^2/n$. Hence by Slutsky's theorem,

$$Q_n = \sqrt{\frac{n}{p(1-p)}} \frac{1}{(X_{(n)} - X_{(1)})} \left(X_{([np]+1)} - a_p(\underline{\theta}) \right) \xrightarrow{L} Z \sim N(0, 1).$$

Thus, Q_n is a pivotal quantity for large n. Given a confidence coefficient $(1 - \alpha)$, adopting the usual procedure we get $100(1 - \alpha)\%$ large sample confidence interval for $a_p(\underline{\theta})$ as

$$\left(X_{([np]+1)} - a_{1-\alpha/2} \frac{\sqrt{p(1-p)}(X_{(n)} - X_{(1)})}{\sqrt{n}}, \; X_{([np]+1)} + a_{1-\alpha/2} \frac{\sqrt{p(1-p)}(X_{(n)} - X_{(1)})}{\sqrt{n}} \right).$$

3.5.30 Suppose $\{X_1, X_2, \ldots, X_n\}$ is a random sample from a Poisson $Poi(\theta)$ distribution, $\theta > 0$. (i) Obtain a CAN estimator of the coefficient of variation $cv(\theta)$ of X when it is defined as $cv(\theta) = $ standard deviation/mean $= 1/\sqrt{\theta}$. (ii) If the estimator of $cv(\theta)$ is proposed as $\tilde{cv}(\theta) = S_n/\overline{X}_n$, where $\overline{X}_n$ is the sample mean and S_n is the sample standard deviation, examine if it is CAN for θ. Compare the two estimators.

Solution: A random variable $X \sim Poi(\theta)$ distribution, hence $E(X) = Var(X) = \theta < \infty$. By the WLLN and the CLT $\overline{X}_n$ is CAN for θ with approximate variance θ/n.
(i) The coefficient of variation $cv(\theta)$ of X is $1/\sqrt{\theta}$. To find its CAN estimator, suppose a function $g : (0, \infty) \to (0, \infty)$ is defined as $g(x) = 1/\sqrt{x}$, then g is a differentiable function with $g'(x) = -1/2x^{3/2} \neq 0$. Hence, by the delta method $g(\overline{X}_n) = 1/\sqrt{\overline{X}_n}$ is CAN for $g(\theta) = 1/\sqrt{\theta}$ with approximate variance $(\theta/n) \times (1/4\theta^3) = 1/4n\theta^2$.
(ii) Suppose the estimator of $cv(\theta)$ is proposed as $\tilde{cv}(\theta) = S_n/\overline{X}_n$. Consistency of $\tilde{cv}(\theta)$ follows from the consistency of S_n for standard deviation of X and consistency of $\overline{X}_n$ for mean of X and the invariance property of consistency. To examine if it CAN, we use the result established in Example 3.3.6 that $\underline{T}_n = (m_1', m_2)'$ is CAN for $\underline{\phi} = (\mu_1', \mu_2)'$ with approximate dispersion matrix Σ/n, where Σ is given by

$$\Sigma = \begin{pmatrix} \mu_2 & \mu_3 \\ \mu_3 & \mu_4 - \mu_2^2 \end{pmatrix}.$$

To find the elements of Σ, we note that the moment generating function $M_X(t)$ of X is $M_X(t) = \exp(\theta\{e^t - 1\})$ and hence the cumulant generating function

$C_X(t)$ of X is $C_X(t) = \theta(e^t - 1) = \theta(t + t^2/2! + t^3/3! + t^4/4! + \cdots)$. It is known that the i-th cumulant is the coefficient of $t^i/i!$. Thus, $k_i = \theta$ for all i. Using the relation between cumulants k_i and moments we have $\mu_1' = k_1 = \theta$, $\mu_i = k_i = \theta, i = 2, 3$ and $\mu_4 = k_4 + 3k_2^2 = \theta + 3\theta^2$. Thus the matrix Σ is given by

$$\Sigma = \begin{pmatrix} \theta & \theta \\ \theta & \theta + 2\theta^2 \end{pmatrix}.$$

To examine whether $\tilde{cv}(\theta) = S_n/\overline{X}_n$ is CAN, we further define a transformation $g : \mathbb{R}^2 \to \mathbb{R}$ such that

$$g(x_1, x_2) = \frac{\sqrt{x_2}}{x_1} \Rightarrow \frac{\partial}{\partial x_1} g(x_1, x_2) = -\frac{\sqrt{x_2}}{x_1^2} \ \& \ \frac{\partial}{\partial x_2} g(x_1, x_2) = \frac{1}{2x_1\sqrt{x_2}}.$$

These partial derivatives are continuous and hence g is a totally differentiable function. The gradient vector Δ evaluated at $(\theta, \theta)'$ is given by $\Delta = [-1/\theta^{3/2}, 1/2\theta^{3/2}]'$. Hence, by Theorem 3.3.4, we get that $g(\overline{X}_n, S_n^2)' = S_n/\overline{X}_n$ is CAN for $g(\theta, \theta) = 1/\sqrt{\theta}$ with approximate variance $\Delta D \Delta'/n$, where $\Delta D \Delta' = (1 + 2\theta)/4\theta^2$. It is to be noted that $(1 + 2\theta)/4\theta^2 > 1/4\theta^2 \ \forall \ \theta > 0$. Hence, $1/\sqrt{\overline{X}_n}$ is a better CAN estimator of $cv(\theta)$ than the CAN estimator $S_n/\overline{X}_n$ of $cv(\theta)$.

3.5.31 Suppose $\{X_1, X_2, \ldots, X_n\}$ is a random sample from a log-normal distribution with parameters μ and σ^2. Find a CAN estimator of $(\mu_1', \mu_2')'$. Hence obtain a CAN estimator for $\underline{\theta} = (\mu, \sigma^2)'$ and its approximate variance-covariance matrix.

Solution: Suppose a random variable X follows log-normal distribution with parameters μ and σ^2. Hence, $Y = \log X \sim N(\mu, \sigma^2)$ distribution. This relation is useful to find moments of $X = e^Y$ from the moment generating function $M_Y(\cdot)$ of Y. Thus,

$$E(X^r) = E(e^{rY}) = M_Y(r) = \exp\{\mu r + \sigma^2 r^2/2\}. \text{ Hence,}$$

$$\mu_1' = \exp\{\mu + \sigma^2/2\} \text{ and } \mu_2' = \exp\{2\mu + 2\sigma^2\}$$
$$\mu_3' = \exp\{3\mu + 9\sigma^2/2\} \text{ and } \mu_4' = \exp\{4\mu + 8\sigma^2\}$$
$$Var(X) = \exp\{2\mu + 2\sigma^2\} - \exp\{2\mu + \sigma^2\}$$
$$= \exp\{2\mu + \sigma^2\}(\exp\{\sigma^2\} - 1)$$
$$Cov(X, X^2) = \exp\{3\mu + 5\sigma^2/2\}(\exp\{2\sigma^2\} - 1)$$
$$\text{and } Var(X^2) = \exp\{4\mu + 4\sigma^2\}(\exp\{4\sigma^2\} - 1).$$

By the WLLN and by the equivalence of marginal and joint consistency, $\underline{T}_n = (m_1', m_2')'$ is consistent for $(\mu_1', \mu_2')'$. Further, by Theorem 3.3.2 it is CAN for $\underline{\phi} = (\mu_1', \mu_2')'$ with approximate dispersion matrix Σ/n, where the dispersion matrix Σ is given by

$$\Sigma = \begin{pmatrix} Var(X) & Cov(X, X^2) \\ Cov(X, X^2) & Var(X^2) \end{pmatrix}$$

and the elements of the matrix are as specified above. We now find a transformation $g : \mathbb{R}^2 \to \mathbb{R}^2$ such that $g(\underline{T}_n)$ is CAN for $g(\underline{\phi}) = \underline{\theta} = (\mu, \sigma^2)'$. Suppose $\underline{g} = (g_1, g_2)'$ is defined as $g_1(x_1, x_2) = 2 \log x_1 - (1/2) \log x_2$ and $g_2(x_1, x_2) = \log x_2 - 2 \log x_1$. Then

$$\frac{\partial}{\partial x_1} g_1(x_1, x_2) = \frac{2}{x_1}, \quad \frac{\partial}{\partial x_2} g_1(x_1, x_2) = -\frac{1}{2x_2}, \quad \frac{\partial}{\partial x_1} g_2(x_1, x_2) = -\frac{2}{x_1}$$

$$\& \quad \frac{\partial}{\partial x_2} g_2(x_1, x_2) = \frac{1}{x_2}.$$

These partial derivatives are continuous and hence g_1 and g_2 are totally differentiable functions. The matrix M of partial derivatives evaluated at $(\mu_1', \mu_2')'$ is given by

$$M = \begin{pmatrix} 2/\mu_1' & -1/(2\mu_2') \\ -2/\mu_1' & 1/\mu_2' \end{pmatrix}.$$

Hence, by Theorem 3.3.4,

$$\begin{aligned} \underline{g}(\underline{T}_n) &= (g_1(m_1', m_2'), g_2(m_1', m_2'))' \\ &= (2 \log m_1' - (1/2) \log m_2', \ \log m_2' - 2 \log m_1')' \end{aligned}$$

is CAN for $g(\mu_1', \mu_2') = \underline{\theta} = (\mu, \sigma^2)'$, with approximate dispersion matrix $M\Sigma M'/n$, where $M\Sigma M'$ is,

$$M\Sigma M' = \begin{pmatrix} 4e^{\sigma^2} - 2e^{2\sigma^2} + e^{4\sigma^2}/4 - 9/4 & -4e^{\sigma^2} + 3e^{2\sigma^2} - e^{4\sigma^2}/2 + 3/2 \\ -4e^{\sigma^2} + 3e^{2\sigma^2} - e^{4\sigma^2}/2 + 3/2 & 4e^{\sigma^2} - 4e^{2\sigma^2} + e^{4\sigma^2} - 1 \end{pmatrix}.$$

It is to be noted that $\sigma^2 > 0 \Rightarrow e^{\sigma^2} > 1$. Hence,

$$4e^{\sigma^2} - 2e^{2\sigma^2} + e^{4\sigma^2}/4 - 9/4 = (1/4)(e^{\sigma^2} - 1)(e^{3\sigma^2} + e^{2\sigma^2} + 9 - 7e^{\sigma^2}) > 0.$$

Similarly,

$$\begin{aligned} 4e^{\sigma^2} - 4e^{2\sigma^2} + e^{4\sigma^2} - 1 &= e^{4\sigma^2} - (2e^{\sigma^2} - 1)^2 \\ &= (e^{2\sigma^2} - 2e^{\sigma^2} + 1)(e^{2\sigma^2} + 2e^{\sigma^2} - 1) \\ &= (e^{\sigma^2} - 1)^2 (e^{2\sigma^2} + e^{\sigma^2} + e^{\sigma^2} - 1) > 0. \end{aligned}$$

3.5.32 Suppose $\{X_1, X_2, \ldots, X_n\}$ is a random sample from a gamma distribution with scale parameter α and shape parameter λ. Find a moment estimator of $(\alpha, \lambda)'$ and examine whether it is CAN. Find its approximate variance-covariance matrix.

Solution: Suppose X follows gamma distribution with scale parameter α and shape parameter λ. Then its probability density function is given by

$$f(x, \alpha, \lambda) = \frac{\alpha^\lambda}{\Gamma(\lambda)} e^{-\alpha x} x^{\lambda - 1}, \quad x > 0, \ \alpha > 0, \ \lambda > 0.$$

Further, $E(X) = \lambda/\alpha$ and $Var(X) = \lambda/\alpha^2$ which implies that $E(X^2) = (\lambda + \lambda^2)/\alpha^2$. Thus, the moment estimator of $(\alpha, \lambda)'$ is a solution of the system of equations given by

$$m_1' = E(X) = \frac{\lambda}{\alpha} \ \& \ m_2' = E(X^2) = \frac{\lambda + \lambda^2}{\alpha^2} \quad \Leftrightarrow \quad m_1' = E(X) = \frac{\lambda}{\alpha}$$

$$\& \ m_2 = Var(X) = \frac{\lambda}{\alpha^2}.$$

The solution is given by $\alpha = m_1'/m_2$ & $\lambda = m_1'^2/m_2$. Thus, the moment estimator of $(\alpha, \lambda)'$ is given by $(\hat{\alpha}_n, \hat{\lambda}_n)' = (m_1'/m_2, \ m_1'^2/m_2)'$. To examine if it is CAN, we use the result established in Example 3.3.6 and the delta method. We find the higher order moments of X from its moment generating function $M_X(t)$. It is given by $M_X(t) = (1 - t/\alpha)^{-\lambda}$, $t < \alpha$. Hence, its cumulant generating function $C_X(t)$ is $C_X(t) = \log M_X(t) = -\lambda \log(1 - t/\alpha)$. Expanding it we have,

$$C_X(t) = \lambda \left(\frac{t}{\alpha} + \frac{t^2}{2\alpha^2} + \frac{t^3}{3\alpha^3} + \frac{t^4}{4\alpha^4} + \cdots \right)$$

$$= \lambda \left(\frac{t}{\alpha} + \frac{t^2}{2!\alpha^2} + \frac{2t^3}{3!\alpha^3} + \frac{6t^4}{4!\alpha^4} + \cdots \right).$$

Using the relation between cumulants k_i and moments we have,

$$\mu_1' = k_1 = \frac{\lambda}{\alpha}, \quad \mu_2 = k_2 = \frac{\lambda}{\alpha^2}, \quad \mu_3 = k_3 = \frac{2\lambda}{\alpha^3}$$

$$\& \ \mu_4 = k_4 + 3k_2^2 = \frac{6\lambda}{\alpha^4} + \frac{3\lambda^2}{\alpha^4}.$$

Thus moments of X up to order 4 are finite and hence from Example 3.3.6, we have $\underline{T}_n = (m_1', m_2)'$ is CAN for $\underline{\phi} = (\mu_1', \mu_2)' = (\lambda/\alpha, \lambda/\alpha^2)'$ with approximate dispersion matrix Σ/n where Σ is

$$\Sigma = \begin{pmatrix} \mu_2 & \mu_3 \\ \mu_3 & \mu_4 - \mu_2^2 \end{pmatrix} = \begin{pmatrix} \frac{\lambda}{\alpha^2} & \frac{2\lambda}{\alpha^3} \\ \frac{2\lambda}{\alpha^3} & \frac{6\lambda+2\lambda^2}{\alpha^4} \end{pmatrix} = \frac{\lambda}{\alpha^2} \begin{pmatrix} 1 & \frac{2}{\alpha} \\ \frac{2}{\alpha} & \frac{6+2\lambda}{\alpha^2} \end{pmatrix}.$$

We now find a transformation $g : \mathbb{R}^2 \to \mathbb{R}^2$ such that $g(\underline{T}_n)$ is CAN for $g(\underline{\phi}) = (\alpha, \lambda)'$. Suppose $\underline{g} = (g_1, g_2)'$ is defined as $g_1(x_1, x_2) = x_1/x_2$ and $g_2(x_1, x_2) = x_1^2/x_2$. Then

$$\frac{\partial}{\partial x_1} g_1(x_1, x_2) = \frac{1}{x_2}, \quad \frac{\partial}{\partial x_2} g_1(x_1, x_2) = -\frac{x_1}{x_2^2}, \quad \frac{\partial}{\partial x_1} g_2(x_1, x_2) = \frac{2x_1}{x_2}$$

$$\& \quad \frac{\partial}{\partial x_2} g_2(x_1, x_2) = -\frac{x_1^2}{x_2^2}.$$

These partial derivatives are continuous and hence g_1 and g_2 are totally differentiable functions. The matrix M of partial derivatives evaluated at $(\mu_1', \mu_2)'$ is given by

$$M = \begin{pmatrix} \frac{\alpha^2}{\lambda} & -\frac{\alpha^3}{\lambda^2} \\ 2\alpha & -\alpha^2 \end{pmatrix} = \alpha \begin{pmatrix} \frac{\alpha}{\lambda} & -\frac{\alpha^2}{\lambda} \\ 2 & -\alpha \end{pmatrix}.$$

Hence, by Theorem 3.3.4, $g(\underline{T}_n) = (g_1(m_1', m_2), \quad g_2(m_1', m_2))' = (m_1'/m_2, \quad m_1'^2/m_2)'$ is CAN for $g(\mu_1', \mu_2) = (\alpha, \lambda)'$, with approximate dispersion matrix $M \Sigma M'/n$, where $M \Sigma M'$ is,

$$M \Sigma M' = \lambda \begin{pmatrix} \frac{3\alpha^2}{\lambda^2} + \frac{2\alpha^2}{\lambda} & \frac{2\alpha}{\lambda} + 2\alpha \\ \frac{2\alpha}{\lambda} + 2\alpha & 2\lambda + 2 \end{pmatrix}.$$

3.5.33 Suppose $X_{ij} = \mu_i + \epsilon_{ij}$ where $\{\epsilon_{ij}, \ i = 1, 2, 3, j = 1, 2, \ldots, n\}$ are independent and identically distributed random variables each having a normal $N(0, \sigma^2)$ distribution.
(i) Obtain a CAN estimator of $\theta = \mu_1 - 2\mu_2 + \mu_3$. (ii) Suppose $\{\epsilon_{ij}\}$ are independent and identically distributed random variables with $E(\epsilon_{ij}) = 0$ and $Var(\epsilon_{ij}) = \sigma^2$. Is the estimator of θ obtained in (i) still a CAN estimator of θ? Justify your answer.

Solution: (i) We have $X_{ij} = \mu_i + \epsilon_{ij} \sim N(\mu_i, \sigma^2), \ i = 1, 2, 3$. Further, $\overline{X}_{in} \sim N(\mu_i, \sigma^2/n), i = 1, 2, 3$ and $\overline{X}_{in}, \ i = 1, 2, 3$ are independent. Hence $T_n = \overline{X}_{1n} - 2\overline{X}_{2n} + \overline{X}_{3n}$ follows $N(\theta, 6\sigma^2/n)$ distribution. Thus, T_n is unbiased for θ and its variance converges to 0 as $n \to \infty$, hence T_n is consistent for θ. Further for each n, $T_n \sim N(\theta, 6\sigma^2/n)$ which implies that its asymptotic distribution is also normal. Thus, T_n is CAN for θ with approximate variance $6\sigma^2/n$.

(ii) It is given that $\{\epsilon_{ij}, \ i = 1, 2, 3, \ j = 1, 2, \ldots, n\}$ are independent and identically distributed random variables. Hence, $\{X_{ij}, \ i = 1, 2, 3, \ j = 1, 2, \ldots, n\}$ are also independent random variables with $E(X_{ij}) = \mu_i$ and $Var(X_{ij}) = \sigma^2$. By the WLLN, $\overline{X}_{in} \overset{P}{\to} \mu_i, \ i = 1, 2, 3$. Hence, $T_n \overset{P}{\to} \theta$, convergence in probability being closed under arithmetic operations. To examine whether T_n is CAN for θ, suppose a random vector $\underline{Z}_j$ is defined as $\underline{Z}_j = (X_{1j}, X_{2j}, X_{3j})'$, $j = 1, 2, \ldots, n$. Observe that $\{\underline{Z}_1, \underline{Z}_2, \ldots, \underline{Z}_n\}$ are independent and identically distributed random vectors with mean vector $\underline{\mu} = (\mu_1, \mu_2, \mu_3)'$ and dispersion matrix $\Sigma = \sigma^2 I_3$, which is positive definite. Hence by the multivariate CLT,

$$\sqrt{n}(\overline{\underline{Z}}_n - \underline{\mu}) \overset{L}{\to} \underline{Z}_1 \sim N_3(\underline{0}, \sigma^2 I_3), \quad \text{where} \quad \overline{\underline{Z}}_n = (\overline{X}_{1n}, \overline{X}_{2n}, \overline{X}_{3n})'.$$

We further define a transformation $g : \mathbb{R}^3 \to \mathbb{R}$ such that $g(x_1, x_2, x_3) = x_1 - 2x_2 + x_3$. Then $\frac{\partial}{\partial x_1} g(x_1, x_2, x_3) = 1$, $\frac{\partial}{\partial x_2} g(x_1, x_2, x_3) = -2$ and $\frac{\partial}{\partial x_3} g(x_1, x_2, x_3) = 1$. These partial derivatives are continuous and hence g is a totally differentiable function. The gradient vector Δ evaluated at μ is given by $\delta = [1, -2, 1]'$. Hence, by Theorem 3.3.4, it follows that $g(\overline{X}_{1n}, \overline{X}_{2n}, \overline{X}_{3n})' = T_n$ is CAN for $g(\mu_1, \mu_2, \mu_3)' = \theta$ with approximate variance $\Delta \Sigma \Delta'/n$, where $\Delta \Sigma \Delta' = 6\sigma^2$. Thus, estimator T_n of θ obtained in (i) is CAN estimator for θ, even if the assumption of normality is relaxed.

3.5.34 Suppose $\underline{X} \equiv \{X_1, X_2, \ldots, X_n\}$ is a random sample from a uniform $U(\theta_1, \theta_2)$ distribution, where $\theta_1 < \theta_2 \in \mathbb{R}$. (i) Find the maximum likelihood estimator of $(\theta_1, \theta_2)'$. Show that it is consistent but not CAN. (ii) Find a CAN estimator of $(\theta_1 + \theta_2)/2$.

Solution: (i) If $X \sim U(\theta_1, \theta_2)$, then its probability density function $f(x, \theta_1, \theta_2)$ and the distribution function $F(x, \theta_1, \theta_2)$ are as follows:

$$f(x, \theta_1, \theta_2) = \begin{cases} \frac{1}{\theta_2 - \theta_1}, & \text{if} \quad \theta_1 < x < \theta_2 \\ 0, & \text{otherwise.} \end{cases}$$

$$F(x, \theta_1, \theta_2) = \begin{cases} 0, & \text{if} \quad x < \theta_1 \\ \frac{x - \theta_1}{\theta_2 - \theta_1}, & \text{if } \theta_1 \leq x < \theta_2 \\ 1, & \text{if} \quad x \geq \theta_2. \end{cases}$$

Corresponding to a random sample $\underline{X}$, the likelihood of θ_1, θ_2 is given by

$$L_n(\theta_1, \theta_2|\underline{X}) = \prod_{i=1}^{n} \frac{1}{\theta_2 - \theta_1} = \frac{1}{(\theta_2 - \theta_1)^n}, \quad \theta_1 \leq X_i \leq \theta_2,$$
$$\forall \ i \ \Leftrightarrow \ X_{(1)} \geq \theta_1, \ X_{(n)} \leq \theta_2.$$

Thus, the likelihood attains maximum when $(\theta_2 - \theta_1)$ is minimum. Now, $\theta_2 \geq X_{(n)}$ and $-\theta_1 \geq -X_{(1)}$. Thus, $(\theta_2 - \theta_1) \geq X_{(n)} - X_{(1)}$ implying that $(\theta_2 - \theta_1)$ is minimum, given data $\underline{X}$, when $\theta_1 = X_{(1)}$ and $\theta_2 = X_{(n)}$. Hence the maximum likelihood estimator $\hat{\theta}_{1n}$ of θ_1 and $\hat{\theta}_{2n}$ of θ_2 are given by $\hat{\theta}_{1n} = X_{(1)}$ and $\hat{\theta}_{2n} = X_{(n)}$. To verify the consistency of these estimators of θ, we define $Y = (X - \theta_1)/(\theta_2 - \theta_1)$ Then $Y \sim U(0, 1)$ and from Example 2.5.2 we have,

$$Y_{(1)} \xrightarrow{P} 0 \ \& \ Y_{(n)} \xrightarrow{P} 1 \ \Rightarrow \ X_{(1)} \xrightarrow{P} \theta_1 \ \& \ X_{(n)} \xrightarrow{P} \theta_2.$$

To derive the asymptotic distribution of $X_{(1)}$ with suitable norming, we define $Y_n = n(X_{(1)} - \theta_1)$ and derive its distribution function $G_{Y_n}(y)$ for $y \in \mathbb{R}$. Since $X_{(1)} \geq \theta_1$, $Y_n \geq 0$, hence for $y < 0$, $G_{Y_n}(y) = 0$. Suppose $y \geq 0$, then

$$G_{Y_n}(y) = P[n(X_{(1)} - \theta_1) \leq y] = P_\theta[X_{(1)} \leq \theta_1 + y/n] = F_{X_{(1)}}(\theta_1 + y/n).$$

Thus,

$$G_{Y_n}(y) \ = \ \begin{cases} 1 - (1-0)^n = 0, & \text{if} & \theta_1 + y/n < \theta_1 \ \Leftrightarrow \ y < 0 \\ 1 - \left(\frac{\theta_2 - \theta_1 - y/n}{\theta_2 - \theta_1}\right)^n, & \text{if } \theta_1 \leq \theta_1 + y/n < \theta_2 \ \Leftrightarrow \ 0 \leq y < n(\theta_2 - \theta_1) \\ 1 - [1-1] = 1, & \text{if} & y \geq n(\theta_2 - \theta_1). \end{cases}$$

Hence,

$$G_{Y_n}(y) \ \rightarrow \ \begin{cases} 0, & \text{if } y < 0 \\ 1 - e^{-\frac{y}{\theta_2 - \theta_1}}, & \text{if } y \geq 0. \end{cases}$$

Thus, the asymptotic distribution of $Y_n = n(X_{(1)} - \theta_1)$ is exponential with location parameter 0 and scale parameter $1/(\theta_2 - \theta_1)$. Thus, with norming factor n, the asymptotic distribution of $X_{(1)}$ is not normal. Proceeding on similar lines as in Example 3.2.1, it follows that there exists no sequence $\{a_n, n \geq 1\}$ of real numbers tending to ∞ as $n \to \infty$, such that the asymptotic distribution of $a_n(X_{(1)} - \theta_1)$ is normal. Hence, we conclude that $X_{(1)}$ is not CAN for θ_1. To derive the asymptotic distribution of $X_{(n)}$ with suitable norming, we define $Y_n = n(\theta_2 - X_{(n)})$ and derive its distribution function $G_{Y_n}(y)$ for $y \in \mathbb{R}$. Since $X_{(n)} \leq \theta_2$, $Y_n \geq 0$, hence for $y < 0$, $G_{Y_n}(y) = 0$. Suppose $y \geq 0$, then

$$G_{Y_n}(y) = P[n(\theta_2 - X_{(n)}) \leq y] = P[X_{(n)} \geq \theta_2 - y/n]$$
$$= 1 - F_{X_{(n)}}(\theta_2 - y/n), \quad \text{hence}$$

$$G_{Y_n}(y) \ = \ \begin{cases} 1 - 0, & \text{if } \theta_2 - y/n < \theta_1 \ \Leftrightarrow \ y \geq n(\theta_2 - \theta_1) \\ 1 - (1 - \frac{y}{n(\theta_2 - \theta_1)})^n, & \text{if} & 0 < y \leq n(\theta_2 - \theta_1) \\ 1 - 1, & \text{if} & y \leq 0. \end{cases}$$

As $n \to \infty$,

$$G_{Y_n}(y) \to \begin{cases} 0, & \text{if } y \leq 0 \\ 1 - e^{-\frac{y}{\theta_2 - \theta_1}}, & \text{if } y \geq 0. \end{cases}$$

Thus, $Y_n = n(\theta_2 - X_{(n)})$ converges in distribution to an exponential distribution with location parameter 0 and scale parameter $1/(\theta_2 - \theta_1)$. Thus, with norming factor n, the asymptotic distribution of $X_{(n)}$ is not normal. Proceeding on similar lines as in Example 3.2.1, it follows that there exists no sequence $\{a_n, n \geq 1\}$ of real numbers tending to ∞ as $n \to \infty$, such that the asymptotic distribution of $a_n(X_{(n)} - \theta)$ is normal. Hence, it is proved that $X_{(n)}$ is not CAN for θ_2. Thus, $(X_{(1)}, X_{(n)})'$ is not CAN for $(\theta_1, \theta_2)'$. It is to be noted that it is enough to prove that one of the two $X_{(1)}$ and $X_{(n)}$ is not CAN to conclude that $(X_{(1)}, X_{(n)})'$ is not CAN for $(\theta_1, \theta_2)'$.

(ii) Since $X \sim U(\theta_1, \theta_2)$, $E(X) = (\theta_1 + \theta_2)/2$ and $Var(X) = (\theta_2 - \theta_1)^2/12 < \infty$. Hence by the WLLN and by the CLT,

$$\overline{X}_n \xrightarrow{P} \frac{\theta_1 + \theta_2}{2} \quad \& \quad \sqrt{n}\left(\overline{X}_n - \frac{\theta_1 + \theta_2}{2}\right) \xrightarrow{L} Z_1' \sim N\left(0, \frac{(\theta_2 - \theta_1)^2}{12}\right).$$

Hence, $\overline{X}_n$ is CAN for $(\theta_1 + \theta_2)/2$ with approximate variance $(\theta_2 - \theta_1)^2/12n$.

3.5.35 Suppose $\{X_1, X_2, \ldots, X_n\}$ are independent and identically distributed random variables with finite fourth order moment. Suppose $E(X_1) = \mu$ and $Var(X_1) = \sigma^2$. Find a CAN estimator of the coefficient of variation σ/μ.

Solution: From Example 3.3.6, we know that $(m_1', m_2)'$ is CAN for $(\mu_1', \mu_2)'$ with approximate dispersion matrix Σ/n where Σ is

$$\Sigma = \begin{pmatrix} \mu_2 & \mu_3 \\ \mu_3 & \mu_4 - \mu_2^2 \end{pmatrix}.$$

To find a CAN estimator of σ/μ, we note that $\sigma/\mu = \sqrt{\mu_2}/\mu_1' = g(\mu_1', \mu_2)$ say, where $g : \mathbb{R}^2 \to \mathbb{R}$ is a function defined as $g(x_1, x_2) = \sqrt{x_2}/x_1$. Then

$$\frac{\partial}{\partial x_1} g(x_1, x_2) = -\frac{x_2}{x_1^2} \quad \& \quad \frac{\partial}{\partial x_2} g(x_1, x_2) = \frac{1}{2x_1\sqrt{x_2}}.$$

These partial derivatives are continuous and hence g is a totally differentiable function. The gradient vector Δ evaluated at $(\mu_1', \mu_2)'$ is $\left[-\sqrt{\mu_2}/(\mu_1')^2, (1/2)\mu_1'\sqrt{\mu_2}\right]' = \left[-\sigma/\mu^2, (1/2)\mu\sigma\right]'$. Hence, by Theorem 3.3.4, $g(m_1', m_2)' = \sqrt{m_2}/m_1'$ is CAN for $g(\mu_1', \mu_2)' = \sigma/\mu$ with approximate variance $\Delta D \Delta'/n$, where $\Delta D \Delta' = \sigma^4/\mu^4 - \mu_3/\mu_1^3 + (\mu_4 - \sigma^4)/4\mu^2\sigma^2.$

3.5.36 Suppose $\{X_1, X_2, \ldots, X_n\}$ is a random sample from a distribution of X with distribution function F. Suppose random variables Z_1 and Z_2 are defined as follows. For $a < b$,

$$Z_1 = \begin{cases} 1, \text{ if } X \leq a \\ 0, \text{ if } X > a. \end{cases}$$

$$Z_2 = \begin{cases} 1, \text{ if } X \leq b \\ 0, \text{ if } X > b. \end{cases}$$

Show that for large n, the distribution of $(\overline{Z}_{1n}, \overline{Z}_{2n})'$ is bivariate normal. Hence obtain a CAN estimator for $(F(a), F(b))'$.

Solution: From the definition of Z_1 and Z_2 we have,

$$E(Z_1) = F(a), \ Var(Z_1) = F(a)(1 - F(a))$$
$$\& \ \ E(Z_2) = F(b), \ Var(Z_2) = F(b)(1 - F(b)).$$

To find $Cov(Z_1, Z_2)$, observe that for $a < b$,

$$E(Z_1 Z_2) = P[Z_1 = 1, Z_2 = 1] = P[X \leq a, X \leq b] = P[X \leq a] = F(a)$$

and hence $Cov(Z_1, Z_2) = F(a) - F(a)F(b) = F(a)(1 - F(b))$. Thus, if we define $\underline{Z} = (Z_1, Z_2)'$ then $E(\underline{Z}) = (F(a), F(b))'$ and dispersion matrix Σ of $\underline{Z}$ is given by

$$\Sigma = \begin{pmatrix} F(a)(1 - F(a)) & F(a)(1 - F(b)) \\ F(a)(1 - F(b)) & F(b)(1 - F(b)) \end{pmatrix}$$

Observe that the first principal minor of Σ is $F(a)(1 - F(a))$ and it is positive. The second principal minor which is the determinant of Σ is $F(a)(1 - F(b))(F(b) - F(a))$ and it is also positive. Hence Σ is a positive definite matrix. Now a random sample $\{X_1, X_2, \ldots, X_n\}$ from the distribution of X gives a random sample $\{\underline{Z}_1, \underline{Z}_2, \ldots, \underline{Z}_n\}$ from the distribution of $\underline{Z}$. Hence by the WLLN and by the multivariate CLT,

$$\underline{Z}_n = (\overline{Z}_{1n}, \overline{Z}_{2n})' \xrightarrow{P} (F(a), F(b))'$$
$$\& \ \sqrt{n}(\underline{Z}_n - (F(a), F(b))') \xrightarrow{L} \underline{Y} \sim N_2(\underline{0}, \Sigma).$$

Thus, we have proved that for large n the distribution of $(\overline{Z}_{1n}, \overline{Z}_{2n})'$ is bivariate normal and $(\overline{Z}_{1n}, \overline{Z}_{2n})'$ is CAN for $(F(a), F(b))'$ with approximate dispersion matrix Σ/n.

3.5.37 Suppose $\underline{X} \equiv \{X_1, X_2, \ldots, X_n\}$ is a random sample from the following distributions (i) Normal $N(\mu, \sigma^2)$ and (ii) exponential distribution with location parameter θ and scale parameter λ. Find the maximum likelihood estimators of the parameters using stepwise maximization procedure and examine whether these are CAN.

Solution: (i) Suppose $X \sim N(\mu, \sigma^2)$. Corresponding to a random sample $\underline{X}$ of size n from normal $N(\mu, \sigma^2)$ distribution, the likelihood of $\underline{\theta} = (\mu, \sigma^2)'$ is given by

$$L_n(\underline{\theta}|\underline{X}) = \prod_{i=1}^{n} \frac{1}{\sqrt{2\pi}\sigma} \exp\{-\frac{1}{2\sigma^2}(X_i - \mu)^2\}$$

$$\Leftrightarrow \log L_n(\underline{\theta}|\underline{X}) = c - \frac{n}{2}\log\sigma^2 - \frac{1}{2\sigma^2}\sum_{i=1}^{n}(X_i - \mu)^2,$$

where c is a constant free from $\underline{\theta}$. Suppose σ^2 is fixed at σ_0^2 and we find maximum of the likelihood with respect to the variations in μ. Thus,

$$\log L_n(\underline{\theta}|\underline{X}) = c - \frac{n}{2}\log\sigma_0^2 - \frac{1}{2\sigma_0^2}\sum_{i=1}^{n}(X_i - \mu)^2.$$

By the usual method, log-likelihood is maximum when $\mu = \overline{X}_n$, hence $\hat{\mu}_n(\sigma_0^2) = \overline{X}_n$. It is to be noted that $\hat{\mu}_n(\sigma_0^2)$ does not depend on the fixed value of σ^2. Now we consider a function $h(\sigma^2) = c - n\log\sigma^2/2 - \sum_{i=1}^{n}(X_i - \overline{X}_n)^2/2\sigma^2$. It is a differentiable function of σ^2 and is maximum when $\sigma^2 = \sum_{i=1}^{n}(X_i - \overline{X}_n)^2/n$. Hence the maximum likelihood estimators of μ and σ^2 are given by $\hat{\mu}_n = \overline{X}_n$ and $\hat{\sigma}_n^2 = \sum_{i=1}^{n}(X_i - \overline{X}_n)^2/n$. In Example 3.3.2, these are shown to be CAN.
(ii) Suppose X follows an exponential distribution with location parameter θ and scale parameter λ. Hence its probability density function is given by

$$f(x, \theta, \lambda) = (1/\lambda)\exp\{-(x - \theta)/\lambda\}, \quad x \geq \theta, \quad \theta \in \mathbb{R}, \quad \lambda > 0.$$

Corresponding to a random sample $\underline{X} \equiv \{X_1, X_2, \ldots, X_n\}$ from this distribution, the likelihood of (θ, λ) is given by

$$L_n(\theta, \lambda|\underline{X}) = \prod_{i=1}^{n}(1/\lambda)\exp\{-(X_i - \theta)/\lambda\}, \quad X_i \geq \theta \ \forall \ i = 1, 2, \ldots, n,$$

which can be expressed as

$$L_n(\theta, \lambda|\underline{X}) = (1/\lambda)^n \exp\left\{-\sum_{i=1}^{n}X_i/\lambda + n\theta/\lambda\right\}, \quad X_{(1)} \geq \theta.$$

Suppose λ is fixed at λ_0, then the likelihood is maximized with respect to variations in θ when θ is maximum given the data. The maximum value of θ given the data is $X_{(1)}$. Note that it does not depend on the fixed value of λ. Now we consider a function $h(\lambda) = (1/\lambda)^n \exp\left\{-\sum_{i=1}^{n}(X_i - X_{(1)})/\lambda\right\}$. It is a differentiable function of λ and $h(\lambda)$ is maximum when $\lambda = \sum_{i=1}^{n}(X_i - X_{(1)})/n$. Hence the maximum likelihood estimators of θ and λ are given by $\hat{\theta}_n = X_{(1)}$ and $\hat{\lambda}_n = \sum_{i=1}^{n}(X_i - X_{(1)})/n$. To verify the consistency of $X_{(1)}$ as an estimator of θ, we find the distribution function of $X_{(1)}$, it is given by $F_{X_{(1)}}(x) = 1 - [1 - F_X(x)]^n$, $x \in \mathbb{R}$. The distribution function $F_X(x)$ is given by

$$F_X(x) \;=\; \begin{cases} 0, & \text{if } x < \theta \\ 1 - \exp\{-\frac{1}{\lambda}(x - \theta)\}, & \text{if } x \geq \theta. \end{cases}$$

Hence, the distribution function of $X_{(1)}$ is given by

$$F_{X_{(1)}}(x) \;=\; \begin{cases} 0, & \text{if } x < \theta \\ 1 - \exp\{-\frac{n}{\lambda}(x - \theta)\}, & \text{if } x \geq \theta. \end{cases}$$

Thus, the distribution of $X_{(1)}$ is again exponential with location parameter θ and scale parameter n/λ. Hence $E(X_{(1)}) = \theta + \lambda/n$ which implies that the bias of $X_{(1)}$ as an estimator of θ is λ/n and it converges to 0 as $n \to \infty$. Further, $Var(X_{(1)}) = \lambda^2/n^2 \to 0$ as $n \to \infty$. Hence, $X_{(1)}$ is consistent for θ. To derive the asymptotic distribution of $X_{(1)}$, with suitable norming, we define $Y_n = n(X_{(1)} - \theta)$ and derive its distribution function $G_{Y_n}(y)$ for $y \in \mathbb{R}$. Since $X_{(1)} \geq \theta$, $Y_n \geq 0$, hence for $y < 0$, $G_{Y_n}(y) = 0$. Suppose $y \geq 0$, then

$$\begin{aligned} G_{Y_n}(y) &= P_\theta[n(X_{(1)} - \theta) \leq y] \\ &= P_\theta[X_{(1)} \leq \theta + y/n] = F_{X_{(1)}}(\theta + y/n) \\ &= 1 - \exp\{-\frac{n}{\lambda}(\theta + y/n - \theta)\} = 1 - \exp\{-y/\lambda\}, \quad y \geq 0. \end{aligned}$$

Thus, for each n, Y_n follows, an exponential distribution with location parameter 0 and scale parameter $1/\lambda$ and hence the asymptotic distribution of $Y_n = n(X_{(1)} - \theta)$ is the same. Thus, with norming factor n, the asymptotic distribution of $X_{(1)}$ is not normal. Proceeding as in Example 3.2.1, we can show that there exists no sequence $\{a_n, n \geq 1\}$ of real numbers tending to ∞ as $n \to \infty$ such that the asymptotic distribution of $a_n(X_{(1)} - \theta)$ is normal, hence we claim that $X_{(1)}$ is consistent but not CAN for θ. Now to examine whether $\hat{\lambda}_n = \sum_{i=1}^{n}(X_i - X_{(1)})/n$ is CAN for λ, observe that $\hat{\lambda}_n$ can be expressed as

$$\hat{\lambda}_n = \frac{1}{n}\left(\sum_{i=1}^{n} X_i - nX_{(1)}\right) = \frac{1}{n}\left(\sum_{i=1}^{n} X_{(i)} - nX_{(1)}\right)$$

$$= \frac{1}{n} \sum_{i=1}^{n} (X_{(i)} - X_{(1)}) = \frac{1}{n} \sum_{i=2}^{n} (X_{(i)} - X_{(1)}).$$

We define random variables Y_i as $Y_i = (n - i + 1)(X_{(i)} - X_{(i-1)})$, $i = 2, 3, \ldots, n$. Then

$$\sum_{i=2}^{n} Y_i = \sum_{i=2}^{n} X_{(i)} - (n-1)X_{(1)} = \sum_{i=2}^{n} (X_{(i)} - X_{(1)}) = n\hat{\lambda}_n.$$

It can be proved that $\{Y_2, Y_3, \ldots, Y_n\}$ are independent and identically distributed random variables each having an exponential distribution with location parameter 0 and scale parameter $1/\lambda$. Thus, $E(Y_2) = \lambda$ and $Var(Y_2) = \lambda^2 < \infty$. Hence by the WLLN,

$$\frac{1}{n-1} \sum_{i=2}^{n} Y_i \xrightarrow{P_{\theta,\lambda}} \lambda \implies \hat{\lambda}_n = \frac{1}{n} \sum_{i=2}^{n} Y_i = \frac{n-1}{n} \frac{1}{n-1} \sum_{i=2}^{n} Y_i \xrightarrow{P} \lambda, \ \forall \ \theta \ \& \ \lambda$$

which proves that $\hat{\lambda}_n$ is consistent. By the CLT

$$\sqrt{n-1} \left(\frac{1}{n-1} \sum_{i=2}^{n} Y_i - \lambda \right) \xrightarrow{L} Z_1 \sim N(0, \lambda^2).$$

From Slutsky's theorem,

$$\sqrt{n} \left(\frac{1}{n-1} \sum_{i=2}^{n} Y_i - \lambda \right) = \frac{\sqrt{n}}{\sqrt{n-1}} \sqrt{n-1} \left(\frac{1}{n-1} \sum_{i=2}^{n} Y_i - \lambda \right)$$

$$\xrightarrow{L} Z_1 \sim N(0, \lambda^2).$$

Now,

$$\sqrt{n}(\hat{\lambda}_n - \lambda) - \sqrt{n} \left(\frac{1}{n-1} \sum_{i=2}^{n} Y_i - \lambda \right) = \sqrt{n} \sum_{i=2}^{n} Y_i \left(\frac{1}{n} - \frac{1}{n-1} \right)$$

$$= \frac{-1}{\sqrt{n}} \frac{1}{n-1} \sum_{i=2}^{n} Y_i$$

converges to 0 in probability, as $\frac{1}{n-1} \sum_{i=2}^{n} Y_i \xrightarrow{P} \lambda$ and hence is bounded in probability. Hence, the limit distribution of $\sqrt{n}(\hat{\lambda}_n - \lambda)$ and $\sqrt{n}(\frac{1}{n-1} \sum_{i=2}^{n} Y_i - \lambda)$ is the same. Thus, $\sqrt{n}(\hat{\lambda}_n - \lambda) \xrightarrow{L} Z_1 \sim N(0, \lambda^2)$. Hence, $\hat{\lambda}_n$ is CAN for λ.

7.3   Chapter 4

4.6.1 Suppose $\underline{X} \equiv \{X_1, X_2, \ldots, X_n\}$ is a random sample from a distribution of X with probability density function $f(x, \theta) = \theta/x^{\theta+1}$ $x > 1$, $\theta > 0$. (i) Examine whether the distribution belongs to a one-parameter exponential family. (ii) On the basis of a random sample of size n from the distribution of X, find the moment estimator of θ based on a sufficient statistic and the maximum likelihood estimator of θ. (iii) Examine whether these are CAN estimators of θ. (iv) Obtain the CAN estimator for $P[X \geq 2]$.

Solution: (i) The probability density function $f(x, \theta)$ of X can be expressed as

$$f(x, \theta) = \frac{\theta}{x^{\theta+1}} = \exp\left\{\log\left(\frac{\theta}{x^{\theta+1}}\right)\right\} = \exp\{\log\theta - (\theta + 1)\log x\}$$
$$= \exp\{U(\theta)K(x) + V(\theta) + W(x)\},$$

where $U(\theta) = -\theta$, $K(x) = \log x$, $V(\theta) = \log\theta$ and $W(x) = -\log x$. Thus, (1) the probability law of X is expressible in the form required in a one-parameter exponential family, (2) support of the probability density function is $(1, \infty)$ and it is free from θ, (3) the parameter space is $(0, \infty)$ which is an open set, (4) $U'(\theta) = -1 \neq 0$ and (5) $K(x)$ and 1 are linearly independent because in the identity $a + b\log x = 0$ if $x = 1$, then $a = 0$, if further in the identity $b\log x = 0$, if $x = 2$ then $b = 0$. Thus, all the requirements of a one-parameter exponential family are satisfied and hence the distribution of X belongs to a one-parameter exponential family.
(ii) To find the maximum likelihood estimator of θ, the log-likelihood of θ corresponding to the data $\underline{X}$ is given by

$$\log L_n(\theta|\underline{X}) = n\log\theta - (\theta + 1)\sum_{i=1}^{n}\log X_i$$

$$\Rightarrow \quad \frac{\partial}{\partial\theta}\log L_n(\theta|\underline{X}) = \frac{n}{\theta} - \sum_{i=1}^{n}\log X_i.$$

Hence, the solution $\hat{\theta}_n$ of the likelihood equation is given by $\hat{\theta}_n = n/\sum_{i=1}^{n}\log X_i$. The second derivative $\frac{\partial^2}{\partial\theta^2}\log L_n(\theta|\underline{X}) = -n/\theta^2 < 0$, $\forall \theta > 0$. Hence, $\hat{\theta}_n$ is the maximum likelihood estimator of θ. From the likelihood it is clear that $\sum_{i=1}^{n}\log X_i$ is a sufficient statistic. Hence the moment estimator of θ based on the sufficient statistic is solution of the equation $\sum_{i=1}^{n}\log X_i/n = E(\log X)$. To find $E(\log X)$, we use the identity $E\left(\frac{\partial}{\partial\theta}\log f(X, \theta)\right) = 0$. Hence, $E(\log X) = 1/\theta$. Hence the moment estimator of θ based on the sufficient statistic is $\tilde{\theta}_n = n/\sum_{i=1}^{n}\log X_i$.

(iii) Since the distribution of X belongs to a one-parameter exponential family, the moment estimator of θ based on the sufficient statistic and the maximum likelihood estimator of θ are the same. These are CAN with approximate variance $1/nI(\theta)$. Now,

$$nI(\theta) = E_\theta \left(-\frac{\partial^2}{\partial\theta^2} \log L_n(\theta|\underline{X}) \right) = \frac{n}{\theta^2}.$$

Thus, $\hat{\theta}_n = \tilde{\theta}_n$ is CAN for θ with approximate variance θ^2/n.

(iv) From the probability density function of X,

$$P[X \geq 2] = \int_2^\infty \frac{\theta}{x^{\theta+1}} dx = 2^{-\theta} = g(\theta), \quad \text{say.}$$

It is clear that g is a differentiable function and $g'(\theta) = -(1/2^\theta)^2 2^\theta \log 2 = -2^{-\theta} \log 2 \neq 0$, $\forall\ \theta > 0$. Hence by the delta method, $2^{-\hat{\theta}_n}$ is CAN for $2^{-\theta}$ with approximate variance $(\theta^2/n) \times (2^{-\theta} \log 2)^2 = \theta^2(\log 2)^2/4^\theta n$.

4.6.2 Suppose $\underline{X} \equiv \{X_1, X_2, \ldots, X_n\}$ is a random sample from a binomial $B(m, \theta)$ distribution, truncated at 0, $0 < \theta < 1$ and m is a known positive integer. Examine whether the distribution belongs to a one-parameter exponential family. Find the moment estimator of θ based on a sufficient statistics and the maximum likelihood estimator of θ. Examine whether the two are the same and whether these are CAN. Find their approximate variances.

Solution: The distribution of X is binomial $B(m, \theta)$, truncated at 0, hence its probability mass function is given by

$$f(x, \theta) = P_\theta[X = x] = \binom{m}{x} \frac{\theta^x(1-\theta)^{m-x}}{(1-(1-\theta)^m)} \Rightarrow \log f(x, \theta)$$
$$= U(\theta)K(x) + V(\theta) + W(x),$$

where $x = 1, 2, \ldots, m$ and $U(\theta) = \log\theta - \log(1-\theta)$, $V(\theta) = m\log(1-\theta) - \log(1-(1-\theta)^m)$, $K(x) = x$ and $W(x) = \binom{m}{x}$. Further, U and V are differentiable functions of θ and can be differentiated any number of times and $U'(\theta) = 1/\theta(1-\theta) \neq 0$. 1 and $K(x)$ are linearly independent. The parameter space is an open set and support of X is free from θ. Thus, binomial $B(m, \theta)$ distribution, truncated at 0, is a member of a one-parameter exponential family. Hence by Theorem 4.2.1, the moment estimator of θ based on a sufficient statistics is the same as the maximum likelihood estimator of θ and it is CAN with approximate variance $1/nI(\theta)$. Corresponding to a random sample of size n from the distribution of X, $\sum_{i=1}^n K(X_i) = \sum_{i=1}^n X_i$ is a sufficient statistics. The moment estimator of θ based on the sufficient statistics is then given by the equation, $\overline{X}_n = E(X) = m\theta(1-(1-\theta)^m)^{-1} = \eta(\theta)$ say. It is to be noted that

$$\eta'(\theta) = \frac{1 - (1-\theta)^m - m\theta(1-\theta)^{m-1}}{(1-(1-\theta)^m)^2}$$

$$= \frac{P_\theta[Y > 1]}{(1-(1-\theta)^m)^2} > 0, \quad \forall\ \theta \in (0, 1)$$

where $Y \sim B(m, \theta)$ distribution with support $\{0, 1, \ldots, m\}$. Hence by the inverse function theorem, η^{-1} exists and by using numerical methods, which are discussed in Sect. 4.4, we get the moment estimator $\tilde{\theta}_n$ of θ based on the sufficient statistics as $\tilde{\theta}_n = \eta^{-1}(\overline{X}_n)$. To find the maximum likelihood estimator, the likelihood and the log-likelihood of θ is given by

$$L_n(\theta|\underline{X}) = \prod_{i=1}^{n} \binom{m}{X_i} \frac{\theta^{X_i}(1-\theta)^{m-X_i}}{(1-(1-\theta)^m)}$$

$$\&\ \ \log L_n(\theta|\underline{X}) = c + \log(\theta/(1-\theta)) \sum_{i=1}^{n} X_i$$

$$+ mn\log(1-\theta) - n\log(1-(1-\theta)^m)$$

where c is a constant free from θ. Thus, the likelihood equation is

$$\frac{\sum_{i=1}^{n} X_i}{\theta(1-\theta)} - \frac{mn}{1-\theta} - \frac{mn(1-\theta)^{m-1}}{1-(1-\theta)^m} = 0 \ \Leftrightarrow\ \overline{X}_n = \frac{m\theta}{(1-(1-\theta)^m)}.$$

Thus, the maximum likelihood estimator $\hat{\theta}_n$ of θ is given by $\hat{\theta}_n = \eta^{-1}(\overline{X}_n)$, which is the same as the moment estimator based on the sufficient statistics. The information function $I(\theta)$ is given by

$$I(\theta) = U'(\theta)\eta'(\theta) = \frac{1 - (1-\theta)^m - m\theta(1-\theta)^{m-1}}{\theta(1-\theta)(1-(1-\theta)^m)^2}.$$

Thus, $\tilde{\theta}_n = \hat{\theta}_n$ is CAN for θ with approximate variance $1/nI(\theta)$.

4.6.3 Suppose $\{X_1, X_2, \ldots, X_n\}$ is a random sample from a distribution of X with probability density function

(i) $f(x, \theta) = (x/\theta) \exp\{-x^2/2\theta\}, \quad x > 0, \ \theta > 0$ and

(ii) $f(x, \theta) = (3x^2/\theta^3) \exp\{-x^3/\theta^3\}, \quad x > 0, \ \theta > 0$.

Examine whether the distribution belongs to a one-parameter exponential family. On the basis of a random sample of size n from these distributions, find the moment estimator based on a sufficient statistic and the maximum likelihood estimator of θ. Examine whether the two are the same and whether these are CAN. Find their approximate variances.

Solution: (i) It is easy to verify that the distribution of X belongs to a one-parameter exponential family. Hence by Theorem 4.2.1, the moment estimator of θ based on a sufficient statistics is the same as the maximum likelihood estimator of θ and it is CAN with approximate variance $1/nI(\theta)$. The maximum likelihood estimator $\hat{\theta}_n$ of θ is given by $\hat{\theta}_n = \sum_{i=1}^{n} X_i^2/2n$, which is same as the moment estimator based on the sufficient statistics. The information function $I(\theta)$ is given by

$$I(\theta) = E\left(-\frac{\partial^2}{\partial\theta^2}\log f(X,\theta)\right) = E\left(-\frac{1}{\theta^2} + \frac{X^2}{\theta^3}\right) = \frac{1}{\theta^2}.$$

Thus, $\tilde{\theta}_n = \hat{\theta}_n$ is CAN for θ with approximate variance $1/nI(\theta) = \theta^2/n$.
(ii) This distribution also belongs to a one-parameter exponential family. Hence by Theorem 4.2.1, the moment estimator of θ based on a sufficient statistic is the same as the maximum likelihood estimator of θ and it is CAN with approximate variance $1/nI(\theta)$. The maximum likelihood estimator $\hat{\theta}_n$ of θ is given by $\hat{\theta}_n = (m_3')^{1/3}$, which is same as the moment estimator based on the sufficient statistic. The information function $I(\theta)$ is given by

$$I(\theta) = E\left(-\frac{\partial^2}{\partial\theta^2}\log f(X,\theta)\right) = E\left(-\frac{3}{\theta^2} + \frac{12X^3}{\theta^5}\right) = \frac{9}{\theta^2}.$$

Thus, $\tilde{\theta}_n = \hat{\theta}_n$ is CAN for θ with approximate variance $1/nI(\theta) = \theta^2/9n$.

4.6.4 Suppose X has a logarithmic series distribution with probability mass function given by

$$p(x,\theta) = \frac{-1}{\log(1-\theta)}\frac{\theta^x}{x}, \quad x = 1, 2, \ldots, \quad 0 < \theta < 1.$$

Show that a logarithmic series distribution is a power series distribution. On the basis of a random sample from a logarithmic series distribution, find the moment estimator of θ based on a sufficient statistic and the maximum likelihood estimator of θ. Examine whether the two are the same and whether these are CAN. Find their approximate variances.

Solution: The probability mass function of X has can be expressed as

$$p(x,\theta) = \frac{-1}{\log(1-\theta)}\frac{\theta^x}{x} = \frac{a_x\theta^x}{A(\theta)}, \quad x = 1, 2, \ldots,$$

where $a_x = 1/x$ for $x = 1, 2, \ldots$, and $A(\theta) = -\log(1-\theta)$, $0 < \theta < 1$. Thus, a logarithmic series distribution is a power series distribution. Suppose

$\{X_1, X_2, \ldots, X_n\}$ is a random sample from a logarithmic series distribution. Hence, as shown in Example 4.2.4, the moment estimator of θ based on the sufficient statistic $\sum_{i=1}^{n} X_i$ is same as the maximum likelihood estimator of θ. It is CAN with approximate variance $1/nI(\theta)$. To find the expression for the estimator and $I(\theta)$, we find

$$E(X) = \frac{-1}{\log(1-\theta)} \sum_{x \geq 1} x \frac{\theta^x}{x} = \frac{-\theta(1-\theta)^{-1}}{\log(1-\theta)}.$$

Thus, the moment estimator of θ based on the sufficient statistic which is same as the maximum likelihood estimator of θ is a solution of the equation $\overline{X}_n = \frac{-\theta(1-\theta)^{-1}}{\log(1-\theta)}$. It exists uniquely and can be obtained by numerical methods discussed in Sect. 4.4. Now to find the information function, we have,

$$\log p(x, \theta) = -\log(-\log(1-\theta)) + x \log\theta - \log x$$

$$\frac{\partial}{\partial\theta} \log p(x, \theta) = \frac{1}{\log(1-\theta)} \frac{1}{1-\theta} + \frac{x}{\theta}$$

$$\frac{\partial^2}{\partial\theta^2} \log p(x, \theta) = \frac{1}{\log(1-\theta)} \frac{1}{(1-\theta)^2} + \frac{1}{(\log(1-\theta))^2} \frac{1}{(1-\theta)^2} - \frac{x}{\theta^2}.$$

Hence the information function $I(\theta)$ is given by

$$I(\theta) = E\left(-\frac{\partial^2}{\partial\theta^2} \log p(X, \theta)\right)$$

$$= \frac{-1}{\log(1-\theta)} \frac{1}{(1-\theta)^2} - \frac{1}{(\log(1-\theta))^2} \frac{1}{(1-\theta)^2} - \frac{1}{\theta^2} \frac{\theta(1-\theta)^{-1}}{\log(1-\theta)}$$

$$= \frac{-(\theta + \log(1-\theta))}{\theta(\log(1-\theta))^2(1-\theta)^2}.$$

Suppose $g(\theta) = \theta + \log(1-\theta)$, then $g(0) = 0$ and $g'(\theta) = 1 - \frac{1}{1-\theta} = \frac{-\theta}{1-\theta} < 0$ which implies that g is a decreasing function. Hence $\theta > 0 \Rightarrow g(\theta) < g(0) = 0$. Thus, $-(\theta + \log(1-\theta)) > 0$ implying that $I(\theta) > 0$.

4.6.5   Suppose $(X, Y)'$ has a bivariate normal distribution with zero mean vector and dispersion matrix Σ given by

$$\Sigma = \sigma^2 \begin{pmatrix} 1 & \rho \\ \rho & 1 \end{pmatrix},$$

$\sigma^2 > 0$, $-1 < \rho < 1$. On the basis of a random sample of size n from the distribution of $(X, Y)'$, find the maximum likelihood estimator of $(\sigma^2, \rho)'$ and examine if it is CAN. Find the approximate dispersion matrix.

Solution: $\underline{Z} = (X, Y)'$ has bivariate normal distribution, hence its probability density function $f(x, y, \sigma^2, \rho)$ is given by

$$f(x, y, \sigma^2, \rho) = \frac{1}{2\pi\sigma^2\sqrt{1-\rho^2}} \exp\left\{-\frac{1}{2\sigma^2(1-\rho^2)}(x^2 - 2\rho xy + y^2)\right\},$$
$$(x, y)' \in \mathbb{R}^2,$$

$\sigma^2 > 0$, $-1 < \rho < 1$. In Example 4.2.5, we have shown that the bivariate normal distribution belongs to a two-parameter exponential family. Hence by Theorem 4.2.3, the moment estimator of $(\sigma^2, \rho)'$ based on a sufficient statistic is the same as the maximum likelihood estimator of $(\sigma^2, \rho)'$ and it is CAN for $(\sigma^2, \rho)'$ with approximate dispersion matrix $I^{-1}(\sigma^2, \rho)/n$. The sufficient statistic for the family is $\left(\sum_{i=1}^{n}(X_i^2 + Y_i^2), \sum_{i=1}^{n} X_i Y_i\right)'$, hence the moment estimator for $(\sigma^2, \rho)'$ based on the sufficient statistic is given by the following system of equations.

$$\frac{1}{n}\sum_{i=1}^{n}(X_i^2 + Y_i^2) = E(K_1(X, Y)) = E(X^2 + Y^2) = 2\sigma^2$$

$$\frac{1}{n}\sum_{i=1}^{n} X_i Y_i = E(K_2(X, Y)) = E(XY) = \sigma^2\rho.$$

Hence, the moment estimator for $(\sigma^2, \rho)'$ based on the sufficient statistic, which is same as the maximum likelihood estimator of $(\sigma^2, \rho)'$, is given by

$$\hat{\sigma}_n^2 = \frac{1}{2n}\sum_{i=1}^{n}(X_i^2 + Y_i^2) \quad \& \quad \hat{\rho}_n = \frac{2\sum_{i=1}^{n} X_i Y_i}{\sum_{i=1}^{n}(X_i^2 + Y_i^2)}.$$

It is to be noted that for this example, it is simpler to find the moment estimator than the maximum likelihood estimator of $(\sigma^2, \rho)'$. The information matrix $I(\sigma^2, \rho)$ is derived in Example 4.2.5 and is given by

$$I(\sigma^2, \rho) = \begin{pmatrix} \frac{1}{\sigma^4} & \frac{-\rho}{\sigma^2(1-\rho^2)} \\ \frac{-\rho}{\sigma^2(1-\rho^2)} & \frac{1+\rho^2}{(1-\rho^2)^2} \end{pmatrix}.$$

Inverse of $I(\sigma^2, \rho)$ is given by

$$I^{-1}(\sigma^2, \rho) = \begin{pmatrix} (1+\rho^2)\sigma^4 & \sigma^2\rho(1-\rho^2) \\ \sigma^2\rho(1-\rho^2) & (1-\rho^2)^2 \end{pmatrix}.$$

Thus, $(\hat{\sigma}_n^2, \hat{\rho}_n)'$ is a CAN estimator of $(\sigma^2, \rho)'$ with approximate dispersion matrix $I^{-1}(\sigma^2, \rho)/n$.

4.6.6 Suppose $(X, Y)'$ is random vector with a joint probability mass function as

$$p_{xy} = P[X = x, Y = y] = \binom{x}{y} e^{-\lambda} \lambda^x p^y (1-p)^{x-y}/x!,$$

$$y = 0, 1, \ldots, x, \quad x = 0, 1, 2, \ldots,$$

where $\lambda > 0$ and $0 < p < 1$. Examine if the distribution belongs to a two-parameter exponential family. Hence, find a CAN estimator for $(\lambda, p)'$ and its approximate dispersion matrix.

Solution: The probability mass function of $(X, Y)'$ is given by

$$p_{xy} = P[X = x, Y = y] = \binom{x}{y} e^{-\lambda} \lambda^x p^y (1-p)^{x-y}/x!,$$

$$y = 0, 1, \ldots, x, \quad x = 0, 1, 2, \ldots,$$

It is to be noted that (i) the parameter space
$\Theta = \{(\lambda, p)' | \lambda > 0, 0 < p < 1\}$ is an open set. (ii) The support S of $(X, Y)'$
is $S = \{(x, y) | y = 0, 1, \ldots, x; \quad x = 0, 1, 2, \ldots, \}$ which does not depend
on the parameters. (iii) The logarithm of the probability mass function of
$(X, Y)'$ is as follows:

$$\log p_{xy} = \log \binom{x}{y} - \log x! - \lambda + x(\log \lambda + \log(1-p))$$

$$+ y(\log p - \log(1-p))$$

$$= W(x, y) + V(\lambda, p) + K_1(x, y)U_1(\lambda, p) + K_2(x, y)U_2(\lambda, p),$$

where $W(x, y) = \log \binom{x}{y} - \log x!$, $V(\lambda, p) = -\lambda$, $K_1(x, y) = x$,
$U_1(\lambda, p) = \log \lambda + \log(1 - p)$ and $K_2(x, y) = y$, $U_2(\lambda, p) = \log p - \log$
$(1 - p)$. (iv) The matrix $J = \left[\frac{dU_i}{d\theta_j}\right]$ of partial derivatives, where $(\theta_1, \theta_2)' =$
$(\lambda, p)'$, is given by

$$J = \begin{pmatrix} \frac{1}{\lambda} & -\frac{1}{(1-p)} \\ 0 & \frac{1}{p(1-p)} \end{pmatrix}.$$

It is clear that U_1 and U_2 have continuous partial derivatives with respect
to λ and p and $|J| = \left|\left[\frac{dU_i}{d\theta_j}\right]\right| = \frac{1}{\lambda p(1-p)} \neq 0$. (v) The functions 1, x and y
are linearly independent. Thus, the joint distribution of $(X, Y)'$ satisfies all
the requirements of a two-parameter exponential family and hence it belongs
to a two-parameter exponential family. Hence by Theorem 4.2.3, based on
a random sample of size n, the moment estimator of $(\lambda, p)'$ based on a
sufficient statistic is the same as the maximum likelihood estimator of $(\lambda, p)'$
and it is CAN for $(\lambda, p)'$ with approximate dispersion matrix $I^{-1}(\lambda, p)/n$.
In Example 3.3.4, we have shown that the moment estimator based on the

sufficient statistic is the same as the maximum likelihood estimator of $(\lambda, p)'$ and it is given by $(\tilde{\lambda}_n, \tilde{p}_n)' = (\overline{X}_n, \overline{Y}_n/\overline{X}_n)'$. Further, the information matrix as obtained in Example 3.3.4 is $I(\lambda, p) = \text{diag}[1/\lambda, \lambda/p(1-p)]$.

4.6.7 Suppose $(X, Y)'$ has a bivariate normal distribution with mean vector $(\mu_1, \mu_2)'$ and dispersion matrix Σ given by

$$\Sigma = \begin{pmatrix} 1 & \rho \\ \rho & 1 \end{pmatrix},$$

where $\rho \neq 0$ and is known and $\mu_1, \mu_2 \in \mathbb{R}$. Show that the distribution belongs to a two-parameter exponential family. Hence, find a CAN estimator $(\mu_1, \mu_2)'$ and its approximate dispersion matrix.

Solution: The probability density function of $(X, Y)'$ is given by

$$f(x, y, \mu_1, \mu_2) = \frac{1}{2\pi\sqrt{1-\rho^2}} \times$$

$$\exp\left\{-\frac{1}{2(1-\rho^2)}((x-\mu_1)^2 - 2\rho(x-\mu_1)(y-\mu_2) + (y-\mu_2)^2)\right\},$$

$(x, y) \in \mathbb{R}^2$, $\mu_1, \mu_2 \in \mathbb{R}$. Thus, the support of the distribution does not depend on the parameters and the parameter space is an open set. Observe that $\log f(x, y, \mu_1, \mu_2)$ can be expressed as follows. Suppose c is a constant free from parameters

$$\log f(x, y, \mu_1, \mu_2) = c - \frac{1}{2(1-\rho^2)}(x^2 + y^2 - 2\rho xy + \mu_1^2 + \mu_2^2 - 2\rho\mu_1\mu_2)$$

$$+ \frac{1}{2(1-\rho^2)}(2\mu_1(x-\rho y) + 2\mu_2(y-\rho x))$$

$$= U_1(\mu_1, \mu_2)K_1(x, y) + U_2(\mu_1, \mu_2)K_2(x, y)$$

$$+ V(\mu_1, \mu_2) + W(x, y),$$

where $U_1(\mu_1, \mu_2) = \mu_1/(1-\rho^2)$,

$K_1(x, y) = x - \rho y$, $\quad U_2(\mu_1, \mu_2) = \mu_2/(1-\rho^2)$

$K_2(x, y) = y - \rho x$,

$V(\mu_1, \mu_2) = -(\mu_1^2 + \mu_2^2 - 2\rho\mu_1\mu_2)/2(1-\rho^2)$ &

$W(x, y) = -(x^2 + y^2 - 2\rho xy)/2(1-\rho^2)$.

The matrix $J = \left[\frac{dU_i}{d\theta_j}\right]$ of partial derivatives, where $(\theta_1, \theta_2)' = (\mu_1, \mu_2)'$, is given by

$$J = \frac{1}{1-\rho^2}\begin{pmatrix} 1 & 0 \\ 0 & 1 \end{pmatrix}.$$

It is clear that U_1 and U_2 have continuous partial derivatives with respect to μ_1 and μ_2 and $|J| = \left| \left[\frac{dU_i}{d\theta_j} \right] \right| = \frac{1}{(1-\rho^2)^2} \neq 0$. (v) It is easy to verify that the functions $1, (x - \rho y)$ and $(y - \rho x)$ are linearly independent. Thus, the joint distribution of $(X, Y)'$ satisfies all the requirements of a two-parameter exponential family and hence it belongs to a two-parameter exponential family. Hence by Theorem 4.2.3, based on a random sample of size n, the moment estimator of $(\mu_1, \mu_2)'$ based on a sufficient statistic is the same as the maximum likelihood estimator of $(\mu_1, \mu_2)'$ and it is CAN for $(\mu_1, \mu_2)'$ with approximate dispersion matrix $I^{-1}(\mu_1, \mu_2)/n$. To find the information matrix, we find the partial derivatives of the $\log f(x, y, \mu_1, \mu_2)$. These are as follows:

$$\frac{\partial}{\partial \mu_1} \log f(x, y, \mu_1, \mu_2) = -\frac{1}{2(1 - \rho^2)} (2\mu_1 - 2\rho\mu_2 - 2(x - \rho y))$$

$$\frac{\partial^2}{\partial \mu_1^2} \log f(x, y, \mu_1, \mu_2) = -\frac{2}{2(1 - \rho^2)} = -\frac{1}{1 - \rho^2}$$

$$\frac{\partial^2}{\partial \mu_2 \partial \mu_1} \log f(x, y, \mu_1, \mu_2) = -\frac{-2\rho}{2(1 - \rho^2)} = \frac{\rho}{1 - \rho^2}$$

$$\frac{\partial}{\partial \mu_2} \log f(x, y, \mu_1, \mu_2) = -\frac{1}{2(1 - \rho^2)} (2\mu_2 - 2\rho\mu_1 - 2(y - \rho x))$$

$$\frac{\partial^2}{\partial \mu_2^2} \log f(x, y, \mu_1, \mu_2) = -\frac{2}{2(1 - \rho^2)} = -\frac{1}{1 - \rho^2}.$$

Hence, the information matrix $I(\mu_1, \mu_2)$ and its inverse are given by

$$I(\mu_1, \mu_2) = \frac{1}{1 - \rho^2} \begin{pmatrix} 1 & -\rho \\ -\rho & 1 \end{pmatrix} \quad \& \quad I^{-1}(\mu_1, \mu_2) = \begin{pmatrix} 1 & \rho \\ \rho & 1 \end{pmatrix}.$$

From the system of likelihood equation, the maximum likelihood estimator of $(\mu_1, \mu_2)'$ is given by $(\overline{X}_n, \overline{Y}_n)'$. It is to be noted that the maximum likelihood estimator of $(\mu_1, \mu_2)'$ derived from the bivariate model and the two marginal univariate models is the same. However, when $\rho \neq 0$,

$$I_{1,1}(\mu_1, \mu_2) = \frac{1}{1 - \rho^2} > 1 = I(\mu_1) \ \& \ I_{2,2}(\mu_1, \mu_2) = \frac{1}{1 - \rho^2} > 1 = I(\mu_2)$$

as observed in Example 4.3.4 and in Example 4.3.5.

4.6.8 Suppose a random variable X has a negative binomial distribution with parameters (k, p) and with the following probability mass function.

$$P[X = x] = \binom{x + k - 1}{k - 1} p^k (1 - p)^x \quad x = 0, 1, 2, \ldots.$$

(i) Show that the distribution belongs to a one-parameter exponential family, if k is known and $p \in (0, 1)$ is unknown. Hence obtain a CAN estimator of p.

(ii) Examine whether the distribution belongs to a one-parameter exponential family, if p is known and k is an unknown positive integer. (iii) Examine whether the distribution belongs to a two-parameter exponential family, if both $p \in (0, 1)$ and k are unknown, where k is a positive integer.

Solution: If (i) k is known and $p \in (0, 1)$ is unknown, then the probability mass function can be expressed as

$$\log p(x, k, p) = \log \binom{x + k - 1}{k - 1} + k \log p + x \log(1 - p)$$
$$= U(p)K(x) + V(p) + W(x),$$

where $U(p) = \log(1 - p)$, $K(x) = x$, $V(p) = k \log p$ and
$W(x) = \log \binom{x+k-1}{k-1}$. Thus,
(i) the probability law of X is expressible in the form required in a one-parameter exponential family, (ii) support of the probability mass function is $\{0, 1, 2, \ldots\}$ and it is free from p, (iii) the parameter space is $(0, 1)$ which is an open set, (iv) $U'(p) = -1/(1 - p) \neq 0$ and (v) x and 1 are linearly independent. Thus, the distribution belongs to a one-parameter exponential family, if k is known and $p \in (0, 1)$ is unknown. Suppose $\underline{X} = \{X_1, X_2, \ldots, X_n\}$ is a random sample from the distribution of X. By Theorem 4.2.1, the moment estimator of p based on a sufficient statistics is the same as the maximum likelihood estimator of p and it is CAN with approximate variance $1/nI(p)$. The moment estimator $\hat{p}_n$ of p based on the sufficient statistics is given by $\hat{p}_n = k/(\overline{X}_n + k)$, which is same as the maximum likelihood estimator of p. The information function $I(p)$ is given by

$$I(p) = E\left(-\frac{\partial^2}{\partial\theta^2} \log p(X, k, p)\right) = E\left(\frac{k}{p^2} + \frac{X}{(1 - p)^2}\right) = \frac{k}{p^2(1 - p)}.$$

If X follows a negative binomial distribution with parameters (k, p) then

$$E(X) = \frac{k(1 - p)}{p} \quad \& \quad Var(X) = \frac{k(1 - p)}{p^2} < \infty.$$

Hence by the WLLN and by the CLT,

$$\overline{X}_n \xrightarrow{P} \frac{k(1 - p)}{p} \quad \& \quad \sqrt{n}\left(\overline{X}_n - \frac{k(1 - p)}{p}\right) \xrightarrow{L} Z_1 \sim N\left(0, \frac{k(1 - p)}{p^2}\right).$$

Thus, $\overline{X}_n$ is CAN for $k(1 - p)/p = \phi$, say, with approximate variance $k(1 - p)/np^2$. To obtain a CAN estimator for p, we find a transformation g such that $g(\phi) = p$. Suppose $g(y) = k/(k + y)$, $y > 0$, then g is a differentiable function with $g'(y) = -k/(k + y)^2$ which is not 0, $\forall \, y > 0$. Hence by the delta method, $g(\overline{X}_n) = k/(k + \overline{X}_n)$ is CAN for $g(\phi) = p$ with approximate variance $(k(1 - p)/np^2) \times (p^4/k^2) = p^2(1 - p)/nk = 1/nI(p)$.

(ii) If p is known and k is an unknown positive integer, the parameter space is not an open set, hence the distribution does not belong to a one-parameter exponential family.

(iii) If both $p \in (0, 1)$ and k are unknown, where k is a positive integer, then again the parameter space is not an open set, hence the distribution does not belong to a two-parameter exponential family.

4.6.9 Examine whether a logistic distribution with probability density function

$$f(x, \theta) = \frac{\exp\{-(x - \theta)\}}{(1 + \exp\{-(x - \theta)\})^2}, \ x \in \mathbb{R}, \ \theta \in \mathbb{R}$$

belongs to a one-parameter exponential family. If not, examine if it belongs to a one-parameter Cramér family. If yes, find a CAN estimator of θ and its approximate variance.

Solution: If X follows a logistic distribution, then $\log f(x, \theta)$ is given by $\log f(x, \theta) = -(x - \theta) - 2\log(1 + \exp\{-(x - \theta)\})$ and it cannot be expressed in the form $U(\theta)K(x) + V(\theta) + W(x)$. Hence, the distribution does not belong to a one-parameter exponential family. To examine whether it belongs to a Cramér family, we note that

(i) support of the probability density function is $\mathbb{R}$ and it is free from θ, (ii) the parameter space is $\mathbb{R}$ which is an open set. The partial derivatives of $\log f(x, \theta)$ up to order 3 are given by

$$\frac{\partial}{\partial \theta} \log f(x, \theta) = 1 - \frac{2\exp\{-(x - \theta)}{(1 + \exp\{-(x - \theta)\})}$$

$$\frac{\partial^2}{\partial \theta^2} \log f(x, \theta) = -\frac{2\exp\{-(x - \theta)\}}{(1 + \exp\{-(x - \theta)\})^2}$$

$$\frac{\partial^3}{\partial \theta^3} \log f(x, \theta) = -\frac{2\exp\{-(x - \theta)\}(1 - \exp\{-(x - \theta)\})}{(1 + \exp\{-(x - \theta)\})^3}.$$

Thus, partial derivatives of $\log f(x, \theta)$ up to order 3 exists. Now observe that

$$\left|\frac{\partial^3}{\partial \theta^3} \log f(x, \theta)\right| \leq \left|\frac{2\exp\{-(x - \theta)\}}{(1 + \exp\{-(x - \theta)\})^3}\right|$$
$$+ \left|\frac{2\exp\{-2(x - \theta)\}}{(1 + \exp\{-(x - \theta)\})^3}\right| = M(x),$$

say. Further,

$$E(M(X)) = \int_{-\infty}^{\infty} M(x) f(x, \theta)\, dx$$

$$= \int_{-\infty}^{\infty} \frac{2 \exp\{-2(x - \theta)\}}{(1 + \exp\{-(x - \theta)\})^5} + \int_{-\infty}^{\infty} \frac{2 \exp\{-3(x - \theta)\}}{(1 + \exp\{-(x - \theta)\})^5}$$

$$= \int_0^{\infty} \frac{2y}{(1 + y)^5} + \int_0^{\infty} \frac{2y^2}{(1 + y)^5} < \infty,$$

with the substitution $\exp\{-(x - \theta)\} = y$. Thus, the third order partial derivative of the log-likelihood is bounded by an integrable random variable. From the second order partial derivative of the log-likelihood we have

$$I(\theta) = E\left(\frac{2 \exp\{-(X - \theta)}{(1 + \exp\{-(X - \theta)\})^2} \right) = \int_{-\infty}^{\infty} \frac{2 \exp\{-2(x - \theta)}{(1 + \exp\{-(x - \theta)\})^4}\, dx$$

$$= 2 \int_0^{\infty} \frac{y}{(1 + y)^4}, \quad \text{by } \exp\{-(x - \theta)\} = y$$

$$= 2B(2, 2) = 2/6 = 1/3.$$

Thus, $0 < I(\theta) < \infty$. Thus, all the Cramér regularity conditions are satisfied and hence the logistic distribution belongs to a one-parameter Cramér family. Hence, the maximum likelihood estimator of θ is CAN for θ with approximate variance $3/n$. The likelihood equation is given by $2 \sum_{i=1}^{n} \frac{\exp\{-(X_i - \theta)\}}{1 + \exp\{-(X_i - \theta)\}} - n = 0$. We can obtain its solution either by the Newton-Raphson procedure or the method of scoring (see Exercise 4.7.10). As an initial iterative value one may take sample median or the sample mean as both are consistent for θ. Method of scoring is easier as the information function $I(\theta) = 1/3$ is free from θ. We have discussed in Example 4.5.3 how to find its solution using `uniroot` function.

4.6.10 Suppose a random variable X follows a Cauchy $C(\theta, \lambda)$ distribution with location parameter θ and shape parameter λ. Examine whether the distribution belongs to a two-parameter Cramér family.

Solution: Part of the solution is given in Example 4.3.1 and in Example 4.5.6. It remains to find out the third order partial derivatives and verify that these are bounded by integrable functions.

4.6.11 Suppose $\{X_1, X_2, \ldots, X_n\}$ is a random sample from a Poisson distribution with parameter $\theta > 0$. An estimator T_n is defined as

$$T_n = \begin{cases} \overline{X}_n, & \text{if } \overline{X}_n > 0 \\ 0.01, & \text{if } \overline{X}_n = 0 \end{cases}$$

Show that $T_{1n} = e^{-T_n}$ is a CAN estimator of $P[X_1 = 0] = e^{-\theta}$. Find its approximate variance. Suppose random variables Y_i, $i = 1, 2, \ldots, n$ are defined as follows:

$$Y_i = \begin{cases} 1, & \text{if } X_i = 0 \\ 0, & \text{otherwise}. \end{cases}$$

Obtain a CAN estimator T_{2n} for $e^{-\theta}$ based on $\{Y_1, Y_2, \ldots, Y_n\}$ and find its approximate variance. Find $ARE(T_{1n}, T_{2n})$.

Solution: Suppose $\{X_1, X_2, \ldots, X_n\}$ is a random sample from a Poisson distribution with parameter $\theta > 0$. An estimator T_n is defined as

$$T_n = \begin{cases} \overline{X}_n, & \text{if } \overline{X}_n > 0 \\ 0.01, & \text{if } \overline{X}_n = 0 \end{cases}$$

For a Poisson distribution with parameter $\theta > 0$, $E(X) = Var(X) = \theta < \infty$. Hence by the WLLN and by the CLT, $\overline{X}_n$ is CAN for θ with approximate variance θ/n. Observe that, for $\epsilon > 0$,

$$P[|T_n - \overline{X}_n| < \epsilon] \geq P[T_n = \overline{X}_n] = P[\overline{X}_n > 0]$$
$$= 1 - \exp(-n\theta) \quad \to \quad 1, \text{ for } \theta > 0.$$

Hence, T_n is also consistent for θ. Similarly,

$$P[\sqrt{n}|T_n - \overline{X}_n| < \epsilon] \geq P[T_n = \overline{X}_n] = P[\overline{X}_n > 0]$$
$$= 1 - \exp(-n\theta) \quad \to \quad 1, \text{ for } \theta > 0.$$

Thus, $\sqrt{n}(T_n - \overline{X}_n) \xrightarrow{P_\theta} 0$, $\forall \; \theta > 0$. Hence, $\sqrt{n}(T_n - \theta)$ and $\sqrt{n}(\overline{X}_n - \theta)$ have the same limiting distribution. By CLT, $\sqrt{n}(\overline{X}_n - \theta) \xrightarrow{L} Z_1 \sim N(0, \theta)$ and hence, $\sqrt{n}(T_n - \theta) \xrightarrow{L} Z_1 \sim N(0, \theta)$, $\forall \; \theta > 0$, which proves that T_n is CAN for θ with approximate variance θ/n. Suppose $g(\theta) = e^{-\theta}$, then it is a differentiable function with $g'(\theta) = -e^{-\theta} \neq 0$. Hence by the delta method, e^{-T_n} is CAN for $e^{-\theta} = P[X_1 = 0]$ with approximate variance $\theta e^{-2\theta}/n$. In the another approach to find a CAN estimator for $P[X_1 = 0]$ we define random variables Y_i, $i = 1, 2, \ldots, n$ as follows:

$$Y_i = \begin{cases} 1, & \text{if } X_i = 0 \\ 0, & \text{otherwise}. \end{cases}$$

Since $\{X_1, X_2, \ldots, X_n\}$ are independent and identically distributed random variables, being Borel functions, $\{Y_1, Y_2, \ldots, Y_n\}$ are also independent and identically distributed random variables each having Bernoulli $B(1, p)$ distribution where $p = E(Y_i) = P[X_i = 0] = e^{-\theta}$. Further, $Var(Y_i) = e^{-\theta}(1 - e^{-\theta}) < \infty$ Hence by the WLLN and the CLT, the sample mean $\overline{Y}_n = T_{2n}$, say, is CAN for $e^{-\theta}$ with approximate variance $e^{-\theta}(1 - e^{-\theta})/n$. Hence,

$$ARE(T_{1n}, T_{2n}) = \frac{e^{-\theta}(1 - e^{-\theta})}{\theta e^{-2\theta}} = \frac{e^{\theta} - 1}{\theta} = 1 + \frac{\sum_{r \geq 2} \frac{\theta^r}{r!}}{\theta} > 1 .$$

Thus, T_{1n} is preferred to T_{2n}.

7.4 Chapter 5

5.4.1 Suppose $\{X_1, X_2, \ldots, X_n\}$ is a random sample from a Laplace distribution with probability density function $f(x, \theta)$ given by

$$f(x, \theta) = \frac{1}{2\theta} \exp\left\{-\frac{|x|}{\theta}\right\}, \quad x \in \mathbb{R}, \quad \theta > 0.$$

Derive a large sample test procedure for testing $H_0 : \theta = \theta_0$ against the alternative $H_1 : \theta < \theta_0$, when the test statistic is a function of $U_n = \sum_{i=1}^{n} |X_i|/n$. Find the power function.

Solution: If $Y = |X|$, then it is easy to verify that X follows an exponential distribution with mean θ. Hence by the WLLN and the CLT, U_n is CAN for θ with approximate variance θ^2/n. Hence we define the test statistic as

$$T_n = \frac{\sqrt{n}}{\theta_0}(U_n - \theta_0) \quad \text{or} \quad S_n = \frac{\sqrt{n}}{U_n}(U_n - \theta_0).$$

The asymptotic null distribution of both T_n and S_n is the standard normal. Hence $H_0 : \theta = \theta_0$ is rejected against the alternative $H_1 : \theta < \theta_0$, if $T_n < -a_{1-\alpha}$ or $S_n < -a_{1-\alpha}$. The power function corresponding to the test statistic T_n is given by

$$\beta(\theta) = P_\theta[T_n < -a_{1-\alpha}] = P_\theta\left[\frac{\sqrt{n}}{\theta_0}(U_n - \theta_0) < -a_{1-\alpha}\right]$$

$$= P_\theta\left[\frac{\sqrt{n}}{\theta}(U_n - \theta) < \frac{\sqrt{n}}{\theta}(\theta_0 - \theta) - \frac{\theta_0}{\theta}a_{1-\alpha}\right]$$

$$= \Phi\left(\frac{\sqrt{n}}{\theta}(\theta_0 - \theta) - \frac{\theta_0}{\theta}a_{1-\alpha}\right).$$

The power function corresponding to the test statistic S_n can be obtained on similar lines.

5.4.2 Suppose $\{X_1, X_2, \ldots, X_n\}$ is a random sample from a Cauchy $C(\theta, 1)$ distribution. Derive a large sample test procedure for testing $H_0 : \theta = 0$ against the alternative $H_1 : \theta \neq 0$. Obtain the power function.

Solution: Suppose $X \sim C(\theta, 1)$. Then given a random sample of size n from the distribution of X, the maximum likelihood estimator $\hat{\theta}_n$ of θ is CAN for θ with approximate variance $2/n$. Hence the test statistic T_n is defined as $T_n = \sqrt{n/2}\, \hat{\theta}_n$. For large n under H_0, $T_n \sim N(0, 1)$ distribution. The null hypothesis H_0 is rejected against H_1 if $|T_n| > c$, where c is determined corresponding to the given level of significance α and the asymptotic null distribution of T_n. Thus, $c = a_{1-\alpha/2}$. The power function $\beta(\theta)$ is derived as follows:

$$\beta(\theta) = P_\theta[|T_n| > c] = 1 - P_\theta\left[-c < \sqrt{n/2}\,\hat{\theta}_n < c\right]$$
$$= 1 - P_\theta\left[-c\sqrt{2}/\sqrt{n} < \hat{\theta}_n < c\sqrt{2}/\sqrt{n}\right]$$
$$= 1 - P_\theta\left[-c - \theta\sqrt{n}/\sqrt{2} < \sqrt{n/2}(\hat{\theta}_n - \theta) < c - \theta\sqrt{n}/\sqrt{2}\right]$$
$$= 1 - \Phi\left(c - \theta\sqrt{n}/\sqrt{2}\right) + \Phi\left(-c - \theta\sqrt{n}/\sqrt{2}\right).$$

It is to be noted that at $\theta = 0$, $\beta(\theta) = \alpha$.

5.4.3 Suppose $\underline{X} \equiv \{X_1, X_2, \ldots, X_n\}$ is a random sample from an exponential distribution with scale parameter 1 and location parameter θ. Develop a test procedure to test $H_0 : \theta = \theta_0$ against the alternative $H_1 : \theta \neq \theta_0$.

Solution: The probability density function of a random variable X having an exponential distribution with location parameter θ and scale parameter 1 is given by

$$f_X(x, \theta) = \exp\{-(x - \theta)\}, \quad x \geq \theta.$$

In Example 3.2.7, it is shown that $X_{(1)}$ is the maximum likelihood estimator of θ and the distribution of $X_{(1)}$ is exponential distribution with location parameter θ and scale parameter n. If we define $Y_n = n\left(X_{(1)} - \theta\right)$ then its distribution is exponential with location parameter 0 and scale parameter 1. To test $H_0 : \theta = \theta_0$ against the alternative $H_1 : \theta \neq \theta_0$, we propose a test statistic T_n as $T_n = n\left(X_{(1)} - \theta_0\right)$. Under $H_0 : \theta = \theta_0$, T_n has the exponential distribution with scale parameter 1. H_0 is rejected against $H_1 : \theta \neq \theta_0$, if $T_n > c$ where c is such that

$$P_{\theta_0}[T_n > c] = \alpha \Leftrightarrow P_{\theta_0}[T_n < c] = 1 - \alpha$$
$$\Leftrightarrow 1 - \exp\{-n(c - \theta_0)\} = 1 - \alpha$$
$$\Leftrightarrow c = \theta_0 - \log\alpha/n.$$

5.4.4 Suppose $\{X_1, X_2, \ldots, X_n\}$ is a random sample from a normal distribution with mean μ and variance σ^2. Develop a test procedure to test the hypothesis $H_0 : P[X < a] = p_0$ against $H_1 : P[X < a] \neq p_0$.

Solution: In Example 3.3.2, it is shown that $\hat{\mu}_n = \overline{X}_n$ and $\hat{\sigma}_n^2 = \frac{1}{n} \sum_{i=1}^n (X_i - \overline{X}_n)^2$ are the maximum likelihood estimators of μ and σ^2 respectively. Further, $(\hat{\mu}_n, \hat{\sigma}_n^2)'$ is CAN for $(\mu, \sigma^2)'$, with approximate dispersion matrix D/n, where

$$D = \begin{pmatrix} \sigma^2 & 0 \\ 0 & 2\sigma^4 \end{pmatrix}.$$

Now $P[X < a] = \Phi((a - \mu)/\sigma) = g(\mu, \sigma^2)$ where $g : \mathbb{R}^2 \to \mathbb{R}$ is such that

$$g(x_1, x_2) = \Phi\left(\frac{a - x_1}{\sqrt{x_2}}\right)$$

$$\Rightarrow \quad \frac{\partial}{\partial x_1} g(x_1, x_2) = -\phi\left(\frac{a - x_1}{\sqrt{x_2}}\right) \frac{1}{\sqrt{x_2}}$$

$$\& \quad \frac{\partial}{\partial x_2} g(x_1, x_2) = -\phi\left(\frac{a - x_1}{\sqrt{x_2}}\right) \frac{a - x_1}{2 x_2^{3/2}}.$$

These partial derivatives are continuous and hence g is a totally differentiable function. The vector Δ of these partial derivatives evaluated at $(\mu, \sigma^2)'$ is $\Delta = \left(-\phi(\frac{a - \mu}{\sigma})\frac{1}{\sigma}, -\phi(\frac{a - \mu}{\sigma})\frac{a - \mu}{2\sigma^3}\right)'$. Hence by the delta method, $\Phi\left(\frac{a - \hat{\mu}_n}{\hat{\sigma}_n}\right)$ is CAN for $P[X < a] = \Phi\left(\frac{a - \mu}{\sigma}\right)$ with approximate variance $v(\mu, \sigma^2) = \Delta' D \Delta / n$ where

$$\Delta' D \Delta = \sigma^2 \left(-\phi\left(\frac{a - \mu}{\sigma}\right) \frac{1}{\sigma}\right)^2 + 2\sigma^4 \left(-\phi\left(\frac{a - \mu}{\sigma}\right)\left(\frac{a - \mu}{2\sigma^3}\right)\right)^2$$

$$= \frac{e^{\frac{-(a - \mu)^2}{\sigma^2}}}{2\pi} \left(1 + \frac{(a - \mu)^2}{2\sigma^2}\right).$$

Consequently,

$$U_n = \frac{\Phi\left(\frac{a - \hat{\mu}_n}{\hat{\sigma}_n}\right) - \Phi\left(\frac{a - \mu}{\sigma}\right)}{\sqrt{v(\mu, \sigma^2)}} \xrightarrow{L} Z \sim N(0, 1)$$

$$\& \quad V_n = \frac{\Phi\left(\frac{a - \hat{\mu}_n}{\hat{\sigma}_n}\right) - \Phi\left(\frac{a - \mu}{\sigma}\right)}{\sqrt{v(\hat{\mu}_n, \hat{\sigma}_n^2)}} \xrightarrow{L} Z \sim N(0, 1),$$

by Slutsky's theorem. Hence we propose the test statistic T_n as

$$T_n = \frac{\Phi\left(\frac{a - \hat{\mu}_n}{\hat{\sigma}_n}\right) - p_0}{\sqrt{v(\hat{\mu}_n, \hat{\sigma}_n^2)}}.$$

H_0 is rejected if $|T_n| > c$ where c is such that $P_{H_0}[|T_n| > c] = \alpha$. For large n, under H_0, $T_n \sim N(0, 1)$ distribution. Hence $c = a_{1-\alpha/2}$.

5.4.5 Suppose $\underline{X} \equiv \{X_1, X_2, \ldots, X_n\}$ is a random sample from a Poisson distribution with parameter θ. Obtain the likelihood ratio test to test $H_0 : P[X = 0] = 1/3$ against the alternative $H_1 : P[X = 0] \neq 1/3$.

Solution: Suppose $X \sim P(\theta)$. To test $H_0 : P[X = 0] = 1/3$ against the alternative $H_1 : P[X = 0] \neq 1/3$ is equivalent to test $H_0 : \theta = \theta_0$ against the alternative $H_1 : \theta \neq \theta_0$, where $\theta_0 = \log_e 3$. Now the entire parameter space Θ is $\Theta = (0, \infty)$ and the null space is $\Theta_0 = \{\theta_0\}$. The maximum likelihood estimator of θ in the entire parameter space is $\overline{X}_n$. Hence, the likelihood ratio test statistic $\lambda(\underline{X})$ is given by

$$\lambda(\underline{X}) = \frac{\sup_{\Theta_0} L_n(\theta|\underline{X})}{\sup_{\Theta} L_n(\theta|\underline{X})} = \frac{e^{-n\theta_0}(\theta_0)^{n\overline{X}_n}}{e^{-n\overline{X}_n}(\overline{X}_n)^{n\overline{X}_n}}.$$

The null hypothesis H_0 against the alternative H_1 is rejected if $\lambda(\underline{X}) < c$, that is, $-2\log\lambda(\underline{X}) > c_1$. If the sample size is large, then $-2\log\lambda(\underline{X}) \sim \chi_1^2$ distribution and H_0 is rejected if $-2\log\lambda(\underline{X}) > \chi_{1,1-\alpha}^2$ where $\chi_{1,1-\alpha}^2$ is $(1 - \alpha)$-th quantile of χ_1^2 distribution.

5.4.6 Suppose $\{X_1, X_2, \ldots, X_n\}$ is a random sample from a lognormal distribution with parameters μ and σ^2. Derive a large sample test procedure to test $H_0 : \mu = \mu_0$, $\sigma^2 = \sigma_0^2$ against $H_1 : \mu \neq \mu_0$, $\sigma^2 \neq \sigma_0^2$.

Solution: In Exercise 3.5.31 we have obtained a CAN estimator of $(\mu, \sigma^2)'$. Using it and the procedure as adopted in Example 5.2.2, we can develop a test procedure to test $H_0 : \mu = \mu_0$, $\sigma^2 = \sigma_0^2$ against $H_1 : \mu \neq \mu_0$, $\sigma^2 \neq \sigma_0^2$.

5.4.7 Suppose $\{X_1, X_2, \ldots, X_n\}$ is a random sample from a Gamma $G(\alpha, \lambda)$ distribution. Derive a large sample test procedure to test $H_0 : \alpha = \alpha_0$ against $H_1 : \alpha \neq \alpha_0$ when λ is (i) known and (ii) unknown.

Solution: Suppose X follows a Gamma $G(\alpha, \lambda)$ distribution. Then $E(X) = \lambda/\alpha$ and $Var(X) = \lambda/\alpha^2$.
(i) If λ is known, then by the WLLN and CLT, $\overline{X}_n$ is CAN for λ/α with approximate variance $\lambda/n\alpha^2$. By the delta method, $U_n = \lambda/\overline{X}_n$ is CAN for α with approximate variance $\alpha^2/n\lambda$. Hence we define a test statistic as

$$T_n = \frac{\sqrt{n\lambda}}{\alpha_0}(U_n - \alpha_0) \quad \text{or} \quad S_n = \frac{\sqrt{n\lambda}}{U_n}(U_n - \alpha_0).$$

The null hypothesis is rejected if $|T_n| > c$ or $|S_n| > c$. The cut-off c is determined using the given level of significance and the asymptotic null distribution. Under H_0, the asymptotic distribution of both the test statistics is standard normal.

(ii) Suppose λ is unknown. In Exercise 3.5.32 we have shown that $(m_1'/m_2, \; m_1'^2/m_2)'$ is CAN for $(\alpha, \lambda)'$ with approximate dispersion matrix D/n, where D is,

$$D = \lambda \begin{pmatrix} \frac{3\alpha^2}{\lambda^2} + \frac{2\alpha^2}{\lambda} & \frac{2\alpha}{\lambda} + 2\alpha \\ \frac{2\alpha}{\lambda} + 2\alpha & 2\lambda + 2 \end{pmatrix}.$$

Hence, $\hat{\alpha}_n = m_1'/m_2$ is CAN for α with approximate variance $v(\alpha, \lambda) = (3\alpha^2/\lambda^2 + 2\alpha^2/\lambda)/n$. Further, $\hat{\lambda}_n = m_1'^2/m_2$ is a consistent estimator of λ. A large sample test procedure to test $H_0 : \alpha = \alpha_0$ against $H_1 : \alpha \neq \alpha_0$ is then based on the following two test statistics, both have the asymptotic null distribution to be the standard normal.

$$T_n = \frac{\sqrt{n}}{\sqrt{v(\alpha_0, \hat{\lambda}_n)}}(\hat{\alpha}_n - \alpha_0) \; \text{ or } \; S_n = \frac{\sqrt{n}}{\sqrt{v(\hat{\alpha}_n, \hat{\lambda}_n)}}(\hat{\alpha}_n - \alpha_0).$$

The null hypothesis is rejected if $|T_n| > c$ or $|S_n| > c$ where $c = a_{1-\alpha/2}$.

5.4.8 Suppose $\{X_1, X_2, \ldots, X_n\}$ is a random sample from a geometric distribution with probability mass function $p_x = P[X = x] = \theta(1 - \theta)^x$, $x = 0, 1, 2, \ldots$. Obtain the likelihood ratio test to test $H_0 : P[X = 0] = 0.3$ against the alternative $H_1 : P[X = 0] \neq 0.3$.

Solution: Observe that $P[X = 0] = \theta$, hence $H_0 : P[X = 0] = 0.3$ implies $\theta = 0.3$. The maximum likelihood estimator of θ in the entire parameter space is $1/(1 + \overline{X}_n)$. Using the routine procedure we can obtain the likelihood ratio test to test H_0 against the alternative H_1.

5.4.9 Suppose $\underline{X} \equiv \{X_1, X_2, \ldots, X_n\}$ are independent and identically distributed random variables with following probability mass function.

$$P[X_1 = 1] = (1 - \theta)/2, \;\; P[X_1 = 2] = 1/2, \;\; P[X_1 = 3] = \theta/2, \;\; 0 < \theta < 1.$$

Derive a likelihood ratio test procedure to test $H_0 : \theta = 1/2$ against the alternative $H_0 : \theta \neq 1/2$. Explain how the critical region will be decided if the sample size is large.

Solution: The likelihood of θ corresponding to the data $\underline{X}$ is given by

$$L_n(\theta | \underline{X}) = \left(\frac{1 - \theta}{2}\right)^{n_1} \left(\frac{1}{2}\right)^{n_2} \left(\frac{\theta}{2}\right)^{n_3}$$

$$= \left(\frac{1}{2}\right)^n (1 - \theta)^{n_1} \theta^{n_3}, \;\; n_1 + n_2 + n_3 = n,$$

where n_i is number of times i occurs in the sample, $i = 1, 2, 3$. Likelihood is a differentiable function of θ, hence the likelihood equation and its solution $\hat{\theta}_n$ are given by

$$-\frac{n_1}{1-\theta} + \frac{n_3}{\theta} = 0 \quad \Rightarrow \quad \hat{\theta}_n = \frac{n_3}{n_1 + n_3}.$$

The second order derivative $\frac{\partial^2}{\partial\theta^2} \log L_n(\theta|\underline{X}) = -\frac{n_1}{(1-\theta)^2} - \frac{n_3}{\theta^2} < 0$, $\forall \ \theta \in (0, 1)$. Hence the maximum likelihood estimator of θ is $\hat{\theta}_n = n_3/(n_1 + n_3)$. Now the entire parameter space Θ is $\Theta = (0, 1)$ and the null space is $\Theta_0 = \{1/2\}$. Hence, the likelihood ratio test statistic $\lambda(\underline{X})$ is given by

$$\lambda(\underline{X}) = \frac{\sup\limits_{\Theta_0} L_n(\theta|\underline{X})}{\sup\limits_{\Theta} L_n(\theta|\underline{X})} = \frac{(\frac{1}{2})^n (\frac{1}{2})^{n_1} (\frac{1}{2})^{n_3}}{(\frac{1}{2})^n (1 - \hat{\theta}_n)^{n_1} (\hat{\theta}_n)^{n_3}} = \frac{(n_1 + n_3)^{(n_1 + n_3)}}{2^{(n_1+n_3)} n_1^{n_1} n_3^{n_3}}.$$

The null hypothesis $H_0 : \theta = 1/2$ against the alternative $H_1 : \theta \neq 1/2$ is rejected if $\lambda(\underline{X}) < c \Leftrightarrow -2\log\lambda(\underline{X}) > c_1$. If sample size is large, then $-2\log\lambda(\underline{X}) \sim \chi_1^2$ distribution and H_0 is rejected if $-2\log\lambda(\underline{X}) > \chi_{1,1-\alpha}^2$ where $\chi_{1,1-\alpha}^2$ is $(1-\alpha)$-th quantile of χ_1^2 distribution.

5.4.10 Suppose X has a discrete distribution with possible values $1, 2, 3, 4$ with probabilities $(2 - \theta_1)/4$, $\theta_1/4$, $\theta_2/4$, $(2 - \theta_2)/4$ respectively. On the basis of a random sample from the distribution of X, derive a likelihood ratio test procedure to test $H_0 : \theta_1 = 1/3$, $\theta_2 = 2/3$ against the alternative $H_1 : \theta \neq 1/3$, $\theta_2 \neq 2/3$.

Solution: Using the procedure similar to that in Example 5.2.5, we get the solution.

5.4.11 Suppose $\underline{X} \equiv \{X_1, X_2, \ldots, X_{n_1}\}$ is a random sample from a Bernoulli $B(1, \theta_1)$ distribution and $\underline{Y} \equiv \{Y_1, Y_2, \ldots, Y_{n_2}\}$ is a random sample from a Bernoulli $B(1, \theta_2)$ distribution. Suppose X and Y are independent random variables. Derive a likelihood ratio test procedure for testing $H_0 : \theta_1 = \theta_2$ against the alternative $H_1 : \theta_1 \neq \theta_2$.

Solution: Suppose $X \sim B(1, \theta_1)$ and $Y \sim B(1, \theta_2)$. In the null setup $\theta_1 = \theta_2 = \theta$, say. Then the likelihood of θ given random samples $\underline{X}$ and $\underline{Y}$, using independence of X and Y is given by

$$L_n(\theta|\underline{X}, \underline{Y}) = \theta^{\left(\sum\limits_{i=1}^{n_1} X_i + \sum\limits_{i=1}^{n_2} Y_i\right)} (1 - \theta)^{\left(n_1 + n_2 - \sum\limits_{i=1}^{n_1} X_i - \sum\limits_{i=1}^{n_2} Y_i\right)}.$$

Then the maximum likelihood estimator of θ is $\hat{\theta}_{n_1+n_2} = (\sum_{i=1}^{n_1} X_i + \sum_{i=1}^{n_2} Y_i)/(n_1 + n_2)$. In the entire parameter space the maximum likelihood estimator $\hat{\theta}_{1n_1}$ of θ_1 is $\hat{\theta}_{1n_1} = \overline{X}_{n_1}$ and the maximum likelihood estimator $\hat{\theta}_{2n_2}$ of θ_2 is $\hat{\theta}_{2n_2} = \overline{Y}_{n_2}$. The likelihood ratio test statistic $\lambda(\underline{X})$ is then given by

$$\lambda(\underline{X}) = \frac{\sup\limits_{\Theta_0} L_n(\underline{\theta}|\underline{X})}{\sup\limits_{\Theta} L_n(\underline{\theta}|\underline{X})} = \frac{\hat{\theta}_{n_1+n_2}^{\left(\sum\limits_{i=1}^{n_1} X_i + \sum\limits_{i=1}^{n_2} Y_i\right)} (1 - \hat{\theta}_{n_1+n_2})^{\left(n_1+n_2 - \sum\limits_{i=1}^{n_1} X_i - \sum\limits_{i=1}^{n_2} Y_i\right)}}{\hat{\theta}_{n_1}^{\left(\sum\limits_{i=1}^{n_1} X_i\right)} \hat{\theta}_{2n_2}^{\left(\sum\limits_{i=1}^{n_2} Y_i\right)} (1 - \hat{\theta}_{n_1})^{\left(n_1 - \sum\limits_{i=1}^{n_1} X_i\right)} (1 - \hat{\theta}_{2n_2})^{\left(n_2 - \sum\limits_{i=1}^{n_2} Y_i\right)}}.$$

The null hypothesis is rejected if $-2\log\lambda(\underline{X}) > c$. If sample size is large, $-2\log\lambda(\underline{X}) \sim \chi_1^2$ distribution, as in the entire parameter space we estimate two parameters and in null space we estimate one parameter. H_0 is rejected if $-2\log\lambda(\underline{X}) > \chi_{1,1-\alpha}^2$ where $\chi_{1,1-\alpha}^2$ is $(1-\alpha)$th quantile of χ_1^2 distribution.

7.5   Chapter 6

6.8.1 In a multinomial distribution with 3 cells, the cell probabilities are

$$p_1(\theta) = p_2(\theta) = (1+\theta)/3 \quad \text{and} \quad p_3(\theta) = (1-2\theta)/3, \quad 0 < \theta < 1/2.$$

(i) Examine whether the distribution belongs to a one-parameter exponential family. On the basis of a random sample of size n from this distribution find the moment estimator based on the sufficient statistic and the maximum likelihood estimator of θ and examine if these are CAN. (ii) Use the result to derive Wald's test and a score test procedure for testing $H_0 : \theta = \theta_0$ against the alternative $H_1 : \theta \neq \theta_0$.

Solution: (i) Suppose $\underline{Y} = (Y_1, Y_2)'$ has a multinomial distribution in three cells with the given cell probabilities. Hence the joint probability mass function of $(Y_1, Y_2)'$ is given by

$$P_\theta[Y_1 = y_1, Y_2 = y_2] = \left(\frac{1+\theta}{3}\right)^{y_1+y_2} \left(\frac{1-2\theta}{3}\right)^{1-y_1-y_2},$$
$$y_1, y_2 = 0, 1 \text{ and } y_1 + y_2 \leq 1.$$

Thus,

$$\log P_\theta[Y_1 = y_1, Y_2 = y_2] = (y_1 + y_2)\log\left(\frac{1+\theta}{3}\right)$$
$$+ (1 - y_1 - y_2)\log\left(\frac{1-2\theta}{3}\right)$$
$$= (y_1 + y_2)\left[\log(1+\theta) - \log(1-2\theta)\right]$$
$$+ \log(1-2\theta) - \log 3$$
$$= U(\theta)K(y_1, y_2) + V(\theta) + W(y_1, y_2),$$

where $U(\theta) = \log(1 + \theta) - \log(1 - 2\theta)$, $K(y_1, y_2) = y_1 + y_2$, $V(\theta) = \log(1 - 2\theta) - \log 3$ and $W(y_1, y_2) = 0$. Thus, (1) the probability law of $(Y_1, Y_2)'$ is expressible in the form required in a one-parameter exponential family, (2) support of the probability mass function is $\{(0, 0), (0, 1), (1, 0)\}$ and it is free from θ, (3) the parameter space is $(0, 1/2)$ which is an open set, (4) $U'(\theta) = 1/(1 + \theta) + 2/(1 - 2\theta) = 3/(1 + \theta)(1 - 2\theta) \neq 0$ and (5) $K(y_1, y_2)$ and 1 are linearly independent because in the identity $a + b(y_1 + y_2) = 0$ if $y_1 = y_2 = 0$, then $a = 0$, if further in the identity $b(y_1 + y_2) = 0$, if either $y_1 = 0$ and $y_2 = 1$ or $y_1 = 1$ and $y_2 = 0$ then $b = 0$. Thus, all the requirements of a one-parameter exponential family are satisfied and hence the joint probability mass function of $(Y_1, Y_2)'$ belongs to a one-parameter exponential family. To find the maximum likelihood estimator of θ, the likelihood of θ corresponding to the data $\underline{X} \equiv \{X_1, X_2, X_3\}$ is given by

$$L_n(\theta|\underline{X}) = ((1 + \theta)/3)^{X_1 + X_2} ((1 - 2\theta)/3)^{X_3}, \quad X_1 + X_2 + X_3 = n,$$

where X_i is the frequency of i-th cell in the sample, $i = 1, 2, 3$. Likelihood is a differentiable function of θ, hence the likelihood equation and its solution $\hat{\theta}_n$ are given by

$$\frac{X_1 + X_2}{1 + \theta} - \frac{2X_3}{1 - 2\theta} = 0 \quad \Rightarrow \quad \hat{\theta}_n = \frac{X_1 + X_2 - 2X_3}{2n} = \frac{1}{2} - \frac{3}{2} \frac{X_3}{n}.$$

The second derivative

$$\frac{\partial^2}{\partial \theta^2} \log L_n(\theta|\underline{X}) = -\frac{X_1 + X_2}{(1 + \theta)^2} - \frac{4X_3}{(1 - 2\theta)^2} < 0, a.s.$$
$$\forall \; \theta \in (0, 1/2) \; \& \; \forall \; X_1, X_2.$$

Hence $\hat{\theta}_n$ is the maximum likelihood estimator of θ. Since the distribution of $(Y_1, Y_2)'$ belongs to a one-parameter exponential family, $\hat{\theta}_n$ is CAN for θ with approximate variance $1/nI(\theta)$. Now,

$$nI(\theta) = E_\theta \left(-\frac{\partial^2}{\partial \theta^2} \log L_n(\theta|\underline{X}) \right) = \frac{np_1(\theta) + np_2(\theta)}{(1 + \theta)^2} + \frac{4np_3(\theta)}{(1 - 2\theta)^2}$$

$$= \frac{n}{3} \left(\frac{2}{1 + \theta} + \frac{4}{1 - 2\theta} \right) = \frac{2n}{(1 + \theta)(1 - 2\theta)}.$$

Thus, $\hat{\theta}_n$ is CAN for θ with approximate variance $(1 + \theta)(1 - 2\theta)/2n$.
(ii) As discussed in Sect. 6.4, in real parameter set up, Wald's test statistic $T_n(W)$ and the score test statistic $T_n(S)$ are defined as

$$T_n(W) = \frac{\hat{\theta}_n - \theta_0}{s.e.(\hat{\theta}_n)} \quad \& \quad T_n(S) = \frac{\hat{\theta}_n - \theta_0}{s.e.(\hat{\theta}_n)|_{\theta_0}},$$

where $s.e.(\hat{\theta}_n)|_{\theta_0}$ is the standard error of $\hat{\theta}_n$ evaluated at θ_0. Here $s.e.(\hat{\theta}_n) = \sqrt{\frac{(1+\hat{\theta}_n)(1-2\hat{\theta}_n)}{2n}}$. In both the procedures under H_0, the asymptotic null distribution of both the test statistics is standard normal. In Wald's test procedure H_0 is rejected if $|T_n(W)| > c$ and in score test procedure H_0 is rejected if $|T_n(S)| > c$, where $c = a_{1-\alpha/2}$.

6.8.2   In a multinomial distribution with four cells, the cell probabilities are

$$p_1(\theta) = p_4(\theta) = (2 - \theta)/4 \quad \text{and} \quad p_2(\theta) = p_3(\theta) = \theta/4 , \quad 0 < \theta < 2.$$

Examine whether the distribution belongs to a one-parameter exponential family. On the basis of a random sample of size n from this distribution find the maximum likelihood estimator of θ and examine if it is CAN. Use the result to derive (i) a likelihood ratio test, (ii) Wald's test, (iii) a score test and (iv) Karl Pearson's chi-square test for testing $H_0 : \theta = \theta_0$ against the alternative $H_1 : \theta \neq \theta_0$.

Solution: Suppose $\underline{Y} = (Y_1, Y_2, Y_3)'$ has a multinomial distribution in four cells with the given cell probabilities. Hence the joint probability mass function $p(y_1, y_2, y_3) = P_\theta[Y_1 = y_1, Y_2 = y_2, Y_3 = y_3]$ of $(Y_1, Y_2, Y_3)'$ is given by

$$p(y_1, y_2, y_3) = \left(\frac{2-\theta}{4}\right)^{y_1 + y_4} \left(\frac{\theta}{4}\right)^{1 - y_1 - y_4},$$
$$y_1, y_2, y_3 = 0, 1 \quad \text{and} \quad y_1 + y_2 + y_3 + y_4 = 1.$$

Thus,

$$\log p(y_1, y_2, y_3) = -\log 4 + (y_1 + y_4)[\log(2 - \theta) - \log \theta] + \log \theta$$
$$= U(\theta)K(y_1, y_2, y_3) + V(\theta) + W(y_1, y_2, y_3),$$

where $U(\theta) = \log(2 - \theta) - \log \theta$, $K(y_1, y_2, y_3) = y_1 + y_4$, $V(\theta) = \log \theta$ and $W(y_1, y_2, y_3) = -\log 4$. Thus, (1) the probability law of $(Y_1, Y_2, Y_3)'$ is expressible in the form required in a one-parameter exponential family, (2) support of Y_i is $\{0, 1\}$, $i = 1, 2, 3$ and it is free from θ, (3) the parameter space is $(0, 2)$ which is an open set, (4) $U'(\theta) = -2/(2 - \theta)\theta \neq 0$ and (5) $K(y_1, y_2, y_3)$ and 1 are linearly independent. Thus, all the requirements of a one-parameter exponential family are satisfied and hence the joint probability mass function of $(Y_1, Y_2, Y_3)'$ belongs to a one-parameter exponential family. By Theorem 4.2.1, the maximum likelihood estimator of θ is CAN for θ with approximate variance $1/nI(\theta)$. To find the maximum likelihood estimator of θ, the log-likelihood of θ corresponding to the data $\underline{X} \equiv \{X_1, X_2, X_3, X_4\}$ is given by

$$\log L_n(\theta|\underline{X}) = -n\log 4 + (X_1 + X_4)[\log(2 - \theta) - \log \theta] + n \log \theta,$$
$$X_1 + X_2 + X_3 + X_4 = n,$$

where X_i is the frequency of ith cell in the sample, $i = 1, 2, 3, 4$. Likelihood is a differentiable function of θ, hence the likelihood equation and its solution $\hat{\theta}_n$ are given by

$$2n - 2(X_1 + X_4) - n\theta = 0 \quad \Rightarrow \quad \hat{\theta}_n = (2X_2 + 2X_3)/n \ .$$

Now,

$$I(\theta) = E_\theta \left(-\frac{\partial^2}{\partial\theta^2} \log p(Y_1, Y_2, Y_3) \right) = \left[\frac{1}{(2-\theta)^2} - \frac{1}{\theta^2} \right] E_\theta(Y_1 + Y_4) + \frac{1}{\theta^2}$$

$$= \left[\frac{1}{(2-\theta)^2} - \frac{1}{\theta^2} \right] \left(\frac{2(2-\theta)}{4} \right) + \frac{1}{\theta^2} = \frac{1}{\theta(2-\theta)} \ .$$

Thus, $\hat{\theta}_n$ is CAN for θ with approximate variance $\theta(2-\theta)/n$.

(i) To test $H_0 : \theta = \theta_0$ against the alternative $H_1 : \theta \neq \theta_0$, the likelihood ratio test statistic $\lambda(\underline{X})$ is given by

$$\lambda(\underline{X}) = \frac{\sup\limits_{\Theta_0} L_n(\theta|\underline{X})}{\sup\limits_{\Theta} L_n(\theta|\underline{X})} = \left(\frac{2 - \theta_0}{2 - \hat{\theta}_n} \right)^{X_1 + X_4} \left(\frac{\theta_0}{\hat{\theta}_n} \right)^{X_2 + X_3} \ .$$

The null hypothesis is rejected if $-2 \log \lambda(\underline{X}) > c$. If the sample size is large, then by Theorem 5.2.1, $-2 \log \lambda(\underline{X}) \sim \chi_1^2$ distribution. H_0 is rejected if $-2 \log \lambda(\underline{X}) > \chi_{1,1-\alpha}^2$.

(ii) and (iii): In real parameter set up, Wald's test statistic $T_n(W)$ and the score test statistic $T_n(S)$ are defined as

$$T_n(W) = \frac{\hat{\theta}_n - \theta_0}{s.e.(\hat{\theta}_n)} \quad \& \quad T_n(S) = \frac{\hat{\theta}_n - \theta_0}{s.e.(\hat{\theta}_n)|_{\theta_0}} \ .$$

Here $s.e.(\hat{\theta}_n) = \sqrt{\frac{\hat{\theta}_n(2-\hat{\theta}_n)}{n}}$. In both the procedures under H_0, the asymptotic null distribution of the test statistics is standard normal. In Wald's test procedure H_0 is rejected if $|T_n(W)| > c$ and in score test procedure H_0 is rejected if $|T_n(S)| > c$, where $c = a_{1-\alpha/2}$.

(iv) It is known that Karl Pearson's chi-square test statistic $T_n(P)$ is same as the score test statistic. Hence, H_0 is rejected if $|T_n(P)| > c$, where $c = a_{1-\alpha/2}$.

6.8.3 In a certain genetic experiment two different varieties of certain species are crossed. A specific characteristic of an offspring can occur at three levels A, B and C. According to the proposed model, probabilities for three levels A, B and C are 1/12, 3/12 and 8/12 respectively. Out of fifty offspring 6, 8 and 36 have levels A, B and C respectively. Test the validity of the proposed model by a score test and by Wald's test.

Solution: The probability model for the given experiment is a trinomial distribution with cell probabilities p_1, p_2, p_3 where $p_i > 0$ for $i = 1, 2, 3$ and $\sum_{i=1}^{3} p_i = 1$. Under the proposed model $p_1 = 1/12$, $p_2 = 3/12$ and $p_3 = 8/12$. Thus, to test the validity of the proposed model, the null hypothesis is $H_0 : p_1 = 1/12$, $p_2 = 3/12$, $p_3 = 8/12$ against the alternative that at least one of the p_i's are not as specified by the model. To find the score test statistic and Wald's test statistic, we find the expected frequencies e_i, expected under the proposed model, as $e_1 = n \times 1/12 = 4.1667$, $e_2 = n \times 3/12 = 12.5$ and $e_3 = n \times 8/12 = 33.3333$ with $n = 50$. Thus, the score test statistic $T_n(S) = \sum_{i=1}^{3} \frac{(o_i - e_i)^2}{e_i} = 2.21$ and Wald's test statistic $T_n(W) = \sum_{i=1}^{3} \frac{(o_i - e_i)^2}{o_i} = 3.03$. Under H_0, asymptotic null distribution of both the statistics is χ_2^2. The null hypothesis H_0 is rejected $T_n(S) > c$ and if $T_n(W) > c$, where c is determined corresponding to the given level of significance $\alpha = 0.05$ and the asymptotic null distribution. Here $c = \chi_{2,0.95}^2 = 5.99$. Thus, value of both the test statistics is less than c. Hence, data do not have sufficient evidence to reject H_0 and the proposed model may be considered to be valid.

6.8.4 On the basis of data in a 3×3 contingency table, derive a likelihood ratio test procedure and Karl Pearson's test procedure to test $H_0 : p_{ij} = p_{ji}, i \neq j = 1, 2, 3$ against the alternative $H_1 : p_{ij} \neq p_{ji}, i \neq j = 1, 2, 3$ for at least one pair.

Solution: To derive a likelihood ratio test procedure and Karl Pearson's test procedure, as a first step we find the maximum likelihood estimators of cell probabilities in the null setup. Under $H_0 : p_{ij} = p_{ji}, i \neq j = 1, 2, 3$, suppose $p_{12} = p_{21} = a$, $p_{13} = p_{31} = b$ and $p_{23} = p_{32} = c$, say. Thus, in the null setup, the parameter to be estimated is

$$\underline{\theta} = (p_{11}, p_{22}, p_{33}, a, b, c) \text{ such that } p_{11}, p_{22}, p_{33}, a, b, c > 0$$
$$\& \ p_{11} + p_{22} + p_{33} + 2a + 2b + 2c = 1.$$

Corresponding to observed cell frequencies $n_{ij}, i, j = 1, 2, 3$, adding to n, the log-likelihood of $\underline{\theta}$ is given by

$$\log L_n(\underline{\theta}|n_{ij}, i, j = 1, 2, 3) = n_{11} \log p_{11} + n_{22} \log p_{22} + n_{33} \log p_{33}$$
$$+ (n_{12} + n_{21}) \log a + (n_{13} + n_{31}) \log b$$
$$+ (n_{23} + n_{32}) \log c.$$

Using Lagranges' method of multipliers, we maximize the log-likelihood with respect to variation in $\underline{\theta}$ subject to the condition that $p_{11} + p_{22} + p_{33} + 2a + 2b + 2c = 1$. Thus, the maximum likelihood estimators of cell probabilities in the null setup are

$$\hat{p}_{iin} = \frac{n_{ii}}{n}, i = 1, 2, 3, \quad \hat{a}_n = \frac{n_{12} + n_{21}}{2n}, \quad \hat{b}_n = \frac{n_{13} + n_{31}}{2n} \ \& \ \hat{c}_n = \frac{n_{23} + n_{32}}{n}.$$

Once we have these maximum likelihood estimators, we can find the expected frequencies and can carry out both the likelihood ratio test procedure and Karl Pearson's test procedure.

6.8.5 On the basis of data in a 2×3 contingency table, derive a likelihood ratio test procedure and Karl Pearson's test procedure to test $H_0 : p_{11} = p_{12} = p_{13}$ against the alternative that there is no restriction as specified in H_0.

Solution: As in the previous example, to derive a likelihood ratio test procedure and Karl Pearson's test procedure, we first find the maximum likelihood estimators of cell probabilities in the null setup. Under $H_0 : p_{11} = p_{12} = p_{13} = a$, say. Thus, in the null setup, the parameter to be estimated is $\underline{\theta} = (a, p_{21}, p_{22}, p_{23})$ such that $a, p_{21}, p_{22}, p_{23} > 0$ and $3a + p_{21} + p_{22} + p_{23} = 1$. Corresponding to observed cell frequencies $n_{ij}, i = 1, 2, \ j = 1, 2, 3$, adding to n, the log-likelihood of $\underline{\theta}$ is given by

$$\log L_n(\underline{\theta}|n_{ij}, i, j = 1, 2, 3) = (n_{11} + n_{12} + n_{13}) \log a$$
$$+ n_{21} \log p_{21} + n_{22} \log p_{22} + n_{23} \log p_{23}.$$

Using Lagranges' method of multipliers, we maximize the log-likelihood with respect to variation in $\underline{\theta}$ subject to the condition that $3a + p_{21} + p_{22} + p_{23} = 1$. Thus, the maximum likelihood estimators of cell probabilities in the null setup are

$$\hat{a}_n = \frac{n_{11} + n_{12} + n_{13}}{3n}, \qquad \hat{p}_{2jn} = \frac{n_{2j}}{n}, j = 1, 2, 3.$$

Once we have these maximum likelihood estimators, we can find the expected frequencies and can carry out both the likelihood ratio test procedure and Karl Pearson's test procedure.

6.8.6 Suppose $\{X_1, X_2, \ldots, X_n\}$ is a random sample from a Laplace distribution with location parameter θ and scale parameter 1. Derive a large sample test procedure to test $H_0 : \theta = \theta_0$ against the alternative $H_0 : \theta > \theta_0$ and examine whether it is a consistent test procedure.

Solution: The probability density function $f(x, \theta)$ of X having Laplace distribution with location parameter θ and scale parameter 1 is given by

$$f(x, \theta) = \frac{1}{2} \exp\{-|x - \theta|\}, \ x \in \mathbb{R}, \ \theta \in \mathbb{R}.$$

For this distribution, the sample median is the maximum likelihood estimator of θ and it is CAN with approximate variance $1/n$. Hence, the test procedure to test $H_0 : \theta = \theta_0$ against $H_1 : \theta > \theta_0$ is based on the sample median $X_{([n/2]+1)}$. The test statistic T_n is given by $T_n = \sqrt{n}(X_{([n/2]+1)} - \theta_0)$. For large n under H_0, $T_n \sim N(0, 1)$ distribution. The null hypothesis $H_0 : \theta = \theta_0$ is rejected against $H_1 : \theta > \theta_0$ if $T_n > c$, where c is determined corresponding to the given level of significance α and the asymptotic null distribution of T_n. Thus, $c = a_{1-\alpha}$. To examine whether it is a consistent test, we find the power function $\beta(\theta)$. Thus,

$$\beta(\theta) = P_\theta[T_n > c] = P_\theta[\sqrt{n}\left(X_{([n/2]+1)} - \theta_0\right) > c]$$

$$= P_\theta\left[\sqrt{n}\left(X_{([n/2]+1)} - \theta\right) > \sqrt{n}\left(\theta_0 - \theta + \frac{c}{\sqrt{n}}\right)\right]$$

$$= 1 - \Phi\left(c + \sqrt{n}(\theta_0 - \theta)\right) \to 1 \quad \forall \quad \theta > \theta_0 .$$

Hence the test is consistent.

7.6   Multiple Choice Questions

In the following multiple choice questions, more than one options may be correct. Answers to all the questions are given in Table 7.1 **at the end of the chapter.**

7.6.1   Chapter 2: Consistency of an Estimator

1. Suppose $X_{ij} \sim N(\mu_i, \sigma^2)$ distribution for $i = 1, 2, \ldots k, j = 1, 2, \ldots n$ and X_{ij}'s are independent random variables. If $\hat{\mu}_{in} = \overline{X}_{in} = \sum_{j=1}^{n} X_{ij}/n$ & $\hat{\sigma}_n^2 = \sum_{i=1}^{k}\sum_{j=1}^{n}(X_{ij} - \overline{X}_{in})^2/nk$, then which of the following statements is/are correct?

(a) $\hat{\mu}_{in}$ is a maximum likelihood estimator of μ_i and is consistent for μ_i.
(b) $\hat{\sigma}_n^2$ is an unbiased estimator of σ^2.
(c) $\hat{\sigma}_n^2$ is a maximum likelihood estimator of σ^2 and is consistent for σ^2.
(d) $\hat{\mu}_{in}$ is not consistent for μ_i.

2. Suppose $\{X_1, X_2 \ldots X_n\}$ are independent and identically distributed random variables each following a uniform $U(0, \theta)$ distribution. Then which of the following statements is/are NOT consistent for θ?

(a) $X_{(1)}$.
(b) $X_{(n)}$.
(c) $X_{(n-1)}$.
(d) $X_{([n/2]+1)}$.

3. Suppose $\{X_1, X_2 \ldots X_n\}$ are independent and identically distributed random variables each following normal $N(\mu, \mu^2)$ distribution. Which of the following statements is/are correct?

(a) $\overline{X}_n$ is a consistent estimator of μ.

(b) $S_n^2 = \frac{1}{n}\sum_{i=1}^n (X_i - \overline{X})^2$ is a consistent estimator of μ^2.

(c) $S_n = \sqrt{S_n^2}$ cannot be a consistent estimator of μ.

(d) Sample median is a consistent estimator of μ.

4. Suppose $\{X_1, X_2 \ldots X_n\}$ are independent random variables, where $X_i \sim U(0, i\theta)$ for $i = 1, 2, \ldots n$. Which of the following statements is/are correct?

(a) $\frac{2}{n}\sum_1^n \frac{X_i}{i}$ is not consistent for θ.

(b) Sample mean is consistent for θ.

(c) $\max\{X_1, X_2/2, X_3/3, \ldots X_n/n\}$ is consistent for θ.

(d) $X_{(n)}$ is consistent for θ.

5. Suppose $\{X_1, X_2 \ldots X_n\}$ are independent and identically distributed random variables with probability density function $f(x, \theta) = \theta x^{\theta-1}$, $0 < x < 1$, $\theta > 0$. Which of the following statements is/are correct?

(a) $\frac{\log(0.5)}{\log(X_{([n/2]+1)})}$ is consistent for θ.

(b) $\frac{\overline{X}_n}{1-\overline{X}_n}$ is consistent for θ.

(c) $\overline{X}_n$ is consistent for θ.

(d) $\frac{\log(X_{([n/2]+1)})}{\log(0.5)}$ is consistent for θ.

6. Suppose $\{X_1, X_2 \ldots X_n\}$ are independent and identically distributed random variables each following Bernoulli $B(1, \theta)$ distribution, where $\theta \in [0.25, 0.75]$. Which of the following is/are NOT consistent for θ?

(a) $\overline{X}_n$.

(b) $\hat{\theta} = \begin{cases} 0.25 & \text{if}\, \overline{X}_n < 0.25 \\ \overline{X}_n & \text{if}\, 0.25 \le \overline{X}_n \le 0.75 \\ 0.75 & if\, \overline{X}_n > 0.75 \end{cases}$

(c) $\min\{0.25, \overline{X}_n\}$.

(d) $\frac{1}{n}\sum X_i^2$.

7. Suppose $\{X_1, X_2 \ldots X_n\}$ are independent and identically distributed random variables with probability density function $f(x, \theta) = 2\theta^2/x^3$ $x \ge \theta$, $\theta > 0$. Then which of the following statements is/are correct?

(a) $\frac{2}{3}X_{([n/3]+1)}^2$ is a consistent estimator of θ^2.

(b) $\frac{1}{3}X_{([n/3]+1)}^2$ is a consistent estimator of θ^2.

(c) $\frac{2}{3}X_{([n/3]+1)}$ is a consistent estimator of θ.

(d) $\sqrt{\frac{2}{3}}X_{([n/3]+1)}$ is a consistent estimator of θ.

8. Suppose $\{X_1, X_2 \ldots X_n\}$ is a random sample from a uniform $U(0, 3\theta)$ distribution, $\theta > 0$. If $T_n = \max\{X_1, \ldots, X_n\}/3$, which of the following statements is/are NOT correct?

(a) T_n is consistent for θ.
(b) T_n is unbiased for θ.
(c) T_n is a sufficient statistic for the family of $U(0, 3\theta)$ distributions.
(d) T_n is a maximum likelihood estimator of θ.

9. Suppose $\{X_1, X_2 \ldots X_n\}$ is a random sample from the distribution with the following probability density function.

$$f(x, \mu, \alpha) = \begin{cases} \frac{1}{\Gamma\alpha}(x - \mu)^{\alpha-1} \exp\{-(x - \mu)\} & x \geq \mu, \ \mu \in \mathbb{R}, \ \alpha > 0 \\ 0 & \text{otherwise} \end{cases}$$

Which of the following statements is/are correct?

(a) $\overline{X}_n$ is a consistent estimator of μ.
(b) $\overline{X}_n - S_n^2$ is a consistent estimator of μ, where S_n^2 is the sample variance.
(c) S_n^2 is a consistent estimator of α.
(d) $\overline{X}_n$ is a consistent estimator of $\mu + \alpha$.

10. Suppose $\{X_1, X_2 \ldots X_n\}$ are independent and identically distributed random variables with probability density function $f(x, \theta) = (2/\theta^2)(x - \theta)$, $\theta < x < 2\theta$, $\theta > 0$. Which of the following statements is/are correct?

(a) $X_{(1)}$ is a consistent estimator of θ.
(b) $X_{(n)}$ is a consistent estimator of θ.
(c) $X_{(n)}/2$ is a consistent estimator of θ.
(d) $\overline{X}_n$ is a consistent estimator of θ.

11. Suppose $(X, Y)'$ is a random vector with joint distribution given by

$$f(x, y) = \frac{e^{-\beta y}(\beta y)^x}{x!}\theta e^{-\theta y} \quad y \geq 0, \ x = 0, 1, 2, \ldots, \ \beta, \theta > 0.$$

Then $\overline{X}_n$ is a consistent estimator of

(a) $\frac{\beta}{\theta}$.
(b) β.
(c) $\frac{\theta}{\beta}$.
(d) $\frac{\beta}{\theta^2}$.

12. Suppose $\{X_1, X_2, \ldots, X_n\}$ is a random sample of size n from a uniform $U(\theta - 1, \theta + 1)$ distribution. Then which of the following statements is/are correct?

    (a) sample mean is consistent for θ.
    (b) sample median is consistent for θ.
    (c) $X_{(n)}$ is consistent for θ.
    (d) $X_{(1)}$ is consistent for θ.

13. Suppose $\{X_1, X_2, \ldots, X_{2n}\}$ is a random sample of size $2n$ from a uniform $U(\theta - 1, \theta + 1)$ distribution. Then which of the following statements is/are correct?

    (a) $X_{(1)}$ is consistent for θ.
    (b) $X_{(1)} + 1$ is consistent for θ.
    (c) $X_{(n+1)}$ is consistent for θ.
    (d) $X_{(2n)} - 1$ is consistent for θ.

14. Following are two statements about an estimator T_n of θ.
    (I) If MSE of T_n as an estimator of θ converges to 0 as $n \to \infty$, then T_n is a consistent estimator of θ.
    (II) If T_n is a consistent estimator of θ, then MSE of T_n as an estimator of θ converges to 0 as $n \to \infty$.
    Which of the following statements is/are correct?

    (a) Both (I) and (II) are true.
    (b) Both (I) and (II) are false.
    (c) (I) is true but (II) is false.
    (d) (I) is false but (II) is true.

15. Suppose $\{X_1, X_2, \ldots, X_n\}$ is a random sample from Cauchy $C(\theta, 1)$ distribution. Then which of the following statements is/are correct?

    (a) Sample mean is consistent for θ.
    (b) Sample median is consistent for θ.
    (c) The maximum likelihood estimator of θ is consistent for θ.
    (d) The first sample quartile is consistent for θ.

16. Suppose $\{X_1, X_2, \ldots, X_n\}$ is a random sample from a Laplace distribution with probability density function $f(x, \theta)$ given by

$$f(x, \theta) = (1/2\theta) \exp\{-|x|/\theta\}, \quad x \in \mathbb{R}, \quad \theta > 0.$$

Which of the following estimators are consistent for θ?

(a) Sample mean.
(b) sample median.
(c) $\sum_{i=1}^{n} |X_i|/n$.
(d) $(\sum_{i=1}^{n} X_i^2/n)^{1/2}$.

17. Following are two statements about an estimator T_n of θ.
(I) If T_n is a strongly consistent estimator of θ, then it is a weakly consistent estimator of θ.
(II) If T_n is a weakly consistent estimator of θ, then it is a strongly consistent estimator of θ.
Which of the following statements is/are correct?

(a) Both (I) and (II) are true.
(b) Both (I) and (II) are false.
(c) (I) is true but (II) is false.
(d) (I) is false but (II) is true.

18. Suppose $\{X_1, X_2, \ldots, X_n\}$ is a random sample from a Bernoulli distribution $B(1, \theta)$. Which of the following statements is/are correct?

(a) Sample mean is a consistent estimator of θ.
(b) Sample mean is a uniformly consistent estimator of θ.
(c) Sample mean is a strongly consistent estimator of θ.
(d) None of a, b, c is true.

19. Suppose $\{X_1, X_2, \ldots, X_n\}$ is a random sample from a distribution with finite second order moment. Which of the following statements is/are correct?

(a) Sample mean is a consistent and unbiased estimator of population mean.
(b) Sample mean is a consistent but a biased estimator of population mean.
(c) Sample variance is a consistent and unbiased estimator of population variance.
(d) Sample variance is a consistent but a biased estimator of population variance.

20. Following are two statements about an estimator T_n of θ.
(I) If T_n is a consistent estimator of θ and g is a continuous function, then $g(T_n)$ is a consistent estimator of $g(\theta)$.
(II) If T_n is a consistent estimator of θ and g is not a continuous function, then $g(T_n)$ is not a consistent estimator of $g(\theta)$.
Which of the following statements is/are correct?

(a) Both (I) and (II) are true.
(b) Both (I) and (II) are false.
(c) (I) is true but (II) is false.
(d) (I) is false but (II) is true.

7.6.2   Chapter 3: Consistent and Asymptotically Normal Estimators

1. Suppose $\{X_1, X_2 \ldots X_n\}$ are independent and identically distributed random variables, each with probability density function $f(x, \theta) = (2/\theta^2)(x - \theta)$, $\theta < x < 2\theta$, $\theta > 0$. Which of the following statements is/are NOT correct?

   (a) $\sqrt{n}(\overline{X}_n - \frac{5\theta}{3}) \xrightarrow{L} N(0, \frac{\theta^2}{9})$ as $n \to \infty$.
   (b) $\sqrt{n}(\frac{3}{5}\overline{X}_n - \theta) \xrightarrow{L} N(0, \frac{\theta^2}{50})$ as $n \to \infty$.
   (c) $\sqrt{n}(\overline{X}_n - \frac{3\theta}{5}) \xrightarrow{L} N(0, \frac{\theta^2}{9})$ as $n \to \infty$.
   (d) $\sqrt{n}(\overline{X}_n - \frac{5\theta}{3}) \xrightarrow{L} N(0, \frac{\theta^2}{18})$ as $n \to \infty$.

2. Suppose $\{X_1, X_2 \ldots X_n\}$ are independent and identically distributed random variables, each following a Poisson distribution with mean μ. Then the asymptotic distribution of $\sqrt{n}(e^{-\overline{X}_n} - e^{-\mu})$ is given by

   (a) $N(0, \mu e^{-\mu})$.
   (b) $N(0, e^{-\mu})$.
   (c) $N(0, \mu^2 e^{-2\mu})$.
   (d) $N(0, \mu e^{-2\mu})$.

3. Suppose $\{X_1, X_2 \ldots X_n\}$ is a random sample from a uniform $U(0, \theta + 1)$ distribution. Then the maximum likelihood estimator of θ is

   (a) $X_{(1)} + 1$.
   (b) $X_{(n)} - 1$.
   (c) $X_{(n)}$.
   (d) any value in the interval $[X_{(1)}, X_{(n)}]$.

4. Suppose $\{X_1, X_2 \ldots X_n\}$ are independent and identically distributed random variables each having an exponential distribution with mean θ. Then the asymptotic distribution of $\frac{\sqrt{n}}{S_n}(\overline{X}_n - \theta)$, where $S_n^2 = \frac{1}{n}\sum_1^n X_i^2 - \overline{X}_n^2$ is

   (a) $N(0, 1)$.
   (b) $N(0, \theta)$.
   (c) $N(1, 1)$.
   (d) exponential with mean θ.

5. Suppose $\{X_1, X_2 \ldots X_n\}$ are independent and identically distributed random variables, each with probability density function

   $$f(x, \alpha, \theta) = \frac{1}{\theta^\alpha \Gamma \alpha} x^{\alpha-1} e^{-x/\theta} \quad x > 0, \ \alpha, \ \theta > 0.$$

   Then which of the following statements is/are correct?

(a) $\sqrt{n}(\overline{X}_n - \alpha\theta) \xrightarrow{L} N(0, \alpha\theta^2)$.

(b) $\sqrt{n}(\log \overline{X}_n - \log \theta) \xrightarrow{L} N(0, \alpha)$.

(c) $\sqrt{n}(\overline{X}_n - \alpha) \xrightarrow{L} N(0, \theta)$.

(d) $\sqrt{n}(\log \overline{X}_n - \log \alpha\theta) \xrightarrow{L} N(0, 1/\alpha)$.

6. Suppose $\{X_1, X_2 \ldots X_n\}$ are independent and identically distributed random variables, each with probability density function $f(x, \theta) = \theta/x^2$, $x \geq \theta$. Then which of the following statements is/are correct?

(a) $\sqrt{n}(X_{([n/4]+1)} - \frac{4\theta}{3}) \xrightarrow{L} N(0, 16\theta^2)$.

(b) $\sqrt{n}(X_{([n/4]+1)} - \frac{4\theta}{3}) \xrightarrow{L} N(0, 16\theta^2/27)$.

(c) $\sqrt{n}(\frac{3}{4}X_{([n/4]+1)} - \theta) \xrightarrow{L} N(0, 9\theta^2)$.

(d) $\sqrt{n}(\frac{3}{4}X_{([n/4]+1)} - \theta) \xrightarrow{L} N(0, \theta^2/3)$.

7. Suppose $\{X_1, X_2 \ldots X_n\}$ are independent and identically distributed random variables, each following an exponential distribution with mean θ. Then $e^{-\overline{X}_n}$ is a CAN estimator of $e^{-\theta}$ with approximate variance

(a) $\theta e^{-2\theta}/n$.

(b) $\theta^2 e^{-\theta}/n$.

(c) $\theta e^{-\theta}/n$.

(d) $\theta^2 e^{-2\theta}/n$.

8. The moment estimator of θ based on a random sample of size n from a distribution with the probability density function $f(x, \theta) = (\theta + 1)x^\theta$, $0 < x < 1$, $\theta > -1$, is given by

(a) $\overline{X}_n/(1 - \overline{X}_n)$.

(b) $\overline{X}_n$.

(c) $\frac{1-2\overline{X}_n}{\overline{X}_n-1}$.

(d) $\frac{4\overline{X}_n-1}{2\overline{X}_n-1}$.

9. Suppose $\{X_1, X_2 \ldots X_n\}$ is a random sample from a uniform $U(0, 2\theta)$ distribution. Then the maximum likelihood estimator of θ

(a) is $X_{(n)}/2$.

(b) is $X_{(n)}$.

(c) is $X_{([n/2]+1)}$.

(d) does not exist.

10. Suppose $\{X_1, X_2 \ldots X_n\}$ is a random sample from the distribution with probability density function

$$f(x, \theta, \lambda) = \frac{\lambda}{\pi} \frac{1}{\lambda^2 + (x - \theta)^2}, \quad x \in \mathbb{R} \ \lambda > 0, \ \theta \in \mathbb{R}.$$

Then which of the following statements is/are correct?

(a) $\sqrt{n}(X_{([n/4]+1)} - (\theta - \lambda)) \xrightarrow{L} N(0, \frac{\pi^2 \lambda^2}{4})$.

(b) $\sqrt{n}(X_{([n/4]+1)} - (\lambda - \theta)) \xrightarrow{L} N(0, \frac{\pi^2 \lambda^2}{4})$.

(c) $\sqrt{n}(X_{([3n/4]+1)} - (\theta + \lambda)) \xrightarrow{L} N(0, \frac{3\pi^2 \lambda^2}{4})$.

(d) $\sqrt{n}(X_{([3n/4]+1)} - (\theta + \lambda)) \xrightarrow{L} N(0, \frac{\pi^2 \lambda^2}{4})$.

11. Suppose $\{X_1, X_2 \ldots X_n\}$ is a random sample from a normal $N(\mu, \sigma^2)$ distribution. Then the approximate dispersion matrix of $\sqrt{n}((\overline{X}_n, S_n^2)' - (\mu, \sigma^2)')$ is given by

(a) $\begin{pmatrix} \sigma^2 & 0 \\ 0 & \sigma^4 \end{pmatrix}$.

(b) $\begin{pmatrix} \sigma^2 & 0 \\ 0 & 2\sigma^4 \end{pmatrix}$.

(c) $\begin{pmatrix} 1/\sigma^2 & 0 \\ 0 & 1/\sigma^4 \end{pmatrix}$.

(d) $\begin{pmatrix} \sigma^2 & 0 \\ 0 & 1/2\sigma^4 \end{pmatrix}$.

12. Suppose $\{X_1, X_2 \ldots X_n\}$ is a random sample from an exponential distribution with mean θ. Then $100(1 - \alpha)\%$ asymptotic confidence interval for θ

(a) is $\left(\frac{\sqrt{n}\overline{X}_n}{a_{1-\alpha/2}+\sqrt{n}}, \frac{\sqrt{n}\overline{X}_n}{a_{\alpha/2}+\sqrt{n}} \right)$.

(b) is $\left(\frac{\sqrt{n}\overline{X}_n}{a_{\alpha/2}+\sqrt{n}}, \frac{\sqrt{n}\overline{X}_n}{a_{1-\alpha/2}+\sqrt{n}} \right)$.

(c) is $\left(\frac{\overline{X}_n}{a_{1-\alpha/2}+\sqrt{n}}, \frac{\overline{X}_n}{a_{\alpha/2}+\sqrt{n}} \right)$.

(d) can not be constructed using $\overline{X}_n$.

13. Suppose $\{X_1, X_2 \ldots X_n\}$ is a random sample from a Poisson $Poi(\theta)$ distribution. The variance stabilizing transformation for constructing an asymptotic confidence interval for θ is

(a) $g(\theta) = \log(\theta)$.

(b) $g(\theta) = 1/\theta$.

(c) $g(\theta) = 1/\sqrt{\theta}$.

(d) $g(\theta) = 2\sqrt{\theta}$.

14. Suppose $\{X_1, X_2 \ldots X_n\}$ is a random sample from a normal $N(\theta, \theta^2)$, $\theta > 0$ distribution. Then, the variance stabilizing transformation for constructing an asymptotic confidence interval for θ based on $\overline{X}_n$ is

   (a) $g(\theta) = \log(\theta)$.
   (b) $g(\theta) = 1/\theta$.
   (c) $g(\theta) = 1/\sqrt{\theta}$.
   (d) $g(\theta) = 2\sqrt{\theta}$.

15. Following are two statements about the estimator of an indexing real parameter θ.
   (I) If T_n is a consistent estimator of θ, then its asymptotic distribution is degenerate at θ.
   (II) If $\sqrt{n}(T_n - \theta) \overset{L}{\to} Z$, then T_n is a consistent estimator of θ.
   Which of the following statements is/are correct?

   (a) Both (I) and (II) are true.
   (b) Both (I) and (II) are false.
   (c) (I) is true but (II) is false.
   (d) (I) is false but (II) is true.

16. Suppose $a_n(T_n - \theta) \overset{L}{\to} Z$, where $a_n \to \infty$ as $n \to \infty$. Which of the following statements is/are correct?

   (a) If g is a continuous function, then $g(a_n(T_n - \theta)) \overset{L}{\to} g(Z)$.
   (b) If g is a differentiable function, then $g(a_n(T_n - \theta)) \overset{L}{\to} g(Z)$.
   (c) Suppose $Z \sim N(0, 1)$. If g is a differentiable function, then the asymptotic distribution of $g(a_n(T_n - \theta))$ is also normal.
   (d) Suppose $Z \sim N(0, 1)$. If g is a differentiable function and the first derivative is always non-zero, then the asymptotic distribution of $g(a_n(T_n - \theta))$ is also normal.

17. Suppose $X \sim C(\theta, 1)$. Which of the following statements is/are correct?

   (a) Approximate variance of the sample median is less than the approximate variance of the sample first quartile.
   (b) Approximate variance of the sample median is less than the approximate variance of the sample third quartile.
   (c) Approximate variance of the sample third quartile is less than the approximate variance of the sample first quartile.
   (d) Approximate variance of the sample third quartile is the same as the approximate variance of the sample first quartile.

18. Suppose $X \sim N(\theta, 1)$. Which of the following statements is/are correct?

   (a) Approximate variance of the sample median is less than the approximate variance of the sample first quartile.
   (b) Approximate variance of the sample median is less than the approximate variance of the sample third quartile.
   (c) Approximate variance of the sample third quartile is less than the approximate variance of the sample first quartile.
   (d) Approximate variance of the sample third quartile is the same as the approximate variance of the sample first quartile.

19. Suppose a distribution of X is indexed by a parameter $\underline{\theta} = (\theta_1, \theta_2)'$. Following are two statements about the estimator of $\underline{\theta}$.
   (I) If T_{in} is a consistent estimator of θ_i, $i = 1, 2$, then $\underline{T}_n = (T_{1n}, T_{2n})'$ is a consistent estimator of $\underline{\theta}$.
   (II) If T_{in} is a CAN estimator of θ_i, $i = 1, 2$ and both have the same norming factor, then $\underline{T}_n = (T_{1n}, T_{2n})'$ is a CAN estimator of $\underline{\theta}$.

   Which of the following statements is/are correct?

   (a) Both (I) and (II) are true.
   (b) Both (I) and (II) are false.
   (c) (I) is true but (II) is false.
   (d) (I) is false but (II) is true.

20. Suppose $(X, Y)' \sim N_2(0, 0, 1, 1, \rho)$ distribution, $-1 < \rho < 1$. To construct a large sample confidence interval for ρ, based on the sample correlation coefficient, the variance stabilizing transformation is

   (a) $g(\rho) = \log \frac{1+\rho}{1-\rho}$.
   (b) $g(\rho) = \frac{1}{2} \log \frac{1-\rho}{1+\rho}$.
   (c) $g(\rho) = \log \frac{1-\rho}{1+\rho}$.
   (d) $g(\rho) = \frac{1}{2} \log \frac{1+\rho}{1-\rho}$.

7.6.3   Chapter 4: CAN Estimators in Exponential and Cramér Families

1. Suppose $\{X_1, X_2 \ldots X_n\}$ is a random sample from the distribution with probability function $f(x, \theta) = \theta x^{\theta-1}, 0 < x < 1, \theta > 0$. Then which of the following statements is/are correct?

   (a) The maximum likelihood estimator of θ is a CAN estimator of θ.
   (b) $\frac{X_n}{X_n - 1}$ is MLE of θ.

(c) $\overline{X}_n$ is a CAN estimator of θ.
(d) There does not exist any CAN estimator of θ which attains the Cramér-Rao lower bound.

2. Suppose $\{X_1, X_2 \ldots X_n\}$ is a random sample from the distribution with probability density function given by $f(x, \theta) = (\theta + 1)x^{\theta}$, $0 < x < 1$, $\theta > -1$. Suppose T_n is a moment estimator of θ based on a sufficient statistic. Which of the following statements is/are correct?

(a) T_n is not CAN for θ.
(b) T_n is not MLE of θ.
(c) $T_n = \frac{1-4\overline{X}_n}{2\overline{X}_n-1}$.
(d) $T_n = -\frac{n}{\sum \log(X_i)} - 1$.

3. In a multinomial distribution with 4 cells, the cell probabilities and the observed cell frequencies are given by

$$1/16 + \theta, \quad 3/16 - \theta, \quad 3/16 - \theta, \quad 9/16 + \theta \ \& \ 31, 37, 35, 187$$

respectively. Then the likelihood equation can be simplified as

(a) $31(1 + 16\theta) - 72(3 - 16\theta) + 187(9 + 16\theta) = 0$.
(b) $(\frac{1}{16} + \theta)^{31}(\frac{3}{16} - \theta)^{72}(\frac{9}{16} + \theta)^{187} = 0$.
(c) $\frac{31}{1+16\theta} - \frac{72}{3-16\theta} + \frac{187}{9+16\theta} = 0$.
(d) $\frac{72}{1+16\theta} - \frac{31}{3-16\theta} + \frac{187}{9+16\theta} = 0$.

4. Suppose $\{X_1, X_2 \ldots X_n\}$ is a random sample from a double exponential distribution with location parameter θ, with probability density function given by

$$f(x) = 0.5e^{-|x-\theta|} \quad x \in \mathbb{R}, \ \theta \in \mathbb{R}$$

Then the maximum likelihood estimator of θ

(a) does not exist.
(b) is the sample median.
(c) is the sample mean.
(d) is the mean of $|X_i|, i = 1, 2, \ldots, n$.

5. Suppose $\{X_1, X_2 \ldots X_n\}$ are independent random variables, where $X_i \sim U(0, i\theta)$ for $i = 1, 2, \ldots, n$. Then the maximum likelihood estimator of θ is

(a) $\overline{X}_n$.
(b) $\max\{X_1, X_2 \ldots X_n\}$.

(c) $\max\{X_1, X_2/2, X_3/3, \ldots X_n/n\}$.

(d) $\frac{1}{n}\sum_{i=1}^{n}\frac{X_i}{i}$.

6. Suppose $(X, Y)'$ is a random vector with joint distribution given by

$$\frac{e^{-\lambda}\lambda^x}{x!}\binom{x}{y}\theta^y(1-\theta)^{x-y} \quad y = 0, 1, 2, \ldots x, \quad x = 0, 1, 2\ldots, \quad 0 < \theta < 1, \quad \lambda > 0.$$

Then a system of likelihood equations based on a random sample of size n are given by

(a) $-n + \frac{\sum_{i=1}^{n}X_i}{\lambda} = 0$ & $\frac{\sum_{i=1}^{n}Y_i}{\theta} - \frac{\sum_{i=1}^{n}X_i-\sum_{i=1}^{n}Y_i}{1-\theta} = 0.$

(b) $-1 + \frac{X_i}{\lambda} = 0$ & $\frac{Y_i}{\theta} - \frac{X_i-Y_i}{1-\theta} = 0.$

(c) $-n + \lambda\sum_{i=1}^{n}X_i = 0$ & $\frac{\sum_{i=1}^{n}Y_i}{\theta} - \frac{\sum_{i=1}^{n}X_i-\sum_{i=1}^{n}Y_i}{1-\theta} = 0.$

(d) none of the above.

7. Suppose $\{X_1, X_2 \ldots X_n\}$ is a random sample from a uniform $U(-\theta, \theta)$ distribution. Then the maximum likelihood estimator of θ is

(a) $X_{(n)}$.

(b) $\max\{|X_1|, \ldots |X_n|\}$.

(c) $-X_{(1)}$.

(d) $\max\{-X_{(1)}, X_{(n)}\}$.

8. Suppose $\{X_1, X_2 \ldots X_n\}$ is a random sample from the distribution with probability density function

$$f(x, \mu, \sigma) = \frac{1}{\sigma}\exp\left(-\frac{x-\mu}{\sigma}\right), \quad x > \mu, \sigma > 0.$$

If $\hat{\mu}_n$ and $\hat{\sigma}_n$ denote the maximum likelihood estimators of μ and σ respectively, then which of the following statements is/are correct?

(a) $(\hat{\mu}_n, \hat{\sigma}_n)' = (X_{(1)}, S_n^2)'$, where S_n^2 is the sample variance.

(b) $(\hat{\mu}_n, \hat{\sigma}_n)' = (X_{(1)}, \overline{X}_n)'$.

(c) $(\hat{\mu}_n, \hat{\sigma}_n)' = (X_{(1)}, \overline{X}_n - X_{(1)})'$.

(d) $(\hat{\mu}_n, \hat{\sigma}_n)' = (\overline{X}_n, S_n^2)'$.

9. On the basis of a random sample of size n from a probability law which is indexed by a real valued parameter θ, which of the following statements is/are correct ?

(a) the maximum likelihood estimator of θ may not exist.

(b) the maximum likelihood estimator of θ, if it exists, is always a function of sufficient statistic.

(c) the maximum likelihood estimator is always consistent for θ.

(d) the maximum likelihood estimator of θ is always asymptotically normally distributed.

10. Following are two statements about the estimator of an indexing parameter $\underline{\theta}$ of a k-parameter exponential family.

(I) The maximum likelihood estimator of $\underline{\theta}$ is a CAN estimator of $\underline{\theta}$.

(II) The moment estimator of $\underline{\theta}$ based on a sufficient statistic is a CAN estimator of $\underline{\theta}$.

Which of the following statements is/are correct?

(a) Both (I) and (II) are true.

(b) Both (I) and (II) are false.

(c) (I) is true but (II) is false.

(d) (I) is false but (II) is true.

11. Suppose X follows an exponential distribution with mean $1/\theta$. Suppose X is discretized to define Y as $Y = k$ if $k < X \le k + 1$ for $k = 0, 1, \ldots$. Suppose a random sample $\{Y_1, Y_2, \ldots, Y_n\}$ is available. Then the moment estimator $\hat{\theta}$ of θ based on a sufficient statistic is

(a) $\hat{\theta} = 1/\overline{Y}_n$.

(b) $\hat{\theta} = 1/\overline{Y}_n + 1$.

(c) $\hat{\theta} = \log(1 + 1/\overline{Y}_n)$.

(d) same as the maximum likelihood estimator of θ.

12. Suppose $\{X_1, X_2 \ldots X_n\}$ is a random sample from the distribution with probability mass function

$$f(x, p) = p(1 - p)^{\sqrt{x}}, \quad x = 0, 1, 4, 9, 16, \ldots, \quad 0 < p < 1.$$

Then which of the following statements is/are correct?

(a) The probability mass function $f(x, p)$ satisfies the Cramér regularity conditions.

(b) The maximum likelihood estimator of p does not exist.

(c) A CAN estimator of p does not exist.

(d) The maximum likelihood estimator of p is not CAN for p.

13. Which one of the following probability laws does NOT fulfill Cramér regularity conditions?

(a) $f(x, \theta) = 1, \quad \theta < x < \theta + 1, \quad \theta > 0$.

(b) $f(x, \theta) = \theta^x (1 - \theta)^{1-x}, \quad x = 0, 1, \quad 0 < \theta < 1$.

(c) $f(x, \theta) = \theta e^{-\theta x}$, $x > 0$, $\theta > 0$.
(d) $f(x, \theta) = (\theta + 1)x^{\theta}$, $0 < x < 1$, $\theta > -1$.

14. Which of the following probability laws is/are NOT a member of exponential family?

(a) $N(\theta, \theta)$, $\theta > 0$.
(b) $N(\theta, \theta^2)$, $\theta > 0$.
(c) $f(x, \theta) = \theta(1 - x)^{\theta-1}$, $0 < x < 1$, $\theta > 0$.
(d) $f(x, \theta) = \theta^x (1 - \theta)^{1-x}$, $x = 0, 1$, $0 < \theta < 1$.

15. Suppose that the joint distribution of a random vector $(X, Y)'$ is,

$$f(x, y, \beta, \theta) = \frac{\theta e^{-(\beta+\theta)y}(\beta y)^x}{x!}, \; y > 0, \; x = 0, 1, \ldots, \; \beta, \theta > 0.$$

Which of the following statements is/are correct?

(a) $f(x, y, \beta, \theta)$ is a member of a two-parameter exponential family.
(b) $1/\overline{Y}_n$ is a maximum likelihood estimator of θ.
(c) $\overline{X}_n/\overline{Y}_n$ is maximum likelihood estimator of β.
(d) $(\overline{X}_n, \overline{Y}_n)$ is consistent for (β, θ).

16. Suppose $\mathbb{E}$ denotes a one-parameter exponential family. Then which of the following statements is/are NOT correct?

(a) Exponential distribution with location parameter $\theta \neq 0$ and scale parameter 1 belongs to $\mathbb{E}$.
(b) Cauchy distribution with location parameter θ belongs to $\mathbb{E}$.
(c) Normal distribution with mean θ and variance θ, $\theta > 0$ belongs to $\mathbb{E}$.
(d) Laplace distribution with location parameter θ and scale parameter 1 belongs to $\mathbb{E}$.

17. Suppose $\mathbb{E}$ denotes a one-parameter exponential family. Then which of the following statements is/are correct?

(a) Uniform distribution $U(0, \theta)$ belongs to $\mathbb{E}$.
(b) Gamma distribution with scale parameter θ and shape parameter 5 belongs to $\mathbb{E}$.
(c) Normal distribution with mean θ and variance θ^2 belongs to $\mathbb{E}$.
(d) Laplace distribution with location parameter 0 and scale parameter θ belongs to $\mathbb{E}$.

18. Suppose $\mathbb{C}$ denotes a Cramér family. Then which of the following statements is/are correct?

    (a) Exponential distribution with location parameter $\theta \neq 0$ and scale parameter 1 belongs to $\mathbb{C}$.
    (b) Cauchy distribution with location parameter θ belongs to $\mathbb{C}$.
    (c) Normal distribution with mean θ and variance θ, $\theta > 0$ belongs to $\mathbb{C}$.
    (d) Laplace distribution with location parameter 0 and scale parameter θ belongs to $\mathbb{C}$.

19. Following are two statements about a multinomial distribution in three cells with cell probabilities as θ^2, $2\theta(1-\theta)$ and $(1-\theta)^2$, $0 < \theta < 1$.
    (I) It belongs to an exponential family.
    (II) It belongs to a Cramér family.
    Which of the following statements is/are correct?

    (a) Both (I) and (II) are true.
    (b) Both (I) and (II) are false.
    (c) (I) is true but (II) is false.
    (d) (I) is false but (II) is true.

20. Following are two statements about the estimator of an indexing parameter $\underline{\theta}$ of a k-parameter exponential family.
    (I) The maximum likelihood estimator of $\underline{\theta}$ is a CAN estimator of $\underline{\theta}$.
    (II) The moment estimator of $\underline{\theta}$ based on a sufficient statistic is a CAN estimator of $\underline{\theta}$.

    Which of the following statements is/are correct?

    (a) Both (I) and (II) are true.
    (b) Both (I) and (II) are false.
    (c) (I) is true but (II) is false.
    (d) (I) is false but (II) is true.

7.6.4 Chapter 5: Large Sample Test Procedures

1. Suppose $\{X_1, X_2 \ldots X_n\}$ is a random sample from a normal $N(\theta, 1)$ distribution. Then the likelihood function attains its maximum under $H_0 : \theta \leq \theta_0$ at

    (a) θ_0.
    (b) $\overline{X_n}$.
    (c) $\min\{\theta_0, \overline{X}\}$.
    (d) $\max\{\theta_0, \overline{X}\}$.

2. Suppose X is a random variable or a random vector with probability law $f(x, \underline{\theta})$ indexed by a parameter $\underline{\theta} \in \Theta \subset \mathbb{R}^k$ and the distribution of X belongs to a Cramér family. Suppose $\lambda(\underline{X})$ is a likelihood ratio test statistic based on a random sample $\underline{X} = \{X_1, X_2, \ldots, X_n\}$ for testing $H_0 : \underline{\theta}^{(1)} = \underline{\theta}_0^{(1)}$ against the alternative $H_1 :$ $\underline{\theta}^{(1)} \neq \underline{\theta}_0^{(1)}$, where $\underline{\theta}^{(1)} = (\theta_1, \theta_2, \ldots, \theta_m)'$ and $\underline{\theta}^{(2)} = (\theta_{m+1}, \theta_{m+2}, \ldots, \theta_k)'$ is partition of $\underline{\theta}$ with $m < k$ and $\underline{\theta}_0^{(1)}$ is a specified vector. Then the asymptotic null distribution of $-2 \log \lambda(\underline{X})$ is

  (a) χ_m^2.
  (b) χ_{k-1}^2.
  (c) χ_{k-m}^2.
  (d) χ_{k-m-1}^2.

3. Suppose $X \sim B(1, p)$ distribution, $0 < p < 1$. On the basis of a random sample of size n from the distribution of X, we want to test $H_0 : p = p_0$ against the alternative $p \neq p_0$, where p_0 is a specified constant. Then asymptotic null distribution of which of the following test statistics is $N(0, 1)$?

  (a) $\dfrac{(\bar{X}_n - p_0)}{\sqrt{\bar{X}_n(1 - \bar{X}_n)/n}}$.

  (b) $\dfrac{(\bar{X}_n - p_0)}{\sqrt{\bar{X}_n(1 - \bar{X}_n)}}$.

  (c) $\dfrac{(\bar{X}_n - p_0)}{\sqrt{p_0(1 - p_0)/n}}$.

  (d) $\dfrac{(\bar{X}_n - p_0)}{\sqrt{p_0(1 - p_0)}}$.

4. Suppose X and Y are independent random variables having Bernoulli $B(1, p_1)$ and $B(1, p_2)$ distributions respectively, $0 < p_1, p_2 < 1$. On the basis of random samples of sizes n_1 and n_2 from the distribution of X and Y respectively, we want to test $H_0 : p_1 = p_2$ against the alternative $p_1 \neq p_2$. Then asymptotic null distribution of which of the following test statistics is $N(0, 1)$?

  (a) $\dfrac{(\bar{X}_{n_1} - \bar{Y}_{n_2})}{\sqrt{\bar{X}_{n_1}(1 - \bar{X}_{n_1})/n_1 + \bar{Y}_{n_2}(1 - \bar{Y}_{n_2})/n_2}}$.

  (b) $\dfrac{(\bar{X}_{n_1} - \bar{Y}_{n_2})}{\sqrt{\bar{X}_{n_1}(1 - \bar{X}_{n_1}) + \bar{Y}_{n_2}(1 - \bar{Y}_{n_2})}}$.

  (c) $\dfrac{(\bar{X}_{n_1} - \bar{Y}_{n_2})}{\sqrt{P_n(1 - P_n)(1/n_1 + 1/n_2)}}$,

  (d) $\dfrac{(\bar{X}_{n_1} - \bar{Y}_{n_2})}{\sqrt{P_n(1 - P_n)}}$,

  where $P_n = (n_1 \bar{X}_{n_1} + n_2 \bar{Y}_{n_2})/(n_1 + n_2)$.

5. Suppose X is a random variable or a random vector with probability law $f(x, \underline{\theta})$ indexed by a parameter $\underline{\theta} \in \Theta \subset \mathbb{R}^k$ and the distribution of X belongs to a Cramér family. Suppose $\lambda(\underline{X})$ is a likelihood ratio test statistic based on a random sample $\underline{X} = \{X_1, X_2, \ldots, X_n\}$ for testing $H_0 : \underline{\theta} \in \Theta_0$ against the alternative $H_1 : \underline{\theta} \in \Theta_1$, where in Θ_0, $\theta_i = g_i(\beta_1, \beta_2, \ldots, \beta_m)'$, $i = 1, 2, \ldots, k$, where $m \leq k$ and $g_1, g_2, \ldots, g_k$ are Borel measurable functions from $\mathbb{R}^m$ to $\mathbb{R}$, having continuous partial derivatives of first order. Then the asymptotic null distribution of $-2 \log \lambda(\underline{X})$ is χ_r^2 where

   (a) $r = k$.
   (b) $r = k - m - 1$.
   (c) $r = m$.
   (d) $r = k - m$.

7.6.5 Chapter 6: Goodness of Fit Test and Tests for Contingency Tables

1. Suppose the distribution of a random variable or a random vector X, indexed by a vector parameter $\underline{\theta}$ belongs to a Cramér family. Suppose $\hat{\underline{\theta}}_n$ is a maximum likelihood estimator of $\underline{\theta}$ based on a random sample of size n from the distribution of X. Then Wald's test statistics for testing $H_0 : \underline{\theta} = \underline{\theta}_0$ is given by

   (a) $n(\hat{\underline{\theta}}_n - \underline{\theta}_0)'(\hat{\underline{\theta}}_n - \underline{\theta}_0)$.
   (b) $n(\hat{\underline{\theta}}_n - \underline{\theta}_0)' M_n (\hat{\underline{\theta}}_n - \underline{\theta}_0)$ where $M_n = \left[-\frac{\partial^2 \log L}{\partial \theta_i \partial \theta_j} \right]$.
   (c) $n(\hat{\underline{\theta}}_n - \underline{\theta}_0)' I(\underline{\theta}_0)(\hat{\underline{\theta}}_n - \underline{\theta}_0)$.
   (d) $n(\hat{\underline{\theta}}_n - \underline{\theta}_0)' I(\hat{\underline{\theta}}_n)(\hat{\underline{\theta}}_n - \underline{\theta}_0)$.

2. Suppose the distribution of a random variable or a random vector X, indexed by a vector parameter $\underline{\theta}$ belongs to a Cramér family. Suppose $\hat{\underline{\theta}}_n$ is a maximum likelihood estimator of $\underline{\theta}$ based on a random sample of size n from the distribution of X. Suppose $\underline{V}_n(\underline{X}, \underline{\theta}_0)$ denote the vector of score functions evaluated at $\underline{\theta}_0$. Then the score test statistics for testing $H_0 : \underline{\theta} = \underline{\theta}_0$ is given by

   (a) $\underline{V}_n'(\underline{X}, \underline{\theta}_0) \underline{V}_n(\underline{X}, \underline{\theta}_0)$.
   (b) $\underline{V}_n'(\underline{X}, \underline{\theta}_0) I^{-1}(\underline{\theta}_0) \underline{V}_n(\underline{X}, \underline{\theta}_0)$.
   (c) $\underline{V}_n'(\underline{X}, \underline{\theta}_0) I(\underline{\theta}_0) \underline{V}_n(\underline{X}, \underline{\theta}_0)$.
   (d) $\underline{V}_n'(\underline{X}, \underline{\theta}_0) I^{-1}(\hat{\underline{\theta}}_n) \underline{V}_n(\underline{X}, \underline{\theta}_0)$.

3. Suppose $\underline{Y}$ follows a trinomial distribution with cell probabilities $p_1, p_2, p_3 > 0$ and $p_1 + p_2 + p_3 = 1$. Suppose $\hat{p}_{in}$ denotes the maximum likelihood estimator of p_i, $i = 1, 2, 3$ and $I \equiv I(p_1, p_2, p_3)$ denotes the information matrix. Which of the following statements is/are correct?

(a) $\sqrt{n}((\hat{p}_{1n}, \hat{p}_{2n}, \hat{p}_{3n}) - (p_1, p_2, p_3))' \xrightarrow{L} N_3(0, I)$.

(b) $\sqrt{n}((\hat{p}_{1n}, \hat{p}_{2n}, \hat{p}_{3n}) - (p_1, p_2, p_3))' \xrightarrow{L} N_3(0, I^{-1})$.

(c) $\sqrt{n}((\hat{p}_{1n}, \hat{p}_{2n}) - (p_1, p_2))' \xrightarrow{L} N_2(0, I)$.

(d) $\sqrt{n}((\hat{p}_{1n}, \hat{p}_{2n}) - (p_1, p_2))' \xrightarrow{L} N_2(0, I^{-1})$.

4. Suppose $\underline{Y}$ follows a multinomial distribution in k cells with cell probabilities $p_i > 0$ and $\sum_{i=1}^{k} p_i = 1$. Suppose $\hat{\underline{p}}_n$ denotes the maximum likelihood estimator of $\underline{p} = (p_1, p_2, \ldots, p_{k-1})$ and $\overline{I}(\underline{p})$ denotes the information matrix. Which of the following statements is/are correct?

(a) $n(\hat{\underline{p}}_n - \underline{p})' I(\underline{p})(\hat{\underline{p}}_n - \underline{p}) \xrightarrow{L} \chi^2_{k-1}$.

(b) $n(\hat{\underline{p}}_n - \underline{p})' I^{-1}(\underline{p})(\hat{\underline{p}}_n - \underline{p}) \xrightarrow{L} \chi^2_{k-1}$.

(c) $n(\hat{\underline{p}}_n - \underline{p})' I(\hat{\underline{p}}_n)(\hat{\underline{p}}_n - \underline{p}) \xrightarrow{L} \chi^2_{k-1}$.

(d) $n(\hat{\underline{p}}_n - \underline{p})' I^{-1}(\hat{\underline{p}}_n)(\hat{\underline{p}}_n - \underline{p}) \xrightarrow{L} \chi^2_{k-1}$.

5. For a multinomial distribution in k cells with cell probabilities $\underline{p} = (p_1, p_2, \ldots, p_k)'$ with $p_i > 0 \ \forall \ i = 1, 2, \ldots, k$ and $\sum_{i=1}^{k} p_i = 1$, suppose we want to test $H_0 : \underline{p} = \underline{p}(\underline{\theta})$ against the alternative $H_1 : \underline{p} \neq \underline{p}(\underline{\theta})$, where $\underline{\theta}$ is an indexing parameter of dimension $l < k$. Suppose $\lambda(\underline{X})$ is a likelihood ratio test statistic based on a random sample of size n. If the multinomial distribution with cell probabilities indexed by $\underline{\theta}$ belongs to a Cramér family, then which of the following statements is/are correct? For large n, $-2 \log \lambda(\underline{X})$ follows

(a) χ^2_{k-1} distribution.

(b) χ^2_{k-l} distribution.

(c) χ^2_{kl-1} distribution.

(d) χ^2_{k-1-l} distribution.

6. Suppose a multinomial distribution with cell probabilities $\underline{p} = (p_1, p_2, \ldots, p_k)'$ where $p_i > 0 \ \forall \ i = 1, 2, \ldots, k$ and $\sum_{i=1}^{k} p_i = 1$, belongs to a Cramér family. For testing $H_0 : \underline{p} = \underline{p}_0$ against the alternative $H_1 : \underline{p} \neq \underline{p}_0$, which of the following statements is/are correct? The asymptotic null distribution of

(a) the likelihood ratio test statistic is χ^2_{k-1}.

(b) Karl Pearson's chi-square test statistic $\sum_{i=1}^{k}(o_i - e_i)^2/e_i$ is χ^2_{k-1}.

(c) Karl Pearson's chi-square test statistic $\sum_{i=1}^{k}(o_i - e_i)^2/e_i$ is χ^2_k.

(d) $\sum_{i=1}^{k}(o_i - e_i)^2/o_i$ is χ^2_k.

7. In a multinomial distribution with k cells having cell probabilities $\underline{p} = (p_1,$ $p_2, \ldots, p_k)'$ where $p_i > 0$ $\forall$ $i = 1, 2, \ldots, k$ and $\sum_{i=1}^{k} p_i = 1$, suppose we want to test $H_0 : \underline{p} = \underline{p}(\underline{\theta})$ against the alternative $H_1 : \underline{p} \neq \underline{p}(\underline{\theta})$, where $\underline{\theta}$ is an indexing parameter of dimension $l < k$. It is assumed that the multinomial distribution when cell probabilities are indexed by $\underline{\theta}$ belongs to a Cramér family. Then which of the following statements is/are correct? The asymptotic null distribution of

(a) the likelihood ratio test statistic is χ^2_{k-l-1}.

(b) Karl Pearson's chi-square test statistic $\sum_{i=1}^{k} (o_i - e_i)^2/e_i$ is χ^2_{k-l-1}.

(c) Karl Pearson's chi-square test statistic $\sum_{i=1}^{k} (o_i - e_i)^2/e_i$ is χ^2_{k-l}.

(d) $\sum_{i=1}^{k} (o_i - e_i)^2/o_i$ is χ^2_{k-l-1}.

8. Suppose $\underline{Y} = \{Y_1, Y_2, \ldots, Y_{k-1}\}$ has a multinomial distribution in k cells with cell probabilities $\underline{p} = \{p_1, p_2, \ldots, p_{k-1}\}$, where $p_i > 0$, $i = 1, 2, \ldots, k$ with $p_k = 1 - \sum_{i=1}^{k-1} p_i$. On the basis of a random sample of size n from the distribution of $\underline{Y}$, suppose we want to test $H_0 : \underline{p} = \underline{p}_0$ against the alternative $H_1 : \underline{p} \neq \underline{p}_0$, where $\underline{p}_0$ is a completely specified vector. Then which of the following statements is/are correct?

(a) Wald's test statistic and $\sum_{i=1}^{k} (o_i - e_i)^2/o_i$ are identically distributed for large n, under H_0.

(b) Wald's test statistic and $\sum_{i=1}^{k} (o_i - e_i)^2/o_i$ are identical random variables, under H_0.

(c) the score test statistic and Karl Pearson's chi-square test statistic $\sum_{i=1}^{k} (o_i - e_i)^2/e_i$ are identically distributed for large n, under H_0.

(d) the score test statistic and Karl Pearson's chi-square test statistic $\sum_{i=1}^{k} (o_i - e_i)^2/e_i$ are identical random variables, under H_0.

9. Suppose $\underline{Y} = \{Y_1, Y_2, \ldots, Y_{k-1}\}$ has a multinomial distribution in k cells with cell probabilities $\underline{p}$ being a function of $\underline{\theta}$, a vector valued parameter of dimension $l \times 1, l < k$. On the basis of a random sample of size n from the distribution of $\underline{Y}$, suppose we want to test $H_0 : \underline{p} = \underline{p}(\underline{\theta})$ against the alternative $H_1 : \underline{p} \neq \underline{p}(\underline{\theta})$. Then which of the following statements is/are correct?

(a) Wald's test statistic and $\sum_{i=1}^{k} (o_i - e_i)^2/o_i$ are identically distributed for large n, under H_0.

(b) Wald's test statistic and $\sum_{i=1}^{k} (o_i - e_i)^2/o_i$ are identical random variables, under H_0.

(c) the score test statistic and Karl Pearson's chi-square test statistic
$\sum_{i=1}^{k} (o_i - e_i)^2 / e_i$ are identically distributed for large n, under H_0.

(d) the score test statistic and Karl Pearson's chi-square test statistic
$\sum_{i=1}^{k} (o_i - e_i)^2 / e_i$ are identical random variables, under H_0.

10. In a $r \times s$ contingency table for testing H_0 : Two attributes A and B are independent against the alternative H_1 : A and B are not independent, which of the following statements is/are correct? The asymptotic null distribution of

(a) the likelihood Ratio test statistic $-2 \log \lambda(\underline{X})$ is $\chi^2_{(r-1)(s-1)}$.

(b) Karl Pearson's chi-square test statistic $\sum_{i=1}^{r} \sum_{j=1}^{s} \frac{(o_{ij} - e_{ij})^2}{e_{ij}}$ is $\chi^2_{(r-1)(s-1)}$.

(c) $\sum_{i=1}^{r} \sum_{j=1}^{s} \frac{(o_{ij} - e_{ij})^2}{o_{ij}}$ is $\chi^2_{(r-1)(s-1)}$.

(d) $\sum_{i=1}^{r} \sum_{j=1}^{s} \frac{(o_{ij} - e_{ij})^2}{o_{ij}}$ is χ^2_{rs-1}.

11. In a $r \times s$ contingency table for testing H_0 : Two attributes A and B are independent against the alternative H_1 : A and B are not independent, which of the following statements is/are correct?

(a) Wald's test statistic and $\sum_{i=1}^{r} \sum_{j=1}^{s} \frac{(o_{ij} - e_{ij})^2}{o_{ij}}$ are identically distributed for large n, under H_0.

(b) Wald's test statistic and $\sum_{i=1}^{r} \sum_{j=1}^{s} \frac{(o_{ij} - e_{ij})^2}{o_{ij}}$ are identical random variables, under H_0.

(c) the score test statistic and Karl Pearson's chi-square test statistic
$\sum_{i=1}^{r} \sum_{j=1}^{s} \frac{(o_{ij} - e_{ij})^2}{e_{ij}}$ are identically distributed for large n, under H_0.

(d) the score test statistic and Karl Pearson's chi-square test statistic
$\sum_{i=1}^{r} \sum_{j=1}^{s} \frac{(o_{ij} - e_{ij})^2}{e_{ij}}$ are identical random variables, under H_0.

12. In a $r \times s$ contingency table for testing H_0 : Two attributes A and B are independent against the alternative H_1 : A and B are not independent, the expected frequency e_{ij} of (i, j)-th cell is given by

(a) $e_{ij} = n_{ij}/n$.
(b) $e_{ij} = n_{i.} n_{.j}$.
(c) $e_{ij} = n_{i.} n_{.j}/n$.
(d) $e_{ij} = n_{i.} n_{.j}/n^2$.

13. In a three way contingency table for testing H_0 : Three attributes A, B and C are mutually independent against the alternative H_1 : A, B and C are not mutually independent, the expected frequency e_{ijk} of (i, j, k)-th cell is given by

(a) $e_{ijk} = n_{ijk}/n$.
(b) $e_{ijk} = (n_{i..}/n)(n_{.j.}/n)(n_{..k})$.
(c) $e_{ijk} = (n_{i..})(n_{.j.})(n_{..k})/n^3$.
(d) $e_{ijk} = (n_{i..})(n_{.j.})(n_{..k})/n$.

14. In a three way contingency table, suppose there are r levels of attribute A, s levels of attribute b and m levels of attribute C. While testing H_0 : Three attributes A, B and C are mutually independent against the alternative H_1 : A, B and C are not mutually independent, the number of parameters to be estimated in the null space is given by

(a) $rsm - 1$.
(b) $r + s + m - 3$.
(c) $r + s + m$.
(d) $r + s + m - 1$.

15. In a three way contingency table for testing H_0 : A and C are conditionally independent given B, the maximum likelihood estimator of p_{ijk}, probability of (i, j, k)-th cell, is given by

(a) $(n_{ij.}n_{.jk})/(nn_{.j.})$.
(b) $(n_{ij.}n_{.jk})/n_{.j.}$.
(c) $(n_{ij.}n_{.jk})/(n^2 n_{.j.})$.
(d) n_{ijk}/n.

16. In a three way contingency table for testing H_0 : A and C are conditionally independent given B, the number of parameters to be estimated in the null space is given by

(a) $rsm - 1$.
(b) $rs + sm - s + 1$.
(c) $rs + sm - rm$.
(d) $rs + sm - s - 1$.

17. In a three way contingency table for testing the null hypothesis that A and (B, C) are independent, the null hypothesis can be expressed as

(a) $H_0 : p_{ijk} = p_{ij.}p_{.jk} \ \forall \ i, j, k$.
(b) $H_0 : p_{ijk} = p_{i..}p_{ij.} \ \forall \ i, j, k$.
(c) $H_0 : p_{ijk} = p_{i..}p_{.jk} \ \forall \ i, j, k$.
(d) $H_0 : p_{ijk} = p_{ij.}p_{i.k} \ \forall \ i, j, k$.

18. In a three way contingency table for testing the null hypothesis that A and (B, C) are independent, the asymptotic null distribution of Karl Pearson's chi-square test statistic is χ_l^2, where l is

   (a) $rsm - 1$.
   (b) $(r - 1)(sm - 1)$.
   (c) $(r - 1)sm$.
   (d) $r(sm - 1)$.

19. Suppose $\{\phi_n(\underline{X}), n \geq 1\}$ is a sequence of test functions based on $\underline{X} = \{X_1, X_2, \dots, X_n\}$ for testing $H_0 : \theta \in \Theta_0$ against the alternative $H_1 : \theta \in \Theta_1$ where $\Theta_0 \cap \Theta_1 = \emptyset$ and $\Theta_0 \cup \Theta_1 = \Theta$. The test procedure governed by a test function ϕ_n is said to be consistent if

   (a) $\sup_{\theta \in \Theta_0} E_\theta(\phi_n(\underline{X})) \to \alpha \in (0, 1)$ and $E_\theta(\phi_n(\underline{X})) \to 1 \ \forall \ \theta \in \Theta_1$.
   (b) $\sup_{\theta \in \Theta_0} E_\theta(\phi_n(\underline{X})) \to \alpha \in (0, 1)$.
   (c) $E_\theta(\phi_n(\underline{X})) \to 1 \ \forall \ \theta \in \Theta_1$.
   (d) $E_\theta(\phi_n(\underline{X})) \to \alpha \in (0, 1)$ for some $\theta \in \Theta_0$ and $E_\theta(\phi_n(\underline{X})) \to 1$ for some $\theta \in \Theta_1$.

20. Suppose $\{X_1, X_2, \dots, X_n\}$ is a random sample from a Cauchy $C(\theta, 1)$ distribution. A test procedure for testing $H_0 : \theta = 0$ against the alternative $H_1 : \theta \neq 0$ is consistent if the test statistic is based on the

   (a) sample median.
   (b) sample first quartile.
   (c) sample third quartile.
   (d) sample mean.

Table 7.1 Answer Key

Q.No.	Chapter 2	Chapter 3	Chapter 4	Chapter 5	Chapter 6
1	a,c	a,c	a	c	d
2	a,d	d	d	a	b
3	a,b,d	b	c	a,c	d
4	c	a	b	a,c	a,c
5	a,b	a,d	c	d	d
6	a,c,d	b,d	a		a,b
7	a,d	d	d		a,b,d
8	b	c	c		a,b,c,d
9	b,c,d	a	a,b		a,c,d
10	a,c	c	a		a,b,c
11	a	b	c,d		a,c,d
12	a,b	a	a		c
13	b,c,d	d	a		b
14	c	a	b		b
15	b,c	a	a,b,c		a
16	c	a,b,d	a,b,d		d
17	c	a,b,d	d		c
18	a,b,c	a,b,d	b,c,d		b
19	a,d	c	a		a
20	c	d	a		a,b,c

Index

© The Author(s), under exclusive license to Springer Nature Singapore Pte Ltd. 2021
S. Deshmukh and M. Kulkarni, *Asymptotic Statistical Inference*,
https://doi.org/10.1007/978-981-15-9003-0

Printed in the United States
by Baker & Taylor Publisher Services